ÆTHEREAL MULTIVERSE

A New Unifying Theoretical Approach to

Cosmology, Particle Physics, and Quantum Gravity

New Delhi | Mumbai | Chennai | Kolkata | Bengaluru | Hyderabad | Kochi | Guwahati

First Indian Edition 2012

VIVA BOOKS PRIVATE LIMITED

- 4737/23 Ansari Road, Daryaganj, New Delhi 110 002
  E-mail: vivadelhi@vivagroupindia.net, Tel. 42242200
- 76, Service Industries, Shirvane, Sector 1, Nerul, Navi Mumbai 400 706
  E-mail: vivamumbai@vivagroupindia.net, Tel. 27721273, 27721274
- Jamals Fazal Chambers, 26 Greams Road, Chennai 600 006
  E-mail: vivachennai@vivagroupindia.net, Tel. 28290304, 28294241
- B-103, Jindal Towers, 21/1A/3 Darga Road, Kolkata 700 017
  E-mail: vivakolkata@vivagroupindia.net, Tel. 22836381, 22816713
- 7, Sovereign Park Apartments, 56-58, K. R. Road, Basavanagudi, Bangalore 560 004
  E-mail: vivabangalore@vivagroupindia.net, Tel. 26607409, 26607410
- 101-102 Mughal Marc Apartments, 3-4-637 to 641, Narayanguda, Hyderabad 500 029
  E-mail: vivahyderabad@vivagroupindia.net, Tel. 27564481, 27564482
- Beevi Towers, First Floor, SRM Road, Kaloor, Kochi 682 018, Kerala
  E-mail: vivakochi@vivagroupindia.net, Tel: 0484-2403055, 2403056
- 232 GNB Road, Beside UCO Bank, Silpukhuri, Guwahati 781 003
  E-mail: vivaguwahati@vivagroupindia.net, Tel: 0361-2666386

www.vivagroupindia.com

Published by arrangement with
Cambridge International Science Publishing
7, Meadow Walk, Great Abington
Cambridge CB1 6AZ, UK

ISBN : 978-81-309-2174-7

Published by Vinod Vasishtha for Viva Books Private Limited, 4737/23, Ansari Road, New Delhi-110 002. Printed and bound by Syndicate Binders, New Delhi.

Dedication

Consolidation of my principal scientific ideas
in the form of this book was an exceptional delight.
The great effort put into the creation process
is solely dedicated to my parents Andrzej and Teresa,
sibling Magdalena, and the friends of mine.

Contents

Preface

A thought is an idea in transit.

Pythagoras

Æthereal Multiverse is my personal attempt to demonstrate that it is possible to make real efficiently the conceptual fusion of two most fundamental concepts of the natural philosophy of Antiquity – Æther and Multiverse – which lay the foundations of the Aristotelian and the Epicurean–Islamic–Hinduistic systems of thought, respectively, and in the present-day state of affairs constitute the basis of human understanding of the world we live in. It should be emphasized that in a standard understanding these concepts are definitely separated from each other. Consequently, presently there are very few scientists whose attempt to understand the actual relationships between Æther and Multiverse and make use of their specific combinations to describe the physical reality productively. This is my research experience-based deep intuitive conviction, however, that such a non-trivial combination of these two ancient ideas is as most natural as possible way to the fruitful and lucid description of physical reality, and leads to the consistent understanding of the anthropic everything, *i.e.* everything what can be observed and detected by a man and all devices produced by humankind. This monograph, in my primary intention, is first of all the proposal of the unifying approach to particle physics, cosmology, and quantum gravity based on the most essential pieces of modern theoretical physics – Quantum Mechanics, Quantum Field Theory, Special Relativity, General Relativity, and Thermodynamics – and creates fruitful methodological background to solve the intriguing problems of high energy physics, cosmology, and gravitational physics.

The content of this monograph consists of Prologue, 16 Chapters, Addendum and Epilogue. The Prologue briefly presents the historical and philosophical development of the concepts of Æther and Multiverse, which lay the foundations of the Aristotelian and Epicurean–Islamic–Hinduistic systems of natural philosophy. We sketch out the evolution of the state of affairs since the origins till the present-day status. Part I is *Cosmology*. Chapter 1 discusses my approach to Multiverse quantum cosmology based on the Einstein–Friedmann Universe. The thermodynamics of space quanta is derived, and a number of related deductions are presented. Chapter 2 discusses my original approach to the inflationary cosmology which is based on the slow-roll approximation. For consistency there are introduced and efficiently employed the concepts of the Higgs–Hubble inflaton and the inflaton constant. Chapter 3 is devoted to the model of massive gravitations, which is based on General Relativity, cosmological perturbation theory, perturbative quantum gravity and the Lorentz force-inspired interpretation. Part II is *Particle Physics*. Chapter 4 presents the two proposals for mass generation mechanisms within the Standard Model. The first mechanism is arranged with using the Higgs potential which is also the foundational element of the accepted mass generation mechanism of the Standard Model, while the second mechanism arises from the modified Higgs potential. In both these models masses of both photon as well as the new neutral scalar field are non-zero. In the first model photon mass depends on the neutral scalar field mass, whereas in the second model these masses are independent variables and the mass of the neutral scalar field remains a free parameter which can be established by analysis of experimental data. Chapter 5 discusses several basic consequences of the Snyder–Sidharth deformation arising from the modification of the Einstein energy-momentum relation due to the non-commutative geometry of the Snyder space-time. There is presented exhaustively, among others, detailed analysis of the Compton effect based on such a non-commutative geometry framework. The Chapters 6 and 7 are focused on an advanced discussion of the model of massive neutrinos arising from the Snyder–Sidharth deformation since the beginning till certain intriguing ramifications. There are presented a number of mathematical details, particularly the mechanism of spontaneous symmetry breaking leading to the chiral condensate of the neutrinos. Other important ramifications of the massive neutrino are discussed in from the points of view based on the both Weyl and Dirac equations. Chapter 8 presents the consistent proposal for extension of Special Relativity to the case of motion with superluminal speeds. There is employed the *ab initio* derived deformation to the Einstein energy-momentum relation. The non-parallelism hypothesis is also investigated which assumes that the angle between momentum and velocity vectors of a superluminally moving object is non-zero. For restoration of the full relativistic invariance, *i.e.* Poincarè invariance, there is implemented the concept of new maximal speed in the superluminal regime of speeds. The superluminal moving neutrinos are consistently explained as the Weyl neutrinos with identically vanishing effective mass. The most recent experimental data of the OPERA Collaboration of CERN Geneva, which give the first evidence for the hypothesis that neutrinos are superluminally moving particles, are briefly discussed in this context. In Chapter 9 we focus on the problem of magnetic monopoles and dyons. First we present certain general outlook on the Yang–Mills gauge field theories, which includes a brief discussion of the most recent results of a number of authors. Then we study the Dirac monopole, the Wu–Yang monopole, the 't Hooft–Polyakov monopole, and the Julia–Zee dyon. In the last case we find generalization of the Bogomol'nyi equations to the case of the non-vanishing scalar field potential. Part III is *Quantum Gravity*. Chapter 10 introduces the reader into the studies of quantum gravity. This chapter presents a certain detailed review of the present state of quantum geometrodynamics: from the Arnowitt–Deser–Misner decomposition and the intrinsic differential geometry, through the classical and the Wheeler–DeWitt quantum geometrodynamics, to the concept of the Wheeler Superspace and a brief survey through a number of modern problems of geometrodynamics, which includes few alternative approaches to the problem of quantization of gravitation recently popularized in the literature. Chapter 11 presents the global one-dimensionality conjecture, the basis of my approach to quantum gravity, and its numerous ramifications to the Wheeler–DeWitt quantum geometrodynamics. The second quantization is performed in the static Fock space using the Bogoliubov transformation suitable for bosons. In Chapter 12 we implement and apply to our considerations the idea of the invariant global dimension to the model of global one-dimensional quantum gravity. There are presented the Cauchy-like wave functionals solving the model of quantum gravity. Chapter 13 is devoted to discussion of a number of the solutions to the four-dimensional Einstein field equations, among others all four-dimensional black holes, the Gödel universe, and the Taub–Newman–Unti–Tamburino solution, in the context of the model of global one-dimensional quantum gravity. In Chapter 14 certain important ramifications of the functional objective geometry arising from the quantum gravity model are discussed in a high detail. It is proposed to study the particular situation within the model given by the Newton–Coulomb potential. Three types of the boundary conditions to the wave functionals are discussed. Chapter 15 presents two approaches to entropy calculation, the analytic and the algebraic one, which lead to distinguished formulations of thermodynamics of space quanta Æther. In Chapter 16 the higher-dimensional generalization of the four-dimensional quantum gravity model is presented. A certain development of the Yang–Mills theories, particularly magnetic monopoles, is also presented in this Chapter. Gauge quantum gravity is investigated in the spirit of the Faddeev–Popov determinant which is the core fundament of quantized non-Abelian gauge field theories. Finally, the relation to the string theory through the dimension of space and the concept of effective space-time is discussed. Part IV is *Addendum*. It presents the extra themes which are: the non-zero photon mass in the framework of wave-particle duality, and one of the Millennium Prize Problems which is existence of smooth solutions to the Navier–Stokes equations. Finally, in the Epilogue, the philosophical summary of a whole content of the book is presented.

In my best intention this monograph presents the advanced analysis between theoretical physics and cosmology arising from a certain fundamental and well-established knowledge. Despite the fact that I mean the level of discussion of this book as rather advanced, in a number of places the reader can find detailed discussion of the presented problems. There are presented many step-by-step explanations which include only the necessary number of mathematical calculations, and for this reason can be useful for the beginners. For this reason, this monograph is easy assimilable for both the beginning as well as professional scientists, researchers, and academics whose are interested in constructive applications of the fundamental knowledge.

Initial motivation for writing an extensive monograph refining my previous research results and enriching them by new developments, was the 2010' book of the mathematical physicist Robert Wayne Carroll [1]. This book expressed an optimistic point of view and, particularly, discussed my research results with a necessary doze of commentaries, what persuaded me to reconsider my results.

In this manner, *Æthereal Multiverse* collects a substantial part of my original non-coauthored research results which in many places are rigorously revisited and enriched by necessary updates and new developments. The main part of these improvements is related to the linguistic typos and the technical mistakes which, fortunately, do not change both the essence of the proposed approaches and the ramifications following from their application. The revision of the results of my previous research was strongly motivated by the fact that not only consistency of any knowledge but also its constructiveness forms the state of affairs. I have supplemented the revised results with the new developments arising from the improved excerpts. There are also presented completely new research results which allowed me to see much more wider horizons of the theory. It should be emphasized that apart from the purely methodological aspects I have attempted to present also my personal purely philosophical standpoint on theoretical physics and cosmology, which arises from the fundamental role of the philosophical interpretation of any mathematical theory of physics.

It should be emphasized that discussions, comments, and remarks benefitted from a number of scientists, researchers, and scholars were valuable. In particular, my as most grateful as possible acknowledgements are directed to: Lubomir Alexandrov, Mikhail Victorovich Altaisky, Andrej Borisovich Arbuzov, Irina Yaroslavovna Aref'eva, Abhay Ashtekar, Boris Mikhailovich Barbashov, Andrey Olegovich Barvinsky, Stanisław Leon Bażański, Andrew Walcott Beckwith, Alexander Alexandrovich Beilinson, Agata Bezubik, Iwo Białynicki-Birula, David Bernhard Blaschke, Paweł Błasiak, Arkadiusz Błaut, Andrzej Zdzisław Borowiec, Stanley Jerome Brodsky, Kirill Alexandrovich Bronnikov, Iosif L'vovich Buchbinder, Čestmir Burdik, Robert Wayne Carroll, Sean Michael Carroll, David Edwin Alvarez Castillo, Marcin Cerkaski, Piotr Henryk Chankowski, Władysław

Jerzy Chmielowski, Tomasz Czyżycki, Ashkbiz Danehkar, Marcin Daszkiewicz, Alessandro De Angelis, Marek Demiański, Jonathan Jay Dickau, Alexander Evgenievich Dorokhov, Emilian Dudas, Remigiusz Durka, Irina Gavrilovna Dymnikova, Gary Vladimirovich Efimov, Anatoli Vasilievich Efremov, Sergey Alexeevich Fedoruk, Ignat Vitalevich Fialkovsky, Mikhail Leonidovich Fil'chenkov, Alexander Tikhonovich Filippov, Andrzej Frydryszak, Dmitry Vladimirovich Fursaev, Oleksii Galan, Philip Edward Gibbs, Ilya Faivilevich Ginzburg, Alexei Valerievich Gladyshev, Anastasia Andreevna Golubtsova, Georgi Georgiev Grahovski, Walter Albin Erhard Greiner, Hovik Alexandrovich Grigorian, John Gideon Hartnett, Thomas Hilger, Furio Honsell, Gerardus 't Hooft, Andrzej Antoni Horzela, Orest Hrycyna, Andrzej Zygmunt Hrynkiewicz, Huping Hu, Alexei Petrovich Isaev, Peter Stepanovich Isaev, Evgeny Alexeevich Ivanov, Vladimir Dmitrievich Ivashchuk, Bernard Jancewicz, Jakub Jankowski, Vladimir Georgievich Kadyshevsky, Jan Henryk Kalinowski, Edward Kapuścik, Vladimir Vsevolodovich Kassandrov, Dmitri Igorevich Kazakov, Marta Krystyna Kicińska-Habior, Roman Konoplya, Ryszard Paweł Kostecki, Jerzy Kowalski-Glikman, Maria Teresa Krawczyk, Harold Walter Kroto, Eduard Alekseevich Kuraev, Burra Subba Lakshmi, Zygmunt Adam Lalak, (Klaas) Nicolaas Pieter Landsman, Peter Gavin Lawrence Leach, Richard Lednický, Anthony James Leggett, Jerzy Krzysztof Lewandowski, Fang-Yu Li, Ricardo Lima, Lev Nikolaevich Lipatov, Jerzy Andrzej Lukierski, Oriana Mansutti, Larry McLerran, Krzysztof Antoni Meissner, Vitaly Nikolaevich Melnikov, Marek Mozrzymas, Viatcheslav Fedorovich Mukhanov, Sergei Nikolaevich Nedelko, Vladimir Vitalievich Nesterenko, Juan Antonio Nieto-Garcia, Anatoly Nikitin, Lech Olgierd Nowicki, Paweł Krzysztof Nurowski, Robert Olkiewicz, Douglas Dean Osheroff, Anna Pachoł, Vladimir Vladimirovich Papoyan, Marek Pawłowski, Victor Nikolaevich Pervushin, Marcin Radosław Piątek, Włodzimierz Zbigniew Piechocki, Stefan Pokorski, Aleksander Antoni Polański, Yuri Andreevich Popov, Roman Popovych, Ziemowit Stanisław Popowicz, Severin Pošta, Frans Pretorius, Vyacheslav Borisovich Priezzhev, Stefan Rauch-Wojciechowski, Katarzyna Rejzner, Jonathan Roberts, Krzysztof Piotr Rolbiecki, Carlo Rovelli, Yuri Grigorevich Rudoy, Alexander Rundel, Nikolai Artemievich Russakovich, Yuri Petrovich

Rybakov, Bijan Saha, Pondichery Kuppuswamy Saiprakash, Artur Sergyeyev, Alexander Alexandrovich Shaideman, Dmitry Vasilievich Shirkov, Burra Gautam Sidharth, Douglas Alexander Singleton, Alexei Norairovich Sissakian, Krystyna Siwek-Wilczyńska, Vladimir Vladimirovich Skokov, Jan Tadeusz Sobczyk, Adam Sobiczewski, Lech Marian Sokołowski, Alexander Savelievich Sorin, Marcin Skórzewski, Michał Spaliński, Simone Speziale, Katepalli Raju Sreenivasan, Alexei Alexandrovich Starobinsky, Nikolaos Stefanis, Gary Van Stephenson, Zbigniew Strycharski, Paulina Suchanek, Aleksander Dmitrievich Sukhanov, Michał Szcząchor, Marek Wojciech Szydłowski, Oleg Valerianovich Teryaev, Andrzej Mariusz Trautman, Yunita Umniyati, Sree Ram Valluri, Sergei Yurovich Vernov, Michel Vittot, Alexei Alexandrovich Vladimirov, Yuri Sergeevich Vladimirov, Victor Vasilevich Voronov, Janusz Wilczyński, Grzegorz Andrzej Wilk, Marek Wolf, Andrzej Woszczyna, Alexander Petrovich Yefremov, Vyacheslav Ivanovich Yukalov, Victor Yulievich Yushankhai, Daniel Zabłocki, Adam Zadrożny, Alexander Fyodorovich Zakharov, Alexander Valerievich Zhiboedov, Antonino Zichichi, Boris Moiseevich Zupnik, Wojciech Hubert Zurek, and Magdalena Zych.

Few of the mentioned colleagues significantly helped. My cordial thanks are directed to Professors: A.B. Arbuzov for initialization of my research in particle physics, A.W. Beckwith for benefitted discussions, K.A. Bronnikov for help to make the consolidation, I.L. Buchbinder for help to make my research lucid, R.W. Carroll for initial motivation, A. De Angelis for invaluable digressions, Sir H.W. Kroto for the matter-of-fact advices, V.N. Pervushin for initialization of my research in particle physics, cosmology and gravitation, and B.G. Sidharth for initialization of my research in non-commutative space-time's aspects.

The front cover picture of the Great Globular Cluster in Hercules (M13, NGC6205) is credited to Robert Lupton, and the Sloan Digital Sky Survey (SDSS) Collaboration, http://www.sdss.org.

Tokens of gratitude are directed to Victor Riecansky, the Publisher at Cambridge International Science Publishing, for generosity of spirit, helpful discussions, kind acceptance of my book and invaluable help in its publication.

Discussions about this book are invited to laglinka@gmail.com.

Łukasz Andrzej Glinka

You cannot hope to build a better world without improving the individuals. To that end, each of us must work for our own improvement and, at the same time, share a general responsibility for all humanity, our particular duty being to aid those to whom we think can be most useful.

Marie Salomea Curie, née Skłodowska

Prologue

> Death is not an event in life: we do not live to experience death. If we take eternity to mean not infinite temporal duration but timelessness, then eternal life belongs to those who live in the present. Our life has no end in the way in which our visual field has no limits.
>
> *Ludwig Wittgenstein*

U-turn to Antiquity

In the second decade of the 21th century the technological progress fully defines all aspects of the evolution of humankind. The present-day civilizations are strictly based on a purely technocratic point of view, which is usually focused on a formal side of the thing and does not accept errors as an inherent feature of the natural state of affairs. Consequently, the resulting state of affairs leaves no doubt that freedom of thought is not welcome, and especially focuses on the creation of new barriers which are called standards and norms. The obstacles for the existence of freedom of thought follow from promotion of feudal submissiveness toward the various traditions, what is as most as possible gross manifestation of deeply-rooted irrationalism. Mostly observed ramifications of such a technocratic fight with rationalism are among others the inhibition of mental growth of individuals and evident effective intellectual poverty. This type of irrationalism is particularly self-destroying because it ruins efforts of the freedom of thought-based rationalists, attempts to establish and stabilize plenty of illusions and delusions, and finally leads to the permanent effect of the damage of an individual mentality. Such a regime has produced recently the intensive and fierce emergence of multiple social disabilities, and above all a regular anti-progressive evolution of humankind. This global situation leads to the crisis in which new aspects are reflected in the everyday life of individuals, and the condensation of these aspects gives the rise to the collective result which is the global crisis in the all regions of the present-day style of life.

On the other hand, however, there are unambiguously detectable symptoms of the vast U-turn to the ideas of the most fundamental epoch in the history of human reasoning, that is *Antiquity*. The necessity of such a non-trivial and fundamental turnabout is fully argued by the globally critical situation of the freedom of thought. In fact, only the Antiquity-based approaches possess the natural ability to reconstruct the most valuable heritage of a human thought. This is caused by the fact that Antiquity is not infected by certain destructive side effects produced in the process of civilization progress. The open question is the fulfillment of the intellectual gap, what will be resulting in amazing healing of the state of affairs. In my opinion such a productive solution, that must be based on the appropriate description of the physical reality, requires fusion of two fundamental concepts of Antiquity: Æther which is the basic idea of the Aristotelian system, and Multiverse which lays the foundations for the Epicurean–Islamic–Hinduistic system.

Æther

In the mid-1950s, when Albert Einstein abandoned the living nature, Cosmology altered manifestly esoteric and religious countenance, and became a respected scientific branch. The satisfactory explanation of observational and experimental data consolidated both the physical and philosophical heritage of General Relativity. Riemannian geometry apparently is able to describe a number of astronomical phenomena like Mercury's perihelion precession and light rays deviation in Sun's vicinity, and offers constructive generalization of the Newton law of universal gravitation. This Einstein's theory described visible objects and predicted existence of new physical beings.

Albeit, Einstein's legacy is wide spread [2] and deserves being called *phenomenology*. Despite the fact that Einstein was established as the specialist in molecular physics and thermodynamics, he gave trailblazing description of fast particles, solids, opalescence, emission and absorption of electromagnetic radiation, and corrected the Zero-Point Energy hypothesis by Max Planck. His pioneering approach to the photoelectric effect used the wave-particle duality which became the main stream of the 20th century theoretical physics. He was the follower of unification of gravitation and electromagnetism. Despite the fact that this heritage incontrovertibly impacted on the mentality and the character of a whole theoretical physics, Einstein's individual intellectual growth was diverse. It is an appearance, because Einstein always attempted to describe physical properties of Æther.

Einstein was, in fact, the most efficient continuator of the more general scientific current focused on Æther. This idea was present in the science of physics since Antiquity and became the dominating contribution in the 19th century, when the first field theoretical approaches to electricity and magnetism were proposed by James Clerk Maxwell. Through the 20th century the idea of Æther remained alive but somewhat hidden under the carpet of alternative concepts, which only seemingly differs from its essence. The crucial theoretical argument against this existence was identification of the nature of Æther with the non-existent nature of Æther drift presented by the various experimental attempts. From the experimental point of view the idea of Æther was replaced by the formal concept of an observational error which is the difference between a measured value of quantity and its true value and, according to the present-day philosophy of a measurement, is not a mistake but such a variability is inherently presented in the nature of the measurement process.

Mythical Nature

The concept of Æther possesses the deepest historical roots within the mythology of Ancient Greece. Ancient Greeks professed Protogenoi, which are the immortal «primordial deities» born in the beginning of Universe. These primeval first born gods were Æther (Mists of Light, Upper Air), Ananke (Inevitability, Necessity, Compulsion), Chaos (Void, Lower Air), Chronos (Time), Erebus (Mists of Darkness), Eros (Love, Generation), Gaea (Earth), Hemera (Day), Hydros (Water), Nesoi

(Islands), Nyx (Night), Oceanus (Ocean), Ourea (Mountains), Phanes (Procreation), Phusis (Nature), Pontus (The Sea), Tartarus (Hell's Pit), Tethys (Fresh Water), Thalassa (Sea Surface), Thesis (Creation), Uranus (Heaven). They subsequently generated Giants, Titans, and Olympian, Oceanic, and Chthonic gods. In Homeric Greek language Æther $\alpha\iota\theta\eta\rho$ symbolizes "pure, fresh air" or "clear sky", which in the Greek mythology is the pure essence where the gods lived and which they breathed. Æther was a personified idea of the cosmogony professed by ancient Greeks, and considered as one of the elementary substances forming the Universe. According to the Orphic hymns Æther is the soul of the world emanating all life.

In alchemy and natural philosophy the concept of Æther was originated in the natural-philosophy works of Aristotle. Aristotle wrote [3]

> (...) That is why the upper part is moved in a circle, while the All is not anywhere. For what is somewhere is itself something, and there must be alongside it some other thing wherein it is and which contains it. But alongside the All or the Whole there is nothing outside the All, and for this reason all things are in the heaven; for the heaven, we may say, is the All. Yet their place is not the same as the heaven. It is part of it, the innermost part of it, which is in contact with the movable body; and for this reason the earth is in water, and this in the air, and the air in the aether, and the aether in heaven, but we cannot go on and say that the heaven is in anything else.

He implemented this idea as the *quinta essentia*, the cognate chemical fifth element of the heavens are made and unifying the sublunary elements, *i.e.* Fire, Earth, Air, and Water. Aristotle claimed that the four sublunary elements move rectilinearly, and in the light of the fact the orbits of the heavenly bodies have a shape of circles and lie on the rotating spheres surfaces he concluded that the physical spheres must be a body. In other words, according to the Aristotelian vision of the world the celestial motions straightforwardly require an existence of the superior element. Aristotle gave several plausible arguments for the existent nature of Æther. One of them, called *argument from incorruptibility*, states that the sublunar elements are easy transiting into each other, but because the eternal heavens they must be made of a different element. Therefore, existence of Æther is the ramification of the necessity for a body endowed with natural circular motion. In the vision of Aristotle the fifth element is unborn, immortal, and invariable, and its name arises from ancient *aei thein* what means eternal motion. Factually, he identified Æther with the mind or the soul, which is divine and does not corrupt with the earthly elements.

From Boyle & Newton to Lorentz

In the 17th century Robert Boyle [4] proposed to consider Æther as the material substance due to what he called subtle particles. In the context of modern physics Isaac Newton, in the corpuscular theory of light presented in his extensive monograph *Opticks* [5], significantly modified the original Boyle's idea. Newton's pioneering proposal was an æthereal medium carrying vibrations traveling faster than light, whose straightforward intervention results in refraction and diffraction of light. In his approach Isaac Newton negated the idea proposed earlier by Christian Huygens, claiming that light travels by Æther medium and, making use of the physically plausible concept of pressure of an atomic Æther impacting upon matter in the abnormal state, presented the physically reasonable explanation of the phenomenon of gravitation. In such a scenario subtle æthereal molecules entered into matter by the pores, and when approached a physical body became less resilient and rarely distributed. A material body find itself under the pressure due to Æther upon all sides. Therefore, between arbitrary two bodies the atomic pressure is less, and rare Æther distribution causes in the phenomenon of gravitation. Newton, however, did not justify the rarefaction mechanism emerging near matter. Consequently, his model of Æther did not find fruitful continuation in the works of subsequent generations of theoretical physicists.

For the historical consistency we should reference also the so-called Le Sage's kinetic theory of gravitation, proposed by Nicolas Fatio de Duillier and Georges-Louis Le Sage, which was Æther-based consistent explanation of the Newton universal gravitation. Le Sage [6] suggested purely mechanical nature of the universal gravitation which in an effect due to streams of ultra-mundane tenuous Æther corpuscules moving at the speed of light and acting on all matter from all directions.

In *Lucrèce Newtonien* Le Sage expressed the following standpoint

> I am well convinced that since the law governing the intensity of universal gravitation is similar to that for light, the thought will have occurred to many physicists that an ethereal substance moving in rectilinear paths may be the cause of gravitation, and that they may have applied to it whatever of skill in the mathematics they have possessed.

William Thomson the 1st Baron Kelvin [7] introduced another impacting concept of Æther. He treated Æther as an elastic solid medium transmitting the electromagnetic waves, and proposed its mechanical model. Such a proposition led him to consistent explanation of the nature of radiation.

Similarly James Clerk Maxwell [8], the father of the theory of electromagnetism which started the field theoretical approach in physics, was a fervent follower of the so-called Luminiferous Æther. Maxwell's Æther is the cosmic medium physically transmitting light. He even applied such a concept to deduce the equations of electrodynamics, known as the Maxwell equations, which show that light is an electromagnetic wave, and thought about the physical lines of forces of electric and magnetic fields as the lines within the Æther. The corresponding idea of vacuum based on the Poisson equation is often referred to as the Maxwell Vacuum.

Another diverse concepts of Æther were widely propagated and developed by a number of eminent scholars, scientists and researchers at the turn of the 19th and 20th centuries (Cf. *e.g.* the contributions in Ref. [9]). Exceptionally detailed analysis of the historical development of Æther till the discovery of quantum mechanics was elegantly performed by Edmund Taylor Whittaker [10]. An excellent example of such a model is the Æther model investigated by Joseph John Thomson, who considered the context of a hypothetical radiation that could be more penetrating than Röntgen's X-rays. Another significant investigation applying a certain Æther model was performed by George Howard Darwin in computation of a geometric deviation from the Newton law of universal gravitation.

The most intriguing idea in the history of Æther was the concept of Æther wind called also Æther drag and Æther drift. In the commonly accepted model this hypothetic phenomenon emerges when Luminiferous Æther is dragged by motion of matter or entrained by matter. Augustin-Jean Fresnel proposed Æther wind with partial entraining, which was empirically confirmed by the 1851 experiment of Armand Hippolyte Louis Fizeau. A different version of the Æther drag hypothesis, founded by George Gabriel Stokes, the 1st Baronet FRS in 1845, was experimentally confirmed by Albert Abraham Michelson and Edward Williams Morley [11] in 1881 and 1887. Fifteen years after interpretation of the results of the Michelson–Morley experiment as the confirmation of Einstein's Special Relativity, Michelson published the book [12] in which one can find several interesting looking reflections

> The standard light waves are not alterable; they depend on the properties of the atoms and upon the universal ether; and these are unalterable. It may be suggested that the whole solar system is moving through space, and that the properties of ether may differ in different portions of space. I would say that such a change, if it occurs, would not produce any material effect in a period of less than twenty millions of years, and by that time we shall probably have less interest in the problem. (...) the vibrations of these particles, or of their electric charges, produce the disturbance in the ether which is propagated in the form of light waves; and that the period of any light wave corresponds to the period of vibration of the electric charge which produces it. (...) the ether itself is electricity; a much more probable one is that electricity is an ether strain - that a displacement of the ether is equivalent to an electric current. If this is true, we are returning to our elastic-solid theory. I may quote a statement which Lord Kelvin made in reply to a rather sceptical question as to the existence of a medium about which so very little is supposed to be known. The reply was: "Yes, ether is the only form of matter about which we know anything at all." In fact, the moment we begin to inquire into the nature of the ultimate particles of ordinary matter, we are at once enveloped in a sea of conjecture and hypotheses - all of great difficulty and complexity. One of the most promising of these hypotheses is the "ether vortex theory,"

which, if true, has the merit of introducing nothing new into the hypotheses already made, but only of specifying the particular form of motion required. The most natural form of such vortex motions with which to deal is that illustrated by ordinary smoke rings, such as are frequently blown from the stack of a locomotive. Such vortex rings may easily be produced by filling with smoke a box which has a circular aperture at one end and a rubber diaphragm at the other, and then tapping the rubber. The friction against the side of the opening, as the puff of smoke passes out, produces a rotary motion, and the result will be smoke rings or vortices. (...) Investigation shows that these smoke rings possess, to a certain degree, the properties which we are accustomed to associate with atoms, notwithstanding the fact that the medium in which these smoke rings exists is far from ideal. If the medium were ideal, it would be devoid of friction, and then the motion, when once started, would continue indefinitely, and that part of the ether which is differentiated by this motion would ever remain so. (...) Another peculiarity of the ring is that it cannot be cut - it simply winds around the knife. Of course, in a very short time the motion in a smoke ring ceases in consequence of the viscosity of the air, but it would continue indefinitely in such a frictionless medium as we suppose the ether to be. (...) Suppose that an ether strain corresponds to an electric charge, an ether displacement to the electric current, these ether vortices to the atoms - if we continue these suppositions, we arrive at what may be one of the grandest generalizations of modern science - of which we are tempted to say that it ought to be true even if it is not - namely, that all the phenomena of the physical universe are only different manifestations of the various modes of motions of one all-pervading substance - the ether. (...) Then the nature of the atoms, and the forces called into play in their chemical union; the interactions between these atoms and the non-differentiated ether as manifested in the phenomena of light and electricity ; the structures of the molecules and molecular systems of which the

> atoms are the units; the explanation of cohesion, elasticity, and gravitation — all these will be marshaled into a single compact and consistent body of scientific knowledge. (. . .) In all probability, it not only exists where ordinary matter does not, but it also permeates all forms of matter. The motion of a medium such as water is found not to add its full value to the velocity of light moving through it, but only such a fraction of it as is perhaps accounted for on the hypothesis that the ether itself does not partake of this motion. (. . .) The phenomenon of the aberration of the fixed stars can be accounted for on the hypothesis that the ether does not partake of the earth's motion in its revolution about the sun. All experiments for testing this hypothesis have, however, given negative results, so that the theory may still be said to be in an unsatisfactory condition.

Hendrik Antoon Lorentz, one of the most eminent theoretical physicists of the turn of the 19th and 20th centuries, also manifestly professed Æther [13]. In his electromagnetic theory, published in 1895, he derived the Fresnel result. In his lectures he straightforwardly supports Æther. In the lectures delivered at Caltech one finds

> Nowadays we are concerned only with the electromagnetic theory of light, in which there is no longer any discussion of a density or elasticity of the ether. In the electromagnetic theory of light attention is fixed on the electric and magnetic fields that can exist in the "ether". (...) the state of the ether is the same at all points of a plane perpendicular to the direction of propagation, and so the waves may be called plane waves.

Similarly, his Leiden's lectures contain the unambiguous opinion

> (. . .)[W]hether there is an aether or not, electromagnetic fields certainly exist, and so also does the energy of electrical oscillations. If we do not like the name of 'aether', we must use another word as a peg to hang all these things upon. It is not certain whether 'space' can be so extended as to take care not only of the geometrical properties but also of

> the electric ones. One cannot deny to the bearer of these properties a certain substantiality, and if so, then one may, in all modesty, call true time the time measured by clocks which are fixed in this medium, and consider simultaneity as a primary concept.

It should be emphasized that also the Lorentz transformations were deduced on the basis of Æther. Lorentz pointed out the inconsistency between results of the Michelson–Morley and the Fizeau experiments. He defended the standpoint, called Lorentz Æther theory, based on his theory of electrons. This theory of stationary Æther removed Luminiferous Æther drag. In 1895, Lorentz predicted the effects of dispersion later measured by Pieter Zeeman in his 1914-27 experiments [14].

George Francis FitzGerald, impressed by Oliver Heaviside's calculations [15] which included deformations of magnetic and electric fields surrounding a moving charge and the effects of it entering a denser medium, today known as the Cherenkov radiation, wrote [16]

> [T]he length of material bodies changes, according as they are moving through the ether or across it, by an amount depending on the square of the ratio of their velocities to that of light.

Another example is the 1915' paper of William Jasper Spillman, the founding father of agricultural economics who independently rediscovered the Mendel laws of genetics, in which reasoning is based on the *Fundamental Assumptions* involving Æther

> **The ether.** - The ether is assumed to exist in the interstices of matter and in open space. It is assumed to be capable of distortion by finite force, and to oppose such distorting force with an equal opposite force. In other words, every point of the ether has a position which it normally occupies, and when removed from that position tends forcibly to return to it, the force being proportional to the distortion. It is further assumed that distortion at a given point in the ether tends to become distributed in the surrounding ether according to the law of inverse squares, and that such distribution occurs at a finite rate (the velocity of light).

The electron. - It is assumed that in the immediate vicinity of the electron there is a region of maximum permanent ether distortion, the distortion at other points in the surrounding ether varying inversely as the square of the distance from the center of the electron, and that pressures are transmitted at a higher velocity through distorted than through non-distorted ether.

The permanent distortion of the ether in the vicinity of the electron may be conceived of as a pushing back of the ether radially from the center of the electron, as if an impenetrable and inelastic body were injected into the midst of an elastic body; or it may be conceived of as being circular, in two hemispheres facing each other, and opposite in direction in the two hemispheres.

In explaining the phenomena of inertia, electric currents, magnetism, chemical affinity, and radiant energy, the above alternative assumptions concerning the character of the distortion lead to essentially similar lines of reasoning, but the treatment on the assumption of radial distortion is very much simpler. For this reason it is used here. In the case of static electricity only the assumption of circular distortion, opposite in direction in the two kinds of electric elements, will explain the facts. The extension of the theory to static electricity is left for future treatment. It must be remembered, however, that the development of the theory on the basis of radial distortion differs only in detail, not in principle, from that on the basis of circular distortion.

The Atom. - The atom is assumed to consist, at least in part, of a Saturnian system of electrons in rapid orbital motion. It will be shown that such orbital motion, with the assumptions here made, would give rise to a pressure in the ether such as Newton showed would account for gravitation.

Physicists are agreed that the phenomena of inertia, the electric current, magnetism, and possibly also chemical affinity are probably related to each other in such manner that when we find the explanation of one of them this explanation will also throw light on the others.

Æther Drag

The most intriguing phenomena related to the concept of Æther, which decided the 20th century fate of this concept, is Æther drag/drift/wind. Both Lorentz Æther theory and Einstein's Special Relativity, the theories of stationary Æther, are interpreted are the theories propagating extraordinary strong antagonism to the Æther wind. The essence of the history of the 20th century physics and the milestone of empirical negating of Æther is the Michelson–Morley experiment which purpose was to detect Earth's motion with respect to Luminiferous Æther by comparison of speed of light in diverse directions with respect to Earth. On the one hand, in the theory of stationary Æther propagated by Hendrik Antoon Lorentz this attempt manifestly and straightforwardly negated existence of Æther drag because contradiction with the result of the Fizeau experiment. On the other hand, the Michelson–Morley experiment is consistently explained and interpreted within Einstein's Special Relativity with no reference to Æther drag.

There were performed a number of other experiments (See Ref. [17]) having the purpose to verify empirically the question of the existence of Æther drift: 1903' Frederick Thomas Trouton and Henry Robertson Noble, 1908' F.T. Trouton and Alexander Oliver Rankine, 1913' Georges Sagnac, 1925' A.A. Michelson and Henry Gordon Gale, 1932' Roy J. Kennedy and Edward M. Thorndike, and 1935' Gustaf Wilhelm Hammar. Their results suffered the fate analogous to the Michelson–Morley experiment, *i.e.* were interpreted as negation of Æther wind and Æther in general. Transparent discussion of some of these results can be found in the book of Wolfgang Pauli [18].

The Trouton–Noble experiment was strict realization of the idea due to G.F. FitzGerald. He concluded that motion of a charged flat condenser through the Æther should result in its perpendicular orientation to the motion. The experimenters detected the lack of the relative motion to the Æther, what was interpreted as the negative result.

The Trouton–Rankine experiment was performed for detection of "preferred frame" which would be the syndrome of existence of the Luminiferous Æther. The measurement assumed that the length contraction produces a measurable effect in the rest frame of the object observed in another frame. Both Special Relativity and Lorentz Æther theory predicted that length contraction is non measurable. F.T. Trouton and A.O. Rankine, however, applying the Ohm law and the Maxwell equations, theoretically predicted that the effect, change of resistance, is measurable in the laboratory frame. The experimenters used the Wheatstone bridge, and the change of resistance was not detected.

Georges Sagnac applied a rotating interferometer, and observed a dependence of the interference fringe position on angular velocity. The result of this experiment was theoretically predicted in 1911 by Max von Laue [19], who proved manifestly its consistence with Special Relativity and numerous models of stationary Æther, including the Lorentz Æther theory. Factually, results of this experiment confirmed existence of the stationary Æther, even in the sense of the Lorentz Æther theory, but Laue's calculations were interpreted as the reflection of Special Relativity correctness. The Sagnac effect is probably the most positive result for the Æther drag, and was also treated as the straightforward demonstration of existence of the Æther. Moreover, the Sagnac effect was constructively explained also within General Relativity which was claimed by Einstein to be the theory of non-physical Æther.

The Michelson–Gale experiment modified the Michelson–Morley experiment by application of enlarged Sagnac's ring interferometer. The purpose was to find out the relation between the Earth rotation motion and the light propagation in the vicinity of the Earth. Similarly as in the case of the Michelson–Morley experiment, the Michelson–Gale version compared the light from a single source after two-directional travel. The difference was replacement of the two of the Michelson–Morley arms with two different rectangles. The obtained results were compatible with both Special Relativity and models of stationary Æther. However, Lorentz Æther theory contradicted the Michelson–Morley experiment and, therefore, the Michelson–Gale version was interpreted also as the confirmation of correctness of Special Relativity.

The investigators of the Kennedy–Thorndike experiment also interpreted their results in terms of relativistic effects. They modified the Michelson–Morley experiment by application of the interferometer's in which one arm is very short in comparison with the second arm. The Michelson–Morley experiment constructively verified the length contraction hypothesis, whereas in the Kennedy–Thorndike version time dilation was straightforwardly examined. The much shorter interferometer's arm and maximal stabilization of the apparatus enabled to verify existence of a specifical fringe shift, which theoretically should change the light frequency. According to theoretical predictions the shift resulted from a change of speed of the Earth with respect to Æther, and this would result in changes of the time of light travel. Albeit, the shift was not detected, and this was interpreted as the confirmation of Special Relativity correctness.

The purpose of the Hammar experiment was to show the emergent nature of the Æther drag asymmetry by making use of massive lead blocks on both sides of only one of the Michelson–Morley interferometer arm. The idea of this investigation was similar to the tests performed by Oliver Joseph Lodge. The blocks absence should result in the equal affect of both arms by Æther, while in the blocks' presence the one arm should be affected. Hammer reported independence of fringe displacements on the blocks' absence/presence, what was the argument against the Æther wind.

The conclusion is unambiguous. Stigmatization of the existence of Æther was based on negative verification of the existence of Æther wind. Moreover the Ockham razor method was applied: the physical interpretation of the experimental data *ad hoc* applied absence of the Æther drag as the crucial symptom for absence of Æther. The purpose of such a strategy was the interpretation of experimental data done in an intentional way supporting elimination of Æther and related phenomena from the description of physical reality. In other words, exclusion of one phenomenon was performed by imposing of absence of another. However, the matter is much more sophisticated because, in general, the Æther drag must not be related to the Æther concept – the idea of Æther can be independent on the idea of the Æther drag.

It is worth stressing that the calculated effects of possible existence of the Æther drag, obtained from diverse Æther models, have been confronted with experimental data many times. Since 1904 Dayton Clarence Miller and collaborators [20] performed over 200,000 perspective empirical investigations devoted to the verification of the Æther drag. Basing on these experimental data, D.C. Miller strongly propagated an existence of the Æther wind whose effects, in his vision, were much smaller than the analogous predictions performed on the basis of the stationary Æther model. Miller's work inspired plenty of scholars, scientists and researchers (Cf. *e.g.* Ref. [21]). In 1955 Robert Sherwood

Shankland et al [22] devaluated the results of Miller's research, and claimed that only the statistical fluctuation due to the local temperature conditions and systematic error generated the effects predicted by D.C. Miller. However, in 1983 W. Broad et al suggested to review with attention the results received by D.C. Miller, and straightforwardly negated the refusal due to R.S. Shankland. Interestingly, there are consistent applications of the Æther drag idea to explain intriguing phenomena. For example, Richard A. Müller [23] constructively described anisotropies of the Cosmic Microwave Background Radiation by application of the concept of the Æther wind.

Einstein's Visions

Detailed discussion of the various Einstein's opinions was recently performed by Ludwik Kostro [24]. It looks like that Einstein, in the name of the innovatory character of Special Relativity, in full inconsistency with his true personal point view expressed in a number of his works, is stigmatized as the principal destructor of the idea of Æther in the 20th century theoretical physics (Cf. *e.g.* the discussion in Ref. [25]). Such an opinion is mostly based on the intentional interpretation of the very small fragment of his 1905' Special Relativity paper [26], which can be found in introduction to this paper

> The introduction of a "luminiferous ether" will prove to be superfluous inasmuch as the view here to be developed will not require an "absolutely stationary space" provided with special properties, nor assign a velocity-vector to a point of the empty space in which electromagnetic processes take place.

It is worth stressing that in 1907 Max von Laue [27] derived the Fresnel drag coefficient from the Einstein velocity-addition formula. Nevertheless, Einstein was also stigmatized as a resurrector of Æther. In famous *Sidelights of Relativity* cited in Ref. [28] one finds

> We may say that according to the general theory of relativity space is endowed with physical qualities; in this sense, therefore, there exists an aether. According to the general theory of relativity space without aether is unthinkable; for

> in such space there not only would be no propagation of light, but also no possibility of existence for standards of space and time (measuring-rods and clocks), nor, therefore, any space-time intervals in the physical sense. But this aether may not be thought of as endowed with the quality characteristic of ponderable media, as consisting of parts which may be tracked through time. The idea of motion may not be applied to it.

Moreover, in the article *Concerning the Aether* Einstein manifestly supported consistency of the idea of Æther

> When we speak here of aether, we are, of course, not referring to the corporeal aether of mechanical wave-theory that underlines Newtonian mechanics, whose individual points each have a velocity assigned to them. (...) Instead of 'aether', one could equally well speak of 'the physical quantities of space'. (...) So we are effectively forced by the current state of things to distinguish between matter and aether, even though we may hope that future generations will transcend this dualistic conception and replace it with a unified theory, as the field theoreticians of our day have tried in vain to accomplish. (...) It is usually believed that aether is foreign to Newtonian physics and that it was only the wave theory of light which introduced the notion of an omnipresent medium influencing, and affected by, physical phenomena. (...) Newtonian mechanics had its 'aether' in the sense indicated, albeit under the name 'absolute space'. To get a clear understanding of this and, at the same time, to explore more fully the concept of aether, we must take a step back. (...) The kinematics, or phoronomy, of classical physics had as little need of an aether as (physically interpreted) Euclidean geometry has. (...) We will call this physical reality which enters the Newtonian law of motion alongside the observable, ponderable real bodies, the aether of mechanics. The occurrence of centrifugal effects with a (rotating) body, whose material points do not change their distances from one another, shows that this aether is

not to be understood as a mere hallucination of the Newtonian theory, but rather that it corresponds to something real that exists in nature. (...) The mechanical aether - which Newton called 'absolute space' - must remain for us a physical reality. Of course, one must not be tempted by the expression aether into thinking that, like the physicists of the 19th century, we have in mind something analogous to ponderable matter. (...) When Newton referred to the space of physics as 'absolute', he was thinking of yet another property of what we call here aether. Every physical thing influences others and is, it its turn, generally influenced by other things. This does not however apply to the aether of Newtonian mechanics. For the inertia-giving property of this aether is, according to classical mechanics, not susceptible to any influence, neither from the configuration of matter nor anything else. Hence the term 'absolute'. (...) Viewed historically, the aether hypothesis has emerged in its present form by a process of sublimation from the mechanical aether hypothesis of optics. After long and fruitless efforts, physicists became convinced that light was not to be understood as the motion of an inertial, elastic medium, that the electromagnetic fields of Maxwell's theory could not be construed as mechanical. So under the pressure of this failure, the electromagnetic fields had gradually come to be regarded as the final, irreducible physical reality, as states of the aether, impervious to further explanation. (...) While at least in Newtonian mechanics all inertial systems were equivalent, it seemed that, in the Maxwell-Lorentz theory, the state of motion of the preferred coordinate system (at rest with respect to the aether) was completely determined. It was accepted implicitly that this preferred coordinate system was also an inertial system, *i.e.* that the principle of inertia [Newton's first law] applied relative to the electromagnetic aether. (...) No longer was a special state of motion to be ascribed to the electromagnetic aether. Now, like the aether of classical mechanics, it resulted not in the favoring of a particular state of motion, only the favoring of a partic-

> ular state of acceleration. Because it was no longer possible to speak, in any absolute sense, of simultaneous states at different locations in the aether, the aether became, as it were, four dimensional, since there was no objective way of ordering its states by time alone. According to special relativity too, the aether was absolute, since its influence on inertia and the propagation of light was thought of as being itself independent of physical influence. (...) Thus geometry, like dynamics, came to depend on the aether. (...) Thus the aether of general relativity differs from those of classical mechanics and special relativity in that it is not 'absolute' but determined, in its locally variable characteristics, by ponderable matter. (...) On the one hand, the metric tensor, which codetermines the phenomena of gravitation and inertia and, on the other, the tensor of the electromagnetic field appear still as different expressions of the state of the aether, whose logical independence one is inclined to attribute rather to the incompleteness of our theoretical edifice than to a complex structure of reality. (...) But even if these possibilities do mature into an actual theory, we will not be able to do without the aether in theoretical physics, that is, a continuum endowed with physical properties; for general relativity, to whose fundamental viewpoints physicists will always hold fast, rules out direct action at a distance. But every theory of local action assumes continuous fields, and thus also the existence of an 'aether'.

In this manner, the actual point of view of Albert Einstein was that Æther is the core fundament of physics. In fact, he never neglected and negated the existence of Æther, and moreover he constructively developed this concept in General Relativity theory. In the famous book coauthored with the Polish scholar Leopold Infeld [29], Einstein presented the history of physics with respect to the development of the concept of Æther. Several significant fragments are cited below

> Our picture of ether might very probably be something like the mechanical picture of a gas that explains the propagation of sound waves. It would be much more difficult to

form a picture of ether carrying transverse waves. To imagine a jelly as a medium made up of particles in such a way that transverse waves are propagated by means of it is no easy task. (...) Yet we know from mechanics that interstellar space does not resist the motion of material bodies. The planets, for example, travel through the ether-jelly without encountering any resistance such as a material medium would offer to their motion. If ether does not disturb matter in its motion, there can be no interaction between particles of ether and particles of matter. Light passes through ether and also through glass and water, but its velocity is changed in the latter substances. How can this fact be explained mechanically? Apparently only by assuming some interaction between ether particles and matter particles. We have just seen that in the case of freely moving bodies such interactions must be assumed not to exist. In other words, there is interaction between ether and matter in optical phenomena, but none in mechanical phenomena! This is certainly a very paradoxical conclusion! (...) We may still use the word ether, but only to express some physical property of space. This word ether has changed its meaning many times in the development of science. At the moment it no longer stands for a medium built up of particles. Its story, by no means finished, is continued by the relativity theory. (...) For the time being, we shall continue to believe that the ether is a medium through which electromagnetic waves, and thus also light waves, are propagated, even though we are fully aware of the many difficulties connected with its mechanical structure.

Theoretical Objections

It is evident from the works of Albert Einstein that all his attempts were strongly concentrated on the consistent explanation of a whole Universe by application of the concept of «non-physical» Æther. In fact, in Special Relativity Einstein implemented to the physical description the concept of the hyperbolic geometry of the Minkowski spacetime which gave the full consistency with the Lorentz transformations and their predictions. Recall that Hendrik Antoon Lorentz deduced these transformations in the rigorous way based on the specific concept of Æther. It seems to be clear that Einstein manifestly replaced the idea of Æther by the concept of hyperbolic space-time geometry, and unexpectedly reinvented the idea of Æther in another formulation, as the manifestation of the dynamics of the Riemannian geometry in General Relativity, when his position in science was consolidated and Special Relativity was accepted as a plausible theory of physics. It is highly probable that separation from the highly impacting scientific personality and authority of H.A. Lorentz was the crucial reason for such a change in Einstein's official point of view. The phrase *non-physical* used in the context of the idea of Æther looks like a diplomatic trick than plausible theoretical physics. Einstein, in fact, understood that rejection of Æther from theoretical physics does not give rise to progress but blocks the development of the theory.

The negating of the existence of Æther, which is definitely irrelevant to General Relativity theory, manifestly supports Special Relativity theory. The empirical results, however, have no absolute physical interpretation and can be easily reinterpreted as the support of the existence of Æther as well as the phenomenon of the Æther drag. Perhaps, the concept of electromagnetic Luminiferous Æther is inadequate, but in general it does not exclude another realization of the idea of Æther. For instance, Æther treated as the primordial cause can exist only initially, say in the very Early Universe, and must not exist in later stages of the evolution of the Universe. An excellent example of the theoretical investigation of Æther was done more than 40 years ago. In mid-1920's Élie Cartan [30] formulated General Relativity in terms of the Newtonian dynamics by application of the language of connections. In 1966 straightforward calculation done by Andrzej Mariusz Trautman [31] showed that the Einstein field equations are the particular case of the equations of the Newtonian gravitation coupled to a thing which Trautman called the Luminiferous Æther. Soon after these results Charles William Misner in *Gravitation*, an excellent textbook of General Relativity coauthored with Kip Stephen Thorne and John Archibald Wheeler [32], axiomatized Trautman's approach to show that the Newtonian dynamics consistently joints General Relativity theory with the idea of Æther realized by Cartan and Trautman.

Paul Charles William Davies [33] interviewed John Stewart Bell, one of founders of quantum physics and originator of Bell's theorem regarding hidden variable theories. J.S. Bell straightforwardly expressed the opinion that the concept of Æther can be a particularly helpful tool in resolving many problems of quantum mechanics, among others the Einstein–Podolsky–Rosen paradox regarding measurements of microscopic systems, by involving a reference frame in which signals go faster than light. In Bell's vision the Lorentz length contraction is correct but inconsistent with Special Relativity, but can emerge in the theory of Æther which is consistent with the results of the Michelson–Morley experiment. Bell manifestly expressed the wrongness of rejection of the concept of Æther from modern physics, and proposed resurrection of Æther emphasizing that a number of presentely unsolvable issues can be straightforwardly and easily solved by taking into account the existence of Æther. A reader interested in more detailed analysis of Æther by J.S. Bell is advised to refer to Ref. [34].

Another attempt, which is a good candidate for the constructive model of Æther, was made by Richard Phillips Feynman [35]. Feynman proposed that the partial-differential equations are able to describe classical macroscopic motion of the so-called X-ons which are certain very small entities. Strictly speaking, the medium formed by such entities can be treated as the model of Æther. Similarly, the approach to electrodynamics based on the action on distance, proposed by J.A. Wheeler and R.P. Feynman [36], gives certain hopes for resurrection of the idea of Æther for description of physical reality.

In this manner, the situation of the concept of Æther is apparently non-established. Moreover, the intentional interpretation of the experimental data, which is widely applied by the antagonists of Æther, enables to consider the existence of Æther as the artefact of physical knowledge. On the other hand, as was suggested by J.S. Bell, Æther is probably one of the most helpful tools for the constructive and consistent explanation of many of presently inexplainable phenomena and effects. A number of heightened attempts, manifestly rejecting Æther from theoretical description of Nature, is based on the methodology which is in itself a certain intentional selection and propagates the preferred interpretation as the absolute interpretation. It should be emphasized that a whole known argumentation supporting the non-existent nature of Æther can be easily reevaluated in the opposite direction, so that the arguments of the antagonists of Æther will become invalid by using both the theoretical as well as the experimental arguments. The question is whether the physical truth should be diplomatic technocracy or rather reformable democracy. Diplomatic technocracy offers only the "rigid body" approach and does not pay attention to the actual state of affairs. Reformable democracy is open for changes but also too weak to establish the truth. Maybe constitutional monarchy is more adequate for the present-day state of theoretical physics. Of course, Nature is governed neither by technocracy nor any other politics, and says the truth with no regard to any system and ideology. The problem of physics is to describe the actual state of affairs with using some preferred background. The concept of Æther, in this manner, has the same rights as any other idea and, from theoretical physics point of view, negation of its existence is unjustified.

Dirac Æther

In 1951 Paul Adrien Maurice Dirac, regarded by Albert Einstein as the founder of relativistic quantum mechanics and quantum field theory, concluded existence of Æther in the light of the nature of four-velocity in the context of theory of electrons following from his new electrodynamics [37]

> It was soon found that the existence of an æther could not be fitted in with relativity, and since relativity was well established, the æther was abandoned. (...) If one reexamines the question in the light of present-day knowledge, one finds that the æther is no longer ruled out by relativity, and good reasons can now be advanced for postulating an æther. (...) at the present time it needs modification, because we have to apply quantum mechanics to the æther. The velocity of the æther, like other physical variables, is subject to uncertainty relations. For a particular physical state the velocity of the æther at a certain point of space-time will not usually be a well-defined quantity, but will be distributed over various possible values according to a probability law obtained by taking the square of the modulus of a wave func-

> tion. We may set up a wave function which makes all values for the velocity of the æther equally probable. Such a wave function may well represent the perfect vacuum state in accordance with the principle of relativity. (...) A thing which cannot be symmetrical in the classical model may very well be symmetrical after quantization. This provides a means of reconciling the disturbance of Lorentz symmetry in space-time produced by the existence of an æther with the principle of relativity. (...) we may very well have an æther, subject to quantum mechanics and conforming to relativity, provided we are willing to consider the perfect vacuum as an idealized state, not attainable in practice. (...) We have now the velocity at all points of space-time, playing a fundamental part in electrodynamics. It is natural to regard it as the velocity of some real physical thing. Thus with the new theory of electrodynamics we are rather forced to have an æther.

One year later Nature magazine published interesting looking polemics between L. Infeld and P.A.M. Dirac [38]. Leopold Infeld combined the formulas, derived by Dirac, in a certain intentional way and applied rather laconic then logical arguments to point out that the new electrodynamics does not need the concept of Æther if "all" its conclusions will be taken into account. Infeld, as the typical representative of the scholastics highly influenced by the Soviet Union's ideology, did not precise what means in his opinion the word "all" applied to such a context. In other words, factually even Leopold Infeld, who required inclusion of "all" aspects of the thing, did not take into account "all" possible conclusions of the Dirac electrodynamics and intentionally selected only the ones which were adequate for elimination of the realization of the idea of Æther proposed by Paul Dirac. In this manner, Infeld used some preferred philosophical interpretation to argue his own beliefs about Æther, which could be influenced by the rules of the People's Republic system, and did not focus attention on the physical aspects of the Dirac Æther. Moreover, such a negative opinion was also straightforwardly opposite to the long-term research of Albert Einstein, which *nota bene* was supported by Infeld several years earlier. It should be emphasized that the deduction process arguing the existent nature of

Æther performed by Paul Dirac applied purely formal nature of the Hamiltonian approach to the Maxwell electrodynamics, *i.e.* the constraints and the action. Recall that over 25 years earlier Paul Dirac discovered the linkage between classical and quantum mechanics by the Poisson brackets correspondence. The Maxwell electrodynamics was unsatisfactorily formulated because the correspondence between quantum and classical based on the Poisson brackets works for only the Hamiltonian version of a classical theory. Dirac performed the Hamiltonian formulation for the case of the Maxwell electrodynamics, and discerned the existence of Æther in such a theory. Applying the typical arguments of the Ockham razor based on personal beliefs, Leopold Infeld manifestly discredited the purposes and efforts due to Paul Dirac.

Such a situation strengthened the strategy of Dirac who was a follower of the revisionist approach with respect to even the established and accepted physical knowledge. The selfish and unnatural selection performed by Infeld quickly obtained an adequate and constructive reply due to Dirac. The reply is brief and can be cited entirely

> Infeld has shown how the field equations of my new electrodynamics can be written so as not to require an æther. This is not sufficient to make a complete dynamical theory. It is necessary to set up an action principle and to get a Hamiltonian formulation of the equations suitable for quantization purposes, and for this the æther velocity is required.
>
> The existence of an æther has not been proved, of course, because my new electrodynamics has not yet justified itself. It will probably have to be modified by the introduction of spin variables before a satisfactory quantum theory of electrons can be obtained from it, and only after this has been accomplished will one to be able to give a definite answer to the æther question.

The method applied by Leopold Infeld was the tip of the iceberg and reflected the true countenance of the regional standpoint based on the fossilized traditional beliefs. At that time the Soviet school of physics dominated European science and the Marxist–Leninist scholastics was one of the most fashionable streams. Such people like Paul Dirac and Albert Einstein were the pioneers whose wanted to change this manifestly irrational state of affairs. In fact, the opinion due to Infeld about the realization of the idea of Æther propagated by Dirac straightforwardly crossed the efforts of both the Nobel laureates. It should be emphasized that for fulfillment of the ideological expectations Infeld decided to negate his own views published several years earlier together with Albert Einstein. Such a controversy was too serious for the science of the region and, in fact, resulted in some kind of hackwork within the Polish physics, which also now possesses a strong impact. Evidently, the lack of the Nobel Prizes in natural sciences in Poland is the most gross manifestation of the fossilized reasoning and approach to science, and labels the stagnation of the civilization. Recall that Leopold Infeld was the only one of 11 signatories to 1955's Russell–Einstein Manifesto who never received a Nobel Prize. His creative efforts with respect to Poland started when in 1950, after a McCarthyite campaign against him organized in the press and in parliament which claimed that he knew atomic secrets, Infeld left Canada where by 12 years worked at the University of Toronto. He came back to communist Poland and decided to help in reconstruction of the Polish science which during World War II lost few generations of scholars. Admittedly his totalitarian approach to science resulted in the tremendous contribution to the Warsaw school of physics, but also transformed this school into the sanctuary of the Soviet scholastics.

Zero Point Energy & Planck Scale

In 1900 Max Planck [39] published the formula for the energy of a single vibrating atom. Several years after, in 1913, Albert Einstein in collaboration with Otto Stern [40] modified the Planck formula by involving of the concept of the cosmic heat bath, directly related to the universal frequency field associated with the Zero Point Energy. From the modern physics point of view one can say that Einstein and Stern renormalized the Planck energy formula at absolute zero temperature by using the residual oscillating energy. In fact, the Einstein–Stern concept of the cosmic bath heat, and the resulting Zero-Point Frequency field, is in itself a certain realization of the idea of Æther and consistently elucidates certain experimental data, like *e.g.* the Casimir effect, the Lamb shift.

Factually, the concept of Æther investigated and developed by James Clerk Maxwell has never been experimentally refuted. In the context of quantum geometrodynamics proposed by J.A. Wheeler [41], which is the pioneering quantization of General Relativity, the Maxwell vacuum can be regarded as quantum foams or, in other words, as a subquantum sea of Zero Point Energy fluctuations. Space-time warps, known as wormholes, follow from tremendous densities of local energy due to the high energetic modes of the universal Zero-Point Frequency field. Wormholes, the notion originally coined by John A. Wheeler, are the tunnels transmitting electricity between two separate places or, in a more general context, between different universes belonging to the configurational space of General Relativity known as the Wheeler superspace. Wheeler proposed to make the reasoning in terms of black and white mini holes, *i.e.* primitive charged particles, provoking the mechanism of electron-positron pairs production. According to his view, the wormholes are related to the local space, and the electricity goes orthogonally through our universe from a fourth dimension.

Interestingly, Harold Aspden [42] proposed the hadronic model based on near-balanced continuum and quons, *i.e.* massless particles of Æther giving a charge and condensing electron-positron pairs. Such a line of thinking was prolonged also by Harold E. Puthoff who showed that by application of the stochastic electrodynamics and the Bohr theory of an atom the ground state of the hydrogen atom results from a dynamic equilibrium between radiation emitted during an accelerated motion of an electron on the ground-state orbit and radiation absorbed from Zero-Point fluctuations of the background vacuum electromagnetic field, and in this way solves the collapse problem of the Bohr atom due to an accelerated orbital motion [43]. In another paper [44] he developed the alternative model of quantum gravity proposed by Andrei Dmitrievich Sakharov [45] in which gravity is induced by the Zero-Point fluctuations of the vacuum similarly to van der Waals and Casimir forces. Puthoff realized this hypothesis by a point-particle Zero-Point fluctuations model in which gravitational mass and the related effects follow from Zitterbewegung, *i.e* the particle motion induced by electromagnetic Zero-Point fluctuations. H.E. Puthoff studied also the possible origins of vacuum electromagnetic Zero-Point energy and showed that generation of them through quantum-fluctuation motion of charged particles that constitutes matter, with the assumption that the Zero-Point field drives particle motion and this motion in turn generates the

Zero-Point energy spectrum, is equivalent to the frequency-cubed spectral distribution and thus indicates dynamic-generated processes for Zero-Point fields [46]. Moreover, Puthoff together with Alfonso Rueda and Bernhard Haisch formulated inertia as a Zero-Point field Lorentz force [47].

Quantum mechanics is also considered as a theory of the Æther formulated in terms of quantum foams. Such a point of view leads to fluctuations of small scales generating quick processes of creation and annihilation of particle pairs. The unifying cosmological fifth element is known as Dark Energy or quintessence and has been considered as the Zero-Point field or quantum vacuum and first identified with Æther by Burra Gautam Sidharth [48]. Recently, also the Æther theory originally proposed by Albert Einstein, which is generally covariant generalization of General Relativity theory describing space-time and endowed both a metric and a unit time-like vector field manifestly violating Lorentz invariance, was developed by several authors (Cf. *e.g.* the papers in Ref. [49]). It should be emphasized that in 2005 Theodore A. Jacobson and Renaud Parentani published the article [50] in which they pointed out that the behaviour of sound waves in a fluid is very similar to the behaviour of light waves in curved space-time, like around black holes which have acoustic counterparts. Such an approach of associating the physical features of space-time to the physical qualities of a fluid leads to consistent solution of certain fundamental problems in the theory of photon radiation, and in itself constitutes a specific realization of the concept of Æther similar to the realizations proposed by William Thomson the 1st Baron Kelvin, and James Clerk Maxwell, *i.e.* the pre-Einsteinian physics.

Multiverse

Epicurus and Eastern Cosmologies

Epicurus [51] was probably the first ancient philosopher who openly propagated the concept of Multiverse. In his Letter to Herodotus one finds manifestly expressed his standpoint

> Moreover, there is an infinite number of worlds, some like this world, others unlike it. For the atoms being in-

> finite in number, as has just been proved, are borne ever further in their course. For the atoms out of which a world might arise, or by which a world might be formed, have not all been expended on one world or a finite number of worlds, whether like or unlike this one. Hence there will be nothing to hinder an infinity of worlds.
>
> (...) After the foregoing we have next to consider that the worlds and every finite aggregate which bears a strong resemblance to things we commonly see have arisen out of the infinite. For all these, whether small or great, have been separated off from special conglomerations of atoms; and all things are again dissolved, some faster, some slower, some through the action of one set of causes, others through the action of another.
>
> And further, we must not suppose that the worlds have necessarily one and the same shape. For nobody can prove that in one sort of world there might not be contained, whereas in another sort of world there could not possibly be, the seeds out of which animals and plants arise and all the rest of the things we see.

Also Al-Qur'an, the holy book of Islam, directly refers to many worlds. Sūratu Al-Fātihah, called *The Seven Verses of Repetition* sounds [52]

> 1 In the name of Allah, Most Gracious, Most Merciful.
> 2 Praise be to Allah, the Cherisher and Sustainer of the worlds!
> 3 Most Gracious, Most Merciful.
> 4 Master of the Day of Judgement.
> 5 Thee we do worship, and Thine aid we seek.
> 6 Show us the straight way,
> 7 The way of those on whom Thou hast bestowed Thy Grace, those whose (portion) is not wrath, nor of those who go astray.

Al-Qur'an straightforwardly inspired the muslim polymath Fakhr Al-Din Al-Razi (Rahmatullah Alayhi), who rejected the Aristotelian-Avicennian single universe revolving around a single world. Adi Setia [53] referred to fragments of the unpublished manuscript *al-Matalib al-'Aliyah* of Imam Razi. In Setia's work one can find the fragment

> It is established by evidence that there exists beyond a void without a terminal limit, and it is established as well by evidence that God Most High has power over all contingent beings. Therefore, he the Most High has the power to create a thousand thousand worlds beyond this world such that each one of those worlds be bigger and more massive than this world as well as having the like of what this world has of the throne, the chair, the heavens,and the earth, and the sun and the moon. The arguments of the philosophers for establishing that the world is one are weak, flimsy arguments founded upon feeble premises.

Multiverse is also present in Puranas [54], the generic texts of Hinduism, Jainism or Buddhism. For example in Bhagavata Purana 9.4.56 one finds the direct reference to multiple universes

> Lord Śiva said: My dear son, I, Lord Brahmā and the other devas, who rotate within this universe under the misconception of our greatness, cannot exhibit any power to compete with the Supreme Personality of Godhead, for innumerable universes and their inhabitants come into existence and are annihilated by the simple direction of the Lord.

According to Hindu scriptures the religious tradition know as Hinduism arises from millions of years experience. Hinduism offers plenty of creation mechanisms, but the only one is consistent with the fundamental Hindu belief – reincarnation. Present and future quantity of both the worlds and multiple universes is higher then the number of drops of the water in the holy river Ganges, and universes are the effect of the work of Lord Brahmā the Creator who is maintained by Lord Vishnu the Preserver and eliminated and recreated by Lord Śiva the Destroyer and the Re-creator. Brahmā, Vishnu and Śiva are both all forms of and the part of the Supreme One which is behind and beyond all. The effect of the annihilation of each universe is a vast ocean in which the great snake Ananta floats and transports Lord Vishnu whose navel is the place of springing out a lotus flower, which is the national flower of India symbolizing divinity, fertility, wealth, knowledge and enlightenment, from which comes Lord Brāhma - the source of all creation. Brāhma grows lonely and splits himself in two to create a male and a female, becomes unified again and creates human beings and the other living things. Some tradition says different parts of Lord Brāhma's body creates everything - animals and people come from his mouth, arms, thighs and feet. Anyway, everything is part of the Supreme One. All universes, whose time of life is beyond imagination, are destroyed every night by Lord Śiva when Lord Brāhma sleeps and are recreated every morning when he awakes. Any universe is completely annihilated by Śiva when Brāhma of a universe has lived a lifetime, and everything disappears into the Supreme One. Chaos and water exist alone for an unimaginable time. Then Lord Vishnu once again emerges floating on the vast ocean, from him comes forth Lord Brāhma of the new universe and such a cycle, known as reincarnation, continues infinite number of times and strongly influences the lifestyle. The important notions of reincarnation are dharma, part of that duty is to respect and care for all what is alive, and karma which emerges from what we do. Good karma is the result of doing of our duty and the next life can be directly related to the actions of our present life. Because everything belongs to the Supreme One thus every living thing is equally important in the reincarnation process.

Modal Realism

In the most general formulation the Multiverse hypothesis takes into account the scenario in which there exists, numerable or innumerable, collection of multiple possible universes. These worlds may include even a whole Nature, including the concepts of space, time, matter, light, and their straightforward psychological impacts related to the concept of mind or soul. Multiverse as the strict scientific concept was introduced by American psychologist and philosopher William James [55], who considered as the essential element of human psychology the influence of divine and mystic experiences. The structure of Multiverse, which in fact defines the nature of a possible universe as well as the various relationships between distinguishable universes, is not rigidly established and manifestly depends on a model of Multiverse.

From the philosophical point of view the concept of Multiverse naturally belongs to the logical system investigated by Ludwig Josef Johann Wittgenstein in his magnum opus *Tractatus Logico-Philosophicus* [56].

In Wittgenstein's logic the logical truth is defined as a statement true in all possible worlds or the quantity which remains unchanged in all possible interpretations, and a fact is only true in this world as it has historically unfolded. This ontological system continues the program originally investigated by Friedrich Ludwig Gottlob Frege [57], but manifestly neglects the Frege axiom whose semantic form is *all true (and, similarly all false) sentences describe the same state of affairs, that is, they have a common referent.* The pioneering formalization of Wittgenstein's *Tractatus* was performed by Polish logician Roman Suszko [58], and resulted in the so-called Non-Fregean Logic in which there are no theorems asserting how many semantic correlates of sentences there can be. Recently this formalization was expanded by his pupil Mieczysław Władysław Omyła [59] onto the logic connecting situations and objects. In his works Omyła many times refers to the works of Bogusław Wolniewicz (Cf. *e.g.* Ref. [60]) studying the ontology of situations which for several years inspired mathematicians working in the theory of conditionally distributive lattices (Cf. *e.g.* Ref. [61]). Wolniewicz's works base on a specific realization of the Multiverse hypothesis, the so-called Logical Atomism, which is characteristic for the Wittgenstein logic. It should be emphasized that Wittgenstein's metaphysics leads to the existence of identical objects in diverse worlds.

The counterpart theory of David Kellogg Lewis [62] showed that identical objects existent in the various worlds should be regarded as similar rather than identical. D.K. Lewis elucidated the role of probability and hypothetical statements in differentiation between similarity and identity. His version of modal realism gives rise to all possible worlds possessing equally realistic character like the actual world. In *Parts of Classes*, Lewis efficiently applied the pluralistic point of view to the foundations of mathematics. He considered such essential issues as set theory, the Peano arithmetic, and the Gödel incompleteness theorems to mereology and plural quantification. In the approach proposed by Lewis such a word like "actual" is a merely indexing procedure, which labels the position within a world. He proposed also the original definition of truth, strictly based on the Multiverse nature of modal realism, which states that things are necessarily true when they are true in all possible worlds. In other words, according to Lewis truth is a kind of invariant quantity with respect to choice of a world.

Lewis was not the first philosopher studying possible worlds, but he contributed comparatively essential ideas about the equal concreteness of all possible worlds, and created the concept of the world in which the existence of the object is no more real than the existence of this object in different possible worlds. Similarly as in the case of the concept of Æther the concept of Multiverse manifestly violates the maxim due to English theologian and a member of the the mendicant Order of Friars Minor (Franciscan) William Ockham, known as the Ockham Razor. This philosophical cutoff says "no" to existence of multiply entities, because the Multiverse hypothesis is manifestly beyond being a necessary explanation of the facts which theories want to describe. It should be emphasized that the straightforward ramification of the Lewis theory is that branching worlds differ in important ways from diverging worlds. This aspect is significant for the so-called Many-Worlds Interpretation of quantum mechanics which we shall discuss below.

Different possible worlds are propagated also by the modern American philosopher and logician Saul Aaron Kripke [63]. He has described modality by application of a metaphysical route, and employed this concept to semantics, what resulted in the so-called Kripke's theory of truth. In this theory a natural language contains its own truth predicate without rising contradiction. Involving the property of partial definition of truth over the set of grammatically well-formed sentences in the language, Kripke recursively showed that a language is able to consistently contain its own truth predicate. In other words, Kripke negated the impossibility of such a situation deduced by Alfred Tarski [64]. In Kripke's view the truth predicate adds new sentences to the language and truth is the union of all the constituents, *i.e.* is in turn defined for all of the elements. The procedure of determination of truth is purely recursive. The infinite number of steps establishes the "fixed point" in the language, which can be treated as the fundamental natural language containing its own truth predicate.

Other pluralistic formulations of the problem of truth involve the notions of correspondence, coherence and constructiveness. The approach proposed by Crispin Wright [65] says that truth must not be a single discourse-invariant analog of identity, and that there are only certain principles of application the truth predicate to a sentence, *i.e.* certain platitudes about true sentences. Wright emphasizes that the crucial role in the problem of truth is played by the context, and defines a truth predicate as superassertible if and only if it is assertible in a certain state of information. In other words, in Wright's approach superassertibility is the rule to find the invariant. C. Wright did not propose or suggest any consistent mechanism for improving or growing of such a state of information. Because arbitrary standards, norms, and habits question the discourse, he associates the fundamental significance with assertiveness. Another approach, proposed by Michael P. Lynch [66], claims that truth is a multiply functional property. In his magnum opus *Truth in Context* he proposed a path where metaphysical pluralism is consistent with robust realism about truth. His studies investigated the so-called relativistic Kantianism, *i.e.* taking of facts and propositions as relative without implications about relativity of an ordinary truth. According to Lynch truths are relative, but individual concepts of truth must not be relative. In his second book *True to Life* Lynch discussed the most fundamental truisms about truth: objectivity, goodness, and arising by worthiness of requesting. He considered the mental origins of cynicism, and presented inadequacy of numerous theories of truth. M.P. Lynch strongly defends caring about truth, and argues that truth has real value for a happy life.

There are also alternatives to modal realism approaches to the Multiverse hypothesis. One of such contributions was created by Henry Nelson Goodman [67]. He exalted artistry in the human-world cognitive relationship, and argued artworks as the symbols referring and constructing diverse worlds. Because arbitrary human activity in itself is an artistry, the level of Goodman's aesthetics is considered as one of the most general ones. According to this approach the interpretation is fundamentally unified to the world by the symbols. The worlds demand interpretation of the symbols they contribute to constructiveness, and with no interpretation of the symbols the worlds are non-existent. Perception, understanding, experience, and discovering always apply symbols as the fundamental method. The interest in symbols is as most natural as possible cognitive, what H.N. Goodman advocates as cognitivism. In his earliest work *Fact, Fiction, and Forecast* Goodman constructed the general problem of projection grounded in the general idea that we project predicates onto reality. In this text he also pointed out how regularities are themselves in a sense problematic - our observations lead to two-folded inductions. The problem arises from involving not just hypotheses but also projection of any predicate onto the world. The new riddle of induction, and also the problem of projection, is consistent explanation of what are the bases for projecting certain predicates onto to the world, and not others. It should be emphasized that there are in principle the predicates which we use and those which we can use are the same, and there is rather a pragmatic difference in habit, or in consolidation of some specific predicates and neglecting of other ones. Combination of the idea of such a consolidation with the idea of projection of certain predicates, and symbols in more general context, modify our observation and perception of reality leads the the Goodman general aesthetics of the cognitive relation to the world in which art plays a basic role. Understanding the worlds of art does not differ from understanding the worlds of science or perception, because interpretation of the various symbols is always required. In another presentation beginning with the dissertation *A Study of Qualities*, through *The Structure of Appearance* and *Ways of Worldmaking* Goodman applied a constructivist style to defend a reality that in itself is formed by such projections. Most of Goodman's aesthetics is expressed in *Languages of Art*. In this book he focused on various issues: art, symbols, linguistic, non-linguistic, science, ordinary life. Goodman's approach is based on the hypothesis that symbols are used for perceiving, understanding, and constructing the worlds of our experience. For Goodman symbols are reference matter.

Many-Worlds Interpretation

Amid a number of philosophical implications of the Multiverse hypothesis to theoretical physics, both the most well known and the most straightforward is the metatheory of quantum theory. In 1956 a doctoral student at Princeton University Hugh Everett III advised by John Archibald Wheeler, proposed his *relative state* (RS) formulation of quantum mechanics [68] which is the accepted form of such a theory. John A. Wheeler was one of the most involved followers of the Copenhagen school of quantum mechanics which at that time was still dominated by the absolute monarch Niels Henrik David Bohr and his devotees, and moreover was the manifestly dominating constituent in the whole theoretical physics. J.A. Wheeler quickly recognized the extraordinary style of reasoning, brilliance and originality of his student, but as a dedicated Bohrian he was forced to support the Copenhagen status quo and denied Everett's original idea leading to the unorthodox theory of measurement. Simultaneously, tried seriously to convince Niels Bohr of these ideas. In 1959 he delegated Everett twice to visit Niels Bohr and discuss the matter with him, but this attempt was apparently ineffective. Everett was not talented in the communication style expected by the Copenhagen school, and he was not able to talk with Bohr about description of the Universe through the objectively existing universal wave function which cannot collapse. Bohr and his devotees regarded Everett as the heretic and the apostate of their sanctuary-like school and applied the classical church-like method of excommunication. Léon Rosenfeld, a Belgian physicist and a fervent devotee of Bohr, described Everett as *undescribably stupid who could not understand the simplest things in quantum mechanics*. In the light of the recently discovered unpublished manuscripts, these attempts resulted in a fervent discussion with the "priests" of the Copenhagen interpretation (for the wide discussion see *e.g.* Ref. [69]). Everett's ideas questioned the rules of the *order*. Effectively, the original version of the doctoral thesis of Hugh Everett III was significantly shorted before publication, and John A. Wheeler with a distinct effort convinced Princeton University to such a version of the dissertation. On the first pages of the introduction to his opus magnum, pp. 8-9, Everett wrote

> Since the universal validity of the state function description is asserted, one can regard the state functions themselves as the fundamental entities, and one can even consider the state function of the entire universe. In this sense this theory can be called the theory of the "universal wave function," since all of physics is presumed to follow from this function alone.

The crucial point of Everett's dissertation was straightforward application of the classical mechanical procedure for consistent defining probability, alternative derivation of the Born rule describing probabilities in quantum mechanics and proving the universality of this rule. Recall that the Born rule was originally derived by Max Born in the context of the wave mechanics of collision processes [70], and was one of the fundamental laws of the Copenhagen interpretation.

In 1970 Bryce Seligman DeWitt attracted the attention of a number of theoreticians to the Everett account [71]. DeWitt reincarnated Everett's relative state formulation of quantum mechanics under the name the Many-Worlds Interpretation [72], and together with his doctoral student R. Neill Graham alternatively derived the Born rule and showed that for an infinite number of worlds, *i.e.* in the situation for which the statistical laws of quantum theory become inadequate for the consistent description, their norm becomes infinite [73]. B.S. DeWitt and R.N. Graham also edited a collective volume devoted to the Many-Worlds Interpretation of quantum mechanics [74], in which the dominating contribution was the Everett's 1956 paper *The Theory of the Universal Wavefunction* which earlier was not known to a wide audience, and also the English translation of the Born's paper and several other articles. In 1976 Wheeler became the director of the Center for Theoretical Physics at the University of Texas at Austin, and in 1977 organized the conference and invited Everett to present a talk. Everett and DeWitt met at this meeting the first and last time. The talk inspired the audience, which included Wheeler's graduate student David Deutsch, who later popularized the Many-Worlds Interpretation to a wider audience. Wheeler wanted to return Everett to a physics career in a new research institute in California, but Everett did not decide to continue this work and eventually worked in game theory. He discovered the revolutionary algorithms which made use in military operations research, performed the pioneering advances in artificial intelligence for the top secret government projects, designed logical systems that modeled rational human and machine behaviours, created the original software for targeting cities in a nuclear war and was one of the first scientists who recognized the danger of nuclear winter. John Wheeler was not satisfied that his name was associated to Everett's ideas and after Everett's death in 1982 he formally renounced the theory. Anyway, in his scientific biography coauthored with Kenneth W. Ford [75] John A. Wheeler expressed enthusiastic views on the Everettian relative state theory. In 1983 Wojciech Hubert Zurek and John A. Wheeler published a more extensive collective volume [76] presenting the papers studying the issues between quantum theory and measurement, in which Everett's paper was also included. An excellent biography with matter-of-fact analysis of the heritage of Hugh Everett

III was recently written by Peter Byrne [77]. It was based on Everett's unpublished papers, recently discovered in his son's basement, and interviews with the people who knew Hugh Everett III personally. The detailed portrait shows the genius afflicted by depression and addictions who, by the personal irrational behaviour, was not able to organize rationally his attempts to constitute the new order in the fossilized professional society in which his role was historically significant.

The Many-Worlds Interpretation, called sometimes a theory of multiple universes, distinctly impacted onto physics and philosophy, possesses a number of consistent applications and references, and is one of the most intriguing themes of the recent years. Andrew Mattei Gleason [78] and James Burkett Hartle [79] independently received the results of the thesis of H. Everett. J.B. Hartle in collaboration with Stephen William Hawking [80] applied certain results of the Many-Worlds Interpretation to the description of initial conditions for Big Bang cosmology by the solution of the Wheeler–DeWitt equation. Andrei Dmitriyevich Linde [81] applied the idea of cold chaotic/eternal inflation to formulate the first cosmology based on the Multiverse hypothesis, where randomly emerging events have independent initial conditions, and partially nucleate in space-time foam as bubbles. John S. Bell in his famous book [34] straightforwardly supported consistency and constructiveness of the Many-Worlds Interpretation.

In Frank Jennings Tipler's article [82], published in the collective volume edited jointly by Roger Penrose & Christopher John Isham, one finds his personal views on the Many-Worlds Interpretation

> I then asked, 'Who does not believe in the many-worlds interpretation?'. About 30 hands went up (including those of Roger Penrose and Bob Wald); clearly Bryce [DeWitt], David [Deutsch] (and I) were in a minority at this meeting. Finally I asked, 'Who is neutral on the many-worlds interpretation?'. The remaining 20 hands went up. I shall do my best in this short paper to persuade both the sceptics and those have not yet formed an opinion as to the validity of the many-worlds interpretation (MWI) that this interpretation is philosophically more beautiful than competing interpretations, and that it can be used in quantum cosmology as a powerful tool not only to interpret the wave function

> of the universe, but also to give us some information about the equation which this wave function obeys. (...) Most sceptics, I've found, have a mistaken idea of what the MWI really means, so it behoves me to review it briefly. The MWI is a theory of measurement, so it is concerned with describing how the universe looks to us qua human beings. At some stage during any measurement, the information is digitalized, and this is true even for the measurement of continuous variables, for example position or momentum. A typical position measurement — of an a-particle nucleus, say — would be carried out by letting the a-particle pass through an array of atoms such as those of a photographic plate. The array cannot make position measurements of unlimited accuracy; at best, the accuracy would be limited by the size of the atom. Even if we were to improve the accuracy of the position measurement at this level of the measuring process, the position measurement would in the end be digitalized when it is transmitted to human beings, for the data corresponding to an arbitrarily precise position measurement would in general exceed the storage capacity of a human brain. Hence we can model any measurement by a measurement of a discrete variable.

John R. Gribbin [83] discussed two important aspects of the Many-Worlds Interpretation: the Schrödinger cat paradox and the Multiverse hypothesis. Michael Lockwood [84], Murray Gell-Mann & James B. Hartle [85], David Z. Albert [86], Roger Penrose [87], David J. Chalmers [88], and David Elieser Deutsch [89] presented the constructive attempts to formulate the theory of evolution straightforwardly based on the Many-Worlds Interpretation. Michio Kaku [90] investigated and developed the idea of parallel worlds, *i.e.* worlds within the Multiverse, in the very speculative context of string theory. Rainer Plaga [91] proposed constructive empirical verification of the Many-Worlds Interpretation, and suggested that there is possible communication between the worlds in the Multiverse. Jeffrey Alan Barrett [92] discussed details of both the Everett and "no collapse" formulations of the Many-Worlds Interpretation of quantum mechanics. David Deutsch [93] suggested to verify the Many-Worlds Interpretation by application of quantum computer, and applying the Many-Worlds Interpretation he derived the information-theoretic version of the Born rule, and determined the concept of Multiverse in terms of an information flow. Don N. Page [94] sees the essential support to the Many-Worlds Interpretation in cosmological observations. Lutz Polley [95] replaced Deutsch's assumptions and derived the Born rule by using the symmetry arguments. The Born rule was derived by Wojciech Hubert Zurek [96] by using the concept of envariance, while entanglement allowed him to deduce probabilities. In other considerations W.H. Zurek discussed also the problem of causality, interaction with environment, and what he calls quantum darwinism describing proliferation, in the environment, of multiple records of selected states of a quantum system [97].

The arguments due to David Deutsch have been defended and significantly improved on by David Wallace [98] and Simon Saunders [99] in their studies on the various aspects of Everett's formulation of the Many-Worlds Interpretation. Interestingly, Saunders and Wallace [100], among others, modified David Lewis's analysis of the fission of identity investigated by Derek Parfit [101]. Parfit constructively applied the Many-Worlds Interpretation to investigate the concept of personal identity which is questioning the fundamental status quo of mental health. Namely, Derek Parfit presented certain sample situations in which a unified person splits into several copies, and justified ambiguousness in the fixation of the state of personality. He concluded that dividing of "I" leads to inadequacy of the concept of personal identity, which is the most celebrated and well-established concept of psychology and psychiatry. Productive introduction of medical norms and social standards based on such a central puzzle on identity is rather far perspective in the present-day state of affairs in psychiatry. Sounders and Wallace used the analysis due to David Lewis to support the idea that prior to undergoing Everett's branching uncertainty is a natural feature of any observation.

Recently, Adrian Kent [102] and Nicolaas Pieter Landsman [103] criticized independently the foundations of the Born rule derived in the context of the Many-Worlds Interpretation. Klaas Landsman's point of view is that the Many-Worlds Interpretation is a desperate answer to a problem that does not exist in reality, namely the measurement problem in quantum mechanics. He believes that measurement outcomes arise in a deterministic way as a consequence of the interaction between the measuring device and the environment (interpreted in a different way from decoherence). On Landsman's views even if the measurement problem really exists, the Many-Worlds Interpretation is a ridiculous and ineffective solution, for reasons explained in the Master's Thesis of his student Hanneke Janssen [104]. In particular, the branching of worlds is ill-defined and all derivations of the Born rule within the Many-Worlds Interpretation have so far turned out to be circular. Similarly, the critical analysis of the Many-Worlds Interpretation was performed by Martin Gardner [105]. Other scholars, Lee Smolin [106] and Brian Greene [107], are strongly supporting the Multiverse hypothesis in the context of string theory only. Some authors consider the only partial aspects of the Many-Worlds Interpretation. Lisa Randall [25] studied the context of the brane worlds, while Colin Bruce [108] focused on the Schrödinger cat paradox.

A number of experienced scientists, scholars, and researchers expressed their opinions about the Multiverse hypothesis in the collective volume edited by Bernard Carr [109]. The context of the Multiverse hypothesis can be found in a brief study of human civilization done by Freeman John Dyson's in his book *Imagined Worlds* [110]. The excellent collective volume edited by Steven J. Dick [111] contains the various opinions of experienced scientists, scholars, and researchers about the Multiverse hypothesis. Max Tegmark [112] provided the consistent classification of multiple universes. Valia Allori, beginning on her doctoral dissertation, independently presented an involving account of the ontology of the Many-Worlds Interpretation [113]. Recently, she presented advanced philosophical studies of the Many-Worlds Interpretation, arising from the De Broglie–Bohm mechanics point of view and other general modal interpretations of quantum mechanics, in collaboration with Sheldon Goldstein, Roderich Tumulka, and Nino Zanghì [114]. In 2009 Michael S. Turner [115] discussed the question of the origin of the Universe in the context of the Many-Worlds Interpretation. Recently Alejandro Jenkins and Gilad Perez [116], and Jonathan L. Feng and Mark Trodden [117] discussed the observational context of the Many-Worlds Interpretation, George Francis Rayner Ellis [118] discussed possibilities of real existence of the Multiverse, and Alexander Vilenkin and Max Tegmark [119] discussed the paradoxical situation of the Multiverse hypothesis in science.

Also recently I supported a certain particular context of the Multiverse hypothesis. I presented my views on quantum cosmology [120] based on the specific realization of the concept of Multiverse, and several ramifications of what I call the Beckwith entropy formula [121] recently suggested by Andrew Walcott Beckwith and his collaborators. In my present view this entropy formula is, however, a manifestly unjustified argument [122], while the impression of Philip Edward Gibbs the Editor of the Prespacetime Journal, which has published my opinion, was rather moderate [123]. Despite the fact that the rigorous analysis seems to show a small but serious mistake in the reasoning, it is clear that A.B. Beckwith's idea can be useful in more general context.

Anthropic Principle

The most non-trivial cosmological ramification arising from application of the Many-Worlds Interpretation is the so-called Cosmological Anthropic Principle, discussed in the book of the theosophist John David Barrow and the mathematical physicist Frank J. Tipler [124], and in the book of Tipler [125]. Barrow and Tipler saw the sense in requiring intelligent life to emerge and formulated the Final Anthropic Principle

> Final anthropic principle (FAP): Intelligent information-processing must come into existence in the Universe, and, once it comes into existence, it will never die out.

This principle was ridiculed by Martin Gardner [126] who quoted the last two sentences of this book as a Completely Ridiculous Anthropic Principle (CRAP)

> At the instant the Omega Point is reached, life will have gained control of all matter and forces not only in a single universe, but in all universes whose existence is logically possible; life will have spread into all spatial regions in all universes which could logically exist, and will have stored an infinite amount of information, including all bits of knowledge which it is logically possible to know. And this is the end.

It should be emphasized that the Cosmological Anthropic Principle is a particular realization, in the context of cosmology, of Generalized Anthropic Principle (GAP) which can be formulated as follows: *any observations are always compatible with the consciousness of an observer*. The inverted statement *the consciousness of an observer determines an observation* is the additional argument which together with the previous one can be treated as a specification of the Generalized Anthropic Principle. Moreover, if consciousness is treated as a purely cognitive result, then Generalized Anthropic Principle can be presented in the cognitive formulation: *any observations are always determined by cognition of an observer*. In other words, according to this principle, the only alive observers are able to detect the true physical state of affairs and, moreover, the physics which is detected by a conscious observer adapts to cognition of such an observer. It is often understood that Everett's interpretation does not solve the problem of the beginning or the problem of the so-called fine-tuned Universe, *i.e.* the Universe in which the conditions leading to life emerge if certain universal fundamental physical constants lie within a very narrow range. This is the reason for elimination of the Many-Worlds Interpretation by a number of modern cosmologists. However, the existence of fine-tuning conditions in the Universe is, in comparison with the Many-Worlds Interpretation, much more abstractive theme and still has not been obtained satisfactory mathematical formulation. Paul Charles William Davies expressed the opinion [127]

> There is now broad agreement among physicists and cosmologists that the universe is in several respects 'fine-tuned' for life. (. . .) the conclusion is not so much that the universe is fine-tuned for life; rather it is fine-tuned for the building blocks and environments that life requires. (. . .) 'anthropic' reasoning fails to distinguish between minimally biophilic universes, in which life is permitted but only marginally possible, and optimally biophilic universes, in which life flourishes because biogenesis occurs frequently.

In his recent book [128] Davies noticed that the Multiverse hypothesis, explaining the observable Universe by using a huge ensemble of unobservable universes, manifestly violates the Occam razor. He concludes that if the Multiverse does exist then there are universes in which other civilizations can be able to simulate another universes on computer, so that our particular entire universe is merely an avatar in some computer simulation.

The supporters of the evidence have suggested the various rational explanations, including among other the Cosmological Anthropic Principle along with the Multiverse hypothesis. The Anthropic Principle is popular in the scientific, philosophic, theological, and creationist societies and with the followers of the intelligent design idea.

Inflationary Cosmology

Recall that plausibility of the Multiverse hypothesis for theoretical physics straightforwardly arises from two cosmological paradoxes discovered in 1960. The first paradox is known as the *flatness problem*. In 1960 it was discovered that the so-called Ω constant, the ratio between the actual mass density of the universe to the critical density which has been governing the expansion of universe since the Big Bang, is very close to one. Detailed calculations have established, in fact, that omega must be 1.0 within one part in 10^{14} [129]. The second problem is known as the *horizon problem*. Different space regions, which have been causally disjointed since the Big Bang, are indistinguishable in Nature. The intensity of the Cosmic Microwave Background Radiation, predicted in 1948 by George Gamow, Ralph Asher Alpher, and Robert Herman [130], recognized as a detectable phenomenon in 1964 by Andrej Georgievich Doroshkevich and Igor Dmitriyevich Novikov [131], discovered in 1965 by Arno Allan Penzias and Robert Woodrow Wilson [132], and interpreted as a signature of the Big Bang in 1965 by Robert Henry Dicke, Phillip James Edwin Peebles, Peter G. Roll, and David Todd Wilkinson [133], was known to be isotropic within one part in 10^5. The question is, however, explanation of the close coordination of space regions on opposite sides of the Universe. There is no physical force, light rays, which could have traversed the distance between the regions since the Big Bang.

In 1981 Alan Harvey Guth [134], however, proposed to take into account an inflationary phase of the very early Universe resulting from false vacuum. According to this scenario, in such a phase the fabric of space increased by the tremendous factor of roughly 10^{30}. Amazingly enough inflation consistently explained the two mentioned paradoxes, and after some 30 years is the strongly established point of view in cosmology. However, one of its paradoxical ramifications is that the whole Universe created at the Big Bang is some 10^{23} times larger than the present-day Universe. The main point of Guth's model was that the cooled early universe is trapped in a false vacuum with a high energy density playing a role of a cosmological constant, while the supercooled very early universe was trapped in a metastable state which can only decay out as the result of the process of bubble nucleation through quantum tunneling. The sea of false vacuum contains spontaneously formed bubbles of true vacuum which are quickly expanding at the speed of light. A.H. Guth found out problems in his model: reheating is inappropriate, *i.e.* during the nucleation process the bubbles do not produce any radiation which could be emerging in the collisions between bubble walls. Moreover, if inflation has sufficiently long life to solve the problems of initial conditions, such collisions are rarely present and in an arbitrary causal patch potentially only one bubble is nucleating.

In 1980 Alexei Alexandrovich Starobinsky studied the exponentially expanding De Sitter phase, and find out that quantum corrections to gravity would replace the initial singularity of the Universe [135]. In the same year, Demosthenes Kazanas suggested that in such a model the particle horizon can be eliminated and perhaps the horizon problem can be solved [136]. In 1981 Katsuhiko Sato suggested that domain walls, which are an exotic relic, can be also removed by exponential expansion [137]. Also, in 1981 K. Sato together with Martin B. Einhorn [138] presented a model similar to the Guth model and showed a way to resolve the riddle of the magnetic monopole abundance in Grand Unification Theories. Similarly to Alan Harvey Guth's model, in their scenario of fine tuning of the cosmological constant, *i.e.* the cosmological constant must be determined very precisely to agree the model with observations, is necessary and, moreover, results in granulation of the Universe resulting from collisions of the bubble walls.

Andrei Dmitriyevich Linde [139] and Andreas Albrecht and Paul Joseph Steinhardt [140] solved the bubble collision problem by the so-called slow-roll inflation model, in which the quantum tunneling out of a false vacuum was replaced by very slow rolling (in comparison with the speed of the expansion of the Universe) of a scalar field on a hill of the potential energy and generated the inflation. The problem is that in this solution inflation ends and reheating can emerge when the slope of the hill becomes steeper. Moreover, as was shown by Viatcheslav Fyodorovich Mukhanov and Gennadiy Vasilevich Chibisov [141], in the analysis of the scenario a quasi-exponential stage of expansion, similar to the Starobinsky model, there are quantum fluctuations in the inflaton which in the slow-roll approximation scenario produce asymmetric universe and create the primordial germs for the large-scale structure of the Universe. The derived perturbative spectrum agrees with measurements of the fluctuations of the Cosmic Microwave Background Radiation in high detail. It should be emphasized that these fluctuations in the strict context of inflation were derived over the course of the 1982 Nuffield Workshop [142] by four groups working separately: Stephen Hawking [143], Alexei Starobinsky [144], Alan Guth and So-Young Pi [145], and James Maxwell Bardeen, Paul Joseph Steinhardt and Michael S. Turner [146].

An essential contribution to the theory of inflation has been provided by Andrei Dmitriyevich Linde and Alexander Vilenkin, who observed that the mechanism of starting and stopping of inflation, which is not explained by the Guth theory, is consistently explainable in terms of quantum fluctuations. Such a strategy implies a process of chaotic eternal inflation which, according Susskind [147], was described by Linde as follows

> The whole process (...) can be considered as an infinite chain reaction of creation and self-reproduction which has no end and which may have no beginning.

Consequently, the Universe is a bubble or a pocket universe amid a rich ensemble of pocket universes. The most intriguing feature of inflation is that its end is a stochastic process: it emerges at different times in various regions of the Universe, and at any time there are regions which are still inflating [81, 148]. If certain constants of Nature are dynamical then inflation randomizes them by quantum fluctuations and, consequently, they have different values in different regions of the Universe. This is, strictly speaking, a specific realization of the Multiverse hypothesis.

String Theory Landscape

The Anthropic Principle leads to straightforward highly non-trivial implications of the Many-Worlds Interpretation and the Multiverse hypothesis for the string theory. The contributions based on or related to the Multiverse hypothesis and the Many-Worlds Interpretation were presented by a number of experienced scholars, including Steven Weinberg [149], Gerardus 't Hooft [150], Stephen W. Hawking [151], Martin John Rees [152], Jacob David Bekenstein [153], and Leonard Susskind [147]. Steven Weinberg, in his famous and intriguing book *Dreams of a Final Theory* [154], unambiguously and manifestly expressed the beliefs which are strictly related to the Many-Worlds Interpretation

> The final approach is to take the Schrodinger equation seriously (...) In this way, a measurement causes the history of the universe for practical purposes to diverge into different non-interacting tracks, one for each possible value of the measured quantity. (...) I prefer this last approach. (...) For what it is worth, I hope that [the multiverse-anthropic view] is not the case. As a theoretical physicist, I would like to see us able to make precise predictions, not vague statements that certain constants have to be in a range that is more or less favorable to life. I hope that string theory really will provide a basis for a final theory and that this theory will turn out to have enough predictive power to be able to prescribe values for all the constants of nature including the cosmological constant. We shall see.

The common account of Leonard Susskind and Gerardus t' Hooft is related to the so-called holographic principle. They suggested that holographic projections of physical processes are applicable to the Universe as a whole. This point of view is consistent with the Generalized Anthropic Principle. A lot of opinions about the Multiverse hypothesis and the Anthropic Principle due to the string theory researchers have been presented by Leonard Susskind in his 2005 book [147]. Let us focus on some of them. David Jonathan Gross cites Winston Churchill's words *Never, ever, ever, ever, ever, ever, ever, ever give up* to urge researchers in looking for a consistent theory removing the necessity for arguments due to the Anthropic Principle and the Multiverse hypothesis. As cited by Susskind, Stephen W. Hawking in his 1999 lecture declared

> I will describe what I see as the framework for quantum cosmology, on the basis of M theory [one formulation of string theory]. I shall adopt the no boundary proposal, and shall argue that the Anthropic Principle is essential, if one is to pick out a solution to represent our universe, from the whole zoo of solutions allowed by M theory.

Andrei Dmitriyevich Linde's opinion is more enthusiastic

> Those who dislike anthropic principles are simply in denial. This principle is not a universal weapon, but a useful tool, which allows us to concentrate on the fundamental problems of physics by separating them from the purely environmental problems, which may have an anthropic solution. One may hate the Anthropic Principle or love it, but I bet that eventually everyone is going to use it.

Juan Martín Maldacena stated *I hope [the multiverse-anthropic principle] isn't true*, but when asked for any hopes in the other direction answered *No, I'm afraid I don't.* According to Susskind, Joseph Polchinski, one of the most influential string theorists, does not see constructive alternative to the view based on the Anthropic Principle and the Multiverse hypothesis. Garardus 't Hooft replied to Susskind's question

> Nobody could really explain to me what it means that string theory has 10^{100} vacuum states. Before you say such a thing you must first give a rigorous definition on what string theory is, and we haven't got such a definition. Or was it 10^{500} vacua, or $10^{10000000000}$? As long as such 'details' are still up in the air, I feel extremely uncomfortable with the anthropic argument. (...) However, some form of anthropic principle I cannot rule out.

Leonard Susskind in his book also expressed his own opinion

> The fact that [the cosmological constant] is not absent is a cataclysm for physicists, and the only way that we know how to make any sense of it is through the reviled and despised Anthropic Principle.

Interestingly, Brian Greene in his 2011 book [155] summarized all the hitherto existing proposals for realization of the Multiverse hypothesis. These are: Quilted Multiverse which describes Davies' pocket universes, Inflationary Multiverse which arises from the Guth–Linde inflation, Brane Multiverse which is the collection of higher-dimensional branes in string theory, Cyclic Multiverse in which the universes are parallel in time and are the result of branes collisions, Landscape Multiverse resulting from the huge number of distinct possible topologies of the fundamental structure of the Universe, Quantum Multiverse in which universes are branching and parallel and straightforwardly arise from the Many-Worlds Interpretation), Holographic Multiverse which is the result of the holographic principle saying that that our universe is mirrored by phenomena taking place on a distant bounding surface, Simulated Multiverse created as simulations running inside futuristic super-powerful computer systems, and Ultimate Multiverse which is the result of Max Tegmark's research saying that every set of mathematical equations describing a possible universe is actually realized.

There are also several scientists who present strongly critical standpoints with respect to the string theory context of the Anthropic Principle and the Multiverse hypothesis. Paul Joseph Steinhardt in his online article *It's a matter of time* [156] stated about

> Desperation has led to a "dangerous" idea: perhaps we live in an anthropically selected universe. According to this view, we live in a multiverse (a multitude of universes) in which the cosmological constant varies randomly from one universe to the next. In most universes, the value is incompatible with the formation of galaxies, planets, and stars. The reason why our cosmological constant has the value it does is because it it is one of the rare examples in which the value happens to lie in the narrow range compatible with life.
>
> This is the ultimate example of "unintelligent design": the multiverse tries every possibility with reckless abandon and only very rarely gets things "right"; that is, consistent with everything we actually observe. It suggests that the creation of unimaginably enormous volumes of uninhabit-

> able space is essential to obtain a few rare habitable spaces.
>
> I consider this approach to be extremely dangerous for two reasons. First, it relies on complex assumptions about physical conditions far beyond the range of conceivable observation so it is not scientifically verifiable. Secondly, I think it leads inevitably to a depressing end to science. What is the point of exploring further the randomly chosen physical properties in our tiny corner of the multiverse if most of the multiverse is so different. I think it is far too early to be so desperate. This is a dangerous idea that I am simply unwilling to contemplate.

Lee Smolin in his controversial book *The Trouble With Physics* [157] also expressed the critical point of view

> The search for quantum gravity is a true quest. The pioneers were explorers in a new landscape of ideas and possible worlds. (...) The scenario of many unobserved universes plays the same logical role as the scenario of an intelligent designer. Each provides an untestable hypothesis that, if true, makes something improbable seem quite probable. (...) The anthropic principle that Susskind refers to is an old idea proposed and explored by cosmologists since the 1970s, dealing with the fact that life can arise only in an extremely narrow range of all possible physical parameters and yet, oddly enough, here we are, as though the universe had been designed to accommodate us (hence the term "anthropic"). The specific version that Susskind invokes is a cosmological scenario that has been advocated by Andrei Linde for some time, called eternal inflation. According to this scenario, the rapidly inflating phase of the early universe gave rise not to one but to an infinite population of universes. You can think of the primordial state of the universe as a phase that is exponentially expanding and never stops. Bubbles appear in it, and in these places the expansion slows dramatically. Our world is one of those bubbles, but there are an infinite number of others. To this scenario, Susskind adds the idea that when a bubble forms, one of the vast number of string theories is chosen by some natural

> process to govern that universe. The result is a vast population of universes, each of which is governed by a string theory randomly chosen from the landscape of theories. Somewhere in the so-called multiverse is every possible theory in the landscape. (...) I find it unfortunate that Susskind and others have embraced the anthropic principle, because it has been understood for some time that it is a very poor basis for doing science. Since every possible theory governs some part of the multiverse, we can make very few predictions. (...) It is not hard to see why. To make a prediction in a theory that posits a vast population of universes satisfying randomly chosen laws, we would first have to write down all the things we know about our own universe. These things would apply to some number of other universes as well, and we can refer to the subset of universes where these facts are true as *possibly true universes.* (...) We physicists need to confront the crisis facing us. A scientific theory [the string theory/multiverse] that makes no predictions and, therefore, is not subject to experiment can never fail, but such a theory can never succeed either, as long as science stands for knowledge gained from rational argument borne out by evidence. There needs to be an honest evaluation of the wisdom of sticking to a research program that has failed after decades to find grounding in either experimental results or precise mathematical formulation. String theorists need to face the possibility that they will turn out to have been wrong and others right.

Peter Woit in his book *Not Even Wrong* [158] is critical too, but his expression is much less radical

> The anthropic principle comes in various versions, but they all involve the fact that the laws of physics must be of a nature that allows the development of intelligent beings such a ourselves. Many scientists believe that this is nothing more than a tautology, which while true, can never be used to create a falsifiable prediction, and thus cannot be part of scientific reasoning. Controversy has arisen as a significant group of superstring theorists have begun to argue

> that superstring theory's inability to make predictions is not a problem with the theory, but a reflection of the true nature of the universe. (...) Weinberg suggested that perhaps the explanation of the problem of the small size of the cosmological constant was the anthropic principle. The idea is that there are huge number of consistent possible universes, and that our universe is part of some larger multiverse or megaverse. Quite naturally, we find ourselves in a part of this multiverse in which galaxies can be produced and thus intelligent life can evolve. If this is the case, there is no hope of ever predicting the value of the cosmological constant, since all one can do is note the tautology that it has a value consistent with one's existence.

There are also various alternative approaches to the Anthropic Principle which are rooted in the ideas of string theory. Adequate examples are the contributions by Juan Antonio Nieto-Gracia and his collaborators [159]. Nieto-Garcia's group have presented extensive studies of several intriguing connections between string theory and the so-called oriented matroid theory, the Chern–Simons theory, p-branes physics, M-theory and certain versions of loop quantum gravity. It should be emphasized that these results look like as a possibly productive background for new constructive realizations of the Multiverse hypothesis.

Æthereal Multiverse

It is clear that in modern physics the concepts of Æther and Multiverse are the ideas inherited from the centuries of human description of Nature. Many generations of scientists, scholars, and researchers studied the various aspects of both these fundamental concepts of Antiquity. Such a situation makes abandoning them impossible, and in any case we are always forced to make use of at least one of them to formulate a consistent description of natural phenomena. The various theoretical realizations of these ideas are, in fact, universally applicable to description of the Universe as a whole.

The essential questions are whether the conceptual fusion of Æther and Multiverse can be done in general, and moreover what is the mathematical level of such an unification. In fact, in the present-day state of affairs theoretical physics does not offer too many diverse variants for the efficient realization of such a construction. The most natural research fields, in which one should seek the possibilities of the unification, are particle physics, cosmology, and quantum gravity. In these branches of modern physics one can find plenty of features of both Æther and Multiverse.

In this book I shall present my personal theoretical standpoint on the relationship between Æther and Multiverse, unifying these concepts into Æthereal Multiverse, consistent and constructive approach to merge a number of physical phenomena, and in itself is a straightforward way of making uniform the various phenomena under the name of one general idea. I have realized that at the present stage of the physical knowledge it is difficult to find the one point of view which is able to unify consistently diverse regions of theoretical reasoning.

For this reason, in this monograph, a certain constructive methodological background, which leads to the consistent formulation of the unifying aesthetics of physical reality and is able to product fruitful advancements, will be applied. This essential background contains pieces of theoretical physics which lay the foundations of our understanding of the world we live in.

The scientific method is based on what I prefer to call the inquiring mindset. It includes all areas of human thoughtful activity that categorically eschew "belief", the enemy of rationality. This mindset is a nebulous mixture of doubt, questioning, observation, experiment and, above all, curiosity, which small children possess in spades. I would argue that it is the most important, intrinsically human quality we possess.

Sir Harold Walter Kroto

Part I

Cosmology

Chapter 1

Quantum Cosmology

A Introduction

Quantum cosmology is one of the most important research tendencies within modern theoretical gravitational physics. Its necessity is defined by problems of formulation of the physics of early and very early stages of our Universe. Its essence, however, is the problem of construction of the model of the Universe, in which quantum theory and statistical mechanics meet and work together. Such a strategy is well known as the crucial aspect of string theory, however, and for this reason an arbitrary model of quantum cosmology which is the linkage between the quantum physics and the statistical physics will have characteristic features and properties of string theory. Moreover, it is possible even that any such a model will be defining new kinds of string theories. Albeit, because of the very abstractive mathematical form of string theory, the problem is how to build the model of quantum cosmology which could be manifestly presenting the value for phenomenology. In other words, the crucial problem in the formulation of quantum cosmology is an experimental verification of any of its model.

One of the most essential steps in the history of quantum cosmology was quantum geometrodynamics, also known as quantum General Relativity or quantum gravity, which had the beginning in the works by John Archibald Wheeler and Bryce Seligman DeWitt. Standardly, in such a strategy the primary canonical quantization procedure is applied to any Hamiltonian formulation of the classical theory of gravitation, General Relativity. The most widely spread canonical formulation of General Relativity is the Hamiltonian approach which started by early works of Paul Adrien Maurice Dirac and obtained a final and commonly accepted appearance in the works by Richard Lewis Arnowitt, Stanley Deser, and Charles William Misner. It is known in the literature as the Arnowitt–Deser–Misner decomposition or $3+1$ splitting.

In this chapter we discuss certain both particular consequences as well as the development of the specific strategy for quantization of gravitational fields arising from the Arnowitt–Deser–Misner decomposition. The main purpose of this chapter is, however, the construction of such a consistent quantum theory of the Universe, which will lead to plausible theoretical predictions possessing both the clearly defined physical sense as well as the possibility of verification by comparison of its results with experimental data obtained in observations of the observed Universe. Applying the well-known epistemological justifications of quantum theory, we propose to describe quantum cosmology as a quantum field theory formulated in terms of the Fock space of creation and annihilation operators, the algebraic system investigated by Vladimir Aleksandrovich Fock to describe quantum states with a variable or unknown number of particles. The particles in the context of quantum cosmology become quantum universes and, consequently, the quantum theory of the Universe is the theory of quantum Multiverse. The Fock space formalism has the exceptionally well established phenomenological meaning for quantum physics. In this manner, the basic problem is the constructive application of the method of secondary quantization, which would result in the phenomenological deductions for quantum cosmology. The natural consequence of the formalism of the Fock repère is the possibility for straightforward construction of statistical mechanics of the quantum states of the Universe. Moreover, the Fock space formalism is the most natural way to realize the Multiverse hypothesis by the collection of quantum universes.

For this reason we employ the quantum field theoretical reasoning for quantum cosmology. As the classical model of the Universe we shall consider the conformal-flat metric first derived by Alexander Alexandrovich Friedmann, and applied in the cosmological context by Georges-Henri Lemaître, Howard Percy Robertson, and Arthur Geoffrey Walker, which lays the foundations of a number of models of the observed Universe. We show that straightforward application of the Hamiltonian approach to General Relativity for such a concrete solution of the Einstein field equations results in the Hamiltonian constraint corresponding to the fundamental state of a bosonic string.

This chapter is not too extensive. First, in section B we sketch very shortly the classical model of Universe described by the conformal-flat Friedmann–Lemaître–Robertson–Walker metric for the case of the finite spatial volume. Such a particular solution, possessing the cylindrical or toroidal topology corresponding to open or closed strings, is the Einstein–Friedmann Universe. We present the Hamiltonian approach which connects both the Dirac approach and the Arnowitt–Deser–Misner Hamiltonian formulation of General Relativity, in which there is splitting of evolution of four-dimensional space-time into the dynamics of three-dimensional space in one-dimensional time. The Hamiltonian constraint is obtained, and the Hubble law is obtained due to its resolution. Moreover, the dynamics of the model is identified with the dynamics of the boson having the negative squared-mass, *i.e.* with the tachyon. In section C we perform the primary canonical quantization procedure with respect to the classical model. By the primary quantization of the Hamiltonian constraint we receive the appropriate Wheeler–DeWitt equation. Applying the separation of variables based on the Hamilton equations of motion we make reduction of order of such an evolution and, therefore, the appropriate one-dimensional Dirac equation for the Universe is deduced. Employing the method of secondary quantization to such a reduced evolution results in the quantum field theory formulated in the Fock space. Application of the appropriate Bogoliubov transformation and the Heisenberg equations of motion leads to the diagonalization of the equations of motion to the canonical form. In this manner, we receive the static Fock repère and quantization of the cosmological model is formulated in terms of the monodromy in the Fock space. In section D we shall construct the statistical mechanics, especially thermodynamics, of quantum states of the Universe. Computations are performed according to the Bose–Einstein statistics in the framework of the method of the density matrix. We apply the approximation in which the quantum Universe is a system with one degenerated state, *i.e.* one-particle approximation. Section E is devoted to discussion of the particular thermodynamical situation of the system of many quantum universes, *i.e.* the early light Multiverse corresponding to cosmological radiation, which is characterized by minimal entropy. In the final section F we summarize briefly the results of this chapter and present the perspective for further development of the proposed approach to quantum cosmology.

B Einstein–Friedmann Universe

General Relativity based on the Einstein field equations [160]

$$R_{\mu\nu} - \frac{1}{2}g_{\mu\nu}R = \kappa T_{\mu\nu}, \tag{1.1}$$

where $\kappa = \dfrac{8\pi G}{c^4}$ is the Einstein constant, $R_{\mu\nu}$ is the Ricci curvature tensor, R is the Ricci scalar curvature, and $T_{\mu\nu}$ is the stress-energy tensor, is commonly accepted as the classical theory of gravitation describing the evolution of metric tensor $g_{\mu\nu}$ of four-dimensional Riemannian space-time manifold M (Cf. *e.g.* Refs. [161, 162]). The Einstein field equations (1.1) can be generated as the Euler–Lagrange equations of motion obtained by the Hilbert–Palatini action principle [163, 164] $\delta S_{EH} = 0$ with respect to the fundamental field $g_{\mu\nu}$ applied to the Einstein–Hilbert action

$$S_{\text{EH}} = \frac{1}{c}\int_M d^4x\sqrt{-g}\left(-\frac{1}{2\kappa}R + \mathscr{L}_M\right). \tag{1.2}$$

In the action (1.2) we denoted $g = \det g_{\mu\nu}$, and $\mathscr{L}_M$ denotes the Lagrangian density of Matter fields. The variational principle allows to express the stress-energy tensor of Matter fields by the Lagrangian density of Matter fields

$$T_{\mu\nu} = \frac{2}{\sqrt{-g}}\frac{\delta\left(\sqrt{-g}\mathscr{L}_M\right)}{\delta g^{\mu\nu}}. \tag{1.3}$$

Let us consider the exact solution of the Einstein field equations (1.1) first derived by A.A. Friedmann, and studied in the extensive cosmological context by G.-H. Lemaître, H.P. Robertson, and A.G. Walker [165], for which the space-time interval has the following form

$$ds^2 = g_{\mu\nu}dx^\mu dx^\nu = -(dx^0)^2 + a^2(x^0)\delta_{ij}dx^i dx^j, \tag{1.4}$$

where $a(x^0)$ is the cosmic scale factor parameter, and x^μ $(\mu = 0,1,2,3)$ is the Cartesian system of space-time coordinates in which the time coordinate x^0 is the object of action of diffeomorphisms [166, 167]

$$x^0 \to x'^0 = x'(x^0). \tag{1.5}$$

When the volume of the three-dimensional space is finite

$$V = \int d^3x < \infty, \tag{1.6}$$

then the interval (1.4) describes the Einstein–Friedmann Universe possessing a topology related to finite space. Such a cosmology is very far from triviality, because the topological structure of finite space relates the model to considerations of string theory. Finite space is associated with cylindrical, toroidal, but also with more general stringy topologies and orbifolds [168]. In other words, the Einstein–Friedmann Universe having finite space looks like to be a topological string. However, such complicated theoretical situations will not be developed in this book. Also, cylindrical topology is far from our present considerations, because the presence of the boundaries in general can result in highly non-trivial physical consequences. The finite space possessing toroidal topology has no boundary because the ends of the cylinder are identified each other. For this reason, for simplicity of the cosmological model, in this book we shall study the case of the Einstein–Friedmann Universe equipped with the toroidal topology. Such a choice in itself is non-trivial, because the toroidal topology can be naturally identified with closed strings. In this manner, we shall consider the cosmological model possessing more general significance, which also can be extended on the other topologies mentioned above.

Let us consider the Friedmann–Lemaître–Robertson–Walker metric (1.4) in the framework of the Hamiltonian approach joining the Dirac approach [169] and the Arnowitt–Deser–Misner Hamiltonian formulation of General Relativity [170]. In this chapter we shall not present the detailed analysis and explanations which can be found in more specialized student textbooks (Cf. *e.g.* Ref. [171]), and shall be discussed in a further part of this book. The Arnowitt–Deser–Misner decomposition of any metric satisfying the Einstein field equation, in a given coordinate system, has the following form

$$g_{\mu\nu} = \begin{bmatrix} -N^2 + N_i N^i & N^i \\ N^j & h_{ij} \end{bmatrix}, \tag{1.7}$$

where N is the lapse function, N_i is the shift vector, h_{ij} is the metric of three-dimensional space embedded in the four-dimensional space-time, and $N^j = h^{ij} N_i$. The classical Universe described by the Friedmann–Lemaître–Robertson–Walker metric (1.4) can be parametrized by

$$N^2 = 1, \tag{1.8}$$

$$N_i = [0,0,0], \tag{1.9}$$

$$h_{ij} = a^2(x^0)\delta_{ij}. \tag{1.10}$$

According to the strategy proposed by Dirac, the lapse function should be preserved explicitly but the shift vector, because its trivialization, becomes absent. Moreover, at the end of calculations the lapse function should be stated explicitly. In this manner, the Hamiltonian approach which involves both the Dirac and the Arnowitt–Deser–Misner formulations of General Relativity allows to present the Friedmann–Lemaître–Robertson–Walker metric in the following form

$$g_{\mu\nu} = \begin{bmatrix} -N^2 & 0 \\ 0 & a^2(x^0)\delta_{ij} \end{bmatrix}, \tag{1.11}$$

which is associated with the space-time interval

$$ds^2 = -N^2(dx^0)^2 + a^2(x^0)dx^i dx^j. \tag{1.12}$$

Application of the diffeomorphism of the coordinate time x^0

$$ct = c\tau + x^0, \tag{1.13}$$

where τ is a reference time constant, allows to make the transformation of cosmic time t to conformal time η on the level of the diffeoinvariant integral measure

$$d\eta = N(x^0)dx^0 \equiv \frac{dt}{a(t)}, \tag{1.14}$$

which allows to rewrite the interval (1.12) in the following form

$$ds^2 = a^2(\eta)\left[-c^2 d\eta^2 + \delta_{ij}dx^i dx^j\right]. \tag{1.15}$$

It means that, in the coordinate system $(c\eta, x^i)$, the conformal-flat Universe is described by the scaled Minkowski space-time

$$g_{\mu\nu} = \Omega(\eta)\eta_{\mu\nu}, \tag{1.16}$$

where the scaling function $\Omega(\eta)$ is

$$\Omega(\eta) = a^2(\eta). \tag{1.17}$$

Derivation of the Christoffel symbols for the metric (1.11),

$$\Gamma^{\rho}_{\mu\nu} = \frac{1}{2}g^{\rho\sigma}\left(\partial_\nu g_{\mu\sigma} + \partial_\mu g_{\sigma\nu} - \partial_\sigma g_{\mu\nu}\right), \tag{1.18}$$

leads to the non-trivial components

$$\Gamma^0_{00} = \frac{\dot{N}}{N}, \tag{1.19}$$

$$\Gamma^0_{ii} = \frac{a\dot{a}}{N^2}, \tag{1.20}$$

$$\Gamma^i_{i0} = \frac{\dot{a}}{a}, \tag{1.21}$$

where dot means differentiation with respect to the time coordinate x^0 and, consequently, the Ricci curvature tensor

$$R_{\mu\nu} = \partial_\alpha \Gamma^\alpha_{\mu\nu} - \partial_\nu \Gamma^\alpha_{\mu\alpha} + \Gamma^\alpha_{\beta\alpha}\Gamma^\beta_{\mu\nu} - \Gamma^\alpha_{\beta\nu}\Gamma^\beta_{\mu\alpha}, \tag{1.22}$$

takes the following form

$$R_{\mu\nu} = \left[\begin{array}{cc} -3\left(\dfrac{\ddot{a}}{a} - \dfrac{\dot{a}}{a}\dfrac{\dot{N}}{N}\right) & \mathbf{0}^{\mathrm{T}} \\ \mathbf{0} & \dfrac{a^2}{N^2}\left(\dfrac{\ddot{a}}{a} - 2\left(\dfrac{\dot{a}}{a}\right)^2 - \dfrac{\dot{a}}{a}\dfrac{\dot{N}}{N}\right)\delta_{ij} \end{array}\right]. \tag{1.23}$$

By straightforward calculation of the contravariant metric components and taking into account the Ricci curvature tensor (1.50), one receives the Ricci scalar curvature

$$R = g^{\mu\nu}R_{\mu\nu} = \frac{6}{N^2}\left[\frac{\ddot{a}}{a} - \left(\frac{\dot{a}}{a}\right)^2 - \frac{\dot{a}}{a}\frac{\dot{N}}{N}\right], \tag{1.24}$$

which allows to evaluate the Einstein–Hilbert action (1.2) for the the case of the classical conformal-flat Universe. Let us describe such a procedure in some detail. First, let us consider the geometric part of the Einstein–Hilbert action, which transforms as follows

$$\int_M d^4x\sqrt{-g}R = \int_M d^3x\int dx^0 Na^3\frac{6}{N^2}\left[\frac{\ddot{a}}{a}-\left(\frac{\dot{a}}{a}\right)^2-\frac{\dot{a}}{a}\frac{\dot{N}}{N}\right]= \tag{1.25}$$

$$= 6V\int dx^0\frac{\ddot{a}a^2N-\dot{a}^2aN-\dot{a}a^2\dot{N}}{N^2}, \tag{1.26}$$

where $V=\int d^3x$ is the volume of the three-dimensional space, which in the case of the Einstein–Friedmann Universe is finite. Applying the total derivative

$$\frac{d}{dx^0}\left(\frac{\dot{a}a^2}{N}\right)=\frac{\ddot{a}a^2N+\dot{a}^2aN+\dot{a}a^2\dot{N}}{N^2}, \tag{1.27}$$

the geometric part of the Einstein–Hilbert action becomes

$$\int_M d^4x\sqrt{-g}R = 6V\int dx^0\left[\frac{d}{dx^0}\left(\frac{\dot{a}a^2}{N}\right)-\frac{3\dot{a}^2a}{N}\right] \tag{1.28}$$

$$= 6V\left(\frac{\dot{a}a^2}{N}\right)\bigg|_{\partial M}-18V\int dx^0\frac{\dot{a}^2a}{N} \tag{1.29}$$

$$= -18V\int dx^0\frac{\dot{a}^2a}{N}, \tag{1.30}$$

where the boundary term has been omitted as identically vanishing

$$\left(\frac{\dot{a}a^2}{N}\right)\bigg|_{\partial M}=0, \tag{1.31}$$

by the choice of the suitable boundary conditions

$$\dot{a}|_{\partial M}=0. \tag{1.32}$$

Let us involve now the relation between the derivative with respect to the time coordinate x^0 and the derivative with respect to conformal time η of the cosmic scale factor parameter

$$\dot{a}=\frac{a'}{a}, \tag{1.33}$$

where the prime means differentiation with respect to the time coordinate $c\eta$. In such a transformation one receives the following result

$$\dot{a}^2 a = \frac{a'^2}{a}, \tag{1.34}$$

and making use of the relation $dx^0 = cdt$ one obtains finally

$$\int_M d^4x\sqrt{-g}R = -18Vc\int dt\frac{1}{N}\frac{a'^2}{a} = -18Vc\int d\eta\frac{a'^2}{N}. \tag{1.35}$$

Let us consider now the action of Matter fields which, in the $3+1$ splitting, transforms in the following way

$$\begin{aligned}\int_M d^4x\sqrt{-g}\mathscr{L}_M &= \int d^3x\int dx^0 Na^3\mathscr{L}_M = c\int d^3x\int dtNa^3\mathscr{L}_M = &(1.36)\\ &= c\int d^3x\int d\eta Na^4\mathscr{L}_M = c\int d\eta Na^4\int d^3x\mathscr{L}_M. &(1.37)\end{aligned}$$

Let us take into account the Legendre transformation

$$\mathscr{L}_M = p_F\dot{F} - \mathscr{H}_M, \tag{1.38}$$

where F is a Matter field (for collection of Matter fields one has a sum over fields), $p_F = \dfrac{\partial\mathscr{L}}{\partial\dot{F}}$ is the conjugate momentum to F, and $\mathscr{H}_M$ is Hamiltonian density of Matter fields. Taking into account the definition of the conjugate momentum, one can write

$$p_F\dot{F} = \dot{F}\frac{\partial\mathscr{L}_M}{\partial\dot{F}} = \frac{\partial}{\partial\dot{F}}\left(\dot{F}\mathscr{L}_M\right) - \mathscr{L}_M, \tag{1.39}$$

and, therefore, one receives the relation

$$\mathscr{L}_M = p_F\dot{F} - \mathscr{H}_M = \frac{\partial}{\partial\dot{F}}\left(\dot{F}\mathscr{L}_M\right) - \mathscr{L}_M - \mathscr{H}_M, \tag{1.40}$$

which allows to establish straightforwardly the Lagrangian density of Matter fields in the following form

$$\mathscr{L}_M = \frac{1}{2}\frac{\partial}{\partial\dot{F}}\left(\dot{F}\mathscr{L}_M\right) - \frac{1}{2}\mathscr{H}_M. \tag{1.41}$$

In this manner, one can determine easily the volume integral of the Lagrangian density of Matter fields

$$\int d^3x \mathscr{L}_M = \frac{1}{2}\int d^3x \frac{\partial}{\partial \dot{F}}\left(\dot{F}\mathscr{L}_M\right) - \frac{1}{2}\int d^3x \mathscr{H}_M. \tag{1.42}$$

Let us consider the first term of this formula. It is evidently seen that the integrand is a total derivative, and applying the chain rule of differentiation one can write the integrand as

$$\frac{\partial}{\partial \dot{F}}\left(\dot{F}\mathscr{L}_M\right) = \frac{\partial V}{\partial \dot{F}}\frac{\partial}{\partial V}\left(\dot{F}\mathscr{L}_M\right). \tag{1.43}$$

However, in the light of the fact $dV = d^3x$ the derivative identically vanishes $\dfrac{\partial V}{\partial \dot{F}} = 0$, so that in such a situation one has

$$\int d^3x \frac{\partial}{\partial \dot{F}}\left(\dot{F}\mathscr{L}_M\right) = \frac{\partial V}{\partial \dot{F}}\int dV \frac{\partial}{\partial V}\left(\dot{F}\mathscr{L}_M\right) = \frac{\partial V}{\partial \dot{F}}\left(\dot{F}\mathscr{L}_M\right) = 0, \tag{1.44}$$

and, therefore, the volume integral of the Lagrangian density of Matter fields is much more simplified

$$\int d^3x \mathscr{L}_M = -\frac{1}{2}\int d^3x \mathscr{H}_M. \tag{1.45}$$

In this manner, finally the action of Matter fields can be entirely expressed through the only Hamiltonian density of Matter fields

$$\int_M d^4x\sqrt{-g}\mathscr{L}_M = -\frac{c}{2}\int d\eta N a^4 \int d^3x \mathscr{H}_M. \tag{1.46}$$

Collecting all these facts one can evaluate the Einstein–Hilbert action on the Friedmann–Lemaître–Robertson–Walker metric. Interestingly, in such a situation the Einstein–Hilbert action has the Hamilton form in conformal time, *i.e.*

$$S_{EH} = \int d\eta \mathrm{L}(a, a', \eta), \tag{1.47}$$

and the Lagrangian $\mathrm{L}(a, a', \eta)$ of the cosmological model has the form

$$\mathrm{L}(a, a', \eta) = \frac{3}{2}E_P \ell_P^2 \frac{V}{V_P}\frac{a'^2}{N} - \frac{1}{2}Na^4 \mathrm{H}_M(\eta). \tag{1.48}$$

where $V_P = \frac{4}{3}\pi\ell_P^3$ is the volume of the Planck sphere, and $\mathrm{H}_M(\eta)$ is the volume integral of the Hamiltonian density of Matter fields

$$\mathrm{H}_M(\eta) = \int d^3x \mathscr{H}_M(x,\eta), \tag{1.49}$$

which has a natural interpretation of the energy of Matter fields.

The momentum conjugated to the cosmic scale factor parameter can be derived straightforwardly from the reduced action (1.47)

$$\mathrm{P}_a = \frac{\partial \mathrm{L}}{\partial v_a} = 3M_P \frac{V}{V_P}\frac{v_a}{N}, \tag{1.50}$$

where we have introduced the speed related to the cosmic scale factor parameter

$$v_a = c\ell_P a'. \tag{1.51}$$

It can be seen straightforwardly that application of the Legendre transformation to the reduced action (1.47) leads to

$$S_{EH} = \int d\eta \left\{\mathrm{P}_a v_a - N\mathrm{H}(\eta)\right\}, \tag{1.52}$$

where $\mathrm{H}(\eta)$ is the Hamilton function of the cosmological model

$$\mathrm{H}(\eta) = \frac{V_P}{3V}\frac{\mathrm{P}_a^2}{2M_P} + \frac{1}{2}a^4\mathrm{H}_M(\eta). \tag{1.53}$$

According to the Arnowitt–Deser–Misner Hamiltonian formulation of General Relativity, the Euler–Lagrange equations of motion obtained by vanishing of functional derivatives of the total action with respect to the parameters N, and N_i are the constraints. In the present situation one has only one constraint – the Hamiltonian constraint

$$\frac{\delta S_{EH}}{\delta N} = -\mathrm{H}(\eta) = 0. \tag{1.54}$$

Such a situation, *i.e.* identically vanishing total energy of the Universe, is the realization of the Feynman formulation of the Mach principle [172]. Consequently, the Hamiltonian constraint to the case of the Friedmann–Lemaître–Robertson–Walker metric is

$$\frac{V_P}{3V}\frac{\mathrm{P}_a^2}{2M_P} + \frac{1}{2}a^4\mathrm{H}_M(\eta) \approx 0. \tag{1.55}$$

Substitution of the explicit form of the conjugated momentum (1.50) to the Hamiltonian constraint (1.55) leads to the equation

$$a'^2 + \frac{N^2}{E_P \ell_P^2} \frac{V_P}{3V} a^4 \mathrm{H}_M(\eta) = 0. \tag{1.56}$$

Now one can apply the value $N^2 = 1$ which is right for the present case. Using the Hubble parameter

$$H(a) \equiv \frac{\dot{a}}{a} = \frac{a'}{a^2}, \tag{1.57}$$

where in this context the dot means t-differentiation, the Hamiltonian constraint can be presented as the equation for the energy of Matter fields

$$\mathrm{H}_M(\eta) = -3E_P \ell_P^2 \frac{V}{V_P} H^2(a), \tag{1.58}$$

which suggests that the energy of Matter fields is negative. In this manner, one can suggest the following general form of this energy

$$\mathrm{H}_M(\eta) = -\left|\mathrm{H}_M(\eta)\right|, \tag{1.59}$$

so that the equation (1.58) can be rewritten in the form

$$H(a) = \pm\sqrt{\frac{1}{E_P \ell_P^2} \frac{V_P}{3V} \left|\mathrm{H}_M\right|}, \tag{1.60}$$

where $\mathrm{H}_M = \mathrm{H}_M(\bullet)$, and $\bullet = t, \eta$. The solution (1.60) is the Hubble law.

The result (1.58) in itself is non-trivial. Namely, it means that the energy of Matter fields explicitly depends not on conformal time η, but on the cosmic scale factor parameter. In other words, the relation $\mathrm{H}_M(\eta) = \mathrm{H}_M(a)$ is always true. Applying the formula (1.58) within the Lagrangian (1.48) and the Hamiltonian (1.53) of the cosmological model one sees that the explicit dependence on conformal time actually transforms into the explicit dependence on the cosmic scale factor parameter a, *i.e.*

$$\mathrm{L}(a, a') = \frac{E_P \ell_P^2}{2} \frac{3V}{V_P} \frac{1}{N} \left[a'^2 + N^2 a^4 H^2(a)\right], \tag{1.61}$$

$$\mathrm{H}(\mathrm{P}_a, a) = \frac{V_P}{3V} \left[\frac{\mathrm{P}_a^2}{2M_P} - \frac{E_P \ell_P^2}{2} \left(\frac{3V}{V_P}\right)^2 a^4 H^2(a)\right]. \tag{1.62}$$

These formulas, however, reflect the unusual tautology within the quantum cosmology because in the light of the Dirac approach to Hamiltonian systems, the Hamiltonian constraint $\mathrm{H}(\mathrm{P}_a, a) \approx 0$ generates the Hubble parameter $H(a) = \frac{1}{N}\frac{a'}{a^2}$ which, after taking into account the value of $N = \pm 1$ proper for the Friedmann–Lemaître–Robertson–Walker metric and the transformation to cosmic time t given by $a' = \dot{a}a$, becomes the usual Hubble parameter, *i.e.* $H(a) = \frac{\dot{a}}{a}$.

The Hamiltonian (1.62), however, in itself is non-trivial in the light of the fact that after taking into account the Hubble parameter expressed in conformal time $H(a) = \frac{a'}{a^2}$ up to the constant multiplier $\frac{V_P}{3V}$ this Hamiltonian takes the form of the Hamiltonian appropriate for a Euclidean one-dimensional oscillator

$$\mathrm{H}(\mathrm{P}_a, a) = \frac{V_P}{3V}\left[T(\mathrm{P}_a) - V(\mathrm{P}_a, a)\right]. \tag{1.63}$$

where the kinetic energy $T(\mathrm{P}_a)$ and the potential energy $V(\mathrm{P}_a, a)$ are

$$T(\mathrm{P}_a) = \frac{\mathrm{P}_a^2}{2M_P}, \tag{1.64}$$

$$V(\mathrm{P}_a, a) = \frac{\mathrm{P}_a^2}{2M_P}a^2. \tag{1.65}$$

This oscillator is the Euclidean one-dimensional harmonic oscillator if one identifies the *cosmological coordinate* with

$$x_C = \ell_P a \tag{1.66}$$

and the potential energy with $V = \frac{kx_C^2}{2}$, where the constant k is

$$k = \frac{\mathrm{P}_a^2}{M_P \ell_P^2}. \tag{1.67}$$

In other words, the Euclidean one-dimensional harmonic oscillator is defined by the constant value of the conjugate momentum

$$\mathrm{P}_a = \pm\sqrt{kM_P}\ell_P, \tag{1.68}$$

or equivalently the constant value of the cosmological speed

$$x'_C = \pm\sqrt{\frac{k}{M_P}}\ell_P\frac{V_P}{3V} = \pm\frac{v_a}{N}. \tag{1.69}$$

This equation can be solved immediately and gives the linear dependence of the cosmological coordinate with respect to conformal time

$$x_C(\eta) = \pm\sqrt{\frac{k}{M_P}}\ell_P\frac{V_P}{3V}(\eta - \eta_I). \tag{1.70}$$

Applying the relation (1.68) and the solution (1.70) to the Hamiltonian (1.63) one obtains the Hamiltonian of the Euclidean one-dimensional harmonic oscillator

$$\mathrm{H}(\eta) = \frac{k\ell_P^2}{2}\frac{V_P}{3V}\left[1 - \frac{k}{M_P}\left(\frac{V_P}{3V}\right)^2(\eta - \eta_I)^2\right]. \tag{1.71}$$

Interestingly, when one identifies the initial value of conformal time as

$$\eta_I = \sqrt{\frac{M_P}{k}\frac{3V}{V_P}}, \tag{1.72}$$

then the Hamiltonian of the Euclidean one-dimensional harmonic oscillator becomes

$$\mathrm{H}(\eta) = -\frac{k^2\ell_P^2}{2M_P}\left(\frac{V_P}{3V}\right)^3\left[\eta^2 - 2\sqrt{\frac{M_P}{k}\frac{3V}{V_P}}\eta\right]. \tag{1.73}$$

Its values are positive for $0 < \eta < 2\eta_I$, negative for $\eta > 2\eta_I$, and zero for $\eta = 2\eta_I$. Interestingly, when one sets *ad hoc* $V = V_P$ and

$$k = k_P = \frac{M_P}{t_P^2} = M_P\omega_P^2 = \frac{E_P}{\ell_P^2} \approx 7.4880571\cdot 10^{78}\frac{\mathrm{kg}}{\mathrm{s}^2}, \tag{1.74}$$

then $\eta_I = 3t_P$, and the Hamiltonian of the Euclidean one-dimensional harmonic oscillator is simplified to

$$\mathrm{H}(\eta) = -\frac{E_P}{54}\left[\left(\frac{\eta}{t_P}\right)^2 - 6\frac{\eta}{t_P}\right], \tag{1.75}$$

so that initially, *i.e.* for $\eta = 3t_P$, one obtains

$$\mathrm{H}(\eta_I) = \frac{E_P}{6}. \tag{1.76}$$

Let us call such a case *the Planckian Euclidean one-dimensional harmonic oscillator*.

In this manner, in the Hamiltonian approach presented above the Einstein–Friedmann Universe is described by the Hamiltonian constraint (1.55), and the Hubble law is obtained due to straightforward integration of this constraint. The Hubble law can be expressed through both the cosmological as well as conformal time

$$\int_{a_I}^{a} \frac{da'}{a'^2 H(a')} = c(\eta_I - \eta), \tag{1.77}$$

$$\int_{a_I}^{a} \frac{da'}{a' H(a')} = c(t_I - t), \tag{1.78}$$

where the subscript I denotes the initial data of given quantities. Let us consider the cosmological redshift z defined by the formula [173],

$$\frac{a}{a_I} \equiv \frac{1}{1+z}, \tag{1.79}$$

where in terms of cosmic time $a = a(t)$, $a_I = a(t_I)$, $z = z(t_I, t)$, while in terms of conformal time $a = a(\eta)$, $a_I = a(\eta_I)$, $z = z(\eta_I, \eta)$. Then the Hubble law (1.77)-(1.78) can be expressed as

$$\int_{z_I}^{z} \frac{dz'}{H(z')} = a(\eta_I) c(\eta - \eta_I), \tag{1.80}$$

$$\int_{z_I}^{z} \frac{dz'}{(1+z')H(z')} = c(t - t_I), \tag{1.81}$$

where $z_I = z(t_I, t_I) = z(\eta_I, \eta_I)$, and by the equation (1.60) and the definition (1.57) the cosmological redshift can be expressed through two distinguishable ways

$$z(\eta_I, \eta) = \pm a(\eta_I) \int_{\eta_I}^{\eta} d\eta' \sqrt{\frac{1}{M_P \ell_P^2} \frac{V_P}{3V} |\mathrm{H}_M(\eta')|} \tag{1.82}$$

$$z(t_I, t) = \exp\left\{ \pm \frac{a(\eta_I)}{a(t_I)} \int_{t_I}^{t} dt' \sqrt{\frac{1}{M_P \ell_P^2} \frac{V_P}{3V} |\mathrm{H}_M(t')|} \right\} - 1. \tag{1.83}$$

The relations (1.77) and (1.78) as well as (1.80) and (1.81) allow to establish the relation between conformal and cosmic time

$$\eta - \eta_I = t - t_I + \frac{1}{c}\int_{a_I}^{a} \frac{a' - 1}{a'^2}\frac{da'}{H(a')}, \tag{1.84}$$

$$a(\eta_I)(\eta - \eta_I) = t - t_I + \frac{1}{c}\int_{z_I}^{z} \frac{z'dz'}{(1+z')H(z')}. \tag{1.85}$$

Interestingly, because the universal definition of the cosmological redshift (1.79) one can suggest *ad hoc* the relation

$$z(\eta_I,\eta) = z(t_I,t), \tag{1.86}$$

which means that cosmological redshift is invariant with respect to the exchange between cosmic time and conformal time. In such a situation, for consistency one should apply the measure (1.14) within the relation (1.82). It can be seen by straightforward calculation that such an invariance is non-trivial, because results in the Riccati equation

$$\frac{dg}{dt} = g + h, \tag{1.87}$$

where we have introduced the notation

$$g = g(t) = \exp[f(t)]\ln[f(t)], \tag{1.88}$$

$$h = h(t) = \frac{a_I}{a(t)}, \tag{1.89}$$

where $a_I = a(t_I)$ and the unknown function $f(t)$ has the form

$$f(t) = \pm\frac{a(\eta_I)}{a(t_I)}\sqrt{\frac{1}{M_P\ell_P^2}\frac{V_P}{3V}\left|\mathrm{H}_M(t')\right|} = H(t), \tag{1.90}$$

where we have applied equality $a(\eta_I) = a(t_I)$. The Riccati equation (1.87) can be solved immediately with the result

$$g(t) = e^{t-t_I}\left(1 + \int_{t_I}^{t} dt' e^{-t'} h(t')\right). \tag{1.91}$$

Differentiation of $g = \exp H \ln H$ leads to

$$\dot{g} = \dot{H}\exp H\ln H + \exp H\frac{\dot{H}}{H}, \tag{1.92}$$

and applying $\exp H = \dfrac{g}{\ln H}$ within the formula (1.92) one obtains

$$\frac{\dot g}{g} = \dot H\left(1+\frac{1}{H\ln H}\right). \tag{1.93}$$

Differentiation of both sides of the equation (1.93) leads to

$$\frac{\ddot g}{g} - \frac{\dot g^2}{g^2} = \ddot H\left(1+\frac{1}{H\ln H}\right) - \frac{\dot H^2}{H^2\ln H}\left(1+\frac{1}{\ln H}\right), \tag{1.94}$$

or after using the equation (1.93)

$$\frac{\ddot g}{g} - \frac{\dot g^2}{g^2} = \frac{\ddot H}{\dot H}\frac{\dot g}{g} - \frac{\dot H^2}{H^2\ln H}\left(1+\frac{1}{\ln H}\right). \tag{1.95}$$

For this reason, one can apply the following identification

$$\frac{\ddot g}{g} = \frac{\ddot H}{\dot H}\frac{\dot g}{g}, \tag{1.96}$$

$$\frac{\dot g^2}{g^2} = \frac{\dot H^2}{H^2\ln H}\left(1+\frac{1}{\ln H}\right). \tag{1.97}$$

The equation (1.96) can be rewritten as

$$\frac{\ddot H}{\dot H} = \frac{\ddot g}{\dot g}. \tag{1.98}$$

Its integration can be performed straightforwardly

$$\ln\dot H = \ln\dot g + \ln C_0, \tag{1.99}$$

where C_0 is the integration constant, or equivalently

$$\dot H = C_0\dot g. \tag{1.100}$$

Integration of the equation (1.100) gives

$$H = C_0 g + C_1, \tag{1.101}$$

where C_1 is the integration constant. In this manner,

$$\frac{\dot H^2}{H^2} = \left(\frac{C_0\dot g}{C_0 g + C_1}\right)^2 = \frac{\dot g^2}{(g+C_1')^2}, \tag{1.102}$$

where $C_1' = \frac{C_1}{C_0}$, and the equation (1.97) can be rewritten as

$$\frac{(g+C_1')^2}{g^2} = \frac{1}{\ln H}\left(1+\frac{1}{\ln H}\right), \tag{1.103}$$

or in equivalent form of the algebraic equation for unknown $\ln H$

$$\left(g+C_1'\right)^2(\ln H)^2 - g^2\ln H - g^2 = 0, \tag{1.104}$$

which has two solutions

$$\ln H = \frac{1}{2}\left(\frac{g}{g+C_1'}\right)^2\left[1\pm\sqrt{1+4\left(\frac{g+C_1'}{g}\right)^2}\right], \tag{1.105}$$

and by $\ln H = \frac{g}{\exp H} > 1$ the correct solution is

$$\ln H = \frac{1}{2}\left(\frac{g}{g+C_1'}\right)^2\left[1+\sqrt{1+4\left(\frac{g+C_1'}{g}\right)^2}\right]. \tag{1.106}$$

Merging the solutions (1.101) and (1.106) allows to establish the algebraic identity for the function g

$$\ln(C_0g+C_1) = \frac{1}{2}\left(\frac{C_0g}{C_0g+C_1}\right)^2\left[1+\sqrt{1+4\left(\frac{C_0g+C_1}{C_0g}\right)^2}\right], \tag{1.107}$$

Equivalently, however, by using $\ln H = \frac{g}{\exp H}$ within the equation (1.104) one obtains the algebraic equation for unknown function $\exp H$

$$(\exp H)^2 + g\exp H - (g+C_1')^2 = 0, \tag{1.108}$$

having also two possible solutions

$$\exp H = \frac{g}{2}\left(1\pm\sqrt{1+4\left(\frac{g+C_1'}{g}\right)^2}\right), \tag{1.109}$$

and because $\exp H > 1$, $g > 1$ the correct solution is

$$\exp H = \frac{g}{2}\left(1+\sqrt{1+4\left(\frac{g+C_1'}{g}\right)^2}\right), \tag{1.110}$$

and by using the solution (1.101) one obtains another identity for g

$$\exp(C_0 g + C_1) = \frac{g}{2}\left(1+\sqrt{1+4\left(\frac{C_0 g+C_1}{C_0 g}\right)^2}\right). \tag{1.111}$$

Anyway, however, employing the solution (1.101) within the definition $g = \exp H \ln H$ allows to establish one more identity for g

$$g = \exp(C_0 g + C_1)\ln(C_0 g + C_1). \tag{1.112}$$

Another identity can be established by application of the equation (1.93) and the relation (1.100) which leads to

$$\ln H = \frac{C_0 g}{1 - C_0 g}\frac{1}{C_0 g + C_1}, \tag{1.113}$$

and when applied to the equation (1.104) this gives the algebraic equation for the function g

$$C_0^2 g^3 + (C_0 C_1 - 3C_0)g^2 + (2 - C_0^4 - 2C_1)g + C_0^3 C_1 - \frac{C_1}{C_0} = 0, \tag{1.114}$$

having three solutions in general. Taking into account the Riccati equation (1.87), *i.e.* non-zero value of $\dot{g} = g + h$, one can differentiate both sides of the equation (1.114) and obtain

$$[3C_0^2 g^2 + 2(C_0 C_1 - 3C_0)g + 2 - C_0^4 - 2C_1]\dot{g} = 0, \tag{1.115}$$

and conclude that the following constraint is fulfilled

$$3C_0^2 g^2 + 2(C_0 C_1 - 3C_0)g + 2 - C_0^4 - 2C_1 = 0. \tag{1.116}$$

It allows to establish the value of function g

$$g = g(C_0, C_1) = -\frac{C_1 - 3}{3C_0} \pm \frac{1}{3C_0}\sqrt{C_1^2 + 3C_0^4 + 3} = const. \tag{1.117}$$

Taking the positive solution as physical

$$g(C_0,C_1) = -\frac{C_1-3}{3C_0} + \frac{1}{3C_0}\sqrt{C_1^2+3C_0^4+3}. \tag{1.118}$$

one can derive the value of the Hubble parameter

$$H = 1+\frac{2}{3}C_1+\frac{1}{3}\sqrt{C_1^2+3C_0^4+3}. \tag{1.119}$$

Therefore, the problem is to establish the constants C_0 and C_1. It can be seen straightforwardly that

$$C_1 = H_I - C_0 g_I, \tag{1.120}$$

$$C_0 = \frac{\dot{H}_I}{\dot{g}_I}, \tag{1.121}$$

where $g_I = g(t_I)$ and $\dot{g}_I = \frac{dg}{dt}(t_I)$. Because the function g and its derivative are determined explicitly by the relation (1.91) and the Riccati equation (1.87), one has

$$g_I = 1, \tag{1.122}$$

$$\dot{g}_I = g_I + h(t_I) = 2, \tag{1.123}$$

and for this reason one obtains finally

$$C_1 = H_I - \frac{\dot{H}_I}{2}, \tag{1.124}$$

$$C_0 = \frac{\dot{H}_I}{2}. \tag{1.125}$$

In other words, in such a situation the Hubble parameter has the form

$$H = 1+\frac{2}{3}\left(H_I-\frac{\dot{H}_I}{2}\right)+\frac{1}{3}\sqrt{H_I^2-H_I\dot{H}_I+\frac{3}{16}\dot{H}_I^4-\frac{1}{4}\dot{H}_I^2+3}. \tag{1.126}$$

Particularly, in the case when $C_1 = 0$ one obtains

$$H_I = H_0\exp\{2(t_I-t_0)\}, \tag{1.127}$$

and the actual value of the Hubble parameter becomes

$$H = 1+\frac{1}{\sqrt{3}}\sqrt{1+H_0^4\exp\{8(t_I-t_0)\}}. \tag{1.128}$$

It can be seen by straightforward computation that if one takes into account the usual definition of the density of the energy of Matter fields

$$\mathcal{H}_M(x,\cdot) = \frac{d\varepsilon_M(\cdot)}{dV(x)}, \tag{1.129}$$

where $\cdot = t, \eta$, $dV(x) = d^3x$ is an infinitesimal volume, and $d\varepsilon_M(\cdot)$ is an infinitesimal energy of Matter fields contained in such a volume, then

$$\varepsilon_M = |\mathrm{H}_M|, \tag{1.130}$$

and, consequently, the relations (1.82) and (1.83) allow to establish dependence of the energy of Matter fields on the cosmological redshift

$$\varepsilon_M(t) = 3M_P\ell_P^2\frac{V}{V_P}\left|\frac{a(t_I)}{a(\eta_I)}\right|^2\left|\frac{1}{1+z(t_I,t)}\frac{dz(t_I,t)}{dt}\right|^2, \tag{1.131}$$

$$\varepsilon_M(\eta) = 3M_P\ell_P^2\frac{V}{V_P}\left|\frac{1}{a(\eta_I)}\right|^2\left|\frac{dz(\eta_I,\eta)}{d\eta}\right|^2. \tag{1.132}$$

By application of the formulas (5.160) and (1.132) and definitions (1.14), (1.60) and (1.79), it can be seen straightforwardly that $\mathcal{H}_M$, ε_M, and H are diffeoinvariants of the coordinate time

$$\mathcal{H}_M = \text{inv.}, \qquad \varepsilon_M = \text{inv.}, \qquad H(a) = \text{inv.}, \tag{1.133}$$

and, moreover, the relation holds

$$\varepsilon_M(t) = 3E_P\ell_P^2\frac{V}{V_P}\left|\frac{a(t_I)}{a(\eta_I)}\right|^2\left|\frac{\dot{a}(t)}{a(t)}\right|^2 = 3E_P\ell_P^2\frac{V}{V_P}\left|\frac{a(t_I)}{a(\eta_I)}\right|^2|H(a)|^2. \tag{1.134}$$

Using the relations (1.60) and (6.75) one obtains

$$\left|\frac{a(t_I)}{a(\eta_I)}\right|^2 = 1. \tag{1.135}$$

Integration of the equation (6.75) leads to the relation

$$\frac{a(t)}{a(t_I)} = \exp\left\{\pm\frac{a(\eta_I)}{a(t_I)}\int_{t_I}^{t} dt'\sqrt{\frac{1}{M_P\ell_P^2}\frac{V_P}{3V}\varepsilon_M(t')}\right\}. \tag{1.136}$$

If one identifies $a(t_I) = a(\eta_I) = a_I$, and $H(a)$ as real and positive then the Hubble parameter is given by the equation

$$H^2(a) = \frac{1}{E_P \ell_P^2} \frac{V_P}{3V} \varepsilon_M, \tag{1.137}$$

and the cosmic scale factor parameter takes the form

$$a(t) = a_I \exp\left\{ \pm \int_{t_I}^{t} dt' \sqrt{\frac{1}{M_P \ell_P^2} \frac{V_P}{3V} \varepsilon_M(t')} \right\}. \tag{1.138}$$

In general, the energy of Matter fields (1.137) defines the evolution of the classical Universe. However, because ε_M is invariant with respect to exchange between cosmic time and conformal time, it can be considered rather as function of the cosmological redshift z or the cosmic scale factor parameter a. Moreover, when there are several kinds of Matter fields then ε_M can be considered as the total energy of Matter fields, and for this reason is an algebraical sum of all contributions

$$\varepsilon_M = \sum_i \varepsilon_i, \tag{1.139}$$

where the subscript i is associated with the type of Matter fields. In this manner, the evolution (1.137) can be considered as the dependence of the cosmological redshift on cosmic as well as conformal time

$$\sqrt{\frac{1}{E_P} \sum_i \varepsilon_i(z)} = -\ell_P \sqrt{\frac{3V}{V_P}} \frac{\dot{z}}{1+z}, \tag{1.140}$$

$$\sqrt{\frac{1}{E_P} \sum_i \varepsilon_i(z)} = -\ell_P \sqrt{\frac{3V}{V_P}} z', \tag{1.141}$$

as well as the dependence of the cosmic scale factor parameter on cosmological as well as conformal time

$$\sqrt{\frac{1}{E_P} \sum_i \varepsilon_i(a)} = -\ell_P \sqrt{\frac{3V}{V_P}} \frac{\dot{a}}{a}, \tag{1.142}$$

$$\sqrt{\frac{1}{E_P} \sum_i \varepsilon_i(a)} = -\ell_P \sqrt{\frac{3V}{V_P}} \frac{a'}{a^2}. \tag{1.143}$$

Interestingly, the Hamiltonian constraint (1.55), describing the classical toroidal Einstein–Friedmann Universe, can be parameterized in the way characteristic for a bosonic string, *i.e.* by using the Einstein energy-momentum relation

$$\mathrm{P}_a^2 c^2 + m^2(a) c^4 = E^2(a), \tag{1.144}$$

where the energy identically vanishes $E^2(a) = 0$, and the squared mass is manifestly negative

$$m^2(a) = -\left(3M_P \ell_P \frac{V}{V_P}\right)^2 a^4 H^2(a) < 0, \tag{1.145}$$

Hence, it is easy to deduce that the Universe can be understood as the tachyon, *i.e.* the groundstate of the classical bosonic string [174]. The mass of the tachyon can be expressed through two ways

$$\sqrt{|m^2(a)|} = 3\frac{V}{V_P} M_P \ell_P a' = \frac{3}{2}\frac{V}{V_P} M_P \ell_P \dot{\Omega}, \tag{1.146}$$

where $\Omega = a^2(t)$ is the scaling function (1.17) and $S_P = 4\pi\ell_P^2$ is the area of the Planck sphere. The Hubble law (1.77)-(1.78) can be interpreted as the relation between the volume of space and the mass of the tachyon

$$M_P \int_{a_I}^{a} \frac{da'}{\sqrt{|m^2(a')|}} = \frac{V_P}{3V}\frac{\eta - \eta_I}{t_P}, \tag{1.147}$$

$$M_P \int_{a_I}^{a} \frac{a' da'}{\sqrt{|m^2(a')|}} = \frac{V_P}{3V}\frac{t - t_I}{t_P}. \tag{1.148}$$

In this manner, we have received the classical point of view on the Einstein–Friedmann Universe. Let us construct the quantum theory of the Universe.

C Quantization of Hamiltonian Constraint

Application of the primary canonical quantization in the form

$$\left[\hat{\mathrm{P}}_a, \hat{a}\right] = -i\frac{\hbar}{\ell_P}, \tag{1.149}$$

where the parameter of quantization $\frac{\hbar}{\ell_P} = M_P c$ is the Planck momentum having the value

$$\frac{\hbar}{\ell_P} = M_P c \approx 6.524806271 \frac{\mathrm{kg}\cdot\mathrm{m}}{\mathrm{s}}. \tag{1.150}$$

In comparison to the standard quantum mechanics, in which the parameter of quantization is the reduced Planck constant $\hbar$, this value of the parameter of quantization is large, *i.e.* about 10^{34} times higher. Such a procedure allows to generate the operator of the momentum conjugated to the cosmic scale factor parameter

$$\hat{\mathrm{P}}_a = -i\frac{\hbar}{\ell_P}\frac{d}{da}. \tag{1.151}$$

Applying such form of the primary canonical quantization within the Hamiltonian constraint (1.144) of the Einstein–Friedmann Universe, one receives the one-dimensional Klein–Gordon equation

$$\left(\frac{d^2}{da^2} + \omega^2(a)\right)\Psi(a) = 0, \tag{1.152}$$

where ω, which plays the role of the frequency, is determined in the following way

$$\omega(a) = \frac{3V}{V_P}\ell_P a^2 H(a), \tag{1.153}$$

where $V_P = \frac{4}{3}\pi\ell_P^3$ is the volume of the Planck sphere, and ℓ_P is the Planck length. In the context of cosmology the one-dimensional Klein–Gordon equation (1.152) is the analog of the Wheeler–DeWitt equation [175, 176] describing more general theory known as the quantum geometrodynamics. In the standard understanding of quantum gravity the Wheeler–DeWitt equation is identified with non-relativistic quantum mechanics, *i.e.* plays the role of the Schrödinger equation. Because the cosmic scale factor parameter a is defined as a function of cosmic or conformal time, in general the wave function $\Psi(a)$ is a functional, and the equation (1.152) is a one-dimensional functional differential equation. In general, by the spirit of the Wheeler superspace, the wave function is a wave functional on the space of three-dimensional metrics characterized by the parameter a. We shall discuss details of this approach in next chapters of this part. Albeit, it should be emphasized that the approach based on the Schrödinger equation, which was proposed more than 50 years ago, since the mid-1980's has not led to any new constructive results.

A different possible interpretation of the quantum theory (1.152) is the Klein–Gordon equation, *i.e.* the relativistic wave equation describing bosons. It is not difficult to see by straightforward computations that the equation (1.152) can be generated as the classical field theoretic Euler–Lagrange equations of motion

$$\left[\frac{\partial}{\partial\Psi} - \frac{\partial}{\partial a}\frac{\partial}{\partial\left(\dfrac{d\Psi}{da}\right)}\right] L\left(\Psi, \frac{d\Psi}{da}\right) = 0, \tag{1.154}$$

by the conventional variational principle

$$\delta S[\Psi] = 0, \tag{1.155}$$

with respect to the action functional $S[\Psi]$ having the following form

$$S[\Psi] = \int da L\left(\Psi, \frac{d\Psi}{da}\right), \tag{1.156}$$

where the Lagrangian of the classical field, *i.e.* one-dimensional wave function $\Psi = \Psi(a)$, is

$$L\left(\Psi, \frac{d\Psi}{da}\right) = \frac{1}{2}\left(\frac{d\Psi}{da}\right)^2 - \frac{1}{2}\omega^2(a)\Psi^2. \tag{1.157}$$

The momentum Π_Ψ, canonically conjugated to the wave function Ψ, can be established immediately from the action (1.156) as

$$\Pi_\Psi = \frac{\partial L\left(\Psi, \dfrac{d\Psi}{da}\right)}{\partial\left(\dfrac{d\Psi}{da}\right)} = \frac{d\Psi}{da}, \tag{1.158}$$

and its straightforward application within the one-dimensional Klein–Gordon equation (1.152) allows to rewrite this equation in the following form

$$\frac{d\Pi_\Psi}{da} + \omega^2(a)\Psi(a) = 0. \tag{1.159}$$

The equations of motion (1.158) and (1.159) can be treated as the system of equations and employed for reduction of the order of the Wheeler–DeWitt equation (1.152). Such a reduction can be performed by introduction of the two-component field

$$\Phi = \begin{bmatrix} \Pi_\Psi \\ \Psi \end{bmatrix}, \tag{1.160}$$

which allows to express the model of quantum cosmology, given by the one-dimensional Klein–Gordon equation (1.152) and describing the Einstein–Friedmann Universe as the Multiverse of quantum universes, in the form of the one-dimensional Dirac equation

$$\left(-i\sigma_2\frac{d}{da} - M\right)\Phi = 0, \tag{1.161}$$

where M plays the role of the mass matrix, and σ_2 is one of the Pauli matrices,

$$M = \begin{bmatrix} -1 & 0 \\ 0 & -\omega^2 \end{bmatrix}, \tag{1.162}$$

$$\sigma_2 = \begin{bmatrix} 0 & -i \\ i & 0 \end{bmatrix}. \tag{1.163}$$

On the other hand, the one-dimensional Dirac equation (1.161) describes the two-component one-dimensional scalar field, which is an axion, and as the particular realization of the one-dimensional quantum mechanics is automatically equipped with the feature of supersymmetry. In this manner, the model of quantum cosmology based on the spatially finite and conformal-flat space-time equipped with the Friedmann–Lemaître–Robertson–Walker metric , *i.e.* the Einstein–Friedmann Universe, is strictly related to one-dimensional type of superstrings which we shall call here the Fermi–Bose superstrings. Such a name arises from the fact that in the dimension one the difference between the Fermi–Dirac and the Bose–Einstein statistics, describing respectively fermions and bosons, is absent as the straightforward ramification of the fact that the spin is identically equal to zero in the dimension one. The appropriate quantum field theory will be constructed relying on the suitable repère in the Fock space of creators and annihilators, the algebraic system of quantum mechanics investigated by Vladimir Alexandrovich Fock [177] and frequently applied to the consistent and constructive description of quantum states in which the number of quantum states in unknown or changeable. Such a specific quantum field theory possesses the natural philosophical interpretation as the quantum theory describing the Multiverse of the Fermi–Bose superstrings, which can be called the superstring Multiverse. In the context of the model of quantum cosmology with the finite volume, the superstring Multiverse contains only the system of multiple quantum Einstein–Friedmann Universes.

Because the model of quantum cosmology (1.152) has manifestly the form of the one-dimensional Klein–Gordon equation, it is natural to perform the canonical secondary quantization of the one-dimensional Dirac equation (1.161) in the way appropriate for bosons [178]

$$\left[\hat{\Pi}_\Psi[a], \hat{\Psi}[a']\right] = -i\delta\left(a - a'\right), \tag{1.164}$$

and other canonical commutation relations are trivial. In the explicit form the bosonic field is

$$\left[\begin{array}{c} \hat{\Psi} \\ \hat{\Pi}_\Psi \end{array}\right] = \left[\begin{array}{cc} \dfrac{1}{\sqrt{2\omega(a)}} & \dfrac{1}{\sqrt{2\omega(a)}} \\ -i\sqrt{\dfrac{\omega(a)}{2}} & i\sqrt{\dfrac{\omega(a)}{2}} \end{array}\right] \left[\begin{array}{c} G[a] \\ G^\dagger[a] \end{array}\right], \tag{1.165}$$

and is consistent with the algebraic approach to canonical commutation relations by John von Neumann, and Huzihiro Araki and E.J. Woods [179]. As a result the Universe is described by the dynamical basis

$$B_a = \left\{\left[\begin{array}{c} G[a] \\ G^\dagger[a] \end{array}\right] : \left[G[a], G^\dagger[a']\right] = \delta\left(a - a'\right), \left[G[a], G[a']\right] = 0\right\}, \tag{1.166}$$

satisfying non-Heisenberg equations of motion

$$\frac{dB_a}{da} = \begin{bmatrix} -i\omega(a) & \frac{1}{2\omega(a)}\frac{d\omega(a)}{da} \\ \frac{1}{2\omega(a)}\frac{d\omega(a)}{da} & i\omega(a) \end{bmatrix} B_a, \tag{1.167}$$

and by perturbation theory holds

$$\left|\frac{1}{2\omega(a)}\frac{d\omega(a)}{da}\right| \ll \frac{1}{a}. \tag{1.168}$$

The equations (1.167) can be diagonalized by taking into account the new basis B'_a

$$B'_a = \left\{ \begin{bmatrix} G'[a] \\ G'^{\dagger}[a] \end{bmatrix} : \left[G'[a], G'^{\dagger}[a']\right] = \delta\left(a - a'\right), \left[G'[a], G'[a']\right] = 0 \right\}, \tag{1.169}$$

obtained by taking together the Bogoliubov transformation and the Heisenberg equations of motion

$$B'_a = \begin{bmatrix} u & v \\ v^* & u^* \end{bmatrix} B_a, \tag{1.170}$$

$$\frac{dB'_a}{da} = \begin{bmatrix} -i\omega' & 0 \\ 0 & i\omega' \end{bmatrix} B'_a, \tag{1.171}$$

where the hyperbolic constraint holds

$$|u|^2 - |v|^2 = 1, \tag{1.172}$$

and ω' is the unknown. It can be seen by straightforward computation that such a procedure generates unambiguously $\omega' = 0$ and, therefore, the new basis is the static Fock repère

$$B_I = \left\{ \begin{bmatrix} \mathrm{w}_I \\ \mathrm{w}_I^{\dagger} \end{bmatrix} : \left[\mathrm{w}_I, \mathrm{w}_I^{\dagger}\right] = 1, \left[\mathrm{w}_I, \mathrm{w}_I\right] = 0 \right\}, \tag{1.173}$$

and for this reason the static vacuum state is obtained

$$\langle 0| = \left\{ \langle 0| : \mathrm{w}_I \langle 0| = 0, \quad 0 = |0\rangle \mathrm{w}_I^{\dagger} \right\}. \tag{1.174}$$

The system of operator equations (1.167) transits to the system of equations for the Bogoliubov coefficients

$$\frac{d}{da}\begin{bmatrix} v(a) \\ u(a) \end{bmatrix} = \begin{bmatrix} -i\omega(a) & \dfrac{1}{2\omega(a)}\dfrac{\partial\omega(a)}{\partial a} \\ \dfrac{1}{2\omega(a)}\dfrac{\partial\omega(a)}{\partial a} & i\omega(a) \end{bmatrix}\begin{bmatrix} v(a) \\ u(a) \end{bmatrix}, \tag{1.175}$$

which is easy to solve in the superfluid parametrization [180]

$$v(a) = \exp(i\theta(a))\sinh\phi(a), \tag{1.176}$$

$$u(a) = \exp(i\theta(a))\cosh\phi(a), \tag{1.177}$$

where θ and ϕ are the angle parameters, which in the present situation are expressed through mass of the tachyon as follows

$$\theta(a) = \pm i\int_{a_I}^{a}\omega(a')da', \tag{1.178}$$

$$\phi(a) = \ln\sqrt{\frac{\omega_I}{\omega(a)}}, \tag{1.179}$$

where $\omega_I = \omega(a_I)$ is the initial datum of $\omega(a)$.

In this manner, *the quantum cosmology is completely determined by the monodromy matrix C between the dynamical and the static Fock repère $B_a = CB_I$ which is explicitly given by*

$$C = \begin{bmatrix} \left(\sqrt{\dfrac{\omega(a)}{\omega_I}} + \sqrt{\dfrac{\omega_I}{\omega(a)}}\right)\dfrac{e^{\lambda}}{2} & \left(\sqrt{\dfrac{\omega(a)}{\omega_I}} - \sqrt{\dfrac{\omega_I}{\omega(a)}}\right)\dfrac{e^{-\lambda}}{2} \\ \left(\sqrt{\dfrac{\omega(a)}{\omega_I}} - \sqrt{\dfrac{\omega_I}{\omega(a)}}\right)\dfrac{e^{\lambda}}{2} & \left(\sqrt{\dfrac{\omega(a)}{\omega_I}} + \sqrt{\dfrac{\omega_I}{\omega(a)}}\right)\dfrac{e^{-\lambda}}{2} \end{bmatrix}, \tag{1.180}$$

where λ is given by the formula

$$\lambda = \lambda(a) = i\theta(a) = \mp\int_{a_I}^{a}\omega(a')da'. \tag{1.181}$$

Now the quantum field $\hat{\Psi}$ can be computed straightforwardly

$$\hat{\Psi}[a] = \frac{1}{\sqrt{2\omega_I}}\left(e^{\lambda(a)}\mathrm{w}_I + e^{-\lambda(a)}\mathrm{w}_I^{\dagger}\right), \tag{1.182}$$

and if one introduces the state

$$|n\rangle \equiv \left(\hat{\Psi}[a]\right)^n \langle 0| = \frac{e^{-n\lambda(a)}}{(2\omega_I)^{n/2}} \mathrm{w}_I^{\dagger\ n} |0\rangle, \tag{1.183}$$

then the interesting relations can be derived

$$\langle m|n\rangle = \frac{e^{(m-n)\lambda(a)}}{(2\omega_I)^{(m+n)/2}} \langle 0| \mathrm{w}_I^m \mathrm{w}_I^{\dagger\ n} |0\rangle, \tag{1.184}$$

$$\langle n|n\rangle = \sum_{p=0}^{n} \frac{e^{i\pi(p+1)}}{(2\omega_I)^n} C_{n-p}^n \langle 0| \left(\mathrm{w}_I^\dagger \mathrm{w}_I\right)^p |0\rangle = \frac{e^{i(2n+1)\pi}}{(2\omega_I)^n}, \tag{1.185}$$

where C_k^n is the Newton binomial symbol, which in the present case is

$$C_{n-p}^n = \frac{n!}{p!(n-p)!}, \tag{1.186}$$

and we have applied the normalization of the stable vacuum state

$$\langle 0|0\rangle = 1. \tag{1.187}$$

Obviously the quantum state $|n\rangle$ is a n-particle state, which in the present cosmological context describes the superstring Multiverse. Its physical significance arises from the straightforward analogy with the quantum theory of many body systems (For basics of the theory and certain applications Cf. *e.g.* Ref. [181]). Namely, in the case of the quantum cosmology one has to deal with the system of many quantum universes. Such a specific quantum many-body system, *i.e.* the secondary-quantized Einstein–Friedmann Universe given by the quantum field (1.182) and the many particle states (1.183), is manifestly a realization of the Multiverse hypothesis. Let us consider now the thermodynamics of such a Multiverse.

D Thermodynamics of Multiverse

Because we have derived the static basis (1.173), there formally also exists the thermal equilibrium for quantum states of the Einstein–Friedmann Universe (1.4). In this section we shall formulate the thermodynamics of the system of many quantum universes. We shall apply the simplest approximation, *i.e.* we shall consider the system with one degenerated state which is defined by the density operator ρ_G having the form

$$\rho_G = G^\dagger G = \tag{1.188}$$

$$= B_a^\dagger \begin{bmatrix} 1 & 0 \\ 0 & 0 \end{bmatrix} B_a = \tag{1.189}$$

$$= B_I^\dagger \begin{bmatrix} |u|^2 & -uv \\ -u^* v^* & |v|^2 \end{bmatrix} B_I \equiv B_I^\dagger \rho_{\rm eq} B_I, \tag{1.190}$$

where $\rho_{\rm eq}$ is the density operator in thermal equilibrium. By straightforward computation of the Boltzmann–von Neumann entropy S

$$\mathrm{S} = \frac{\mathrm{tr}\left(\rho_{\rm eq}\ln\rho_{\rm eq}\right)}{\mathrm{tr}\rho_{\rm eq}} = \tag{1.191}$$

$$= \ln\left(2|u|^2 - 1\right) \equiv -\ln\Omega_{\rm eq}, \tag{1.192}$$

it is easy to deduce the distribution function $\Omega_{\rm eq}$:

$$\Omega_{\rm eq} = \frac{1}{2|u|^2 - 1} = \frac{1}{2|v|^2 + 1}. \tag{1.193}$$

On the other side, one can use the occupation number n

$$n \equiv \langle 0|G^\dagger[a]G[a]|0\rangle = \tag{1.194}$$

$$= \frac{1}{4}\left(\frac{\omega(a)}{\omega_I} + \frac{\omega_I}{\omega(a)}\right) - \frac{1}{2} = |v|^2, \tag{1.195}$$

where $m_I = m(a_I)$, to derivation of the entropy

$$\mathrm{S} = \ln\langle \mathrm{n}\rangle, \tag{1.196}$$

where $\langle \mathrm{n}\rangle$ is the averaged occupation number

$$\langle \mathrm{n}\rangle = 2n + 1 = \frac{1}{2}\left(\frac{\omega(a)}{\omega_I} + \frac{\omega_I}{\omega(a)}\right). \tag{1.197}$$

By combination of the formulas (1.192), (1.195) and (1.196), with the natural conditions $n \geq 0$ and $\omega(a) \geq \omega_I$, one receives the mass spectrum of the system of many Einstein–Friedmann Universes

$$\frac{\omega(a)}{\omega_I} = \langle \mathrm{n} \rangle + \sqrt{\langle \mathrm{n} \rangle^2 - 1}. \tag{1.198}$$

Similarly, by application of the averaging method one can derive the internal energy U

$$\begin{aligned} \mathrm{U} &\equiv \frac{\mathrm{tr}(\rho_{\mathrm{eq}}\mathrm{H}_{\mathrm{eq}})}{\mathrm{tr}\rho_{\mathrm{eq}}} = & (1.199)\\ &= \left(\langle \mathrm{n} \rangle + \frac{1}{2}\right)\left(\langle \mathrm{n} \rangle + \sqrt{\langle \mathrm{n} \rangle^2 - 1}\right)\omega_I, & (1.200) \end{aligned}$$

and chemical potential μ

$$\begin{aligned} \mu &\equiv \frac{\partial \mathrm{U}}{\partial \mathrm{n}} = & (1.201)\\ &= 2\left(\frac{\langle \mathrm{n} \rangle + \frac{1}{2}}{\sqrt{\langle \mathrm{n} \rangle^2 - 1}} + 1\right)\left(\langle \mathrm{n} \rangle + \sqrt{\langle \mathrm{n} \rangle^2 - 1}\right)\omega_I, & (1.202) \end{aligned}$$

where we have applied the Hamiltonian of the classical theory expressed in the static Fock repère

$$H = \left(G^\dagger[a]G[a] + G[a]G^\dagger[a]\right)\frac{\omega(a)}{2} = B_I^\dagger \mathrm{H}_{\mathrm{eq}} B_I, \tag{1.203}$$

where H_{eq} is the Hamiltonian in equilibrium

$$H_{\mathrm{eq}} = \begin{bmatrix} |u|^2 + |v|^2 & -2uv \\ -2u^*v^* & |u|^2 + |v|^2 \end{bmatrix} \frac{\omega(a)}{2}. \tag{1.204}$$

Let us establish the temperature of the many universe system by application of the method of quantum statistics. For this one must take into account the concrete form of the quantum statistics of the system, which can arise from the nature of the system or some other deductions. The bosonic character of the model of quantum cosmology (1.152) manifestly suggests application of the Bose–Einstein statistics which naturally describes bosonic systems. In this manner, one can *ad hoc* take into account the following statistics

$$\Omega_{\mathrm{eq}} \equiv \frac{1}{\exp\left\{\frac{\mathrm{U}-\mu n}{\mathrm{T}}\right\}-1}. \tag{1.205}$$

Because the quantum statistics derived in this section has the form

$$\Omega_{\mathrm{eq}} = \frac{1}{|u|^2+|v|^2} = \frac{1}{2|u|^2-1}, \tag{1.206}$$

after taking into account the obvious relation $2|u|^2 = \langle n\rangle + 1$ one can establish the relation

$$\frac{\mathrm{U}-\mu n}{\mathrm{T}} = \ln\left(\langle n\rangle + 1\right). \tag{1.207}$$

Consequently, such a procedure allows to determine the temperature of the system of the quantum states of the Universe as the fixed parameter

$$\mathrm{T} = \mathrm{T}_0\frac{\ln 2}{\ln(\langle \mathrm{n}\rangle+1)}\left(\langle \mathrm{n}\rangle + \sqrt{\langle \mathrm{n}\rangle^2-1}\right)\left(1+\frac{2\langle \mathrm{n}\rangle+1}{3}\sqrt{\frac{\langle \mathrm{n}\rangle-1}{\langle \mathrm{n}\rangle+1}}\right), \tag{1.208}$$

where T_0 is the temperature corresponding with $n = 0$ is

$$\mathrm{T}_0 = \frac{3\omega_I}{2\ln 2}. \tag{1.209}$$

E Early Light Multiverse

Interestingly, there is a certain particular thermodynamical situation of the system of many quantum universes. Namely, these are the quantum universes possessing minimal entropy

$$\mathrm{S}_{\min} = 0. \tag{1.210}$$

It can be deduced straightforwardly that such a Multiverse is characterized by the following conditions

$$n = 0, \quad \langle \mathrm{n}\rangle = 1, \quad \omega(a) = \omega_I, \tag{1.211}$$

and, therefore, its thermodynamical parameters have the values

$$\mathrm{U}_{\min} = \frac{3}{2}\omega_I, \tag{1.212}$$

$$\mu_{\min} = \infty, \tag{1.213}$$

$$\mathrm{T}_{\min} = \frac{\mathrm{U}_{\max}}{\ln 2}. \tag{1.214}$$

The relations (1.212), (1.213) and (1.214) shows that such a specific collection of quantum universes is determined by only the initial datum, *i.e.* the parameter ω_I. The relation (1.214) nontrivially connects temperature and internal energy of these quantum states and, therefore, in the general spirit of statistical mechanics it is the equation of state of the collection of quantum universes. The infinite value of the chemical potential is the only characteristic syndrome of openness of the system, and shows that at the point $n = 0$ a phase transition happens. Such a point can be understood as the point at which quantum universes start their existence. There is a question about the physical meaning of the quantum states of Universe characterized by the minimal entropy. It is not difficult to see that for such Universes holds the relation

$$\omega_I = \pm\frac{3V}{V_P}\ell_P a_I^2 H(a_I), \tag{1.215}$$

which defines the initial datum of the Hubble parameter

$$H(a_I) = \frac{Q}{a_I^2}, \tag{1.216}$$

where Q is a constant which depends on two free parameters: the initial datum of the mass and the volume of space

$$Q = \pm\frac{V_P}{3V}\frac{\omega_I}{\ell_P} = const. \tag{1.217}$$

Interestingly, when ω_I is finite and non-zero and the volume of space is infinite, then identically $Q \equiv 0$ and, therefore, the initial datum of the Hubble parameter also vanishes. It proves that for consistency the volume of space must be finite and non-zero. In such a situation, the value of the Hubble parameter (1.216) corresponds with radiation, and this is the physical sense of this collection of quantum universes. In this manner, the Multiverse of such quantum universes expresses the cosmological nature of light, and for this reason we propose to call this *the light Multiverse*. It can be seen by straightforward derivation from the Hubble parameter (1.216) and its definition

$$H(a_I) = \frac{\dot{a}_I}{a_I} = \frac{a'_I}{a_I^2}, \tag{1.218}$$

that the light Multiverse is described by the following initial values of the cosmic scale factor parameter

$$a_I(t_I, t_0) = \sqrt{a_0^2 + 2Qc(t_I - t_0)}, \tag{1.219}$$

$$a_I(\eta_I, \eta_0) = a_0 + Qc(\eta_I - \eta_0), \tag{1.220}$$

where t_0, η_0, $a_0 = a_I(t_0, t_0) = a_I(\eta_0, \eta_0)$ are the the integration constants, and

$$a_I(t_I, t_0) \geqslant a_0, \tag{1.221}$$

$$a_I(\eta_I, \eta_0) \geqslant a_0. \tag{1.222}$$

However, because both the scale factor parameters $a_I(t_I, t_0)$ and $a_I(\eta_I, \eta_0)$ are the only constants, one can suggest *ad hoc* that they are equal

$$a_I(t_I, t_0) = a_I(\eta_I, \eta_0) \equiv a_I. \tag{1.223}$$

Such a conjecture allows to establish the linkage between the cosmological and the conformal times for the early evolution of the Multiverse

$$\eta_I - \eta_0 = \frac{a_0}{Qc}\left(\sqrt{1 + \frac{2Qc}{a_0^2}(t_I - t_0)} - 1\right). \tag{1.224}$$

The Taylor series expansion of this relation is also interesting

$$\eta_I - \eta_0 = \sqrt{\pi}\sum_{n=0}^{\infty} \frac{(2Qc/a_0)^n}{\Gamma(n+2)\Gamma\left(\frac{1}{2} - n\right)}(t_I - t_0)^{n+1}. \tag{1.225}$$

In the sense of perturbation theory the proposed quantum cosmology is consistent if and only if the condition (1.168) is satisfied. This inequality can be expressed in terms of the Hubble parameter

$$H(a) \ll H(a_I), \tag{1.226}$$

and straightforwardly integrated. Application of cosmic time allows to rewrite the inequality (1.226) as

$$\frac{\dot{a}}{a} \ll \frac{\dot{a}_I}{a_I}, \tag{1.227}$$

and straightforward integration in the region $a_0 \leqslant a_I \leqslant a$ gives

$$\ln\left|\frac{a}{a_I}\right| \ll \ln\left|\frac{a_I}{a_0}\right|, \tag{1.228}$$

what leads to the bound for cosmic scale factor parameter

$$a \ll \frac{a_I^2}{a_0}. \tag{1.229}$$

In the light of the inequalities (1.221) and (1.222) one has

$$\frac{a_I^2}{a_0} = \frac{a_I}{a_0} a_I \geqslant a_I, \tag{1.230}$$

which gives the following bound for the cosmic scale factor parameter

$$a \geqslant a_I, \tag{1.231}$$

which can be expressed equivalently as the bound for the cosmological redshift

$$z(t_I, t) \leqslant 0, \tag{1.232}$$

and defines the early Universe. Similarly, the inequality (1.226) can be rewritten in terms of conformal time

$$\frac{a'}{a^2} \ll \frac{a'_I}{a_I^2}, \tag{1.233}$$

and straightforwardly integrated with the result

$$-\frac{1}{a} + \frac{1}{a_I} \ll -\frac{1}{a_I} + \frac{1}{a_0}, \tag{1.234}$$

which finally gives the bound for the cosmic scale factor parameter

$$a \ll \frac{a_0 a_I}{2a_0 - a_I}. \tag{1.235}$$

Interestingly, the results (1.229) and (1.235) coincide in only one case, *i.e.* $a_I = a_0$.

For this reason, using the formulas (1.219) and (1.220), the early light Multiverse in itself expresses the applicability conditions for the model of quantum cosmology

$$a \ll a_0 + \frac{2Qc}{a_0}(t_I - t_0), \tag{1.236}$$

$$a \ll a_0 + \frac{Qc(\eta_I - \eta_0)}{1 - \frac{Qc}{a_0}(\eta_I - \eta_0)}. \tag{1.237}$$

However, because one can suggest also that the cosmic scale factor parameter expressed through cosmological and conformal time is the same, particularly in the case of early Universe, one obtains the equation

$$\frac{2Qc}{a_0}(t_I - t_0) = \frac{Qc(\eta_I - \eta_0)}{1 - \frac{Qc}{a_0}(\eta_I - \eta_0)}, \tag{1.238}$$

having the solution

$$\eta_I - \eta_0 = \frac{2(t_I - t_0)}{a_0 + 2\frac{Qc}{a_0}(t_I - t_0)}, \tag{1.239}$$

which compared with the relation (1.224)

$$\frac{2(t_I - t_0)}{a_0 + 2\frac{Qc}{a_0}(t_I - t_0)} = \frac{a_0}{Qc}\left(\sqrt{1 + \frac{2Qc}{a_0^2}(t_I - t_0)} - 1\right), \tag{1.240}$$

leads to the following equation

$$2\frac{Qc}{a_0}(t_I - t_0)\left[\left(2\frac{Qc}{a_0}(t_I - t_0)\right)^2 - 2\frac{Qc}{a_0}(t_I - t_0) - 1\right] = 0. \tag{1.241}$$

Interestingly, despite the fact that the equation (1.241) is satisfied for $t_I = t_0$, *i.e.* if $a_I = a_0$, it also possesses two other solutions

$$2\frac{Qc}{a_0}(t_I - t_0) = \varphi_{\pm}, \tag{1.242}$$

where $\varphi_{\pm}$ are the irrational constants

$$\varphi_{\pm} = \frac{1 \pm \sqrt{5}}{2}, \tag{1.243}$$

which in the case $\varphi_{+} = \varphi \approx 1.6180339887$ is the Fibonacci golden ratio, and in the case $\varphi_{-} = 1 - \varphi = \varphi - \sqrt{5}$. Employing the difference $t_I - t_0$ established by (1.242) within the equation (1.239), one receives

$$\eta_I - \eta_0 = \frac{a_0}{Qc} \frac{\varphi_{\pm}}{a_0 + \varphi_{\pm}}. \tag{1.244}$$

First let us consider the situation based on cosmic time. When one knows the value of the difference

$$\tau = t_I - t_0, \tag{1.245}$$

the relation (1.242) can be used for determination of the constant Q

$$Q = a_0 \frac{\varphi_{\pm}}{2c\tau}. \tag{1.246}$$

Because $Q = H_I a_I^2$ one has

$$\frac{a_I^2}{a_0} = \frac{\varphi_{\pm}}{2H_I c\tau}, \tag{1.247}$$

i.e. when one knows the initial datum of the Hubble parameter H_I and the difference (1.245) then by the condition (1.229) one obtains

$$a \ll \frac{\varphi_{\pm}}{2H_I c\tau}. \tag{1.248}$$

It is easy to see from (1.247) that the initial datum of the cosmic scale factor parameter is given by the initial value of this parameter

$$a_I = \sqrt{\frac{\varphi_{\pm}}{2H_I c\tau} a_0}, \tag{1.249}$$

which means that for consistency it must be $a_0 \neq 0$. The relations (1.247) and (1.249) allow to derive the ratio

$$\frac{a_I}{a_0} = \frac{\varphi_{\pm}}{2H_I c\tau} \frac{1}{a_I} = \sqrt{\frac{\varphi_{\pm}}{2H_I c\tau} \frac{1}{a_0}} \tag{1.250}$$

and, consequently, the initial datum of the cosmological redshift $z_I = z(t_0, t_I)$ is

$$\frac{a_I}{a_0} = \frac{1}{1+z_I}, \tag{1.251}$$

where

$$z(t_0, t) = \exp\left\{\pm\frac{a(\eta_0)}{a(t_0)}\int_{t_0}^{t} dt' \sqrt{\frac{1}{M_P \ell_P^2}\frac{V_P}{3V}|\mathrm{H}_M(t')|}\right\} - 1, \tag{1.252}$$

can be obtained straightforwardly

$$z_I = \sqrt{\frac{2a_0}{\varphi_\pm} H_I c\tau} - 1, \tag{1.253}$$

which after taking account that $a(\eta_0) = a(t_0) = a_0$ in the relation (1.252) leads to the result

$$\pm\int_{t_0}^{t_I} dt' \sqrt{\frac{1}{M_P \ell_P^2}\frac{V_P}{3V}|\mathrm{H}_M(t')|} = \frac{1}{2}\ln\left|\frac{2a_0}{\varphi_\pm} H_I c(t_I - t_0)\right|. \tag{1.254}$$

After differentiating of both sides of the equation (1.254) with respect to t_I one obtains

$$H_I = \sqrt{\frac{1}{M_P \ell_P^2}\frac{V_P}{3V}|\mathrm{H}_M(t_I)|} = \frac{1}{2c(t_I - t_0)}, \tag{1.255}$$

which after straightforward integration

$$\ln\frac{a_I}{a_0} = \frac{1}{2}\int_0^{t_I}\frac{dt'}{t' - t_0}, \tag{1.256}$$

on the one hand allows to establish in an unambiguous way the value of the initial datum for the cosmic scale factor parameter

$$a_I = a_0\sqrt{\frac{\tau}{t_0}}, \tag{1.257}$$

and, on the other hand, allows to derive unambiguously the energy of Matter fields

$$\varepsilon_M(t_I) = \frac{3}{4} E_P \left(\frac{t_P}{\tau}\right)^2 \frac{V}{V_P}. \tag{1.258}$$

Taking into account the Planck sphere $V = V_P$ and $\tau = t_P$, one receives

$$\varepsilon_M(t_I) = \frac{3}{4}E_P. \tag{1.259}$$

Similarly, for the Planck volume $V = \ell_P^3$ and $\tau = t_P$ one obtains

$$\varepsilon_M(t_I) = \frac{9}{16\pi}E_P. \tag{1.260}$$

Applying the formula (1.249) one receives both the initial value and the initial datum of the cosmic scale factor parameter

$$a_0 = \frac{\varphi_\pm}{2H_I c\tau}\frac{t_0}{\tau}, \tag{1.261}$$

$$a_I = \frac{\varphi_\pm}{2H_I c\tau}\sqrt{\frac{t_0}{\tau}}. \tag{1.262}$$

Moreover, because $Q = \pm\frac{V_P}{3V}\omega_I$ one can establish the initial datum

$$\omega_I = \pm\frac{3}{2}a_0\varphi_\pm\frac{t_P}{\tau}\frac{V}{V_P}, \tag{1.263}$$

which for the case of plus sign becomes

$$\omega_I^+ = \frac{3}{2}a_0\varphi\frac{t_P}{\tau}\frac{V}{V_P}, \tag{1.264}$$

while for the case of the minus sign is

$$\omega_I^- = \frac{3}{2}a_0(\varphi-1)\frac{t_P}{\tau}\frac{V}{V_P} = \frac{\varphi-1}{\varphi}\omega_I^+. \tag{1.265}$$

Interestingly, when one considers the Einstein–Friedmann Universe having the volume equals the Planck volume $V = \ell_P^3$ and the time difference (1.245) equal to the Planck time $\tau = t_P = \sqrt{\frac{\hbar G}{c^5}}$, then one obtains approximatively

$$\omega_I^+ = \frac{9\varphi}{8\pi}a_0, \tag{1.266}$$

$$\omega_I^- = \frac{9}{8\pi}(\varphi-1)a_0. \tag{1.267}$$

Similarly, for the Planck sphere $V = \frac{4}{3}\pi\ell_P^3$ one receives

$$\omega_I^+ = \frac{3}{2}\varphi a_0, \tag{1.268}$$

$$\omega_I^- = \frac{3}{2}(\varphi - 1)a_0. \tag{1.269}$$

Anyway, however, the relations (1.219) and (1.220) can be used to eliminate the parameter Q by the following combination

$$\frac{a_I^2(t_I,t_0) - a_0^2}{2(t_I - t_0)} = \frac{a_I(\eta_I,\eta_0) - a_0}{\eta_I - \eta_0}, \tag{1.270}$$

which after taking into account that $a_I(t_I,t_0) = a_I(\eta_I,\eta_0) = a_I$ becomes

$$\frac{a_I + a_0}{2} = \frac{t_I - t_0}{\eta_I - \eta_0}, \tag{1.271}$$

and leads to another relation between cosmological and conformal times

$$\eta_I - \eta_0 = \frac{2}{a_I + a_0}(t_I - t_0), \tag{1.272}$$

which in the case $a_0 = 0$ becomes

$$\eta_I - \eta_0 = \frac{2}{a_I}(t_I - t_0). \tag{1.273}$$

Applying the result (1.272) to the series (1.225) one obtains the rule

$$\frac{1}{\sqrt{\pi}} = \frac{a_I + a_0}{2} \sum_{n=0}^{\infty} \frac{(2Qc/a_0)^n}{\Gamma(n+2)\Gamma\left(\frac{1}{2} - n\right)}(t_I - t_0)^n. \tag{1.274}$$

Let us consider the situation $\eta = \eta_0$ and $t = t_0$. Then the cosmological redshift (1.82) and (1.83) must be redefined as follows

$$z(\eta_0,\eta) = \pm a(\eta_0) \int_{\eta_0}^{\eta} d\eta' \sqrt{\frac{1}{M_P\ell_P^2} \frac{V_P}{3V} |\mathrm{H}_M(\eta')|}, \tag{1.275}$$

Let us denote by z_0 the value of the cosmological redshift at the beginning of evolution of the classical Universe. Then using the definition (1.275), one can establish

$$z_0 = z(t_0,t_0) \equiv 0, \tag{1.276}$$

which is independent of the value of the cosmic scale factor parameter a_0 at the beginning of evolution of the Universe. The relations (1.84) and (1.85) computed for $\eta = \eta_0$ and $t = t_0$ are

$$\eta_I - \eta_0 = t_I - t_0 + \frac{1}{c}\int_{a_0}^{a_I} \frac{a'-1}{a'^2}\frac{da'}{H(a')}, \qquad (1.277)$$

$$a_0(\eta_I - \eta_0) = t_I - t_0 + \frac{1}{c}\int_0^{z_I} \frac{z'dz'}{(1+z')H(z')}, \qquad (1.278)$$

and by application of the result (1.272) become

$$\left(\frac{2}{a_I + a_0} - 1\right)(t_I - t_0) = \frac{1}{c}\int_{a_0}^{a_I} \frac{a'-1}{a'^2}\frac{da'}{H(a')}, \qquad (1.279)$$

$$\frac{a_I - a_0}{a_I + a_0}(t_I - t_0) = \frac{1}{c}\int_0^{z_I} \frac{z'dz'}{(1+z')H(z')}. \qquad (1.280)$$

Interestingly, the left hand side of the formula (1.279) vanishes identically when the initial datum a_I and the initial value a_0 of the cosmic scale factor parameter are constrained by the equation

$$a_I + a_0 = 2. \qquad (1.281)$$

If one expresses the integral on the right hand side of (1.279) by time variables, then this equation says that

$$0 = \int_{t_0}^{t_I} dt' - \int_{\eta_0}^{\eta_I} d\eta', \qquad (1.282)$$

which leads to the equality between the differences

$$\eta_0 - \eta_I = t_0 - t_I, \qquad (1.283)$$

which is consistent with the relation (1.272). Such an equality means that if the condition (1.281) holds then in such a region of evolution of the Universe conformal time and cosmic time flow in such a way that their difference is constant

$$\eta - t = const, \qquad (1.284)$$

and for convenience can be taken equal to zero. The equation (1.284) expresses the law of conservation for the difference $\eta - t$. In such a situation, the second relation (1.280) becomes

$$(1 - a_0)(t_I - t_0) = \frac{1}{c}\int_0^{z_I} \frac{z'dz'}{(1+z')H(z')}, \qquad (1.285)$$

where the initial datum of the cosmological redshift is

$$z_I = 2\frac{a_0 - 1}{2 - a_0} = 2\frac{1 - a_I}{a_I}. \tag{1.286}$$

Interestingly, when $a_0 = 1$ then $a_I = 1$ the initial datum of the cosmological redshift is trivial $z_I = 0$, and the left hand side of the formula (1.285) identically vanishes. After expressing the integral on the right hand side of (1.285) by time variables one obtains

$$0 = \int_{t_0}^{t_I} z(t_0, t')dt' = \int_{t_0}^{t_I} \left(\frac{a(t_0)}{a(t')} - 1\right) dt' = a_0 \int_{\eta_0}^{\eta_I} d\eta' - \int_{t_0}^{t_I} dt', \tag{1.287}$$

which can be computed straightforwardly

$$a_0(\eta_I - \eta_0) = t_I - t_0, \tag{1.288}$$

and for $a_0 = 1$ leads once again to the law of conservation (1.283).

Let us see what happens in the particular situation for which $a_0 = 0$. First, let us establish the initial datum of the cosmological redshift z_I

$$z_I = -1. \tag{1.289}$$

Because the constraint (1.281) one has $a_I = 2$ and, therefore, the relations (1.279) and (1.280) take the form

$$0 = \frac{1}{c}\int_0^2 \frac{a' - 1}{a'^2}\frac{da'}{H(a')}, \tag{1.290}$$

$$t_0 - t_I = \frac{1}{c}\int_{-1}^0 \frac{z'dz'}{(1 + z')H(z')}. \tag{1.291}$$

Interestingly, in such a case the equation (1.238) is simplified

$$\frac{Qc}{2}(\eta_I - \eta_0)^2 - (t_I - t_0) = 0, \tag{1.292}$$

and in the light of the relation (1.283) allows to establish the value of the parameter Q crucial for the initial datum of the Hubble parameter (1.216)

$$Q = 2\frac{t_I - t_0}{c(\eta_I - \eta_0)^2} = \pm\frac{2}{c\tau}, \tag{1.293}$$

where we have denoted $\pm\tau = \eta_I - \eta_0 = t_I - t_0$. In the light of the definition (1.217) one can determine the initial datum ω_I

$$\omega_I = \frac{6V}{V_P}\frac{t_P}{\tau}. \tag{1.294}$$

For the finite volume of space V and finite time τ the initial datum ω_I is established consistently. The time τ can be interpreted as the time between the beginning of the Universe and creation of initial data. If its value is taken *ad hoc* as identical to the Planck time $\tau = t_P$ then

$$\omega_I = \frac{6V}{V_P}, \tag{1.295}$$

and, therefore, if the value of the initial datum ω_I is small, $\omega_I \sim 1$ say, then the volume of space is

$$V \sim \frac{V_P}{6}. \tag{1.296}$$

Such a volume can be treated as the definition of the early Multiverse.

In the light of the result (1.291) one obtains another definition of the time τ

$$\tau = -\frac{1}{c}\int_{-1}^{0} \frac{z' dz'}{(1+z')H(z')}, \tag{1.297}$$

which by using the fact

$$H(z') = -\frac{1}{c}\frac{1}{1+z'}\frac{dz'}{dt'}, \tag{1.298}$$

and the definition $z' = z(t_0, t')$ becomes

$$\tau = \int_{t_0}^{t_I} z(t_0, t')dt', \tag{1.299}$$

or after expressing by the energy of Matter fields

$$\tau = \int_{t_0}^{t_I}\left[\exp\left[\pm\int_{t_0}^{t'} dt''\sqrt{\frac{1}{M_P\ell_P^2}\frac{V_P}{3V}\varepsilon_M(t'')}\right] - 1\right]dt'. \tag{1.300}$$

finally results in the relation

$$\tau = \frac{1}{2}\int_{t_0}^{t_I}\exp\left[\pm\int_{t_0}^{t'} dt''\sqrt{\frac{1}{M_P\ell_P^2}\frac{V_P}{3V}\varepsilon_M(t'')}\right]dt'. \tag{1.301}$$

The natural generalization of the equation (1.301) is

$$t_2 - t_1 = \frac{1}{2}\int_{t_1}^{t_2} \exp\left[\pm\int_{t_1}^{t'} dt''\sqrt{\frac{1}{M_P\ell_P^2}\frac{V_P}{3V}\varepsilon_M(t'')}\right] dt', \tag{1.302}$$

which in the case $t_1 = 0$ and $t_2 = t$ can be used for determination of cosmic time

$$t - t_0 = \frac{1}{2}\int_{t_0}^{t} \exp\left[\pm\int_{t_0}^{t'} dt''\sqrt{\frac{1}{M_P\ell_P^2}\frac{V_P}{3V}\varepsilon_M(t'')}\right] dt', \tag{1.303}$$

In the light of the relations (1.60) and (1.129) one has

$$H(z') = \sqrt{\frac{1}{E_P\ell_P^2}\frac{V_P}{3V}\varepsilon_M(z')}, \tag{1.304}$$

where $\varepsilon_M(z')$ is the energy of Matter fields. For the given energy of Matter fields dependence on the cosmological redshift z', the equation (1.304) can be solved as a differential equation for the cosmological redshift as an unknown function of cosmic or conformal time. Application of the explicit form of $\varepsilon_M(z')$ to the formula (1.297) results in the exact evaluation of the parameter τ. Equivalently, the cosmological redshift derived from the evolution (1.304) can be applied to the formula (1.299) for determination of time τ.

Interestingly, one can apply the equation (1.304) to the generalized relation (1.303)

$$\begin{aligned} t - t_0 &= \frac{1}{2}\int_{t_0}^{t} \exp\left[\pm\int_{t_0}^{t'} dt''\frac{1}{a(t'')}\frac{da(t'')}{dt''}\right] dt' = \\ &= \frac{1}{2}\int_{t_0}^{t} \exp\left[\pm\int_{a_0}^{a(t')}\frac{da}{a}\right] dt' = \frac{1}{2}\int_{t_0}^{t} \exp\left[\pm\ln\frac{a(t')}{a_0}\right] dt' = \\ &= \frac{1}{2}\int_{t_0}^{t}\left(\frac{a(t')}{a_0}\right)^{\pm 1} dt', \end{aligned} \tag{1.305}$$

and it is evident that the case of the plus sign, *i.e.* an expanding Universe, is distinguishable from the case of the minus sign, *i.e.* a collapsing Universe. The case of plus sign gives

$$\begin{aligned} (t - t_0)^+ &= \frac{1}{2a_0}\int_{t_0}^{t} a(t')dt' = \frac{1}{2a_0}\int_{t_0}^{t} a^2(t')\frac{dt'}{a(t')} = \\ &= \frac{1}{2a_0}\int_{\eta_0}^{\eta} a^2(\eta')d\eta', \end{aligned} \tag{1.306}$$

which after application of the definition of the Hubble parameter expressed in terms of conformal time

$$a^2(\eta)d\eta = \frac{1}{c}\frac{da}{H(a)}, \tag{1.307}$$

allows to establish the result

$$(t-t_0)^+ = \frac{1}{2ca_0}\int_{a_0}^{a}\frac{da'}{H(a')}. \tag{1.308}$$

The case of the minus sign leads to a completely different result

$$(t-t_0)^- = \frac{a_0}{2}\int_{t_0}^{t}\frac{dt'}{a(t')} = \frac{a_0}{2c}\int_{a_0}^{a}\frac{da'}{a'^2H(a')}, \tag{1.309}$$

or after straightforward computation

$$(t-t_0)^- = \frac{a_0}{2}\int_{\eta_0}^{\eta}d\eta' = \frac{a_0}{2}(\eta-\eta_0), \tag{1.310}$$

which allows to determine the non-trivial relation between cosmological and conformal time

$$t = \frac{1}{2}a_0\eta. \tag{1.311}$$

Because the plus sign is related to expansion of the Multiverse and the minus sign describes the opposite situation, the relations (1.306) and (1.309) for cosmic time mean that the flow of cosmic time in an expanding Multiverse is different than such flow in a collapsing Multiverse. Strictly speaking, the equation is satisfied

$$a_0^2d(t-t_0)^+ = a^2d(t-t_0)^-, \tag{1.312}$$

and because $dt_0^- = dt_0^+ = 0$ one has the condition

$$a_0^2dt^+ - a^2dt^- = 0. \tag{1.313}$$

In other words, cosmic time flows in the same way if and only if for an arbitrary value of conformal or cosmic time the relation $a^2 \equiv a_0^2$ holds identically. In such a situation, the relation (1.311) is automatically satisfied, while the Hubble parameter identically vanishes $H = \frac{1}{a^2}\frac{da}{d\eta} =$

0. In this manner, such an approach suggests that the Multiverse is non-dynamical or is dynamical but considered at the fixed moment of time.

On the other side, however, it can be seen straightforwardly that application of the trivial value $H = 0$ of the Hubble parameter within the definitions (1.308) and (1.309) of cosmic time leads to manifestly divergent integrands. Such a singular behaviour suggests that the non-dynamical Multiverse has non-physical nature or, in other words, that the fixation of time has purely non-physical nature. However, from the point of view of modern physics fixation of time is the standard tool in the theoretical explanations of quantum field theory.

Interestingly, there is another possible interpretation of the relations (1.312) and (1.313). Namely, one can say that during the expansion of the Multiverse the initial value of the cosmic scale factor parameter is *ad hoc* established to a certain value a_0, while during a collapse of the Multiverse this initial value is undetermined. In such a situation, the identification $a^2 = a_0^2$ means exactly that Multiverse is non expanding and is not collapsing, *i.e.* is non dynamical.

F Summary

In this chapter we proposed the specific realization of the Multiverse hypothesis as the collection of multiple quantum universes obtained by the method of quantization in the static Fock space of the creator and annihilator operators applied to the classical field theory of the Einstein–Friedmann Universe. Making straightforward elementary calculations we received the thermodynamics of the system of many quantum universes, which in itself is the result possessing both the most natural and strong value for phenomenology and empirical verification of the theoretical results of the proposed model of quantum cosmology. This elegant feature allows to conclude that using the simple solution of the Einstein field equations allowed to formulate the physical model of Multiverse, and make real the proposed programme of quantum cosmology. Such line of reasoning opens the door to straightforward consistent generalization of the constructive physical scenario of the observed Universe.

There are many vitally important philosophical ramifications arising from the Multiverse hypothesis-based formulation of quantum cosmology. Let us mention the several ones which were established in this chapter:

1. Constructive realization of the Multiverse hypothesis. The rational and simple scheme of the natural emergence of the observed Universe as the system of multiple quantum universes.

2. Methodological linkage between quantum cosmology and theoretical physics. Formulation of quantum cosmology by using various models of theoretical physics having a well-established value for phenomenology.

3. Non-trivial coherence between the quantum cosmology and string theory. Theoretical description of (very) early Universe as the static Multiverse of the superfluid Fermi–Bose superstrings which, in general, can be open or closed.

4. Consistent conceptual way which leads to understanding the physical role of quantum gravity, string theory, and supersymmetry for the birth and early evolution of the observed Universe.

5. Inherent cosmological role of the superfluidity phenomenon and light for the Multiverse hypothesis.

6. Syncretism of quantum cosmology and the fundamental natural sciences, like organic chemistry and evolutionary biology, on the ideological level of discussion. Establishment of the role of the Multiverse hypothesis as the philosophical landscape for consistent comprehension of the observed Universe with no referring to interference of supernatural forces in creation and development of the observed Universe.

All these findings need further development. It is worth stressing that the realization of the Multiverse hypothesis is fully consistent with the R.P. Feynman's understanding of the Mach principle [172]. In this manner, the model of quantum cosmology is strictly related to the Machian type of Multiverse.

To see the sense of the proposed strategy, let us sketch briefly the crucial elements of the constructed model of quantum cosmology

1. Description of a certain selected solution of the Einstein field equations using the framework of the Hamiltonian approach merging the Dirac and the Arnowitt–Deser–Misner approach to the Hamiltonian formulation of General Relativity.

2. Application of methods of the primary and the secondary canonical quantization to the received Hamiltonian constraint, and formulation of the quantum cosmology through the one-dimensional Dirac equation.

3. Elementary derivation of the thermodynamics of quantum states for the case of the selected metric by application of the fundamental methods of statistical mechanics involving the method of the static Fock repère.

There is an obvious conjecture immediately following from such a programme of quantum cosmology. Namely, it is intuitively clear that the proposed strategy of construction of quantum cosmology can be straightforwardly generalized for another solutions of the Einstein field equations, and as a result the appropriate models of quantum gravity can be easily generated. In other words, the realization of the Multiverse hypothesis based on the one-dimensional Dirac equation can be straightforwardly generalized to the case of the class of solutions to the Einstein field equations which are parameterized by the Arnowitt–Deser–Misner decomposition. Such a strategy would result in the elegant transition between quantum cosmology and quantum gravity, and establish new non-trivial connections between quantum gravity and string theory. In the next chapters of this book we shall present a certain idea for constructive realization of such a generalized strategy.

Chapter 2

Inflationary Multiverse

A Outlook on Inflationary Cosmology

Let us consider the Friedmann–Lemaître–Robertson–Walker metric $g_{\mu\nu} = \text{diag}[-1, a^2, a^2, a^2]$, for which the space-time interval is

$$ds^2 = -c^2dt^2 + a^2(t)dx^i dx_i, \tag{2.1}$$

where t is cosmic time, a is the cosmic scale factor parameter, and

$$dx^i dx_i = \frac{dr^2}{1-kr^2} + r^2(d\theta^2 + \sin^2\theta d\phi^2), \tag{2.2}$$

where k is: 0 m^{-2} for flat space, $+1$ m^{-2} for sphere of constant positive curvature, and -1 m^{-2} for hyperbolic space having constant negative curvature. Taking into account an isotropic fluid characterized by energy density ρ, pressure p, and four-velocity $u_\mu = [\pm 1, 0, 0, 0]$, for the one can write the stress-energy tensor as

$$T_{\mu\nu} = pg_{\mu\nu} + (p+\rho)u_\mu u_\nu = \text{diag}[\rho, a^2p, a^2p, a^2p], \tag{2.3}$$

and, consequently, one can establish the Einstein field equations for such situation. The 00-component is the Friedmann equation

$$H^2 + \frac{k}{a^2} = \frac{\kappa}{3}\rho, \tag{2.4}$$

while ij-component is the equation

$$2\dot{H} + 3H^2 + \frac{k}{a^2} = -\kappa p, \tag{2.5}$$

where $\kappa = \dfrac{8\pi G}{c^4}$ is the Einstein constant, $H = \dfrac{\dot{a}}{a}$ is the Hubble parameter, and dot means differentiation with respect to the time coordinate ct. Combining the equations (2.4) and (2.5) one obtains

$$\dot{\rho} = -3H(p+\rho), \tag{2.6}$$

$$\frac{\ddot{a}}{a} = -\frac{\kappa}{6}(\rho+3p). \tag{2.7}$$

Introducing conformal time η as $ad\eta = dt$, the interval (2.1) takes the form

$$ds^2 = a^2(\eta)\left(-c^2 d\eta^2 + \delta_{ij}dx^i dx^j\right), \tag{2.8}$$

while the Hubble parameter becomes $H = \dfrac{\mathscr{H}}{a}$, where $\mathscr{H} = \dfrac{a'}{a}$ is the conformal Hubble parameter, and prime means differentiation with respect to the time coordinate $c\eta$.

Inflation is defined by positive acceleration of the Universe $\ddot{a} > 0$, *i.e.* $\rho + 3p < 0$. Inflationary cosmology (Cf. *e.g.* Ref. [182]) arises from quantum-mechanical cosmological perturbations of isotropic and homogeneous space-time given by the Friedmann–Lemaître–Robertson–Walker metric. The question is how small quantum fluctuations around this solution of the Einstein field equations behave during inflation, the hypothetical phase of accelerated expansion that took place in the early Universe. Inflation takes place in the very early universe, at very high energies. Then fluid description of matter is inappropriate, but quantum field theory is appropriate. Usually, a scalar field φ, called *the inflaton*, is considered as the reason of the very early Universe evolution. We shall consider the conformal time dependence $\varphi = \varphi(\eta)$.

The perturbed space-time metric is [183]

$$ds^2 = a^2(\eta)\left(-c^2(1-2\phi)d\eta^2 + 2cB_{,i}dx^i d\eta + [(1-2\psi)\delta_{ij} + 2E_{,ij} + h_{ij}]\,dx^i dx^j\right), \tag{2.9}$$

where ϕ, ψ (both dimensionless), B (dimension m), and E (dimension m^2) are scalar functions. The traceless tensor h_{ij}, satisfying $h_{ij}{}^{;j} = 0$, describes the gravitational waves. Vector perturbations are absent because a single scalar field cannot seed rotational perturbations. At the linear level scalar and tensor perturbations can be treated separately.

Let us consider a single scalar field σ in an arbitrary curved space.

Then the action functional is

$$S = \frac{1}{c}\int d^4x\sqrt{-g}\left(\frac{E_P\ell_P}{2}g^{\mu\nu}\partial_\mu\sigma\partial_\nu\sigma + \frac{1}{\ell_P^3}\mathrm{V}(\sigma)\right), \tag{2.10}$$

and, therefore, σ satisfies the Klein–Gordon equation

$$\frac{1}{\sqrt{-g}}\partial_\mu\left(\sqrt{-g}g^{\mu\nu}\partial_\nu\sigma\right) - \frac{1}{\ell_P^4E_P}\frac{d\mathrm{V}(\sigma)}{d\sigma} = 0. \tag{2.11}$$

If one performs the following perturbation

$$\sigma(x,\eta) = \varphi(\eta) + \delta\varphi(x,\eta), \tag{2.12}$$

then, using the unperturbed space-time metric $g_{\mu\nu} = a^2(\eta)\eta_{\mu\nu}$, one obtains the Klein–Gordon equation for the unperturbed homogeneous scalar field $\varphi = \varphi(\eta)$

$$\varphi'' + 2\mathscr{H}\varphi' + \frac{a^2}{E_P\ell_P^4}\frac{d\mathrm{V}(\varphi)}{d\varphi} = 0, \tag{2.13}$$

where the prime denotes differentiation with respect to the time coordinate $c\eta$.

The action of Matter fields is considered as the action of the unperturbed scalar field

$$S = \frac{1}{c}\int_M d^4x\sqrt{-g}\mathscr{L}_M, \tag{2.14}$$

where the Lagrangian density is

$$\mathscr{L}_M = -\frac{E_P\ell_P}{2}g^{\mu\nu}\partial_\mu\varphi\partial_\nu\varphi - \frac{1}{\ell_P^3}\mathrm{V}(\varphi), \tag{2.15}$$

and, consequently, one can determine the stress-energy tensor

$$T_{\mu\nu} = g_{\mu\nu}\mathscr{L}_M - 2\frac{\delta\mathscr{L}_M}{\delta g^{\mu\nu}} = E_P\ell_P\partial_\mu\varphi\partial_\nu\varphi - g_{\mu\nu}\left(\frac{E_P\ell_P}{2}g^{\alpha\beta}\partial_\alpha\varphi\partial_\beta\varphi + \frac{1}{\ell_P^3}\mathrm{V}(\varphi)\right). \tag{2.16}$$

Now one can rewrite this stress-energy tensor in the form (2.3) which leads to the system of equations

$$E_P\ell_P\frac{(\varphi')^2}{a^2} - g_{00}\left(\frac{E_P\ell_P}{2}g^{\alpha\beta}\partial_\alpha\varphi\partial_\beta\varphi + \frac{1}{\ell_P^3}\mathrm{V}(\varphi)\right) = pg_{00} + (p+\rho)u_0u_0, \quad (2.17)$$

$$E_P\ell_P\frac{\varphi'}{a}\partial_i\varphi - g_{0i}\left(\frac{E_P\ell_P}{2}g^{\alpha\beta}\partial_\alpha\varphi\partial_\beta\varphi + \frac{1}{\ell_P^3}\mathrm{V}(\varphi)\right) = pg_{0i} + (p+\rho)u_0u_i, \quad (2.18)$$

$$E_P\ell_P\partial_i\varphi\partial_j\varphi - g_{ij}\left(\frac{E_P\ell_P}{2}g^{\alpha\beta}\partial_\alpha\varphi\partial_\beta\varphi + \frac{1}{\ell_P^3}\mathrm{V}(\varphi)\right) = pg_{ij} + (p+\rho)u_iu_j, \quad (2.19)$$

where prime means differentiation with respect to the time coordinate $c\eta$. Taking into account the normalization condition

$$g^{\mu\nu}u_\mu u_\nu = -1, \quad (2.20)$$

one can easily find the solutions

$$p = -\frac{E_P\ell_P}{2}g^{\alpha\beta}\partial_\alpha\varphi\partial_\beta\varphi - \frac{1}{\ell_P^3}\mathrm{V}(\varphi), \quad (2.21)$$

$$\rho = -\frac{E_P\ell_P}{2}g^{\alpha\beta}\partial_\alpha\varphi\partial_\beta\varphi + \frac{1}{\ell_P^3}\mathrm{V}(\varphi), \quad (2.22)$$

$$u_\mu = \sqrt{\frac{E_P\ell_P}{p+\rho}}\partial_\mu\varphi = \frac{\partial_\mu\varphi}{\sqrt{-g^{\alpha\beta}\partial_\alpha\varphi\partial_\beta\varphi}}. \quad (2.23)$$

so that inflation takes the place when

$$E_P\ell_P g^{\alpha\beta}\partial_\alpha\varphi\partial_\beta\varphi + \frac{1}{\ell_P^3}\mathrm{V}(\varphi) > 0. \quad (2.24)$$

According to the perturbation theory, calculation should be made for the case of the unperturbed metric, *i.e.* $g_{\mu\nu} = \mathrm{diag}[-1, a^2, a^2, a^2]$. Then, for $\varphi = \varphi(\eta)$, one obtains

$$p = \frac{E_P\ell_P}{2}\dot{\varphi}^2 - \frac{1}{\ell_P^3}\mathrm{V}(\varphi), \quad (2.25)$$

$$\rho = \frac{E_P\ell_P}{2}\dot{\varphi}^2 + \frac{1}{\ell_P^3}\mathrm{V}(\varphi), \quad (2.26)$$

$$u_\mu = \sqrt{\frac{E_P\ell_P}{p+\rho}}\left[\frac{\varphi'}{a}, 0, 0, 0\right] = [\pm 1, 0, 0, 0], \quad (2.27)$$

while inflation takes the place when

$$\dot{\varphi}^2 < \frac{1}{\ell_P^4}\frac{V}{E_P}. \tag{2.28}$$

The slower situation

$$\dot{\varphi}^2 \ll \frac{1}{\ell_P^4}\frac{V}{E_P}, \tag{2.29}$$

is known as the slow-roll approximation, and the resulting inflation is called the slow-roll inflation. Using the solutions (2.25) and (2.26) within the equation (2.6) one receives the result

$$\ddot{\varphi} + 3H\dot{\varphi} + \frac{1}{E_P\ell_P^4}\frac{dV}{d\varphi} = 0, \tag{2.30}$$

which, after transformation to conformal time coordinate $c\eta$, coincides with the Klein–Gordon equation (2.13).

In inflationary cosmology the background space-time is conformally flat, *i.e.* $k = 0$. By taking time differentiation of the Friedmann equation (2.4) and then using the Klein–Gordon equation (2.30) one obtains

$$\dot{H} = -S_P\dot{\varphi}^2, \tag{2.31}$$

where $S_P = 4\pi\ell_P^2$ is the area of the Planck sphere.

In the case of scalar perturbations of the geometry evoked above, due to freedom to choose the coordinate system the four functions are in fact redundant, and the scalar fluctuations of the geometry can be characterized by the gauge-invariant Bardeen potentials [184]

$$\Phi = \phi + \frac{1}{a}\left[a\left(B - E'\right)\right]', \tag{2.32}$$

$$\Psi = \psi - \frac{a'}{a}(B - E'). \tag{2.33}$$

In general, the Einstein field equations for the space-time metric (2.9) are solely expressed in terms of these gauge-invariant quantities. Let us start with density perturbations, for which tedious calculations using the Fourier transforms

$$\Phi(\eta,x) = \int d^3\mathrm{k}\tilde{\Phi}(\eta,\mathrm{k})e^{i\mathrm{kx}}, \tag{2.34}$$

$$\Psi(\eta,x) = \int d^3\mathrm{k}\tilde{\Psi}(\eta,\mathrm{k})e^{i\mathrm{kx}}, \tag{2.35}$$

for a fixed mode k lead to the system of equations

$$-3\mathscr{H}\left(\mathscr{H}\tilde{\Phi}-\tilde{\Psi}'\right)=S_P\left[-\varphi'^2\tilde{\Phi}+\varphi'(\delta\varphi^g)'+\frac{a^2}{E_P\ell_P^4}\frac{dV}{d\phi}\delta\varphi^g\right], \quad (2.36)$$

$$\mathscr{H}\tilde{\Phi}+\tilde{\Psi}'=S_P\varphi'\delta\varphi^g, \quad (2.37)$$

$$\tilde{\Phi}-\tilde{\Psi}=0, \quad (2.38)$$

$$(2\mathscr{H}'+\mathscr{H}^2)\tilde{\Phi}+\mathscr{H}\tilde{\Phi}'+\tilde{\Psi}''+2\mathscr{H}\tilde{\Psi}'-\frac{k^2}{3}(\tilde{\Phi}-\tilde{\Psi})=$$

$$=S_P\left[-\varphi'^2\tilde{\Phi}+\varphi'(\delta\varphi^g)'-\frac{a^2}{E_P\ell_P^4}\frac{dV}{d\phi}\delta\varphi^g\right], \quad (2.39)$$

where $\delta\varphi^g$ is the gauge-invariant perturbation

$$\delta\varphi^g(\eta,x)=\delta\varphi+\varphi'\left(B-E'\right), \quad (2.40)$$

which characterizes the fluctuations in the inflaton scalar field. In this manner, in the simple case of a single scalar field the scalar perturbations are reduced to evolution of Φ, and it is true as long as there is no anisotropic stress. Straightforward reduction leads to

$$\Phi''+2\left(\mathscr{H}-\frac{\varphi''}{\varphi'}\right)\Phi'+\left[k^2+2\left(\mathscr{H}'-\mathscr{H}\frac{\varphi''}{\varphi'}\right)\right]\Phi=0, \quad (2.41)$$

and using the variable

$$u=\sqrt{\frac{8}{3}}\frac{a\Phi}{\mathscr{H}\sqrt{\gamma}}, \quad (2.42)$$

where γ is the background function

$$\gamma=-\frac{\dot{H}}{H}=1-\frac{\mathscr{H}'}{\mathscr{H}^2}, \quad (2.43)$$

transforms the equation (2.41) to

$$u''+\frac{\omega^2}{c^2}u=0, \quad (2.44)$$

where

$$\frac{\omega^2}{c^2} = k^2-\frac{(a\sqrt{\gamma})''}{a\sqrt{\gamma}}+2\left(\frac{(a\sqrt{\gamma})'}{a\sqrt{\gamma}}\right)^2 \quad (2.45)$$

$$= k^2-\mathscr{H}'+\mathscr{H}^2+\frac{1}{2}\left(\gamma''-\frac{3}{2}\frac{\gamma'^2}{\gamma}\right). \quad (2.46)$$

Because the Einstein field equations couple the Bardeen potential and the gauge-invariant perturbation, one has one degree of freedom. The scalar perturbations are reduced to study the quantity

$$\mu_S(\eta,x) = -\sqrt{2S_P}v(\eta,x) = -2a\sqrt{\gamma}\zeta(\eta,x), \tag{2.47}$$

where $v(\eta,x)$ is the Mukhanov–Sasaki variable

$$v(\eta,x) = a\left(\delta\varphi^g + \varphi'\frac{\Phi}{\mathscr{H}}\right). \tag{2.48}$$

and

$$\zeta = \frac{2}{3}\frac{\mathscr{H}^{-1}\Phi' + \Phi}{1+w} + \Phi, \tag{2.49}$$

where $w = \dfrac{p}{\rho}$, is the first integral of the equation (2.41) for the Fourier modes that are much larger than the Hubble scale $\mathrm{k} \ll \mathscr{H}$, *i.e.* for superhorizon modes. The automatically gauge-invariant tensor sector is described by the quantity $\mu_T(\eta,x)$ which is defined according to the following formula

$$h_{ij} = \frac{\mu_T}{a}Q_{ij}^{TT}, \tag{2.50}$$

where Q_{ij}^{TT} are the transverse and traceless eigentensors of the Laplace operator on the space-like sections, and by the Fourier transform

$$h_{ij}(\eta,\mathrm{x}) = \int d^3\mathrm{k} \sum_{s=1,2} p_{ij}^s(\mathrm{k})h_{ij}^s(\eta,\mathrm{k})e^{i\mathrm{kx}}, \tag{2.51}$$

where $p_{ij}^s(\mathrm{k})$ is the transverse and traceless polarization tensor satisfying the following property $p_{ij}^s(\mathrm{k})p^{ijs'}(\mathrm{k}) = 2\delta^{ss'}$, and $s = 1,2$ denote the two possible states of polarization of the gravitational waves. The first order Euler–Lagrange equations of motion can be obtained from the Einstein–Hilbert action pertubatively expanded up to second order in the metric and scalar field perturbations. The perturbations can be studied mode by mode by using the Fourier transforms

$$\widetilde{\mu}_S(\eta,\mathrm{k}) = \int d^3\mathrm{x}\mu_S(\eta,\mathrm{x})e^{-i\mathrm{kx}}, \tag{2.52}$$

$$\widetilde{\mu}_T(\eta,\mathrm{k}) = \int d^3\mathrm{x}\mu_T(\eta,\mathrm{x})e^{-i\mathrm{kx}}, \tag{2.53}$$

obeying the following Euler–Lagrange equations of motion

$$\frac{d^2}{d\eta^2}\widetilde{\mu}_S(\eta,\mathrm{k})+\omega_S(\eta,\mathrm{k})\widetilde{\mu}_S(\eta,\mathrm{k}) = 0, \tag{2.54}$$

$$\frac{d^2}{d\eta^2}\widetilde{\mu}_T(\eta,\mathrm{k})+\omega_T(\eta,\mathrm{k})\widetilde{\mu}_T(\eta,\mathrm{k}) = 0, \tag{2.55}$$

where $\omega_S(\eta,\mathrm{k})$ and $\omega_T(\eta,\mathrm{k})$ are the frequencies

$$\omega_S^2(\eta,\mathrm{k}) = c^2\left[\mathrm{k}^2-\frac{(a\sqrt{\gamma})''}{a\sqrt{\gamma}}\right], \tag{2.56}$$

$$\omega_T^2(\eta,\mathrm{k}) = c^2\left[\mathrm{k}^2-\frac{a''}{a}\right]. \tag{2.57}$$

The cosmological perturbations obey exactly the same type of equation as scalar field $\Phi(t,x)$ interacting with a classical electric field in the Schwinger effect [185], *i.e.* the equation of a parametric oscillator

$$\ddot{\widetilde{\Phi}}(t,\mathrm{k})+\omega^2(t,\mathrm{k})\widetilde{\Phi}(t,\mathrm{k})=0, \tag{2.58}$$

where the frequency has the form

$$\omega^2(t,\mathrm{k})=\mathrm{k}^2c^2+\frac{m^2c^4}{\hbar^2}-2\frac{c^2}{\hbar}eEk_zt+\frac{1}{\hbar^2}e^2E^2t^2, \tag{2.59}$$

where e is the elementary charge, and E is the electric field. The only difference is the classical source which, in the case of inflationary cosmological perturbations, is the background gravitational field determined by the Friedmann–Lemaître–Robertson–Walker metric. Also, the time dependence of the frequencies is qualitatively different. The primary canonical quantization of the theory proceeds as in the usual Schwinger effect, *i.e.* with using of the squeezed states technique and the Wigner functions. The consequence of the interaction between the quantum cosmological perturbations and the classical background is the creation of particles which, in the context of inflationary cosmology, are interpreted as gravitons. Classically, this corresponds to the amplification growing mode of the fluctuations. In this manner, the Schwinger effect is essential phenomenon for inflation. Detailed analysis of the situation was performed by Jérôme Martin [186].

The slow-roll approximation leads to the specific situation within the scheme. First of all, in the light of this approximation the Friedmann equation (2.4) is reduced to the form

$$H^2 = \frac{8\pi}{3E_P\ell_P^2}\mathrm{V}(\varphi), \tag{2.60}$$

and the Klein–Gordon equation (2.30) to

$$\dot{\varphi} = -\frac{1}{3HE_P\ell_P^4}\frac{d\mathrm{V}}{d\varphi} = -\frac{1}{\sqrt{24\pi E_P\ell_P^3}}\frac{\mathrm{V}_\varphi}{\sqrt{\mathrm{V}}}, \tag{2.61}$$

where we have applied the shortened notation $\mathrm{V}_\varphi = \frac{d\mathrm{V}}{d\varphi}$. The third equation is the equation (2.31). At leading order, the slow-roll approximation is controlled by the three slow-roll parameters [187]

$$\varepsilon = -\frac{\dot{H}}{H^2} = \gamma = \frac{1}{4S_P}\left(\frac{\mathrm{V}_\varphi}{\mathrm{V}}\right)^2, \tag{2.62}$$

$$\delta = -\frac{\ddot{H}}{H\dot{H}} = -\frac{\ddot{\varphi}}{H\dot{\varphi}} = \frac{1}{4S_P}\left[2\frac{\mathrm{V}_{\varphi\varphi}}{\mathrm{V}} - \left(\frac{\mathrm{V}_\varphi}{\mathrm{V}}\right)^2\right], \tag{2.63}$$

$$\xi = \frac{\dot{\varepsilon} - \dot{\delta}}{H} = \frac{1}{4S_P^2}\left(\frac{\mathrm{V}_\varphi}{\mathrm{V}}\right)^2\left[3\frac{\mathrm{V}_{\varphi\varphi}}{\mathrm{V}} - \frac{\mathrm{V}_{\varphi\varphi\varphi}}{\mathrm{V}_\varphi} - 2\left(\frac{\mathrm{V}_\varphi}{\mathrm{V}}\right)^2\right], \tag{2.64}$$

obeying the equations of motion

$$\frac{\dot{\varepsilon}}{H} = 2\varepsilon(\varepsilon - \delta), \tag{2.65}$$

$$\frac{\dot{\delta}}{H} = 2\varepsilon(\varepsilon - \delta) - \xi. \tag{2.66}$$

In such approximation, the frequencies of the gauge-invariant cosmological perturbations (2.56) and (2.57), at linear order can be rewritten by using the slow-roll parameters

$$\omega_S^2(\eta,\mathrm{k}) = c^2\left[\mathrm{k}^2 - \frac{2 + 6\varepsilon - 3\delta}{c^2\eta^2}\right], \tag{2.67}$$

$$\omega_T^2(\eta,\mathrm{k}) = c^2\left[\mathrm{k}^2 - \frac{2 + 3\varepsilon}{c^2\eta^2}\right]. \tag{2.68}$$

It is convenient to express the slow-roll parameters by using the horizon flow functions ε_1, ε_2, and ε_3

$$\varepsilon = \varepsilon_1, \tag{2.69}$$

$$\delta = \varepsilon_1 - \frac{1}{2}\varepsilon_2, \tag{2.70}$$

$$\xi = \frac{1}{2}\varepsilon_2\varepsilon_3. \tag{2.71}$$

It can be seen straightforwardly that the quotient of the first and third horizon flow functions $\frac{\varepsilon_1}{\varepsilon_3}$ measures the ratio of the kinetic energy to the total energy, whereas the horizon flow function ε_2 represents a model where the kinetic energy, with respect to the total energy, itself increases when $\varepsilon_2 > 0$ or decreases when $\varepsilon_2 < 0$. Provided the slow-roll conditions are valid if and only if $\varepsilon_{1,2} \ll 1$. One can express the slow-roll parameters by using the inflaton potential V

$$\varepsilon_1 = \frac{1}{4S_P}\left(\frac{V_\varphi}{V}\right)^2, \tag{2.72}$$

$$\varepsilon_2 = \frac{1}{S_P}\left[\left(\frac{V_\varphi}{V}\right)^2 - \frac{V_{\varphi\varphi}}{V}\right], \tag{2.73}$$

$$\varepsilon_3 = \frac{1}{2S_P}\left(\frac{V_\varphi}{V}\right)^2 \frac{3\frac{V_{\varphi\varphi}}{V} - \frac{V_{\varphi\varphi\varphi}}{V_\varphi} - 2\left(\frac{V_\varphi}{V}\right)^2}{\left(\frac{V_\varphi}{V}\right)^2 - \frac{V_{\varphi\varphi}}{V}}. \tag{2.74}$$

B Power Law Inflaton

Let us consider first the inflaton potential in the form of the power law

$$\mathrm{V}(\varphi) = -\frac{\ell_P^p}{E_P}\frac{m^2c^4}{2}\varphi^p, \tag{2.75}$$

where m is the mass of the power law inflaton φ.

In such a situation, the horizon flow functions are easy to derive

$$\varepsilon_1 = \frac{1}{4S_P}\frac{p^2}{\varphi^2}, \tag{2.76}$$

$$\varepsilon_2 = \frac{1}{S_P}\frac{p}{\varphi^2}, \tag{2.77}$$

$$\varepsilon_3 = -\frac{1}{2S_P}\frac{p^2(p^2-4p+5)}{\varphi^2}, \tag{2.78}$$

so that the slow roll parameters have the form

$$\varepsilon = \frac{1}{4S_P}\frac{p^2}{\varphi^2}, \tag{2.79}$$

$$\delta = \frac{1}{4S_P}\frac{p(p-2)}{\varphi^2}, \tag{2.80}$$

$$\xi = -\frac{1}{4S_P^2}\frac{p^3(p^2-4p+5)}{\varphi^4}. \tag{2.81}$$

In this manner, one can establish straightforwardly the frequencies of the scalar and tensor cosmological perturbations

$$\omega_S^2(\mathrm{k},\eta) = \mathrm{k}^2c^2 - \left(2+\frac{3}{4S_P}\frac{p(p+2)}{\varphi^2}\right)\frac{1}{\eta^2}, \tag{2.82}$$

$$\omega_T^2(\mathrm{k},\eta) = \mathrm{k}^2c^2 - \left(2+\frac{3}{4S_P}\frac{p^2}{\varphi^2}\right)\frac{1}{\eta^2}. \tag{2.83}$$

The problem is to establish the total frequency of the inflationary cosmological perturbations of the power law inflaton. Let us postulate such an effective frequency by the Pythagorean theorem

$$\omega_{\text{eff}}^2 = \omega_S^2(\mathrm{k},\eta) + \omega_T^2(\mathrm{k},\eta), \tag{2.84}$$

which we shall call the *Pythagorean frequency of scalar fluctuations*, which has the following explicit form

$$\omega_{\text{eff}} = \sqrt{2\mathrm{k}^2c^2 - 2\left(2 + \frac{3}{4S_P}\frac{p(p+1)}{\varphi^2}\right)\frac{1}{\eta^2}}. \tag{2.85}$$

One can consider the effective energy of the cosmological perturbations of the power law inflaton. Let us propose *ad hoc* that such energy of the inflationary cosmological perturbations is simply given by the Planck energy-frequency relation expressing the wave-particle duality

$$E_{\text{eff}} = \hbar\omega_{\text{eff}}. \tag{2.86}$$

C Higgs–Hubble Inflaton

Let us discuss first the non-trivial linkage of the power law inflaton with the Multiverse model presented in the previous chapter of this part. For this let us introduce the auxiliary field $\varphi = \varphi(\eta)$, which expressed in terms of the Planck units is

$$\varphi(\eta) = \varphi_0 a(\eta), \tag{2.87}$$

where $a(\eta)$ is the cosmic scale factor parameter, η is conformal time describing the classical general relativistic evolution of the Friedmann–Lemaître–Robertson–Walker metric, and $\varphi_0 = \dfrac{\alpha}{\ell_P}$ is the initial datum of the auxiliary field $\varphi(\eta)$ (with $a(\eta_0) = 1$, and dimensionless constant α).

The Einstein–Hilbert action of General Relativity evaluated on the Friedmann–Lemaître–Robertson–Walker metric $g_{\mu\nu} = \mathrm{diag}[-1, a^2(t)\delta_{ij}]$, which we established in the previous section as (1.47), can be presented in the following form

$$S[\varphi] = \int d\eta \left[E_P\ell_P V\frac{3}{2}\frac{\alpha^2\ell_P^3}{V_P}\varphi'^2 + \frac{\varphi^4}{\varphi_0^4}\varepsilon_M(\eta)\right], \tag{2.88}$$

where $V = \int d^3x < \infty$ is the spatial volume, η is conformal time, the prime denotes differentiation with respect to η, $d\eta = \dfrac{dt}{a(t)}$ is conformal time integral measure, and $\varepsilon_M(\eta) = \int d^3x \mathscr{H}_M(x,\eta)$ is the energy of Matter fields. In this manner, the most convenient choice of the constant dimensionless parameter α can be done in such a way that

$$\alpha = \frac{2}{3}\sqrt{\pi}, \tag{2.89}$$

so that in such a case the auxiliary field has the form

$$\varphi(\eta) = \varphi_0 a(\eta), \tag{2.90}$$

where φ_0 is the constant

$$\varphi_0 = \frac{2\sqrt{\pi}}{3\ell_P} \approx 7.3109596 \cdot 10^{34} \frac{1}{\mathrm{m}}, \tag{2.91}$$

and the Einstein–Hilbert action becomes

$$S[\varphi] = \int d\eta \left[\frac{E_P \ell_P}{2} V \varphi'^2 + \frac{\varphi^4}{\varphi_0^4} \varepsilon_M(\eta) \right]. \tag{2.92}$$

If one introduces the speed

$$v_\varphi = c\ell_P^2 \varphi' \tag{2.93}$$

then the conjugate momentum can be calculated as

$$P_\varphi = \frac{\delta S[\varphi]}{\delta v_\varphi} = M_P \frac{V}{\ell_P^3} v_\varphi, \tag{2.94}$$

and it is easily noticeable that the quantity

$$m = M_P \frac{V}{\ell_P^3}, \tag{2.95}$$

has the natural interpretation of the mass of the Higgs–Hubble inflaton. Applying the Markov hypothesis, *i.e.* $m \leqslant M_P$, one obtains the consistency condition for the theory

$$V \leqslant \ell_P^3, \tag{2.96}$$

which says that the upper bound for the volume of space is given by the Planck volume. In other words, in the light of the Markov hypothesis, considered theory is physically consistent when the volume of space is at most equal to the Planck volume.

By application of the Legendre transformation, one can present the action (2.92) in the Hamilton form

$$S[\varphi] = \int d\eta \left\{P_\varphi v_\varphi - H(\eta)\right\}, \tag{2.97}$$

where $H(\eta)$ is the Hamiltonian

$$H(\eta) = \frac{P_\varphi^2}{2m} - \frac{\varphi^4}{\varphi_0^4}\varepsilon_M(\eta) \approx 0, \tag{2.98}$$

which vanishes automatically by virtue of the Dirac method of canonical primary quantization. In such a situation, the Hamiltonian constraint can be straightforwardly resolved

$$P_\varphi = \pm\sqrt{\frac{V}{\ell_P^3}\frac{\varphi^2}{\varphi_0^2}}\sqrt{2M_P\varepsilon_M(\eta)}, \tag{2.99}$$

which generates the solution in the form of the Hubble law

$$\frac{\varphi(\eta)}{\varphi_0} = \frac{1}{1+z(\eta_0,\eta)}, \tag{2.100}$$

where z is the cosmological redshift

$$z(\eta_0,\eta) = \pm\frac{1}{\varphi_0}\int_{\eta_0}^{\eta}\sqrt{\frac{2\varepsilon_M(\eta')}{M_P\ell_P V}}d\eta'. \tag{2.101}$$

Application of the Dirac method of canonical primary quantization

$$[\hat{P}_\varphi, \varphi] = -i\frac{\hbar}{\ell_P^2}, \tag{2.102}$$

leads to the momentum operator

$$\hat{P}_\varphi = -i\frac{\hbar}{\ell_P^2}\frac{d}{d\varphi}, \tag{2.103}$$

which, applied to the Hamiltonian constraint, leads to the Wheeler–DeWitt equation which in this case has the form of the Klein–Gordon equation governing the Multiverse

$$\left(\frac{d^2}{d\varphi^2}+\Omega_\varphi^2\right)\Psi(\varphi)=0, \tag{2.104}$$

where Ω_φ, which plays the role of the frequency, has the form

$$\Omega_\varphi=\ell_P\sqrt{\frac{2V}{\ell_P^3}\frac{\varphi^2}{\varphi_0^2}}\sqrt{\frac{\varepsilon_M(\eta)}{E_P}}. \tag{2.105}$$

and can be presented in the equivalent form

$$\Omega_\varphi=\ell_P\sqrt{\frac{2V}{\ell_P^3}}\left(\frac{3E_P\mathrm{V}(\varphi)}{\sqrt{\pi}m^2c^4}\right)^{2/p}\sqrt{\frac{\varepsilon_M(\eta)}{E_P}}, \tag{2.106}$$

which for only $p=2$ becomes proportional to the inflaton potential

$$\mathrm{V}(\varphi)=\mathrm{V}_H(\varphi)=-\ell_P^2\frac{m^2c^4}{2}\varphi^2. \tag{2.107}$$

In such a situation, the meaning of the frequency (2.106) becomes much more unambiguous. In other words, such a Multiverse emerges due to *the Higgs–Hubble inflaton*, and for this reason we shall call it *the Higgs–Hubble Multiverse* In general, the auxiliary field (2.87) satisfies the Hubble law (2.100) and, therefore, we shall call it *the Hubble auxiliary field*, which after identification with the power law inflaton becomes *the Hubble inflaton*.

D Cosmology of Higgs–Hubble Inflaton

In this section we shall discuss in certain detail the cosmological context of the Higgs–Hubble inflaton. Our proposal is to interpret the Higgs–Hubble inflaton (2.87) as the homogeneous unperturbed scalar field φ. In other words, we shall take into account the situation

$$\varphi = \frac{\varphi_0}{1+z} = \varphi_0 a, \tag{2.108}$$

where $\varphi_0 = \dfrac{2\sqrt{\pi}}{3\ell_P}$ is the initial datum of the Higgs–Hubble inflaton established in the previous section. In such a situation, one has

$$\begin{aligned}
\varphi' &= -\frac{z'}{\varphi_0}\varphi^2, && (2.109)\\
\varphi'' &= 2\left(\frac{z'}{\varphi_0}\right)^2\varphi^3, && (2.110)\\
a &= \frac{\varphi}{\varphi_0}, && (2.111)\\
\mathscr{H} &= \frac{\varphi'}{\varphi}, && (2.112)\\
\mathrm{V}(\varphi) &= -\ell_P^2\frac{m^2c^4}{2E_P}\varphi^2. && (2.113)
\end{aligned}$$

Therefore, the Klein–Gordon equation (2.13) becomes

$$\varphi'' + 2\frac{\varphi'^2}{\varphi} - \frac{1}{\varphi_0^2}\left(\frac{mc}{\hbar}\right)^2\varphi^3 = 0, \tag{2.114}$$

which after taking into account the explicit form of the derivatives of the scalar field φ, and excluding the trivial solution $\varphi = 0$, becomes the differential equation for the cosmological redshift

$$z'^2 - \left(\frac{mc}{2\hbar}\right)^2 = 0. \tag{2.115}$$

The equation (2.115) can be solved straightforwardly

$$z(\eta,\eta_0) = \pm\frac{mc^2}{2\hbar}(\eta - \eta_0). \tag{2.116}$$

In this manner, in general the Higgs–Hubble inflaton has the form

$$\varphi = \frac{\varphi_0}{1 \pm \dfrac{mc^2}{2\hbar}(\eta - \eta_0)}, \tag{2.117}$$

so, consequently, the conformal Hubble parameter is

$$\mathscr{H} = -\frac{\pm\dfrac{mc}{2\hbar}}{1 \pm \dfrac{mc^2}{2\hbar}(\eta - \eta_0)}, \tag{2.118}$$

while the Hubble parameter $H = \mathscr{H}(1+z)$ becomes constant

$$H = \mp\frac{mc}{2\hbar}. \tag{2.119}$$

Because in the light of the definition (2.101) one has

$$z' = \pm\frac{1}{\varphi_0}\sqrt{\frac{2\varepsilon_M(\eta)}{E_P \ell_P V}}, \tag{2.120}$$

which compared to the result of the equation (2.115)

$$z' = \pm\frac{1}{\sqrt{2}}\frac{mc}{\hbar}, \tag{2.121}$$

allows to establish the energy of Matter fields in the form

$$\varepsilon_M(\eta) = \frac{4}{27}\pi^2 \frac{V}{V_P}\frac{m^2c^4}{E_P}, \tag{2.122}$$

where $V_P = \dfrac{4}{3}\pi\ell_P^3$ is the volume of the Planck sphere. Therefore, the parameter (2.105) can be rewritten in the form

$$\Omega_\phi = \frac{3}{4\sqrt{\pi}} V \frac{m}{M_P}\varphi^2, \tag{2.123}$$

and using the definition (2.95), which gives $V = \dfrac{m}{M_P}\ell_P^3$, one obtains

$$\Omega_\phi = \frac{3\ell_P^3}{4\sqrt{\pi}}\left(\frac{mc^2}{E_P}\right)^2 \varphi^2. \tag{2.124}$$

The energy of Matter fields (2.122) can be rewritten in the form

$$\varepsilon_M(\eta) = \frac{\pi}{9}\left(\frac{mc^2}{E_P}\right)^3 E_P = \hbar\omega_M(\eta), \tag{2.125}$$

where ω_M is the frequency of Matter fields, which in the light of the formula (2.122) can be rewritten in the form

$$\omega_M = \frac{\pi}{9}\left(\frac{mc^2}{E_P}\right)^3 \omega_P, \tag{2.126}$$

where $\omega_P = \dfrac{M_P c^2}{\hbar}$ is the Planck frequency.

In such a situation, the Einstein–Hilbert action calculated for the case of the Einstein–Friedmann Universe, given by the formula (2.92), takes the form of the action describing the one-dimensional φ^4-theory evolving in conformal time

$$S[\varphi] = \int d\eta \left(\frac{mv_\varphi^2}{2} + \frac{9}{16\pi}\left(\frac{mc^2}{E_P}\right)^3 E_P \ell_P^4 \varphi^4\right). \tag{2.127}$$

If one takes into account the usual action of φ^4-theory expressed in terms of conformal time and complemented by the Planck units

$$S[\varphi] = \int d\eta \left(\frac{mv_\varphi^2}{2} - \frac{g}{4!} E_P \ell_P^4 \varphi^4\right), \tag{2.128}$$

where g is the coupling constant, then one obtains the coupling constant in the form

$$g = -\frac{27}{2\pi}\left(\frac{mc^2}{E_P}\right)^3 = -\frac{27}{2\pi}\left(\frac{m}{M_P}\right)^3. \tag{2.129}$$

In such a situation, one can establish the beta function

$$\beta(g) = \frac{dg}{d\ln m} \tag{2.130}$$

which for the Higgs–Hubble inflaton is

$$\beta(g) = m\frac{dg}{dm} = 2g. \tag{2.131}$$

Interestingly, this value of the beta function coincides with the asymptotics $g \to \infty$ of $\beta(g)$ calculated from the duality relation for the two-dimensional Ising model [188], where the dimension D following from the asymptotic relation $\beta(g) = Dg$ is $D = 2$.

However, the situation is rather strange, because one has to deal with the dimension 1. There is, however, a different definition of the beta function (Cf. *e.g.* Ref. [189])

$$\beta(g) = \frac{dg}{d \ln m^2}, \tag{2.132}$$

which for the Higgs–Hubble inflaton is

$$\beta(g) = m^2 \frac{dg}{dm^2} = g. \tag{2.133}$$

This value of the beta function coincides with the asymptotics $g \to \infty$ of $\beta(g)$ in quantum electrodynamics. If one takes into account the asymptotics $\beta(g) = Dg$, then $D = 1$ coincides with the situation of the Higgs–Hubble inflaton. In this manner, the Einstein–Friedmann Universe as well as the Higgs–Hubble inflaton obtained the non-trivial physical significance.

In our situation, however, the potential can be straightforwardly deduced from the Einstein–Hilbert action (2.127) as

$$V(\varphi, g, \ell_P) = \frac{g(\ell_P)}{4!} E_P \ell_P^4 \varphi^4, \tag{2.134}$$

where the coupling constant is

$$g(\ell_P) = -\frac{27}{2\pi} \left(\frac{mc}{\hbar}\right)^3 \ell_P^3. \tag{2.135}$$

Taking into account the scaling

$$\varphi \to \varphi_\lambda = \frac{\varphi}{\lambda}, \tag{2.136}$$

for $\lambda = \ell_P$ one receives the property

$$V(\varphi, g(\lambda), \lambda) = \hat{V}(\varphi_\lambda, g(\lambda)), \tag{2.137}$$

where the scaled potential is

$$\hat{V}(\varphi_\lambda, g(\lambda)) = \frac{g(\lambda)}{4!} \varphi^4. \tag{2.138}$$

In general, one can compute the Callan–Symanzik beta function by application of the Callan–Symanzik function $\psi(g)$

$$\frac{d\ln g}{d\ln\lambda} = \psi(g) = \frac{\beta(g)}{g}. \tag{2.139}$$

It is easy to see that in our case $\ln g = \ln\lambda + \ln C$, where C is a certain constant, and for this reason one receives

$$\psi(g) = 1. \tag{2.140}$$

In this manner, one can establish the Callan–Symanzik beta function of the Higgs–Hubble inflaton as follows

$$\beta(g) = g. \tag{2.141}$$

It can be seen by straightforward analysis that such a simple beta function amazingly coincides with the beta function (2.133) obtained due to the asymptotics $g \to \infty$ applied to quantum electrodynamics, *i.e.* corresponds exactly with the case of the dimension one. The problem is, therefore, to construct the appropriate renormalization group equation. Its construction, however, can be easily performed by the deformation of the usual Callan–Symanzik equation applied to the scalar field potential

$$\left(\lambda\frac{\partial}{\partial\lambda} - \beta(g)\frac{\partial}{\partial g} + \mu\right)V(\varphi, g, \lambda) = 0, \tag{2.142}$$

where μ is a certain deformation parameter which can be established by straightforward calculation. Taking into account the derivatives

$$\frac{\partial}{\partial\lambda}V(\varphi, g, \lambda) = \frac{4}{\lambda}V(\varphi, g, \lambda), \tag{2.143}$$

$$\frac{\partial}{\partial g}V(\varphi, g, \lambda) = \frac{1}{g}V(\varphi, g, \lambda), \tag{2.144}$$

and the Callan–Symanzik beta function of the Higgs–Hubble inflaton in the form (2.141) established above, one obtains the following value of the deformation parameter

$$\mu = 3. \tag{2.145}$$

Therefore, the renormalization group equation related to the Higgs–Hubble inflaton has the form

$$\left(\lambda\frac{\partial}{\partial\lambda}-\beta(g)\frac{\partial}{\partial g}+3\right)V(\varphi,g,\lambda)=0. \tag{2.146}$$

E Phononic Hubble Inflaton

Let us see in some detail what happens in the Hubble Multiverse and Higgs–Hubble Multiverse, *i.e.* the Multiverse generated by the Hubble inflaton and the Higgs–Hubble inflaton, respectively. Applying the explicit form of the inflaton (2.87) to the inflaton energy (2.86) one can straightforwardly express the inflaton energy in terms of the cosmic scale factor parameter. The result is as follows

$$E_H \equiv E_{\text{eff}} = \hbar\sqrt{2\mathrm{k}^2c^2-2\left(2+\frac{27}{64\pi^2}\frac{p(p+1)}{a^2}\right)\frac{1}{\eta^2}}. \tag{2.147}$$

First, let us analyse this formula from the point of view of Special Relativity, *i.e.* the Einstein energy-momentum relation

$$E=\sqrt{\mathrm{p}^2c^2+m^2c^4}. \tag{2.148}$$

It can be seen by direct computation from such a point of view that the equation (2.147) describes the particle-universe possessing the following values of momentum and mass

$$\mathrm{p} = \sqrt{2}\hbar\mathrm{k}, \tag{2.149}$$

$$m^2 = -2M_P^2\left(2+27\frac{p(p+1)}{(8\pi a)^2}\right)\frac{t_P^2}{\eta^2}. \tag{2.150}$$

It is easy to see that the squared mass (2.150) is positive if and only if the power p satisfies the following inequality

$$p^2+p+\frac{2}{3}\left(\frac{8\pi a}{9}\right)^2<0, \tag{2.151}$$

which can be solved immediately

$$p\in\left[-\frac{1}{2}-\frac{1}{2}\sqrt{1-\frac{1}{6}\left(\frac{32\pi a}{9}\right)^2},-\frac{1}{2}+\frac{1}{2}\sqrt{1-\frac{1}{6}\left(\frac{32\pi a}{9}\right)^2}\right]. \tag{2.152}$$

The power p must be a real, which means that the cosmic scale factor parameter satisfies the inequality

$$1-\frac{1}{6}\left(\frac{32\pi a}{9}\right)^2 \geqslant 0, \tag{2.153}$$

which can be easily solved

$$a \in [-a_s, a_s], \tag{2.154}$$

where

$$a_s = \frac{9\sqrt{6}}{32\pi} \approx 0.21929. \tag{2.155}$$

Taking into account the definition $a = 1/(1+z)$ one can determine the appropriate regime of the cosmological redshift

$$z \in [z_-, z_+], \tag{2.156}$$

where

$$z_- = -1-\frac{16\sqrt{6}}{27}\pi \approx -5.560177, \tag{2.157}$$

$$z_+ = -1+\frac{16\sqrt{6}}{27}\pi \approx 3.560177. \tag{2.158}$$

In this regime of the cosmological redshift z the model based on the energy (2.147) is physically consistent.

The squared mass (2.150) is negative if and only if

$$p^2+p+\frac{2}{3}\left(\frac{8\pi a}{9}\right)^2 > 0, \tag{2.159}$$

which can be solved immediately

$$p \in \left(-\infty, -\frac{1}{2}-\frac{1}{2}\sqrt{1-\frac{1}{6}\left(\frac{32\pi a}{9}\right)^2}\right] \cup \left[-\frac{1}{2}+\frac{1}{2}\sqrt{1-\frac{1}{6}\left(\frac{32\pi a}{9}\right)^2}, \infty\right). \tag{2.160}$$

And as previously the power p is real when $a \in [-a_s, a_s]$. For this reason one has to deal with *tachyon* equipped with the mass

$$m_t = im. \tag{2.161}$$

Equivalently, the equation (2.150) can be understood as the expression for the cosmic scale factor parameter by the mass of a particle, *i.e.*

$$a^2(\eta) = \frac{27}{128\pi^2}\frac{p(p+1)}{\left(\frac{m_t c^2}{2\hbar}\eta\right)^2 - 1}. \tag{2.162}$$

In such a context, the Hubble inflaton becomes bosonic equipped with the negative value of the squared mass. Applying the simple identification $E_H = \hbar\omega_H$

$$\omega_H = \sqrt{2\mathrm{k}^2c^2 - 2\left(2 + \frac{27}{64\pi^2}\frac{p(p+1)}{a^2}\right)\frac{1}{\eta^2}}, \tag{2.163}$$

one can derive straightforwardly the group velocity of the particle-universe

$$v_g = \frac{d\omega_H}{d\mathrm{k}} = \frac{c}{\sqrt{1+\left(\frac{mc}{\mathrm{p}}\right)^2}} \tag{2.164}$$

$$= \frac{c}{\sqrt{1 - \frac{1}{(\mathrm{k}\ell_P a)^2}\left(2a^2 + \frac{27}{64\pi^2}p(p+1)\right)\frac{1}{\eta^2}}}, \tag{2.165}$$

and, similarly, the phase velocity can be obtained

$$v_{ph} = \frac{\omega_H}{\mathrm{k}} = c\sqrt{1+\left(\frac{mc}{\mathrm{p}}\right)^2} \tag{2.166}$$

$$= c\sqrt{1 - \frac{1}{(\mathrm{k}\ell_P a)^2}\left(2a^2 + \frac{27}{64\pi^2}p(p+1)\right)\frac{t_P^2}{\eta^2}}. \tag{2.167}$$

On the other hand, the inflaton energy (2.147) can be analyzed by the point of view suitable for phonons, *i.e.* quanta of sound in solids. In such a context, an application of the dispersion relation for phonons

$$\omega_k = \sqrt{2\omega^2(k)(1-\cos(k\ell_P a))}, \tag{2.168}$$

leads to the following identification

$$\omega(k) = \mathrm{k}c, \tag{2.169}$$

and the cosmic scale factor parameter a becomes the lattice spacing. Then, the Hubble inflaton becomes phononic, and straightforward comparison of the relations (2.147) and (2.168) leads to the following non-algebraic equation

$$(k\ell_P a)^2 \cos(k\ell_P a) = \left(2a^2 + \frac{27}{64\pi^2}p(p+1)\right)\frac{t_P^2}{\eta^2}, \tag{2.170}$$

where for the known value of conformal time η the unknown is the lattice spacing a. Solutions of the equation (2.170) can be found only by the numerical way. Let us apply the principles of quantum solid state physics (For basics, advances, and applications Cf. *e.g.* Ref. [190]).

According to the quantization rule for phonons, which we shall call *the phononic quantization*, the wave vector and the lattice spacing are nontrivially joined by the relation

$$k\ell_P a = \frac{n}{N}\pi, \tag{2.171}$$

where N is the number of identical atoms, $n = 0, \pm 1, \ldots, \pm N$, *i.e.* the product $k\ell_P a$ takes integer values in the range $\left[-\frac{\pi}{N}, \frac{\pi}{N}\right]$. These integers, however, cannot be chosen arbitrarily, because they must solve the non-algebraic equation (2.170). Possibly, there is no integer solution of this equation. In such a situation, one must reinterpret the equation (2.170) as the equation for conformal time η while the product $k\ell_P a$ is determined according to phononic quantization (2.171).

Similarly, as in the case of the particle-universe, one can derive straightforwardly the group velocity $v_g = \dfrac{d\omega_k}{dk}$ and the phase velocity $v_{ph} = \dfrac{\omega_k}{k}$ of the phonon-universe

$$v_g = v_{ph}\left(2 + k\ell_P a\sqrt{1 + \cos(k\ell_P a)}\right), \tag{2.172}$$

$$v_{ph} = c\sqrt{1 - \cos(k\ell_P a)}, \tag{2.173}$$

where the following relation

$$\cos(k\ell_P a) = \frac{1}{k^2\ell_P^2}\left(2 + \frac{27}{64\pi^2}\frac{p(p+1)}{a^2}\right)\frac{t_P^2}{\eta^2}, \tag{2.174}$$

holds. Applying the constraint (2.170) within the formulas (2.172) and (2.173), one receives

$$v_{ph} = c\sqrt{1-\frac{1}{(k\ell_P a)^2}\left(2a^2+\frac{27}{64\pi^2}p(p+1)\right)\frac{t_P^2}{\eta^2}}, \tag{2.175}$$

$$v_g = v_{ph}\left(2+\sqrt{(k\ell_P a)^2+\left(2a^2+\frac{27}{64\pi^2}p(p+1)\right)\frac{t_P^2}{\eta^2}}\right). \tag{2.176}$$

Applying phononic quantization (2.171) lead to the following quantization of the velocities

$$v_{ph} = c\sqrt{1-\left(\frac{N}{\pi n}\right)^2\left(2a^2+\frac{27}{64\pi^2}p(p+1)\right)\frac{t_P^2}{\eta^2}}, \tag{2.177}$$

$$v_g = v_{ph}\left(2+\sqrt{\left(\pi\frac{n}{N}\right)^2+\left(2a^2+\frac{27}{64\pi^2}p(p+1)\right)\frac{t_P^2}{\eta^2}}\right). \tag{2.178}$$

The phase velocity, however, is real if and only if

$$k\ell_P a \geqslant \sqrt{2a^2+\frac{27}{64\pi^2}p(p-1)}\frac{t_P}{\eta}, \tag{2.179}$$

and, therefore, the quantization is not chosen arbitrarily, but its choice is restricted by the following inequality

$$n \geqslant \frac{N}{\pi}\sqrt{2a^2+\frac{27}{64\pi^2}p(p+1)}\frac{t_P}{\eta}. \tag{2.180}$$

It can be seen straightforwardly that the following restriction

$$n \geqslant \frac{N}{\sqrt{2}\pi}\frac{v_g-2v_{ph}}{v_{ph}}, \tag{2.181}$$

is also essential. In the limiting situation

$$k\ell_P a = \sqrt{2a^2+\frac{27}{64\pi^2}p(p-1)}\frac{t_P}{\eta}, \tag{2.182}$$

both the group velocity and the phase velocity vanish identically, and by the constraint (2.170) such a situation corresponds to the equation

$$\cos\left[\sqrt{2a^2+\frac{27}{64\pi^2}p(p-1)}\frac{t_P}{\eta}\right]=1, \tag{2.183}$$

or, in other words, with the following quantization of the cosmic scale factor parameter

$$a_n=\sqrt{\frac{1}{2}\left(2\pi n\omega_P\eta\right)^2-\frac{27}{128\pi^2}p(p-1)}, \tag{2.184}$$

where $n\in\mathbf{Z}$, which corresponds with the quantization of cosmological redshift

$$z_n=\left(\frac{1}{2}\left(2\pi n\omega_P\eta\right)^2-\frac{27}{128\pi^2}p(p-1)\right)^{-1/2}-1. \tag{2.185}$$

Interestingly, $a_n\equiv 0$, *i.e.* the Multiverse evolution starts if and only if conformal time is quantized as follows

$$\eta_n=\frac{3t_P}{8\pi^2}\frac{\sqrt{3p(p+1)}}{2n}. \tag{2.186}$$

This situation means that such a Multiverse is cyclic, *i.e.* its evolution begins several times.

It can be seen by straightforward comparison, however, that the phase velocities of the particle (2.175) and the phonon (2.177) are identical, while the group velocities (2.176) and (2.178) have a blatantly different form. For the full *bosonic-phononic duality* of the Hubble inflaton, the most natural way is to take into account the equality between the group velocities of the bosonic universe and the phononic universe. We shall call *bonons* the Hubble inflatons satisfying the bononic (bosonic-phononic) duality. Such a kind of duality is definitely non-trivial because, in fact, it establishes the duality between sound (phonons) and matter (bosons). In this manner, bonons are the Hubble inflatons following from *the matter-sound duality*.

It is easy to see that such a duality condition can be presented as the non-algebraic equation

$$\begin{aligned}&(k\ell_P a)^2\cos^3(k\ell_P a)-(k\ell_P a)^2\cos^2(k\ell_P a)-(k\ell_P a)^2\cos(k\ell_P a)-\\&2\cos(k\ell_P a)+(k\ell_P a)^2+1=0,\end{aligned} \tag{2.187}$$

which with using the constraint (2.170) can be presented as the algebraic equation of degree 3

$$x^3+(1-f)x^2-f(f+2)x+f^3=0, \tag{2.188}$$

where we have introduced the notation

$$x = (k\ell_P a)^2 \geqslant 0, \tag{2.189}$$

$$f = \left(2a^2+\frac{27}{64\pi^2}p(p+1)\right)\frac{t_P^2}{\eta^2}. \tag{2.190}$$

If one wishes to use phononic quantization (2.171), then the equation (2.188) can be used to determine the quantization of the cosmic scale factor parameter. This quantization can be obtained as the solution of the following discrete equation

$$f_n^3-x_nf_n^2-x_n(x_n+2)f_n+x_n^2+x_n^3=0, \tag{2.191}$$

where we have used the notation

$$x_n = \left(\pi\frac{n}{N}\right)^2 \geqslant 0, \tag{2.192}$$

$$f_n = \left(2a_n^2+\frac{27}{64\pi^2}p(p+1)\right)\frac{t_P^2}{\eta^2}. \tag{2.193}$$

The equation (2.191) does not possess real roots. It means the bononic duality has non-physical nature.

Let us take into account the phononic Hubble inflaton. Interestingly, after application of the phonon quantization, the inflaton energy (2.147) in general is nontrivially quantized

$$E_H^{(n)} = \sqrt{2\left(\pi\frac{n}{N}\right)^2\left[1-\cos\left(\pi\frac{n}{N}\right)\right]}\frac{E_P}{a}, \tag{2.194}$$

where E_P is the Planck energy. Interestingly, in the most general situation this inflaton energy is an even function with respect to n, and behaves as $E_H \sim \frac{1}{a}$. Therefore, the total energy of the Hubble Multiverse

$$E_H^{\mathrm{TOT}}(a) = \int_{a_I}^{a} da' E_H^{(n)}(a') = \sqrt{2\left(\pi\frac{n}{N}\right)^2\left[1-\cos\left(\pi\frac{n}{N}\right)\right]}E_P\ln\frac{a}{a_I}, \tag{2.195}$$

has a divergent behaviour in the limit $a \to \infty$

$$\lim_{a\to\infty} E_H^{\mathrm{TOT}}(a) = \infty. \tag{2.196}$$

Another interesting quantity is the inflaton energy summarized with respect to the number of excitations n. The general formula for such energy can be constructed as follows

$$\langle E_H \rangle_N = 2\sum_{n=0}^{N} E_H^{(n)}, \tag{2.197}$$

where the multiplier 2 arises from inclusion of the states with negative n. In this manner, one can establish the mean inflaton energy in the Multiverse. The result can be presented in the form

$$\langle E_H \rangle_N (a) = \Lambda_N \frac{E_P}{a}, \tag{2.198}$$

where Λ_N is given by the formula

$$\Lambda_N = 2\sum_{n=0}^{N} \sqrt{2\left(\pi\frac{n}{N}\right)^2 \left[1-\cos\left(\pi\frac{n}{N}\right)\right]}. \tag{2.199}$$

For consistency, one can also consider the mean inflaton energy averaged over values of the cosmic scale factor parameter a

$$\overline{\langle E_H \rangle_N} = \frac{1}{a-a_I}\int_{a_I}^{a} da' \langle E_H \rangle_N (a') = E_P \Lambda_N \frac{\ln\dfrac{a}{a_I}}{a-a_I}, \tag{2.200}$$

which gives the physical interpretation of the constant Λ_N

$$\Lambda_N = \lim_{a\to a_I} \frac{\overline{\langle E_H \rangle_N}}{E_P}. \tag{2.201}$$

Interestingly, when the Multiverse becomes infinite, *i.e.* $a \to \infty$, then the mean inflaton energy of the Hubble inflaton averaged over a (2.200) tends to zero.

F Inflaton Constant

The question is, however, the convergence of quantity Λ_N for the huge N limit, *i.e.* the value of Λ_∞ describing the Multiverse which is full of the inflatons. In such a situation, the averaged energy of the Higgs–Hubble inflaton takes the following form

$$\langle E_H \rangle_\infty = \Lambda_\infty \frac{E_P}{a}, \tag{2.202}$$

where formally Λ_∞ is the limit value of Λ_N

$$\Lambda_\infty = \lim_{N\to\infty} \Lambda_N. \tag{2.203}$$

In other words, the problem is to establish the value of Λ_∞ as well as give the answer to the question whether Λ_∞ is an universal constant. One can identify the identical atoms following from phononic quantization with the spatial dimensions $N \equiv D$, and treat the Multiverse model presented above as the multidimensional Universe. Another interpretation of this situation is that N is the number of *atoms of space*, and then the limit $N \to \infty$ would correspond to the classical space, *i.e.* the space which is solid medium of the atoms of space which in this case are *the phononic Hubble inflatons*. Naturally, the most stable mean inflaton energy is obtained for infinite number of the identical atoms $N \to \infty$, *i.e.* when the space-time solid medium creates specific **Æther model**. For simplicity we propose to call Λ_N *the inflaton N-atomic constant*, and Λ_∞ *the inflaton constant*.

In such a situation, the value of the inflaton constant is essential for further discussion. Namely, one can consider the frequency ω_I

$$\omega_I = \Lambda_\infty \omega_P, \tag{2.204}$$

where $\omega_P = \dfrac{E_P}{\hbar}$ is the Planck frequency, which can be interpreted as the Zero-Point Frequency field. Such an idea allows to introduce the characteristic time of the inflation

$$t_I = \frac{2\pi}{\omega_I} = \frac{2\pi}{\Lambda_\infty} t_P, \tag{2.205}$$

where t_P is the Planck time. If one defines the cosmological potential energy in the following simple way

$$V_C(x_C) = \langle E_H \rangle_\infty, \tag{2.206}$$

where $x_C = \ell_P a$ is the cosmological coordinate, then one can determine the related force according to the standard formula

$$F_H = -\frac{dV_C(x_C)}{dx_C}. \tag{2.207}$$

The result can be presented in several ways

$$F_H = -\frac{\Lambda_\infty \hbar c}{x_C^2} = -\frac{\Lambda_\infty}{4\pi}\frac{8}{\kappa}\left(\frac{p_C c}{E_P}\right)^2 = -\frac{G\Lambda_\infty M_P^2}{x_C^2}, \tag{2.208}$$

where $\kappa = \frac{8\pi G}{c^4} = \frac{8\pi\ell_P}{E_P}$ is the Einstein constant, and we have introduced the cosmological equivalent of the De Broglie relation

$$p_C = \frac{h}{x_C}. \tag{2.209}$$

It is evident that this force is similar to the Newton law of universal gravitation for $\Lambda_\infty M_P^2 = m_1 m_2$, where $m_{1,2}$ are masses of two interacting bodies. Interestingly, however, this universal gravitation has the opposite sign to the standard Newton universal gravitation, *i.e.* in fact is anti-gravitation.

Let us try to establish the value of the inflaton constant, which is defined as the sum of the series

$$\Lambda_N = 2\sum_{n=0}^{N}\sqrt{2\left(\pi\frac{n}{N}\right)^2\left[1-\cos\left(\pi\frac{n}{N}\right)\right]}. \tag{2.210}$$

It is convenient to change the summation method by introduction of the summation index $k = \frac{n}{N} = \{0,1\}$. Applying such a modification to the formula (2.210) one obtains the result

$$\Lambda_N = 2\sum_{k=0}^{1}\sqrt{2(\pi k)^2\left[1-\cos(\pi k)\right]}, \tag{2.211}$$

which is very easy to derive straightforwardly

$$\Lambda_N = 4\pi, \tag{2.212}$$

and its independency on N is evident. In the light of such a result the N-atomic inflaton constant is exactly the same as the inflaton constant, *i.e.* the number of atoms of space does not play role for this constant.

Such a situation allows to write out explicitly the averaged energy

$$\langle E_H \rangle_N = \langle E_H \rangle_\infty = 4\pi \frac{E_P}{a}, \tag{2.213}$$

and, consequently, one can evaluate straightforwardly the Zero-Point Frequency field (2.204), the characteristic time of the inflation (2.205) and derive the law of anti-gravitation

$$\omega_I = 4\pi\omega_P \approx 2.3308857 \cdot 10^{44}\text{Hz}, \tag{2.214}$$

$$t_I = \frac{t_P}{2} \approx 2.9656213 \cdot 10^{-44}\text{s}, \tag{2.215}$$

$$F_H = -\frac{32\pi^2}{\kappa}\left(\frac{p_C c}{E_P}\right)^2 = -\frac{4\pi G M_P^2}{x_C^2}. \tag{2.216}$$

G Cosmological Uncertainty Principle

Let us consider the most general uncertainty principle for the uncertainties σ_A and σ_B of any operators A and B

$$\sigma_A \sigma_B \geqslant \sqrt{\left(\frac{1}{2}\langle\{A,B\}\rangle - \langle A\rangle\langle B\rangle\right)^2 + \left(\frac{1}{2i}\langle[A,B]\rangle\right)^2}, \tag{2.217}$$

which was first derived by Erwin Schrödinger [191]. We shall consider this principle in the case of the cosmological operators $A = x_C$ and $B = p_C = \frac{h}{x_C}$. We take into account the standard definition of the quantum mechanical expectation value of an operator A

$$\langle A \rangle = \int_{a_I}^{a_F} \Psi^*(a) A \Psi(a) da, \tag{2.218}$$

where a is the cosmic scale factor parameter, a_I and a_F are the initial and the final value of a, and $\Psi(a)$ is the wave function of the Universe governed by the Klein–Gordon equation. Moreover, we shall apply the usual quantum mechanical normalization condition

$$\int_{a_I}^{a_F} \Psi^*(a)\Psi(a) da = 1. \tag{2.219}$$

It is easy to see that the commutator and the anticommutator are

$$[x_C, p_C] = 0, \tag{2.220}$$
$$\{x_C, p_C\} = 2h. \tag{2.221}$$

The quantum expectation values $\langle x_C\rangle$ and $\langle p_C\rangle$ are

$$\langle x_C\rangle = \ell_P \int_{a_I}^{a_F} a|\Psi(a)|^2 da, \tag{2.222}$$
$$\langle p_C\rangle = \frac{h}{\ell_P}\int_{a_I}^{a_F} \frac{1}{a}|\Psi(a)|^2 da, \tag{2.223}$$

where both the integrals can be calculated straightforwardly by the method of integration by parts. In a general case one obtains

$$\begin{aligned}\int_{a_I}^{a_F} a^n|\Psi(a)|^2 da &= \left(a^n|\Psi(a)|^2\right)\Big|_{a_I}^{a_F} - \int_{a_I}^{a_F} na^{n-1}\left(\int_{a_I}^{a}|\Psi(a')|^2 da'\right)da \\ &= -\int_{a_I}^{a_F} na^{n-1}\left(\int_{a_I}^{a}|\Psi(a')|^2 da'\right)da. \end{aligned}\tag{2.224}$$

In this manner, one can establish easy the product of the expectation values

$$\langle x_C\rangle\langle p_C\rangle = -h\left[\int_{a_I}^{a_F}\left(\int_{a_I}^{a}|\Psi(a')|^2 da'\right)da\right]\left[\int_{a_I}^{a_F}\frac{1}{a^2}\left(\int_{a_I}^{a}|\Psi(a')|^2 da'\right)da\right], \tag{2.225}$$

so that the Schrödinger uncertainty principle in such a situation becomes

$$\sigma_{x_C}\sigma_{p_C} \geqslant h\left|1+\left[\int_{a_I}^{a_F}\left(\int_{a_I}^{a}|\Psi(a')|^2 da'\right)da\right]\left[\int_{a_I}^{a_F}\frac{1}{a^2}\left(\int_{a_I}^{a}|\Psi(a')|^2 da'\right)da\right]\right|. \tag{2.226}$$

Making use of the Hölder inequality

$$\left(\int_S |f(x)|^p dx\right)^{1/p}\left(\int_S |f(x)|^q dx\right)^{1/q} \geqslant \int_S |f(x)g(x)|dx, \tag{2.227}$$

where $\frac{1}{p}+\frac{1}{q}\leqslant 1$, which in the case of $p=q=2$ is called the Cauchy–Bunyakovsky–Schwarz inequality. This inequality is consistent for the case of $p=q$ with $q\geqslant 2$. Let us consider such a case with the functions

$$|f(a)| = \left(\int_{a_I}^{a}|\Psi(a')|^2da'\right)^{1/q}, \tag{2.228}$$

$$|g(a)| = \left(\frac{1}{a^2}\int_{a_I}^{a}|\Psi(a')|^2da'\right)^{1/q}. \tag{2.229}$$

Then one receives

$$\begin{aligned}&\left[\int_{a_I}^{a_F}\left(\int_{a_I}^{a}|\Psi(a')|^2da'\right)da\right]\left[\int_{a_I}^{a_F}\frac{1}{a^2}\left(\int_{a_I}^{a}|\Psi(a')|^2da'\right)da\right]\geqslant\\ &\geqslant\left[\int_{a_I}^{a_F}\left|\left(\int_{a_I}^{a}|\Psi(a')|^2da'\right)^{1/q}\left(\frac{1}{a^2}\int_{a_I}^{a}|\Psi(a')|^2da'\right)^{1/q}\right|da\right]^q,\end{aligned} \tag{2.230}$$

or after minor algebraic manipulations

$$\begin{aligned}&\left[\int_{a_I}^{a_F}\left(\int_{a_I}^{a}|\Psi(a')|^2da'\right)da\right]\left[\int_{a_I}^{a_F}\frac{1}{a^2}\left(\int_{a_I}^{a}|\Psi(a')|^2da'\right)da\right]\geqslant\\ &\geqslant\left[\int_{a_I}^{a_F}\left|\frac{1}{a}\left(\int_{a_I}^{a}|\Psi(a')|^2da'\right)\right|^{2/q}da\right]^q,\end{aligned} \tag{2.231}$$

and in this manner, one obtains

$$\sigma_{x_C}\sigma_{p_C}\geqslant h\left|1+\left[\int_{a_I}^{a_F}\left|\frac{1}{a}\left(\int_{a_I}^{a}|\Psi(a')|^2da'\right)\right|^{2/q}da\right]^q\right|. \tag{2.232}$$

Now one sees that for any value of q and independently on the choice of the wave function $\Psi(a)$ one has

$$\sigma_{x_C}\sigma_{p_C}\geqslant h. \tag{2.233}$$

This result links the Higgs–Hubble inflaton to quantum cosmology, and expresses an universal character of inflation.

Chapter 3

Massive Gravitons

A Outlook on Gravitational Waves

Gravitational waves are one of the most intriguing questions of modern physics. The earliest approach was studied in 1776 by Pierre Simon de Laplace [192] who, to explain the observed decrease of the Moon's orbital period with respect to ancient eclipse observations, considered an orbital damping force arising from a finite speed of propagation of gravity. In 1905 and 1908 Jules Henri Poincaré [193] suggested that planetary orbits slowly lose energy to wave emission in the gravitational field and indicated that any such effect would be too small to explain the perihelion shift of Mercury, which was the first time gravitational waves were mentioned.

In the modern fashion gravitational waves were predicted in the early General Relativity's papers of Albert Einstein. In the 1916 paper [194] Einstein proposed gravitational radiation as one of the consequences of General Relativity, whereas in the 1918 paper [195] he improved mistakes of the previous paper, and calculated the flux of energy carried by the waves far from their source on the basis of the law of conservation of energy. He calculated the equivalent loss of energy to the source system, an effect known from electromagnetism as radiation/back reaction or radiation damping in the situations involving the decay of periodic motion. Applying the linearized field equations, the basis of study of weak gravitational fields arising from strong analogy with electromagnetism, Einstein found out the rate of energy transport by gravitational radiation and, in general, showed that any material configuration characterized by a time-dependent quadrupole or higher multipole moment should emit gravitational waves. The quadrupole formula was, therefore, similar to the multipole radiation formulas of electromagnetism, which as the lowest order of emission has the dipole radiation.

In 1917, Tullio Levi–Civita, in response to the incorrect results derived by Einstein in his 1916 paper on gravitational waves, proposed an alternative theory [196]. In 1922, the British astrophysicist Arthur Stanley Eddington did not support Einstein vision about reality of gravitational waves, and sceptically commented that these waves are *not objective, and (like absolute velocity) are not detectable by any conceivable experiment. They are merely sinuosities in the co-ordinate system, and the only speed of propagation relevant to them is "the speed of thought"*. [197]. In other words, Eddington drastically criticized Einstein's vision saying that gravitational waves are the only undetectable ripples in space-time. The Eddington metaphor was adopted as the title of the book by Daniel Kennefick [198] in which one can find the history of gravitational waves clearly explained.

However, in 1937 Einstein together with American-Israeli physicist Nathan Rosen studied the motion of test particles in the gravitational waves space-times [199]. Their analysis showed that the focusing properties of geodesic lines yield the instability resulting in collapse of plane gravitational waves. Gravitational instabilities can be the reason for the concentration of energy density into black holes, but the instabilities are under control for the case of plane waves. After a brief time Einstein, in the letter to Max Born [200], interpreted this result as the theoretical evidence for the non-existent nature of gravitational waves, *i.e.* that gravitational radiation is the only a mathematical artifact resulting from the linearization of the Einstein field equations. In fact, Einstein and Rosen rediscovered the so-called Beck vacuums, named after Guido Beck [201], first discovered in 1925 which describe the family of exact gravitational wave solutions having a cylindrical symmetry which are known as the Einstein–Rosen waves.

This paper has enlightening history [198]. It is worth stressing that in the early summer of 1936 Einstein and Rosen submitted their manuscript, first entitled *Do gravitational waves exist?*, to Physical Review. After the papers about the Einstein–Podolsky–Rosen paradox and the Einstein–Rosen bridge, this was their third joint paper submitted to this journal. In this article they described the waves and suggested their non-physical nature, infeasibility of energy transport and yielding detectable phenomena. An anonymous referee found the interpretation wrong and the Managing Editor John Torrence Tate sent the article to the authors for revision, with a brief letter and the referee comments. Einstein did not accept this criticism and critically replied in a rather angry way. The manuscript was then sent to a minor journal, the Journal of the Franklin Institute where it was eventually published. However, in the published version Einstein added the note in which he pointed out the wrongness of the initial interpretation of their formal results, and thanked Howard Percy Robertson for help.

In early October 1936, Leopold Infeld arrived at Princeton as Einstein's new assistant after Rosen's depart for the Soviet Union in mid-to-late July [202]. Einstein convinced him that gravitational waves are non-existent, and Infeld even constructed his own proof. Infeld become friendly with Howard Percy Robertson, the established cosmologist, for whom the proof was not convincing. Infeld quickly communicated Robertson with Einstein who, thankfully to the communication, understood that the space-time metric describing the plane gravitational waves can be straightforwardly transformed to the metric appropriate for cylindrical waves. After such a transformation the new insight emerged: the cylindrical waves were generated by the troublesome space-time singularities. From the modern point of view the entire argument about the singularities was wrong, because the singularities are seeming. Einstein independently found out the mistake in Infeld's proof. Robertson noticed that locally the Einstein–Rosen waves are gravitational plane waves, which was earlier studied by Hans W. Brinkmann [203] and then by O. R. Baldwin and George Barker Jeffery [204]. Einstein and Rosen showed that test particles form caustics in sinusoidal plane waves, and transformation using the Brinkmann coordinates does not lead to a contradiction.

In mid-November Einstein made changes within the proof of the paper for the Journal of the Franklin Institute in Philadelphia, and in late 1936/early 1937 Rosen also understood the error. The third version of the manuscript, entitled *On gravitational waves*, strongly influenced by Robertson's suggestion of taking into account cylindrical coordinates, was eventually published. The Einstein–Rosen cylindrical gravitational waves are locally isometric to plane waves. Rosen eventually published his own version of the theory, which continued to neglect of gravitational radiation [205]. The crucial documents were made available to Daniel Kennefick by Martin Blume, the current Editor-in-Chief of The American Physical Society with responsibility for all of the Physical Review journals, Physical Review Letters, and Reviews of Modern Physics, and clearly showed that the referee was H.P. Robertson [198]. In letter to Tate, Robertson reported that his counterarguments were eventually accepted by Einstein, but actually he never published in the Physical Review again.

In 1938, Einstein together with Infeld and Banesh Hoffmann investigated the approximation method [206] which did not contain the terms corresponding to gravitational radiation. They formulated the problem of motion within the post-Newtonian theory which makes corrections to the Newtonian motion of the system. Such a method arises from the highly non-linear nature of the Einstein field equations which blocks finding out of exact solutions. The Einstein–Infeld–Hoffmann method is the most influential approximation method. The first post-Newtonian approaches to gravitational radiation based on the problem of motion emerged after the war. In 1946, Ning Hu studied an equal mass binary system in a circular orbit by the Einstein–Infeld–Hoffmann method, and reported energy loss disagreeing with the Einstein quadrupole formula [207]. However, before publication Hu corrected the result and obtained anti-damping instead of damping, so that the system emitting radiation should receive rather than lose energy and should be slowly increasing. After the war, both Infeld and Rosen still claimed that gravitational waves are non-physical.

In 1951, Infeld and his student Adrian Scheidegger made use of the Einstein–Infeld–Hoffmann method and the standing wave boundary condition [208]. They showed that this approximation leads to a no-radiation-reaction result and argued that all back-reaction terms can be removed by choice of an appropriate chart. In the years 1954 and 1955 Aleksei Zinovyevich Petrov published a fundamental classification of the Riemann–Christoffel curvature tensor [209], which inspired next generations of theoreticians. In 1955, at the conference at Bern, Rosen showed that both the non-covariant descriptions of energy of a gravitational field given by the Einstein pseudotensor and the Landau–Lifshitz pseudotensor vanish for the Einstein–Rosen waves, and convicted himself that such waves do not exist [210]. Rosen realized that this result supports the Infeld–Scheidegger point of view. In 1955 George Cunliffe McVittie, a doctoral student of Eddington, presented certain arguments for the non-existence of the plane gravitational waves following from the gas dynamics [211]. This paper made the reviewer Felix Arnold Edward Pirani highly dissatisfied [212].

In 1955, Joshua Goldberg, a student of Peter Bergmann who had criticized the Infeld–Scheidegger views, examined the reaction problem in the formalism following from the Einstein–Infeld–Hoffmann method [213]. On the one hand, Goldberg denied that the slow motion approach tended to exclude the possibility of damping but, on the other hand, he determined that it was poorly suited to the back reaction problem, principally because the restriction to slow motions of the source. Also in 1955, Vladimir Alexandrovich Fock considered the orbital damping problem in his book (Fock's book is cited in Ref. [171]). He made use of a slow-motion expansion, an alternative to the Einstein–Infeld–Hoffmann method, coupled with "no-ingoing wave" boundary conditions in the past of the system. Fock regarded his back-reaction result as the syndrome of inconsistency of wave phenomena in the problem of motion in General Relativity, which was argued by smallness of the effect for certain astronomical systems. In 1956, G. C. McVittie published a book, which did not support the existence of gravitational waves [214]. At the same year F. A. E. Pirani introduced the Petrov classification of algebra of the Weyl tensor [215].

In 1957 a conference devoted to the methods developed by John Lighton Synge, Aleksei Zinovyevich Petrov [209] and André Lichnerowicz [887] was organized at Chapel Hill. F. A. E. Pirani, influenced by J. L. Synge during a year spent in Dublin, made use of the Petrov classification to understand gravitational radiation [216]. He considered the vacuum configurations and characterized gravitational waves by the algebraic structure of the Riemann–Christoffel curvature tensor. Pirani took into account propagation of the waves with the fundamental velocity, and gravitational wave fronts emerged as discontinuities in the Riemann–Christoffel curvature tensor across null three-surfaces.

Interestingly, he did not exclude the possibility of a propagation with a speed less than the speed of light. During the conference, Pirani, applying the geodesic deviation, the favorite method of Synge, showed how initially mutually static test particles in the path of a sinusoidal gravitational wave move about relative to each other by the space-time curvature of the passing wave. According to this point of view, gravitational waves were depicted as ripples in the fabric of space-time itself, which was the realization of the Eddington metaphor, and physical effects were detectable by observation of the relative motion of two adjacent particles during the passage of a wave.

At the same conference Richard Phillips Feynman, using Pirani's arguments, produced the so-called sticky bead argument. He argued that a particle lying beside a stick would be rubbed back and forth against the stick when a wave is passing, and the friction would generate heat, so that energy would be extracted from the wave. Moreover, he stated that any absorber of waves could also be an emitter. In such a situation, Feynman made gravitational waves naturally existent and for this reason showed a simple counterargument to Rosen's 1955 argument. Also in 1957, Bondi [217] made use of a somewhat modified version of Pirani' argument to refute Rosen's argument. Independently Joseph Weber and John Archibald Wheeler [218] in the same year used the sticky bead argument to demolish Rosen's argument on energy transport. In 1958, Nathan Rosen noticed that using the Cartesian coordinates produces energy in the cylindrical waves and revised his 1955 opinion [219]. Also in 1958, Peter Gabriel Bergmann together with his doctoral student Rainer Kurt Sachs introduced definition of multipoles in General Relativity [220], which helped to understand gravitational radiation. In the same year Andrzej Mariusz Trautman, representative of Infeld's school of Warsaw, in contrast to his mentor adopted "outgoing wave only" boundary conditions within the Einstein–Infeld–Hoffmann approximation. Trautman confirmed Goldberg' 1955 supposition that the net back-reaction effect could not be simply transformed away, and found the positive damping which slightly differs from the Einstein quadrupole formula [221].

F. A. E. Pirani, dissatisfied by G. C. McVittie's results, decided to join efforts with Hermann Bondi and Ivor Robinson. Both Pirani and Robinson believed that gravitational radiation would indeed be produced, while Bondi studied the problem of determinacy of the energy and momentum for a system radiating gravitational waves. In 1959 Bondi, Pirani, and Robinson noted that Rosen's regularity conditions for the plane wave metric were unnecessarily severe, and made use of the Lichnerowicz approach in imposing regularity conditions [222]. They defined the plane wave metrics as solutions of the vacuum Einstein field equations that permit the $G5$ isometry group acting multiply-transitively on null three-surfaces, *i.e.* two spacelike, one null translation, and two null rotations. Bondi, Pirani, and Robinson used the Lichnerowicz junction conditions to show permissibility of "sandwich waves" propagating through the Minkowski space-time - this is possible when the regularity conditions do not require analyticity of the metric components. They noticed that the definitions in 1957 paper of Pirani are suitable to pure gravitational radiation (Petrov type N) only, and do not apply to more general cases, including the presence of the Matter fields. They proposed an invariant definition of gravitational radiation which is less severe than the one of Pirani 1957, in order to include the Matter fields, and possibly of Petrov type I.

Also in 1959, Roy Patrick Kerr investigated the alternative approach to the fast-motion problem and applied this formulation to the electromagnetic field in the curved space-time [223]. In the same year 1959, John Stachel's Master thesis showed certain study in the energy problem for gravitational waves [224]. Another attempt was in 1959-1960 by Asher Peres [225], who initially found anti-damping, similarly to the earlier finding of Ning Hu, but he corrected calculations shortly after so that the new result for circular binary orbits agreed with the 1941 result of Landau and Lifshitz published in their well-known book cited in Ref. [173].

Infeld was rather dissatisfied by the state of affair within his group, but in the 1960 book coauthored with Jerzy Franciszek Plebański he included a detailed argument against the existence of back reaction in freely falling systems without the knowledge or agreement of Plebański [226]. In 1960, Weber showed consistency of the physical nature of gravitational waves by starting within a few years an experimental program to detect them, using large resonant metal bars as antennae [227]. Also in 1960 Plebański together with Bruno Bertotti presented the theory of cosmological perturbations employing the alternative fast-motion approach [228]. In 1961, Rainer Kurt Sachs, considered vacuum gravitational fields on the base of quasi-algebraic approach involving the Riemann–Christoffel curvature tensor and the second Bianchi identities [229]. Sachs introduced optical parameters, shear, divergence, twist, and asymptotic structure of the curvature tensor. He gave a covariant definition of vacuum gravitational fields with "asymptotically geodesic rays", derived the propagation equations for the optical scalars of a geodesic null congruence, and proposed the outgoing radiation condition: a bounded source field has no mixed large-scale radiation if and only if it has asymptotically geodesic rays.

1962 was the year of the significant progress. Goldberg and Sachs [230] introduced an important theorem linking the vacuum solution to the Einstein field equations with algebraic special property of the Weyl tensor. Goldberg together with Peter Havas [231] formulated the alternative fast-motion approach based on Lorentz-invariant equations of motion. Ezra Ted Newman and Roger Penrose presented the approach to gravitational waves by the method of spin coefficients based on the Goldberg–Sachs theorem and peeling, which is presently the well-established classical null frame formalism [232]. Bondi together with M.G.J. van der Burg and A.W.K. Metzner [233] improved the understanding of gravitational waves far from the source by introduction of null coordinates and asymptotic solutions of the Einstein field equations, and the mass loss theorem. Sachs [234] generalized this work to gravitational waves in asymptotically flat space-time. Bondi, van der Burg and Metzner, and Sachs showed that an isolated system would lose mass to the emission of gravitational waves when a certain function, called the Bondi news function, is present. The resulting algebraic structure of asymptotic symmetry group of General Relativity is known as the Bondi–Metzner–Sachs group. Newman together with Theodore W.J. Unti developed this theme [235]. At the 1962 Infeld's conference, Bondi in the discussion with Bergmann and Feynman, emphasized crucial role of specific equations of state in the components of the binary system, because according to his views binaries composed solely of pressure free dust do not emit gravitational radiation [236]. He argued this statement by the supposition that all particles follow geodesics and there is no possibility a departure from geodesic motion, *i.e.* "news". For any physical system, even if there is no deviation from geodesic motion, this is news, since no news, if not good news, is still news, if news was expected. It is known that Feynman had rather sceptical impression about this meeting. Algebraically special metrics inspired Robinson and Trautman whose considered the twist-free condition [237]. Also in 1962, Mikhail Evgeniyevich Gertsenshtein first concerned high-frequency gravitational waves [238].

In 1963, Penrose introduced null infinity and conformal compactification of space-time. In the same year Freeman John Dyson, inspired at first by Weber's work, began to suggest supernovae and binary neutron stars as possible astrophysical sources [240]. In 1964, William Alfred Fowler, later 1983 Nobel Laureate in Physics for *his theoretical and experimental studies of the nuclear reactions of importance in the formation of the chemical elements in the universe*, showed that radio galaxy emissions are an excellent astrophysical application of gravitational waves [241]. In the 1965 book edited by Ivor Robinson, Alfred Schild and Engelbert Lothar Schücking, one can find more such applications emerging with the discovery of the quasar radio sources [242]. In 1965 Subrahmanyan Chandrasekhar, later 1983 Nobel Laureate in Physics for *his theoretical studies of the physical processes of importance to the structure and evolution of the star*, formulated hydrodynamics based on the post-Newtonian approach within General Relativity [243], and by the end of the decade he was able to describe reaction effects. In 1966, Peter Szekeres implemented the concepts of gravitational inert and gravitational active parts of a matter fluid [244]. In 1967, Fred Isaac Cooperstock presented a paper about energy transported by gravitational waves in quasars [245].

In 1968, Richard Allen Isaacson first investigated the limit of high frequency for gravitational waves and considered the linear approximation, geometric optics, nonlinear terms, and effective stress tensor in such a limit [246]. In 1969, William Lionel Burke removed the ambiguous nature of the slow-motion approach following from imposing of the boundary conditions [247]. He applied the method of matched asymptotic expansions, which determined the solution to the problem of motion in the zone near the source, by matching it through an intermediate zone, to Trautman's "outgoing wave only" boundary conditions in the far zone of the waves. Consequently, the solution of the near zone problem was correctly defined by unambiguous application of the chosen boundary condition. Employing the Burke approach, in 1969 Burke together with his supervisor Kip Stephen Thorne derived the quadrupole formula for emission from binary systems [248]. Also in 1969, Weber surprisingly announced detection of gravitational waves [249], which confused the theoretical front of research and was apparently a false result.

In 1970, S. Chandrasekhar together with F. Paul Esposito [250] received the Einstein quadrupole formula using Chandrasekhar's early worked out methods. In 1973, Havas expressed doubts about the usefulness of the multipole formalism in the context of astrophysical applications [251]. In 1974, Thibault Damour and Remo Jacopo Ruffini expressed doubts that the binary system would exhibit measurable orbital damping effects [252]. In 1975, Russell Alan Hulse and Joseph Hooton Taylor, Jr., discovered object Pulsar B1913+16, also known as PSR J1915+1606 or PSR 1913+16, which is formed by a radiating neutron star orbiting around a common center of mass together with another neutron star and called now the Hulse–Taylor binary system. The analysis performed by R.A. Hulse and J.H. Taylor [253] manifestly showed that the experimental data are consistently explainable if one assumes that the system loses energy through emission of gravitational radiation in strict accordance with General Relativity. This discovery crystalized the excitement in the field of relativity, providing the first test bed for strong field effects of General Relativity. Hulse and Taylor were jointly awarded 1993 Nobel Prize in Physics for *new type of pulsar, a discovery that has opened up new possibilities for the study of gravitation*. In 1976, Jürgen Ehlers together with Arnold Rosenblum, Joshua Goldberg and Peter Havas noticed that there is no rigorous derivation of the quadrupole formula for the binary pulsar, inadequacy of the previous attempts to the back-reaction problem, and advanced an outline of the pioneering program based on the higher-order terms [588]. In 1977, Beverly Kobre Berger, Douglas Matthew Eardley, and Donald Wallace Olson found that the "quasispherical" space-times investigated by Peter Szekeres considered in comoving coordinates have conformally flat space slices, so that these space-times can represent the pressureless non-spherical gravitational collapse with no gravitational radiation. In 1979, Ehlers organized the summer school in Varenna devoted to formulating a research program going to the rigorous solution of the problem [256].

In 1980, J.H. Taylor together with Peter M. McCulloch, after years of careful observation of the system, announced an orbital period decrease in line with the predictions of the quadrupole formula with the accuracy of measurement of about 20 [257]. Also in 1980, Thorne suggested that the multipole formalism is able to give correct approximate estimates of source strength [258], which met criticism of Ehlers's school. However, in 1980 Martin Walker and Clifford Martin Will in 1980 noticed that a basic iterative algorithm, applicable for both fast and slow motion, leads to the Einstein quadrupole formula through reaction calculations [259]. They analyzed the cross section of well known calculations and argued that a certain sufficient number of steps in the iteration recovers the quadrupole formula. This kills the sense of the research program proposed by Ehlers. In 1981, Rosenblum brought the physically unclear significance of the quadrupole formula, and derived the equations of motion necessary to compute the emission of gravitational radiation during small-angle scattering by making use of a fast motion approximation [260]. In 1982, Cooperstock together with David Wesley Hobill reviewed the problem of the general relativistic prediction for the gravitational radiation emitted by gravitationally interacting systems with particular emphasis on the two-body problem [261]. They refused to set forward a general scheme or advocate a particular result and instead argued against preconceived notions.

In 1983, Rosenblum, making use of a 'fast motion approach' and first principles, worked out explicitly the equations of motion to second order [262]. Also in 1983, Damour calculated evolution of the orbit of a binary pulsar under the action of a gravitational radiation reaction with no approximation of weak gravity inside the individual stars to certain order, and showed that the result reveals no acceleration of the center of mass of the system, which quantitatively agreed with both the quadrupole formula and with observations [263]. In the same year Toshifumi Futamase applied, to the gravitational radiation-reaction problem, the asymptotic approximation scheme based on the theory of the Newtonian limit which allowed to remove all divergences of the previous approaches. The asymptotic approximation was finite to all orders even beyond the radiation-reaction order, it proved the prediction derived from the standard quadrupole formula applies in the

Newtonian limit, and showed that random data for the initial gravitational wave field do not affect the calculation of the radiation reaction, even if their amplitude is of first post-Newtonian order [264]. In 1987, James L. Anderson returned to the Einstein–Infeld–Hoffmann method complete with its surface integral method, and combined this with Burke's matched asymptotic expansions, to derive the quadrupole formula [265]. Also in 1987, Ehlers together with Aragam Ramarao Prasanna and Reinhard A. Breuer investigated the propagation of the small-amplitude gravitational waves through the pressureless matter (dust) [266]. In 1989 there were two important papers. Ivor Robinson [267] applied conditions for radiation from a bounded source to the Robinson–Trautman metrics, for the axially symmetric case showed simplification of the field equations. Bernard Frederick Schutz [268] discussed detectability of gravitational waves: estimation of strength of sources, possible sources, dependence on its distance, total energy, and the dominant frequency, and certain physical consequences.

In 1991, Jerry B. Griffiths published book establishing his approach to collisions of plane gravitational waves [269]. In 1992, F.I. Cooperstock making use of the hypothesis that the preferred reference the framework can be used for elimination of the Landau–Lifshitz pseudotensor, suggested that energy of the gravitational field should be described only by an invariant tensor. This led him to the conclusion that the suitable law of conservation requires localization of energy only in the regions of non-vanishing energy–momentum tensor, which blocks propagation of gravitational waves through empty space [270] and is incompatible with Feynman's sticky bead argument. Also in 1992, the group headed by Alex Abramovici [271] announced the Laser Interferometer Gravitational-Wave Observatory (LIGO) Project whose purpose was to detect and study astrophysical gravitational waves and use data from them for research in physics and astronomy. LIGO was planned to study the nature and nonlinear dynamics of gravity, the structures of black holes, the equation of state of nuclear matter, the masses, birth rates, collisions, and distributions of black holes and neutron stars in the universe, the cores of supernovae, the very early universe. LIGO's project was completed after 20 years, its construction was planned on 1992, and gravitational-wave searches were planned to begin in 1998. In the same year, Roberto O. Gómez and Jeffrey Winicour investigated the discrepancies that arise if wave forms are observed at a finite distance as opposed to infinity and, on the basis of the spherically symmetric Einstein–Klein–Gordon system, applied this method to isolate the effects of backscattering and redshifting while avoiding more complicated effects that arise in nonspherical systems with gravitational radiation [272]. Also in 1992, Michael Efroimsky [273] derived and studied the stress-energy tensor of gravitational waves in arbitrary media and on an arbitrary curved background, introduced the natural low-frequency cut-off, and discussed arising nonlinear wave equation for weak gravity oscillations.

In 1993, Nathan Rosen together with Kumar Shwetketu Virbhadra presented new calculations on the energy content of cylindrical waves which did not appear until after some delay [274]. In 1993 there were several other important contributions to gravitational waves research. Luc Blanchet and Gerhard Schäfer [275] investigated and discussed some applications of the spectral (or Fourier) decomposition of gravitational wave tails, produced by back-scattering of the outgoing gravitational radiation (emitted by an isolated system) off the curved space-time associated with the total mass of the system, at large distances from the system. Donald E. Neville [276] constructed physical observables for plane gravitational waves, defined to be gauge-invariant quantities which commute with all the constraints, from the amplitudes for the two polarization states obtained through canonical transformation of the variables describing plane gravitational waves. Schutz [277] reviewed the most likely sources of gravitational radiation that may be seen by the large-scale laser-interferometric detectors that will soon be built around the world.

In 1994, Blanchet together with Bangalore S. Sathyaprakash [278] investigated the level at which the tail of a gravitational wave, resulting from the non-linear interaction between the usual quadrupole radiation generated by an isolated system (with total mass-energy Mc^2) and the static monopole field associated with Mc^2, could be detected in future laser interferometric detectors. In the same year, Damour and Balasubramanian R. Iyer [279] described the generation of gravitational waves by the post-Newtonian spin octupole moment. In 1995, Bondi, interviewed by Daniel Keenefick [198], eventually removed his opinion that binary systems consisting of ideal dust do not generate gravitational waves. In 1996, Georgy Andreevich Alekseev and Jerry B. Griffiths [280] presented a large family of exact vacuum solutions of the Einstein field equations, with fixed background Minkowski spacetime, which describe gravitational waves distinguishable by the spherical wavefront. Also in 1996, Blanchet, Iyer, Will together with Alan G. Wiseman [281] presented two independent 'plus' and 'cross' polarization waveforms associated with the gravitational waves emitted by inspiralling, non-spinning, and compact binaries, which are based on the second-post-Newtonian approximation of General Relativity beyond the dominant (Newtonian) quadrupolar field.

In 1997, Roy Maartens, George Francis Rayner Ellis, and Stephen Theodore Chesmer Siklos [282] showed that the electric and magnetic parts of the Weyl curvature tensor are crucial for the existence of gravitational waves. In the same year, Richard Alfred Matzner and collaborators [283] used the complete set of $3+1$ ADM equations for linearized waves to show the properties of the initial data and of the specific evolution of a wave problem: arising of the gauge modes and constraint error modes which are of the form which will be controlled in any well specified convergent computational discretization. Also in 1997, Liliana E. Simone, Stephen W. Leonard, Eric Poisson and Clifford M. Will [284] for the case of circular and apart from the radiation-induced inspiral orbits, considered a truncated model for gravitational radiation from binary systems in which each symmetric trace-free mass and current moment of order l is given by the lowest-order, Newtonian-like l-pole radiative moment of the orbiting masses, which summed over all l give gravitational radiation. In 1999, Henk van Elst and G.F.R. Ellis [285] using the extended $3+1$ orthonormal frame formalism for fluid spacetimes containing the Bianchi equations for the Weyl curvature, derived a first-order 44-dimensional symmetric hyperbolic system. Taking into account the baryotropic perfect fluid, they obtained the propagation velocities of disturbances in the matter and the gravitational field.

It is worth stressing that there have been published a number of papers devoted to the subject which can be probably verified by the experimental front, mainly LIGO and VIRGO. There is also a lot of intriguing ideas in the theoretical front, which are more or less based on the knowledge established before 1980. The most recent specialistic monograph in the theme of gravitational waves was presented in 2004 by J. Weber [286], in 2007 by Michele Maggiore [287], in 2009 by Piotr Jaranowski and Andrzej Królak [288], and in 2011 by Jolien D. E. Creighton and Warren G. Anderson [289]. We are not able to discuss all the results here, because the SPIRES database includes more than 3600 bibliographical items related to the phrases "gravitational waves" and "gravitational radiation".

One of such directions, which in the author's opinion is promising for further fruitful development, are high-frequency gravitational waves, which were first considered by M.E. Gertsenshtein [238]. In this field the greatest hope gives the group of the Li–Baker HFGW Detector project, co-founded by the research groups of Robert M. L. Baker, Jr. and Fang-Yu Li which among others include Andrew Walcott Beckwith and Gary Van Stephenson. We shall not discuss details of this research line, but certain significant references of the Li–Baker HFGW Detector are collected in Ref. [290].

At the 2011 conference *Quantum Theory and Gravitation* which held in the Eidgenössische Technische Hochschule Zürich, Viatcheslav Fyodorovich Mukhanov of Ludwig-Maximilians-Universität München presented the the lecture devoted to the idea of massive gravity [291] following from Einstein's gravitational waves, cosmological perturbation theory, and perturbative quantum gravity. Paying no heed to absence of both details as well as crucial references, and his manifestly ignorant attitude of the lecturer who did not respond to my questions, he suggested that his idea leads to the conclusion that gravitation acting on a massive graviton is given by the Newton law of universal gravitation multiplied by $4/3$. This result was intriguing, but the misleading lecture was not convincing. This section is straightforwardly motivated by this lecture, and shows that the situation is really much more complicated than V.F. Mukhanov indicated. We show that the idea propagated by the Mukhanov lecture is generally realizable and plausible, but the results are generally different and much more intriguing than the Mukhanov's ones.

B Einstein's Gravitational Waves

Let us first focus our attention on gravitational waves formulated in the framework of General Relativity. Here we shall present the only basic formulas with no detailed derivation and discussion, which can be found in standard handbooks (Cf. *e.g.* book of Steven Weinberg in Ref. [559], and books of Sean Carroll, and Landau–Lifshitz's course in Ref. [173]). In General Relativity, gravitational waves are defined through the weak field condition

$$g_{\mu\nu} = \eta_{\mu\nu} + h_{\mu\nu}, \tag{3.1}$$

where $\eta_{\mu\nu} = \mathrm{diag}(-1,1,1,1)$ is the metric of the Minkowski space-time, and $h_{\mu\nu}$ is a perturbation such that

$$\det h_{\mu\nu} \ll 1, \tag{3.2}$$

everywhere in space-time. The smallness leads to the inverted metric $g^{\mu\nu} = \eta^{\mu\nu} - h^{\mu\nu}$, where $h^{\mu\nu} = \eta^{\mu\alpha}\eta^{\nu\beta}h_{\alpha\beta}$. The weak symmetric field $h_{\mu\nu}$ can be treated as propagating on the background Minkowski space-time.

The Lorentz invariance for gravitational waves can be proven easily and straightforwardly. Namely, let us assume that space-time coordinates transform like $x'_\mu = \Lambda^\nu_\mu x_\nu$, where Λ^ν_μ is the Lorentz transformation known from Special Relativity. Then the usual conditions of the Lorentz invariance

$$g_{\mu\nu} = \Lambda^\alpha_\mu \Lambda^\beta_\nu g_{\alpha\beta}, \tag{3.3}$$

$$\eta_{\mu\nu} = \Lambda^\alpha_\mu \Lambda^\beta_\nu \eta_{\alpha\beta}, \tag{3.4}$$

lead to the following conclusion

$$g_{\alpha\beta} = \eta_{\alpha\beta} + h_{\alpha\beta}, \tag{3.5}$$

$$h_{\alpha\beta} = \Lambda^\mu_\alpha \Lambda^\nu_\beta h_{\mu\nu}. \tag{3.6}$$

There is also the specific gauge transformation for Einstein's gravitational waves. Namely, the change of coordinates

$$x^\alpha = x^\alpha + \xi^\alpha(x^\beta), \tag{3.7}$$

generates the Lorentz transformation

$$\Lambda^{\alpha}_{\beta} = \frac{\partial x'^{\alpha}}{\partial x^{\beta}} = \delta^{\alpha}_{\beta} + \partial_{\beta}\xi^{\alpha}, \tag{3.8}$$

which leads to the following equation

$$g'_{\alpha\beta} = \eta_{\alpha\beta} + h'_{\alpha\beta}, \tag{3.9}$$

where $h'_{\alpha\beta}$ is the transformation of the metric $h_{\alpha\beta}$

$$h'_{\alpha\beta} = h_{\alpha\beta} - 2\partial_{(\beta}\xi_{\alpha)}, \tag{3.10}$$

and $\xi_{\beta} = \eta_{\alpha\beta}\xi^{\alpha}$.

In such an approximation one obtains the fundamental geometric characteristics. The Christoffel symbols are

$$\Gamma^{\alpha}_{\mu\nu} = \frac{1}{2}\eta^{\alpha\beta}\left(\partial_{\mu}h_{\nu\beta} + \partial_{\nu}h_{\beta\mu} - \partial_{\beta}h_{\mu\nu}\right), \tag{3.11}$$

the Riemann–Christoffel curvature tensor, the Ricci curvature tensor, and the Ricci scalar curvature are

$$R_{\mu\nu\alpha\beta} = 2\eta_{\mu\lambda}\partial_{[\alpha}\Gamma^{\lambda}_{\nu\beta]} = \partial_{\alpha}\partial_{[\nu}h_{\mu]\beta} + \partial_{\beta}\partial_{[\mu}h_{\nu]\alpha}, \tag{3.12}$$

$$R_{\mu\nu} = \eta^{\alpha\beta}R_{\mu\nu\alpha\beta} = \partial_{\alpha}\partial_{(\nu}h^{\alpha}_{\mu)} - \frac{1}{2}\left(\partial_{\mu}\partial_{\nu}h + \Box h_{\mu\nu}\right), \tag{3.13}$$

$$R = \eta^{\mu\nu}R_{\mu\nu} = \partial_{\alpha}\partial^{\mu}h^{\alpha}_{\mu} - \Box h, \tag{3.14}$$

where $h = \eta^{\mu\nu}h_{\mu\nu} = h^{\mu}_{\mu}$ is the trace, $\Box = \eta^{\mu\nu}\partial_{\mu}\partial_{\nu}$ is the D'Alembert operator, and we have used the usual notation for symmetrization () and antisymmetrization [] near the indexes. In this way, the Einstein tensor for gravitational waves is

$$\begin{aligned} G_{\mu\nu} &= R_{\mu\nu} - \frac{1}{2}\eta_{\mu\nu}R \\ &= \partial_{\alpha}\partial_{(\nu}h^{\alpha}_{\mu)} - \frac{1}{2}\left(\partial_{\mu}\partial_{\nu}h + \Box h_{\mu\nu} + \eta_{\mu\nu}\partial_{\alpha}\partial_{\beta}h^{\alpha\beta} + \eta_{\mu\nu}\Box h\right). \end{aligned} \tag{3.15}$$

One sees that the Einstein field equations for gravitational waves are

$$\partial_{\mu}\partial_{\nu}h + \Box h_{\mu\nu} + \eta_{\mu\nu}\partial_{\alpha}\partial_{\beta}h^{\alpha\beta} + \eta_{\mu\nu}\Box h - 2\partial_{\alpha}\partial_{(\nu}h^{\alpha}_{\mu)} = -2\kappa T_{\mu\nu}, \tag{3.16}$$

where $T_{\mu\nu}$ is the energy-momentum tensor. To generate the wave equation, one should use the tensor

$$\bar{h}_{\mu\nu} = h_{\mu\nu} - \frac{1}{2}\eta_{\mu\nu}h, \tag{3.17}$$

and require the appropriate Lorenz-like gauge

$$\partial^{\mu}\bar{h}_{\mu\nu} = 0, \tag{3.18}$$

in which the Einstein field equations for gravitational waves (3.16) take the form of the usual wave equation

$$\Box\bar{h}_{\mu\nu} = -2\kappa T_{\mu\nu}. \tag{3.19}$$

In such a situation, there exist the limiting procedure which leads to the Newton law of universal gravitation. Taking into account the 00-component of the wave equation (3.22 and the spherical symmetry

$$\bar{h}_{00} = \bar{h}_{00}(r), \tag{3.20}$$

together with the energy-momentum tensor of the form

$$T_{00} = \rho c^2, \tag{3.21}$$

one receives the Poisson equation

$$\triangle\phi = \frac{4\pi G}{c^2}\rho, \tag{3.22}$$

where the gravitational potential is

$$\phi = -\frac{1}{4}\bar{h}_{00}. \tag{3.23}$$

In such an approximation, one can neglect other components of $\bar{h}_{\mu\nu}$, so that

$$h = \bar{h}_{00}, \tag{3.24}$$

which gives the metric components

$$h^{00} = h^{11} = h^{22} = h^{33} = -2\phi, \tag{3.25}$$

and, consequently, the space-time is characterized by the interval

$$ds^2 = -(1+2\phi)dx^0dx_0 + (1-2\phi)dx^idx_i. \tag{3.26}$$

C Cosmological Perturbations

Let us look on an alternative point of view given by the cosmological perturbation theory, which is one of foundational theories of relativistic cosmology and theoretical astrophysics. The present state of cosmological perturbation theory arises from the work of Evgeny Mikhailovich Lifshitz [292]. The reference suitable for the application of this section is [293]. The most suitable handbooks are Sean Carroll's and Steven Weinberg's books cited in Ref. [173].

In the cosmological perturbation theory, the fundamental role is played by the background Minkowski space-time perturbed by scalar, vector, and tensor perturbations

$$ds^2 = -(1+2\Phi)(dx^0)^2 + 2\partial_i B dx^i dx^0 + [(1-2\Psi)\delta_{ij} + 2s_{ij}]dx^i dx^j, \tag{3.27}$$

where the scalar and the tensor perturbations are determined as

$$\Psi = -\frac{1}{6}h^k_k, \tag{3.28}$$

$$s_{ij} = \frac{1}{2}\left(h_{ij} + 2\Psi\delta_{ij}\right), \tag{3.29}$$

Φ is the gravitational potential which is responsible for the appropriate law of gravitation, and $\partial_i B$ is the vector perturbation. Such a situation corresponds with the weak field in the form

$$h_{00} = -2\Phi, \tag{3.30}$$

$$h_{0i} = \partial_i B, \tag{3.31}$$

$$h_{ij} = 2s_{ij} - 2\Psi\delta_{ij}. \tag{3.32}$$

The smallness condition $\det h_{\mu\nu} \ll 1$ gives

$$\det h_{ij} \ll \frac{1+|\partial_i B|^2}{2|\Phi|}. \tag{3.33}$$

Such a situation gives slightly different results than General Relativity. This is caused by the fact that the cosmological perturbation theory corresponds with fixation of an inertial reference frame in the background Minkowski space-time and decomposition of the perturbation metric components is made according to their transformation properties under spatial rotations: the 00 component is a scalar, the *Oi* components form a three-vector, and the *ij* components form a two-index symmetric spatial tensor. This decomposition, known as the scalar-vector-tensor decomposition, is similar to decomposition of the electromagnetic field into electric and magnetic fields. The $(0,2)$ spatial tensor can be further decomposed into a trace and a trace-free part, which corresponds with looking for an irreducible representations of the rotation group.

Let us present briefly the geometric characteristics for the cosmological perturbations. The non-zero Christoffel symbols are

$$\Gamma^0_{00} = \partial_0\Phi, \quad (3.34)$$
$$\Gamma^i_{00} = \partial^i(\Phi+\partial_0 B), \quad (3.35)$$
$$\Gamma^0_{i0} = \partial_i\Phi, \quad (3.36)$$
$$\Gamma^i_{j0} = \frac{1}{2}\partial_0 h_{ij}, \quad (3.37)$$
$$\Gamma^0_{ij} = -\partial_i\partial_j B + \frac{1}{2}\partial_0 h_{ij}, \quad (3.38)$$
$$\Gamma^i_{jk} = \partial_{(j}h^i_{k)} - \partial^i h_{jk}, \quad (3.39)$$

so that non-vanishing components of the Riemann–Christoffel curvature tensor are

$$R_{0i0j} = \partial_i\partial_j(\Phi+\partial_0 B) - \frac{1}{2}\partial_0^2 h_{ij}, \quad (3.40)$$
$$R_{0ijk} = -\partial_0\partial_{[j}h_{k]i}, \quad (3.41)$$
$$R_{ijkl} = 2\partial_{[j}\partial_{[k}h_{l]i]}, \quad (3.42)$$

and, consequently, the Ricci curvature tensor has the components

$$R_{00} = \triangle(\Phi+\partial_0 B) + 3\partial_0^2\Psi, \quad (3.43)$$
$$R_{0i} = \partial_0\partial_k\left(3\delta^k_i\Psi + \frac{1}{2}h^k_i\right), \quad (3.44)$$
$$R_{ij} = -\partial_i\partial_j(\Phi - 3\Psi + \partial_0 B) - \frac{1}{2}\Box h_{ij} + \partial_k\partial_{(i}h^k_{j)}. \quad (3.45)$$

while the Ricci scalar curvature is

$$R = 6\Box\Psi - 2\triangle(\Phi+\partial_0 B) + \partial_i\partial_j h^{ij}. \quad (3.46)$$

In this way, components of the Einstein tensor are

$$
\begin{aligned}
G_{00} &= 3\triangle\Psi + \frac{1}{2}\partial_k\partial_l h^{kl}, &(3.47)\\
G_{0i} &= \partial_0\partial_k\left(3\delta_i^k\Psi + \frac{1}{2}h_i^k\right), &(3.48)\\
G_{ij} &= (\delta_{ij}\triangle - \partial_i\partial_j)(\Phi + \partial_0 B) - 3(\delta_{ij}\Box - \partial_i\partial_j)\Psi\\
&- \frac{1}{2}\Box h_{ij} + \partial_k\partial_{(i}h_{j)}^k - \frac{1}{2}\delta_{ij}\partial_k\partial_l h^{kl}, &(3.49)
\end{aligned}
$$

so consequently the Einstein field equations for the cosmological perturbations theory are

$$
\begin{aligned}
&3\triangle\Psi + \frac{1}{2}\partial_k\partial_l h^{kl} = \kappa T_{00}, &(3.50)\\
&\partial_0\partial_k\left(3\delta_i^k\Psi + \frac{1}{2}h_i^k\right) = \kappa T_{0i}, &(3.51)\\
&(\delta_{ij}\triangle - \partial_i\partial_j)(\Phi + \partial_0 B) - 3(\delta_{ij}\Box - \partial_i\partial_j)\Psi\\
&-\frac{1}{2}\Box h_{ij} + \partial_k\partial_{(i}h_{j)}^k - \frac{1}{2}\delta_{ij}\partial_k\partial_l h^{kl} = \kappa T_{ij}, &(3.52)
\end{aligned}
$$

and, amazingly, can be solved straightforwardly.

Fist of all, it is easy to see that the 00 component and $0i$ component can be used to establish the relation for the energy-momentum tensor

$$\partial_0 T_{00} = \partial^i T_{0i}. \tag{3.53}$$

The $0j$-component of the Einstein field equations can be solved through straightforward elementary integration

$$h_j^k = -6\delta_j^k\Psi + 2\kappa\int\int T_{0j}dx^0 dx^k. \tag{3.54}$$

Contraction of both sides of this solution with δ_k^j and application of the relation (3.28) yields the scalar perturbations

$$\Psi = \frac{\kappa}{6}\int\int T_{0k}dx^0 dx^k, \tag{3.55}$$

and, consequently, the solution (3.54) can be presented in the following form

$$h_j^i = -\kappa\delta_j^i\int\int T_{0k}dx^0 dx^k + 2\kappa\int\int T_{0j}dx^0 dx^i. \tag{3.56}$$

Let us notice that now one can deduce straightforwardly and easily the following relationships

$$\begin{aligned}
\partial_k\partial_l h^{kl} &= \kappa T_{00}, &(3.57)\\
\partial_k\partial_{(i}h^k_{j)} &= 2\kappa\int\partial_{(i}T_{0j)}dx^0-\kappa\int\int\partial_i\partial_jT_{0k}dx^0dx^k, &(3.58)\\
\Box h_{ij} &= \kappa\left(\int\partial^0T_{0k}dx^k-T_{00}\right)\delta_{ij}\\
&+ 2\kappa\left(\int\partial_iT_{0j}dx^0-\int\partial^0T_{0j}dx_i\right), &(3.59)\\
\Box\Psi &= \frac{\kappa}{6}\left(T_{00}-\int\partial^0T_{0k}dx^k\right), &(3.60)\\
\partial_i\partial_j\Psi &= \frac{\kappa}{6}\int\int\partial_i\partial_jT_{0k}dx^0dx^k &(3.61)
\end{aligned}$$

Now it can be seen by simple manipulation that the 00-component of the Einstein field equations takes the form

$$\triangle\Psi=\frac{\kappa}{6}T_{00}. \qquad (3.62)$$

Making use of the relation (9.182) after certain manipulations, one receives the following relationships

$$\begin{aligned}
\int\partial^0T_{0j}dx_i &= \int\int\partial_0^2T_{00}dx_idx_j, &(3.63)\\
\int\partial^0T_{0k}dx^k &= \int\int\partial_0^2T_{00}dx^kdx_k, &(3.64)\\
\int\int\partial_i\partial_jT_{0k}dx^0dx^k &= \int\int\partial_i\partial_jT_{00}dx^kdx_k, &(3.65)\\
\int\partial_jT_{0i}dx^0 &= \int\partial_jT_{00}dx_i &(3.66)
\end{aligned}$$

which can be applied to reduce the ij component of the Einstein field equations to the following form

$$\begin{aligned}
(\delta_{ij}\triangle-\partial_i\partial_j)(\Phi+\partial_0B) &= \kappa T_{ij}+\frac{\kappa}{2}\left(T_{00}+\frac{2}{3}\int\int\partial_0^2T_{00}dx^kdx_k\right)\delta_{ij}\\
&+\frac{\kappa}{2}\int\int\partial_i\partial_jT_{00}dx^kdx_k-\kappa\int\int\partial_0^2T_{00}dx_idx_j-\kappa\int\partial_jT_{00}dx_i, \qquad (3.67)
\end{aligned}$$

which after contraction with δ^{ij} yields the equation

$$\triangle(\Phi+\partial_0 B)=\frac{\kappa}{2}\left(T+T_{00}\right), \tag{3.68}$$

where $T=\delta^{ij}T_{ij}$ is the trace of the spatial part of the energy-momentum tensor. Finally, one can easy establish

$$h_{ij} = 2\kappa\int\int T_{00}dx_i dx_j-\kappa\delta_{ij}\int\int T_{00}dx_k dx^k, \tag{3.69}$$

$$\Psi = \frac{\kappa}{6}\int\int T_{00}dx_k dx^k, \tag{3.70}$$

and the condition (3.33) for non-zero T_{00} can be rewritten in the following form

$$\frac{\kappa^3\left(\int\int T_{00}dx_k dx^k\right)^3}{\det\left(\delta_{ij}-2\frac{\int\int T_{00}dx_i dx_j}{\int\int T_{00}dx_k dx^k}\right)}\ll\frac{1+|\partial_i B|^2}{2|\Phi|}. \tag{3.71}$$

The visible problem is, however, the equation (3.68) in which evolution of the field B can not be separated from the evolution of the potential Φ. However, for full consistency with the Newton law of universal gravitation, one can propose that the Poisson equation

$$\triangle\Phi=\frac{4\pi G}{c^2}\rho, \tag{3.72}$$

where ρ is the energy density of the Matter fields

$$T_{00}=\rho c^2, \tag{3.73}$$

is satisfied. In such a situation, the field B satisfies the equation

$$\triangle B=\frac{\kappa}{2}\int T dx^0. \tag{3.74}$$

In this manner, the whole problem is to take into account the suitable energy-momentum tensor $T_{\mu\nu}$, and consider its 00 component to solve it consistently.

D Perturbative Graviton

Let us consider the perturbative point of view on graviton [294], i.e. the gauge field theory suitable for spin-2 particle. In such a situation, the energy-momentum tensor $T_{\mu\nu}$ of graviton, which in gauge field theories is the coefficient in the functional Taylor series expansion of the Lagrangian density describing the Matter fields

$$\mathscr{L}_M = \mathscr{L}_M(\eta_{\mu\nu}) + \frac{1}{2}T^{\mu\nu}\left(h_{\mu\nu} - \eta_{\mu\nu}\right) + \ldots, \tag{3.75}$$

is given by the explicit formula

$$T^{\mu\nu} = 2\left.\frac{\delta\mathscr{L}_M}{\delta h_{\mu\nu}}\right|_{h^{\mu\nu}=\eta^{\mu\nu}}. \tag{3.76}$$

In such a situation, the Lagrangian of Matter fields has the form

$$\mathscr{L}_M = \alpha\frac{m^2}{4}\left(h^\mu_\mu h^\nu_\nu - h_{\mu\nu}h^{\mu\nu}\right), \tag{3.77}$$

where $h^\mu_\mu = h^{\mu\nu}\eta_{\mu\nu}$, m is graviton mass, and α is some constant coefficient which guarantees proper dimension of the Lagrangian, i.e. $[\mathscr{L}_M] = E\cdot L^{-3}$. To receive the dimensional correctness one can make use of the Planck units. Then

$$\alpha = \frac{c^2}{M_P\ell_P^3} = \frac{c^6}{\hbar^2 G} = \frac{G}{\ell_P^4} = 2\pi\frac{\kappa c^4}{S_P^2}, \tag{3.78}$$

where $M_P = \dfrac{\hbar}{c\ell_P}$, $\ell_P = \sqrt{\dfrac{\hbar G}{c^3}}$, and $S_P = 4\pi\ell_P^2$ are the Planck mass, the Planck length, and the area of the Planck sphere, respectively.

In such a situation, straightforward calculations lead to the following energy-momentum tensor

$$T^{\mu\nu} = \frac{m^2c^2}{M_P\ell_P^3}\eta^{\mu\nu}, \tag{3.79}$$

for which one receives

$$T = \delta^{ij}T_{ij} = 3\frac{m^2c^2}{M_P\ell_P^3}, \tag{3.80}$$

$$T_{00} = -\frac{m^2c^2}{M_P\ell_P^3}. \tag{3.81}$$

In this manner, implementation of the massless limit to the energy-momentum tensor of a massive graviton is not difficult to perform and shows that exactly *all its components* identically vanish. Applying the massless limit to the basic equations obtained in the previous subsection, one receives

$$\Psi = 0, \tag{3.82}$$
$$h_{ij} = 0, \tag{3.83}$$
$$\triangle(\Phi + \partial_0 B) = 0. \tag{3.84}$$

In such a situation, the metric (3.27) becomes

$$ds^2 = -(1+2\Phi)(dx^0)^2 + 2\partial_i B dx^i dx^0 + dx^i dx^j, \tag{3.85}$$

and this state of affairs is fully consistent in the light of the condition (3.33)

$$\frac{1+|\partial_i B|^2}{2|\Phi|} \gg 0. \tag{3.86}$$

E Massless Limit

The Laplace equation (3.84) can be rewritten in the form

$$\triangle\Phi = -\triangle\partial_0 B, \tag{3.87}$$

and can be solved by the most general separation of variables

$$\triangle\Phi = C, \tag{3.88}$$
$$\triangle\partial_0 B = -C, \tag{3.89}$$

where C is a separation constant. In such a situation, one obtains the following solutions

$$\Phi(r) = C_0 - \frac{C_1}{r} + \frac{C}{6}r^2, \tag{3.90}$$
$$B(r,t) = \left(C_0' - \frac{C_1'}{r} - \frac{C}{6}r^2\right)(t - t_0). \tag{3.91}$$

Let us focus on the gravitational potential Φ, because a similar procedure can be repeated for the field B. Taking into account the boundary conditions

$$\Phi(r_0) = \Phi_0, \tag{3.92}$$
$$\partial_i\Phi(r_0) = \partial_i\Phi_0, \tag{3.93}$$

one receives the general solution in the form

$$\Phi(r) = \Phi_0 + \left(1 - \frac{r_0}{r}\right) r_0^i \partial_i \Phi_0 + \frac{C}{6}\left(r^2 - 3r_0^2 + \frac{2r_0^3}{r}\right), \tag{3.94}$$

so that one constant remains undetermined. The separation constant, however, can be established simply. Namely, one sees that the constant in the gravitational potential

$$\Phi_0 - r_0^i \partial_i \Phi_0 - \frac{C}{2} r_0^2, \tag{3.95}$$

does not play an essential role for the force law of gravitation because it vanishes when differentiation, $F = -m\partial_i\Phi$, is performed. In this manner, one can suggest identical vanishing of this constant contribution to the gravitational potential. Such a condition allows to determine unambiguously the separation constant

$$C = \frac{2}{r_0^2}\left(\Phi_0 - r_0^i \partial_i \Phi_0\right), \tag{3.96}$$

and, for this reason, one obtains finally

$$\Phi(r) = \frac{2}{3}\left(\Phi_0 - \frac{5}{2} r_0^i \partial_i \Phi_0\right)\frac{r_0}{r} + \frac{\Phi_0 - r_0^i \partial_i \Phi_0}{3}\frac{r^2}{r_0^2}. \tag{3.97}$$

Let us consider the general situation

$$\frac{2}{3}\left(\Phi_0 - \frac{5}{2} r_0^i \partial_i \Phi_0\right) = -\frac{\Omega_0^2}{2} r_0^2, \tag{3.98}$$
$$\frac{\Phi_0 - r_0^i \partial_i \Phi_0}{3} = \frac{\omega_0^2}{2} r_0^2, \tag{3.99}$$

where Ω_0 and ω_0 are two constant angular frequency parameters, for which

$$\Phi_0 = \frac{1}{2}(5\omega_0^2 + \Omega_0^2)r_0^2, \tag{3.100}$$

$$r_0^i \partial_i \Phi_0 = \frac{1}{2}(2\omega_0^2 + \Omega_0^2)r_0^2, \tag{3.101}$$

and the force law of gravitation is

$$F_i = \lim_{m \to 0} \left(-\frac{m\Omega_0^2}{2} \frac{r_0^3}{r^3} r_i - m\omega_0^2 r_i \right). \tag{3.102}$$

One sees now that the force law is, in general, non-trivial if and only if

$$\Omega_0^2 = \frac{K_0}{m}, \tag{3.103}$$

$$\omega_0^2 = \frac{k_0}{m}, \tag{3.104}$$

where K_0 and k_0 are positive constants.

F Massive Gravitons

One sees now that there are two non-trivial important cases within the general situation given by the gravitational potential. Let m be a positive graviton mass parameter. Then:

1. The Newton law of universal gravitation is restored if and only if

$$\omega_0 = 0, \tag{3.105}$$

$$\Omega_0 = \frac{GM}{r_0^3}, \tag{3.106}$$

where M is a positive mass parameter describing a spherically-symmetric body of radius r_0.

2. The *oscillatory law of gravitation* is obtained when

$$\omega_0 \neq 0, \tag{3.107}$$

$$\Omega_0 = 0. \tag{3.108}$$

3. The force law of gravitation $\sim \frac{1}{r^2}$ can be obtained when

$$r_0^i \partial_i \Phi_0 = \Phi_0, \tag{3.109}$$

$$\frac{2}{3}\left(\Phi_0 - \frac{5}{2} r_0^i \partial_i \Phi_0\right) = \frac{\alpha}{m}, \tag{3.110}$$

where α is some reference constant. Let us select $\alpha = M\Phi_1$, where M is the mass of spherically symmetric body and Φ_1 is some constant reference value of the potential Φ. In such a situation, one obtains

$$\Phi_0 = -\frac{M}{m}\Phi_1, \tag{3.111}$$

$$r_0^i \partial_i \Phi_0 = -\frac{M}{m}\Phi_1. \tag{3.112}$$

However, for consistency it must be

$$M = \beta m = const, \tag{3.113}$$

where β is a dimensionless constant, which means that the mass of a body generating gravitation is multiplication of the graviton mass. In such a situation, the potential is

$$\Phi(r) = \beta \Phi_1 \frac{r_0}{r}. \tag{3.114}$$

where the constant $\beta \Phi_1 r_0$ can be established by suitable experimental data.

4. The most general force law of gravitation, which is not trivial in the massless limit,

$$\Phi(r) = \frac{M_1 \Phi_1 r_0}{m} \frac{1}{r} + \frac{M_2 \Phi_2}{m r_0^2} r^2, \tag{3.115}$$

where $M_1\Phi_1$ and $M_2\Phi_2$ are some reference constants, can be received from the conditions

$$\Phi_0 = \frac{5M_2\Phi_2 - M_1\Phi_1}{m}, \tag{3.116}$$

$$r_0^i \partial_i \Phi_0 = \frac{2M_2\Phi_2 - M_1\Phi_1}{m}. \tag{3.117}$$

calculation of Φ_0 from the definition (3.115) and comparison with the result (3.116) leads to the conclusion

$$M_2\Phi_2 = \frac{1}{2}M_1\Phi_1, \tag{3.118}$$

and, consequently,

$$\Phi_0 = \frac{3}{2}\frac{M_1\Phi_1}{m}, \tag{3.119}$$

$$r_0^i\partial_i\Phi_0 = 0. \tag{3.120}$$

The relations (3.118) and (3.119) allows to write

$$\Phi(r) = \frac{2\Phi_0}{3}\frac{r_0}{r} + \frac{\Phi_0}{3}\frac{r^2}{r_0^2}, \tag{3.121}$$

The question is how to interpret the relation (3.119). One can consider $M_1 = M$ as the mass of the spherically symmetric body generating the gravitational field, and $\Phi_1 = -\frac{GM}{r_0}$ so that

$$\Phi_0 = -\frac{3}{2}\frac{GM^2}{mr_0}. \tag{3.122}$$

In such a situation, one receives

$$\Phi(r) = \frac{\omega_0^2 r_0^3}{r} - \frac{\omega_0^2}{2}r^2, \tag{3.123}$$

where the frequency of the oscillatory part of gravitation becomes

$$\omega_0 = \sqrt{\frac{GM^2}{mr_0^3}}, \tag{3.124}$$

and after taking into account the relation $M = \beta m$ takes the form

$$\omega_0 = \sqrt{\beta\frac{GM}{r_0^3}}. \tag{3.125}$$

The parameter r_0 can be chosen arbitrarily. Let us check $r_0 = R$, where R is the radius of the spherical body. In such a situation,

$$\omega_0 = \sqrt{\beta}\Omega_0, \tag{3.126}$$

where $\Omega_0 = \frac{GM}{R^3}$. In other words, one can write the graviton mass

$$m = \left(\frac{\Omega_0}{\omega_0}\right)^2 M. \tag{3.127}$$

The formula (3.127) suggests that there the various types of gravitons emitted from any massive body.

G Electromagnetic Gravity

The obtained force law of gravitation

$$F_i = -\frac{m\Omega_0^2}{2}\frac{r_0^3}{r^3}r_i - m\omega_0^2 r_i. \tag{3.128}$$

can be considered as the analogue of the Lorentz force

$$F_i = mg_i + m\varepsilon_i^{jk} v_j d_k, \tag{3.129}$$

where the field d_k plays a role analogous to magnetic field, gravitational acceleration g_i plays a role of electric field, and mass m of a moving particle plays a role of electric charge.

Let us analyze the situation in the straightforward analogy to the Maxwell electrodynamics point of view. The field tensor $\mathbb{G}_{\mu\nu}$ analogous to the Faraday tensor should have the following form

$$\mathbb{G}_{\mu\nu} = \alpha \begin{bmatrix} 0 & -\frac{g_i}{c} \\ \frac{g_j}{c} & \varepsilon_{ij}^k d_k \end{bmatrix}, \tag{3.130}$$

where α is a constant which should give the field tensor the dimension of the Faraday tensor, i.e. $\frac{\mathrm{N}}{\mathrm{A}\cdot\mathrm{m}}$. Straightforward dimensional analysis shows that, $[\alpha] = \frac{\mathrm{kg}}{\mathrm{A}\cdot\mathrm{s}}$, so one can make use of the Planck units to construct this constant as follows

$$\alpha = \frac{M_P}{q_P} = \frac{1}{\sqrt{4\pi\varepsilon_0 G}}, \tag{3.131}$$

where $M_P = \sqrt{\frac{\hbar c}{G}}$ is the Planck mass and $q_P = \sqrt{4\pi\varepsilon_0\hbar c}$ is the Planck charge. Consequently, the field tensor is

$$\mathbb{G}_{\mu\nu} = \frac{1}{\sqrt{4\pi\varepsilon_0 G}}\begin{bmatrix} 0 & \frac{g_i}{c} \\ -\frac{g_j}{c} & \varepsilon_{ij}^k d_k \end{bmatrix}. \tag{3.132}$$

so that its dual tensor has the form

$$\tilde{\mathbb{G}}_{\mu\nu} = \varepsilon_{\mu\nu\kappa\lambda}\mathbb{G}^{\kappa\lambda} = \frac{1}{\sqrt{4\pi\varepsilon_0 G}}\begin{bmatrix} 0 & d_i \\ -d_j & -\varepsilon_{ij}^k \frac{g_k}{c} \end{bmatrix}. \tag{3.133}$$

We shall take into account an analogue of electromagnetic four-current

$$J_\nu = [c\rho, J_i]. \tag{3.134}$$

Now one can construct immediately the analogue of the Maxwell field equations. The first pair of the field equations is

$$\partial^\mu \mathbb{G}_{\mu\nu} = \mu_0 J_\nu, \tag{3.135}$$

and lead to the analogue of the Gauss law for electricity

$$\partial^i g_i = -\sqrt{\frac{4\pi G}{\varepsilon_0}}\rho, \tag{3.136}$$

and to the analogue of the Ampère law

$$\varepsilon_{ij}^k \partial^i d_k = \sqrt{\frac{4\pi G}{\varepsilon_0}}\frac{1}{c^2}J_i - \frac{1}{c^2}\dot{g}_i. \tag{3.137}$$

Similarly, the Bianchi identities

$$\partial^\mu \tilde{\mathbb{G}}_{\mu\nu} = 0, \tag{3.138}$$

give the second pair of the field equations: the analogue of the Gauss law for magnetism

$$\partial^k d_k = 0, \tag{3.139}$$

and the analogue of the Maxwell–Faraday law

$$\dot{d}_j = \varepsilon_{ij}^k \partial^i g_k. \tag{3.140}$$

The field tensor can be expressed by the analogue of the electromagnetic four-potential A_μ

$$\mathbb{G}_{\mu\nu} = \partial_\mu A_\nu - \partial_\nu A_\mu, \tag{3.141}$$

where straightforward analysis shows that

$$A_\mu = \frac{1}{\sqrt{4\pi\varepsilon_0 G}} v_\mu, \tag{3.142}$$

where $v_\mu = [c, v_i]$ is the velocity four-vector. In such a situation, by making use of the definition (3.141), one receives

$$G_{0i} = -\frac{1}{\sqrt{4\pi\varepsilon_0 G}}\frac{g_i}{c} = \partial_0 A_i - \partial_i A_0 = \frac{1}{\sqrt{4\pi\varepsilon_0 G}}\frac{1}{c}\dot{v}_i, \tag{3.143}$$

$$G_{ij} = \frac{1}{\sqrt{4\pi\varepsilon_0 G}}\varepsilon_{ij}^{k} d_k = \partial_i A_j - \partial_j A_i = \frac{1}{\sqrt{4\pi\varepsilon_0 G}}\left(\partial_i v_j - \partial_j v_i\right), \tag{3.144}$$

which after minor manipulations give the relations

$$g_i = -\dot{v}_i, \tag{3.145}$$

$$d_i = \varepsilon_i^{kl}\partial_k v_l. \tag{3.146}$$

Also, the stress-energy tensor is the analogue of the electromagnetic one

$$T_{\mu\nu} = \frac{1}{\mu_0}\left(\eta^{\alpha\beta}\mathbb{G}_{\mu\alpha}\mathbb{G}_{\beta\nu} - \frac{1}{4}\eta^{\mu\nu}\mathbb{G}_{\alpha\beta}\mathbb{G}^{\alpha\beta}\right) = \begin{bmatrix} \varepsilon & \frac{S_i}{c} \\ \frac{S_j}{c} & -\sigma_{ij} \end{bmatrix}, \tag{3.147}$$

where ε is the field energy density

$$\varepsilon = \frac{g^2 + c^2 d^2}{8\pi G}, \tag{3.148}$$

S_i is the analogue of the Poynting vector

$$S_i = \frac{c}{4\pi G}\varepsilon_i^{kl} g_k d_l \tag{3.149}$$

and σ_{ij} is the analogue of the Maxwell stress tensor

$$\sigma_{ij} = \frac{1}{4\pi G}\left[g_i g_j + c^2 d_i d_j - \frac{g^2 + c^2 d^2}{2}\delta_{ij}\right]. \tag{3.150}$$

Making use of the definitions (3.145) and (3.146) within the equation (3.137), one receives

$$\Box v_j - \partial_j(\partial^i v_i) = \sqrt{\frac{4\pi G}{\varepsilon_0}}\frac{1}{c^2}J_j, \tag{3.151}$$

and, similarly, the equation (3.136) leads to

$$\partial^i \dot{v}_i = \sqrt{\frac{4\pi G}{\varepsilon_0}}\rho, \tag{3.152}$$

so, consequently, one can establish the four-current

$$J_\mu = c^2\sqrt{\frac{\varepsilon_0}{4\pi G}}\left[\frac{1}{c}\partial^i \dot{v}_i, \Box v_j - \partial_j(\partial^i v_i)\right]. \tag{3.153}$$

If one consider a graviton possessing non-zero mass m, then the four-current should have the form

$$J_\mu = -\frac{1}{\mu_0}\left(\frac{mc}{\hbar}\right)^2 A_\mu, \tag{3.154}$$

which after taking into account of the definition (3.142) becomes

$$J_\mu = c^2\sqrt{\frac{\varepsilon_0}{4\pi G}}\left[-c\left(\frac{mc}{\hbar}\right)^2, -\left(\frac{mc}{\hbar}\right)^2 v_j\right]. \tag{3.155}$$

Consequently, comparison of the relations (3.153) and (3.155) gives

$$\partial^i \dot{v}_i = -\left(\frac{mc^2}{\hbar}\right)^2, \tag{3.156}$$

$$\Box v_j - \partial_j(\partial^i v_i) = -\left(\frac{mc}{\hbar}\right)^2 v_j, \tag{3.157}$$

which lead to the equations

$$m = \frac{\hbar}{c^2}\sqrt{\partial^i g_i}, \tag{3.158}$$

$$\left(\Box + \left(\frac{mc}{\hbar}\right)^2\right) v_j = 0. \tag{3.159}$$

The equation (3.158) is the definition of the graviton mass, while the equation (3.159) is the analogue of the Proca equation describing the massive graviton.

The straightforward analogy with the covariant formulation of the Maxwell electrodynamics allows to construct the action

$$\mathscr{S} = \frac{1}{c} \int d^4x \mathscr{L}, \qquad (3.160)$$

of the field theory leading to the field equations (3.135) and (3.138) through the appropriate variational principle. The Lagrangian density $\mathscr{L}$ is

$$\mathscr{L} = -\frac{1}{4\mu_0} \mathbb{G}_{\mu\nu}\mathbb{G}^{\mu\nu} - A_\mu J^\mu. \qquad (3.161)$$

Consequently, on the quantum level this field theory is also the straightforward analogue of quantum electrodynamics, where the concept of photon is replaced by the concept of the graviton – the vector four-potential A_μ, which in the present case up to the constant multiplier is the velocity four-vector v_μ. In such a situation, the Feynman diagrams of the corresponding quantum field theory are fully analogous to the Feynman diagrams of QED. Making use of the definition (3.154) and doing certain algebraic manipulations, one sees that the Lagrangian density (3.161) takes the form

$$\mathscr{L} = -\frac{c^2}{4\pi G} \left[\frac{1}{2} \left(\partial_\mu v_\nu \partial^\mu v^\nu - \partial_\mu v_\nu \partial^\nu v^\mu \right) + \left(\frac{mc}{\hbar} \right)^2 v_\mu v^\mu \right], \qquad (3.162)$$

which after using the definition $v_\mu = [c, v_i]$ becomes

$$\mathscr{L} = -\frac{c^2}{4\pi G} \left[\frac{1}{2} \left(\partial_\mu v_i \partial^\mu v^i - \partial_i v_j \partial^j v^i \right) + \left(\frac{mc}{\hbar} \right)^2 v_i v^i \right], \qquad (3.163)$$

where we have omitted the constant Lagrangian which does not contribute to the field equations.

Let us notice that in the light of the straightforward analogy with the Maxwell electrodynamics, the continuity equation

$$\partial^\mu J_\mu = 0, \qquad (3.164)$$

should be satisfied. In such a situation, the equation (3.154) leads to the condition

$$\partial^\mu A_\mu = 0, \qquad (3.165)$$

which gives the divergence-free velocity field

$$\partial^\mu v_\mu = \partial^i v_i = 0. \tag{3.166}$$

This condition is highly non-trivial, because it implies that $\partial^i \dot{v}_i = 0$ and, consequently, the graviton mass identically vanishes $m = 0$. However, in the most general physical situation one must not take this condition into account, because this is an additional relation which can be broken and which is not needed for consistency of the theory. In other words, when the continuity equation is non-valid then the non-zero graviton mass and also the non-zero photon mass in the case of massive electrodynamics, is existent and fully argued. We shall consider such a situation here.

Let us return to initial considerations of this section. The equations (3.128) and (3.129) can be used to find the explicit form of the fields g_i and d_i. There are two fundamental identifications. The first one is

$$g_i = -\frac{\Omega_0^2}{2}\frac{r_0^3}{r^3} r_i, \tag{3.167}$$

$$\varepsilon_i^{jk} v_j d_k = -\omega_0^2 r_i. \tag{3.168}$$

while the second one is

$$g_i = -\omega_0^2 r_i, \tag{3.169}$$

$$\varepsilon_i^{jk} v_j d_k = -\frac{\Omega_0^2}{2}\frac{r_0^3}{r^3} r_i. \tag{3.170}$$

Let us consider the first identification. In such a situation,

$$\partial^i g_i = 0, \tag{3.171}$$

and, consequently, graviton mass is identically vanishing, and the continuity equation is satisfied automatically. However, the second identification gives slightly different result

$$\partial^i g_i = -3\omega_0^2, \tag{3.172}$$

and, consequently, the graviton mass is

$$m = \sqrt{3}\frac{\hbar}{c^2}\sqrt{-\omega_0^2}. \tag{3.173}$$

One sees now that if $\omega_0^2 > 0$ then the graviton is a tachyon. However, if $\omega_0^2 = -|\omega_0^2| < 0$, which corresponds to the so-called inverted harmonic oscillator, then the graviton mass becomes a real quantity

$$m = \sqrt{3}\frac{\hbar|\omega_0|}{c^2}. \tag{3.174}$$

Let us consider this situation in the light of the formula (3.127). It can be seen by straightforward comparison of this relation with the result (3.174) that

$$M\Omega_0^2 = -\sqrt{3}\frac{\hbar|\omega_0|^3}{c^2}, \tag{3.175}$$

so for non-tachyonic mass M one should have $\Omega_0^2 = -\frac{GM}{R^3} = -|\Omega_0|^2$, which corresponds to repulsive gravitation. In such a situation, one obtains

$$|\omega_0| = \frac{1}{R}\left(\frac{GM^2c^2}{\sqrt{3}\hbar}\right)^{1/3}. \tag{3.176}$$

Let us consider the case of the Planck object, *i.e.* a sphere of radius equal to the Planck length $R = \ell_P$ and mass equal to the Planck mass $M = M_P$. In such a situation, one receives

$$|\omega_0| = \frac{\omega_P}{3^{1/6}} \approx 1.544 \cdot 10^{19}\text{YHz}, \tag{3.177}$$

$$|\Omega_0| = \omega_P \approx 1.855 \cdot 10^{19}\text{YHz}, \tag{3.178}$$

$$m = 3^{1/3}M_P \approx 1.7610 \cdot 10^4\frac{\text{YeV}}{c^2}, \tag{3.179}$$

where ω_P is the Planck frequency, and the force law of gravitation takes the form

$$F_i = M_P\omega_P^2\left(1 + \frac{3^{1/3}}{2}\frac{\ell_P^3}{r^3}\right)r_i. \tag{3.180}$$

Such a situation is fully justified when the object of mass $M = M_P$ is the so-called Planckian particle, which is the hypothetical particle existing in the Planck scale energetic regime. This means that, in particular, we have described the graviton emitted by such Planckian particle. One sees that in such a situation the graviton mass is higher than the Planck mass. In the light of the Markov hypothesis [337], masses of measurable particles are less than the Planck mass, so this suggests that this type of graviton is nonmeasurable. However, for consistency this question should be verified by ultra-high energy experiments.

Let us finally focus on the analogue of the magnetic field. The equation (3.146) allows to calculate

$$\varepsilon_i^{jk} v_j d_k = v^l \partial_i v_l - v_l \partial^l v_i, \tag{3.181}$$

so when the velocity field v_i is known, for example as the solution to the equation (3.159), each of the equations (3.168) and (3.170) is the equation for the distance r_i. These equations are

$$r_i = R^2 \left(\frac{GM^2c^2}{\sqrt{3}\hbar} \right)^{-2/3} \left(v^j \partial_i v_j - v_j \partial^j v_i \right), \tag{3.182}$$

$$\frac{r_i}{r^3} = \frac{2}{GM} \left(v^l \partial_i v_l - v_l \partial^l v_i \right), \tag{3.183}$$

in the case of the first and the second identification, respectively.

H Summary

From the theoretical point of view gravitational waves are responsible for transport energy in the form of the specific kind of radiation emitted from any body of any system of bodies. This radiation is still hypothetical because its existence has not been confirmed by the suitable experimental data. There is still no straightforward proof of existence of gravitational radiation, but there are certain plausible experimental data indirectly requiring gravitational waves for physical consistency, like *e.g.* the Hulse–Taylor binary system. At present, the focus in the field of gravitational waves is on a higher order calculations level, but the state of affairs has been stagnating.

The results of this section are based on various approaches to gravity: General Relativity, the cosmological perturbation theory, perturbative quantum gravity, and the Lorentz force-inspired model. Taken together, they present a qualitatively new approach to massless and massive gravitons. This theory is, strictly speaking, non general relativistic, but is plausible and its applicability should be confronted with experimental data.

Part II

Particle Physics

Chapter 4

Two Scalar Field-Based Mass Generation Mechanisms

A Introduction

In modern physics, from both theoretical and experimental points of view, the most fruitful and prospective theory is the Standard Model, a commonly accepted and plausible physical theory of elementary particles and fundamental interactions. It has been shown in recent years that various extensions and modifications of the Standard Model create the most natural possibilities for new physics. Albeit, the problematic question is the presence within the Standard Model of experimentally unverified mathematical physics. This part of particle physics is usually believed to be the part responsible for the invisible or undetectable part of particle physics. The Higgs particle, called also Higgs boson or God particle, naturally belongs to such a collection of ideas. This status of the Higgs particle is confirmed by negative verification of its hypothetical existence in confrontation with the rigorous experimental high-statistics data of particle physics. In spite of such a state of affairs, still the greatest success and possibly the most intriguing and attracting part of the Standard Model is the consistent mass generation mechanism predicting the masses of various particles, and in itself leading to existence of the hypothetical Higgs particle.

However, despite of the fact that the Standard Model has had a number of successful predictions, among which is the application of the idea of spontaneous symmetry breaking to the beautiful explanation of why the $W^{\pm}$ and Z^{0} bosons are massive while the photon is massless [295], the acceptable mass generation mechanism leaves a lot to be desired from the empirical point of view. Namely, the commonly propagated explanation demands the existence of at least one additional particle, and since 1964, when the mass generation mechanism based on spontaneous symmetry breaking was proposed by Jeffrey Goldstone and later laureate of the 2008 Nobel Prize in Physics Yoichiro Nambu [296], the existence of the Higgs particle has never been satisfactorily confirmed by numerous experiments of particle physics. Recently, the experiments especially directed to finding out the Higgs particle were performed at CERN by using the Large Hadron Collide (LHC) and at Fermi National Accelerator Laboratory by using the Tevatron until its closure in late 2011. However, in spite of these essential efforts the mass generation mechanism based on existent nature of the Higgs particle, leading to probably the most beautiful picture of modern physics, possesses certain singular points that, in the light of experimental data, result in the inconsistency between the theory and the experiment. One of the glaring points is the question about the mass of the photon. According to the accepted electroweak mass generation mechanism the photon is a massless gauge field, while there are optical experimental data, for example the Goos–Hänchen effect [297, 298, 299], which conclude something different about this important property of the photon. In the light of these results, namely, the photon is manifestly massive particle, and its mass is very small but detectable. In the light of such a situation, the present mass generation mechanism of the Standard Model is inconsistent.

In this chapter we shall present two formulations of the mass generation mechanism for any Abelian gauge field of the Standard Model, including the photon, which can be straightforwardly generalized to the case of non-Abelian fields described in the framework of the Yang–Mills theories. Our proposal, however, manifestly differs from the mass generation mechanism of the Standard Model by including the non-zero mass of the photon. We shall express all our calculations by application of the Planck units, which creates the possibility of application of the model for new physics at the Planck scale.

The content of the chapter is as follows. Section B discusses concisely the mass generation mechanism for the Abelian sector of the Standard Model based on the Higgs potential and the charged scalar field. In Section C we present the construction of the alternative mass generation mechanism to Abelian fields which, however, still remains within the Standard Model. This model also arises from a charged scalar field, but in this case the Higgs potential is specifically modified. Both the models are manifestly invariant under action of the $O(2)$ symmetry group and based on the existence of a new scalar neutral boson χ and an auxiliary scalar neutral field φ which can be interpreted as a dilaton. It should be emphasized that both mass generation mechanisms are only seemingly similar to the Higgs mechanism, but the difference is that they lead to non-zero value of the photon mass and predict distinguishable scenarios for the mass of the new neutral scalar boson. In the first model, a new scalar particle is massive, and the value of its mass can be estimated by the present day experimental limits on the photon mass. In the second model, dilaton is massless and the mass of a new scalar particle remains a free parameter of the theory and can be estimated by experimental data. The mass of the photon in this model does not depend on the mass of the new scalar particle. Finally, in Section D the results of the whole chapter are concisely summarized.

B Higgs Potential

The consistent approach to the mass generation mechanism of elementary particles within the Standard Model, well-known as the Higgs mechanism (For detailed discussion Cf. *e.g.* Ref. [300]), is based on the Higgs potential and, in fact, presently lays the foundations of the theoretical high energy particle physics. In this chapter we shall focus our attention on the generation of mass to an Abelian gauge field A_μ, for example the photon, which interacts with a charged scalar field Φ. From the theoretical point of view, this is the theory of a gauge field coupled to a charged spinless boson, which is usually called scalar quantum electrodynamics (Cf. *e.g.* Refs. [301]), described by the Lagrangian invariant under action of the $U(1)$ symmetry group. In such a particular situation the algorithm of the spontaneous symmetry breaking is realized by the so-called Abelian Higgs mechanism.

It should be emphasized that both scalar quantum electrodynamics and the Higgs potential possess a number of essential applications, which have the established status in theoretical physics. The first context is the pioneering model of superconductor, proposed in 1950 by Lev Davidovich Landau and Vitaly Lazarevich Ginzburg [302] whose received the 1962 and 2003 Nobel Prizes in Physics, which formulates superconductivity as a charged Bose–Einstein condensate. In this model a complex order field is applied for description of fluctuations in the order parameter by adding a gradient to the Gorter–Casimir two-fluid model of superconductors, which lead them to the theory of superconductivity applicable near the critical temperature. As was shown in 1957 by Alexei Alexeyevich Abrikosov [303], co-laureate of the 2003 Nobel Prize in Physics, and in another context by Holger Bech Nielsen and Poul Olesen in 1973 [304], for the $2+1$-dimensional situation within the Ginzburg–Landau model there are vortices carrying magnetic flux. The Abrikosov vortex, called a fluxon, is a vortex of supercurrent in type-II superconductors. The non-superconducting core of fluxon, whose size is the parameter of the Ginzburg–Landau model, is the center of circulations of the supercurrent. The London penetration depth λ_L is the approximate distance where the supercurrents decay, and in type-II superconductors this length is always higher than the size of the core. Circulation of the supercurrents results in induced magnetic fields with the total flux equal to a single flux quantum Φ_0. Defects in a type-II superconductor are able to trap the Abrikosov vortices. The Nielsen–Olesen vortex has a purely formal and far from phenomenology nature. This is a configuration of scalar fields with the non-trivial topological property of non-contractible circles. This is a solution of $2+1$-dimensional classical field theory – a point-like object living in two-dimensional space.

Many authors studied numerous applications of the Higgs potential in particle physics including the charged scalar field (Cf. *e.g.* the papers in Ref. [305]). The phenomenological context, which manifestly lies in the borderland of cosmology and particle physics, has been discussed recently by Andrej Borisovich Arbuzov, Lukasz Andrzej Glinka, and Victor Nikolaevich Pervushin [306].

Let us focus our attention on the consistent formulation of the mass generation mechanism, which is based on the Higgs potential and lead to the non-zero photon mass. The Standard Model is described in the Minkowski space-time by the action of the form

$$S=\frac{1}{c}\int d^4x\mathscr{L}, \tag{4.1}$$

in which $\mathscr{L}$ is the total Lagrangian density of the theory

$$\mathscr{L} = \mathscr{L}_0+\mathscr{L}_I, \tag{4.2}$$

where $\mathscr{L}_0$ is the Lagrangian density of free fields, and $\mathscr{L}_I$ is the Lagrangian density describing interaction in the theory. In the case of scalar quantum electrodynamics free fields and interaction Lagrangian densities have the following form

$$\mathscr{L}_0 = E_P\ell_P\partial_\mu\Phi^\dagger\partial^\mu\Phi-\frac{1}{\ell_P^3}V(|\Phi|)-\frac{1}{4\mu_0}F_{\mu\nu}F^{\mu\nu}, \tag{4.3}$$

$$\mathscr{L}_I = -j^\mu A_\mu+\frac{e^2}{M_P\ell_P}\Phi^\dagger\Phi A_\mu A^\mu, \tag{4.4}$$

where we have made use of the Planck units, *i.e.* the Planck energy $E_P=M_Pc^2$, the Planck mass $M_P=\dfrac{\hbar}{c\ell_P}$ and the Planck length $\ell_P=\sqrt{\dfrac{\hbar G}{c^3}}$, immediately as the language which allows to express the theory in the dimensionally correct form. When one applies the standard unit system of particle physics, then it must be taken $\hbar=1$, $c=1$ and $G=1$ so that the Planck units also are equal to 1. The field $\Phi(x)$ is a complex (charged) scalar field, and $\Phi^\dagger(x)$ is the complex conjugate of $\Phi(x)$

$$\Phi = \frac{\varphi_1(x)+i\varphi_2(x)}{\sqrt{2}}, \tag{4.5}$$

$$\Phi^\dagger = \frac{\varphi_1(x)-i\varphi_2(x)}{\sqrt{2}}, \tag{4.6}$$

both having physical dimension L^{-1}, characterized by mass parameter m and *dimensionless* coupling constant g. The potential $V(|\Phi|)$ is an effective $O(2)$-symmetric potential of such a scalar field theory which can be written in the standard form of the Higgs potential,

$$V(|\Phi|)=\ell_P^2\frac{m^2c^2}{M_P}|\Phi|^2+gE_P\ell_P^4|\Phi|^4, \tag{4.7}$$

where $|\Phi|^2 = \Phi^\dagger\Phi$. In the case of scalar quantum electrodynamics the total Lagrangian density (4.2) remains invariant under action of the $U(1)$ symmetry group transformations

$$\Phi' = \exp(-i\theta)\Phi, \tag{4.8}$$
$$\Phi^{\dagger\prime} = \exp(i\theta)\Phi^\dagger, \tag{4.9}$$
$$A_\mu{}' = A_\mu + \frac{\hbar}{e}\partial_\mu\theta, \tag{4.10}$$
$$A^{\mu\prime} = A^\mu - \frac{\hbar}{e}\partial^\mu\theta, \tag{4.11}$$

where $\theta(x)$ is a local phase. In such a situation, the conserved Noether current j_μ has the following form

$$j^\mu = iec\left(\Phi^\dagger\partial^\mu\Phi - \left(\partial^\mu\Phi^\dagger\right)\Phi\right). \tag{4.12}$$

The electromagnetic interaction is described by the Abelian gauge field which is electromagnetic four-potential A_μ known as the photon in the Maxwell electrodynamics. In the standard theory the photon is neutral and massless, and its description is characterized by the Faraday tensor $F_{\mu\nu} = \partial_\mu A_\nu - \partial_\nu A_\mu$. We shall use here the standard conventions $A_\mu = \eta_{\mu\nu}A^\nu$, $\partial_\mu = \eta_{\mu\nu}\partial^\nu$, where $\eta_{\mu\nu} = \mathrm{diag}[-1,1,1,1]$ is the metric of the Minkowski space-time.

We take into account the following vacuum expectation values of the real scalar fields building the charged scalar field

$$\langle 0|\varphi_1(x)|0\rangle = \varphi_0, \tag{4.13}$$
$$\langle 0|\varphi_2(x)|0\rangle = 0, \tag{4.14}$$

where φ_0 is a real constant, and the vacuum state $|0\rangle$ is assumed to belong to the static Fock space of the theory. Such a notation arises from the traditions of particle physics. The scalar fields, which in our mechanism provoke existence of mass, have the following decomposition in terms of the Fourier harmonics

$$\varphi_1(x) = \varphi_0 + \chi(x), \tag{4.15}$$
$$\varphi_2(x) = \varphi(x), \tag{4.16}$$

where both the Higgs particle $\chi(x)$ and auxiliary scalar field φ have identically vanishing vacuum expectation values

$$\begin{aligned} \langle 0|\chi(x)|0\rangle &= 0, && (4.17)\\ \langle 0|\varphi(x)|0\rangle &= 0. && (4.18)\end{aligned}$$

In such a situation, the vacuum expectation value of the Higgs potential (4.7) can be derived straightforwardly as

$$\langle 0|V(|\Phi|)|0\rangle = \ell_P^2\frac{m^2c^2}{2M_P}\varphi_0^2 + \ell_P^4 E_P\frac{g}{4}\varphi_0^4 \equiv \mathscr{E}(\varphi_0), \tag{4.19}$$

and extremal values of such energy are established by identical vanishing of the corresponding force, *i.e.* the derivative

$$-\frac{d\mathscr{E}(\varphi_0)}{d\varphi_0} = -\left(\ell_P^2\frac{m^2c^2}{M_P}\varphi_0 + gE_P\ell_P^4\varphi_0^3\right) = 0. \tag{4.20}$$

This equation possesses two solutions

$$\begin{aligned} \varphi_0 &= 0, && (4.21)\\ \varphi_0^2 &= -\frac{1}{g}\frac{m^2}{M_P^2}\frac{1}{\ell_P^2}. && (4.22)\end{aligned}$$

The values of the energy at these points are respectively

$$\begin{aligned} \mathscr{E}(\varphi_0=0) &= 0, && (4.23)\\ \mathscr{E}\left(\varphi_0^2=-\frac{1}{g}\frac{m^2}{M_P^2}\frac{1}{\ell_P^2}\right) &= -\frac{E_P}{4g}\frac{m^4}{M_P^4}. && (4.24)\end{aligned}$$

The first solution is trivial and does not give an essential contribution to the theory. The second one is essential for the theory if and only if the following stability conditions are satisfied

$$\begin{aligned} g &> 0, && (4.25)\\ m^2 &= -m_0^2, && (4.26)\\ m_0^2 &> 0, && (4.27)\end{aligned}$$

which means that the scalar Higgs particle must be a *tachyon*.

By application of all the conventions presented above, the total Lagrangian density of the scalar quantum electrodynamics (4.2) can be rewritten in the following form

$$\mathscr{L} = \mathscr{L}_C + \mathscr{L}_\chi + \mathscr{L}_\varphi + \mathscr{L}_A + \mathscr{L}_{\chi\varphi A}, \tag{4.28}$$

where the parts of the Lagrangian density describe respectively:

- constant contribution $\mathscr{L}_C$

$$\mathscr{L}_C = \frac{E_P \ell_P}{2}\left(\frac{m_0^2}{M_P^2 \ell_P^2} - \frac{g}{2}\varphi_0^2\right)\varphi_0^2, \tag{4.29}$$

- free real massive scalar Higgs field $\mathscr{L}_\chi$

$$\begin{aligned} \mathscr{L}_\chi &= \frac{E_P\ell_P}{2}\partial_\mu\chi\partial^\mu\chi + E_P\ell_P\varphi_0\left(\frac{m_0^2}{M_P^2\ell_P^2} - g\varphi_0^2\right)\chi & (4.30)\\ &+ \frac{E_P\ell_P}{2}\left(\frac{m_0^2}{M_P^2\ell_P^2} - 3g\varphi_0^2\right)\chi^2 - E_P\ell_P\varphi_0 g\chi^3 - E_P\ell_P\frac{g}{4}\chi^4, & (4.31)\end{aligned}$$

- free real massive auxiliary scalar field $\mathscr{L}_\varphi$

$$\mathscr{L}_\varphi = \frac{E_P\ell_P}{2}\partial_\mu\varphi\partial^\mu\varphi + \frac{E_P\ell_P}{2}\left(\frac{m_0^2}{M_P^2\ell_P^2} - g\varphi_0^2\right)\varphi^2 - E_P\ell_P\frac{g}{4}\varphi^4, \tag{4.32}$$

- free massive photon $\mathscr{L}_A$

$$\mathscr{L}_A = -\frac{1}{4\mu_0}F_{\mu\nu}F^{\mu\nu} + \frac{e^2\varphi_0^2}{2M_P\ell_P}A_\mu A^\mu, \tag{4.33}$$

- effective interaction $\mathscr{L}_{\chi\varphi A}$

$$\begin{aligned} \mathscr{L}_{\chi\varphi A} &= -E_P\ell_P\varphi_0 g\chi\varphi^2 - E_P\ell_P\frac{g}{2}\chi^2\varphi^2 + \frac{e^2}{2M_P\ell_P}\eta_{\mu\nu}\chi^2A^\mu A^\nu & (4.34)\\ &+ \frac{e^2}{2M_P\ell_P}\eta_{\mu\nu}\varphi^2A^\mu A^\nu + \frac{e^2\varphi_0}{M_P\ell_P}\eta_{\mu\nu}\chi A^\mu A^\nu & (4.35)\\ &+ ec\eta_{\mu\nu}\chi\partial^\mu\varphi A^\nu - ec\eta_{\mu\nu}\varphi\partial^\mu\chi A^\nu. & (4.36)\end{aligned}$$

The quadratic in the fields and the derivatives part of the total Lagrangian density (4.28)

$$\mathscr{L}_Q = -\frac{1}{4\mu_0}F_{\mu\nu}F^{\mu\nu} + \frac{e^2 m_A^2}{2M_P^3\ell_P^3}A_\mu A^\mu + \frac{E_P\ell_P}{2}\partial_\mu\chi\partial^\mu\chi - \frac{m_\chi^2 c^2}{2M_P\ell_P}\chi^2 \quad (4.37)$$

$$+ \frac{E_P\ell_P}{2}\partial_\mu\varphi\partial^\mu\varphi - \frac{m_\varphi^2 c^2}{2M_P\ell_P}\varphi^2, \quad (4.38)$$

allows to establish the masses of the particles

$$m_A = M_P\ell_P\varphi_0, \quad (4.39)$$

$$m_\varphi = \sqrt{gm_A^2 - m_0^2}, \quad (4.40)$$

$$m_\chi = \sqrt{3gm_A^2 - m_0^2}. \quad (4.41)$$

In this manner, one receives the value of the constant φ_0

$$\varphi_0 = \frac{m_A}{M_P}\frac{1}{\ell_P} = \frac{m_A c}{\hbar}. \quad (4.42)$$

Applying the formulas (4.39), (4.40) and (4.41), one can express the coupling constant g by the masses of the particles

$$g = \frac{m_\chi^2 - m_\varphi^2}{2m_A^2}. \quad (4.43)$$

Since only the masses which are real numbers have any physical meaning, one obtains the system of inequalities

$$\begin{cases} gm_A^2 - m_0^2 \geqslant 0, \\ 3gm_A^2 - m_0^2 \geqslant 0, \end{cases} \quad (4.44)$$

which lead to the lower bound for the coupling constant

$$g \geqslant \frac{m_0^2}{m_A^2}, \quad (4.45)$$

which applied to the basic relation (4.43) gives the upper bound for the mass of the charged scalar field Φ

$$m_0 \leqslant \sqrt{\frac{m_\chi^2 - m_\varphi^2}{2}}. \quad (4.46)$$

Applying the Markov hypothesis [337] $m_0 \leqslant M_P$, one obtains the condition for the squared-mass difference $\Delta m^2_{\chi\varphi} = m^2_\chi - m^2_\varphi$

$$\Delta m^2_{\chi\varphi} \leqslant 2M^2_P. \tag{4.47}$$

The special case arises by application of the solution (4.22), *i.e.* $g\varphi_0^2 = \dfrac{m_0^2}{M_P^2\ell_P^2}$ to the total Lagrangian density of scalar quantum electrodynamics (4.28). Then the particular parts of this Lagrangian become

- constant contribution $\mathscr{L}_C$

$$\mathscr{L}_C = \frac{m_0^4 c^2}{8M_P^3\ell_P^3}, \tag{4.48}$$

- free real massive scalar Higgs field $\mathscr{L}_\chi$

$$\mathscr{L}_\chi = \frac{E_P\ell_P}{2}\partial_\mu\chi\partial^\mu\chi - \frac{m_0^2c^2}{M_P\ell_P}\chi^2 - \sqrt{g}m_0c^2\chi^3 - E_P\ell_P\frac{g}{4}\chi^4, \tag{4.49}$$

- free real massive auxiliary scalar field $\mathscr{L}_\varphi$

$$\mathscr{L}_\varphi = \frac{E_P\ell_P}{2}\partial_\mu\varphi\partial^\mu\varphi - E_P\ell_P\frac{g}{4}\varphi^4, \tag{4.50}$$

- free massive photon $\mathscr{L}_A$

$$\mathscr{L}_A = -\frac{1}{4\mu_0}F_{\mu\nu}F^{\mu\nu} + \frac{e^2m_0^2}{2gM_P^3\ell_P^3}A_\mu A^\mu, \tag{4.51}$$

- effective interaction $\mathscr{L}_{\chi\varphi A}$

$$\mathscr{L}_{\chi\varphi A} = -\sqrt{g}m_0c^2\chi\varphi^2 - E_P\ell_P\frac{g}{2}\chi^2\varphi^2 + \frac{e^2}{2M_P\ell_P}\eta_{\mu\nu}\chi^2A^\mu A^\nu \tag{4.52}$$

$$+ \frac{e^2}{2M_P\ell_P}\eta_{\mu\nu}\varphi^2A^\mu A^\nu + \frac{e^2m_0}{\sqrt{g}M_P^2\ell_P^2}\eta_{\mu\nu}\chi A^\mu A^\nu \tag{4.53}$$

$$+ ec\eta_{\mu\nu}\chi\partial^\mu\varphi A^\nu - ec\eta_{\mu\nu}\varphi\partial^\mu\chi A^\nu. \tag{4.54}$$

The masses derived in the maximal point have the values

$$m_A = \frac{m_0}{\sqrt{g}}, \tag{4.55}$$

$$m_\varphi = 0, \tag{4.56}$$

$$m_\chi = \sqrt{2}m_0, \tag{4.57}$$

and possess natural interpretation as the ground state masses of the elementary particles – the photon $A_\mu(x)$, the auxiliary scalar field $\varphi(x)$, the $\chi(x)$ boson. In this manner, in the model presented above the scalar auxiliary field $\varphi(x)$ is massless, whereas both the Higgs particle $\chi(x)$ and the photon A^μ are massive and their masses are determined by the free parameter m_0. Let us mention only that for the trivial solution $\varphi_0 = 0$ one obtains

$$m_A = 0, \tag{4.58}$$

$$m_\varphi = im_0, \tag{4.59}$$

$$m_\chi = im_0, \tag{4.60}$$

what means that in such a situation the photon is massless and both the Higgs particle and the auxiliary scalar field are tachyons, *i.e.* are non-physical.

There is a highly non-trivial case within the scalar quantum electrodynamics. Namely, when the constant contribution Lagrangian (4.48) identically vanishes, which is fully acceptable because constant term in the total Lagrangian does not provide a contribution to the appropriate Euler–Lagrange equations of motion, then one has the relation

$$2\frac{m_0^2}{M_P^2\ell_P^2} = g\varphi_0^2. \tag{4.61}$$

In such a situation, another contributions to the total Lagrangian density (4.28) are

- free real massive scalar Higgs field $\mathscr{L}_\chi$

$$\mathscr{L}_\chi = \frac{E_P\ell_P}{2}\partial_\mu\chi\partial^\mu\chi - \sqrt{\frac{2}{g}}\frac{m_0^3c^2}{M_P^2\ell_P^2}\chi - \frac{5m_0^2c^2}{2M_P\ell_P}\chi^2 - \sqrt{2g}m_0c^2\chi^3 - E_P\ell_P\frac{g}{4}\chi^4, \tag{4.62}$$

- free real massive auxiliary scalar field $\mathscr{L}_\varphi$

$$\mathscr{L}_\varphi = \frac{E_P \ell_P}{2} \partial_\mu \varphi \partial^\mu \varphi - \frac{m_0^2 c^2}{2 M_P \ell_P} \varphi^2 - E_P \ell_P \frac{g}{4} \varphi^4, \tag{4.63}$$

- free massive photon $\mathscr{L}_A$

$$\mathscr{L}_A = -\frac{1}{4\mu_0} F_{\mu\nu} F^{\mu\nu} + \frac{e^2 m_0^2}{g M_P^3 \ell_P^3} A_\mu A^\mu, \tag{4.64}$$

- the effective interaction $\mathscr{L}_{\chi\varphi A}$

$$\mathscr{L}_{\chi\varphi A} = -\sqrt{2g} m_0 c^2 \chi \varphi^2 - E_P \ell_P \frac{g}{2} \chi^2 \varphi^2 + \frac{e^2}{2 M_P \ell_P} \eta_{\mu\nu} \chi^2 A^\mu A^\nu \tag{4.65}$$

$$+ \frac{e^2}{2 M_P \ell_P} \eta_{\mu\nu} \varphi^2 A^\mu A^\nu + \sqrt{\frac{2}{g}} \frac{e^2 m_0}{M_P^2 \ell_P^2} \eta_{\mu\nu} \chi A^\mu A^\nu \tag{4.66}$$

$$+ ec \eta_{\mu\nu} \chi \partial^\mu \varphi A^\nu - ec \eta_{\mu\nu} \varphi \partial^\mu \chi A^\nu. \tag{4.67}$$

In this light the photon mass (4.39) becomes

$$m_A = \sqrt{\frac{2}{g}} m_0, \tag{4.68}$$

and the masses of the auxiliary field (4.40) and the Higgs particle (4.41) take the following form

$$m_\varphi = m_0, \tag{4.69}$$

$$m_\chi = \sqrt{5} m_0. \tag{4.70}$$

In this manner, in such a situation all masses in the theory are described by only the free parameter m_0. Let us call this particularly non-trivial situation the *one-parameter mass generation mechanism*. In this interesting case the relation (4.43) leads to the simple identity $g = g$, and other obtained relations are also fully consistent. In this especial, but important, case the Markov hypothesis (4.47) gives the upper bound for the mass parameter m_0 by the Planck mass

$$m_0 \leqslant \frac{M_P}{2}, \tag{4.71}$$

so that one can take *ad hoc* the maximal value of this parameter

$$m_0 = \frac{M_P}{2}. \tag{4.72}$$

Such a choice implies highly non-trivial ramifications. The parts of the total Lagrangian (4.28) become much simpler

- free real massive scalar Higgs field $\mathscr{L}_\chi$

$$\mathscr{L}_\chi = \frac{E_P \ell_P}{2} \partial_\mu \chi \partial^\mu \chi - \frac{1}{\sqrt{2g}} \frac{E_P}{4\ell_P^2} \chi - \frac{5E_P}{8\ell_P} \chi^2 - \sqrt{\frac{g}{2}} E_P \chi^3 - E_P \ell_P \frac{g}{4} \chi^4, \tag{4.73}$$

- free real massive auxiliary scalar field $\mathscr{L}_\varphi$

$$\mathscr{L}_\varphi = \frac{E_P \ell_P}{2} \partial_\mu \varphi \partial^\mu \varphi - \frac{E_P}{8\ell_P} \varphi^2 - E_P \ell_P \frac{g}{4} \varphi^4, \tag{4.74}$$

- free massive photon $\mathscr{L}_A$

$$\mathscr{L}_A = -\frac{1}{4\mu_0} F_{\mu\nu} F^{\mu\nu} + \frac{e^2}{4g M_P \ell_P^3} A_\mu A^\mu, \tag{4.75}$$

- effective interaction $\mathscr{L}_{\chi\varphi A}$

$$\begin{aligned} \mathscr{L}_{\chi\varphi A} &= -\sqrt{\frac{g}{2}} E_P \chi \varphi^2 - E_P \ell_P \frac{g}{2} \chi^2 \varphi^2 + \frac{e^2}{2M_P \ell_P} \eta_{\mu\nu} \chi^2 A^\mu A^\nu && (4.76) \\ &+ \frac{e^2}{2M_P \ell_P} \eta_{\mu\nu} \varphi^2 A^\mu A^\nu + \frac{1}{\sqrt{2g}} \frac{e^2}{M_P \ell_P^2} \eta_{\mu\nu} \chi A^\mu A^\nu && (4.77) \\ &+ ec\eta_{\mu\nu} \chi \partial^\mu \varphi A^\nu - ec\eta_{\mu\nu} \varphi \partial^\mu \chi A^\nu, && (4.78) \end{aligned}$$

while the generated masses of the photon, the Higgs particle, and the auxiliary field are expressed only through the Planck mass

$$\begin{aligned} m_A &= \frac{M_P}{\sqrt{2g}}, && (4.79) \\ m_\varphi &= \frac{M_P}{2}, && (4.80) \\ m_\chi &= \frac{\sqrt{5}}{2} M_P. && (4.81) \end{aligned}$$

One sees, however, that in such a situation $m_\chi > M_P$ breaks the Markov hypothesis, *i.e.* the χ-boson can be regarded as non-detectable, while the φ field is physical and $m_A \leqslant M_P$ means that $g \geqslant \frac{1}{2}$, so then φ can be treated as the Higgs particle.

Let us discuss now the Euler–Lagrange equations of motion following from the total Lagrangian (4.28) in the most general case. The equations of motion for the photon are

$$\Box A^\mu - \partial^\mu(\partial_\nu A^\nu) + \frac{\mu_0 e^2 m_A^2}{M_P^3 \ell_P^3} A^\mu = \mu_0 j^\mu - \frac{\mu_0 e^2}{M_P \ell_P}\left(\chi^2 + \varphi^2 + \frac{2m_A}{M_P \ell_P}\chi\right) A^\mu, \tag{4.82}$$

where $\Box = \partial^\nu \partial_\nu$ and j^μ is the conserved Noether current

$$j^\mu = ec\left(\varphi \partial^\mu \chi - \chi \partial^\mu \varphi\right). \tag{4.83}$$

The equations for motion for the auxiliary scalar field are

$$\Box\varphi + \frac{m_\varphi^2}{M_P^2 \ell_P^2}\varphi + g\varphi^3 = -2\frac{g m_A}{M_P \ell_P}\chi\varphi - g\chi^2\varphi + \frac{e^2}{M_P \ell_P}\varphi A^\mu A_\mu \tag{4.84}$$

$$- 2\frac{e}{\hbar}\partial^\mu \chi A_\mu - \frac{e}{\hbar}\chi \partial^\mu A_\mu, \tag{4.85}$$

and the equations for motion for the Higgs particle in scalar quantum electrodynamics are

$$\Box\chi + \frac{m_\chi^2}{M_P^2 \ell_P^2}\chi + 3\frac{g m_A}{M_P \ell_P}\chi^2 + g\chi^3 = -\frac{m_\varphi^2}{M_P^2 \ell_P^2}\varphi_0 - \frac{g m_A}{M_P \ell_P}\varphi^2 - g\chi\varphi^2 \tag{4.86}$$

$$+ \frac{e^2}{M_P \ell_P}\chi A^\mu A_\mu + \frac{e^2 m_A}{M_P^2 \ell_P^2} A^\mu A_\mu \tag{4.87}$$

$$+ 2\frac{e}{\hbar}\partial^\mu \varphi A_\mu + \frac{e}{\hbar}\chi \partial^\mu A_\mu. \tag{4.88}$$

It can be seen by straightforward and easy calculation that if one takes into consideration the Markov hypothesis [337], stating that the mass of any elementary particle is not more than the Planck mass, applied to the mass of the charged scalar field m_0

$$m_0 \leqslant M_P, \tag{4.89}$$

then one obtains the upper bound for the initial datum φ_0

$$\varphi_0 \leqslant \frac{1}{\sqrt{g}}\frac{1}{\ell_P}. \tag{4.90}$$

The deductions for the Higgs–Hubble inflaton presented in the previous chapter, *i.e.* the case of the Higgs particle obeying the Hubble law, give the most adequate inflationary cosmology on the Planck scale. In such a situation, the initial datum of the scalar field is

$$\varphi_0 = \frac{2}{3}\sqrt{\pi}\frac{1}{\ell_P}. \tag{4.91}$$

Applying this value to the inequality (4.90), one receives the upper bound to the coupling constant in such a Planck-scale situation

$$g \leqslant \frac{9}{4\pi} \approx 0.716, \tag{4.92}$$

and applying the maximal initial datum value $\varphi_0^2 = \frac{1}{g}\frac{m_0^2}{M_P^2}\frac{1}{\ell_P^2}$ to the equation (4.91), one receives the following result

$$m_0 = \frac{2}{3}\sqrt{g\pi}M_P, \tag{4.93}$$

which can be rewritten in the equivalent form

$$g = \frac{9}{4\pi}\left(\frac{m_0}{M_P}\right)^2. \tag{4.94}$$

Using this relation together with the formula (4.43) one receives

$$m_0 = \sqrt{2\pi\Delta m_{\chi\varphi}^2\frac{M_P}{3m_A}}, \tag{4.95}$$

where $\Delta m_{\chi\varphi}^2 = m_\chi^2 - m_\varphi^2$. This formula allows to deduce that

$$\Delta m_{\chi\varphi}^2 = \frac{9}{2\pi}\left(\frac{m_0 m_A}{M_P}\right)^2, \tag{4.96}$$

so that involving the bound (4.47) one obtains the inequality

$$m_0 m_A \leqslant \frac{2}{3}\sqrt{\pi}M_P^2. \tag{4.97}$$

Another consequence of taking as the initial datum of the scalar field φ the initial datum of the Higgs–Hubble inflaton (4.91) are the masses of the particles. In the maximum of the Higgs potential they have the following values

$$m_A = \frac{2}{3}\sqrt{\pi}M_P, \tag{4.98}$$
$$m_\varphi = 0, \tag{4.99}$$
$$m_\chi = \sqrt{2g}m_A. \tag{4.100}$$

Using the inequalities (4.45) and (4.92), one can straightforwardly deduce the upper bound for the free mass parameter m_0

$$m_0 \leqslant \frac{3}{2\sqrt{\pi}}m_A, \tag{4.101}$$

which after application of the formulas (4.93) and (4.39) leads to the lower bound for the initial datum φ_0

$$\varphi_0 \geqslant \frac{4\pi}{9}\sqrt{g}\frac{1}{\ell_P}. \tag{4.102}$$

The scenario presented in this section in itself is a certain algorithm for generation of the masses of the Abelian field A^μ within the Standard Model. This mechanism involves two neutral scalar fields - the massive Higgs particle χ, and the auxiliary scalar field φ which is massless at the maximum of the vacuum expectation value of the Higgs potential. The photon A^μ is always massive in such a scenario. This mechanism can be straightforwardly generalized to the case of non-Abelian gauge fields, but we will not discuss here such generalization.

C Modified Higgs Potential

In the present section we shall take into account the alternative mass generation mechanism for Abelian fields within the framework of the Standard Model. This model will be based on the Higgs potential modified due to the constant shift $|\Phi_0|$ of the variable $|\Phi|$, where Φ is the charged scalar field considered in the previous section. This mechanism can be also straightforwardly generalized to the situations involving non-Abelian gauge fields theories, and we shall not discuss such generalization in this book.

In our discussion we shall consider the mass generation mechanism arising from the following modification of the Higgs potential of the real scalar field $|\Phi|$

$$V(|\Phi|) = \ell_P^2 \frac{m^2c^2}{M_P} (|\Phi| - |\Phi_0|)^2 + gE_P\ell_P^4 (|\Phi| - |\Phi_0|)^4 . \tag{4.103}$$

This potential is essentially is the standard scalar field potential, discussed in the previous section, rotated around the point $|\Phi| = 0$. The modified Higgs potential (4.103) is still invariant under action of the $O(2)$ symmetry group. The constant shift Φ_0 of the real scalar field can be simply associated with the zero Fourier harmonic of the charged scalar field Φ. Let us presume that there is the following decomposition

$$\sigma(x) := |\Phi(x)| \equiv \sqrt{\frac{\varphi_1^2(x) + \varphi_2^2(x)}{2}} = \frac{\sigma_0 + \chi(x)}{\sqrt{2}}, \tag{4.104}$$

and that the neutral scalar field χ is the new neutral scalar boson, *i.e.* its vacuum expectation value vanishes identically

$$\langle 0|\chi(x)|0\rangle = 0. \tag{4.105}$$

Consequently, the vacuum expectation value of the real scalar field σ is exactly the zeroth Fourier harmonic of the real scalar field $|\Phi(x)|$

$$\langle 0|\sigma(x)|0\rangle = \frac{\sigma_0}{\sqrt{2}}. \tag{4.106}$$

If one takes into account *ad hoc* the following relation

$$\langle 0|\Phi|0\rangle = \Phi_0, \tag{4.107}$$

then one receives the identification between the constants

$$\frac{\sigma_0}{\sqrt{2}} \equiv |\Phi_0|. \tag{4.108}$$

Consequently, the modified Higgs potential (4.103) can be expressed entirely by using the neutral scalar field χ

$$V(|\Phi|) = V(\chi) = \ell_P^2 \frac{m_\chi^2 c^2}{2M_P} \chi^2 + g\frac{E_P\ell_P^4}{4}\chi^4, \tag{4.109}$$

and it is easy to see that the mass of such a new scalar boson coincides with the mass parameter m of the charged scalar field Φ

$$m_\chi = m. \tag{4.110}$$

Let us present the charged scalar field $\Phi(x)$ in the form resulting from application of the polar decomposition

$$\Phi(x) = \sigma(x)\exp\{i\theta(x)\}, \tag{4.111}$$

where the local phase $\theta(x)$ can be established straightforwardly

$$\theta(x) = \arg\Phi(x) = \arctan\frac{\Im\Phi(x)}{\Re\Phi(x)} = \arctan\frac{\varphi_2(x)}{\varphi_1(x)} + 2\pi n, \tag{4.112}$$

where $n \in \mathbb{Z}$. According to the proposed decomposition (4.104) the neutral scalar fields φ_1 and φ_2 cannot be taken arbitrarily but in such a way that they must satisfy the following constraint

$$\varphi_1^2(x) + \varphi_2^2(x) = \sigma_0^2 + 2\sigma_0\chi(x) + \chi^2(x), \tag{4.113}$$

which means that the dependence of the neutral scalar fields φ_1 and φ_2 on the neutral scalar field χ is not unambiguous.

Let us consider the local phase $\theta(x)$ in the decomposition (4.111) of the charged scalar field. In the most general situation such a local phase can be also expressed through an integral dependence on the wave four-vector $k_\mu(x)$

$$\theta(x) = \theta(x^\mu) = \alpha\int^{x^\mu} dx'^\mu k_\mu(x'), \tag{4.114}$$

where $x^\mu = [ct, x^i]$ is the position four-vector, and $k^\mu = \eta^{\mu\nu}k_\nu$ is the wave four-vector given by the formula

$$k^\mu(x) = \left[\frac{1}{c}\omega(k^i(x)), k^i(x)\right], \tag{4.115}$$

where $\omega(k^i)$ is the oscillation frequency. The length of the wave vector $k^i(x)$ is linked to wavelength $\lambda(x)$

$$k(x) = \frac{2\pi}{\lambda(x)}, \tag{4.116}$$

of the wave identified with the charged scalar field $\Phi(x)$. The momentum of the particle represented by the charged scalar field $\Phi(x)$ shall be determined by quantum theory as $p^i = \hbar k^i$, so that the De Broglie relation $p = \frac{2\pi\hbar}{\lambda}$ of the wave-particle duality is still valid. In the formula (4.114) the quantity α is a certain dimensionless parameter, which we shall establish straightforwardly below. In this manner, one can connect the formulas (4.112) and (4.114) to receive the unambiguous expressions for the neutral scalar fields φ_1 and φ_2

$$\varphi_1(x) = \frac{\sigma_0 + \chi(x)}{\sqrt{1 + \tan\left(\alpha \int^{x^\mu} dx'^\mu k_\mu(x') - 2\pi n\right)}}, \tag{4.117}$$

$$\varphi_2(x) = (\sigma_0 + \chi(x)) \frac{\tan\left(\alpha \int^{x^\mu} dx'^\mu k_\mu(x') - 2\pi n\right)}{\sqrt{1 + \tan\left(\alpha \int^{x^\mu} dx'^\mu k_\mu(x') - 2\pi n\right)}}. \tag{4.118}$$

The problem is, however, to establish the wave four-vector $k_\mu(x)$. Note that in the case when the space-time is the Minkowski space-time one has the following identity

$$\int^{x^\mu} dx'^\mu k_\mu(x') = \int^{x_\mu} dx'_\mu k^\mu(x'), \tag{4.119}$$

which does not hold for the general case of space-times characterized by non-flat metric tensors $g^{\mu\nu}(x)$.

Let us consider the total Lagrangian of the scalar quantum electrodynamics

$$\mathscr{L} = E_P \ell_P \partial_\mu \Phi^\dagger \partial^\mu \Phi - \frac{1}{\ell_P^3} V(|\Phi|) - \frac{1}{4\mu_0} F_{\mu\nu} F^{\mu\nu} - j^\mu A_\mu + \frac{e^2}{M_P \ell_P} |\Phi|^2 A_\mu A^\mu, \tag{4.120}$$

and express such a theory in terms of the neutral scalar boson χ following from the decomposition (4.104), and by the wave four-vector $k_\mu(x)$ following from the definition of the local phase (4.114).

By straightforward calculation one can find that the following relations between derivatives of the complex scalar field $\Phi(x)$ and the real scalar field $\sigma(x) = |\Phi(x)|$

$$\partial^\mu \Phi = \exp\left\{ i\alpha \int^{x^\mu} dx'^\mu k_\mu(x') \right\} [\partial^\mu \sigma(x) + i\alpha\sigma(x) k^\mu(x)], \quad (4.121)$$

$$\partial_\mu \Phi^\dagger = \exp\left\{ -i\alpha \int^{x^\mu} dx'^\mu k_\mu(x') \right\} [\partial_\mu \sigma(x) - i\alpha\sigma(x) k_\mu(x)], \quad (4.122)$$

hold and, therefore, the kinetic Lagrangian of the charged scalar field has the form

$$\partial_\mu \Phi^\dagger \partial^\mu \Phi = \partial_\mu \sigma \partial^\mu \sigma + \alpha^2 \sigma^2 k_\mu k^\mu = \quad (4.123)$$

$$= \frac{1}{2} \partial_\mu \chi \partial^\mu \chi + \frac{\alpha^2 \sigma_0^2}{2} k_\mu k^\mu + \alpha^2 \sigma_0 \chi k_\mu k^\mu + \frac{\alpha^2}{2} \chi^2 k_\mu k^\mu. \quad (4.124)$$

Similarly, one can determine the Noether current

$$j^\mu = -2ec\alpha\sigma^2 k^\mu = -\alpha ec \left(\sigma_0^2 k^\mu + 2\sigma_0 \chi k^\mu + \chi^2 k^\mu \right), \quad (4.125)$$

and the Lagrangian quadratic in the Abelian gauge field

$$|\Phi|^2 A_\mu A^\mu = \frac{\sigma_0^2}{2} A_\mu A^\mu + \sigma_0 \chi A_\mu A^\mu + \frac{1}{2} \chi^2 A_\mu A^\mu. \quad (4.126)$$

Collecting all these relations together one receives the total Lagrangian of scalar quantum electrodynamics in the following form

$$\begin{aligned}
\mathscr{L} &= \frac{E_P \ell_P}{2} \partial_\mu \chi \partial^\mu \chi + \alpha^2 \sigma_0^2 \frac{E_P \ell_P}{2} k_\mu k^\mu + \alpha^2 \sigma_0 E_P \ell_P \chi k_\mu k^\mu + \alpha^2 \frac{E_P \ell_P}{2} \chi^2 k_\mu k^\mu \\
&- \frac{m_\chi^2 c^2}{2 M_P \ell_P} \chi^2 - g \frac{E_P \ell_P}{4} \chi^4 - \frac{1}{4\mu_0} F_{\mu\nu} F^{\mu\nu} + ec\alpha\sigma_0^2 k^\mu A_\mu + 2ec\alpha\sigma_0 \chi k^\mu A_\mu \\
&+ ec\alpha\chi^2 k^\mu A_\mu + \frac{e^2 \sigma_0^2}{2 M_P \ell_P} A_\mu A^\mu + \frac{e^2 \sigma_0}{M_P \ell_P} \chi A_\mu A^\mu + \frac{e^2}{2 M_P \ell_P} \chi^2 A_\mu A^\mu. \quad (4.127)
\end{aligned}$$

The Lagrangian density (4.127) describes the classical theory of three interacting fields – the photon A_μ, the neutral scalar field χ, and the non-dynamical massive wave four-vector field k_μ. On the one hand, such a theory is rather complicated if there is no given restriction for the relation between the wave four-vector k_μ and the photon A_μ. On the other hand, such a non-trivial relation is necessary because the non-kinetic nature of the wave four-vector k_μ. Anyway, the choice of the suitable restraint is, in fact, the choice of the appropriate gauge and straightforwardly imparts the physical significance to the massive gauge vector field k_μ.

Let us consider first the case of the generic gauge

$$k_\mu = \frac{ec}{E_P \ell_P} A_\mu = \frac{e}{\hbar} A_\mu. \tag{4.128}$$

In such a situation, the total Lagrangian od the scalar quantum electrodynamics (4.127) simplifies to the theory of interacting photon A_μ and neutral scalar field χ

$$\mathscr{L} = \frac{E_P \ell_P}{2} \partial_\mu \chi \partial^\mu \chi - \frac{m_\chi^2 c^2}{2 M_P \ell_P} \chi^2 - g \frac{E_P \ell_P}{4} \chi^4 \tag{4.129}$$

$$- \frac{1}{4\mu_0} F_{\mu\nu} F^{\mu\nu} + \frac{e^2 \sigma_0^2}{2 M_P \ell_P} (\alpha+1)^2 A_\mu A^\mu \tag{4.130}$$

$$+ \frac{e^2 \sigma_0}{M_P \ell_P} (\alpha+1)^2 \chi A_\mu A^\mu \tag{4.131}$$

$$+ \frac{e^2}{2 M_P \ell_P} (\alpha+1)^2 \chi^2 A_\mu A^\mu. \tag{4.132}$$

Writing out the part of the Lagrangian which is quadratic in fields and its derivatives in the following form

$$\mathscr{L}_Q = \frac{E_P \ell_P}{2} \partial_\mu \chi \partial^\mu \chi - \frac{m_\chi^2 c^2}{2 M_P \ell_P} \chi^2 - \frac{1}{4\mu_0} F_{\mu\nu} F^{\mu\nu} + \frac{e^2 m_A^2}{2 M_P^3 \ell_P^3} A_\mu A^\mu, \tag{4.133}$$

one receives immediately the mass of the photon

$$m_A = M_P \ell_P \sigma_0 |\alpha + 1|. \tag{4.134}$$

The mass (4.134) of the photon does not vanish for all the situations different from $\alpha \neq -1$. This mass formula is completely different from the analogical relation obtained in the previous section from the mass generation mechanism applying the Higgs potential. The crucial and the most important difference is that in the present mechanism the mass of photon is independent of the mass of the neutral scalar field χ of the theory. The mass of the neutral scalar field χ is a free parameter and can be established by analysis of suitable experimental data. Meanwhile, the mass of the photon is solely determined only by two independent constants, the initial datum σ_0 and the dimensionless parameter α. The parameter α shall be established below, while the initial datum σ_0 possesses the freedom of choice. One can study the concrete physical situations which predict the appropriate values of this number. For example, one can select *ad hoc* the value of the initial datum $\sigma_0 = \frac{2}{3}\sqrt{\pi}\frac{1}{\ell_P}$ and study the scalar quantum electrodynamics in the context of the Higgs–Hubble inflaton.

Recall that non-zero value of the mass of the photon is deeply rooted in the physical interpretation of the experimental data. Such a belief was expressed in early 1970s by Aline Mazet, Christian Imbert, and Serge Huard [297] on the basis of the observations of total internal reflection to test the Goos–Hänchen effect of the beam shift [298], named after Hermann Fritz Gustav Goos and Hilda Hänchen. Soon after this essential analysis Louis De Broglie and Jean-Pierre Vigier [299] discussed this contradiction in the context of the quantum theory of radiation and generalized it to the quantum theory of massive spin-1 photons. Soon after this important paper Gordon John Troup et al [307] suggested the De Broglie–Vigier proposal as the untenable. Albeit, still the question of the non-zero mass of the photon belongs to the most important and most actual problems of modern physics. One of the best studies of this problem is the discussion by Roderic Lakes [308], in which the photon mass is analyzed in the framework of the Proca equation. Moreover, the possibility of the non-zero mass of the photon was studied also by Ludvik Bass and Erwin Schrödinger [309], and Richard Phillips Feynman [310]. Howard Mason Georgi III, Paul Ginsparg, and Sheldon Lee Glashow [311] have suggested the non-zero photon mass to justify existence and properties of Cosmic Microwave Background Radiation. It should be emphasized that since at least three decades there has been non successful verification of existence of the Higgs particle arising within the well-accepted mass generation mechanism offered by the Standard Model. Such a state of affairs suggests that the Higgs particle does not exist. In this manner, the model presented in this section is consistent with the collection of the empirical facts.

The problem is to establish the value of the dimensionless parameter α. Let us employ for this the Euler–Lagrange equations of motion arising from the total Lagrangian (4.127) of scalar quantum electrodynamics. The equations of motion for the neutral scalar field χ are

$$\Box\chi + g\chi^3 + \frac{m_\chi^2}{M_P^2\ell_P^2}\chi = \left(\alpha k_\mu + \frac{e}{\hbar}A_\mu\right)\left(\alpha k^\mu + \frac{e}{\hbar}A^\mu\right)(\sigma_0 + \chi), \tag{4.135}$$

while the photon A_μ obeys the equations of motion

$$\Box A^\mu - \partial^\mu(\partial_\nu A^\nu) + \frac{\mu_0 e^2\sigma_0^2}{M_P\ell_P}A^\mu = \mu_0 j^\mu - \frac{\mu_0 e^2}{M_P\ell_P}\left(\chi^2 + 2\sigma_0\chi\right)A^\mu, \tag{4.136}$$

and the non-dynamical vector field k_μ satisfies the equations of motion

$$(\sigma_0 + \chi)^2\left(\frac{1}{2}\alpha k_\mu + \frac{e}{\hbar}A_\mu\right) = 0. \tag{4.137}$$

The equations (4.137) are non-dynamical and for this reason are the constraints. They can be solved straightforwardly with the result

$$\alpha k_\mu = -2\frac{e}{\hbar}A_\mu, \tag{4.138}$$

which suggests that $\alpha = -2$. Involving this solution into the equations of motion (14.177) one receives

$$\Box\chi + \frac{m_\chi^2}{M_P^2\ell_P^2}\chi + g\chi^3 = \frac{e^2\sigma_0}{\hbar^2}A_\mu A^\mu + \frac{e^2}{\hbar^2}\chi A_\mu A^\mu, \tag{4.139}$$

while the conserved current (4.125) simplifies to

$$j^\mu = \frac{2e^2}{M_P\ell_P}\left(\sigma_0^2 + 2\sigma_0\chi + \chi^2\right)A^\mu, \tag{4.140}$$

so that the equations of motion for the photon (14.178) become

$$\Box A^\mu - \partial^\mu(\partial_\nu A^\nu) - \frac{\mu_0 e^2\sigma_0^2}{M_P\ell_P}A^\mu = \frac{2\mu_0 e^2}{M_P\ell_P}\left(\chi^2 + \sigma_0\chi\right)A^\mu. \tag{4.141}$$

Applying, however, the gauge (4.138) to the total Lagrangian (4.127) one receives

$$\begin{aligned}\mathscr{L} &= \frac{E_P\ell_P}{2}\partial_\mu\chi\partial^\mu\chi - \frac{m_\chi^2c^2}{2M_P\ell_P}\chi^2 - g\frac{E_P\ell_P}{4}\chi^4 \\ &- \frac{1}{4\mu_0}F_{\mu\nu}F^{\mu\nu} + \frac{e^2\sigma_0^2}{2M_P\ell_P}A_\mu A^\mu + \frac{e^2\sigma_0}{M_P\ell_P}\chi A_\mu A^\mu + \frac{e^2}{2M_P\ell_P}\chi^2 A_\mu A^\mu,\end{aligned} \tag{4.142}$$

which means that the mass of the photon has the following form

$$m_A = M_P \ell_P \sigma_0, \tag{4.143}$$

and exactly corresponds to the value $\alpha = -2$ applied *ad hoc* within the formula (4.134).

The photon mass formula (4.143) obtained in this section is principally the same as the photon mass formula (4.39) calculated in the previous section. The non-essential difference are the constants φ_0 and σ_0 which, in general, can be different but can be also the same constant. Similarly, as in the previous model, in the present model the free parameter σ_0 can be also expressed via the photon mass only

$$\sigma_0 = \frac{m_A}{M_P \ell_P} = \frac{m_A c}{\hbar}, \tag{4.144}$$

and all deductions related to the Higgs–Hubble inflaton, performed in the previous section, can be performed in the present case in the analogous way. The difference between the present and the previous model of the mass generation mechanism is, however, essential because in the present situation the mass of the neutral scalar field χ remains a free parameter which can be determined by experimental data, and this parameter is independent of the photon mass. Recall that in the case of the previous model the mass of the scalar field, *i.e.* the Higgs particle, is in general determined by the photon mass and the free parameter m_0.

Interestingly, in the present mass generation mechanism the neutral scalar fields which create the charged scalar field possess the nontrivial decomposition

$$\varphi_1(x) = \frac{\sigma_0 + \chi(x)}{\sqrt{1 - \tan\left(2\frac{e}{\hbar}\int^{x^\mu} dx'^\mu A_\mu(x') + 2\pi n\right)}}, \tag{4.145}$$

$$\varphi_2(x) = -(\sigma_0 + \chi(x)) \frac{\tan\left(2\frac{e}{\hbar}\int^{x^\mu} dx'^\mu A_\mu(x') + 2\pi n\right)}{\sqrt{1 - \tan\left(2\frac{e}{\hbar}\int^{x^\mu} dx'^\mu A_\mu(x') + 2\pi n\right)}}. \tag{4.146}$$

It is evident that the charged scalar field $\Phi = \frac{1}{\sqrt{2}}(\varphi_1 + i\varphi_2)$ becomes a neutral scalar field if and only if the neutrality condition is satisfied

$$\int^{x^\mu} dx'^\mu A_\mu(x') = \left(\frac{p}{2} - n\right)\frac{\pi\hbar}{e} = \left(\frac{p}{2} - n\right)\frac{h}{2e}, \quad p, n \in \mathbb{Z} \tag{4.147}$$

where the photon A_μ obeys the system of equations (14.63) and (4.141).

The states of the photon defined by the condition (4.147) are highly non-trivial. In fact, such a non-triviality can be seen straightforwardly from the equations of motion (14.63) and (4.141). The equations (14.63) can be rewritten in the following form

$$A_\mu A^\mu = \frac{\hbar^2}{e^2}\frac{\Box\chi + \dfrac{m_\chi^2}{M_P^2\ell_P^2}\chi + g\chi^3}{\sigma_0 + \chi}. \tag{4.148}$$

Applying the convention $A_\mu = \left[\dfrac{\varphi(x)}{c}, \mathbf{A}(x)\right]$, where $\varphi(x)$ is the electric potential and $\mathbf{A}$ is the magnetic potential, one receives

$$\varphi(x) = \frac{\hbar c}{e}\sqrt{\frac{\Box\chi + \dfrac{m_\chi^2}{M_P^2\ell_P^2}\chi + g\chi^3}{\sigma_0 + \chi} - \left(\frac{e\mathbf{A}}{\hbar}\right)^2}. \tag{4.149}$$

In this manner, the neutrality condition (4.147) says that

$$c\int dt\sqrt{\frac{\Box\chi + \dfrac{m_\chi^2}{M_P^2\ell_P^2}\chi + g\chi^3}{\sigma_0 + \chi} - \left(\frac{e\mathbf{A}}{\hbar}\right)^2} + \frac{e}{\hbar}\int d\mathbf{x}\mathbf{A} = \left(\frac{p}{2} - n\right)\pi. \tag{4.150}$$

It can be seen straightforwardly that the Aharonov–Bohm effect [312] is the particular case in such a situation. Namely, the magnetic flux in the Aharonov–Bohm effect is quantized by the rule

$$\int d\mathbf{x}\mathbf{A} = 4\pi n\Phi_0, \tag{4.151}$$

where $\Phi_0 = \dfrac{\hbar}{2e}$ is the magnetic flux quantum. In this manner, in the present scenario the Goldstone field $\varphi_2(x)$ identically vanishes if and only if the generalized Aharonov–Bohm effect for the photon is present. It should be also emphasized that both the neutral scalar field χ and the magnetic potential $\mathbf{A}$ cannot be chosen arbitrarily, but in such a way that the Euler–Lagrange equations of motion (4.141) must be satisfied.

D Summary

In this chapter we have considered two models of the mass generation mechanism strictly based on the Higgs mechanism, which is the well-accepted paradigm within the Standard Model, but still remains empirically the half-true model for the consistent explanation the nature of the masses of elementary particles. Both the models presented in this chapter were based on the concept of charged scalar field Φ, which lay the foundations of the Higgs mechanism, and in each of these models a neutral scalar field was implemented for the description of mass generation. We have considered the case of Abelian gauge fields, but all the deductions can be straightforwardly generalized to non-Abelian Yang–Mills theories.

The first model was strictly based on the Higgs potential and led us to consistent and constructive explanation of the non-zero values of the photon mass and the mass of the neutral scalar field χ which breaks the symmetry spontaneously. The result of such a model is the dependence between the masses of the photon and neutral scalar field χ, which is similar to the usual Higgs mechanism of the Standard Model.

The second model is based on the modified Higgs potential, where the modification is due to a constant shift field from the real scalar field $|\Phi|$. We have also employed the idea of the wave four-vector k_μ which allowed to look in a new way on the structure of scalar quantum electrodynamics. The concept of the neutral scalar field χ was also determined in an alternative way. This model resulted in the formula for the photon mass, which is practically the same as the formula obtained within the first model. The crucial difference, however, is contained in the constants of the models. Moreover, in the model based on the modified Higgs potential there is no dependence between the mass of the neutral scalar field χ and the mass of the photon, and the mass of the neutral scalar field χ remains a free parameter which can be established by analysis of suitable experimental data. In this model we have established and discussed the presence of generalization of the Aharonov–Bohm effect.

Both models presented in this chapter are in themselves certain scenarios for the consistent explanation of the mass of the photon , and there are different points of view on the idea of the neutral scalar field which breaks the symmetry spontaneously. Both models are seemingly very similar to the standard Higgs mechanism. However, in the Higgs mechanism the photon is explained as massless and, moreover, the Higgs field is a charged four-real-component complex spinor. In both presented scenarios we have not considered *ad hoc* that there is the Higgs field. The scenarios have been based on the *ad hoc* assumption of the existence of the new neutral scalar field, which plays the role of the Higgs particle, and an auxiliary scalar neutral field φ which can be consistently interpreted as a dilaton. The phenomenological status of both the new scalar particle and the dilaton is not clear in the light of the present-day experimental data. The second model, however, is much more promising from this point of view, because in this model the dilaton is massless and the mass of the neutral scalar particle, which is a free parameter, can be determined solely by the suitable experimental data. Moreover, in the second model the mass of the photon does not depend on the mass of the neutral scalar field χ, while in the first model the mass of the photon strongly depends on the mass of the neutral scalar field. Within the framework of the Standard Model estimation of the mass of the photon seems to be a senseless idea, because in the light of the Higgs mechanism the photon is automatically massless. In the first model presented in this chapter the new scalar particle is massive, and the value of its mass can be estimated by the present-day experimental limits on the photon mass.

Chapter 5

Snyder–Sidharth Deformation

Special Relativity can be formulated on the basis of the momentum space, in which the Einstein energy-momentum relation holds

$$E^2 = m^2c^4 + p^2c^2, \tag{5.1}$$

where c is the speed of light in vacuum, p the momentum value of a relativistic particle possessing mass m. Factually, the relation (5.1) can be rewritten in the more conventional form

$$E^2 - p^2c^2 - m^2c^4 = 0, \tag{5.2}$$

which defines the Hamiltonian constraint of Special Relativity. Solving this constraint with respect to particle energy E leads to fixation of the energy as the approximation. For Special Relativity the Lorentz symmetry holds and (5.1), as a quadratic form on the Minkowski energy-momentum space of a particle $p^\mu = [E, p^i c]$, is the Lorentz invariant.

From the modal realism point of view, the Lorentz invariance of the constraint (5.1) leads to the fundamental problem. Namely, the equation (5.1) is not the only such a quadratic form possible. Naturally, the constructive generalization of Special Relativity is

$$E^2 = m^2c^4 + p^2c^2 + \Delta(E, p), \tag{5.3}$$

where the deformation $\Delta(E, p)$ contains also a set of free parameters, which I shall call deformation parameters. Such an extension, however, determines the Multiverse in which the possible worlds are diverse Æther theories, and such an Æther theory defines new physical beings and effects. For Lorentz invariance a whole expression (5.3) must be a quadratic form defined on the Minkowski energy-momentum space. This chapter presents the updated results of the author's paper [313]. Some of these results are removed and replaced by more adequate ideas.

A Linear Deformation

Let us consider first the deformation of Special Relativity due to a simple linear term in a particle momentum p

$$E^2 = m^2c^4 + p^2c^2 + \mathscr{P}^i p_i c^2, \tag{5.4}$$

where $\mathscr{P}^i$, $i = 1,2,3$ is a three-vector of the deformation parameters, which in fact is a certain constant reference momentum three-vector distinguishing the axis related to its direction. Let us introduce also the length of the 3-momentum vector $\mathscr{P}^i$ as

$$\mathscr{P} = \sqrt{\mathscr{P}^i \mathscr{P}_i}. \tag{5.5}$$

It is not difficult to deduce that the deformation momentum three-vector $\mathscr{P}^i$ can be physically understood as the Æther momentum vector. Therefore, the linear deformation of Special Relativity (5.4) describes the simplest type interaction between a particle and the Æther in the Minkowski energy-momentum space. In such a situation the Æther is characterized by a constant momentum vector field $\mathscr{P}^i$ directed according to the direction of the momentum three-vector p^i.

From the Minkowski energy-momentum space point of view the modified Special Relativity (5.4) corresponds to the nontrivially deformed invariant hyperboloid. The theory determined by the energy-momentum relation (5.4) can be described by a quadratic form which by elementary algebraic manipulation can be obtained from (5.4) as its canonical form

$$E^2 + \frac{1}{4}\mathscr{P}^2 c^2 = \left(\frac{\mathscr{P}^i}{\mathscr{P}} p_i c + \frac{\mathscr{P} c}{2}\right)^2 + m^2c^4, \tag{5.6}$$

which can be worked out by multiple ways, *i.e.* in itself determines the Multiverse of possible physical theories. The identification method creates a lot of routes for possible interpretations between all the parts of the equation (5.6). Let us consider here the three basic interpretations.

A1 Dirac Equation and New Algebra

The first case is the following identification

$$\begin{cases} m^2c^4 = \dfrac{\mathscr{P}^2c^2}{4} \\ E^2 = \left(\dfrac{\mathscr{P}^i}{\mathscr{P}}p_ic + \dfrac{\mathscr{P}c}{2}\right)^2 \end{cases}, \tag{5.7}$$

which leads to the mass values

$$\pm mc = \frac{\mathscr{P}}{2}, \tag{5.8}$$

where we have included negative sign mass as physical. The second equation is not difficult to solve. The solution determines energy as

$$\gamma^0 E = \gamma^i p_i c \pm mc^2, \tag{5.9}$$

where γ's are defined by the following relations

$$\gamma^{0^2} = 1, \tag{5.10}$$

$$\gamma^i = \frac{\mathscr{P}^i}{\mathscr{P}}. \tag{5.11}$$

where the equality (5.11) for the present case (5.8) takes the form

$$\gamma^i = \frac{\mathscr{P}^i}{2mc}. \tag{5.12}$$

In other words, from Eq. (5.11) one has $\mathscr{P}^i = \mathscr{P}\gamma^i$ which, after application of the usual raising of the lower index $\gamma_i = \delta_{ij}\gamma^j$, allows to write

$$\mathscr{P}^i\mathscr{P}_i = \mathscr{P}^2\gamma^i\gamma_i, \tag{5.13}$$

i.e. for correctness with the definition (5.5) the identity $\gamma^i\gamma_i \equiv 1$ must hold in an arbitrary case. It is, albeit blatantly incorrect if one treats γ^i as the Dirac gamma matrices obeying the Clifford algebra

$$\{\gamma^i, \gamma^j\}_C = 2\delta^{ij}, \tag{5.14}$$

where δ_{ij} is the $D \times D$ unit matrix, for $\gamma^i\gamma_i = \delta^i_i = D$ and, therefore, in the case of three-dimensional space $\gamma^i\gamma_i = 3$. In fact, such an algebraical treatment (5.14) was investigated by Wolfgang Pauli [314], and is commonly used in both relativistic quantum mechanics as well as quantum field theory [315]. Moreover, the Clifford algebra is the fundamental computational rule in particle physics, for instance for cross sections for reactions. Alternatively, the relations (5.5) and (5.13) can be understood as the suggestion that the linear deformation is true for only $D = 1$ dimensional space, *i.e.* that the Æther exists only in one-dimensional space. It looks like, however, that such an explanation is logically inconsistent and the Clifford algebra (5.14) must be exchanged for the more adequate one.

However, because the definition (5.11) imparts the Æther momentum non-commutative nature, there is an essential difference between expressions $\mathscr{P}^i\mathscr{P}_i$ and $\mathscr{P}_i\mathscr{P}^i$ which in the Abelian case are identical. For this reason let us introduce the definition of the Æther momentum square including the non-commutative nature of $\mathscr{P}^i$

$$\mathscr{P}^2 = \frac{1}{2}\left(\mathscr{P}^i\mathscr{P}_i + \mathscr{P}_i\mathscr{P}^i\right), \tag{5.15}$$

which in the commutative situation leads to (5.5). Thus, one obtains

$$\mathscr{P}^2 = \frac{1}{2}\left(\mathscr{P}^2\gamma^i\gamma_i + \mathscr{P}^2\gamma_i\gamma^i\right), \tag{5.16}$$

which after using the identity $\gamma_i = \delta_{ij}\gamma^j$ gives

$$\mathscr{P}^2 = \frac{\mathscr{P}^2}{2}\delta_{ij}\left(\gamma^i\gamma^j + \gamma^j\gamma^i\right), \tag{5.17}$$

and results in the basic relation

$$1 = \frac{1}{2}\delta_{ij}\left\{\gamma^i, \gamma^j\right\}, \tag{5.18}$$

which can be expressed in more conventionally

$$\left\{\gamma^i, \gamma^j\right\} = \frac{2}{\delta_{ij}}. \tag{5.19}$$

The Clifford algebra (5.14) can be reconstructed from (5.18) if and only if one puts *ad hoc* the identity $\delta^{ij} = \dfrac{1}{\delta_{ij}}$, *i.e.* $\delta^{ij}\delta_{ij} = \delta^i_i = 1$. However, such an algebraical strategy is, in general, incorrect because in the D dimensional space case $\frac{1}{\delta_{ij}} = \frac{\delta^{ij}}{\delta_{ij}\delta^{ij}} = \frac{\delta^{ij}}{\delta^i_i} = \frac{1}{D}\delta^{ij}$. Strictly speaking, it means that the spatial gamma matrices (5.11) do not belong to the Clifford algebra for $D \neq 1$.

In itself, however, the obtained problem is solved by introduction of the new algebra, which can be established straightforwardly and rather easily. Namely, application of the relation $\frac{1}{\delta_{ij}} = \frac{1}{D}\delta^{ij}$ within the equation (5.19) results in the following anticommutator

$$\{\gamma^i, \gamma^j\} = \frac{2}{D}\delta^{ij}, \tag{5.20}$$

which for the three-dimensional case gives the rule

$$\{\gamma^i, \gamma^j\} = \frac{2}{3}\delta^{ij}. \tag{5.21}$$

It is not difficult to see straightforwardly that the extension of spatial gamma matrices to the space-time version

$$\gamma^i \to \gamma^\mu = \left(-\gamma^0, \gamma^i\right), \tag{5.22}$$

leads to simple generalization of the basic formula (5.18)

$$1 = \frac{1}{2}\eta_{\mu\nu}\{\gamma^\mu, \gamma^\nu\}, \tag{5.23}$$

where $\eta_{\mu\nu} = \mathrm{diag}(-1,1,1,1)$ is the metric of the Minkowski space-time. Therefore, one obtains the new four-dimensional gamma matrix algebra

$$\{\gamma^\mu, \gamma^\nu\} = \frac{2}{4}\eta^{\mu\nu} = \frac{1}{2}\eta^{\mu\nu}, \tag{5.24}$$

which is distinguishable from the space-time Clifford algebra

$$\{\gamma^\mu, \gamma^\nu\}_C = 2\eta^{\mu\nu}. \tag{5.25}$$

It is not difficult to prove that for $D+1$ dimensional space-time, where D is the spatial dimension, the new algebra is

$$\{\gamma^\mu, \gamma^\nu\} = \frac{2}{D+1}\eta^{\mu\nu}. \tag{5.26}$$

Let us consider now the constraint (5.9), and apply to this the canonical relativistic quantization procedure

$$(E, p_i c) \to i\hbar\partial_\mu = i\hbar(-\partial_0, c\partial_i). \tag{5.27}$$

The resulting equation

$$-i\hbar\gamma^0\partial_0\psi = (ic\hbar\gamma^i\partial_i \pm mc^2)\psi, \tag{5.28}$$

in fact is the Dirac equation

$$(i\hbar\gamma^\mu\partial_\mu \pm mc^2)\psi = 0, \tag{5.29}$$

for which the Lorentz symmetry is fully validated. In other words, we have obtained the Dirac equation, where however, the gamma matrices do not belong to the Clifford algebra but obey the new algebra (5.24). Interestingly, factually we have generated the Dirac equation independently of the gamma matrix algebra and this suggests that the Dirac equation must not be obtained as "square root taking" of the Klein–Gordon equation, like Dirac originally deduced and applied [316]. Dirac's computations did not involve relations between gamma matrices manifestly and, therefore, his deductions are true.

Interestingly, the general relation between the new algebra (5.20) and the Clifford algebra (5.14) can be established straightforwardly

$$\{\gamma^i, \gamma^j\} = \frac{1}{D}\{\gamma^i, \gamma^j\}_C. \tag{5.30}$$

In other words, because the Clifford algebra we have the unit $D \times D$ matrix and the limit $D \to \infty$ gives trivially

$$\lim_{D\to\infty}\{\gamma^i, \gamma^j\} = 0, \tag{5.31}$$

while the Clifford algebra is conserved in such a limit situation. Albeit, in the light of the general relation (5.18) this case generates the blatantly incorrect equality $1 = 0$. It suggests that $D \to \infty$ is a non-physical situation when the spatial metric is a constant $D \times D$ unit matrix δ_{ij}. Possibly, such an infinite limit holds if and only if the space/space-time metric depends on the number of dimensions D. However, we shall not discuss such examples in this book. It is easy to see that a similar situation is present for the space-time new algebra (5.26) and the space-time Clifford algebra (5.25)

$$\{\gamma^{\mu},\gamma^{\nu}\} = \frac{1}{D+1}\{\gamma^{\mu},\gamma^{\nu}\}_{C}, \tag{5.32}$$

$$\lim_{D\to\infty}\{\gamma^{\mu},\gamma^{\nu}\} = 0. \tag{5.33}$$

Let us establish a certain identity important for this chapter. Namely, let us compute $\gamma^{i}\gamma_{i}$ in the light of the basic relation (5.18)

$$\begin{aligned} \gamma^{i}\gamma_{i} &= \gamma^{i}\delta_{ij}\gamma^{j} = \delta_{ij}\gamma^{i}\gamma^{j} = \frac{1}{2}(\delta_{ij}+\delta_{ji})\gamma^{i}\gamma^{j} \\ &= \frac{1}{2}(\delta_{ij}\gamma^{i}\gamma^{j}+\delta_{ji}\gamma^{i}\gamma^{j}) = \frac{1}{2}(\delta_{ij}\gamma^{i}\gamma^{j}+\delta_{ij}\gamma^{j}\gamma^{i}) \\ &= \frac{1}{2}\delta_{ij}(\gamma^{i}\gamma^{j}+\gamma^{j}\gamma^{i}) = \frac{1}{2}\delta_{ij}\{\gamma^{i},\gamma^{j}\} = 1, \end{aligned} \tag{5.34}$$

which differs from the D-dimensional Clifford algebra result $\gamma^{i}\gamma_{i} = D$. The result (5.34) holds for any D and, first of all, also for the space-time version of the new algebra $\gamma^{\mu}\gamma_{\mu} = 1$, and is independent of the spatial dimension. This is an important elucidation.

Because the presented approach, based on *identification method*, is essentially new and evidently changes deductions and explanations related to the Clifford algebra, we shall call the new algebra *the Æther algebra* . The presented way of reasoning shows that the Dirac equation is not related to the Clifford algebra only, but factually can be deduced by a way very far from the Dirac "square root" technique and produce other algebras of gamma matrices. We show here that, in general, gamma matrices can be deduced by techniques different from the methods of relativistic physics, propagated by Dirac in his contributions to quantum mechanics. Albeit, we regard Dirac's results and, particularly, their diverse consequences for particle physics as an the inspiration. The Æther algebra proposed above can be related to other, possibly unknown, particles and forces. Possibly, the Dirac equation with non-Dirac gamma matrices defines an effective theory.

A2 Alternative Identifications

The second possible identification is

$$\begin{cases} m^2c^4 = E^2 \\ \dfrac{\mathscr{P}^2c^2}{4} = \left(\dfrac{\mathscr{P}^i}{\mathscr{P}}p_ic + \dfrac{\mathscr{P}c}{2}\right)^2 \end{cases}. \tag{5.35}$$

In such a situation, the first equality leads to the relation

$$\gamma^0 E = \pm mc^2, \tag{5.36}$$

where γ^0 is defined by the relation (5.10). The formula (5.36) looks like the Einstein mass-energy relation. Similarly, the second equality in (5.35) leads to the following non-trivial and manifestly distinguishable physical situations

$$\frac{\mathscr{P}^i}{\mathscr{P}}p_ic = 0, \tag{5.37}$$

$$\frac{\mathscr{P}^i}{\mathscr{P}}p_ic = -\mathscr{P}c. \tag{5.38}$$

Similarly, as in the previous case, one can introduce the Clifford algebra of spatial gamma matrices given by the relations (5.10) and (5.11), and applying the canonical relativistic quantization procedure (5.27) one obtains the appropriate projections conditions. The first such a condition arises from (5.36) and has the form

$$\left(i\hbar\gamma^0\partial_0 \mp mc^2\right)\psi = 0, \tag{5.39}$$

while the second one, following from (5.37) and (5.38), has the form

$$ic\hbar\gamma^i\partial_i\psi = 0, \tag{5.40}$$

$$\left(ic\hbar\gamma^i\partial_i + \mathscr{P}c\right)\psi = 0. \tag{5.41}$$

Interestingly, the equation (5.39) added to the condition (5.40) leads to the usual Dirac equation (5.29), while addition of the condition (5.40) to the equation (5.39) allows to establish the new quantum relativistic equation joining a Dirac particle and the Æther

$$\left(i\hbar\gamma^\mu\partial_\mu + Mc^2\right)\psi = 0, \tag{5.42}$$

where M is the effective mass of the particle-Æther system

$$M = \mp m + \frac{1}{c}\mathscr{P}, \tag{5.43}$$

i.e. for $[\mathscr{P}] \sim [c] = 3 \cdot 10^8$ the correction due to the Æther plays an essential physical role. Interestingly, for the positive sign near particle mass m in (9.193) the effective mass M is always positive, while for the negative sign the effective mass M is positive for $mc < \mathscr{P}$, negative for $mc > \mathscr{P}$, and vanishes when $mc = \mathscr{P}$.

On the one hand, the effective mass (9.193) manifestly contains the correction to particle mass due to the Æther momentum value $\mathscr{P}$ but not due to $\mathscr{P}_i$. Such a property involves a situation when the Æther momentum vector $\mathscr{P}_i$ is non-trivial but its length $\mathscr{P}$ vanishes. In such instance the arbitrary component of the Æther momentum vector is determined by the two remaining components which are still arbitrary. It can be seen that the classical theory is then deformed Special Relativity (5.4) while, because $M = \pm m$ according to (9.193), quantum theory (5.42) is the Dirac relativistic quantum mechanics. In other words, the Dirac theory is related not only to the Einstein theory, but possesses a wider sense.

On the other hand, the quantum theory given by the projections (5.39), and (5.40) and (5.41) carries a different content than the usual Dirac relativistic quantum mechanics. In the Dirac theory there is only one cumulative projection condition (5.29), while in our theory the conditions (5.39), and (5.40) and (5.41) in general are not cumulative. We mean that we have obtained the condition for time evolution (5.39) and two alternative conditions for spatial evolution (5.40) and (5.41), while in the Dirac theory there is unified space-time evolution (5.29). For this reason our situation is physically distinguished from the theory based on the four-dimensional Dirac equation . However, the unification obtained by the simple algebraic sum of the time projection and the spatial projection leads us to the usual Dirac theory and the Dirac theory with the effective mass (9.193). This particular case is within the general theory given by the projection conditions (5.39), and (5.40) and (5.41).

Interestingly, also the classical physics context of the conditions (5.37) and (5.38) is non-trivial. Namely, these relations establish two possible constraints for a particle momentum components, which allows to express an arbitrary one component of a particle momentum by the Æther momentum. In other words, the constraints (5.37) and (5.38) join a classical particle with the Æther. In fact, the first relation is

$$\mathscr{P}^1 p_1 + \mathscr{P}^2 p_2 + \mathscr{P}^3 p_3 = 0, \tag{5.44}$$

and the latter one is

$$\mathscr{P}^1(p_1 + \mathscr{P}_1) + \mathscr{P}^2(p_2 + \mathscr{P}_2) + \mathscr{P}^3(p_3 + \mathscr{P}_3) = 0. \tag{5.45}$$

There are, in general, three types of solutions for each of these constraints, in which one component of a particle momentum is dependent on two other (arbitrary) components of particle momentum and all components of the reference momentum three-vector. The constraint (16.308) can be solved by

$$p_i = \left(-\frac{\mathscr{P}^2}{\mathscr{P}^1}p_2 - \frac{\mathscr{P}^3}{\mathscr{P}^1}p_3, p_2, p_3\right), \tag{5.46}$$

$$p_i = \left(p_1, -\frac{\mathscr{P}^1}{\mathscr{P}^2}p_1 - \frac{\mathscr{P}^3}{\mathscr{P}^2}p_3, p_3\right), \tag{5.47}$$

$$p_i = \left(p_1, p_2, -\frac{\mathscr{P}^1}{\mathscr{P}^3}p_1 - \frac{\mathscr{P}^2}{\mathscr{P}^3}p_2\right). \tag{5.48}$$

Similarly, the constraint given by (5.45) possesses in general three possible solutions

$$p_i = \left(-\mathscr{P}_1 - \frac{\mathscr{P}^2}{\mathscr{P}^1}(p_2 + \mathscr{P}_2) - \frac{\mathscr{P}^3}{\mathscr{P}^1}(p_3 + \mathscr{P}_3), p_2, p_3\right), \tag{5.49}$$

$$p_i = \left(p_1, -\mathscr{P}_2 - \frac{\mathscr{P}^1}{\mathscr{P}^2}(p_1 + \mathscr{P}_1) - \frac{\mathscr{P}^3}{\mathscr{P}^2}(p_3 + \mathscr{P}_3), p_3\right), \tag{5.50}$$

$$p_i = \left(p_1, p_2, -\mathscr{P}_3 - \frac{\mathscr{P}^1}{\mathscr{P}^3}(p_1 + \mathscr{P}_1) - \frac{\mathscr{P}^2}{\mathscr{P}^3}(p_2 + \mathscr{P}_2)\right). \tag{5.51}$$

Particularly, the solution of the first constraint can be trivial, *i.e.* $p_i = 0$, and the second constraint can be solved simply by $p_i = -\mathscr{P}_i$. Both these cases have a physical interpretation of an inertial reference frame of a particle: either rest frame or motion of a particle under the constant momentum opposite to the Æther momentum. For both these particular solutions the Lorentz symmetry also holds.

The third interesting identification is

$$\left\{\begin{array}{l} -E^2 = \dfrac{\mathscr{P}^2 c^2}{4} \\ -m^2 c^4 = \left(\dfrac{\mathscr{P}^i}{\mathscr{P}} p_i c + \dfrac{\mathscr{P} c}{2}\right)^2 \end{array}\right. . \tag{5.52}$$

The first constraint can be resolved straightforwardly with the result

$$\pm i\gamma^0 E = \frac{\mathscr{P} c}{2}. \tag{5.53}$$

The solution of the second equation also can be easily established

$$\pm imc^2 = \frac{\mathscr{P}^i}{\mathscr{P}} p_i c + \frac{\mathscr{P} c}{2}. \tag{5.54}$$

Employing (5.53) this solution can be written as

$$\pm imc^2 = \gamma^i p_i c \pm i\gamma^0 E. \tag{5.55}$$

After application of the canonical relativistic quantization to the solution (5.55) one receives the equation

$$\left(ic\hbar\gamma^i\partial_i \mp \hbar\gamma^0\partial_0 \mp imc^2\right)\psi = 0, \tag{5.56}$$

having blatantly real and imaginary parts which are

$$\mp\hbar\gamma^0\partial_0\psi = 0, \tag{5.57}$$

$$\left(c\hbar\gamma^i\partial_i \mp mc^2\right)\psi = 0, \tag{5.58}$$

and must be treated as the system of equations.

With using (5.53) the solution (5.54), however, can be rewritten in another form and understood alternatively. Namely, because one has determined the Æther momentum value by a particle energy and non-determined the Æther momentum vector, the equation (5.54) can be presented in an equivalent form

$$\pm imc^2 = \frac{\mathscr{P}^i c}{\pm 2i\gamma^0 E} p_i c \pm i\gamma^0 E. \tag{5.59}$$

One sees straightforwardly, however, that the equation (5.59) after elementary algebraic manipulations can be presented in the form of a quadratic equation

$$(\pm\gamma^0 E)^2 \mp mc^2(\pm\gamma^0 E) - \frac{\mathscr{P}^i c}{2} p_i c = 0. \tag{5.60}$$

Now it is easy to conclude that essentially for the arbitrary sign of m this equation can be rewritten as

$$(\gamma^0 E)^2 \pm mc^2(\gamma^0 E) - \frac{\mathscr{P}^i c}{2} p_i c = 0, \tag{5.61}$$

while one can treat the positive mass case in the constraint (5.59) as the physical situation. The canonical relativistic quantization applied to the constraint (5.61) results in the following projection condition

$$\left(-\hbar^2 \partial_0^2 \pm i\hbar mc^2 \gamma^0 \partial_0 - ic\hbar \frac{\mathscr{P}^i c}{2} \partial_i\right) \psi = 0, \tag{5.62}$$

which after multiplication by $-1/mc^2$ and taking into account *ad hoc* the following identification of the spatial gamma matrices

$$\gamma^i = \frac{\mathscr{P}^i}{2mc}. \tag{5.63}$$

can easily lead to a more convenient form

$$\left(\frac{\hbar^2}{mc^2} \partial_0^2 \mp i\hbar \gamma^0 \partial_0 + ic\hbar \gamma^i \partial_i\right) \psi = 0. \tag{5.64}$$

Interestingly, when one takes the plus sign in (5.64) as the physical case then this condition can be written as

$$\left(i\hbar \gamma^\mu \partial_\mu + \frac{\hbar^2}{mc^2} \partial_0^2\right) \psi = 0, \tag{5.65}$$

and then the Dirac equation is recovered

$$(i\hbar \gamma^\mu \partial_\mu + Mc^2) \psi = 0, \tag{5.66}$$

where M is an effective mass term

$$Mc^2 \psi \equiv \frac{\hbar^2}{mc^2} \partial_0^2 \psi. \tag{5.67}$$

It is easy to see that the Lorentz symmetry is validated in such mass generation mechanism.

The case of the minus sign in the equation (5.64) can be also considered in terms of the Dirac equation (5.66), but then the mass term is determined by one of two equivalent conditions

$$Mc^2\psi \equiv \left(-\frac{\hbar^2}{mc^2}\partial_0^2 - 2ic\hbar\gamma^i\partial_i\right)\psi, \tag{5.68}$$

$$Mc^2\psi \equiv \left(\frac{\hbar^2}{mc^2}\partial_0^2 - 2i\hbar\gamma^0\partial_0\right)\psi. \tag{5.69}$$

From the classical physics point of view, however, the mass term conditions (5.68) and (5.69) are respectively

$$Mc^2 = -\frac{E^2}{mc^2} - 2\frac{\mathscr{P}^i}{2mc}p_i c, \tag{5.70}$$

$$Mc^2 = \frac{E^2}{mc^2} - 2\gamma^0 E, \tag{5.71}$$

where we used ${\gamma_0}^2 = 1$, and their equivalence leads to the constraint

$$\frac{E^2}{mc^2} - \gamma^0 E + \frac{\mathscr{P}^i}{2mc}p_i c = 0, \tag{5.72}$$

which can be rewritten in a more conventional form

$$\gamma^0 E = \frac{E^2}{mc^2} + \frac{\mathscr{P}^i}{2mc}p_i c. \tag{5.73}$$

Applying the basic constraint (5.4) to the right hand side of (5.73) one receives the relation

$$\gamma^0 E = mc^2 + \frac{3\mathscr{P}^i c p_i c}{2mc^2} + \frac{p^i c p_i c}{mc^2}, \tag{5.74}$$

that defines the resolution of the constraint (5.4) in this case. Employing once again the canonical relativistic quantization to (5.74) and identifying the spatial gamma matrices as (5.63) one obtains the following equation

$$i\hbar\gamma^0\partial_0\psi = \left(mc^2 + 3ic\hbar\gamma^i\partial_i - \frac{\hbar^2}{mc^2}\triangle\right)\psi, \tag{5.75}$$

where $\triangle = \gamma^i\gamma_i\partial^i\partial_i = \partial^i\partial_i$ is the Laplace operator, which can be presented as the Dirac equation (5.66) with the effective mass given by two equivalent mass generation rules

$$Mc^2\psi = \left(-mc^2 - 4ic\hbar\gamma^i\partial_i + \frac{\hbar^2}{mc^2}\triangle\right)\psi, \tag{5.76}$$

$$Mc^2\psi = \frac{1}{3}\left(mc^2 - 4ic\hbar\gamma^0\partial_0 - \frac{\hbar^2}{mc^2}\triangle\right)\psi. \tag{5.77}$$

It should be emphasized that the Laplace operator $\triangle = \gamma^i\gamma_i\partial^i\partial_i$ has the form $\triangle = \partial^i\partial_i$ for the new algebra, while for the Clifford algebra it is manifestly different $\triangle = D\partial^i\partial_i$. It is the confirmation of the correctness of the new algebra.

Another route to a new equation in the case under consideration, can be obtained *e.g.* by application of the first relation (5.52) within the constraint (5.73). First one obtains

$$\gamma^0 E = -\frac{\mathscr{P}^i}{2mc}\frac{\mathscr{P}_i}{2mc}mc^2 + \frac{\mathscr{P}^i}{2mc}p_i c, \tag{5.78}$$

then by using (5.63) and the identity $\mathscr{P}_i = \delta_{ij}\mathscr{P}^j$ one has

$$\gamma^0 E = -\delta_{ij}\gamma^i\gamma^j mc^2 + \gamma^i p_i c, \tag{5.79}$$

and by the new algebra result (5.34) one receives finally

$$\gamma^0 E - \gamma^i p_i c + mc^2 = 0. \tag{5.80}$$

In this manner, applying once again the canonical relativistic quantization one obtains the equation

$$\left(i\hbar\gamma^0\partial_0 - ic\hbar\gamma^i\partial_i + mc^2\right)\psi = 0, \tag{5.81}$$

which differs from the Dirac equation for a particle with mass m by the minus sign presence near the spatial derivative. However, one sees straightforwardly that the equation (5.81) can be interpreted as the Dirac equation (5.66) with the effective mass M, if and only if the effective mass term is given by one of the two equivalent relations

$$Mc^2\psi \equiv \left(-2ic\hbar\gamma^i\partial_i + mc^2\right)\psi, \tag{5.82}$$

$$Mc^2\psi \equiv \left(-2i\hbar\gamma^0\partial_0 - mc^2\right)\psi. \tag{5.83}$$

The obtained Dirac-like quantum theories can be always presented in the form of the Schrödinger equation, *i.e.*

$$i\hbar\partial_0\psi = H\psi, \tag{5.84}$$

where $H = H(\partial_0, \partial_i)$ is the Hamilton operator describing a quantum system. In comparison to the standard Dirac theory in our case the operator H, however, in general must not be always a hermitian operator. The equation (5.84) should be solved with the usual spatial normalization condition for the wave function ψ

$$\int d^D x |\psi(x)|^2 = 1, \tag{5.85}$$

where 1 is the D-dimensional unit matrix. The question of solvability of the systems deduced above, however, is not the main theme of this book. It is a good exercise for a reader.

As we proposed initially, the linear deformation of Special Relativity constraint in (5.4) couples a particle and the Æther in the Minkowski energy-momentum space. The reasoning done in the spirit of Dirac's relativistic quantum mechanics led us to the linkage (5.11) between the new algebra of the spatial gamma matrices, and the Æther momentum vector, and its particular case (5.63) was also discussed. Straightforward application of the Æther algebra (5.20) results in the following non-commutative algebra of the Æther momentum

$$\left\{\frac{\mathscr{P}^i}{\mathscr{P}}, \frac{\mathscr{P}^j}{\mathscr{P}}\right\} = \frac{2}{D}\delta^{ij}. \tag{5.86}$$

Let us generalize this algebra to the Minkowski space-time case

$$\left\{\frac{\mathscr{P}^\mu}{\mathfrak{P}}, \frac{\mathscr{P}^\nu}{\mathfrak{P}}\right\} = \frac{2}{D+1}\eta^{\mu\nu}, \tag{5.87}$$

where $\mathfrak{P}$ is the length of the Æther four-momentum

$$\mathfrak{P}^2 = \frac{1}{2}\left(\mathscr{P}^\mu \mathscr{P}_\mu + \mathscr{P}_\mu \mathscr{P}^\mu\right), \tag{5.88}$$

which can be initially postulated as

$$\mathscr{P}^\mu = \left(\mathscr{P}^0, \mathscr{P}\gamma^i\right), \tag{5.89}$$

where $\mathscr{P}^0$ is the time component of $\mathscr{P}^\mu$, and $\mathscr{P}$ is the length of the Æther three-momentum. Let us compute (5.88) straightforwardly

$$\mathfrak{P}^2 = \frac{1}{2}\left(\mathscr{P}^0\mathscr{P}_0 + \mathscr{P}_0\mathscr{P}^0 + \mathscr{P}^2\delta_{ij}\left\{\gamma^i,\gamma^j\right\}\right), \tag{5.90}$$

and postulate $\mathscr{P}^0 = -\alpha\gamma^0$, where α is an unknown multiplier. Then, by definition one has $\mathscr{P}_0 = \eta_{00}\mathscr{P}^0 = \alpha\gamma_0$, one receives

$$\mathfrak{P}^2 - \frac{1}{2}\mathscr{P}^2\delta_{ij}\left\{\gamma^i,\gamma^j\right\} = -\alpha^2\left(1 - \frac{1}{2}\delta_{ij}\left\{\gamma^i,\gamma^j\right\}\right), \tag{5.91}$$

what in the light of the spatial algebra (5.18) is satisfied if and only if

$$\mathfrak{P}^2 = \mathscr{P}^2, \quad \mathfrak{P}^2 = -\alpha^2, \tag{5.92}$$

and, consequently, the time component of the Æther four-momentum is

$$\mathscr{P}^0 = \pm i\mathfrak{P}\gamma^0 = \pm i\mathscr{P}\gamma^0. \tag{5.93}$$

Therefore, $\mathscr{P}^0 = 0$ if and only if $\mathfrak{P} = \mathscr{P} = 0$.

For the particular situation $\gamma^i = \dfrac{\mathscr{P}^i}{2mc}$ the particle mass can be expressed through the Æther momentum vector

$$m = \frac{1}{2}\frac{\mathscr{P}}{c}, \tag{5.94}$$

so in fact the energy-momentum relation (5.4) can be expressed through the Æther and particle momenta only

$$E^2 = \left(\frac{1}{4}\mathscr{P}^i\mathscr{P}_i + p^i p_i + \mathscr{P}^i p_i\right)c^2. \tag{5.95}$$

The fixed value of the square of the Æther momentum vector allows to establish three equivalent forms of the the momentum three-vector

$$\mathscr{P}^i = \left[\sqrt{(2mc)^2 - \mathscr{P}^2\mathscr{P}_2 - \mathscr{P}^3\mathscr{P}_3}, \mathscr{P}^2, \mathscr{P}^3\right] = \tag{5.96}$$

$$= \left[\mathscr{P}^1, \sqrt{(2mc)^2 - \mathscr{P}^1\mathscr{P}_1 - \mathscr{P}^3\mathscr{P}_3}, \mathscr{P}^3\right] = \tag{5.97}$$

$$= \left[\mathscr{P}^1, \mathscr{P}^2, \sqrt{(2mc)^2 - \mathscr{P}^1\mathscr{P}_1 - \mathscr{P}^2\mathscr{P}_2}\right], \tag{5.98}$$

which allows also to derive easily the normalized Æther momentum vector $\frac{\mathscr{P}^i}{\mathscr{P}}$. Moreover, one can reconsider the situations defined by the constraints (5.37) and (5.38). In the first case the energetic constraint (9.182) takes the following form

$$E^2 = \left(\frac{1}{4}\mathscr{P}^i\mathscr{P}_i + p^ip_i\right)c^2, \tag{5.99}$$

with the particle momentum defined by the solutions (5.46), (5.47) or (5.48), while in the second situation one obtains

$$E^2 = \left(-\frac{3}{4}\mathscr{P}^i\mathscr{P}_i + p^ip_i\right)c^2. \tag{5.100}$$

where the particle momentum is given by (5.49), (5.50) or (5.51). It is visible that in the case (9.189) the energy is always non-zero. However, in the case (9.192) the energy can be trivially vanishing if and only if the particle momentum is constrained by

$$p_i = \sqrt{\frac{3}{4}}\mathscr{P}_i. \tag{5.101}$$

It is easy to see for (5.49), (5.50) or (5.51) that in the case (5.101) holds

$$\mathscr{P}^i\mathscr{P}_i = 0. \tag{5.102}$$

Particularly, the constraint (5.102) is satisfied when the Æther momentum vanishes identically $\mathscr{P}^i = 0$, but by (5.101) such a situation implies

$$m = \frac{1}{\sqrt{3}}\frac{p}{c}, \tag{5.103}$$

where $p = \sqrt{p^ip_i}$ is the momentum value of a particle. Moreover, the vanishing Æther momentum implies that the deformation vanishes, *i.e.* Special Relativity should be reconstructed. Using both $\mathscr{P}^i = 0$ and (5.103) in the light of the Hamiltonian constraint (5.4) one obtains

$$E^2 = 4m^2c^4, \tag{5.104}$$

but by (5.101) the energy square (9.192) vanishes, so reconstruction of Special Relativity gives finally $m^2 = 0$. Application of the constraint

(5.102) within the solutions (5.49)-(5.51) reconstructs the case (5.46)-(5.48), so (5.102) is equivalent to $\mathscr{P}^i p_i = 0$ in this case.

Interestingly, one can also consider other types of linear deformations of Special Relativity. For example

$$\Delta_1 = \mathscr{P}_i p^i c^2, \tag{5.105}$$

$$\Delta_2 = \frac{1}{2}\left(\mathscr{P}_i p^i c^2 + \mathscr{P}^i p_i c^2\right), \tag{5.106}$$

$$\Delta_3 = \frac{1}{2}\left(\mathscr{P}_i p^i c^2 + \mathscr{P}^i p_i c^2\right) \pm \mathscr{P}^2 c^2, \tag{5.107}$$

and so on, and perform analogous considerations. The deformations (5.105) and (5.106) are equivalent to (5.4) if and only if $\mathscr{P}_i p^i = \mathscr{P}^i p_i$, *i.e.*

$$\delta_{ik}\gamma^k p^i = \gamma^j p_j \tag{5.108}$$

what, after multiplication of both sides by $\delta^{ik}\gamma_k = \gamma^i$ and taking into account the identity $\delta_{ik}\delta^{ik} = D$, takes the form

$$p^i = \frac{1}{D}\gamma^i \gamma^j p_j. \tag{5.109}$$

In this section we have presented the approach based on the linear deformation of the Hamiltonian constraint of Special Relativity which generates new equations in the framework of relativistic quantum mechanics. The linear deformation can be generalized for other deformations, and similar strategy can be used. All these new models can be treated as Dark Matter and/or Dark Energy models.

B Snyder–Sidharth Hamiltonian

Let us consider now a more complex situation. Namely, we shall focus our attention on the following deformation of the Hamiltonian constraint of Special Relativity

$$E^2 = m^2c^4 + c^2p^2 + \alpha\left(\frac{\ell}{\hbar}\right)^2 c^2p^4, \tag{5.110}$$

where ℓ is any physical minimal scale, and $\alpha \sim 1$ is a dimensionless constant. Factually, this deformation has been investigated by Burra

Gautam Sidharth who considered such a Hamiltonian constraint compatible with the quantized space-time investigated by Hartland Sweet Snyder [317] in the context of the infrared catastrophe of soft photons in the Compton scatterin, and in general to renormalize quantum field theory by application of the non-commutative quantum space-time, as it is widely studied by numerous authors and scholars [318]. Sidharth have considered the following model of discrete space-time

$$\frac{i}{\hbar}[p,x] = 1+\frac{\ell^2}{\hbar^2}p^2, \tag{5.111}$$

$$[x,y] = O(\ell^2), \tag{5.112}$$

in which one has to deal with the structure of a non-differentiable manifold, or lattice model of space-time. We shall discuss wider mathematical details of such a space-time model (5.111) in next two chapters. It should be emphasized that the deformation (5.111) reveals Lorentz invariance. In fact Burra Gautam Sidharth (See the books in Ref. [48]) was one of these scholars who first accepted the space-time non-commutative geometry as the serious argument for new physics. He has considered the energy-momentum relation (8.20) in the particle astrophysics context, and proposed to treat such a deformation in a more general sense as a kind of the perturbational series in the minimum scale ℓ, that can be *e.g.* the Planck scale or the Compton scale. For this reason we shall call the Hamiltonian constraint (8.20) by the name *the Snyder–Sidharth Hamiltonian constraint* or briefly the Snyder–Sidharth Hamiltonian.

It should be emphasized that a number of distinguished scholars like Sheldon Lee Glashow, Sidney Richard Coleman, and others have considered the various schemes which manifestly depart from the Einstein Special Relativity. It must be stressed here that these all scenarios are purely *ad hoc*. Recently, observations of ultra-high-energy cosmic rays and rays from gamma bursts [319] suggested violation of the Lorentz (CP) symmetry, and in this context the Snyder–Sidharth Hamiltonian constraint (8.20) is applicable in the light of the fact that this energy-momentum relation violates the Lorentz (CP) symmetry manifestly. It should be emphasized that, in general, situation the Hamiltonian constraint is not a Hamiltonian, *i.e.* energy. Recall that in the general scheme of Hamiltonian systems energy can be obtained as a solution of a Hamiltonian constraint, as it was shown in the previous section, and will be continued in the next chapters of this book. Interestingly, the discrete space-time model possesses a certain context in string theory, where can be referred as a stringy-like space-time.

Similarly, as in the case of the linear deformation discussed in the previous section, the deformed Hamiltonian constraint (8.20) can be treated as a quadratic form and can be straightforwardly lead to the canonical form

$$E^2 + \frac{\hbar^2 c^2}{4\alpha\ell^2} = \alpha\left(\frac{c\ell}{\hbar}\right)^2\left(p^2 + \frac{\hbar^2}{2\alpha\ell^2}\right)^2 + m^2c^4, \tag{5.113}$$

and as previously there are basically three possible mathematical interpretations of this relation. However, in the case of the fourth-order deformation (8.20) one has no linear terms in particle momentum p, but there are actually the powers of p^2. In the light of the principles of relativistic quantum theory (Cf. *e.g.* Refs. [320] and [321]) it means that in such a situation the structure of the Clifford algebra must be at least hidden if not hidden manifestly. Let us consider the three basic identifications to the case of fermions and the case of bosons.

B1 Fermions

1. First, we can interpret the Hamiltonian constraint (5.113) as the system of two equations

$$\begin{cases} m^2c^4 = \dfrac{\hbar^2 c^2}{4\alpha\ell^2} \\ E^2 = \alpha\left(\dfrac{c\ell}{\hbar}\right)^2\left(p^2 + \dfrac{\hbar^2}{2\alpha\ell^2}\right)^2 \end{cases} \tag{5.114}$$

The first equation in this system leads to solution that looks like formally as the bosonic string tension

$$m = \frac{\hbar}{2\sqrt{\alpha}c\ell}. \tag{5.115}$$

Expressing minimal scale ℓ in terms of mass m and the constants, one can write the solution of the second equality in the system

(5.114) in the following way

$$E = \frac{p^2}{2m} + mc^2, \tag{5.116}$$

where we have omitted the solution with the minus sign as a manifestly non-physical situation. This Hamiltonian is exactly the energy of a free point particle in semi-classical mechanics, *i.e.* the Newtonian kinetic energy supplemented by the contribution describing the Einstein–Poincarè rest energy. Interestingly, both the relations for mass (5.115) as well as for energy (5.116) do not lose consistency in the case when mass m is the Planck mass, *i.e.* $m = M_P = \sqrt{\frac{\hbar c}{G}}$ and minimal scale ℓ is the Planck length, *i.e.* $\ell_P = \sqrt{\frac{\hbar G}{c^3}}$. Recall that the Planck mass in itself determines a unifying scale, which is known as the Planck scale, where the classical and the quantum meet and collaborate. The Schwarzschild radius $r_S(m) = \frac{Gm}{c^2}$ evaluated for the case of the Planck mass equals the Compton wavelength of a hypothetical particle possessing such a value of mass [322]. In comparison with Special Relativity, higher relativistic corrections to the semi-classical case are absent. After application of the primary canonical quantization to the Hamiltonian (5.116) one receives exactly the Schrödinger equation describing free quarks in quantum chromodynamics, because quarks are non-relativistic and massive [323]. For the case of the identically vanishing minimal scale and non-zero constant α as well as for identically vanishing α and a fixed non-zero value of minimal scale ℓ, both mass and energy are formally divergent and, consequently, such a situation describes a non-physical singularity which looks like as a black-hole. For the large scale limit and non-vanishing value of α, the mass spectrum is trivial point-like $m = 0$. For the case of non-zero momentum energy remains infinite, *i.e.* this is a non-physical situation. Anyway, such an analysis shows that the Hamiltonian (5.116) is manifestly compatible with the Snyder–Sidharth deformation (5.111).

2. Let us consider the second identification which exchanges the roles of energy and mass

$$\begin{cases} -m^2c^4 = \alpha\left(\dfrac{c\ell}{\hbar}\right)^2\left(p^2+\dfrac{\hbar^2}{2\alpha\ell^2}\right)^2 \\ -E^2 = \dfrac{\hbar^2c^2}{4\alpha\ell^2} \end{cases}, \tag{5.117}$$

and leads to the discrete energy spectrum for a fixed value of minimal scale ℓ. In the light of the fact that the Snyder–Sidharth Hamiltonian constraint (8.20) in the case of positive α is true for fermions, and in the case of negative α is validated for bosons (See the paper of B.G. Sidharth in Ref. [319]), for the case of fermions energy is manifestly imaginary

$$iE = \frac{\hbar c}{2\sqrt{\alpha}\ell}, \tag{5.118}$$

similarly as mass

$$imc^2 = \sqrt{\alpha}\frac{c\ell}{\hbar}p^2 + \frac{\hbar c}{2\sqrt{\alpha}\ell}, \tag{5.119}$$

where we have omitted the negative values as non-physical. However, one can eliminate minimal scale by using the energy relation (5.118) with the following result

$$mc^2 = -\frac{p^2c^2}{2E} + E, \tag{5.120}$$

which after minor algebraic manipulations can be rewritten in the form of the quadratic equation

$$E^2 = mc^2E + \frac{p^2c^2}{2}, \tag{5.121}$$

and by application of the Snyder–Sidharth Hamiltonian constraint (8.20) it yields

$$m^2c^4 + c^2p^2 + \alpha\left(\frac{\ell}{\hbar}\right)^2 c^2p^4 = mc^2E + \frac{p^2c^2}{2}. \tag{5.122}$$

One can find now the energy (not square of energy!), that has the form of the $4th$ order polynomial in the momentum

$$E = mc^2 + \frac{p^2}{2m} + \alpha\left(\frac{\ell}{\hbar}\right)^2 \frac{p^4}{m}. \tag{5.123}$$

Once again one can treat this result as a quadratic form and apply suitable factorization

$$E + \frac{\hbar^2}{16\alpha\ell^2}\frac{1}{m} = \alpha\left(\frac{\ell}{\hbar}\right)^2 \frac{1}{m}\left(p^2 + \frac{\hbar^2}{4\alpha\ell^2}\right)^2 + mc^2, \tag{5.124}$$

and consider the identification method for the mass-energy relation.

(a) The first obvious interpretation yields

$$\begin{cases} mc^2 = \dfrac{\hbar^2}{16\alpha\ell^2}\dfrac{1}{m} \\ E = \alpha\left(\dfrac{\ell}{\hbar}\right)^2 \dfrac{1}{m}\left(p^2 + \dfrac{\hbar^2}{4\alpha\ell^2}\right)^2 \end{cases} \tag{5.125}$$

and once again one can extract easily and straightforwardly the solution to the first equation

$$m = \frac{\hbar}{4\sqrt{\alpha}c\ell}, \tag{5.126}$$

and solution of the latter equality can be written in the form of the Pauli Hamiltonian constraint

$$E = mc^2 + \frac{p^2}{2m} + \frac{p^4}{16m^3c^2}, \tag{5.127}$$

or after elimination of mass by using (5.126) one receives

$$E = \frac{\hbar c}{4\sqrt{\alpha}\ell} + 2c\sqrt{\alpha}\frac{\ell}{\hbar}p^2 + 4c\left(\sqrt{\alpha}\frac{\ell}{\hbar}\right)^3 p^4. \tag{5.128}$$

However, it can be easily seen that the relation (5.118) combined with (5.126) determines energy as

$$iE = 2mc^2, \tag{5.129}$$

or equivalently the mass square value

$$m^2 = -\frac{E^2}{2c^4} < 0, \tag{5.130}$$

which means that momentum values are non-hermitian, so one has to deal with a tachyon. Moreover, by straightforward application of the formula (5.126) together with the relation (5.119) one establishes that

$$p^2 = \frac{\hbar^2}{2\alpha\ell^2}\left(-1+\frac{1}{2}i\right). \tag{5.131}$$

Using the polar form of the complex number in brackets

$$-1+\frac{1}{2}i = \frac{\sqrt{5}}{2}\exp\left(-i\arctan\frac{1}{2}+2ni\pi\right), \tag{5.132}$$

where $n \in \mathbb{Z}$, one can take its square root

$$\sqrt{-1+\frac{1}{2}i} = \frac{1}{\sqrt{2}}\left(\sqrt{\frac{\sqrt{5}}{2}-1}+i\sqrt{\frac{\sqrt{5}}{2}+1}\right)e^{ni\pi}, \tag{5.133}$$

and obtains the momentum spectrum in dependence on ℓ

$$p_n = \pm\frac{(-1)^n}{2}\left(\sqrt{\frac{\sqrt{5}}{2}-1}+i\sqrt{\frac{\sqrt{5}}{2}+1}\right)\frac{\hbar}{\sqrt{\alpha}\ell}. \tag{5.134}$$

(b) The latter subcase is

$$\begin{cases} mc^2 = E \\ \dfrac{\hbar^2}{16\alpha\ell^2}\dfrac{1}{m} = \alpha\left(\dfrac{\ell}{\hbar}\right)^2\dfrac{1}{m}\left(p^2+\dfrac{\hbar^2}{4\alpha\ell^2}\right)^2 \end{cases} \tag{5.135}$$

Once again, solution to the first equation by using the relation (5.118) is straightforward

$$m = -i\frac{\hbar}{2\sqrt{\alpha}c\ell}, \quad m^2 < 0, \tag{5.136}$$

and once again the tachyon is obtained - there are particles with the momentum spectrum

$$p = \left\{0, \pm\frac{\hbar}{\sqrt{2\alpha}\ell}\right\}. \qquad (5.137)$$

This is a discrete momentum spectrum for fixed scale ℓ. For the running scale this is non-compact spectrum, but compactification to the point is done in the large scale limit

$$\lim_{\ell\to\infty} p = 0, \qquad (5.138)$$

and it is the rest. For $\alpha = 0$ and fixed scale ℓ there are two singular values of the momentum p. For all $\ell \neq 0$ and $\alpha \neq 0$, the case of non-zero p is related to the existence of the tachyon.

(c) The third interpretation yields

$$\begin{cases} -E = \dfrac{\hbar^2}{16\alpha\ell^2}\dfrac{1}{m} \\ -mc^2 = \alpha\left(\dfrac{\ell}{\hbar}\right)^2\dfrac{1}{m}\left(p^2+\dfrac{\hbar^2}{4\alpha\ell^2}\right)^2 \end{cases}. \qquad (5.139)$$

Similarly, as in the previous subcase, employing the relation (5.118) one obtains from the first equation the mass

$$m = -i\frac{\hbar}{8\sqrt{\alpha}c\ell}, \qquad (5.140)$$

and again one has a tachyon, *i.e.* $m^2 < 0$. Consequently, the momentum spectrum can be established as

$$p = \left\{\pm i\sqrt{\frac{1}{8\alpha}}\frac{\hbar}{\ell}, \pm i\sqrt{\frac{3}{8\alpha}}\frac{\hbar}{\ell}\right\}, \qquad (5.141)$$

and is non-hermitian, as one has expected. For this reason this particular case, that is the Pauli Hamiltonian constraint with mass related to minimum scale, describes a tachyon, the hypothetical particle moving with a speed faster than light.

3. The third possible solution of the Snyder–Sidharth Hamiltonian constraint can be constructed by the system of equations

$$\begin{cases} E^2 = m^2c^4 \\ \dfrac{\hbar^2 c^2}{4\alpha\ell^2} = \alpha\left(\dfrac{c\ell}{\hbar}\right)^2\left(p^2 + \dfrac{\hbar^2}{2\alpha\ell^2}\right)^2 \end{cases} . \tag{5.142}$$

The first equation gives the standard Einstein mass-energy relation

$$E = mc^2, \tag{5.143}$$

and the latter equality gives the discrete momentum spectrum

$$p = \left\{0, \pm\frac{1}{\sqrt{\alpha}}\frac{\hbar}{\ell}\right\}, \tag{5.144}$$

for a fixed scale ℓ value. For the case of a running scale this spectrum losses compactness, but in the limit of large scales the spectrum is manifestly compactified to the fixed point

$$\lim_{\ell\to\infty} p = 0, \tag{5.145}$$

and it is the rest. For $\alpha = 0$ and fixed scale ℓ there are two singular values of the momentum p. For all $\ell \neq 0$ and $\alpha \neq 0$, the case of non-zero p is related to the existence of a relativistic particle.

A different situation arises when particle mass vanishes identically, *i.e.* $m \equiv 0$. In such instance one sees that the constraint (5.113) takes the form

$$E^2 + \frac{\hbar^2 c^2}{4\alpha\ell^2} = \alpha\left(\frac{c\ell}{\hbar}\right)^2\left(p^2 + \frac{\hbar^2}{2\alpha\ell^2}\right)^2, \tag{5.146}$$

and applying the non-trivial identification

$$E^2 + \frac{\hbar^2 c^2}{4\alpha\ell^2} = 0, \tag{5.147}$$

$$p^2 + \frac{\hbar^2}{2\alpha\ell^2} = 0, \tag{5.148}$$

can be solved as the tachyonic case

$$E^2 = -\frac{\hbar^2 c^2}{4\alpha\ell^2}, \tag{5.149}$$

$$p^2 = -\frac{\hbar^2}{2\alpha\ell^2}, \tag{5.150}$$

which can be combined into a one constraint

$$E^2 = \frac{1}{2}p^2c^2, \tag{5.151}$$

which can be solved immediately

$$E = \pm\frac{1}{\sqrt{2}}pc, \tag{5.152}$$

The equation (5.152) differs from the usual Special Relativity condition for massless particles, *i.e.* $E = pc$. Such a difference is the reflection of the fact that the Snyder–Sidharth deformation is an algebraic deformation in the Minkowski energy-momentum space. It can be seen straightforwardly from the relations (5.149) and (5.150) that the energy and momentum of the massless particles are purely imaginary quantities and, therefore, such a situation corresponds to a massless fermionic tachyon.

B2 Bosons

Fermions obey the Dirac equation which is the square root of the Klein–Gordon equation governing bosons. Therefore, fermions are approximation and bosons are fundamental. Previous section results work for fermions when $\alpha > 0$ and for $\alpha < 0$ are true for boson, *i.e.* the case of bosons can be received by a simple exchange

$$\alpha \longrightarrow -|\alpha|, \tag{5.153}$$

within mass and energy formulas. The basic relation (5.113) then is

$$E^2 - \frac{\hbar^2 c^2}{4|\alpha|\ell^2} = -|\alpha|\left(\frac{c\ell}{\hbar}\right)^2\left(p^2 - \frac{\hbar^2}{2|\alpha|\ell^2}\right)^2 + m^2c^4. \tag{5.154}$$

1. In the first case one has the system of equations

$$\begin{cases} m^2c^4 = -\dfrac{\hbar^2 c^2}{4|\alpha|\ell^2} \\ E^2 = -|\alpha|\left(\dfrac{c\ell}{\hbar}\right)^2\left(p^2 - \dfrac{\hbar^2}{2|\alpha|\ell^2}\right)^2 \end{cases} \tag{5.155}$$

which defines the tachyon with mass and energy

$$m = i\frac{\hbar}{2\sqrt{|\alpha|}c\ell}, \tag{5.156}$$

$$E = \frac{p^2}{2m} + mc^2. \tag{5.157}$$

2. The second case is the system of equations

$$\begin{cases} m^2c^4 = |\alpha|\left(\dfrac{c\ell}{\hbar}\right)^2\left(p^2 - \dfrac{\hbar^2}{2|\alpha|\ell^2}\right)^2 \\ E^2 = \dfrac{\hbar^2 c^2}{4|\alpha|\ell^2} \end{cases}, \tag{5.158}$$

which can be solved by

$$mc^2 = \frac{p^2c^2}{2E} - E, \tag{5.159}$$

$$E = \frac{\hbar c}{2\sqrt{|\alpha|}\ell}. \tag{5.160}$$

However, the relation (5.159) in itself is the quadratic equation

$$E^2 = -mc^2E + \frac{p^2c^2}{2}, \tag{5.161}$$

which, by using (8.20) and (5.153), is the constraint

$$E = -mc^2 - \frac{p^2}{2m} + |\alpha|\left(\frac{\ell}{\hbar}\right)^2\frac{p^4}{m}, \tag{5.162}$$

similar to the Pauli Hamiltonian constraint, and can be easily factorized

$$E - \frac{\hbar^2}{16|\alpha|\ell^2}\frac{1}{m} = -|\alpha|\left(\frac{\ell}{\hbar}\right)^2\frac{1}{m}\left(p^2 - \frac{\hbar^2}{4|\alpha|\ell^2}\right)^2 + mc^2, \tag{5.163}$$

and considered by three ways.

(a) In the first one we have the system

$$\begin{cases} mc^2 = -\dfrac{\hbar^2}{16|\alpha|\ell^2}\dfrac{1}{m} \\ E = -|\alpha|\left(\dfrac{\ell}{\hbar}\right)^2\dfrac{1}{m}\left(p^2 - \dfrac{\hbar^2}{4|\alpha|\ell^2}\right)^2 \end{cases} \tag{5.164}$$

which can be solved straightforwardly

$$m = i\frac{\hbar}{4\sqrt{|\alpha|}c\ell}, \tag{5.165}$$

$$E = mc^2 + \frac{p^2}{2m} + \frac{p^4}{16m^3c^2}. \tag{5.166}$$

and by application of the mass (5.165) the energy (5.166) becomes

$$E = i\frac{\hbar c}{4\sqrt{|\alpha|}\ell} - i2c\sqrt{|\alpha|}\frac{\ell}{\hbar}p^2 + i4c\left(\sqrt{|\alpha|}\frac{\ell}{\hbar}\right)^3 p^4. \tag{5.167}$$

Combining of the energy (5.160) and the mass (5.165) leads to the simple relation

$$iE = 2mc^2, \tag{5.168}$$

which means that the square of mass has manifestly negative value for the real value of energy

$$m^2 = -\frac{E^2}{2c^4} < 0, \tag{5.169}$$

i.e. such a situation describes a tachyon. Applying the first equation in the system (5.158) together with the mass relation (5.165) allows to establish the values of the monentum

$$p^2 = \frac{\hbar^2}{2|\alpha|\ell^2}\left(1 + \frac{1}{2}i\right), \tag{5.170}$$

which, after using the fact that

$$\sqrt{-1 + \frac{1}{2}i} = \frac{1}{\sqrt{2}}\left(\sqrt{\frac{\sqrt{5}}{2} + 1} + i\sqrt{\frac{\sqrt{5}}{2} - 1}\right)e^{ni\pi}, \tag{5.171}$$

where $n \in \mathbb{Z}$, leads to the momentum spectrum

$$p_n = \pm\frac{(-1)^n}{2}\left(\sqrt{\frac{\sqrt{5}}{2}+1}+i\sqrt{\frac{\sqrt{5}}{2}-1}\right)\frac{\hbar}{\sqrt{|\alpha|}\ell}. \tag{5.172}$$

Interestingly, writing (5.134) as $p_n(|\alpha|)$ and (5.172) as $p_n(-|\alpha|)$, one can define the spectral mean

$$\langle p_n(|\alpha|)p_n(-|\alpha|)\rangle = \frac{1}{|\alpha|-|\alpha_0|}\int_{|\alpha_0|}^{|\alpha|} p_n(x)p_n(-x)dx, \tag{5.173}$$

which is easy to derive

$$\langle p_n(|\alpha|)p_n(-|\alpha|)\rangle = \pm i\frac{\sqrt{5}}{4}\frac{\hbar^2}{\ell^2}\frac{\ln\left|\dfrac{\alpha}{\alpha_0}\right|}{|\alpha|-|\alpha_0|}, \tag{5.174}$$

and it is not difficult to see that

$$\lim_{|\alpha|\to|\alpha_0|}\langle p_n(|\alpha|)p_n(-|\alpha|)\rangle = \pm\sqrt{5}\frac{\hbar}{2\sqrt{|\alpha_0|}\ell}\frac{\hbar}{2\sqrt{-|\alpha_0|}\ell}. \tag{5.175}$$

Now it is not difficult to establish also the spectral means

$$\langle |p_n(|\alpha|)|^2\rangle = \frac{1}{|\alpha|-|\alpha_0|}\int_{|\alpha_0|}^{|\alpha|} |p_n(x)|^2 dx, \tag{5.176}$$

$$\langle |p_n(-|\alpha|)|^2\rangle = \frac{1}{|\alpha|-|\alpha_0|}\int_{|\alpha_0|}^{|\alpha|} |p_n(-x)|^2 dx, \tag{5.177}$$

which are equal to

$$\langle |p_n(|\alpha|)|^2\rangle = \frac{\sqrt{5}}{4}\frac{\hbar^2}{\ell^2}\frac{\ln\left|\dfrac{\alpha}{\alpha_0}\right|}{|\alpha|-|\alpha_0|}, \tag{5.178}$$

$$\langle |p_n(-|\alpha|)|^2\rangle = \frac{\sqrt{5}}{4}\frac{\hbar^2}{\ell^2}\frac{\ln\left|\dfrac{\alpha}{\alpha_0}\right|}{|\alpha|-|\alpha_0|}, \tag{5.179}$$

and have the limiting values

$$\lim_{|\alpha|\to|\alpha_0|}\langle |p_n(|\alpha|)|^2\rangle = \frac{\sqrt{5}}{4}\frac{\hbar^2}{|\alpha_0|\ell^2}, \tag{5.180}$$

$$\lim_{|\alpha|\to|\alpha_0|}\langle |p_n(-|\alpha|)|^2\rangle = \frac{\sqrt{5}}{4}\frac{\hbar^2}{|\alpha_0|\ell^2}. \tag{5.181}$$

The formulas (5.178)-(5.179) and (5.174) allow to construct the variance

$$\sigma^2 = \langle |p_n(|\alpha|)|^2\rangle\langle |p_n(-|\alpha|)|^2\rangle - |\langle p_n(|\alpha|)p_n(-|\alpha|)\rangle|^2, \quad (5.182)$$

which, in general, vanishes

$$\sigma^2 = 0, \quad (5.183)$$

as well as in the limiting case

$$\lim_{|\alpha|\to|\alpha_0|} \sigma^2 = 0. \quad (5.184)$$

One can compute also the following means

$$\langle p_n(|\alpha|)\rangle = \frac{1}{|\alpha|-|\alpha_0|}\int_{|\alpha_0|}^{|\alpha|} p_n(x)dx, \quad (5.185)$$

$$\langle p_n(-|\alpha|)\rangle = \frac{1}{|\alpha|-|\alpha_0|}\int_{|\alpha_0|}^{|\alpha|} p_n(-x)dx, \quad (5.186)$$

and obtain the quantity

$$\langle p_n(|\alpha|)\rangle\langle p_n(-|\alpha|)\rangle = \pm i\sqrt{5}\frac{\hbar^2}{\ell^2}\frac{1}{\left(\sqrt{|\alpha|}+\sqrt{|\alpha_0|}\right)^2}, \quad (5.187)$$

which, together with (5.174), allows to establish the variance

$$\sigma^2 = \langle p_n(|\alpha|)\rangle\langle p_n(-|\alpha|)\rangle - \langle p_n(|\alpha|)p_n(-|\alpha|)\rangle, \quad (5.188)$$

with the final result

$$\sigma^2 = \pm i\frac{\sqrt{5}}{4}\frac{\hbar^2}{\ell^2}\left[\frac{4}{\left(\sqrt{|\alpha|}+\sqrt{|\alpha_0|}\right)^2} - \frac{\ln\left|\frac{\alpha}{\alpha_0}\right|}{|\alpha|-|\alpha_0|}\right]. \quad (5.189)$$

It is easy to derive now the limiting case of (5.190)

$$\lim_{|\alpha|\to|\alpha_0|} \sigma^2 = i0. \quad (5.190)$$

(b) The latter subcase is

$$\begin{cases} mc^2 = E \\ \dfrac{\hbar^2}{16\alpha\ell^2}\dfrac{1}{m} = \alpha\left(\dfrac{\ell}{\hbar}\right)^2\dfrac{1}{m}\left(p^2+\dfrac{\hbar^2}{4\alpha\ell^2}\right)^2 \end{cases} \tag{5.191}$$

Again, solution of the first equation by using (5.118) is rather simple

$$m = -i\frac{\hbar}{2\sqrt{\alpha}c\ell}, \quad m^2 < 0, \tag{5.192}$$

and once again the tachyon is obtained - there are particles characterize by the momentum spectrum

$$p = \left\{0, \pm\frac{\hbar}{\sqrt{2\alpha}\ell}\right\}. \tag{5.193}$$

This is a discrete momentum spectrum for a fixed scale ℓ. For the running scale this is non-compact spectrum, but compactification to the point is done in the large scale limit

$$\lim_{\ell\to\infty} p = 0, \tag{5.194}$$

and it is the rest. For $\alpha = 0$ and fixed scale ℓ there are two singular values of the momentum p. For all $\ell \neq 0$ and $\alpha \neq 0$, the case of non-zero p is related to the existence of a tachyon.

(c) The third interpretation yields

$$\begin{cases} -E = \dfrac{\hbar^2}{16\alpha\ell^2}\dfrac{1}{m} \\ -mc^2 = \alpha\left(\dfrac{\ell}{\hbar}\right)^2\dfrac{1}{m}\left(p^2+\dfrac{\hbar^2}{4\alpha\ell^2}\right)^2 \end{cases}. \tag{5.195}$$

Similarly, as in the previous subcase, employing the relation (5.118) one obtains from the first equation the mass

$$m = -i\frac{\hbar}{8\sqrt{\alpha}c\ell}, \tag{5.196}$$

and again one has a tachyon, *i.e.* $m^2 < 0$. Consequently, the momentum spectrum can be established as

$$p = \left\{ \pm i\sqrt{\frac{1}{8\alpha}}\frac{\hbar}{\ell}, \pm i\sqrt{\frac{3}{8\alpha}}\frac{\hbar}{\ell} \right\}, \tag{5.197}$$

and is non-hermitian, as one has expected. For this reason this particular case, that is the Pauli Hamiltonian constraint with mass related to the minimum scale, describes the tachyon, the hypothetical particles with velocity faster then light.

3. The third possible solution of the Snyder–Sidharth Hamiltonian constraint can be constructed by the system of equations

$$\begin{cases} E^2 = m^2c^4 \\ \dfrac{\hbar^2c^2}{4\alpha\ell^2} = \alpha\left(\dfrac{c\ell}{\hbar}\right)^2\left(p^2 + \dfrac{\hbar^2}{2\alpha\ell^2}\right)^2 \end{cases}. \tag{5.198}$$

The first equation gives the standard Einstein mass-energy relation

$$E = mc^2, \tag{5.199}$$

and the latter equality results in the discrete momentum spectrum

$$p = \left\{0, \pm\frac{1}{\sqrt{\alpha}}\frac{\hbar}{\ell}\right\}, \tag{5.200}$$

for the fixed scale ℓ value. For the running scale, this spectrum is non-compact, but in the large scale limit the spectrum is manifestly compactified to the point

$$\lim_{\ell\to\infty} p = 0, \tag{5.201}$$

and it is the rest. For $\alpha = 0$ and fixed scale ℓ there are two singular values of the momentum p. For all $\ell \neq 0$ and $\alpha \neq 0$, the case of non-zero p is related to the existence of a relativistic particle.

C Modified Compton Effect

In particle astrophysics there is great interest in the various situations between light and particles, and actually plenty of intriguing conclusions arise from an analysis of this type of phenomenon from both the theoretical and the experimental points of view (Cf. *e.g.* general books in Ref. [324]), which give physical information about the various aspects of the observations. One such phenomenon is the Compton scattering, discovered by Arthur Holly Compton in 1923 [325], which is usually classified as a mid-energy interaction of light and matter. The Compton scattering is realized by the electromagnetic radiation, X-rays and gamma (γ) rays, coming through a material objects which consist of electrons. A decrease in photon energy or photon wavelength due to the inelastic scattering is the phenomenon known as the Compton effect. We shall not discuss in this book the detailed classical analysis of the Compton scattering, because actually the analysis is based on the framework involving the law of conservation of energy, the law of conservation of momentum, and certain concepts of Einstein's Special Relativity. Let us consider the case of the Compton effect within the framework employing the Snyder–Sidharth Hamiltonian constraint (6.65). Such an analysis can be performed, however, in three different routes minimally. In the case of lack of deformation all the ways lead to the same result, called the Compton equation. However, in the presence of deformation due to the non-commutative geometry of space-time one obtains three different results. It is not clear which of the formulations is correct and also, in general, other alternative formulations are not known. Let us consider the three approaches step by step.

C1 Relativistic Approach

Our approach is based on the standard Special Relativity formulation (Cf. *e.g.* Ref. [326]). Let us consider a photon and an electron having the initial momentum four-vectors

$$p_\gamma^\mu = [E_\gamma \,,\, p_\gamma^i c], \tag{5.202}$$

$$p_e^\mu = [E_0 \,,\, 0], \tag{5.203}$$

where $E_0 = m_e c^2$ is the electron rest energy. An electron at rest is scattered by an incoming photon, and an outgoing photon is observed under the scattering angle θ relatively to the incident direction of an incoming photon. The final momentum four-vectors are

$$p^{\mu}_{\gamma'} = [E_{\gamma'}\ ,\ p^{i}_{\gamma'}c], \tag{5.204}$$

$$p^{\mu}_{e'} = [E_{e}\ ,\ p^{i}_{e}c]. \tag{5.205}$$

Let us preserve unchanged the Planck–Einstein relations of the wave-particle duality. Thus, for a photon one has

$$E_{\gamma} = \hbar\omega, \tag{5.206}$$

$$p^{i}_{\gamma} = \hbar k^{i}, \tag{5.207}$$

where ω and k_{γ} are angular frequency and wave vector, and the value of wave vector of a photon is

$$k = \sqrt{k_i k^i} = \frac{2\pi}{\lambda} = \frac{1}{\lambda\!\!\!^{-}}, \tag{5.208}$$

where $\lambda\!\!\!^{-}$ is the reduced wavelength of a photon. Similarly, for a matter particle, including electrons, one has

$$E_{e} = \hbar\omega_{e}, \tag{5.209}$$

$$p^{i}_{e} = \hbar k^{i}_{e}, \tag{5.210}$$

where ω_e and k_e are angular frequency and wave vector of an electron, and the value of the wave vector of an electron is

$$k_e = \sqrt{k_{e i} k^{i}_{e}} = \frac{2\pi}{\lambda_e} = \frac{1}{\lambda\!\!\!^{-}_e}, \tag{5.211}$$

where $\lambda\!\!\!^{-}_e$ is the reduced wavelength of an electron.

In addition, we assume that Special Relativity is deformed due to the non-commutative geometry of the Snyder quantized space-time, *i.e.* that for any photon γ and any electron e the Hamiltonian constraints following from the Snyder–Sidharth deformation are

$$E^2_{\gamma} = p^2_{\gamma}c^2 + \frac{1}{\varepsilon^2}\left(p^2_{\gamma}c^2\right)^2, \tag{5.212}$$

$$E^2_{e} = E^2_0 + p^2_{e}c^2 + \frac{1}{\varepsilon^2}\left(p^2_{e}c^2\right)^2, \tag{5.213}$$

where $p_e = \sqrt{p_{ei}p_e^i}$ and $p_\gamma = \sqrt{p_{\gamma_i}p_\gamma{}^i}$ are values of momenta of an electron and a photon, respectively, and for shortened notation we have introduced the following energy parameter

$$\varepsilon = \frac{\hbar c}{\sqrt{\alpha}\ell} = \frac{\hbar c}{\ell_P}\frac{1}{\sqrt{\alpha}}\frac{\ell_P}{\ell} = E_P\frac{1}{\sqrt{\alpha}}\frac{\ell_P}{\ell}, \tag{5.214}$$

where $E_P = \sqrt{\frac{\hbar c^5}{G}}$ is the Planck energy. By application of the wave-particle duality the Snyder–Sidharth Hamiltonian constraints for a photon and an electron can be presented as

$$\frac{\omega^2}{c^2} = k^2 + \frac{k^4}{\kappa^2}, \tag{5.215}$$

$$\frac{\omega_e^2}{c^2} = k_C^2 + k_e^2 + \frac{k_e^4}{\kappa^2}, \tag{5.216}$$

where k_C is the wave vector

$$k_C = \frac{2\pi}{\lambda_C} = \frac{1}{\lambda\!\!\!^{-}_C}, \tag{5.217}$$

related to the Compton wavelength λ_C of an electron, and its reduced form

$$\lambda\!\!\!^{-}_C = \frac{\hbar c}{E_0} = \frac{\hbar}{m_e c}, \tag{5.218}$$

and κ is the parameter

$$\kappa = \frac{\varepsilon}{\hbar c} = \frac{1}{\sqrt{\alpha}\ell}, \tag{5.219}$$

which can be interpreted as a wave vector value associated to the (non-reduced!) wavelength $\lambda_\kappa = 2\pi\sqrt{\alpha}\ell$, which for

$$\alpha = \frac{1}{(2\pi)^2}, \tag{5.220}$$

becomes $\lambda_\kappa \equiv \ell$. The parameter (5.214) can be physically interpreted as the Æther energy, so κ is the wave vector value of the Æther wave-particle.

We are interested in the relation for the angle θ of the scattered photon, *i.e.* the angle between the wave vectors k_i and k_i'. Let us apply first the law of conservation of energy-momentum, *i.e.*

$$p_\gamma^\mu + p_e^\mu = p_{\gamma'}^\mu + p_{e'}^\mu. \tag{5.221}$$

For the momentum this principle gives

$$p_\gamma^i c = p_{\gamma'}^i c + p_e^i c, \tag{5.222}$$

or in terms of the wave vectors

$$k^i = k'^i + k_e^i, \tag{5.223}$$

and by elementary algebraic operations leads to the result

$$k_e^2 = k^2 + k'^2 - 2kk'\cos\theta. \tag{5.224}$$

Next one can use the law of conservation of energy which, in the light of (5.221), for the present case has the following form

$$E_\gamma + E_0 = E_{\gamma'} + E_e, \tag{5.225}$$

and for this reason one obtains

$$\frac{\omega_e}{c} = k_C + \frac{\omega}{c} - \frac{\omega'}{c}. \tag{5.226}$$

Taking the square of both sides of the equation (5.226)

$$\frac{\omega_e^2}{c^2} = k_C^2 + \frac{\omega^2}{c^2} + \frac{\omega'^2}{c^2} - 2\frac{\omega}{c}\frac{\omega'}{c} + 2k_C\left(\frac{\omega}{c} - \frac{\omega'}{c}\right), \tag{5.227}$$

and applying the Snyder–Sidharth Hamiltonian constraint for photons and an electron, one obtains the relation

$$\begin{aligned} k_C^2 + k_e^2 + \frac{k_e^4}{\kappa^2} \quad &= \quad k_C^2 + k^2 + \frac{k^4}{\kappa^2} + k'^2 + \frac{k'^4}{\kappa^2} - 2\sqrt{k^2 + \frac{k^4}{\kappa^2}}\sqrt{k'^2 + \frac{k'^4}{\kappa^2}} + \\ & 2k_C\left(\sqrt{k^2 + \frac{k^4}{\kappa^2}} - \sqrt{k'^2 + \frac{k'^4}{\kappa^2}}\right), \end{aligned} \tag{5.228}$$

or after a small reduction

$$k_e^2+\frac{k_e^4}{\kappa^2} = k^2+k'^2+\frac{k^4}{\kappa^2}+\frac{k'^4}{\kappa^2}-2\sqrt{k^2+\frac{k^4}{\kappa^2}}\sqrt{k'^2+\frac{k'^4}{\kappa^2}}+ 2k_C\left(\sqrt{k^2+\frac{k^4}{\kappa^2}}-\sqrt{k'^2+\frac{k'^4}{\kappa^2}}\right), \tag{5.229}$$

Applying the relation (5.224) within the left hand side of the equation (5.229), some algebraic identities, and factorization one obtains

$$-\cos\theta+\frac{\kappa^2}{2kk'}\frac{\left(k^2+k'^2\right)^2}{\kappa^4}-2\frac{k^2+k'^2}{\kappa^2}\cos\theta+\frac{2kk'}{\kappa^2}\cos^2\theta = \frac{\kappa^2}{2kk'}\frac{k^4+k'^4}{\kappa^4}-\sqrt{1+\frac{k^2}{\kappa^2}}\sqrt{1+\frac{k'^2}{\kappa^2}}+k_C\left(\frac{1}{k'}\sqrt{1+\frac{k^2}{\kappa^2}}-\frac{1}{k}\sqrt{1+\frac{k'^2}{\kappa^2}}\right). \tag{5.230}$$

Now by ordering in $\cos\theta$ powers one receives the quadratic equation

$$\beta_\kappa\cos^2\theta-\delta_\kappa\cos\theta+\eta_\kappa=0, \tag{5.231}$$

where we have introduced the notation

$$\beta_\kappa=\frac{2kk'}{\kappa^2}, \tag{5.232}$$

$$\delta_\kappa=1+2\frac{k^2+k'^2}{\kappa^2}, \tag{5.233}$$

$$\eta_\kappa=\frac{kk'}{\kappa^2}+\sqrt{1+\frac{k^2}{\kappa^2}}\sqrt{1+\frac{k'^2}{\kappa^2}}-k_C\left(\frac{1}{k'}\sqrt{1+\frac{k^2}{\kappa^2}}-\frac{1}{k}\sqrt{1+\frac{k'^2}{\kappa^2}}\right). \tag{5.234}$$

First of all, one sees straightforwardly that in the Special Relativity limit $\ell\to0$, *i.e.* $\kappa\to\infty$, the coefficients (5.232), (5.233), and (5.234) read

$$\beta_\infty=0, \tag{5.235}$$

$$\delta_\infty=1, \tag{5.236}$$

$$\eta_\infty=1-k_C\left(\frac{1}{k'}-\frac{1}{k}\right), \tag{5.237}$$

so that the modified Compton equation (5.231) becomes linear in the cosinus

$$-\cos\theta + 1 - k_C\left(\frac{1}{k'} - \frac{1}{k}\right) = 0, \tag{5.238}$$

and its solution can be presented in the form

$$\lambda\!\!\!^{-}_C(1-\cos\theta) = \frac{1}{k'} - \frac{1}{k}, \tag{5.239}$$

called the Compton equation, describing the photon-electron scattering in the framework of Special Relativity, as originally deduced by Arthur Holly Compton. Because the boundaries $-1 \leq \cos\theta \leq 1$ hold, one obtains that in the Compton effect the non-trivial restriction

$$\lambda \leq \lambda' \leq \lambda + 2\lambda_C, \tag{5.240}$$

for the outgoing photon wavelength, holds.

C2 Relativistic Limit. Lensing Hypothesis

However, in general the Snyder–Sidharth deformation results in the modified Compton equation (5.231) which has solutions different from the Compton equation (5.239). By straightforward easy computation one obtains formally two mathematically and physically different solutions of the equation (5.231)

$$\cos\theta = \frac{1}{2\beta_\kappa}\left(\delta_\kappa \pm \sqrt{\delta_\kappa^2 - 4\beta_\kappa\eta_\kappa}\right), \tag{5.241}$$

and in this way one can determine the exact formulas for the cosinus

$$\cos\theta = \left[1 \pm \sqrt{g^2\left(\frac{k}{\kappa}, \frac{k'}{\kappa}\right) + \frac{2k_C}{\kappa} g\left(\frac{k}{\kappa}, \frac{k'}{\kappa}\right) + 1}\,\right]\frac{\kappa^2}{4kk'} + \frac{1}{2}\left(\frac{k}{k'} + \frac{k'}{k}\right), \tag{5.242}$$

where we have introduced the dimensionless quantity

$$g\left(\frac{k}{\kappa}, \frac{k'}{\kappa}\right) = \frac{2E_\gamma}{\kappa} - \frac{2E_{\gamma'}}{\kappa}, \tag{5.243}$$

where we have used the Snyder–Sidharth Hamiltonian constraint for photons (5.212). In the light of the law of conservation of energy (5.225) one sees that the function (5.243) in fact can be completely expressed through energies of rest and scattered electrons

$$g\left(\frac{k}{\kappa},\frac{k'}{\kappa}\right)=\frac{2(E_{\gamma}-E_{\gamma'})}{\kappa}=\frac{2(E_e-E_0)}{\kappa}, \tag{5.244}$$

where E_e is given by the Snyder–Sidharth Hamiltonian constraint for electrons (5.213). It means that the function (5.243) is always positively defined $g\left(\frac{k}{\kappa},\frac{k'}{\kappa}\right)\geq 0$. It can be seen straightforwardly that if one exchanges momenta of photons $k'\leftrightarrow k$ then the function (5.243) changes the sign $g\left(\frac{k}{\kappa},\frac{k'}{\kappa}\right)\to -g\left(\frac{k}{\kappa},\frac{k'}{\kappa}\right)$, *i.e.* $g\left(\frac{k}{\kappa},\frac{k'}{\kappa}\right)$ is an odd function with respect to such an exchange. Therefore, the presence of the square root term in the cosinus formula (5.242) breaks crossing symmetry manifestly. This fact contradicts the nature of cosinus which is an even function. It results in the conclusion that the presence of minimal scale breaks crossing symmetry. Moreover, in general, the cosinus (5.242) does not belong to the range of cosinus $[-1,1]$, but we shall discuss and solve this problematic point later.

In the relativistic limit $\kappa\to\infty$ the Eqs. (5.242) behave like

$$\cos\theta=(1\pm 1)\frac{\kappa^2}{4kk'}+\frac{1}{2}\left(\frac{k}{k'}+\frac{k'}{k}\right). \tag{5.245}$$

For the minus sign case the second term in (5.245) vanishes as $\kappa\to\infty$, while the plus sign case leads to non renormalizable divergence. In other words, the physical solution is the solution with the minus sign, and when one applies the limit $\kappa\to\infty$ it gives the finite result

$$\lambda_C(1-\cos\theta)=\frac{1}{k_C}\left(1-\frac{1}{2}\left(\frac{k}{k'}+\frac{k'}{k}\right)\right), \tag{5.246}$$

which blatantly differs from the Compton equation (5.239). It means that despite using the relativistic limit to the deformed scattering the received result does not reconstruct the relativistic scattering.

Anyway, for conceptual correctness one can compare *ad hoc* the equation (5.246) with the Compton equation (5.239)

$$\frac{1}{k'}-\frac{1}{k}=\frac{1}{k_C}\left(1-\frac{1}{2}\left(\frac{k}{k'}+\frac{k'}{k}\right)\right), \tag{5.247}$$

and easily obtain the condition for which the relativistic limit agrees with the Compton equation

$$k'^2 - 2(k_C + k)k' + k^2 + 2k_C k = 0. \tag{5.248}$$

This condition is satisfied for two cases $k' = k_C + k \pm k_C$, *i.e.*

$$k' - k = 2k_C, \tag{5.249}$$

$$k' - k = 0, \tag{5.250}$$

which can be expressed through incoming and outgoing photons wavelengths

$$\frac{1}{\lambda'} - \frac{1}{\lambda} = \frac{2}{\lambda_C}, \tag{5.251}$$

$$\frac{1}{\lambda'} - \frac{1}{\lambda} = 0. \tag{5.252}$$

Interestingly, the first solution (5.251) looks like a particular case of the lensmaker's equation,

$$\frac{1}{r_1} - \frac{1}{r_2} + \left(1 - \frac{1}{n}\right)\frac{\delta}{r_1 r_2} = \frac{1}{(n-1)f}, \tag{5.253}$$

for the refractive index $n = 3/2 = c/v$, focal length of the lens $f = \lambda_C$, and lens radii of curvature $r_1 = \lambda'$ and $r_2 = \lambda$, and the thickness of the lens $\delta = 0$, *i.e.* the thin lens created by the medium in which speed of light is $v = 2c/3$. The second solution (5.252) corresponds with the case of telescopic lens case $f = \infty$ in such a medium, but because photon wavelength does not change in such a situation it is justified to consider this solution as a non-physical one.

In this way one can take into account the *ad hoc* generalization

$$\frac{\omega'}{c} - \frac{\omega}{c} + \delta\frac{n-1}{n}\frac{\omega}{c}\frac{\omega'}{c} = \frac{k_C}{n-1}, \tag{5.254}$$

which in the relativistic limit $\kappa \to \infty$ leads to

$$k' - k + \delta\frac{n-1}{n}kk' = \frac{k_C}{n-1}, \tag{5.255}$$

i.e. with $k = 1/\lambda\!\!\!^{-} = \dfrac{2\pi}{\lambda}$, $\delta = 0$, and $n = 3/2$ reconstructs the first solution (5.251). The problem is to establish the linkage between the thickness d and minimal scale ℓ and, in general, $n > 1$ as a numerical coefficient. We shall call (5.254) *the lensing hypothesis*.

C3 Modified Compton Equation

Let us construct straightforwardly the general solution of the modified Compton equation (5.231). Because $\delta_\kappa \neq 0$ always holds, let us extract $\cos\theta$ manifestly from the Eq. (5.231)

$$\cos\theta = \frac{\beta_\kappa}{\delta_\kappa}\cos^2\theta + \frac{\eta_\kappa}{\delta_\kappa}. \tag{5.256}$$

Then however, one must involve the fact that $\cos\theta$ is a bounded function, *i.e.* $-1 \leq \cos\theta \leq 1$. Application of these boundaries to (5.256) leads to the following restrictions for $\cos^2\theta$

$$-\frac{\delta_\kappa + \eta_\kappa}{\beta_\kappa} \leq \cos^2\theta \leq \frac{\delta_\kappa - \eta_\kappa}{\beta_\kappa}. \tag{5.257}$$

The equation (5.257), however, must be considered with taking into account the fact $\cos^2\theta$ is naturally bounded to $0 \leq \cos^2\theta \leq 1$. Employment of these boundary values results in the system of constraints

$$-\frac{\delta_\kappa + \eta_\kappa}{\beta_\kappa} \equiv 0, \tag{5.258}$$

$$\frac{\delta_\kappa - \eta_\kappa}{\beta_\kappa} \equiv 1, \tag{5.259}$$

which factually reduces the momentum space of outgoing and incoming photons. In other words, we consider the equation (5.231) as the basic relation for the scattering, but not its solution which is a secondary result. Let us see the consequences of such an approach.

For $\beta_\kappa \neq 0$, *i.e.* factually for non-zero values of incoming and outgoing photon momenta, the reduction is given by the equivalent conditions

$$\delta_\kappa + \eta_\kappa = 0, \tag{5.260}$$

$$\delta_\kappa - \eta_\kappa = \beta_\kappa, \tag{5.261}$$

which are not difficult to solve. The results are

$$\delta_\kappa = \frac{1}{2}\beta_\kappa, \tag{5.262}$$

$$\eta_\kappa = -\frac{1}{2}\beta_\kappa. \tag{5.263}$$

It is easy to see that the first condition says that

$$2\frac{k'^2+k^2}{\kappa^2}-\frac{kk'}{\kappa^2}+1=0, \tag{5.264}$$

while the second constraint can be written as

$$2\frac{kk'}{\kappa^2}+\sqrt{1+\frac{k^2}{\kappa^2}}\sqrt{1+\frac{k'^2}{\kappa^2}}-k_C\left(\frac{1}{k'}\sqrt{1+\frac{k^2}{\kappa^2}}-\frac{1}{k}\sqrt{1+\frac{k'^2}{\kappa^2}}\right)=0. \tag{5.265}$$

Application of (5.264) and (5.265) allows to reduce the equation (5.231)

$$\cos^2\theta-\frac{1}{2}\cos\theta-\frac{1}{2}=0. \tag{5.266}$$

Because (5.238) does not contain any wave vectors, therefore, its solutions are constant and independent of photon wavelengths. In the usual Compton effect the cosinus depends on the photon wavelengths. In the modified case the cosinus (5.242), in general, does not satisfy the cosinus variability range $[-1,1]$, and application of these limits to the equation (5.231) led us to the bounds (5.264)-(5.265), and this resulted in the equation (5.266). In this way the Compton effect modified due to the Snyder–Sidharth Hamiltonian constraint is solved by a constant scattering angle which solves the equation (5.266), and the values of the wave vectors k and k' following from the bounds (5.264)-(5.265). The constant angle values are easy to derive

$$\cos\theta \;=\; 1\longrightarrow\theta=2n\pi, \tag{5.267}$$

$$\cos\theta \;=\; -\frac{1}{2}\longrightarrow\theta=\pm\frac{\pi}{3}+2n\pi, \tag{5.268}$$

where $n\in\mathbb{Z}$. The first solution (5.267) means no scattering or backward scattering. Solutions of the system (5.264)-(5.265) are not easy to extract. With no additional constraint(s) between k and k' this system leads to a polynomial equation of more than 40 degree which must be treated by complicated methods of the Galois group.

However, one can use the lensing hypothesis (5.254) as the additional constraint which written out explicitly is

$$\sqrt{k'^2+\frac{k'^4}{\kappa^2}}-\sqrt{k^2+\frac{k^4}{\kappa^2}}+\delta\frac{n-1}{n}\sqrt{k^2+\frac{k^4}{\kappa^2}}\sqrt{k'^2+\frac{k'^4}{\kappa^2}}=\frac{k_C}{n-1}, \tag{5.269}$$

which after elementary algebraic manipulations takes the form

$$k_C\left(\frac{1}{k'}\sqrt{1+\frac{k^2}{\kappa^2}}-\frac{1}{k}\sqrt{1+\frac{k'^2}{\kappa^2}}\right)=k_C\delta\frac{n-1}{n}\sqrt{1+\frac{k^2}{\kappa^2}}\sqrt{1+\frac{k'^2}{\kappa^2}}-\frac{\kappa^2}{kk'}\frac{1}{n-1}\frac{k_C^2}{\kappa^2}. \tag{5.270}$$

In other words, using the additional constraint (5.270) within the constraint (5.265) gives the result

$$2\frac{kk'}{\kappa^2}+\left(1+k_C\delta\frac{n-1}{n}\right)\sqrt{1+\frac{k^2}{\kappa^2}}\sqrt{1+\frac{k'^2}{\kappa^2}}-\frac{\kappa^2}{kk'}\frac{1}{n-1}\frac{k_C^2}{\kappa^2}=0, \tag{5.271}$$

which can be presented as

$$\sqrt{1+\frac{k^2+k'^2}{\kappa^2}+\left(\frac{kk'}{\kappa^2}\right)^2}=\frac{n}{(1+k_C\delta)n-k_C\delta}\left(\frac{1}{n-1}\frac{k_C^2}{\kappa^2}\frac{\kappa^2}{kk'}-2\frac{kk'}{\kappa^2}\right). \tag{5.272}$$

Now one can apply the constraint (5.264) to the left hand side of the equation (5.272). It gives

$$\sqrt{\frac{1}{2}+\frac{1}{2}\frac{kk'}{\kappa^2}+\left(\frac{kk'}{\kappa^2}\right)^2}=\frac{n}{(1+k_C\delta)n-k_C\delta}\left(\frac{1}{n-1}\frac{k_C^2}{\kappa^2}\frac{\kappa^2}{kk'}-2\frac{kk'}{\kappa^2}\right). \tag{5.273}$$

By squaring both sides of the equation (5.273), doing very few elementary algebraic manipulations, and grouping the result with respect to powers of $x=\frac{kk'}{\kappa^2}$ one obtains finally the condition

$$Ax^4+Bx^3+Cx^2-D=0 \tag{5.274}$$

where we have introduced the coefficients

$$A = 2\frac{(n-1)^3}{n}k_C\delta\left(2+\frac{n-1}{n}k_C\delta\right), \tag{5.275}$$

$$B = (n-1)^2\left(1+\frac{n-1}{n}k_C\delta\right)^2, \tag{5.276}$$

$$C = (n-1)^2\left(1+\frac{n-1}{n}k_C\delta\right)^2+4(n-1)\frac{k_C^2}{\kappa^2}, \tag{5.277}$$

$$D = 2\frac{k_C^4}{\kappa^4}, \tag{5.278}$$

which are always positive and non-zero. Naturally, in general the equation (5.274) can be solved and possesses the solution

$$k' = x\frac{\kappa^2}{k}, \tag{5.279}$$

where x is a coefficient. Because $k' \geqslant 0$ and $k > 0$, the real and positive values of x are physical. In this way there are two physical x

$$x = \frac{1}{4A}\left(\sqrt{2a-b-c \pm \frac{d}{\sqrt{a+b+c}}} - \sqrt{a+b+c} - B\right), \tag{5.280}$$

with the conditions

$$a+b+c > 0, \tag{5.281}$$

$$2a \pm \frac{d}{\sqrt{a+b+c}} \geqslant b+c, \tag{5.282}$$

$$\sqrt{2a-b-c \pm \frac{d}{\sqrt{a+b+c}}} \geqslant B + \sqrt{a+b+c}, \tag{5.283}$$

where we have introduced the shortened notation

$$a = B^2 - \frac{8}{3}CA, \tag{5.284}$$

$$b = \frac{4}{3}2^{1/3}A\beta\left(\alpha + \sqrt{\alpha^2 - 4\beta^3}\right)^{-\frac{1}{3}}, \tag{5.285}$$

$$c = 2^{1/3}A\left(\alpha + \sqrt{\alpha^2 - 4\beta^3}\right)^{\frac{1}{3}}, \tag{5.286}$$

$$d = \frac{2B}{A}\left(-B^2 + 4CA\right), \tag{5.287}$$

and $\alpha = 2C^3 - 27B^2D + 72ACD$ and $\beta = C^2 - 12AD$.

Interestingly, one can see easily that the case $n = 1$ applied to (5.274) leads to $D = 0$, what is true for the only relativistic limit $\kappa \to \infty$. However, the result of the relativistic limit applied to the equation (5.274) has also different countenance. Before taking the limit one must reduce this equation maximally with respect to powers of $1/\kappa \to 0$, *i.e.*

$$A\frac{(kk')^4}{\kappa^4} + B\frac{(kk')^3}{\kappa^2} + C\left(kk'\right)^2 - 2k_C^4 = 0, \tag{5.288}$$

and for this reason in such a limit one obtains the condition

$$(n-1)^2\left(1+\frac{n-1}{n}k_C\delta\right)^2(kk')^2-2k_C^4=0, \tag{5.289}$$

which is easy to solve

$$k'=\frac{\dfrac{\sqrt{2}}{n-1}}{1+\dfrac{n-1}{n}k_C\delta}\frac{k_C^2}{k}\sim\frac{1}{k}, \tag{5.290}$$

where we neglected the negative solution as non-physical. It is seen that $n=1$ is not appropriate for such a solution.

D Dispersional Generalization

Let us construct finally the other approach to the modified Compton effect argued by the generalization due to the specific form of the dispersion relations. For this let us consider first the deduction due to the unmodified case, *i.e.* Special Relativity. In such a situation, the dispersion relations for a photon and an electron have the form

$$\frac{\omega^2}{c^2} = k^2, \tag{5.291}$$

$$\frac{\omega_e^2}{c^2} = k_C^2+k_e^2. \tag{5.292}$$

For this reason the laws of conservation of momentum (5.224) and energy (5.226) expressed through the dispersion relations (5.291) and (5.292) have the following form

$$\frac{\omega_e^2}{c^2}-k_C^2 = \frac{\omega^2}{c^2}+\frac{\omega'^2}{c^2}-\frac{\omega}{c}\frac{\omega'}{c}\cos\theta, \tag{5.293}$$

$$\frac{\omega_e}{c} = k_C+\frac{\omega}{c}-\frac{\omega'}{c}. \tag{5.294}$$

and after elementary algebraic manipulations lead to the result

$$\lambda_C(1-\cos\theta)=\frac{c}{\omega'}-\frac{c}{\omega}, \tag{5.295}$$

which after application of the explicit form of the dispersion relation (5.291) gives the Compton equation (5.239). In this manner, we shall call this equation *the generalized Compton equation*.

Similar reasoning can be performed to the modified dispersion relations (5.215 and (5.216). Then the laws of conservation of momentum and energy expressed through the dispersion relations are preserved in the form (5.293)-(5.294) obtained by Special Relativity. Therefore, also the Compton equation (5.295) is preserved. However, the change is due to the specific dispersion relations (5.215) and (5.216), so that the modified Compton effect is described by the equation

$$\lambda\!\!\!^{-}_C(1-\cos\theta) = \frac{1}{\sqrt{k'^2 + \dfrac{k'^4}{\kappa^2}}} - \frac{1}{\sqrt{k^2 + \dfrac{k^4}{\kappa^2}}}. \tag{5.296}$$

Interestingly, in the relativistic limit $\kappa \to \infty$ the modified Compton equation (5.296) coincides with the usual Compton equation (5.295) with no additional presumptions. We call such an approach *the dispersional generalization*. It is clear that this generalization should also work for elementary processes other than the Compton scattering. For this reason let us express the proposition

Proposition (The Dispersional Generalization). *Let us presume that there is an initial theory of elementary precesses having established results, which is characterized by certain dispersion relations and the laws of conservation. Let us consider the theory due to a modification of the initial theory. The consistent analysis of an arbitrary elementary process within the modified theory is based on the three-step procedure:*

1. *Derivation of the dispersion relations due of the modified theory,*
2. *Application of the dispersion relations to the laws of conservation of the modified theory,*
3. *Using the explicit form of the dispersion relations within the results obtained due to the laws of conservation in the framework of the modified theory.*

The analysis is consistent if and only if the results of the modified theory coincide with the results of the initial theory for lack of the modification.

Because $-1 \leqslant \cos\theta \leqslant 1$, the relation (5.296) allows to establish the bounds for the wave vector of an outgoing photon $k'_{min} \leqslant k \leqslant k'_{max}$, where

$$k'_{min} = \frac{\kappa}{\sqrt{2}}\sqrt{\sqrt{1+\frac{4}{\kappa^2}\left(k^2+\frac{k^4}{\kappa^2}\right)}-1}, \tag{5.297}$$

$$k'_{max} = \frac{\kappa}{\sqrt{2}}\sqrt{\sqrt{1+\frac{4}{\kappa^2}\frac{k^2+\dfrac{k^4}{\kappa^2}}{\left(1+2\lambda_C\sqrt{k^2+\dfrac{k^4}{\kappa^2}}\right)^2}}-1}. \tag{5.298}$$

Let us apply the lensing hypothesis to the obtained general result (5.295). The formula (5.254) can be rewritten in the form

$$\frac{c}{\omega'}-\frac{c}{\omega}=\delta\frac{n-1}{n}-\frac{c}{\omega}\frac{c}{\omega'}\frac{k_C}{n-1}, \tag{5.299}$$

what means that the generalized Compton equation (5.295) becomes

$$\lambda_C(1-\cos\theta)=\delta\frac{n-1}{n}-\frac{c}{\omega}\frac{c}{\omega'}\frac{k_C}{n-1}. \tag{5.300}$$

In this manner, by application of the identification method, one can establish the following relations

$$\delta = \frac{n}{n-1}\lambda_C, \tag{5.301}$$

$$\cos\theta = \frac{c}{\omega}\frac{c}{\omega'}\frac{k_C^2}{n-1}, \tag{5.302}$$

which allows to derive

$$\cos\theta=\frac{c}{\omega}\frac{c}{\omega'}\frac{\delta-\lambda_C}{\lambda_C^3}. \tag{5.303}$$

However, because $-1 \leq \cos\theta \leq 1$ one has the bound

$$\frac{\omega'}{c} \geqslant \frac{\delta-\lambda_C}{\lambda_C^3}\frac{c}{\omega}, \tag{5.304}$$

which for the Compton effect means that

$$k' \geqslant \frac{\delta - \lambda_C}{\lambda_C^3} \frac{1}{k}, \tag{5.305}$$

while for the modified Compton effect gives

$$k' \geqslant \frac{\kappa}{\sqrt{2}} \left[\sqrt{1 + \frac{4}{\kappa^2} \left(\frac{\delta - \lambda_C}{\lambda_C^3} \right)^2 \frac{1}{k^2 + \frac{k^4}{\kappa^2}}} - 1 \right]. \tag{5.306}$$

In the case of the usual Compton effect the formula (5.303) gives

$$\cos\theta = \frac{1}{k} \frac{1}{k'} \frac{\delta - \lambda_C}{\lambda_C^3}, \tag{5.307}$$

while in the modified case one receives

$$\cos\theta = \frac{1}{\sqrt{k'^2 + \frac{k'^4}{\kappa^2}}} \frac{1}{\sqrt{k^2 + \frac{k^4}{\kappa^2}}} \frac{\delta - \lambda_C}{\lambda_C^3}. \tag{5.308}$$

The lensing hypothesis (5.299) together with (5.301) leads to

$$\frac{\omega'}{c} = \frac{1 + \frac{c}{\omega} \frac{\delta - \lambda_C}{\lambda_C^2}}{\lambda_C + \frac{c}{\omega}}, \tag{5.309}$$

what together with (5.304) leads to the another bound

$$\frac{\omega}{c} \geqslant \frac{1}{\lambda_C} \sqrt{\frac{\delta}{\lambda_C} - 1} = \frac{\sqrt{n-1}}{\lambda_C}, \tag{5.310}$$

which for the usual Compton effect means that

$$k \geqslant \frac{1}{\lambda_C} \sqrt{\frac{\delta}{\lambda_C} - 1}, \tag{5.311}$$

while for the modified Compton effect leads to

$$k \geqslant \frac{\kappa}{\sqrt{2}}\sqrt{\sqrt{1+\frac{4}{\kappa^2}\frac{1}{\lambda_C}\sqrt{\frac{\delta}{\lambda_C}-1}}-1}. \tag{5.312}$$

The formula (5.309) can be rewritten in the form

$$\frac{\omega'}{c} = \frac{1}{\lambda_C} + \left(\frac{1}{\lambda_C}\right)^2 \frac{\delta - 2\lambda_C}{1+\lambda_C\dfrac{\omega}{c}}, \tag{5.313}$$

and studied approximatively with respect to the point $\frac{\omega}{c} = \frac{1}{\lambda_C}$, which defines the following value of the wave vector of the incoming photon.

$$k = \frac{\kappa}{\sqrt{2}}\sqrt{\sqrt{1+4\frac{k_C^2}{\kappa^2}}-1}. \tag{5.314}$$

It is easy to see that, because the formula (5.309), in such a situation

$$\frac{\omega'}{c} = \frac{1+\dfrac{\delta-\lambda_C}{\lambda_C}}{2\lambda_C} = \frac{\delta}{2\lambda_C^2}, \tag{5.315}$$

and for this reason the cosinus formula (5.303) takes the form

$$\cos\theta = 2\left(1-\frac{\lambda_C}{\delta}\right) = \frac{2}{n}, \tag{5.316}$$

i.e. the scattering is possible when the thickness of the lens is

$$\frac{2}{3}\lambda_C \leqslant \delta \leqslant 2\lambda_C, \tag{5.317}$$

or equivalently the refraction index of the medium is

$$n \geqslant 2. \tag{5.318}$$

For $\frac{\omega}{c} < \frac{1}{\lambda_C}$ the correct expansion of the formula (5.313) is given by

$$\frac{\omega'}{c} = \frac{1}{\lambda_C} + (\delta - 2\lambda_C)\left(\frac{1}{\lambda_C}\right)^2 \sum_{n=0}^{\infty}(-1)^n \lambda_C^n \left(\frac{\omega}{c}\right)^n, \tag{5.319}$$

while for $\frac{\omega}{c} > \frac{1}{\lambda_C}$ the appropriate expansion has the following form

$$\frac{\omega'}{c} = \frac{1}{\lambda_C} + (\delta - 2\lambda_C)\left(\frac{1}{\lambda_C}\right)^2 \sum_{n=1}^{\infty} (-1)^n \left(\frac{1}{\lambda_C}\right)^n \left(\frac{c}{\omega}\right)^n. \tag{5.320}$$

In the neighborhood of the point $\frac{\omega}{c} = \frac{1}{\lambda_C}$ the approximations (5.319) and (5.320)

$$\frac{\omega'}{c} \approx \frac{\delta - 2\lambda_C}{\lambda_C^2} - \frac{\delta - 2\lambda_C}{\lambda_C}\frac{\omega}{c}, \tag{5.321}$$

$$\frac{\omega'}{c} \approx \frac{1}{\lambda_C} - \frac{\delta - 2\lambda_C}{\lambda_C^3}\frac{c}{\omega}, \tag{5.322}$$

must coincide. For this reason in such a situation one obtains

$$\delta = 3\lambda_C, \tag{5.323}$$

i.e. the refraction index n near $\frac{\omega}{c} = \frac{1}{\lambda_C}$ has the value

$$n = 1.5. \tag{5.324}$$

In the light of the relation (5.323) the cosinus (5.316) becomes

$$\cos\theta = \frac{4}{3}, \tag{5.325}$$

i.e. is non-physical. It means that near the point $\frac{\omega}{c} = \frac{1}{\lambda_C}$ the modified Compton effect, considered in the framework of the lensing hypothesis, has no place because lack of scattering.

If θ, ω, and ω' are established, *e.g.* by experimental data, then the thickness of the lens d and the refractive index n are

$$\delta = \lambda_C\left(1 + \lambda_C^2 \frac{\omega}{c}\frac{\omega'}{c}\cos\theta\right), \tag{5.326}$$

$$n = 1 + \frac{1}{\lambda_C^2}\frac{c}{\omega}\frac{c}{\omega'}\sec\theta. \tag{5.327}$$

However, because $n = c/v$, where v is velocity of light in the medium in which the refraction takes place, one can see that

$$v = c\frac{\lambda_C^2 \frac{\omega}{c}\frac{\omega'}{c}\cos\theta}{1+\lambda_C^2 \frac{\omega}{c}\frac{\omega'}{c}\cos\theta}. \tag{5.328}$$

In this section we presented the approach to the Compton effect modified due to the non-commutative geometry of the Snyder quantized space-time. First, we carried out the relativistic analysis which produced the modified Compton equation coinciding with the Compton equation in the case of lack of the modification due to the Snyder–Sidharth deformation. Then, we solved the modified Compton equation straightforwardly and showed that in the case of lack of the modification the result differs from deductions of Special Relativity. For this reason we proposed to take into account the lensing hypothesis. Then, we showed that the scattered angle values are independent of particle energies, and derived the equation combining the wave vectors of incoming and outgoing photons. Finally, we carried out the dispersional generalization which led us to the correct result in the case of lack of the modification. Application of the lensing hypothesis and the identification method together with the dispersional generalization produced a number of new results arising from the modified Compton effect. There are two different routes for treatment of the results of this section. First they could be confronted with suitable experimental data. Second they could be used for prediction of new experiments. Anyway, with no consistent verification with observations, it is pointless to determine their physical value.

Chapter 6

Neutrinos: Masses & Chiral Condensate

A Snyder Quantized Space-Time

In 1947 the American physicist Hartland Sweet Snyder [317] proposed the non-commutative geometry model of quantized space-time, and employed such a model in the analysis of the electromagnetic field. In the present-day conventions the Snyder quantized space-time can be presented in the compact form

$$\frac{i}{\hbar c}[p_\mu, p_\nu] = 0, \tag{6.1}$$

$$\frac{i}{\hbar c}[x_\mu, p_\nu] = -\delta_{\mu\nu} + \left(\frac{a}{\hbar c}\right)^2 p_\mu p_\nu, \tag{6.2}$$

$$\frac{i}{\hbar c}[x_\mu, x_\nu] = c\left(\frac{a}{\hbar c}\right)^2 \begin{bmatrix} 0 & -M_i \\ M_i & \varepsilon_{ij}^k L_k \end{bmatrix}, \tag{6.3}$$

where $p_\mu = [E, p_i c]$ is the momentum four-vector and $x_\mu = [ct, x_i]$ is the position four-vector, and

$$M_i = -c\left(\frac{\hbar}{a}\right)^2 \frac{i}{\hbar}[t, x_i] = x_i \frac{E}{c} + ct p_i, \tag{6.4}$$

$$L_i = \frac{1}{2}\left(\frac{\hbar}{a}\right)^2 \varepsilon_i^{jk} \frac{i}{\hbar}[x_j, x_k] = \varepsilon_i^{jk} x_j p_k, \tag{6.5}$$

are the infinitesimal generators of the Lorentz group: M_i is the three-vector of the Lorentz boosts and L_i is the three-vector of angular momentum

$$\frac{i}{\hbar}[L_i,x_j] = -\varepsilon_{ij}^k x_k, \quad (6.6)$$

$$\frac{i}{\hbar}[L_i,t] = 0, \quad (6.7)$$

$$\frac{i}{\hbar}[L_i,p_j] = -\varepsilon_{ij}^k p_k, \quad (6.8)$$

$$\frac{i}{\hbar}[L_i,E] = 0, \quad (6.9)$$

$$\frac{i}{\hbar}[L_i,L_j] = -\varepsilon_{ij}^k L_k, \quad (6.10)$$

$$\frac{i}{\hbar}[L_i,M_j] = -\varepsilon_{ij}^k M_k, \quad (6.11)$$

$$\frac{i}{\hbar}[M_i,M_j] = -\varepsilon_{ij}^k L_k, \quad (6.12)$$

$$\frac{i}{\hbar}[M_i,x_j] = \delta_{ij}ct, \quad (6.13)$$

$$\frac{i}{\hbar}[M_i,p_j] = -\delta_{ij}\frac{E}{c}, \quad (6.14)$$

$$\frac{i}{\hbar}[M_i,E] = -cp_i. \quad (6.15)$$

Introducing the matrix

$$M_{\mu\nu} = x_\mu p_\nu - x_\nu p_\mu, \quad (6.16)$$

and using the property of a commutator $[AB,C] = A[B,C] + [A,C]B$, one can determine the Lie brackets

$$\frac{i}{\hbar}[M_{\mu\nu},p_\rho] = \frac{i}{\hbar}[x_\mu,p_\rho]p_\nu - \frac{i}{\hbar}[x_\nu,p_\rho]p_\mu, \quad (6.17)$$

$$\frac{i}{\hbar}[M_{\mu\nu},M_{\rho\sigma}] = x_\nu\frac{i}{\hbar}[M_{\rho\sigma},p_\mu] - x_\mu\frac{i}{\hbar}[M_{\rho\sigma},p_\nu] \quad (6.18)$$

$$+ \frac{i}{\hbar}[M_{\rho\sigma},x_\mu]p_\mu - \frac{i}{\hbar}[M_{\rho\sigma},x_\mu]p_\nu. \quad (6.19)$$

For calculation of the second bracket it is necessary to establish the commutator

$$[M_{\mu\nu},x_\sigma] = x_\nu[x_\sigma,p_\mu] - x_\mu[x_\sigma,p_\nu] + [x_\mu,x_\sigma]p_\nu - [x_\nu,x_\sigma]p_\mu, \quad (6.20)$$

which after using the fact

$$[x_\nu, x_\sigma]p_\mu = [\delta^\mu_\nu x_\mu, x_\sigma]p_\mu = [x_\mu, x_\sigma]\delta^\mu_\nu p_\mu = [x_\mu, x_\sigma]p_\nu, \tag{6.21}$$

and the relation (6.2) gives

$$\frac{i}{\hbar c}[M_{\mu\nu}, x_\sigma] = -x_\nu \delta_{\sigma\mu} + x_\mu \delta_{\sigma\nu} - \left(\frac{a}{\hbar c}\right)^2 M_{\mu\nu} p_\sigma, \tag{6.22}$$

so that consequently one receives

$$\frac{i}{\hbar}[M_{\mu\nu}, p_\rho] = -\delta_{\mu\rho} p_\nu + \delta_{\nu\rho} p_\mu, \tag{6.23}$$

$$\frac{i}{\hbar}[M_{\mu\nu}, M_{\rho\sigma}] = \delta_{\mu\sigma} M_{\nu\rho} + \delta_{\nu\rho} M_{\mu\sigma} - \delta_{\mu\rho} M_{\sigma\nu} - \delta_{\nu\sigma} M_{\rho\mu}. \tag{6.24}$$

The relations (6.1), (6.23) and (6.24) create the algebra which is the Euclidean equivalent of the Poincaré algebra, *i.e.* the group of isometries of the Euclidean space-time. In this manner, the algebra of the Snyder quantized space-time is the $SO(4)$ algebra. However, such a situation does not mean that the Snyder quantized space-time arises from the Euclidean space-time. Moreover, H.S. Snyder did not consider such a situation and based his model on the De Sitter space. Strictly speaking, for both the Lorentz invariance and the Poincaré invariance modified due to the natural unit length a, Snyder considered a momentum space of a constant curvature isometry group, *i.e.* the Poincaré algebra deformation in the De Sitter space.

There are some differences between the Snyder quantized space-time and quantum mechanics. One of such dissimilarities is entering time in the formulation. In the case of the Snyder quantized space-time time is an operator quantity, what is the straightforward ramification of the Lorentz invariance adopted to the concept of space-time, because in the light of the Lorentz transformations space and time play equivalent roles. Meanwhile, the concept of time in quantum mechanics is inherited from classical mechanics, where time is an affine parameter, and for this reason there are problems with implementation of a time operator within standard quantum mechanics. In spite of this difference between Snyder quantized space-time and quantum mechanics, both the formulations possess some common points. For example, there are unmodified commutation relations of the angular momentum operators and defined by the position and momentum operators, so that the orbital angular momentum spectrum remains unchanged. In his model, H.S. Snyder did not take into account the concept of spin, and it is still an unsolved question whether spin quantum-mechanical effects remain unchanged in the Snyder quantized space-time. The most important and intriguing aspects of the thing is that many of quantum mechanical features are present in the case of the Snyder quantized space-time [327].

Let us consider the Minkowski space-time projection of the Snyder space-time. In such a situation, the formula (6.2) can be contracted with the Minkowski metric $\eta^{\mu\nu}$

$$\frac{i}{\hbar c}[x^{\mu}, p_{\mu}] = -2 + \left(\frac{a}{\hbar c}\right)^2 p_{\mu}p^{\mu}, \tag{6.25}$$

where $p^{\mu} = \eta^{\mu\nu}p_{\nu}$ and $p_{\mu}p^{\mu} = E^2 - p^2c^2$, and used for definition of the four-momentum square

$$p_{\mu}p^{\mu} = \left(\frac{\hbar c}{a}\right)^2 \left(2 + \frac{i}{\hbar c}[x^{\mu}, p_{\mu}]\right), \tag{6.26}$$

so that one can derive the Hamiltonian constraint

$$E^2 = p^2c^2 + \left(\frac{\hbar c}{a}\right)^2 \left(2 + \frac{i}{\hbar c}[x^{\mu}, p_{\mu}]\right). \tag{6.27}$$

Taking into account the identification

$$m = \frac{\sqrt{2}\hbar}{ac} = \sqrt{2}\frac{\ell_P}{a}M_P, \tag{6.28}$$

where ℓ_P is the Planck length and $M_P = \dfrac{\hbar}{\ell_P c}$ is the Planck mass, one can rewrite the Hamiltonian constraint (6.29) in the following form

$$E^2 = p^2c^2 + m^2c^4 + \Delta, \tag{6.29}$$

where the deformation term Δ is

$$\Delta = \left(\frac{\hbar c}{a}\right)^2 \frac{i}{\hbar c}[x^{\mu}, p_{\mu}]. \tag{6.30}$$

The Snyder model is a quantum mechanical scenario for space-time. In his model operators of space-time coordinates do not commute and gave discrete spectrum similarly to the position and momentum operators quantum mechanics. The Snyder quantized space-time is Lorentz invariant, but translation invariance is violated so that the full relativistic symmetry – the Poincaré invariance – is not preserved. Some authors deduced that this aspect of the problem leads to the manifestly non-physical character of the Snyder quantized space-time [328]. Recently, however, there has been a visible certain trend towards the Snyder quantized space-time, in which particularly studied are the quantum groups and non-commutative geometry aspects of the problem.

The physical significance and developments of the Snyder quantized space-time were considered seriously by very few scholars, scientists, and researches including Burra Gautam Sidharth (Cf. *e.g.* the papers in Ref. [329]), Lukasz Andrzej Glinka [313, 330], Burra Subba Lakshmi [331], Venkata Aditya Cherikuri and B.G. Sidharth [332], and Raoelina Andriambololona and Christian Rakotonirina [333]. Sidharth applied the model of discrete space-time

$$\frac{i}{\hbar}[p,x] = 1+\left(\frac{\ell}{\hbar}\right)^2 p^2, \tag{6.31}$$

$$\frac{i}{\hbar}[x,y] = O(\ell^2) \quad , \tag{6.32}$$

where ℓ is minimal scale. Sidharth proposed to consider the Hamiltonian constraint

$$E^2 = m^2c^4 + p^2c^2 + \alpha\left(\frac{\ell}{\hbar}\right)^2 c^2p^4, \tag{6.33}$$

where $\alpha \sim 1$ is a dimensionless constant. This energy-momentum relation, named in the previous chapter the Snyder–Sidharth Hamiltonian constraint, describes a particular case of the formula (6.29) in which the deformation is

$$\left(\frac{\hbar c}{\ell}\right)^2 \frac{i}{\hbar c}[x^\mu, p_\mu] = \alpha\left(\frac{\ell}{\hbar}\right)^2 c^2p^4, \tag{6.34}$$

and is compatible with the Snyder model of quantized space-time if and only if the following Lie bracket

$$\frac{i}{\hbar c}[x^\mu, p_\mu] = \alpha\left(\frac{\ell}{\hbar}\right)^4 p^4, \tag{6.35}$$

is satisfied. The model (6.31)-(6.32) studied by Sidharth is manifestly based on the Snyder model but also manifestly differs from this model of quantized space-time. In this manner, we shall call the model (6.31)-(6.32) the Snyder–Sidharth space-time.

B Outlook on Non-Commutative Geometry

The Snyder–Sidharth space-time (6.31)–(6.32) in itself defines a specific non-commutative geometry and an algebra deformation. Let us study this model in some detail. First, for better insight, let us sketch concisely the basic rules of non-commutative geometry in a certain general outlook. The reader interested in more basic explanations and more advanced applications is referred to the contributions collected in Ref. [334].

Let us consider associative Lie algebra A for which $\tilde{A} = A[[\lambda]]$ is the module due to the ring of formal series $\mathbb{K}[[\lambda]]$ in a parameter λ. Let us call $\tilde{A}$ deformation of A, *i.e.* $\mathbb{K}[[\lambda]]$-algebra such that $\tilde{A}/\lambda\tilde{A} \approx A$. If A is endowed with a locally convex topology with continuous laws, *i.e.* is topological algebra, then $\tilde{A}$ is called topologically free. We presume that in the Lie algebra A the law of composition is determined by an ordinary product and the related bracket is $[\cdot,\cdot]$. In such a situation, $\tilde{A}$ is an associative Lie algebra if and only if for arbitrary two elements of the algebra $f, g \in A$ a new product $\star$ and the related bracket $[\cdot,\cdot]_\star$ are defined as follows

$$f \star g = fg + \sum_{n=1}^{\infty} \lambda^n C_n(f,g), \qquad (6.36)$$

$$[f,g]_\star \equiv f \star g - g \star f = [f,g] + \sum_{n=1}^{\infty} \lambda^n B_n(f,g), \qquad (6.37)$$

where C_n and B_n are the Hochschild and the Chevalley 2-cochains, and for arbitrary three elements of the algebra $f, g, h \in A$ two conditions are satisfied: the Jacobi identity

$$[[f,g]_\star,h]_\star + [[h,f]_\star,g]_\star + [[g,h]_\star,f]_\star = 0 \qquad (6.38)$$

and the law of associativity

$$(f \star g) \star h = f \star (g \star h). \qquad (6.39)$$

If b and ∂ are the Hochschild and the Chevalley coboundary operators, *i.e.* such that $b^2 = 0$ and $\partial^2 = 0$, then for each n and $j,k \geqslant 1$ such that $j+k = n$ the following relations hold

$$bC_n(f,g,h) = \sum_{j,k}[C_j(C_k(f,g),h) - C_j(f,C_k(g,h))], \tag{6.40}$$

$$\partial B_n(f,g,h) = \sum_{j,k}[B_j(B_k(f,g),h) + B_j(B_k(h,f),g) + B_j(B_k(g,h),f)], \tag{6.41}$$

Let $C^\infty(M)$ be an algebra of smooth functions on a differentiable manifold M. The law of associativity yields the Hochschild cohomologies. An antisymmetric contravariant 2-tensor θ, which trivialize the Schouten–Nijenhuis bracket $[\theta,\theta]_{SN} = 0$ on M, determines the Poisson brackets $\{f,g\} = i\theta df \wedge dg$ satisfying the Jacobi identity and the Leibniz rule. Then $(M,\{\cdot,\cdot\})$ is called the Poisson manifold.

In 1997 the Russian mathematician Maxim L'vovich Kontsevich [335] defined deformation quantization of a general Poisson differentiable manifold. Let $\mathbb{R}^d$ be endowed with the Poisson brackets

$$\alpha(f,g) = \sum_{1\leqslant i,j\leqslant n} \alpha^{ij}\frac{\partial f}{\partial x^i}\frac{\partial g}{\partial x^j}, \tag{6.42}$$

where $1 \leqslant k \leqslant d$. For $\star$-product and $n \geqslant 0$, exists a family $G_{n,2}$ of $(n(n+1))^n$ oriented graphs Γ. Let V_Γ be the set of vertices of Γ. This set has $n+2$ elements collected in two subsets: the first type $\{1,\ldots,n\}$ and the second type $\{\bar{1},\bar{2}\}$. Let E_Γ denotes the set of oriented edges of Γ, having $2n$ elements. The rule is that there is no edge starting at a second type vertex. Let $Star(k)$ denotes the set of oriented edges starting at a first type vertex k with cardinality $\sharp k = 2$, $\sum_{1\leqslant k\leqslant n}\sharp k = 2n$. Then $\{e_k^1,\ldots,e_k^{\sharp k}\}$ are the edges of Γ starting at vertex k. Vortices starting and ending in the edge v are $v = (s(v),e(v))$ where $s(v) \in \{1,\ldots,n\}$ and $e(v) \in \{1,\ldots,n;\bar{1},\bar{2}\}$. Γ has no loop and no parallel multiple edges. For arbitrary two elements $f,g \in C^\infty(\mathbb{R}^d)$ a bidifferential operator $(f,g) \mapsto B_\Gamma(f,g)$ is associated to Γ. The symbols $\alpha^{e_k^1 e_k^2}$ are associated to each first type vertex k from where the edges $\{e_k^1,e_k^2\}$ start; f is the vertex 1, and g is the vertex $\bar{2}$. An edge e_k^1 acts like differentiation operator $\partial/\partial x^{e_k^1}$ on its ending vertex. Than B_Γ is a sum over all maps $I : E_\Gamma \to \{1,\ldots,d\}$

$$B_\Gamma(f,g) = \sum_I\left(\prod_{k=1}^n\prod_{k'=1}^n \partial_{I(k',k)}\alpha^{I(e_k^1)I(e_k^2)}\right)\left(\prod_{k_1=1}^n \partial_{I(k_1,\bar{1})}f\right)\left(\prod_{k_2=1}^n \partial_{I(k_2,\bar{2})}g\right). \tag{6.43}$$

Let us denote by $\mathscr{H}_n$ the configuration space of n distinct points in upper half-plane $\mathscr{H} = \{z \in \mathbb{C} | \Im(z) > 0\}$ with the Lobachevsky hyperbolic metric, which is an open submanifold of $\mathbb{C}^n$. Let for the vertex k such that $1 \leqslant k \leqslant n$, $z_k \in \mathscr{H}$ denotes a variable associated to Γ. The vertices $\bar{1}$ and $\bar{2}$ are associated to $0 \in \mathbb{R}$ and $1 \in \mathbb{R}$, respectively. If $\tilde{\phi}_v = \phi(s(v), e(v))$ is a function on $\mathscr{H}_n$, associated to v, determined by the angle function $\phi : \mathscr{H}_2 \to \mathbb{R}/2\pi\mathbb{Z}$ having the following form

$$\phi(z_1, z_2) = \operatorname{Arg}\frac{z_2 - z_1}{z_2 - \bar{z}_1} = \frac{1}{2i}\operatorname{Log}\frac{\bar{z}_2 - z_1}{z_2 - \bar{z}_1}\frac{z_2 - z_1}{\bar{z}_2 - \bar{z}_1}, \tag{6.44}$$

then the integral of $2n$-form $w(\Gamma) \in \mathbb{R}$ is a weight associated to $\Gamma \in G_{n,2}$

$$w(\Gamma) = \frac{1}{n!(2\pi)^{2n}} \int_{\mathscr{H}_n} \bigwedge_{1\leqslant k\leqslant n} \left(d\tilde{\phi}_{e_k^1} \wedge d\tilde{\phi}_{e_k^2} \right), \tag{6.45}$$

which does not depend on the Poisson structure or the dimension d. On $(\mathbb{R}^d, \alpha)$ the Kontsevich $\star$-product maps $C^\infty(\mathbb{R}) \times C^\infty(\mathbb{R}) \to C^\infty(\mathbb{R})[[\lambda]]$

$$(f, g) \mapsto f \star g = \sum_{n\geqslant 0} \lambda^n C_n(f, g), \tag{6.46}$$

where $C_0(f,g) = fg$, $C_1(f,g) = \{f,g\}_\alpha = \alpha df \wedge dg$ and in general

$$C_n(f,g) = \sum_{\Gamma \in G_{n,2}} w(\Gamma) B_\Gamma(f,g). \tag{6.47}$$

Equivalence classes of (6.46) are bijective to the equivalence classes of the Poisson brackets $\alpha_\lambda = \sum_{k\geqslant 0} \lambda^k \alpha_k$. For linear Poisson structures, *i.e.* on coalgebra $A^\star$, the weight (6.45) of all even wheel graphs vanishes, and the Kontsevich star product (6.46) coincides with the $\star$-product determined by the Duflo isomorphism. This case allows to quantize the class of quadratic Poisson brackets belonging to the image of the Drinfeld map which associates a quadratic to a linear bracket.

Let us consider the deformations of phase-space and space given by the parameters λ_{ph}, λ_s being

$$\lambda_{ph} = \frac{i\alpha}{2}, \tag{6.48}$$

$$\lambda_s = \frac{i\beta}{2}, \tag{6.49}$$

where $\alpha \sim 1$ and $\beta \sim 1$ are dimensionless constants, and resulting in the appropriate star product (6.36), or equivalently to the Kontsevich star-product (6.46), on the phase space (x,p) and between two distinct space points x and y

$$x \star p = px + \sum_{n=1}^{\infty} \left(\frac{i\alpha}{2}\right)^n C_n(x,p), \tag{6.50}$$

$$x \star y = xy + \sum_{n=1}^{\infty} \left(\frac{i\beta}{2}\right)^n C_n(x,y), \tag{6.51}$$

where $C_n(x,p)$ and $C_n(x,y)$ are the Hochschild cochains in the Kontsevich formula (6.46) related to the phase space and the space deformations, respectively. The Lie brackets arising from the star products (6.50) and (6.51) have the following form

$$[x,p]_\star = [x,p] + \sum_{n=1}^{\infty} \left(\frac{i\alpha}{2}\right)^n B_n(x,p), \tag{6.52}$$

$$[x,y]_\star = [x,y] + \sum_{n=1}^{\infty} \left(\frac{i\beta}{2}\right)^n B_n(x,y), \tag{6.53}$$

where $B_n(x,p)$ and $B_n(x,y)$ are the Chevalley cochains related to phase-space and space, respectively. Taking the first approximation in the formulas (6.52) and (6.53), and application of the non-deformed commutators $[x,p] = -i\hbar$ and $[x,y] = 0$ leads to

$$[x,p]_\star = -i\hbar + \frac{i\alpha}{2} B_1(x,p), \tag{6.54}$$

$$[x,y]_\star = \frac{i\beta}{2} B_1(x,y). \tag{6.55}$$

or after using the Dirac "method of classical analogy" [336]

$$\frac{1}{i\hbar}[p,x]_\star = 1 - \frac{\alpha}{2\hbar} B_1(x,p), \tag{6.56}$$

$$\frac{1}{i\hbar}[x,y]_\star = \frac{\beta}{2\hbar} B_1(x,y). \tag{6.57}$$

However, since for two arbitrary elements $f,g \in C^\infty(M)$ one has $B_1(f,g) = 2\theta(df \wedge dg)$, therefore, the formulas (6.56) and (6.57) can be rewritten in the following form

$$\frac{1}{i\hbar}[p,x]_\star = 1-\frac{\alpha}{\hbar}\left(\frac{dx\wedge dp}{\alpha}\right), \tag{6.58}$$

$$\frac{1}{i\hbar}[x,y]_\star = \frac{\beta}{\hbar}dx\wedge dy. \tag{6.59}$$

Let us consider the space lattice characterized by an infinitesimal growth of a space coordinate identified with the fundamental scale ℓ

$$dx = \ell, \tag{6.60}$$

and presume that the momentum of a particle is related to the coordinate of a particle by the De Broglie relation known from the wave-particle duality

$$p = \frac{\hbar}{x}. \tag{6.61}$$

Application of the model (6.60)-(6.61) allows to derive straightforwardly an infinitesimal growth of the momentum

$$dp = -\frac{\hbar}{x^2}dx = -\frac{p^2}{\hbar}\ell. \tag{6.62}$$

Therefore, the infinitesimal growths (6.60) and (6.62) applied to the deformed brackets (6.58) and (6.59) allows to establish finally

$$\frac{i}{\hbar}[p,x]_\star = 1+\frac{1}{\hbar^2}\ell^2p^2, \tag{6.63}$$

$$\frac{i}{\hbar}[x,y]_\star = -\frac{\beta}{\hbar}\ell^2. \tag{6.64}$$

The relations (6.63) and (6.64) prove that the Snyder–Sidharth space-time (6.31)-(6.32) possesses a structure of an non-commutative geometry obtained by the first approximation of the Kontsevich deformation quantization.

In the 1960s the Soviet physicist Moisey Alexandrovich Markov [337] proposed to take into account a fundamental length scale as minimal scale and determined such a scale through the Planck length, *i.e.* $\ell = \ell_P = \sqrt{\frac{\hbar c}{G}}$. M.A. Markov also expressed the hypothesis, called sometimes the Markov hypothesis, that the upper bound for mass m of any elementary particle is given by the maximal mass identified with the Planck mass, *i.e.* $m \leqslant M_P = \dfrac{\hbar}{c\ell_P} = \sqrt{\dfrac{G\hbar}{c^3}}$. Applying this fundamental idea, since 1978 the Soviet–Russian theoretician Vladimir Georgievich Kadyshevsky and his collaborators [338] studied the various aspects of the Snyder quantized space-time in the contexts strictly related to certain problems of particle physics. Recently, Vasilii Nikolaevich Rodionov developed independently the stream of Kadyshevsky [339]. The problems discussed in this chapter seem to be more related to a general current [340], where particularly the Snyder–Sidharth space-time (6.31)-(6.32) has been found a number of applications.

Beginning 2000 the Indian scholar and philosopher Burra Gautam Sidharth has been studying the model of discrete space-time strictly based on the Snyder quantized space-time which breaks the two fundamental paradigms celebrated in relativistic physics: the Einstein energy-momentum relation as well as the Lorentz invariance. In his considerations B.G. Sidharth took into account the Hamiltonian constraint of Special Relativity with the deformation which is proportional to the fourth power of three-momentum of a relativistic particle and the second power of a fundamental scale ℓ (Cf. Ref. [341])

$$E^2 = m^2c^4 + c^2p^2 + \alpha\left(\frac{c}{\hbar}\right)^2 \ell^2 p^4. \tag{6.65}$$

Sidharth has identified a fundamental scale with minimal scale, *i.e.* the Planck length or the Compton wavelength of an electron. Neglecting negative mass states as non-physical, he established a number of intriguing new ramifications [342, 343]. Particularly, by straightforward application of the Dirac "square root" technique to the modified Hamiltonian constraint (6.65) B.G. Sidharth concluded that the corresponding modified Dirac equation should have the following form

$$\left(\gamma^\mu \hat{p}_\mu + mc^2 + \sqrt{\alpha}\frac{c}{\hbar}\ell\gamma^5\hat{p}^2\right)\psi = 0. \tag{6.66}$$

which differs from the conventional Dirac relativistic quantum mechanics by a correction due to the γ^5-term proportional to the second power of the three momentum of a particle and to minimal scale ℓ.

However, it looks like that Sidharth has neglected the fact that the modified Special Relativity (6.65) leads to one additional possibility which is physically nonequivalent to the modified Dirac equation (6.66) considered by him as the physical quantum theory. Namely, there is also possible the modification by the γ^5-term with negative sign

$$\left(\gamma^\mu \hat{p}_\mu + mc^2 - \sqrt{\alpha}\frac{c}{\hbar}\ell\gamma^5\hat{p}^2\right)\psi = 0. \tag{6.67}$$

Fortunately, however, such an issue seems to be easily solvable because the possible physical results of the quantum theory (6.67) can be straightforwardly concluded from the results arising from the Dirac equation modified by the positive γ^5-term (6.66) by application of the mirror reflection in minimal scale $\ell \to -\ell$. We are not going to neglect also the negative mass states in the modified Dirac theory as non-physical, because this situation strictly corresponds with the results obtained from the equation (6.66) transformed by a mirror reflection in mass of a relativistic particle $m \to -m$. Moreover, it should be emphasized that in the standard relativistic quantum mechanics the negative sign corresponds to the antimatter, which recently has been considered as the element of Reality (Cf. *e.g.* Ref. [344]).

Therefore, we propose to consider the result of the canonical relativistic quantization

$$p_\mu = (E, p_i c) \to \hat{p}_\mu = i\hbar(\partial_0, c\partial_i) \tag{6.68}$$

applied to the deformed Special Relativity (6.65) linearized by the Dirac "square root" technique. In general, such a procedure leads to the four possible physically nonequivalent quantum theories which can be presented in the form of one compact equation

$$\left(\gamma^\mu \hat{p}_\mu \pm mc^2 \pm \sqrt{\alpha}\frac{c}{\hbar}\ell\gamma^5\hat{p}^2\right)\psi = 0. \tag{6.69}$$

We shall presume that by analogy with the standard Dirac relativistic quantum mechanics, the wave function ψ of the modified Dirac equation (6.69) is a four-component spinor $\psi = [\phi_0, \phi_1, \phi_2, \phi_3]^T$, and that the Dirac gamma matrices satisfy the four-dimensional Æther algebra $\{\gamma^\mu, \gamma^\nu\} = \frac{1}{2}\eta_{\mu\nu}$, introduced in the previous chapter, rather than the usual Dirac algebra. It should be emphasized that presence of the γ^5-term in the modified Dirac equation (6.69) results in a manifest violation of the parity symmetry and, therefore, also the Lorentz symmetry is violated due to such a correction. For simplicity, however, we shall consider one of the four cases (6.69) given by the Sidharth's Dirac equation (6.66). The results due to the three remained situations can be described by straightforward application of the mentioned mirror transformations in the mass of a particle and minimal scale to the results due to the generic theory (6.66).

Both this and next chapter are strictly based on the recent results of the author [330] enriched by necessary minor updates. The research value of these two chapters is justified by the fact that the approach based on deformed Special Relativity, including the Snyder–Sidharth deformation (6.65), recently has became one of the most intensively developing and fruitful research direction in astrophysics of gamma rays, especially in the context of CP violation [345]. Interestingly, the modified Dirac equation (6.66) was originally proposed as the idea for ultra-high energy physics (See Refs. [342, 343]). Sidharth, however, has not presented computations based on this idea, which could result in experimentally verifiable physical predictions. In communication with the author [346], Sidharth presented a number of intriguing and interesting looking speculations about the extra mass terms deforming Special Relativity and the corresponding Dirac equations. Also, we discussed a lot of philosophical issues and suggestions about the foundational role of non-commutative geometry for new physics based on the Lorentz symmetry (CP) violation. The discussions did not establish a physical truth and, therefore, derivation of the modified Dirac equation and the reasoning performed by Sidharth, in general, possesses philosophical countenance. Albeit, the physical role of deformation theory and non-commutative geometry is still a great riddle to the same degree as it is an amazing hope, and factually nobody established the real significance of such abstractive mathematics for theoretical physics. In author's opinion the most hopeful observational research region for verification of the theories (6.69) is high-energy and ultra-high-energy astrophysics. The astrophysical phenomena are probably the best test for the Planck scale. Particularly, ultra-high-energy cosmic rays coming from gamma bursts sources, neutrinos coming from supernovas, and other effects observed in this energy region, are the most fruitful research material for tests of the modified theories. This cognitive aspect of the problem is both the most logical and rational justification for considering the equation (6.66), arising due to the the non-commutative geometry of the Snyder quantized space-time (6.31)-(6.32), and trying pull out possibly novel valuable extensions of the well-grounded physical knowledge. It should be emphasized that an arbitrary abstractive mathematics creates potentially new physical theories, but extraction of physics is usually a heroic work. By its simplicity the non-commutative geometry of the Snyder quantized space-time is one of themes of this book, but other models are not forbidden. Possibly, however, the only non-commutative geometry of the Snyder quantized space-time possesses clear and unambiguous physical meaning. Such a hypothesis is also the good point for experimental verification.

C Massive Neutrinos

In fact, the Sidharth γ^5-term, emerging from the non-commutative geometry of the Snyder quantized space-time of phase space (6.31), is the shift of the conventional Dirac relativistic quantum mechanics. Let us presume that new physics arises from the physical picture in which the modified Dirac equation holds, but Special Relativity stays *non-modified*. In other words, we shall preserve Einstein's Special Relativity unchanged, but change the Dirac relativistic quantum mechanics by the non-commutative geometry of the Snyder quantized space-time. Such a modification is an algebra deformation. It is easy to deduce that such a deformation can be realized by preservation of the hyperbolic geometry of both the Minkowski energy-momentum space as well as the space-time. While the physical foundation of the Einstein theory is dynamics of a relativistic particle, the physical foundations of an algebra deformation are based on a non-dynamical justification. For example a deformation can be due to finite sizes of a particle. In this manner, in our view while the Snyder–Sidharth deformation of Special Relativity (6.65) can be interpreted as a dynamical result, the corresponding modification of the Dirac equation (6.66) is due to the non-dynamical γ^5-term despite the fact that this term is explicitly dependent on a particle three-momentum. To make such a constructive strategy evident, we propose to apply the formalism of the Minkowski energy-momentum space, despite the presence of the γ^5-term, straightforwardly to both the modified Special Relativity (6.65) and the modified Dirac equation (6.66).

The standard identity of the Minkowski energy-momentum space

$$p_\mu p^\mu = \left(\gamma^\mu p_\mu\right)^2 = E^2 - c^2 p^2, \tag{6.70}$$

allows to extract the square of the momentum three-vector which, together with the mass shell condition $p_\mu p^\mu = mc^2$, applied to the modified Special Relativity (6.65) results in the equation

$$\alpha\left(\frac{c}{\hbar}\right)^2 \ell^2 p^4 = 0, \tag{6.71}$$

which for non-zero momentum has the unique and unambiguous solution $\ell = 0$. In other words, the hyperbolic identity (6.70) on the mass shell leads to direct reconstruction of Special Relativity.

Let us now apply the identity (6.70) to the modified Dirac equation (6.66). The square of three-momentum can be extracted by the identity (6.70) and applied within the equation (6.66)

$$\left[\gamma^\mu \hat{p}_\mu + mc^2 + \frac{\sqrt{\alpha}}{\hbar c}\ell\gamma^5\left[E^2 - \left(\gamma^\mu \hat{p}_\mu\right)^2\right]\right]\psi = 0, \tag{6.72}$$

what leads to the quadratic equation

$$\left[-\frac{\sqrt{\alpha}}{\hbar c}\ell\gamma^5\left(\gamma^\mu \hat{p}_\mu\right)^2 + \gamma^\mu \hat{p}_\mu + mc^2 + \frac{\sqrt{\alpha}}{\hbar c}\ell\gamma^5 E^2\right]\psi = 0, \tag{6.73}$$

which after multiplication of both sides by γ^5 and using the combination $\gamma^5\gamma^\mu p_\mu$ can be rewritten in the following form

$$\left[\left(\gamma^5\gamma^\mu \hat{p}_\mu\right)^2 - \varepsilon\left(\gamma^5\gamma^\mu \hat{p}_\mu\right) + E^2 - \varepsilon mc^2\gamma^5\right]\psi = 0, \tag{6.74}$$

where ε is maximal energy due to minimal scale ℓ

$$\varepsilon = \frac{\hbar c}{\sqrt{\alpha}\ell}. \tag{6.75}$$

The equation (6.74) expresses projection of the operator

$$\left(\gamma^5\gamma^\mu \hat{p}_\mu\right)^2 - \varepsilon\left(\gamma^5\gamma^\mu \hat{p}_\mu\right) + E^2 - \varepsilon mc^2\gamma^5, \tag{6.76}$$

on the spinor field ψ. Elementary algebraic manipulations applied to the quadratic operator (6.76) lead to the factorized form of this operator

$$\left(\gamma^5\gamma^\mu \hat{p}_\mu - \varepsilon_+\right)\left(\gamma^5\gamma^\mu \hat{p}_\mu - \varepsilon_-\right), \tag{6.77}$$

where $\varepsilon_\pm$ are the manifestly non-hermitian energies

$$\varepsilon_\pm = \frac{\varepsilon}{2}\left(1 \pm \sqrt{1-\frac{4E^2}{\varepsilon^2}}\sqrt{1+\frac{4\varepsilon mc^2}{\varepsilon^2-4E^2}\gamma^5}\right). \tag{6.78}$$

Principally, the quantities (6.78) were obtained by the order reduction, and for this reason lead to the linearization similar to the original Dirac's reduction of the Klein–Gordon equation.

We shall treat energy E of a particle, mass m of a particle, and maximal energy ε (or equivalently minimal scale ℓ) in the formula (6.78) as free parameters. It can be observed straightforwardly that the modified Dirac equation (6.66) is equivalent to the two nonequivalent relativistic quantum theories

$$\left(\gamma^\mu \hat{p}_\mu - M_+c^2\right)\psi = 0, \tag{6.79}$$

$$\left(\gamma^\mu \hat{p}_\mu - M_-c^2\right)\psi = 0, \tag{6.80}$$

where $M_\pm$ are the generated mass matrices

$$M_\pm = \frac{\varepsilon}{2c^2}\left(-1 \mp \sqrt{1-\frac{4E^2}{\varepsilon^2}+\frac{4mc^2}{\varepsilon}\gamma^5}\right)\gamma^5. \tag{6.81}$$

The formula (6.81) in itself is a highly non-trivial result. In fact, by application of the Minkowski energy-momentum space, the Dirac equation modified due to the non-commutative geometry of the Snyder quantized space-time has been reduced to two distinguishable standard Dirac equations which, in the present situation, describe a kind of effective particles characterized by manifestly non-hermitian mass matrices $M_\pm$. Both these received theories are all the more so intriguing because an entire information about minimal scale ℓ and, therefore, about the non-commutative geometry of the Snyder quantized space-time, was placed in the mass matrices $M_\pm$ only, while the relativistic space-time and the Minkowski energy-momentum space formalisms are exactly the same as in the standard relativistic quantum mechanics which arises from Einstein's Special Relativity.

It is worth stressing that such a procedure is also consistent from the methodological point of view. We have applied the tensor formalism of the hyperbolic geometry of the four-dimensional Minkowski energy-momentum space within the Dirac equation modified due to the extra γ^5-term. This important point has not been noticed or has been omitted in the analysis made by B.G. Sidharth as well as many authors who treat the Snyder quantized space-time as a seemingly relativistic construction and do not consider the relativistic formalism in such a case. In this manner, we have developed a completely new realization of the idea of the *mass generation mechanism*, which cannot be received within the framework involving only Special Relativity, *i.e.* in the situation when the minimal scale is identically vanishing $\ell = 0$ or equivalently the maximal energy goes to infinite limit $\varepsilon = \infty$.

In fact, we have shown that it is possible to consistently apply the non-commutative geometry in such a way that a finite value of maximal energy exists, and results in the natural renormalization of the Einstein Special Relativity at ultra-high energies. Therefore, the mass generation mechanism presented above in itself is the effect of such a non-trivial renormalization procedure, possesses purely kinetic nature and, above all, arises due to the Dirac-like factorization applied to the operator (6.76) projecting onto the spinor wave function ψ. It should be emphasized that such a kinetic effect is the result arising due to the abstractive mathematics of the non-commutative geometry and the algebra deformation of the Snyder quantized space-time.

The key problem is, however, the straightforward generalization of such a realization of the idea of the mass generation mechanism to more general situations described in the framework of the Kontsevich deformation quantization. In other words, the crucial and presently unsolved problem is a reply to the question: *does a physical contribution from any non-commutative geometry of space-time, in general, essentially results in only a specific realization of mass generation mechanism?*. The necessity of reply to such a query is fully justified above all by the problem of the experimental verification of the theoretical results obtained within a framework involving any non-commutative geometry of space-time, and possesses fundamental significance for the philosophical foundations of the new physics obtained through the constructive applications of such abstractive mathematics.

Let us continue discussion of the ramifications of the mass generation mechanism following from the Snyder–Sidharth space-time. The mass matrices $M_{\pm}$ can be presented in a more convenient form employing a linear dependence of the γ^5 matrix. Fist, let us apply the Taylor series expansion to the square root part of the mass matrices (6.81). The expansion can be performed in the following way

$$\sqrt{1-\frac{4E^2}{\varepsilon^2}+\frac{4mc^2}{\varepsilon}\gamma^5} = \sqrt{1-\frac{4E^2}{\varepsilon^2}}\sqrt{1+\frac{\frac{4mc^2}{\varepsilon}}{1-\frac{4E^2}{\varepsilon^2}}\gamma^5} =$$
$$= \sqrt{1-\frac{4E^2}{\varepsilon^2}}\sum_{n=0}^{\infty}\binom{1/2}{n}\left(\frac{\frac{4mc^2}{\varepsilon}}{1-\frac{4E^2}{\varepsilon^2}}\gamma^5\right)^n, \quad (6.82)$$

where the generalized Newton binomial symbol was used

$$\binom{n}{k} = \frac{\Gamma(n+1)}{\Gamma(k+1)\Gamma(n+1-k)}.$$

By application of the basic properties of γ^5 matrix, *i.e.* $(\gamma^5)^{2n} = -1$ and $(\gamma^5)^{2n+1} = -\gamma^5$, one can decompose the sum present in the last term of the formula (6.82) into the two components related to the odd and even powers of the γ^5 matrix

$$\sum_{n=0}^{\infty}\binom{1/2}{n}\left(\frac{\frac{4mc^2}{\varepsilon}}{1-\frac{4E^2}{\varepsilon^2}}\gamma^5\right)^n =$$
$$= -\sum_{n=0}^{\infty}\binom{1/2}{2n}\left(\frac{\frac{4mc^2}{\varepsilon}}{1-\frac{4E^2}{\varepsilon^2}}\right)^{2n} - \sum_{n=0}^{\infty}\binom{1/2}{2n+1}\left(\frac{\frac{4mc^2}{\varepsilon}}{1-\frac{4E^2}{\varepsilon^2}}\right)^{2n+1}\gamma^5. \quad (6.83)$$

Straightforward application of the standard summation procedure allows to establish the sums presented in both the components of the decomposition (6.83)

$$\sum_{n=0}^{\infty}\binom{1/2}{2n}\left(\frac{\frac{4mc^2}{\varepsilon}}{1-\frac{4E^2}{\varepsilon^2}}\right)^{2n} = \sqrt{1+\frac{\frac{4mc^2}{\varepsilon}}{1-\frac{4E^2}{\varepsilon^2}}}+\sqrt{1-\frac{\frac{4mc^2}{\varepsilon}}{1-\frac{4E^2}{\varepsilon^2}}}, \quad (6.84)$$

$$\sum_{n=0}^{\infty}\binom{1/2}{2n+1}\left(\frac{\frac{4mc^2}{\varepsilon}}{1-\frac{4E^2}{\varepsilon^2}}\right)^{2n+1} = \sqrt{1+\frac{\frac{4mc^2}{\varepsilon}}{1-\frac{4E^2}{\varepsilon^2}}}-\sqrt{1-\frac{\frac{4mc^2}{\varepsilon}}{1-\frac{4E^2}{\varepsilon^2}}}. \quad (6.85)$$

In this manner, it can be seen straightforwardly that both mass matrices $M_{\pm}$ possess the following decomposition onto two components: the hermitian $\mathfrak{H}(M_{\pm})$ and the antihermitian $\mathfrak{A}(M_{\pm})$

$$M_{\pm} = \mathfrak{H}(M_{\pm}) + \mathfrak{A}(M_{\pm}), \quad (6.86)$$

where both parts can be presented in a compact form

$$\mathfrak{H}(M_{\pm}) = \pm\frac{\varepsilon}{2c^2}\left[\sqrt{1-\frac{4E^2}{\varepsilon^2}}\left(\sqrt{1+\frac{\frac{4mc^2}{\varepsilon}}{1-\frac{4E^2}{\varepsilon^2}}}-\sqrt{1-\frac{\frac{4mc^2}{\varepsilon}}{1-\frac{4E^2}{\varepsilon^2}}}\right)\right], \quad (6.87)$$

$$\mathfrak{A}(M_{\pm}) = -\frac{\varepsilon}{2c^2}\left[1\pm\sqrt{1-\frac{4E^2}{\varepsilon^2}}\left(\sqrt{1+\frac{\frac{4mc^2}{\varepsilon}}{1-\frac{4E^2}{\varepsilon^2}}}+\sqrt{1-\frac{\frac{4mc^2}{\varepsilon}}{1-\frac{4E^2}{\varepsilon^2}}}\right)\right]\gamma^5 \quad (6.88)$$

By application of certain elementary algebraic manipulations one obtains the equivalent decomposition of the mass matrices $M_{\pm}$ into the basis of the commutating projectors $\left\{\Pi_i : \frac{1+\gamma^5}{2}, \frac{1-\gamma^5}{2}\right\}$,

$$M_{\pm} = \sum_i \mu_i^{\pm}\Pi_i = \mu_R^{\pm}\frac{1+\gamma^5}{2} + \mu_L^{\pm}\frac{1-\gamma^5}{2}, \quad (6.89)$$

where $\mu_R^{\pm}$ and $\mu_L^{\pm}$ are the projected masses related to the Dirac theories with signs $\pm$ in agreement with the mass matrix signs. The values of the projected masses can be established easily

$$\mu_R^{\pm} = -\frac{1}{c^2}\left(\frac{\varepsilon}{2} \pm \sqrt{\varepsilon^2 - 4\varepsilon mc^2 - 4E^2}\right), \quad (6.90)$$

$$\mu_L^{\pm} = \frac{1}{c^2}\left(\frac{\varepsilon}{2} \pm \sqrt{\varepsilon^2 + 4\varepsilon mc^2 - 4E^2}\right). \quad (6.91)$$

For physical consistency, we shall presume that the masses of right-handed neutrinos (6.90) and the left-handed neutrinos (6.91) are real numbers. It is a non-trivial condition, which allows to express the range of maximal energy ε through the energy and the mass of an original quantum state obeying the Dirac equation. In the case of massive states one obtains

$$\begin{aligned} \varepsilon \in & \left(-\infty, -2mc^2\left(1+\sqrt{1+\left(\frac{E}{mc^2}\right)^2}\right)\right] \cup \\ \cup & \left[-2mc^2\left(1-\sqrt{1+\left(\frac{E}{mc^2}\right)^2}\right), 2mc^2\left(1+\sqrt{1+\left(\frac{E}{mc^2}\right)^2}\right)\right] \cup \\ \cup & \left[2mc^2\left(1+\sqrt{1+\left(\frac{E}{mc^2}\right)^2}\right), \infty\right), \end{aligned} \quad (6.92)$$

while in the case of massless quantum states possessing energy E

$$\varepsilon \in (-\infty, -2|E|] \cup [2|E|, \infty). \quad (6.93)$$

The properties of the projectors $\Pi_i^{\dagger}\Pi_i = \mathbf{1}_4$, $\Pi_1\Pi_2 = \frac{1}{2}\mathbf{1}_4$, $\Pi_1^{\dagger} = \Pi_2$ and $\Pi_1 + \Pi_2 = \mathbf{1}_4$ allow to derive the relation

$$M_{\pm}M_{\pm}^{\dagger} = \frac{(\mu_R^{\pm})^2 + (\mu_L^{\pm})^2}{2}\mathbf{1}_4. \quad (6.94)$$

By introducing the right- and left-handed chiral Weyl fields

$$\psi_R = \frac{1+\gamma^5}{2}\psi, \quad (6.95)$$

$$\psi_L = \frac{1-\gamma^5}{2}\psi, \quad (6.96)$$

where wave function ψ is a solution of the appropriate Dirac equations (6.79) and (6.80), both theories can be rewritten as the system of two equations

$$\left(\gamma^\mu \hat{p}_\mu + \mu^+ c^2\right) \begin{bmatrix} \psi_R^+ \\ \psi_L^+ \end{bmatrix} = 0, \tag{6.97}$$

$$\left(\gamma^\mu \hat{p}_\mu + \mu^- c^2\right) \begin{bmatrix} \psi_R^- \\ \psi_L^- \end{bmatrix} = 0, \tag{6.98}$$

where now the mass matrices $\mu^\pm$, related to the chiral fields $\psi_{R,L}^\pm$, are manifestly hermitian quantities

$$\mu^\pm = \begin{bmatrix} \mu_R^\pm & 0 \\ 0 & \mu_L^\pm \end{bmatrix} = \begin{bmatrix} \mu_R^\pm & 0 \\ 0 & \mu_L^\pm \end{bmatrix}^\dagger . \tag{6.99}$$

Note that the masses (6.90) and (6.91) are invariant with respect to a choice of representation of the Dirac matrices γ^μ. For this reason these quantities have physical character. It is interesting that for the mirror reflection in minimal scale $\ell \to -\ell$, or equivalently for the mirror reflection $\varepsilon \to -\varepsilon$, one has the exchange between the masses $\mu_R^\pm \leftrightarrow \mu_L^\pm$ while the chiral Weyl fields stay unchanged. Similarly, in the situation of the mirror reflection in the mass of an original quantum state $m \to -m$ one has the exchange between the projected masses $\mu_R^\pm \leftrightarrow -\mu_L^\pm$. The case of originally massless quantum states $m = 0$ is also intriguing from a theoretical point of view.

From the mass formulas (6.90) and (6.91) one can see straightforwardly that in such a case the equality between masses of the neutrinos $\mu_R = -\mu_L$ holds. In the limiting case of the generic Einstein's Special Relativity $\ell = 0$, the projected masses have interesting properties

$$\mu_R^\pm = \begin{cases} -\infty & \text{for } + \\ \infty & \text{for } - \end{cases}, \tag{6.100}$$

$$\mu_L^\pm = \begin{cases} \infty & \text{for } + \\ -\infty & \text{for } - \end{cases}. \tag{6.101}$$

Such a situation means that in the case of Special Relativity the massive Weyl theories are undetermined, *i.e.* the particles described by the Weyl equations (6.97) and (6.98) are non-existent. In general, however, for correctness of the projection splitting (6.89) both projected masses

(6.90) and (6.91) must be real numbers. Strictly speaking, when the projected masses are complex numbers the decomposition (6.89) does not yield hermitian mass matrices (6.99). Therefore, in such a situation the presented construction losses the physical sense, and for this reason should be replaced by another theory based on different arguments.

In the approach propagated by the Standard Model neutrinos are massless quantum particles, *i.e.* are the Weyl neutrinos. In this manner, it is evident that application of the non-commutative geometry of the Snyder quantized space-time has generated the non-triviality which is *the kinetic mass generation mechanism of the neutrino mass*. It should be emphasized that in all the cited contributions Sidharth mentioned a possibility of neutrino mass "due to the mass term", where by the mass term he understands the dynamical extra γ^5-term in the modified Dirac equation (6.66). Albeit, regarding the Standard Model which is well-accepted theory of elementary particles and fundamental interactions, mass term cannot have a dynamical nature, and for this reason the nomenclature of Sidharth is misleading in further analysis of the philosophical ideas about the neutrino. Sidharth's statement has even more serious ramifications: the dynamical mass term breaks the Poincaré invariance of the model, and in this manner interpretation of the extra γ^5-term as the term leads to inconsistencies.

We have obtained the massive neutrinos due to the two-step mass generation mechanism. The first one is factorization of the modified Dirac equation (6.66). The second one is the decomposition of the mass matrices (6.81) into the projectors basis and introducing the chiral Weyl fields (6.95) and (6.96). Calling such an unique procedure as the result "due to the mass term" is at least inaccurate, creates a number of inequivalent interpretations, and in itself is escaping from physics to pure philosophy. It should be emphasized that an arbitrary mass generation mechanism has not been presented or suggested within Sidharth's contributions and the reasoning presented there completely differs from our analysis, manifestly omits a number of important physical and mathematical details (Cf. *e.g.* Refs. [343, 347]).

The interpretation proposed in this section, *i.e.* the unique treatment of the extra γ^5-term in the modified Dirac equation (6.66) and preservation the hyperbolic geometry of Minkowski energy-momentum space (6.70) of the Einstein Special Relativity, has resulted in generation of the Weyl equations (6.97)-(6.98) describing two left- $\psi_L^{\pm}$ and two right- $\psi_R^{\pm}$ chiral fields possessing manifestly non-zero masses. In other words, we have established the constructive and full consistent with Special Relativity theory of massive neutrinos, related to both originally massive $m \neq 0$ as well as massless $m = 0$ quantum states described by relativistic quantum mechanics.

For this reason our approach changes the physical essence of the concepts *neutrino* and *neutrino mass*. In our understanding the neutrino is a chiral field due to any originally massive and massless quantum particle, which in itself is also a quantum particle. Meanwhile, neutrino mass is generated due to the mass of the originally quantum particle described by the Dirac equation. Furthermore, it is easy to see that the Weyl theories (6.97) and (6.98) associate four massive neutrinos with a one original quantum state. Because these are four possible theories (6.69), the maximal number of the neutrinos obtained through the non-commutative geometry of the Snyder quantized space-time is sixteen. However, this number can be reduced due to experimental verification of the results of these theories.

D Compton–Planck Scale

The Planck scale, defined by taking into account minimal scale ℓ identical with the Planck length $\ell_P = \sqrt{\dfrac{\hbar G}{c^3}}$, in modern cosmology and particle physics is the energetic region at which quantum physics meets classical physics. At this scale the standard methodology of particle physics, *i.e.* the Standard Model, is manifestly an inadequate tool for constructive description because the necessity for a theory of quantum gravity. There are few approaches to the adequate formalism of the Planck scale physics, including string theory, M-theory, loop quantum gravity, and non-commutative geometry (Cf. *e.g.* Ref. [348]). A theory of quantum gravity, however, is the main requirement for the consistent description of the physics at the Planck scale. We shall present a certain proposal for such a theory in the next chapters of this book.

Let us look on few coincidences at the Planck scale, which are informative for general understanding. In such a situation, maximal energy

(6.75) coincides with the Planck energy divided by $\sqrt{\alpha}$

$$\varepsilon(\ell_P) = \frac{1}{\sqrt{\alpha}}\sqrt{\frac{\hbar c^5}{G}} = \frac{1}{\sqrt{\alpha}} M_P c^2 = \frac{E_P}{\sqrt{\alpha}}. \tag{6.102}$$

Let us take into account the Compton wavelength of a particle possessing the rest mass m, *i.e.* $\lambda_C(m) = 2\pi\frac{\hbar}{mc}$. Maximal energy (6.75) computed at the scale identical to the Compton wavelength $\lambda_C(m)$ is proportional to rest energy of a particle

$$\varepsilon(\lambda_C(m)) = \frac{1}{2\pi\sqrt{\alpha}} mc^2, \tag{6.103}$$

and, if α has the proposed value (5.220), becomes simply the rest energy of a particle $\varepsilon = mc^2$.

Interestingly, in the situation when the rest mass of a particle equals the Planck mass $m \equiv M_P = \sqrt{\frac{\hbar c}{G}}$ or, equivalently, the rest energy of a particle equals the Planck energy $mc^2 = E_P$, there is a number of nontrivialities. Let us denote the Compton wavelength of such a Planckian particle by $\ell_C = \lambda_C(M_P)$. In fact, this wavelength defines the mixed scale which we propose to call *the Compton–Planck (CP) scale*. It can be seen straightforwardly that if the CP scale holds

$$\varepsilon(\ell_C) = \frac{\varepsilon(\ell_P)}{2\pi}. \tag{6.104}$$

Moreover, when one regards (5.220), *i.e.* $\alpha = 1/(2\pi)^2$, then also

$$\alpha = \left(\frac{\varepsilon(\ell_C)}{\varepsilon(\ell_P)}\right)^2 = \left(\frac{\ell_P}{\ell_C}\right)^2. \tag{6.105}$$

Moreover, in the CP scale the doubled Compton wavelength equals the circumference of a circle with a radius of the Schwarzschild radius $r_S(m) = \frac{2Gm}{c^2}$ of the Planck mass (Cf. Ref. [322])

$$2\ell_C = 2\pi r_S(M_P). \tag{6.106}$$

Straightforward and easy computation shows that the Compton wavelength of the Planck mass is identified with the circumference of the circle with a radius of the Planck length, *i.e.*

$$\ell_C = 2\pi\ell_P, \tag{6.107}$$

and for this reason the doubled Planck length equals the Schwarzschild radius of the Planck mass

$$2\ell_P = r_S(M_P). \tag{6.108}$$

In general, the ratio of the Planck length and the Compton wavelength of a particle with mass m is

$$\frac{\ell_P}{\lambda_C(m)} = \frac{1}{2\pi}\frac{m}{M_P}. \tag{6.109}$$

Let us generalize the last relation in the equation (6.105) as follows

$$\alpha \equiv \left(\frac{1}{2\pi}\frac{m}{M_P}\right)^2. \tag{6.110}$$

Taking α according to the choice (5.220) together with the presumption $\alpha \sim 1$, which guarantees correctness of the Kontsevich deformation quantization. Such a reasoning establishes the mass of a particle for which the non-commutative geometry of the Snyder quantized space-time is adequate. It is not difficult to see that the mass of a particle must be of the order of the Planck mass, *i.e.*

$$m \sim M_P. \tag{6.111}$$

If, however, one wishes to neglect (5.220) but preserve (6.110) together with the condition $\alpha \sim 1$, then

$$m \sim 2\pi M_P, \tag{6.112}$$

i.e. the mass of a particle described by the non-commutative geometry is of the order $m \sim (10^{22} - 10^{23})\mathrm{MeV}/\mathrm{c}^2$.

E Global Effective Chiral Condensate

Let us consider the meaning of the massive Weyl equations (6.97)-(6.97) in the spirit of the gauge field theories [349], which are the base of the Standard Model. The problem is to construct the Lagrangian $\mathscr{L}$, revealing Lorentz invariance, of the gauge field theory characterized by the massive Weyl equations treated as the Euler–Lagrange equations of motion for the chiral fields $\psi_L^{\pm}$ and $\psi_R^{\pm}$. In general, such a construction is not easy to perform, but because the massive Weyl equations are in fact the Dirac theories, it can be seen by straightforward computations that the following four Dirac-like Lagrangians

$$\mathscr{L}_R^{\pm} = \bar{\psi}_R^{\pm}\left(\gamma^{\mu}\hat{p}_{\mu} + \mu_R^{\pm}c^2\right)\psi_R^{\pm}, \tag{6.113}$$

$$\mathscr{L}_L^{\pm} = \bar{\psi}_L^{\pm}\left(\gamma^{\mu}\hat{p}_{\mu} + \mu_L^{\pm}c^2\right)\psi_L^{\pm}, \tag{6.114}$$

where $\bar{\psi}_{R,L}^{\pm} = \left(\psi_{R,L}^{\pm}\right)^{\dagger}\gamma^0$ are the Dirac adjoint of $\psi_{R,L}^{\pm}$, lead to the massive Weyl equations by the appropriate principle of the least action. In other words, the Euler–Lagrange equations of motion

$$\frac{\partial \mathscr{L}_{R,L}^{\pm}}{\partial \psi_{R,L}^{\pm}} - \partial_{\mu}\frac{\partial \mathscr{L}_{R,L}^{\pm}}{\partial\left(\partial_{\mu}\psi_{R,L}^{\pm}\right)} = 0, \tag{6.115}$$

coincide with the massive Weyl equations (6.97)-(6.97). In this manner, the appropriate full gauge field theory of massive neutrinos can be directly constructed by using the Lagrangian which is an algebraical sum of the four gauge field theories (6.113)-(6.114)

$$\mathscr{L} = \mathscr{L}_R^{+} + \mathscr{L}_R^{-} + \mathscr{L}_L^{+} + \mathscr{L}_L^{-}, \tag{6.116}$$

i.e. the massive Weyl equations are obtained from the Euler–Lagrange equations of motion (6.115) with the exchange $\mathscr{L}_{R,L}^{\pm} \to \mathscr{L}$. It should be emphasized that the choice of the Lagrangians in the form (6.113)-(6.114) is due to straightforward analogy between the massive Weyl equation and the Dirac equation. The essential difference between these theories is only the number of spinor components which, however, has no influence on the form of the Lagrangian and the principle of the least action. The Lagrangians (6.113) and (6.114) describe the two components of the Weyl spinor. In other words, this choice is both most intuitive and simplest. However, it does not mean that there is no another, possibly more complicated, choice dictated by another justification.

One can see easily that the gauge field theories (6.113) and (6.114) exhibit several well-known gauge symmetries. Namely, the (local) chiral symmetry $SU(2)_R^{\pm} \otimes SU(2)_L^{\pm}$ is expressed through one of the transformations

$$\left\{ \begin{array}{c} \psi_R^{\pm} \to \exp\left\{i\theta_R^{\pm}\right\} \psi_R^{\pm} \\ \psi_L^{\pm} \to \psi_L^{\pm} \end{array} \right. , \tag{6.117}$$

$$\left\{ \begin{array}{c} \psi_R^{\pm} \to \psi_R^{\pm} \\ \psi_L^{\pm} \to \exp\left\{i\theta_L^{\pm}\right\} \psi_L^{\pm} \end{array} \right. , \tag{6.118}$$

the vector symmetry $U(1)_V^{\pm}$

$$\left\{ \begin{array}{c} \psi_R^{\pm} \to \exp\{i\theta^{\pm}\} \psi_R^{\pm} \\ \psi_L^{\pm} \to \exp\{i\theta^{\pm}\} \psi_L^{\pm} \end{array} \right. , \tag{6.119}$$

and the axial symmetry $U(1)_A^{\pm}$

$$\left\{ \begin{array}{c} \psi_R^{\pm} \to \exp\{-i\theta^{\pm}\} \psi_R^{\pm} \\ \psi_L^{\pm} \to \exp\{i\theta^{\pm}\} \psi_L^{\pm} \end{array} \right. . \tag{6.120}$$

In this manner, the total symmetry group is

$$SU(3)_C^{+} \oplus SU(3)_C^{-}, \tag{6.121}$$

where $SU(3)_C^{\pm}$ are the global (chiral) 3-flavor gauge symmetries related to each of the gauge theories (6.113) and (6.114), *i.e.*

$$SU(2)_R^{+} \otimes SU(2)_L^{+} \otimes U(1)_V^{+} \otimes U(1)_A^{+} \equiv SU(3)^{+} \otimes SU(3)^{+} = SU(3)_C^{+}, \tag{6.122}$$
$$SU(2)_R^{-} \otimes SU(2)_L^{-} \otimes U(1)_V^{-} \otimes U(1)_A^{-} \equiv SU(3)^{-} \otimes SU(3)^{-} = SU(3)_C^{-}, \tag{6.123}$$

describing 2-flavor massive free quarks - *the neutrinos* in our proposition. Because, the group (6.121) does not possess a name in literature, we shall call the group *the composite symmetry* $SU(3)_C^{TOT}$.

By application of the definitions for the chiral Weyl fields (6.95) and (6.96) one obtains

$$\bar{\psi}^{\pm}_{R,L}\gamma^{\mu}p_{\mu}\psi^{\pm}_{R,L}=\bar{\psi}^{\pm}\frac{1\pm\gamma^5}{2}\gamma^{\mu}p_{\mu}\frac{1\pm\gamma^5}{2}\psi^{\pm}=\bar{\psi}^{\pm}\left(\frac{1\pm\gamma^5}{2}\gamma^{\mu}\frac{1\pm\gamma^5}{2}\right)p_{\mu}\psi^{\pm}, \tag{6.124}$$

where $\bar{\psi}^{\pm}=(\psi^{\pm})^{\dagger}\gamma^0$ is the Dirac adjoint of the Dirac fields $\psi^{\pm}$ related to the chiral Weyl fields by the transformations (6.95) and (6.96). Because the identity

$$\frac{1\pm\gamma^5}{2}\gamma^{\mu}\frac{1\pm\gamma^5}{2}=\frac{\gamma^{\mu}\pm\{\gamma^{\mu},\gamma^5\}+\gamma^5\gamma^{\mu}\gamma^5}{4}=\frac{1-(\gamma^5)^2}{4}\gamma^{\mu}=\frac{1}{2}\gamma^{\mu}, \tag{6.125}$$

where we have applied the basic properties of γ^5 matrix $\{\gamma^{\mu},\gamma^5\}=0$ and $(\gamma^5)^2=-1$, one obtains finally

$$\bar{\psi}^{\pm}_{R,L}\gamma^{\mu}p_{\mu}\psi^{\pm}_{R,L}=\frac{1}{2}\bar{\psi}^{\pm}\gamma^{\mu}p_{\mu}\psi^{\pm}. \tag{6.126}$$

Similarly, applying the identity

$$\frac{1\pm\gamma^5}{2}\frac{1\pm\gamma^5}{2} \;=\; \frac{1}{4}\left(1\pm2\gamma^5+(\gamma^5)^2\right)=\pm\frac{1}{2}\gamma^5, \tag{6.127}$$

one can establish the quantity

$$\mu^{\pm}_{R,L}c^2\bar{\psi}^{\pm}_{R,L}\psi^{\pm}_{R,L}=\mu^{\pm}_{R,L}c^2\bar{\psi}^{\pm}\frac{1\pm\gamma^5}{2}\frac{1\pm\gamma^5}{2}\psi^{\pm}, \tag{6.128}$$

with the result

$$\mu^{\pm}_{R,L}c^2\bar{\psi}^{\pm}_{R,L}\psi^{\pm}_{R,L}=\frac{1}{2}\bar{\psi}^{\pm}\left(\pm\mu^{\pm}_{R,L}c^2\gamma^5\right)\psi^{\pm}, \tag{6.129}$$

where the plus sign of the mass is appropriate for the right-handed neutrinos, while the minus sign is appropriate for the left-handed neutrinos. In this way one obtains finally the partial Lagrangians

$$\begin{aligned}\mathscr{L}^{\pm}_{R,L} &= \bar{\psi}^{\pm}_{R,L}\left(\gamma^{\mu}p_{\mu}+\mu^{\pm}_{R,L}c^2\right)\psi^{\pm}_{R,L}=\\ &= \bar{\psi}^{\pm}_{R,L}\gamma^{\mu}p_{\mu}\psi^{\pm}_{R,L}+\mu^{\pm}_{R,L}c^2\bar{\psi}^{\pm}_{R,L}\psi^{\pm}_{R,L}=\\ &= \frac{1}{2}\bar{\psi}^{\pm}\gamma^{\mu}p_{\mu}\psi^{\pm}+\frac{1}{2}\bar{\psi}^{\pm}\left(\pm\mu^{\pm}_{R,L}c^2\gamma^5\right)\psi^{\pm}=\\ &= \frac{1}{2}\bar{\psi}^{\pm}\left(\gamma^{\mu}p_{\mu}\pm\mu^{\pm}_{R,L}c^2\gamma^5\right)\psi^{\pm}.\end{aligned} \tag{6.130}$$

which reveal the Lorentz invariance. In this manner, it can be seen straightforwardly that using the partial Lagrangians (6.130) the global chiral Lagrangian (6.116) takes the following form

$$\begin{aligned}\mathscr{L} &= \frac{1}{2}\bar{\psi}^{+}\left(\gamma^{\mu}p_{\mu}+\mu_{R}^{+}c^{2}\gamma^{5}\right)\psi^{+}+\frac{1}{2}\bar{\psi}^{+}\left(\gamma^{\mu}p_{\mu}-\mu_{L}^{+}c^{2}\gamma^{5}\right)\psi^{+}+\\ &+ \frac{1}{2}\bar{\psi}^{-}\left(\gamma^{\mu}p_{\mu}+\mu_{R}^{-}c^{2}\gamma^{5}\right)\psi^{-}+\frac{1}{2}\bar{\psi}^{-}\left(\gamma^{\mu}p_{\mu}-\mu_{L}^{-}c^{2}\gamma^{5}\right)\psi^{-},\end{aligned} \quad (6.131)$$

or after summation

$$\mathscr{L} = \bar{\psi}^{+}\left(\gamma^{\mu}\hat{p}_{\mu}+\mu_{eff}^{+}c^{2}\right)\psi^{+}+\bar{\psi}^{-}\left(\gamma^{\mu}\hat{p}_{\mu}+\mu_{eff}^{-}c^{2}\right)\psi^{-}, \quad (6.132)$$

where $\mu_{eff}^{\pm}$ are the effective mass matrices of the gauge fields $\psi^{\pm}$,

$$\mu_{eff}^{\pm} = \frac{\mu_{R}^{\pm}-\mu_{L}^{\pm}}{2}\gamma^{5}. \quad (6.133)$$

After introduction of the global effective 8-component field

$$\Psi = \begin{bmatrix} \psi^{+} \\ \psi^{-} \end{bmatrix}, \quad (6.134)$$

the theory (6.132) becomes

$$\mathscr{L} = \bar{\Psi}\left(\gamma^{\mu}\hat{p}_{\mu}+M_{eff}c^{2}\right)\Psi, \quad (6.135)$$

where M_{eff} is the mass matrix given by

$$M_{eff} = \begin{bmatrix} \mu_{eff}^{+} & 0 \\ 0 & \mu_{eff}^{-} \end{bmatrix}, \quad (6.136)$$

γ^{μ} are 8×8 matrices, and

$$\hat{p}_{\mu} = i\hbar\begin{bmatrix} \partial_{\mu} \\ \partial_{\mu} \end{bmatrix}, \quad (6.137)$$

is the 8-component momentum operator. Therefore, (6.135) can be associated with octonions. Hermiticity of both the effective mass matrices $\mu_{eff}^{\pm}$ and, therefore, also of the global effective mass matrix M_{eff}, depends on the choice of representation of the γ^{5} matrix. For consistency, the preferred representation of γ^{5} matrix must be hermitian. It means that the effective global gauge field theory (6.135) is physical for only such representation.

Obviously, the global effective gauge field theory (6.135) manifestly demonstrates an invariance with respect to action of the composite gauge vector symmetry $SU(2)_V^{TOT}$

$$SU(2)_V^{TOT} = SU(2)_V^{+} \oplus SU(2)_V^{-}, \tag{6.138}$$

where $SU(2)_V^{\pm}$ are the $SU(2) \otimes SU(2)$ group transformations applied separately to each of the gauge fields $\psi^{\pm}$

$$\begin{cases} \psi^{\pm} \to \exp\{i\theta^{\pm}\}\, \psi^{\pm} \\ \bar{\psi}^{\pm} \to \bar{\psi}^{\pm} \exp\{-i\theta^{\pm}\} \end{cases} . \tag{6.139}$$

Such a situation means that there is realized the mechanism of spontaneous symmetry breaking for the global effective gauge field theory (6.135). The broken symmetry is the composite global chiral symmetry $SU(3)_C^{TOT}$, and the result of the symmetry breaking is breakdown to its subgroup which is the composite isospin symmetry group $SU(2)_V^{TOT}$

$$SU(3)_C^{TOT} \longrightarrow SU(2)_V^{TOT}. \tag{6.140}$$

In such a situation, the physical interpretation is clear and unambiguous. Namely, it is the syndrome of the existence of the global effective chiral condensate of the massive neutrinos, being the composition of two independent chiral condensates, which is the global effective gauge field theory invariant under action of the gauge symmetry (See, *e.g.* the book of Steven Weinberg in Ref. [349])

$$SU(2)_V^{TOT} = (SU(2)^{+} \otimes SU(2)^{+}) \oplus (SU(2)^{-} \otimes SU(2)^{-}). \tag{6.141}$$

However, because action of the composite global chiral gauge symmetry $SU(3)_C^{TOT}$ (6.121), the gauge field theories (6.113) and (6.114) look like formally as the theories of non-interacting massive free quarks. Such a situation is very similar to the formalism of quantum chromodynamics [350], but factually in the presented physical scenario one has to deal with the composition of two independent copies of quantum chromodynamics. For each of these theories the space of fields is distinguishable for the space of fields of the single quantum chromodynamics. The difference is contained, namely, in the fact that there are only two massive chiral fields - the left- and the right-handed Weyl fields, which are the neutrinos in our proposition.

The global effective chiral condensate of massive neutrinos (6.135) is the result manifestly beyond the Standard Model, but essentially can be included in the fundamental theory of particle physics as the new contribution due to the non-commutative geometry of the Snyder quantized space-time. Usually, the mechanism of spontaneous symmetry breaking results in generation of the Goldstone bosons. However, in the situation presented above such a mechanism results in the chiral condensate, *i.e.* does not generate new particles. In the context of the massive neutrinos, the result of the mechanism of the spontaneous symmetry breaking is the global effective chiral condensate of the massive neutrinos. For this reason such a situation is manifestly distinguishable, and is beyond the present state of the Standard Model.

F Conclusion

The specific case of the deformation of Special Relativity given by the Snyder–Sidharth Hamiltonian constraint (6.65) obtained due to the non-commutative geometry of the Snyder quantized space-time (6.31)-(6.31) is essentially different from the Einstein energy-momentum relation which lays the foundations of Special Relativity's ramifications. In particular, it is self-evident from the specific form of the Snyder–Sidharth Hamiltonian constraint(6.65) that the non-commutative geometry of the Snyder quantized space-time is compatible with the deformation to the Einstein energy-momentum relation considered by Burra Gautam Sidharth. This is brought out very clearly in the crucial ramifications of the Snyder–Sidharth space-time – the modified Dirac equations (6.79)-(6.80) which are manifestly non-hermitian, as well as the massive Weyl equations (6.97)-(6.98) which are blatantly hermitian and responsible for description of the neutrinos in our proposition.

A massless neutrino, characteristic for the conventional Weyl theory and the Standard Model which presently makes use of the massless neutrino theory, now becomes manifestly massive particle, and further, it can be straightforwardly noticed from the mass formulas (6.86) and (6.89) that this mass possesses two left-handed components and two right-handed components. Once this is recognized, the mass matrix, which otherwise appears non-hermitian, turns out to be actually hermitian, as seen in the formula (6.99), but if and only if the masses (6.86) and (6.89) of the neutrinos are real numbers. There are not any restrictions, however, for the sign of these masses. In other words, the masses can have positive as well as negative values. In this manner, the underlying non-commutative geometry of the Snyder quantized space-time (6.31)-(6.31) is reflected in the modified Dirac equation (6.66) and, naturally, and highly nontrivially, gives rise to the mass of the neutrino.

As we have mentioned in the previous discussions within this chapter, in analogy with the Standard Model, B.G. Sidharth [347] has suggested that such a situation can be a result "due to mass term", however, with no concrete calculations and the proposals for the realization of the idea of the mass generation mechanism for neutrinos. The mass generation mechanism, proposed in this chapter for such a constructive and consistent realization of the general idea of Sidharth, is of purely kinetic nature, and moreover it is formally the ramification arising from the first approximation of more general non-commutative geometry framework determined by the Kontsevich deformation quantization. In this manner, we have shown that Sidharth's general idea of the mass generation mechanism "due to mass term" can be elegantly formulated in the framework of the non-commutative geometry, which particular case is the Snyder–Sidharth space-time.

We have shown also that the model of massive neutrinos can be understood and consistently described from the point of view of gauge field theories, which naturally includes both the Poincaré invariance and the Lorentz invariance. Such a formulation leads to interesting construction involving two independent copies of quantum chromodynamics and non-interacting massive free quarks, which also employs the effective composite isospin group resulting in the global effective chiral condensate of the massive neutrinos. The mechanism of spontaneous symmetry breaking presented in this chapter, which is the tool to receiving the composite isospin group, does not require existence of the Goldstone bosons whereas the role of the Goldstone bosons is played by the chiral condensate of massive neutrinos. In itself this is a new type of mass generation mechanism.

The results are consistent with some recent experimental data. It must be remembered that presently in the Standard Model the neutrino is considered as massless, but in 1998 the Super-Kamiokande neutrino detector experiments determined that neutrinos can oscillate from one flavor to another, which requires that neutrino must have a non-zero mass [351]. This is the leading motivation to taking into account the models beyond the Standard Model. The model of massive neutrinos presented in this chapter is the best possible example of such a proposal. In this connection, it is also relevant to mention that currently the Standard Model requires the Higgs mechanism for the generation of mass in general, though the Higgs particle has been undetected for forty five years and it is hoped will be detected by researchers of the Fermi National Accelerator Laboratory or the Large Hadron Collider.

Chapter 7

Neutrinos: Energy Renormalization & Integrability

A Introduction

In the previous chapter we have established that the modified Dirac equation, arising due to the non-commutative geometry of the Snyder quantized space-time, yields the conventional Dirac theory with non-hermitian mass, or equivalently the Weyl equation with a diagonal and hermitian mass matrices which describes the massive neutrinos. The obtained model of massive neutrinos involves four massive chiral fields related to any originally massive or massless quantum state obeying the usual Dirac equation. By application of the mechanism of spontaneous symmetry breaking with respect to the global chiral symmetry the model was converted into the form of the isospin-symmetric global effective gauge field theory of the eight-component field Ψ which is associated with the composed chiral condensate of massive neutrinos.

All these ramifications violate the Lorentz (CP) symmetry manifestly, albeit the possible physical application arising from them can be considered in various ways. On the one hand, the global effective gauge field theory is beyond the Standard Model, yet can be considered as the new contributory part due to the non-commutative geometry of the Snyder quantized space-time. On the other hand, in the model of massive neutrinos the masses of the two left-handed and two right-handed chiral Weyl fields arise due to mass and energy of an original state and minimal scale, *e.g.* the Planck scale. Therefore, its quantum mechanical countenance becomes almost a mystical riddle. In fact, possible existence of the massive neutrinos would be the logically consistent justification of the physical consistency of the non-commutative geometry of the Snyder quantized space-time.

This chapter is mostly focused on the quantum mechanical aspect of the model of massive neutrinos presented on the previous chapter. We shall present manifestly that the model in itself yields the consistent physical explanation of the non-commutative geometry of the Snyder quantized space-time and, consequently, leads to energy renormalization of an original quantum relativistic particle. We shall perform calculations arising directly from the Schrödinger equation formulation of both the modified Dirac equation and the massive Weyl equation.

First of all, the essential problem for discussion is the manifestly non-hermitian modified Dirac Hamiltonian. Its integrability is formulated by straightforward application of the Zassenhaus formula for exponentiation of the sum of two non commuting operators. It is shown straightforwardly, however, that this approach does not lead to well-defined solutions, because for such a formulation the Zassenhaus exponents are still sums of two non commuting operators. Therefore, such an integrability procedure possesses a cyclic problem which cannot be removed, and for this reason is not a good algorithm. In such instance, the only approximations can be studied, but extraction of the full solution is an extremely difficult problem.

For solving the problem we shall change the integrability strategy, *i.e.* instead of the modified Dirac equation we shall employ the Schrödinger equation form of the massive Weyl equation with the pure hermitian mass matrix. Integration of this equation can be performed straightforwardly, elementary, and analogously to the integration of the usual Dirac equation. Calculations will be presented in both the Dirac and the Weyl representations of the Dirac gamma matrices.

It is worth stressing that we perform calculations in the Clifford algebra because the representations of gamma matrices obeying the Æther algebra are not established and are a very good problem for future research. This is caused by the fact, that the Æther algebra was proposed first in this book, and was not considered in earlier literature, and we would like to establish first the ramifications of the usual Dirac algebra.

B Energy Renormalization

Let us focus our attention on the masses of left-handed and right-handed chiral Weyl fields (6.90) and (6.91). By straightforward elementary algebraic manipulations these two relations can be rewritten as the following system of equations

$$\begin{cases} \left(\mu_R^{\pm}c^2 + \frac{\varepsilon}{2}\right)^2 = \varepsilon^2 - 4\varepsilon mc^2 - 4E^2 \\ \left(\mu_L^{\pm}c^2 - \frac{\varepsilon}{2}\right)^2 = \varepsilon^2 + 4\varepsilon mc^2 - 4E^2 \end{cases} \tag{7.1}$$

which allows to study dependence between the deformation parameter, *i.e.* maximal energy ε, and energy E and mass m of a particle and the masses $\mu_R^{\pm}$ and $\mu_L^{\pm}$ of neutrinos treated as physically measurable quantities which can be established by experimental data. Subtraction of the second equation from the first one in the system of equations (7.1) allows to obtain

$$\left(\mu_L^{\pm}c^2 - \frac{\varepsilon}{2}\right)^2 - \left(\mu_R^{\pm}c^2 + \frac{\varepsilon}{2}\right)^2 = 8\varepsilon mc^2, \tag{7.2}$$

or applying the difference of two squares $a^2 - b^2 = (a-b)(a+b)$

$$\left[\left(\mu_L^{\pm} - \mu_R^{\pm}\right)c^2 - \varepsilon\right]\left(\mu_L^{\pm} + \mu_R^{\pm}\right)c^2 = 8\varepsilon mc^2, \tag{7.3}$$

which allows to derive maximal energy as

$$\varepsilon = \frac{\left(\mu_L^{\pm} - \mu_R^{\pm}\right)c^2}{1 + \dfrac{8m}{\mu_L^{\pm} + \mu_R^{\pm}}}. \tag{7.4}$$

Because $\varepsilon \geqslant 0$ one has the condition of masses of neutrinos

$$\mu_L^{\pm} \geqslant \mu_R^{\pm}, \tag{7.5}$$

which after using the explicit formulas (6.90) and (6.91) leads to two alternative conditions for ε

$$\varepsilon \geqslant \mp\left(\sqrt{\varepsilon^2+4\varepsilon mc^2-4E^2}+\sqrt{\varepsilon^2-4\varepsilon mc^2-4E^2}\right). \tag{7.6}$$

The first condition leads to

$$\varepsilon \in \left(-\infty, -\frac{8}{3}mc^2-\frac{4}{3}\sqrt{3E^2+4m^2c^4}\right] \cup \left[-2mc^2+2\sqrt{E^2+m^2c^4}, \infty\right), \tag{7.7}$$

while the second one gives

$$\varepsilon \in \left[-2mc^2+2\sqrt{E^2+m^2c^4}, -\frac{8}{3}mc^2+\frac{4}{3}\sqrt{3E^2+4m^2c^4}\right], \tag{7.8}$$

and by taking these results together one obtains finally

$$-2mc^2+2\sqrt{E^2+m^2c^4} \leqslant \varepsilon \leqslant -\frac{8}{3}mc^2+\frac{4}{3}\sqrt{3E^2+4m^2c^4}. \tag{7.9}$$

However, because $\varepsilon \geqslant 0$ one can receive the conditions for the mass and energy of a particle

$$-2mc^2+2\sqrt{E^2+m^2c^4} \geqslant 0, \tag{7.10}$$

$$-\frac{8}{3}mc^2+\frac{4}{3}\sqrt{3E^2+4m^2c^4} \geqslant 0, \tag{7.11}$$

$$-\frac{8}{3}mc^2+\frac{4}{3}\sqrt{3E^2+4m^2c^4} \geqslant -2mc^2+2\sqrt{E^2+m^2c^4}. \tag{7.12}$$

The first and the second conditions lead to the trivial relation $E^2 \geqslant 0$, while the third one states that

$$\frac{9}{32}E^4+m^2c^4E^2+m^4c^8 \geqslant 0, \tag{7.13}$$

which is satisfied if and only if mass m and energy E of a particle are real numbers. By application of the bounds (7.9) to the relation (7.4) one obtains the condition for masses

$$m \leqslant \frac{\mu_L^{\pm}+\mu_R^{\pm}}{8}, \tag{7.14}$$

or equivalently the inequality for the ratio

$$\frac{8m}{\mu_L^{\pm}+\mu_R^{\pm}} \leqslant 1. \tag{7.15}$$

By the definition (7.4) one has

$$\frac{8m}{\mu_L^{\pm}+\mu_R^{\pm}}=\frac{\left(\mu_L^{\pm}-\mu_R^{\pm}\right)c^2}{\varepsilon}-1, \tag{7.16}$$

what in the light of the inequality (7.15) leads to the bound

$$\varepsilon \geqslant \frac{\mu_L^{\pm}-\mu_R^{\pm}}{2}c^2, \tag{7.17}$$

which can be translated into the language of minimal scale

$$\ell \leqslant \frac{\hbar}{\sqrt{\alpha}c}\frac{2}{\mu_L^{\pm}-\mu_R^{\pm}}. \tag{7.18}$$

Using the Compton wavelength of a neutrino

$$\lambda_C(\mu_{R,L}^{\pm})=2\pi\frac{\hbar}{\mu_{R,L}^{\pm}c}, \tag{7.19}$$

the bound (7.25) can be written in the form

$$\ell \leqslant \frac{1}{2\pi\sqrt{\alpha}}\frac{2\lambda_C(\mu_L^{\pm})\lambda_C(\mu_R^{\pm})}{\lambda_C(\mu_R^{\pm})-\lambda_C(\mu_L^{\pm})}, \tag{7.20}$$

which, if one wishes to apply $\alpha=\dfrac{1}{2\pi}$ established by (5.220), becomes

$$\ell \leqslant \frac{2\lambda_C(\mu_L^{\pm})\lambda_C(\mu_R^{\pm})}{\lambda_C(\mu_R^{\pm})-\lambda_C(\mu_L^{\pm})}. \tag{7.21}$$

Equivalently, the bound (7.25) with the condition (5.220) can be presented in the most conventional form

$$\ell \leqslant \lambda_C\left(\frac{\mu_R^{\pm}-\mu_L^{\pm}}{2}\right). \tag{7.22}$$

If one wishes not to preserve (5.220), then the bound (7.20) for the fixed scale value establishes the following inequality for α

$$\alpha \leqslant \left(\frac{1}{2\pi}\frac{1}{\ell}\frac{2\lambda_C(\mu_L^{\pm})\lambda_C(\mu_R^{\pm})}{\lambda_C(\mu_R^{\pm})-\lambda_C(\mu_L^{\pm})}\right)^2, \tag{7.23}$$

which in the light of the generalization (6.110) leads to

$$\frac{m}{M_P} \leqslant \frac{1}{\ell} \frac{2\lambda_C(\mu_L^{\pm})\lambda_C(\mu_R^{\pm})}{\lambda_C(\mu_R^{\pm}) - \lambda_C(\mu_L^{\pm})}, \tag{7.24}$$

or after application of the relation (7.25)

$$m \leqslant \left(\frac{\hbar}{c}\right)^2 \frac{2\pi}{\ell\ell_P} \frac{2}{\mu_L^{\pm} - \mu_R^{\pm}}. \tag{7.25}$$

This result in the light of the bound (7.14), however, leads to

$$\left(\frac{\hbar}{c}\right)^2 \frac{2\pi}{\ell\ell_P} \frac{2}{\mu_L^{\pm} - \mu_R^{\pm}} = \frac{\mu_L^{\pm} + \mu_R^{\pm}}{8}, \tag{7.26}$$

which results in the squared-mass difference

$$\Delta\mu_{LR}^2 = \left(\mu_L^{\pm}\right)^2 - \left(\mu_R^{\pm}\right)^2 = \left(\frac{\hbar}{c}\right)^2 \frac{32\pi}{\ell\ell_P} = 32\pi\frac{\ell_P}{\ell}M_P^2. \tag{7.27}$$

In this manner, if $\Delta\mu_{LR}^2$ is fixed by experimental data, then by the equation (7.27) establishes minimal scale

$$\ell = 32\pi\frac{M_P^2}{\Delta\mu_{LR}^2}\ell_P, \tag{7.28}$$

which has the approximative value

$$\ell \approx 2.4220 \cdot 10^{23} m \frac{1\frac{eV^2}{c^4}}{\Delta\mu_{LR}^2}. \tag{7.29}$$

Interestingly, at the Planck scale $\ell = \ell_P$, (7.28) then generates

$$\Delta\mu_{LR}^2 = 32\pi M_P^2 \approx 1.4985 \cdot 10^{10}\frac{YeV^2}{c^4}, \tag{7.30}$$

where $1YeV = 10^{24}eV$, while at the Compton scale $\ell = \lambda_C(m_e)$

$$\Delta\mu_{LR}^2 \approx 10^5\frac{PeV^2}{c^4}, \tag{7.31}$$

where $1PeV = 10^{15}eV$. Similarly, at the Compton–Planck scale $\ell = \lambda_C(M_P)$

$$\Delta\mu_{LR}^2 = 16M_P^2 \approx 2.3850\cdot 10^9\frac{YeV^2}{c^4}. \tag{7.32}$$

In the light of the mass formulas (6.90) and (6.91) one can deduce the squared-mass difference

$$\Delta\mu_{LR}^2 = 16\pi\frac{\ell_P}{\ell}M_P^2\left[1 \pm \frac{\sqrt{\varepsilon^2 - 4E^2 + 8\pi\frac{\ell_P}{\ell}M_P^2c^4} - \sqrt{\varepsilon^2 - 4E^2 - 8\pi\frac{\ell_P}{\ell}M_P^2c^4}}{8mc^2}\right], \tag{7.33}$$

where we have applied the generalization (6.110). Comparison of the equations (7.27) and (7.33) gives

$$1 = \pm\frac{\sqrt{\varepsilon^2 - 4E^2 + 8\pi\frac{\ell_P}{\ell}M_P^2c^4} - \sqrt{\varepsilon^2 - 4E^2 - 8\pi\frac{\ell_P}{\ell}M_P^2c^4}}{8mc^2}, \tag{7.34}$$

which is the equation for minimal scale ℓ as a function of mass m and energy E of a particle. The solution of this equation is easy to establish

$$\alpha\ell = \frac{\sqrt{3}}{16\pi}\frac{\ell_P}{\sqrt{1 + \left(\frac{E}{2mc^2}\right)^2}}, \tag{7.35}$$

and can be equivalently presented as the appropriate formula for the energy of a particle

$$\frac{E^2}{(2mc^2)^2} = \frac{3}{256\pi^2}\frac{\ell_P^2}{\alpha^2\ell^2} - 1, \tag{7.36}$$

and application of the generalization (6.110) leads to

$$E^2 = \frac{3}{16}\frac{\ell_P^2}{\ell^2}E_P^2 - (2mc^2)^2 = \frac{3}{16}\varepsilon^2 - (2mc^2)^2, \tag{7.37}$$

where $E_P = M_Pc^2$ is the Planck energy. However, because the square of energy is positive one obtains the following inequality

$$m\ell \leqslant \frac{\sqrt{3}}{8}M_P\ell_P = \frac{\sqrt{3}}{8}\frac{\hbar}{c}, \tag{7.38}$$

which for the fixed value of mass of a particle leads to the upper bound for minimal scale

$$\ell \leqslant \frac{2\pi\sqrt{3}}{8}\lambda_C(m), \tag{7.39}$$

or the lower bound for maximal energy

$$\varepsilon \geqslant \frac{8}{\sqrt{3}}E_P, \tag{7.40}$$

while for fixed minimal scale leads to the upper bound for the of a particle.

Applying minimal scale (7.35) within the definition (6.75) allows to eliminate minimal scale dependence

$$\varepsilon = \frac{16\pi}{\sqrt{3}}E_P\sqrt{\alpha}\sqrt{1+\left(\frac{E}{2mc^2}\right)^2}, \tag{7.41}$$

which after application of the generalization (6.110) takes the form

$$\varepsilon = \frac{4}{\sqrt{3}}\sqrt{E^2+(2mc^2)^2}, \tag{7.42}$$

and in the light of (7.40) leads to the bound for the energy of a particle

$$E^2 \geqslant 4\left(E_P^2 - m^2c^4\right), \tag{7.43}$$

which by $E^2 \geqslant 0$ gives the upper bound for the mass of a particle

$$m \leqslant M_P, \tag{7.44}$$

explicitly proving the Markov hypothesis expressing the supposition that the upper bound for mass of an observable particle is given by the Planck mass. Recall that, as we have deduced earlier, the formula (6.111) says that the only case $m \sim M_P$ is consistent with the Kontsevich deformation quantization. In the light of the inequality (7.44) such a situation suggests that a particle described by the non-commutative geometry of the Snyder quantized space-time is the Planckian particle, *i.e.* the particle equipped with the mass identical to the Planck mass $m = M_P$. In other words, possible existence of the Planckian particle will establish the physical sense of both the non-commutative geometry of the Snyder quantized space-time and the Kontsevich deformation quantization. It suggests also that physics at the Planck scale is the physics of the Planckian particle.

Maximal energy (7.4) does not vanish for all $\mu_L^{\pm} \neq \mu_R^{\pm} \neq 0$, and is finite if and only if $\mu_{L,R}^{\pm} < \infty$. The mass m of an original quantum state as well as the masses of neutrinos $\mu_R^{\pm}$ and $\mu_L^{\pm}$ are presumed to be physical quantities, which can be established by experimental data. In the case when an original quantum state is massless, maximal energy has the maximal value which equals

$$\varepsilon(m=0) = \left(\mu_L^{\pm} - \mu_R^{\pm}\right) c^2 \equiv \varepsilon_0, \tag{7.45}$$

and is finite and non-vanishing for finite $\mu_R^{\pm} \neq 0$ and $\mu_L^{\pm} \neq 0$. In this manner, one can study approximation around such defined state. For $|\mu_R^{\pm} + \mu_L^{\pm}| > 8m$ the appropriate Taylor series expansion is

$$\varepsilon = \varepsilon_0 \left[1 - \frac{8m}{\mu_R^{\pm} + \mu_L^{\pm}} + O\left(\left(\frac{8m}{\mu_R^{\pm} + \mu_L^{\pm}} \right)^2 \right) \right], \tag{7.46}$$

while for $|\mu_R^{\pm} + \mu_L^{\pm}| < 8m$ one has the following expansion

$$\varepsilon = \varepsilon_0 \left[\frac{\mu_R^{\pm} + \mu_L^{\pm}}{8m} + O\left(\left(\frac{\mu_R^{\pm} + \mu_L^{\pm}}{8m} \right)^2 \right) \right]. \tag{7.47}$$

For the case $|\mu_R^{\pm} + \mu_L^{\pm}| = 8m$ both these series coincide and

$$\varepsilon = \frac{\varepsilon_0}{2}. \tag{7.48}$$

The established inequality (7.15), however, allows to remove from considerations the case $|\mu_R^{\pm} + \mu_L^{\pm}| > 8m$ given by the Taylor power series expansion (7.46).

On the other hand, addition of the second equation to the first one in the system of equations (7.1) gives the relation

$$\left(\mu_L^{\pm} c^2 - \frac{\varepsilon}{2} \right)^2 + \left(\mu_R^{\pm} c^2 + \frac{\varepsilon}{2} \right)^2 = 2\left(\varepsilon^2 - 4E^2 \right). \tag{7.49}$$

The left hand side of the equation (7.49) is always positive as a sum of two squares of real numbers and, therefore, the right hand side of this equation is always positive. In this manner, one obtains the renormalization of energy of a particle expressed through maximal energy

$$-\frac{\varepsilon}{2} \leqslant E \leqslant \frac{\varepsilon}{2}. \tag{7.50}$$

Naturally, for the generic case of Special Relativity we have $\varepsilon \equiv \infty$ and for this reason values of energy E of a particle are not bounded. Therefore, taking into account the relation (7.50) it is evident that the noncommutative geometry of the Snyder quantized space-time results in renormalization of energy of a particle.

The relation (7.49) can be treated as the constraint for the energy E of a particle, and immediately solved with respect to E. The solution is a quadratic form which can be presented in the canonical form with respect to maximal energy ε

$$E^2 = \frac{3}{16}\left\{\left[\varepsilon + \frac{\mu_L^\pm - \mu_R^\pm}{3}c^2\right]^2 - \left[\frac{\mu_L^\pm - \mu_R^\pm}{3}c^2\right]^2\left[7 + \frac{12\mu_L^\pm\mu_R^\pm}{(\mu_L^\pm - \mu_R^\pm)^2}\right]\right\}. \tag{7.51}$$

By explicit application of maximal energy (7.4) within the energetic constraint (7.51) one can present the particle energy only by the masses of the neutrinos related to this particle

$$E^2 = \frac{\left[(\mu_L^\pm - \mu_R^\pm)\,c^2\right]^2}{48}\left\{\left(\frac{4 + \dfrac{8m}{\mu_L^\pm + \mu_R^\pm}}{1 + \dfrac{8m}{\mu_L^\pm + \mu_R^\pm}}\right)^2 - \left[7 + \frac{12\mu_L^\pm\mu_R^\pm}{(\mu_L^\pm - \mu_R^\pm)^2}\right]\right\}, \tag{7.52}$$

which for the case of the originally massless state has the value

$$E^2(m=0) = \frac{1}{16}\left[(\mu_L^\pm - \mu_R^\pm)\,c^2\right]^2\left[3 - 4\frac{\mu_L^\pm\mu_R^\pm}{(\mu_L^\pm - \mu_R^\pm)^2}\right] \equiv E_0^2. \tag{7.53}$$

For this reason the energy of a particle is

$$E^2 = E_0^2 + \Delta E^2, \tag{7.54}$$

where ΔE^2 is the correction generated due to the non-zero mass of the original quantum state

$$\Delta E^2 = -\frac{5}{16}\left[\left(\mu_L^{\pm} - \mu_R^{\pm}\right)c^2\right]^2 \frac{\dfrac{8m}{\mu_L^{\pm}+\mu_R^{\pm}}}{1+\dfrac{8m}{\mu_L^{\pm}+\mu_R^{\pm}}} \frac{\dfrac{8}{5}+\dfrac{8m}{\mu_L^{\pm}+\mu_R^{\pm}}}{1+\dfrac{8m}{\mu_L^{\pm}+\mu_R^{\pm}}}, \tag{7.55}$$

which can be expanded into the Taylor series around the massless state

$$\Delta E^2 = \frac{1}{16}\left[\left(\mu_L^{\pm} - \mu_R^{\pm}\right)c^2\right]^2 \sum_{n=1}^{\infty}(-1)^n(3n+5)\left(\frac{8m}{\mu_L^{\pm}+\mu_R^{\pm}}\right)^n. \tag{7.56}$$

In the light of the relation (7.15) it is more convenient to consider the Taylor power series expansion of the squared energy around the point $\mu_L^{\pm}+\mu_R^{\pm} = 8m$ in the situation $|\mu_L^{\pm}+\mu_R^{\pm}| > 8m$. In such a situation, the decomposition has a somewhat different form

$$E^2 = E_{8m}^2 + \Delta E_{8m}^2, \tag{7.57}$$

where

$$E_{8m}^2 = E_0^2 + \frac{13}{64}\left[\left(\mu_L^{\pm} - \mu_R^{\pm}\right)c^2\right]^2, \tag{7.58}$$

$$\Delta E_{8m}^2 = \frac{\left[\left(\mu_L^{\pm} - \mu_R^{\pm}\right)c^2\right]^2}{64}\sum_{n=1}^{\infty}\left(-\frac{1}{2}\right)^n(3n+7)\left(\frac{8m}{\mu_L^{\pm}+\mu_R^{\pm}}-1\right)^n. \tag{7.59}$$

In this way, in such expansion the leading correction to the square of the energy of a particle is

$$\Delta E_{8m}^{2\ (1)} = -\frac{5}{64}\left[\left(\mu_L^{\pm} - \mu_R^{\pm}\right)c^2\right]^2\left(\frac{8m}{\mu_L^{\pm}+\mu_R^{\pm}}-1\right). \tag{7.60}$$

For defined E_{8m}^2 the equation (7.58) establishes the relation between the masses of the neutrinos

$$\mu_L^{\pm} = \left(\frac{33}{25} \pm \sqrt{\frac{464}{625} + \left(\frac{8}{5}\frac{E_{8m}}{\mu_R^{\pm}c^2}\right)^2}\right)\mu_R^{\pm}, \tag{7.61}$$

and because of the condition $\mu_L^{\pm} - \mu_R^{\pm} \geqslant 0$

$$\mu_L^{\pm} - \mu_R^{\pm} = \left(\frac{8}{25} \pm \sqrt{\frac{464}{625} + \left(\frac{8}{5} \frac{E_{8m}}{\mu_R^{\pm} c^2} \right)^2} \right) \mu_R^{\pm} \geqslant 0. \tag{7.62}$$

the case of the minus sign leads to the inequality

$$\left(\frac{E_{8m}}{\mu_R^{\pm} c^2} \right)^2 + \frac{1}{4} \leqslant 0, \tag{7.63}$$

which has no solutions for real values of the energy E_{8m} and the masses $\mu_R^{\pm}$. Therefore, the physical solution is

$$\mu_L^{\pm} = \left(\frac{33}{25} + \frac{4}{25} \sqrt{29 + 4 \left(\frac{E_{8m}}{\mu_R^{\pm} c^2} \right)^2} \right) \mu_R^{\pm}, \tag{7.64}$$

and has the minimal value for $E_{8m} = 0$ with the value

$$\mu_L^{\pm} = \left(\frac{33}{25} + \frac{4}{25} \sqrt{29} \right) \mu_R^{\pm} \approx 2.1816 \mu_R^{\pm}. \tag{7.65}$$

In fact, for the given value of energy of massless state E_0 the equation (7.53) can be used for establishment of the relation between the masses of the neutrinos. As a result one receives two possible solutions

$$\mu_L^{\pm} = \left(\frac{5}{3} \pm \frac{4}{3} \sqrt{1 + \frac{1}{3} \left(\frac{E_0}{\mu_L^{\pm} c^2} \right)^2} \right) \mu_R^{\pm}, \tag{7.66}$$

and the physical solution is established by the condition (7.5)

$$\mu_L^{\pm} - \mu_R^{\pm} = \left(\frac{2}{3} \pm \frac{4}{3} \sqrt{1 + \frac{1}{3} \left(\frac{E_0}{\mu_L^{\pm} c^2} \right)^2} \right) \mu_R^{\pm} \geqslant 0, \tag{7.67}$$

which in the case of the minus sign states that

$$\left(\frac{E_0}{\mu_L^{\pm} c^2} \right)^2 + \frac{9}{4} \leqslant 0, \tag{7.68}$$

which is not satisfied for real E_0 and $\mu_L^{\pm}$. This argument allows to generate the physical solution

$$\mu_L^{\pm} = \left(\frac{5}{3} + \frac{4}{3}\sqrt{1 + \frac{1}{3}\left(\frac{E_0}{\mu_L^{\pm} c^2}\right)^2}\right)\mu_R^{\pm}. \tag{7.69}$$

which is minimized by $E_0 = 0$ with the value

$$\mu_L^{\pm} = 3\mu_R^{\pm}. \tag{7.70}$$

The relation (7.52) for the squared energy E^2 of a particle, *i.e.* factually the Hamiltonian constraint, is naturally helpful in the analysis of certain specific situations. Albeit, in general, if one knows the approximate value of this Hamiltonian constraint the value of energy E of a particle cannot be established by taking the square root of E^2. The square root method in itself establishes an approximation, and if one wishes to consider approximations of the energy E then this energy should be determined separately by the appropriate Taylor power series expansion. The problem is that the obtained series for E^2 and for E will have, in general, a different behaviour in the same regimes, and taking the square root of an arbitrary fixed order of an approximation of the square energy E^2 will not be compatible with the same order of an approximation of energy E.

In this manner, one can perform the deductions for energy E of a particle which are analogous to the deductions for its square. Let us write out the formula for energy explicitly

$$E = \frac{\left(\mu_L^{\pm} - \mu_R^{\pm}\right)c^2}{4}\sqrt{\frac{1}{3}\left(\frac{4 + \dfrac{8m}{\mu_L^{\pm} + \mu_R^{\pm}}}{1 + \dfrac{8m}{\mu_L^{\pm} + \mu_R^{\pm}}}\right)^2 - \frac{7}{3} - \frac{4\mu_L^{\pm}\mu_R^{\pm}}{\left(\mu_L^{\pm} - \mu_R^{\pm}\right)^2}}. \tag{7.71}$$

It is easy to see that for the originally massless case the value of energy is $E = E_0$, where

$$E_0 = \frac{\left(\mu_L^{\pm} - \mu_R^{\pm}\right)c^2}{4}\sqrt{3 - \frac{4\mu_L^{\pm}\mu_R^{\pm}}{\left(\mu_L^{\pm} - \mu_R^{\pm}\right)^2}}, \tag{7.72}$$

and in this way

$$E - E_0 = \frac{(\mu_L^\pm - \mu_R^\pm)\, c^2}{4} \left\{ \sqrt{\frac{1}{3}\left(\frac{4+\dfrac{8m}{\mu_L^\pm + \mu_R^\pm}}{1+\dfrac{8m}{\mu_L^\pm + \mu_R^\pm}}\right)^2 - \left[\frac{7}{3} + \frac{4\mu_L^\pm \mu_R^\pm}{(\mu_L^\pm - \mu_R^\pm)^2}\right]} - \sqrt{3 - \frac{4\mu_L^\pm \mu_R^\pm}{(\mu_L^\pm - \mu_R^\pm)^2}} \right\}. \tag{7.73}$$

Now one can apply the expansion (7.57) to the relation (7.73). In this instance, *i.e.* E not E^2, the expansion is more difficult to apply, because one has to deal with square roots. The expansion has the form

$$E = \frac{(\mu_L^\pm - \mu_R^\pm)\, c^2}{4} \sum_{n=0}^{\infty} A_n \left(\frac{8m}{\mu_L^\pm + \mu_R^\pm} - 1\right)^n. \tag{7.74}$$

The explicit formula for the coefficients A_n is not easy to extract, because for $n \geqslant 0$ they satisfy the recurrence equation

$$\begin{aligned}&(66+22n+8(3+n)a^2)A_{n+3} + (51+23n+12(2+n)a^2)A_{n+2} +\\ &+(9+8n+6(1+n)a^2)A_{n+1} + n(1+a^2)A_n = 0,\end{aligned} \tag{7.75}$$

with the initial conditions

$$A_0 = \sqrt{\frac{11}{4} + a^2}, \tag{7.76}$$

$$A_1 = -\frac{5}{8\sqrt{\dfrac{11}{4} + a^2}}, \tag{7.77}$$

$$A_2 = \frac{59 + 26a^2}{64\left(\dfrac{11}{4} + a^2\right)^{3/2}}, \tag{7.78}$$

where we have introduced the parameter

$$a = \frac{4E_0/c^2}{\mu_L^\pm - \mu_R^\pm}. \tag{7.79}$$

A similar situation occurs for expansion around the massless state. In this instance the expansion has the form

$$E = \frac{(\mu_L^{\pm} - \mu_R^{\pm})\, c^2}{4} \sum_{n=0}^{\infty} B_n \left(\frac{8m}{\mu_L^{\pm} + \mu_R^{\pm}} \right)^n, \tag{7.80}$$

where, in general for $n \geqslant 0$, the expansion coefficients B_n satisfy the following recurrence equation

$$\begin{aligned} &(18+6n+(3+n)a^2)B_{n+3} + (24+10n+3(2+n)a^2)B_{n+2} + \\ &+(6+5n+3(1+n)a^2)B_{n+1} + n(1+a^2)B_n = 0, \end{aligned} \tag{7.81}$$

having the following initial conditions

$$B_0 = \sqrt{6+a^2}, \tag{7.82}$$

$$B_1 = -\frac{4}{\sqrt{6+a^2}}, \tag{7.83}$$

$$B_2 = -\frac{50+11a^2}{2(6+a^2)^{3/2}}. \tag{7.84}$$

C Integrability Problem for Dirac Equations

The derived Dirac equations

$$\left(i\hbar c \gamma^{\mu} \partial_{\mu} - M_{\pm} c^2\right) \psi = 0, \tag{7.85}$$

where the mass matrices $M_{\pm}$ is given by the formula (6.89), can be rewritten in the form of the Schrödinger equation

$$i\hbar \partial_0 \psi^{\pm} = \hat{H} \psi^{\pm}, \tag{7.86}$$

where in the present case the Hamilton operator $\hat{H}$ has the form

$$\hat{H} = -i\hbar c \gamma^0 \gamma^i \partial_i - \frac{\mu_L^{\pm} + \mu_R^{\pm}}{2} c^2 \gamma^0 + \frac{\mu_L^{\pm} - \mu_R^{\pm}}{2} c^2 \gamma^0 \gamma^5, \tag{7.87}$$

which can be split into the components: the hermitian $\mathfrak{H}(\hat{H})$ and the antihermitian $\mathfrak{A}(\hat{H})$ one

$$\hat{H} = \mathfrak{H}(\hat{H}) + \mathfrak{A}(\hat{H}), \tag{7.88}$$

$$\mathfrak{H}(\hat{H}) = -i\hbar c \gamma^0 \gamma^i \partial_i - \frac{\mu_L^{\pm} + \mu_R^{\pm}}{2} c^2 \gamma^0, \tag{7.89}$$

$$\mathfrak{A}(\hat{H}) = \frac{\mu_L^{\pm} - \mu_R^{\pm}}{2} c^2 \gamma^0 \gamma^5, \tag{7.90}$$

with (anti)hermiticity defined standardly

$$\int d^3x \bar{\psi}^{\pm} \mathfrak{H}(\hat{H}) \psi^{\pm} = \int d^3x \overline{\mathfrak{H}(\hat{H}) \psi^{\pm}} \psi^{\pm}, \tag{7.91}$$

$$\int d^3x \bar{\psi}^{\pm} \mathfrak{A}(\hat{H}) \psi^{\pm} = -\int d^3x \overline{\mathfrak{A}(\hat{H}) \psi^{\pm}} \psi^{\pm}. \tag{7.92}$$

Let us consider the situation when the masses of the neutrinos have the same value, which we shall call μ

$$\mu_R^{\pm} = \mu_L^{\pm} \equiv \mu. \tag{7.93}$$

It is easy to see that the antihermitian component (7.90) then vanishes identically. However, the hermitian component (7.89) is still nontrivial. Consequently, the Hamilton operator (7.88) takes the form of the conventional Dirac Hamiltonian

$$\hat{H}_D = -\gamma^0 \left(i\hbar c \gamma^i \partial_i + \mu c^2 \right). \tag{7.94}$$

In such a situation, however, in the light of the relation (7.52) the energy of a particle vanishes identically

$$E = 0, \tag{7.95}$$

and, moreover, taking into account the bounds (7.50) one obtains also that maximal energy is identically vanishing

$$\varepsilon \equiv 0, \tag{7.96}$$

and, therefore, the minimal scale has an infinite value $\ell = \infty$. In the light of the formulas (6.90)-(6.91) for the masses of the neutrinos, one obtains

$$\mu_R^{\pm} = \mu_L^{\pm} \equiv 0, \tag{7.97}$$

i.e. in the light of the definition (7.93)

$$\mu \equiv 0. \tag{7.98}$$

Therefore, the Dirac Hamiltonian (7.94) becomes massless

$$\hat{H}_D = -i\hbar c \gamma^0 \gamma^i \partial_i. \tag{7.99}$$

and, consequently, the Dirac equations (7.85) take the form of the Weyl equation

$$i\hbar c\gamma^{\mu}\partial_{\mu}\psi = 0, \tag{7.100}$$

describing the massless particle - the Weyl neutrino. It means that the equality between the masses of the neutrinos (7.93) defines the massless particle obeying the Weyl equation. However, in the light of the Super-Kamiokande results the neutrino is equipped with non-zero mass. In this manner, such an embarrassing situation created by the Weyl equation (7.100), and laying the foundations of the Standard Model, is manifestly non-physical. In other words, in the light of the non-commutative geometry of the Snyder quantized space-time the theory of massive neutrinos is consistent if and only if the difference between masses of the left-handed and the right-handed neutrinos is non-zero and positive.

The full modified Hamiltonian operator (7.87) possesses evidently non-hermitian nature. Therefore, consequently the time evolution in the form of the Schrödinger equation (7.86) is manifestly non-unitary. Its formal integration, however, can be carried out by the standard method of quantum mechanics

$$\psi^{\pm}(x,t) = G(t,t_0)\psi^{\pm}(x,t_0), \tag{7.101}$$

involving the following time evolution operator

$$G(t,t_0) \equiv \exp\left\{-\frac{i}{\hbar}\int_{t_0}^{t} d\tau \hat{H}(\tau)\right\}. \tag{7.102}$$

For this reason, the integrability problem for the wave equation (7.86) with the Hamilton operator (7.87) can be formulated in terms of the appropriate Zassenhaus exponents

$$\exp\{A+B\} = \exp(A)\exp(B)\prod_{n=2}^{\infty}\exp C_n, \tag{7.103}$$

$$C_2 = -\frac{1}{2}C, \tag{7.104}$$

$$C_3 = -\frac{1}{6}(2[C,B]+[C,A]), \tag{7.105}$$

$$C_4 = -\frac{1}{24}([[C,A],A]+3[[C,A],B]+3[[C,B],B]), \tag{7.106}$$

$$\ldots$$

where $C = [A, B]$. In the light of the definition (7.102) one can establish the following identification

$$A \equiv A(t) = -\frac{i}{\hbar}\int_{t_0}^{t} d\tau \mathfrak{H}(\hat{H})(\tau), \tag{7.107}$$

$$B \equiv B(t) = -\frac{i}{\hbar}\int_{t_0}^{t} d\tau \mathfrak{A}(\hat{H})(\tau), \tag{7.108}$$

and, therefore, the commutator C can be derived straightforwardly and rather easily. The result is

$$C = -\frac{1}{\hbar^2}\int_{t_0}^{t} d\tau' \int_{t_0}^{t} d\tau'' \mathfrak{C}\left(\tau', \tau''\right), \tag{7.109}$$

where we have introduced the quantity

$$\mathfrak{C}\left(\tau', \tau''\right) \equiv \left[\mathfrak{H}(\hat{H})(\tau'), \mathfrak{A}(\hat{H})(\tau'')\right], \tag{7.110}$$

which can be computed with using elementary algebra

$$\begin{aligned} \mathfrak{C} &= \left(i\hbar\frac{\mu_R^{\pm} - \mu_L^{\pm}}{2}c^3\partial_i\right)\gamma^0\gamma^i\gamma^0\gamma^5 + \left(\frac{(\mu_R^{\pm})^2 - (\mu_L^{\pm})^2}{4}c^4\right)\gamma^0\gamma^0\gamma^5 - \\ &- \left(i\hbar\frac{\mu_R^{\pm} - \mu_L^{\pm}}{2}c^3\partial_i\right)\gamma^0\gamma^5\gamma^0\gamma^i - \left(\frac{(\mu_R^{\pm})^2 - (\mu_L^{\pm})^2}{4}c^4\right)\gamma^0\gamma^5\gamma^0 = \\ &= 2\left(i\hbar\frac{\mu_R^{\pm} - \mu_L^{\pm}}{2}c^3\partial_i\right)\gamma^0\gamma^i\gamma^0\gamma^5 + 2\left(\frac{(\mu_R^{\pm})^2 - (\mu_L^{\pm})^2}{4}c^4\right)\gamma^0\gamma^0\gamma^5, \end{aligned} \tag{7.111}$$

where we have applied the relations

$$\gamma^0\gamma^5\gamma^0\gamma^i = -\gamma^0\gamma^i\gamma^0\gamma^5, \tag{7.112}$$

$$\gamma^0\gamma^5\gamma^0 = -\gamma^0\gamma^0\gamma^5, \tag{7.113}$$

arising from the property of the γ^5 matrix $\{\gamma^5, \gamma^\mu\} = 0$. Therefore, consequently one obtains finally the result

$$\mathfrak{C}(\tau', \tau'') = 2\mathfrak{H}(\hat{H})(\tau')\mathfrak{A}(\hat{H})(\tau''), \tag{7.114}$$

that leads to the equivalent statement - for arbitrary two times τ' and τ'' the Poisson brackets of the hermitian $\mathfrak{H}(\hat{H})(\tau')$ and the antihermitian $\mathfrak{A}(\hat{H})(\tau'')$ components of the full Hamiltonian (7.88) is trivial

$$\left\{\mathfrak{H}(\hat{H})(\tau'), \mathfrak{A}(\hat{H})(\tau'')\right\} = 0. \tag{7.115}$$

Naturally, by simple factorization one obtains also

$$C = 2AB, \quad \{A,B\} = 0, \tag{7.116}$$

and, consequently,

$$\begin{aligned}
[C,A] &= CA, && (7.117)\\
[C,B] &= CB, && (7.118)\\
[[C,A],A] &= 2[C,A]A, && (7.119)\\
[[C,A],B] &= 2[C,A]B, && (7.120)\\
[[C,B],A] &= 2[C,B]A, && (7.121)
\end{aligned}$$

and so on. In this manner, the 4th order approximation of the formula (7.103) in the present case has the form

$$\begin{aligned}
\exp\{A+B\} &\approx \exp(A)\exp(B)\exp C_2 \exp C_3 \exp C_4, && (7.122)\\
C_2 &= -\frac{1}{2}C, && (7.123)\\
C_3 &= -\frac{1}{6}(CA+2CB), && (7.124)\\
C_4 &= -\frac{1}{12}\left(CA^2+3CB^2+\frac{3}{2}C^2\right). && (7.125)
\end{aligned}$$

For the case of constant in time masses $\mu_R^{\pm}$ and $\mu_L^{\pm}$ one can determine the relations

$$\begin{aligned}
A &= \frac{i}{\hbar}(t-t_0)\left(-i\hbar c\gamma^i\partial_i+\frac{\mu_L^{\pm}+\mu_R^{\pm}}{2}c^2\right)\gamma^0, && (7.126)\\
B &= \frac{i(\mu_L^{\pm}-\mu_R^{\pm})c^2}{2\hbar}(t-t_0)\gamma^5\gamma^0, && (7.127)\\
C &= \frac{(\mu_L^{\pm}-\mu_R^{\pm})c^2}{\hbar^2}(t-t_0)^2\left(-i\hbar c\gamma^i\partial_i+\frac{\mu_L^{\pm}+\mu_R^{\pm}}{2}c^2\right)\gamma^5, && (7.128)
\end{aligned}$$

and, consequently, by elementary algebraic manipulations one can establish the Zassenhaus exponents as

$$C_2 = -\frac{(\mu_L^\pm - \mu_R^\pm)c^2}{2\hbar^2}(t-t_0)^2\left(-i\hbar c\gamma^i\partial_i + \frac{\mu_L^\pm + \mu_R^\pm}{2}c^2\right)\gamma^5, \tag{7.129}$$

$$\begin{aligned} C_3 &= -\frac{i}{6\hbar^3}(\mu_L^\pm - \mu_R^\pm)c^2(t-t_0)^3\left(-i\hbar c\gamma^i\partial_i + \frac{\mu_L^\pm + \mu_R^\pm}{2}c^2\right)\times \\ &\times \left[\left(-i\hbar c\gamma^i\partial_i + \frac{\mu_L^\pm + \mu_R^\pm}{2}c^2\right)\gamma^5 + (\mu_L^\pm - \mu_R^\pm)c^2\right]\gamma^0, \end{aligned} \tag{7.130}$$

$$\begin{aligned} C_4 &= \frac{(\mu_L^\pm - \mu_R^\pm)c^2}{12\hbar^4}(t-t_0)^4\left(-i\hbar c\gamma^i\partial_i + \frac{\mu_L^\pm + \mu_R^\pm}{2}c^2\right)\times \\ &\times \left\{\left[\left(-i\hbar c\gamma^i\partial_i + \frac{\mu_L^\pm + \mu_R^\pm}{2}c^2\right)^2 + 3\left(\frac{\mu_L^\pm - \mu_R^\pm}{2}c^2\right)^2\right]\gamma^5 + \right. \\ &+ \left. 3\frac{\mu_L^\pm - \mu_R^\pm}{2}c^2\left(-i\hbar c\gamma^i\partial_i + \frac{\mu_L^\pm + \mu_R^\pm}{2}c^2\right)\right\}. \end{aligned} \tag{7.131}$$

This approximation is sufficient to make conclusion about the general properties of the procedure and the ramifications arising from these features. The explicit form of the Zassenhaus exponents C_n shows manifestly that the integrability problem formulated in terms of the Zassenhaus formula is not well defined in the case studied in this section. Namely, the problem is that when the Hamiltonian operator is a sum of non-commuting antihermitian and hermitian components then also the Zassenhaus exponents C_n inherit analogous decomposition, *i.e.* are sums of two non-commuting operators. The fundamental stage, *i.e.* the exponentiation procedure, must be applied once again and, therefore, consequently in the next step of the procedure one meets the same property, *i.e.* sums of two non-commuting operators. In this manner, the problem has a cyclic nature and cannot be solved in any approximation, while estimation of full integration formula (7.103) becomes the tremendous computational problem and its convergence is not well-defined. Therefore, for the case the Schrödinger equation (7.86) with the Hamiltonian operator (7.87), such a recurrence integrability procedure based on the Zassenhaus exponents is not an algorithm which results in the conclusion that the quantum system is non integrable. For this reason one must construct any different integrability procedure having a finite number of steps and being an appropriate algorithm.

For realization of such a construction let us formulate the integrability problem in a certain different form.

D Integrability Problem for Weyl Equations

For constructive solving of the problem, let us consider another integrability procedure. Instead of the Dirac equation let us focus on the massive Weyl equations (6.97)-(6.98), which define the model of massive neutrinos. These two equations can be straightforwardly rewritten in the form of the effective two-component time evolution described by the Schrödinger equation

$$i\hbar\partial_0 \left[\begin{array}{c} \psi_R^{\pm}(x,t) \\ \psi_L^{\pm}(x,t) \end{array} \right] = \hat{H}\left(\partial_i\right) \left[\begin{array}{c} \psi_R^{\pm}(x,t) \\ \psi_L^{\pm}(x,t) \end{array} \right], \qquad (7.132)$$

where the Hamilton operator $\hat{H}$

$$\hat{H} = -\gamma^0 \left(i\hbar c\gamma^i\partial_i + \left[\begin{array}{cc} \mu_R^{\pm}c^2 & 0 \\ 0 & \mu_L^{\pm}c^2 \end{array} \right] \right), \qquad (7.133)$$

is manifestly hermitian and, therefore, the Schrödinger time evolution (7.132) is unitary. In this manner, the integration procedure can be performed in the usual way well known from quantum mechanics.

Integrability of (7.132) is well defined. The solutions are

$$\left[\begin{array}{c} \psi_R^{\pm}(x,t) \\ \psi_L^{\pm}(x,t) \end{array} \right] = U(t,t_0) \left[\begin{array}{c} \psi_R^{\pm}(x,t_0) \\ \psi_L^{\pm}(x,t_0) \end{array} \right], \qquad (7.134)$$

where $U(t,t_0)$ is the unitary time-evolution operator which for the constant masses is explicitly given by

$$U(t,t_0) = \exp\left\{ -\frac{i}{\hbar}(t-t_0)\hat{H} \right\}, \qquad (7.135)$$

and $\psi_{R,L}^{\pm}(x,t_0)$ are the initial time t_0 eigenstates with defined momenta

$$i\hbar\sigma^i\partial_i\psi_{R,L}^{\pm}(x,t_0) = p_{R,L}^{\pm\,0}\psi_{R,L}^{\pm}(x,t_0), \qquad (7.136)$$

where the initial momenta $p_R^{\pm 0}$ and $p_L^{\pm 0}$ are related to the right-handed $\psi_R^{\pm}(x,t_0)$ and the left-handed $\psi_L^{\pm}(x,t_0)$ chiral fields, respectively. The eigenequation (7.136), however, can be straightforwardly integrated and the result will be determined by the spatial part of the evolution. The result can be presented in the symbolic form

$$\psi^{\pm}_{R,L}(x,t_0) = \exp\left\{-\frac{i}{\hbar}p^{\pm\,0}_{R,L}(x-x_0)_i\sigma^i\right\}\psi^{\pm}_{R,L}(x_0,t_0), \tag{7.137}$$

which after direct exponentiation leads to

$$\psi^{\pm}_{R,L}(x,t_0) = \left(\mathbf{1}_2\cos\eta - i\eta_i\sigma^i\frac{\sin\eta}{\eta}\right)\psi^{\pm}_{R,L}(x_0,t_0), \tag{7.138}$$

where $\eta = |\eta_i|$ and η_i is the three dimensionless vector

$$\eta_i = \frac{p^{\pm\,0}_{R,L}}{\hbar}(x-x_0)_i. \tag{7.139}$$

In the present situation the embarrassing problem which emerges in the integration procedure for the Dirac equation, discussed in the previous section, is absent. Now the Zassenhaus exponents are not troublesome because, by definition, the γ^5 matrix is included into the chiral Weyl fields. Therefore, the Hamilton operator (7.133) is manifestly hermitian and, consequently, the exponentiation (7.135) can be straightforwardly performed by the standard method of quantum mechanics. At first glance, however, the mass matrix presence in the Hamilton operator (7.133) causes that one can choose between at least two nonequivalent representations of the Dirac γ matrices. On the one hand, the straightforward analogy to the Weyl equation suggests that the appropriate choice is the Weyl basis. Albeit, on the other hand, the Hamilton operator (7.133) can be treated as the usual hermitian Dirac Hamiltonian and, consequently, the Dirac basis would be the right representation for the Dirac γ matrices. Other choices can be also applied, but they have no unambiguous justification because the mass matrix presence in the Weyl equation. For example the Majorana basis, which is an adequate choice for the case of massless neutrino, is not an adequate choice for the case of massive neutrino. In this manner, in fact, one should not prefer the representation but rather consider both the chiral fields and the time evolution operator (7.135) in both the Weyl and Dirac representations. Let us denote by superscript r the chosen representation. For this reason the adequate labeling procedure is

$$U(t,t_0) \rightarrow U^r(t,t_0), \tag{7.140}$$

$$\psi^{\pm}_{R,L}(x,t_0) \rightarrow (\psi^{\pm}_{R,L})^r(x,t_0), \tag{7.141}$$

$$\psi^{\pm}_{R,L}(x_0,t_0) \rightarrow (\psi^{\pm}_{R,L})^r(x_0,t_0) \tag{7.142}$$

where in our considerations the superscript has two values $r = D,W$ corresponding with the quantities taken in the Dirac and Weyl basis, respectively. Interestingly, the eigenequation (7.136) is independent of the representation choice and, therefore, the initial momenta $p^{\pm\,0}_{R,L}$ of the chiral Weyl fields are measurable. For full consistency, let us test both the representations.

D1 Dirac Basis

The Dirac basis of the gamma matrices is defined as

$$\gamma^0 = \begin{bmatrix} I & 0 \\ 0 & -I \end{bmatrix}, \tag{7.143}$$

$$\gamma^i = \begin{bmatrix} 0 & \sigma^i \\ -\sigma^i & 0 \end{bmatrix}, \tag{7.144}$$

$$\gamma^5 = \begin{bmatrix} 0 & I \\ I & 0 \end{bmatrix}, \tag{7.145}$$

where I is the 2×2 unit matrix, and $\sigma^i = [\sigma_x, \sigma_y, \sigma_z]$ is a vector of the 2×2 Pauli matrices

$$\sigma_x = \begin{bmatrix} 0 & 1 \\ 1 & 0 \end{bmatrix}, \tag{7.146}$$

$$\sigma_y = \begin{bmatrix} 0 & -i \\ i & 0 \end{bmatrix}, \tag{7.147}$$

$$\sigma_z = \begin{bmatrix} 1 & 0 \\ 0 & -1 \end{bmatrix}. \tag{7.148}$$

Application of the Dirac basis allows to express the Hamilton operator (7.133) as follows

$$\hat{H} = \begin{bmatrix} \mu^{\pm}_R & i\frac{\hbar}{c}\sigma^i\partial_i \\ i\frac{\hbar}{c}\sigma^i\partial_i & -\mu^{\pm}_L \end{bmatrix} c^2, \tag{7.149}$$

and for the case of constant in time neutrinos masses yields a solution (7.134) with the unitary time evolution operator U

$$U^D = \exp\left\{ -i\frac{c^2}{\hbar}(t-t_0) \begin{bmatrix} \mu_R^\pm & i\frac{\hbar}{c}\sigma^i\partial_i \\ i\frac{\hbar}{c}\sigma^i\partial_i & -\mu_L^\pm \end{bmatrix} \right\}. \tag{7.150}$$

Straightforward exponentiation in (7.150) leads to the result

$$\begin{aligned} U^D &= \left\{ \begin{bmatrix} I & 0 \\ 0 & I \end{bmatrix} \cos\left[\frac{t-t_0}{\hbar} c^2 \sqrt{\left(\frac{\mu_R^\pm + \mu_L^\pm}{2}\right)^2 + \left(i\frac{\hbar}{c}\sigma^i\partial_i\right)^2} \right] - \right. \\ &- i \begin{bmatrix} \frac{\mu_L^\pm + \mu_R^\pm}{2} & i\frac{\hbar}{c}\sigma^i\partial_i \\ i\frac{\hbar}{c}\sigma^i\partial_i & -\frac{\mu_L^\pm + \mu_R^\pm}{2} \end{bmatrix} \times \\ &\times \left. \frac{\sin\left[\frac{t-t_0}{\hbar} c^2 \sqrt{\left(\frac{\mu_R^\pm + \mu_L^\pm}{2}\right)^2 + \left(i\frac{\hbar}{c}\sigma^i\partial_i\right)^2} \right]}{\sqrt{\left(\frac{\mu_R^\pm + \mu_L^\pm}{2}\right)^2 + \left(i\frac{\hbar}{c}\sigma^i\partial_i\right)^2}} \right\} \times \\ &\times \exp\left\{ -i\frac{(\mu_R^\pm - \mu_L^\pm)c^2}{2\hbar}(t-t_0) \right\}, \end{aligned} \tag{7.151}$$

where we understand that all the functions are treated by the appropriate Taylor series expansions.

D2 Weyl Basis

As we have mentioned, however, application of the Weyl representation of the Dirac γ matrices is also justified by theoretical reasons. Such a basis is defined as follows

$$\gamma^0 = \begin{bmatrix} 0 & I \\ I & 0 \end{bmatrix}, \tag{7.152}$$

$$\gamma^i = \begin{bmatrix} 0 & \sigma^i \\ -\sigma^i & 0 \end{bmatrix}, \tag{7.153}$$

$$\gamma^5 = \begin{bmatrix} -I & 0 \\ 0 & I \end{bmatrix}, \tag{7.154}$$

and the Hamilton operator (7.133) in this representation has the form

$$\hat{H} = \begin{bmatrix} i\frac{\hbar}{c}\sigma^i\partial_i & -\mu_L^{\pm} \\ -\mu_R^{\pm} & -i\frac{\hbar}{c}\sigma^i\partial_i \end{bmatrix} c^2. \tag{7.155}$$

Consequently, for the case of constant in time neutrinos masses one establishes the unitary time evolution operator

$$U^W = \exp\left\{ -i\frac{c^2}{\hbar}(t-t_0)\begin{bmatrix} i\frac{\hbar}{c}\sigma^i\partial_i & -\mu_L^{\pm} \\ -\mu_R^{\pm} & -i\frac{\hbar}{c}\sigma^i\partial_i \end{bmatrix} \right\}, \tag{7.156}$$

which after straightforward exponentiation becomes

$$U^W = \begin{bmatrix} I & 0 \\ 0 & I \end{bmatrix} \cos\left[\frac{t-t_0}{\hbar}c^2\sqrt{\mu_L^{\pm}\mu_R^{\pm} + \left(i\frac{\hbar}{c}\sigma^i\partial_i\right)^2}\right] - $$
$$- i\begin{bmatrix} i\frac{\hbar}{c}\sigma^i\partial_i & -\mu_L^{\pm} \\ -\mu_R^{\pm} & -i\frac{\hbar}{c}\sigma^i\partial_i \end{bmatrix} \frac{\sin\left[\frac{t-t_0}{\hbar}c^2\sqrt{\mu_L^{\pm}\mu_R^{\pm} + \left(i\frac{\hbar}{c}\sigma^i\partial_i\right)^2}\right]}{\sqrt{\mu_L^{\pm}\mu_R^{\pm} + \left(i\frac{\hbar}{c}\sigma^i\partial_i\right)^2}}. \tag{7.157}$$

Evidently, the time evolution operator evaluated in the Weyl representation (7.157) has a meaningfully simpler form than the result of the evaluation performed in the Dirac basis (7.151). In this manner, the results of the choices are manifestly distinguishable and, consequently, are not physically equivalent, *i.e.* yield different solutions of the same equation. Albeit, it is not the strangest feature of the problem. Namely, both the choices can be related to physics in different energetic regions and, therefore, it is naturally necessary to find the solutions to the massive Weyl equation in both the Weyl and Dirac representations of the gamma matrices.

D3 Space-Time Evolution

It should be emphasized that the results obtained in both previous subsections are strictly related to the massive Weyl equations presented in the form of the Schrödinger equation (7.132). Presently, one can straightforwardly apply these key results, *i.e.* the momentum eigenequations (7.136), the spatial evolutions (7.137), and the evaluations of the unitary time evolution operators (7.151) and (7.157), to exact determination of the corresponding wave functions of the massive Weyl equation (7.132) in both the Dirac and Weyl bases of the Dirac gamma matrices.

Dirac-like Solutions

Let us derive first the wave functions in the Dirac basis. For this we shall employ the shortened notation

$$E^D(p_R^{\pm 0}) \equiv c^2\sqrt{(\mu_\pm^D)^2 + \left(\frac{p_R^{\pm 0}}{c}\right)^2}, \tag{7.158}$$

where $\mu_\pm^D$ is the arithmetic mean of the masses of the neutrinos

$$\mu_\pm^D = \frac{\mu_R^\pm + \mu_L^\pm}{2}, \tag{7.159}$$

by elementary algebraic manipulations one receives the right-handed chiral Weyl fields

$$\begin{aligned}(\psi_R^\pm)^D(x,t) = &\left\{\left[\cos\left[\frac{t-t_0}{\hbar}E^D(p_R^{\pm 0})\right] - \right.\right.\\ - &\left. i\mu_\pm^D c^2 \frac{\sin\left[\frac{t-t_0}{\hbar}E^D(p_R^{\pm 0})\right]}{E^D(p_R^{\pm 0})}\right]\exp\left\{-\frac{i}{\hbar}p_R^{\pm 0}(x-x_0)_i\sigma^i\right\}(\psi_R^\pm)_0^D - \\ - &\left. ip_L^{\pm 0}c\frac{\sin\left[\frac{t-t_0}{\hbar}E^D(p_L^{\pm 0})\right]}{E^D(p_L^{\pm 0})}\exp\left\{-\frac{i}{\hbar}p_L^{\pm 0}(x-x_0)_i\sigma^i\right\}(\psi_L^\pm)_0^D\right\}\times\\ \times\ &\exp\left\{-i\frac{(\mu_R^\pm-\mu_L^\pm)c^2}{2\hbar}(t-t_0)\right\},\end{aligned} \tag{7.160}$$

where $(\psi^{\pm}_{R,L})^D_0 = (\psi^{\pm}_{R,L})^D(x_0,t_0)$. Similarly, the left-handed chiral Weyl fields also can be also established in an exact way

$$
\begin{aligned}
(\psi^{\pm}_L)^D(x,t) = & \left\{\left[\cos\left[\frac{t-t_0}{\hbar}E^D(p^{\pm 0}_L)\right] + \right.\right.\\
+ & \left. i\mu^D_{\pm}c^2\frac{\sin\left[\frac{t-t_0}{\hbar}E^D(p^{\pm 0}_L)\right]}{E^D(p^{\pm 0}_L)}\right]\exp\left\{-\frac{i}{\hbar}p^{\pm 0}_L(x-x_0)_i\sigma^i\right\}(\psi^{\pm}_L)^D_0 - \\
- & \left. ip^{\pm 0}_R c\frac{\sin\left[\frac{t-t_0}{\hbar}E^D(p^{\pm 0}_R)\right]}{E^D(p^{\pm 0}_R)}\exp\left\{-\frac{i}{\hbar}p^{\pm 0}_R(x-x_0)_i\sigma^i\right\}(\psi^{\pm}_R)^D_0\right\}\times \\
\times & \exp\left\{-i\frac{(\mu^{\pm}_R-\mu^{\pm}_L)c^2}{2\hbar}(t-t_0)\right\}. \qquad (7.161)
\end{aligned}
$$

Weyl-like Solutions

A similar line of reasoning can be carried out for derivation of the wave functions in the Weyl basis. As in the previous case we shall employ the following shortened notation

$$
E^W(p^{\pm 0}_R) \equiv c^2\sqrt{(\mu^W_{\pm})^2 + \left(\frac{p^{\pm 0}_R}{c}\right)^2}, \qquad (7.162)
$$

where $\mu^W_{\pm}$ is the geometric mean of the masses of the neutrinos

$$
\mu^W_{\pm} = \sqrt{\mu^{\pm}_R\mu^{\pm}_L}, \qquad (7.163)
$$

and performing elementary calculation one can deduce the right-handed chiral Weyl fields

$$
\begin{aligned}
(\psi_R^{\pm})^W(x,t) = & \left\{ \cos\left[\frac{t-t_0}{\hbar} E^W(p_R^{\pm 0})\right] - \right. \\
- & \left. i p_R^{\pm 0} c \frac{\sin\left[\frac{t-t_0}{\hbar} E^W(p_R^{\pm 0})\right]}{E^W(p_R^{\pm 0})} \right\} \exp\left\{-\frac{i}{\hbar} p_R^{\pm 0}(x-x_0)_i \sigma^i\right\} (\psi_R^{\pm})_0^W + \\
+ & i\mu_L^{\pm} c^2 \frac{\sin\left[\frac{t-t_0}{\hbar} E^W(p_L^{\pm 0})\right]}{E^W(p_L^{\pm 0})} \exp\left\{-\frac{i}{\hbar} p_L^{\pm 0}(x-x_0)_i \sigma^i\right\} (\psi_L^{\pm})_0^W, \quad (7.164)
\end{aligned}
$$

where as in the case of the Dirac-like solutions we have introduced the shortened notation $(\psi_{R,L}^{\pm})_0^W = (\psi_{R,L}^{\pm})^W(x_0,t_0)$. For the left-handed chiral Weyl fields one obtains the formula

$$
\begin{aligned}
(\psi_L^{\pm})^W(x,t) = & \left\{ \cos\left[\frac{t-t_0}{\hbar} E^W(p_L^{\pm 0})\right] - \right. \\
+ & \left. i p_L^{\pm 0} c \frac{\sin\left[\frac{t-t_0}{\hbar} E^W(p_L^{\pm 0})\right]}{E^W(p_L^{\pm 0})} \right\} \exp\left\{-\frac{i}{\hbar} p_L^{\pm 0}(x-x_0)_i \sigma^i\right\} (\psi_L^{\pm})_0^W + \\
+ & i\mu_R^{\pm} c^2 \frac{\sin\left[\frac{t-t_0}{\hbar} E^W(p_R^{\pm 0})\right]}{E^W(p_R^{\pm 0})} \exp\left\{-\frac{i}{\hbar} p_R^{\pm 0}(x-x_0)_i \sigma^i\right\} (\psi_R^{\pm})_0^W. \quad (7.165)
\end{aligned}
$$

In this manner, one sees that the difference between the obtained wave functions is crucial. Straightforward comparison of the Weyl-like solutions (7.164) and (7.165) with the Dirac-like solutions (7.160) and (7.161) shows that in the case of the Dirac basis there are different coefficients of cosinuses and sinuses and there is an additional time-exponent. Moreover, the functions $M^D(p_R^{\pm 0})$ and $M^W(p_R^{\pm 0})$ having the basic status for both received solutions also have different form which manifestly depends on the choice of the Dirac representation of the γ matrices. As we have suggested earlier, the difference is not a problem, because the solutions can be related to different regions of energy. Anyway, however, the validation of both representations, and also other ones, should be confronted with suitable experimental data.

D4 Probability Density. Normalization

If one knows the explicit form of the chiral Weyl fields then, applying the Dirac basis, one can derive the usual Dirac fields by the following procedure

$$(\psi^{\pm})^D = \begin{bmatrix} \frac{(\psi_R^{\pm})^D + (\psi_L^{\pm})^D}{2}\mathbf{1}_2 & \frac{(\psi_R^{\pm})^D - (\psi_L^{\pm})^D}{2}\mathbf{1}_2 \\ \frac{(\psi_R^{\pm})^D - (\psi_L^{\pm})^D}{2}\mathbf{1}_2 & \frac{(\psi_R^{\pm})^D + (\psi_L^{\pm})^D}{2}\mathbf{1}_2 \end{bmatrix}, \tag{7.166}$$

where we have used the shortened notation $(\psi^{\pm})^D = (\psi^{\pm})^D(x,t)$, and $(\psi_{R,L}^{\pm})^D = (\psi_{R,L}^{\pm})^D(x,t)$. Similarly, employing the Weyl basis, the Dirac fields can be determined as follows

$$(\psi^{\pm})^W = \begin{bmatrix} (\psi_L^{\pm})^W \mathbf{1}_2 & \mathbf{0}_2 \\ \mathbf{0}_2 & (\psi_R^{\pm})^W \mathbf{1}_2 \end{bmatrix}, \tag{7.167}$$

where, like in the case of the Dirac basis, we have applied the shortened notation $(\psi^{\pm})^W = (\psi^{\pm})^W(x,t)$, and $(\psi_{R,L}^{\pm})^W = (\psi_{R,L}^{\pm})^W(x,t)$. It is evident now, that in general these two cases are different from physical, mathematical, and computational points of view. In this manner, if we consider the quantum mechanical probability density and its normalization, we are forced to relate the probability density revealing Lorentz invariance to the chosen representation

$$\Omega^{D,W} \equiv (\bar{\psi}^{\pm})^{D,W} (\psi^{\pm})^{D,W}, \tag{7.168}$$

$$\int d^3x \Omega^{D,W} = \mathbf{1}_4. \tag{7.169}$$

Applying the Dirac field in the Dirac basis (7.166) by elementary derivation, one can obtain

$$\Omega^D = \begin{bmatrix} \frac{(\bar{\psi}_R^{\pm})^D(\psi_R^{\pm})^D + (\bar{\psi}_L^{\pm})^D(\psi_L^{\pm})^D}{2}\mathbf{1}_2 & \frac{(\bar{\psi}_R^{\pm})^D(\psi_R^{\pm})^D - (\bar{\psi}_L^{\pm})^D(\psi_L^{\pm})^D}{2}\mathbf{1}_2 \\ \frac{(\bar{\psi}_R^{\pm})^D(\psi_R^{\pm})^D - (\bar{\psi}_L^{\pm})^D(\psi_L^{\pm})^D}{2}\mathbf{1}_2 & \frac{(\bar{\psi}_R^{\pm})^D(\psi_R^{\pm})^D + (\bar{\psi}_L^{\pm})^D(\psi_L^{\pm})^D}{2}\mathbf{1}_2 \end{bmatrix}, \tag{7.170}$$

and, similarly, the probability density (7.168) computed for the Dirac field in the Weyl basis (7.167) has the form

$$\Omega^W = \begin{bmatrix} (\bar{\psi}_R^{\pm})^W (\psi_R^{\pm})^W \mathbf{1}_2 & \mathbf{0}_2 \\ \mathbf{0}_2 & (\bar{\psi}_L^{\pm})^W (\psi_L^{\pm})^W \mathbf{1}_2 \end{bmatrix}. \tag{7.171}$$

Employing the normalization condition (7.169) in the Dirac representation, one obtains the system of equations

$$\frac{1}{2}\left(\int d^3x(\bar{\psi}_R^{\pm})^D(\psi_R^{\pm})^D + \int d^3x(\bar{\psi}_L^{\pm})^D(\psi_L^{\pm})^D\right) = 1, \tag{7.172}$$

$$\frac{1}{2}\left(\int d^3x(\bar{\psi}_R^{\pm})^D(\psi_R^{\pm})^D - \int d^3x(\bar{\psi}_L^{\pm})^D(\psi_L^{\pm})^D\right) = 0, \tag{7.173}$$

which leads to the particular normalization conditions

$$\int d^3x(\bar{\psi}_R^{\pm})^D(\psi_R^{\pm})^D = 1, \tag{7.174}$$

$$\int d^3x(\bar{\psi}_L^{\pm})^D(\psi_L^{\pm})^D = 1. \tag{7.175}$$

In the case of Weyl representation one receives

$$\int d^3x(\bar{\psi}_R^{\pm})^W(\psi_R^{\pm})^W = 1, \tag{7.176}$$

$$\int d^3x(\bar{\psi}_L^{\pm})^W(\psi_L^{\pm})^W = 1. \tag{7.177}$$

In this manner, one sees straightforwardly that the normalization conditions (7.174), (7.175) and (7.176), (7.177)) are the same

$$\int d^3x(\bar{\psi}_{R,L}^{\pm})^{D,W}(x,t)(\psi_{R,L}^{\pm})^{D,W}(x,t) = 1, \tag{7.178}$$

i.e. are invariant with respect to the choice of the gamma matrices representations, which means that they are physical conditions. Using the fact that full space-time evolution is determined as

$$(\psi_{R,L}^{\pm})^{D,W}(x,t) = U^{D,W}(t,t_0)(\psi_{R,L}^{\pm})^{D,W}(x,t_0), \tag{7.179}$$

$$\left[U^{D,W}(t,t_0)\right]^{\dagger} U^{D,W}(t,t_0) = \mathbf{1}_2, \tag{7.180}$$

one finds easily that

$$\int d^3x(\bar{\psi}_{R,L}^{\pm})^{D,W}(x,t_0)(\psi_{R,L}^{\pm})^{D,W}(x,t_0) = 1. \tag{7.181}$$

Using the spatial evolution (7.138) one obtains the relation

$$\left|(\psi_{R,L}^{\pm})^{D,W}(x_0,t_0)\right|^2 \int d^3x\left(\mathbf{1}_2 + \frac{(x-x_0)_i}{|x-x_0|}\Im\sigma^i \sin\left|2\frac{p_{R,L}^{\pm\,0}}{\hbar}(x-x_0)_i\right|\right) = 1, \tag{7.182}$$

where $\Im\sigma^i = \dfrac{\sigma^i - \sigma^{i\dagger}}{2i}$ is the imaginary part of the vector σ^i. The decomposition $\sigma_i = [\sigma_x, 0, \sigma_z] + i[0, -i\sigma_y, 0]$ yields $\Im\sigma^i = [0, -i\sigma_y, 0]$ and, consequently, the equation (7.182) becomes

$$\left|(\psi^{\pm}_{R,L})^{D,W}(x_0,t_0)\right|^2 \int d^3x \left(\mathbf{1}_2 - i\frac{(x-x_0)_y}{|x-x_0|}\sigma_y \sin\left|2\frac{p^{\pm\,0}_{R,L}}{\hbar}(x-x_0)_i\right|\right) = 1. \tag{7.183}$$

Introducing the dimensionless variable

$$a_i \equiv 2\frac{p^{\pm\,0}_{R,L}}{\hbar}(x-x_0)_i, \tag{7.184}$$

and the effective volume V' due to the vector a_i

$$V'\mathbf{1}_2 = V\mathbf{1}_2 - i\sigma_y V_y, \tag{7.185}$$

or explicitly

$$\begin{bmatrix} V' & 0 \\ 0 & V' \end{bmatrix} = \begin{bmatrix} V & V_y \\ V_y & V \end{bmatrix} \tag{7.186}$$

where $V = \int d^3x$ is the usual volume, and V_y is the axial volume

$$V_y = \int d^3x a_y \frac{\sin|a|}{|a|}. \tag{7.187}$$

In such a situation, equation (7.183) can be rewritten in the form

$$\left|(\psi^{\pm}_{R,L})^{D,W}(x_0,t_0)\right|^2 V'\mathbf{1}_2 = \mathbf{1}_2, \tag{7.188}$$

and, therefore, one obtains

$$(\psi^{\pm}_{R,L})^{D,W}(x_0,t_0) = \frac{1}{\sqrt{V'}}\exp i\theta_{\pm}, \tag{7.189}$$

where $\theta_{\pm}$ are constant.

The relation (7.186) suggest that $V_y = 0$, but this is not generally true. Let us consider the axial volume (7.187) which by using the relation (7.184) takes the following form

$$V_y = \left(\frac{2\hbar}{p^{\pm\,0}_{R,L}}\right)^3 \int d^3a a_y \frac{\sin|a|}{|a|}, \tag{7.190}$$

and in spherical coordinate system (a,θ,φ) one has

$$d^3a = |a|^2 \sin\theta d|a| d\theta d\varphi, \tag{7.191}$$
$$a_y = |a| \sin\theta \sin\varphi, \tag{7.192}$$

so, taking into account the region of integration $a \in [0,A]$, $\theta \in [0,\Theta]$, $\varphi \in [0,\Phi]$, one obtains

$$V_y = \left(\frac{2\hbar}{p_{R,L}^{\pm 0}}\right)^3 (\Theta - \sin\Theta\cos\Theta)\sin\Phi \left(\frac{A^2}{2}\cos A - A\sin A - \cos A + 1\right). \tag{7.193}$$

One sees that the axial volume identically vanishes for $A < \infty$, $\Theta = \pi$, $\Phi = 2\pi$, is indefinite for $A \to \infty$, $\Theta = \pi$, $\Phi = 2\pi$, and is finite for any finite values of the constants A, Θ, Φ, or in other words

$$V_y = \begin{cases} 0 & \text{for finite spherical space} \\ \infty & \text{for infinite spherical space} \\ < \infty & \text{for finite spherical sections} \end{cases}, \tag{7.194}$$

and, consequently, selection of the normalization region as the finite sphere of the radius

$$R = \frac{2\hbar}{p_{R,L}^{\pm 0}} A, \tag{7.195}$$

gives $V_y = 0$. In this manner, only such choice of the normalization region is consistent. In general, however, for finite values of the angles Θ and Φ the axial volume vanishes if and only if the following non-algebraic equation for the constant parameter A

$$\frac{A^2}{2}\cos A - A\sin A - \cos A + 1 = 0, \tag{7.196}$$

is satisfied. The equation (7.196) has infinite number of solutions, and can solved only numerically. However, here we shall focus on the solution (7.195) which, in fact, is usually applied.

Now one sees straightforwardly that the normalization is, strictly speaking, dependent on the choice of an appropriate region of integrability. For infinite symmetric spatial regions such a normalization procedure is not well defined, because the axial volume effect is infinite. However, one can study certain reasonable situations which consider solutions of the quantum theory on finite symmetric spatial regions. Moreover, the problem of integrability is defined with respect to the choice of the initial momentum of the chiral Weyl fields $p_{R,L}^{\pm\,0}$. In fact, there are many possible nonequivalent physical situations connected with the choice of a concrete initial momentum eigenvalue. In the next section we are going to discuss one such situation related to a finite symmetric spatial region, the concrete example of the model of massive neutrinos which, in general, was solved in the present section.

E Ultra-Relativistic Massive Neutrinos

As the final piece of this chapter let us consider the concrete application of the general model presented in the previous section, which is based on the normalization in a finite symmetrical box and putting *ad hoc* the eigenvalue of the initial momenta of the chiral Weyl fields according to the ultra-relativistic limit of Special Relativity

$$p_{R,L}^{\pm\,0} = \mu_{R,L}^{\pm} c. \tag{7.197}$$

Then the formula (7.195) gives

$$R = 2\lambda_C(\mu_{R,L}^{\pm})A, \tag{7.198}$$

where $\lambda_C(\mu_{R,L}^{\pm}) = \dfrac{\hbar}{\mu_{R,L}^{\pm} c}$ is the Compton wavelength of the right- or left-handed neutrino. In such a situation,

$$V' = V = \frac{4}{3}\pi R^3 = \frac{32}{3}\pi A^3 \lambda_C^3(\mu_{R,L}^{\pm}), \tag{7.199}$$

and the normalization discussed in the previous section leads to the following initial data condition

$$(\psi_{R,L}^{\pm})^{D,W}(x_0, t_0) = \sqrt{\frac{3\pi}{32A^3\lambda_C^3(\mu_{R,L}^{\pm})}} \exp i\theta_{\pm}, \tag{7.200}$$

When the theory is normalized in the region of the volume

$$V = \frac{4}{3}\pi\lambda_C^3(\mu_{R,L}^{\pm}), \tag{7.201}$$

then $A = \frac{1}{2}$ and, consequently, the initial data of the chiral Weyl fields are defined as follows

$$(\psi_{R,L}^{\pm})^{D,W}(x_0, t_0) = \sqrt{\frac{3\pi}{4\lambda_C^3(\mu_{R,L}^{\pm})}} \exp i\theta_{\pm}. \tag{7.202}$$

Interestingly, when one considers normalization of the left- and right-handed neutrino on the finite spheres of radii $R_{R,L}^{\pm}$ then the normalization radii are equal to the Compton wavelength

$$R_{R,L}^{\pm} = \lambda_C(\mu_{R,L}^{\pm}). \tag{7.203}$$

In this manner, if one can measure the normalization radius $R_{R,L}^{\pm}$, then the masses of the neutrinos can be established as

$$\mu_{R,L}^{\pm} = \frac{\hbar}{c} \frac{1}{R_{R,L}^{\pm}} = M_P \frac{\ell_P}{R_{R,L}^{\pm}}, \tag{7.204}$$

where ℓ_P is the Planck length and M_P is the Planck mass, and using the Markov hypothesis, *i.e.* $\mu_{R,L}^{\pm} \leqslant M_P$, one obtains

$$R_{R,L}^{\pm} \geqslant \ell_P. \tag{7.205}$$

Consequently, the squared-mass difference is

$$\Delta\mu_{LR}^2 = M_P^2 \left[\left(\frac{\ell_P}{R_L^{\pm}} \right)^2 - \left(\frac{\ell_P}{R_R^{\pm}} \right)^2 \right]. \tag{7.206}$$

Because the squared-mass difference is positive $\Delta\mu_{LR}^2 > 0$, one has

$$R_R^{\pm} > R_L^{\pm}. \tag{7.207}$$

Moreover, application of the relation (7.27) leads to the conclusion

$$\frac{1}{32\pi} \left[\frac{1}{\left(R_L^{\pm}\right)^2 / \ell_P} - \frac{1}{\left(R_R^{\pm}\right)^2 / \ell_P} \right] = \frac{1}{\ell}, \tag{7.208}$$

which can be treated as the lensmaker's equation

$$(n-1) \left[\frac{1}{R_1} - \frac{1}{R_2} + \frac{n-1}{n} \frac{d}{R_1 R_2} \right] = \frac{1}{f}, \tag{7.209}$$

for the convergent lens of thickness d small compared with the radii of curvature R_1 and R_2. In the case of the lens described by the lens-maker's equation (7.208), one sees that the thickness is identically zero

$$d = 0, \tag{7.210}$$

the focal length is identical to the minimal scale

$$f = \ell, \tag{7.211}$$

the radii of curvature are given by the normalization radii

$$R_1 = \frac{\left(R_L^{\pm}\right)^2}{\ell_P}, \tag{7.212}$$

$$R_2 = \frac{\left(R_R^{\pm}\right)^2}{\ell_P}, \tag{7.213}$$

and the value of the refractive index is

$$n = 1 + \frac{1}{32\pi} \approx 1.00995. \tag{7.214}$$

Let us call such lens by the name *the neutrino lens*. The equation (7.208) expresses the minimal scale by the normalization radii, *i.e.* if and only if $R_{R,L}^{\pm}$ are established by experimental data then ℓ is also established.

Let us consider such a normalization, *i.e.* the particular case of the general solutions which describes the situation of the neutrino lens. First let us derive the appropriate wave functions in the Dirac representation. Introducing the function

$$E^D(x,y) \equiv c^2 \sqrt{\left(\frac{x+y}{2}\right)^2 + x^2}, \tag{7.215}$$

the right-handed chiral Weyl fields are

$$
\begin{aligned}
(\psi_R^{\pm})^D(x,t) &= \frac{1}{\sqrt{V}}\left\{\left[\cos\left[\frac{t-t_0}{\hbar}E^D(\mu_R^{\pm},\mu_L^{\pm})\right]-\right.\right.\\
&- i\frac{\mu_{\pm}^D c^2}{E^D(\mu_R^{\pm},\mu_L^{\pm})}\sin\left[\frac{t-t_0}{\hbar}E^D(\mu_R^{\pm},\mu_L^{\pm})\right]\Bigg]\exp\left\{-\frac{ic}{\hbar}\mu_R^{\pm}(x-x_0)_i\sigma^i\right\}-\\
&- i\frac{\mu_L^{\pm}c^2}{E^D(\mu_L^{\pm},\mu_R^{\pm})}\sin\left[\frac{t-t_0}{\hbar}E^D(\mu_L^{\pm},\mu_R^{\pm})\right]\exp\left\{-\frac{ic}{\hbar}\mu_L^{\pm}(x-x_0)_i\sigma^i\right\}\Bigg\}\times\\
&\times\ \exp\left\{i\left[\theta_{\pm}-\frac{(\mu_R^{\pm}-\mu_L^{\pm})c^2}{2\hbar}(t-t_0)\right]\right\}.
\end{aligned}
\tag{7.216}
$$

while the left-handed chiral Weyl fields are

$$
\begin{aligned}
(\psi_L^{\pm})^D(x,t) &= \frac{1}{\sqrt{V}}\left\{\left[\cos\left[\frac{t-t_0}{\hbar}E^D(\mu_L^{\pm},\mu_R^{\pm})\right]+\right.\right.\\
&+ i\frac{\mu_{\pm}^D c^2}{E^D(\mu_L^{\pm},\mu_R^{\pm})}\sin\left[\frac{t-t_0}{\hbar}E^D(\mu_L^{\pm},\mu_R^{\pm})\right]\Bigg]\exp\left\{-\frac{ic}{\hbar}\mu_L^{\pm}(x-x_0)_i\sigma^i\right\}-\\
&- i\frac{\mu_R^{\pm}c^2}{E^D(\mu_R^{\pm},\mu_L^{\pm})}\sin\left[\frac{t-t_0}{\hbar}E^D(\mu_R^{\pm},\mu_L^{\pm})\right]\exp\left\{-\frac{ic}{\hbar}\mu_R^{\pm}(x-x_0)_i\sigma^i\right\}\Bigg\}\times\\
&\times\ \exp\left\{i\left[\theta_{\pm}-\frac{(\mu_R^{\pm}-\mu_L^{\pm})c^2}{2\hbar}(t-t_0)\right]\right\}.
\end{aligned}
\tag{7.217}
$$

Similarly, one can derive the appropriate wave functions in the Weyl representation. Introducing the function

$$
E^W(x,y) \equiv c^2\sqrt{xy+x^2},
\tag{7.218}
$$

the right-handed chiral Weyl fields are

$$
\begin{aligned}
(\psi_R^{\pm})^W(x,t) &= \frac{\exp i\theta_{\pm}}{\sqrt{V}}\left\{\left[\cos\left[\frac{t-t_0}{\hbar}E^W(\mu_R^{\pm},\mu_L^{\pm})\right]-\right.\right.\\
&- \frac{i\mu_R^{\pm}c^2}{E^W(\mu_R^{\pm},\mu_L^{\pm})}\sin\left[\frac{t-t_0}{\hbar}E^W(\mu_R^{\pm},\mu_L^{\pm})\right]\Bigg]\exp\left\{-\frac{ic}{\hbar}\mu_R^{\pm}(x-x_0)_i\sigma^i\right\}+\\
&+ \frac{i\mu_L^{\pm}c^2}{E^W(\mu_L^{\pm},\mu_R^{\pm})}\sin\left[\frac{t-t_0}{\hbar}E^W(\mu_L^{\pm},\mu_R^{\pm})\right]\exp\left\{-\frac{ic}{\hbar}\mu_L^{\pm}(x-x_0)_i\sigma^i\right\}\Bigg\},
\end{aligned}
\tag{7.219}
$$

while the left-handed chiral Weyl fields are

$$(\psi_L^\pm)^W(x,t) = \frac{\exp i\theta_\pm}{\sqrt{V}}\left\{\left[\cos\left[\frac{t-t_0}{\hbar}E^W(\mu_L^\pm,\mu_R^\pm)\right] - \right.\right.$$
$$\left. - \frac{i\mu_L^\pm c^2}{E^W(\mu_L^\pm,\mu_R^\pm)}\sin\left[\frac{t-t_0}{\hbar}E^W(\mu_L^\pm,\mu_R^\pm)\right]\right]\exp\left\{-\frac{ic}{\hbar}\mu_L^\pm(x-x_0)_i\sigma^i\right\} +$$
$$\left. + \frac{i\mu_R^\pm c^2}{E^W(\mu_R^\pm,\mu_L^\pm)}\sin\left[\frac{t-t_0}{\hbar}E^W(\mu_R^\pm,\mu_L^\pm)\right]\exp\left\{-\frac{ic}{\hbar}\mu_R^\pm(x-x_0)_i\sigma^i\right\}\right\}. \quad (7.220)$$

The situation considered above is only one of many examples arising from the model of massive neutrinos which can be based on the massive Weyl equations (6.97)-(6.98) obtained by application of the non-commutative geometry of the Snyder quantized space-time (6.31)-(6.32). Because such a physical situation is straightforwardly related to the ultra-relativistic limit, we shall call this type of massive neutrinos by the name *ultra-relativistic massive neutrinos*. There is plenty of alternative possibilities for determination of the relation between the initial values of eigenmomentum $p_{R,L}^{\pm\,0}$ and the masses $\mu_{R,L}^\pm$ of the right- and left-handed chiral Weyl fields $\psi_{R,L}^\pm$. However, the choice studied in this section (12.14) presents the crucial reasonability which is the straightforward ramification of its special-relativistic nature. Such a situation expresses validation of the Einstein equivalence principle for the initial data of the space-time evolution of the massive neutrinos, *i.e.* $E_{R,L}^\pm = \mu_{R,L}^\pm c^2 = p_{R,L}^{\pm\,0} c$. This case, however, is also non-trivial from the high energy physics point of view [352]. Namely, it is strictly related to the ultra-high energy region, widely studied in the modern astrophysics (Cf. *e.g.* Refs. [345] and [353] and suitable references therein). The presented particular space-time evolution describes physics of the massive neutrinos in such an energetic region and, therefore, its results should be verified by experimental data due to ultra-high energy astrophysics.

Chapter 8

Ab Initio Deformed Relativity: Superluminal Neutrinos

A Introduction

22nd September 2011 was a truly revolutionary day for high energy particle physics. The OPERA Collaboration, international experimental project of CERN Geneva, announced that their results arising from the high-statistics experimental data evidently demonstrate that neutrinos travel with superluminal speeds [354]. The next five months showed many various speculations about this result, and Special Relativity was considered to be wrong. On 22nd February 2012 the OPERA Collaboration pointed out two GPS-based issues which could impact on the measurement process [355]. The first one is linked to the oscillator producing the events time-stamps in between the GPS synchronizations, whereas the second one is the cable of the optical fiber bringing the external GPS signal to the OPERA master clock. Both these possibilities could give the anomaly leading to the superluminal speed of a neutrino. In February 2012 the second reason was considered more seriously than the first one, and OPERA Collaboration was going to verify the situation in 2012. Anyway, it leads to the essential question about the foundations of Special Relativity, i.e. the limits of the Einstein postulate about the maximal character of the speed of light c.

It is worth stressing that the analogous state of affairs is well-known in modern astronomy since the early 1980s, when faster-than-light motion was suggested to contradict the quasars having cosmological distances. The present day experimental data show that superluminal travels are very often phenomena in radio galaxies, quasars and microquasars. Usually, it is believed that existence of a black hole within these sources provokes the high speed mass ejection. It should be emphasized that the optical illusions are mostly believed to be theoretical justifications for superluminal speeds, and such arguments do not destroy Special Relativity. For this reason the speculations based on the conclusions within Special Relativity about time travels and other hypothetical possibilities, arising from the exceeding of the speed of light, are cordially propagated.

It should be emphasized that even Albert Einstein never excluded the possibility of superluminal speeds. His account of the theory of motion is the postulate of Special Relativity saying that the speed of light c is the maximal speed, which does not mean that the measured speed of a moving object must be less than c. It is very probable that Special Relativity is the well-working approximation for the speeds less than the speed of light, and for the superluminal speeds the Einstein theory should be complemented by new arguments. In fact, the speed of light postulate is the only internal restriction of Special Relativity, but not the limitation of Nature, so the sceptics are forced to accept the state of affairs in which Special Relativity is a consistent physical theory.

In this chapter we present the complement of Special Relativity which consistently describes motion with faster-than-light speeds. We show that superluminally traveling particles can be efficiently theorized using Special Relativity complemented by the additional arguments which are specific for the superluminal speeds only. In this manner, a whole mathematical structure of the Minkowski energy-momentum space, the four-vector formalism, as well as the Einstein postulates remain unchanged, while the energy-momentum relation expressing the Einstein equivalence principle is deformed due to certain additional assumptions. It should be emphasized that the postulates of Special Relativity can be taken into account separately from the original Einstein's energy-momentum relation $E^2 = m^2c^4 + p^2c^2$ and, for this reason, its modification does not lead to inconsistencies. The only one problem is, however, to complement Special Relativity in such a way that the Poincaré invariance, and its special case the Lorentz invariance, remains preserved. We show here that, in general, such a restoration is possible. For creation of such an appropriate theory we shall discuss in some detail general consequences for Special Relativity arising from deformed energy-momentum relations, which of course include existence of faster-than-light particles.

For the consistent form of the theory we propose to take into account the non-parallelism hypothesis, expressing the lack of parallelism between momentum and velocity vectors of a moving body. The parallelism is usual for both Newton's and Einstein's theories of motion. Moreover, the deformation of Einstein's energy-momentum relation is not given *ad hoc*, like it is done in the case of the Snyder–Sidharth deformation discussed in previous chapters, but is derived *ab initio* from the definition of velocity. The third essential change is introduction of the new speed of light, which in some particular situations can exceed the speed of light c. These collection of assumptions taken together result in the theory which is able to describe motion with superluminal speeds within Special Relativity, and above all preserve both Poincaré and Lorentz invariance of the Minkowski energy-momentum space. Finally, we discuss our theory in the context of the superluminal neutrinos detected by OPERA Collaboration, and demonstrate that both exceeding of the speed of light and high-speed mass ejection of neutrinos are consistently described within our framework.

Let us consider Special Relativity, in which the Hamiltonian constraint called the Einstein energy-momentum relation, is modified by the constituent Δ arising due to certain additional arguments. This extra-term has the physical dimensionality of squared energy and, in general, is a function of momentum vector and energy of a moving object, and also other non-dynamical parameters

$$E^2 = m^2c^4 + p^2c^2 + \Delta(E, p_i; m, \ldots). \tag{8.1}$$

We assume that energy is a function of momentum value, *i.e.* $E = E(p)$. It is obvious that for the trivial case $\Delta \equiv 0$ one has to deal with the situation described by Special Relativity and its all consequences among which the key one is the Standard Model of particles and fundamental interactions. In this manner, existence of deformation Δ leads to new physics which is manifestly beyond the Standard Model. The origins of the deformation $\Delta(E,p;m,\dots)$ can have diverse nature. All well-known models, including the Snyder–Sidharth deformation, preserve the structure of the Minkowski energy-momentum space and the Einstein postulates, but manifestly violate both the Poincaré and Lorentz (CP) symmetries.

In this manner, it is reasonable to suspect that the most general Hamiltonian constraint (8.1) also leads to consistent description of the new physics, including superluminal speeds. We shall call the extra-term Δ by *superluminal deformation*. The problem is to establish the concrete form of superluminal deformation, and for the derived formula of Δ to deduce the physical content of the corresponding deformed Special Relativity.

B Superluminal Speeds

Let us examine the velocity and speed of a particle characterized by any energy-momentum relation of the form (8.1). According to the usual definition, which is particularly valid for Special Relativity, classical mechanics and other well-known cases, the velocity vector is

$$v^i = \frac{\partial E}{\partial p_i}. \tag{8.2}$$

Using (8.1) leads to the relation

$$\frac{\partial E}{\partial p_i} = \frac{1}{2E}\left(2p^i c^2 + \frac{\partial \Delta}{\partial p_i} + \frac{\partial \Delta}{\partial E} v^i\right). \tag{8.3}$$

For consistency Δ should be a scalar quantity and, therefore, one can suggest that Δ depends on the powers of momentum value $p = \sqrt{p^i p_i}$. Hence the p_i-derivative can be transformed into the p-derivative

$$\frac{\partial \Delta}{\partial p_i} = \frac{\partial \Delta}{\partial p}\frac{\partial p}{\partial p_i} + \frac{\partial \Delta}{\partial E}\frac{\partial E}{\partial p_i} = \frac{\partial \Delta}{\partial p}\frac{p^i}{p} + \frac{\partial \Delta}{\partial E} v^i, \tag{8.4}$$

applied to the relation (8.3) leads to

$$v^i\left(1 - \frac{1}{E}\frac{\partial \Delta}{\partial E}\right) = \frac{1}{2E}\left(2p^i c^2 + \frac{\partial \Delta}{\partial p}\frac{p^i}{p}\right). \tag{8.5}$$

If $\frac{1}{E}\frac{\partial\Delta}{\partial E} \neq 1$, then the velocity of a moving object has the form

$$v^i = v\frac{p^i}{p}, \tag{8.6}$$

where $v = \sqrt{v^i v_i}$ is the speed of a moving object

$$v = \frac{pc^2 + \frac{1}{2}\frac{\partial\Delta}{\partial p}}{E - \frac{\partial\Delta}{\partial E}}. \tag{8.7}$$

Both sides of the relation (8.6) can be multiplied by v_i or by p_i which gives

$$vp = v_i p^i = v^i p_i. \tag{8.8}$$

Applying the definition of the scalar product one has $v^i p_i = vp\cos\alpha$, where $\alpha = (v^i, p_i)$ is the angle between the vectors v^i and p^i and, consequently, (8.8) gives identically $\cos\alpha = 1$, *i.e.* $\alpha = 0$.

The speed formula (8.7) can be presented in the equivalent form

$$v = v_r + \delta v, \tag{8.9}$$

where v_r is the speed, which does not contain the derivatives of superluminal deformation Δ and in the trivial situation $\Delta \equiv 0$ leads to the speed computed from standard Special Relativity

$$v_r = \frac{pc^2}{E}, \tag{8.10}$$

and δv is the speed correction due to superluminal deformation

$$\delta v = v - v_r = \frac{v_r\frac{\partial\Delta}{\partial E} + \frac{1}{2}\frac{\partial\Delta}{\partial p}}{E - \frac{\partial\Delta}{\partial E}}. \tag{8.11}$$

In the usual sense of smallness of the speed correction must be

$$\left|\frac{\delta v}{v}\right| \ll 1, \tag{8.12}$$

which means that

$$-\frac{1}{2pc^2}\frac{\partial\Delta}{\partial p} \leqslant \frac{1}{E}\frac{\partial\Delta}{\partial E} \ll 1, \quad \text{for} \quad -\frac{1}{2pc^2}\frac{\partial\Delta}{\partial p} \leqslant 1, \tag{8.13}$$

$$1 \ll \frac{1}{E}\frac{\partial\Delta}{\partial E} \ll -1 - \frac{1}{pc^2}\frac{\partial\Delta}{\partial p}, \quad \text{for} \quad -\frac{1}{2pc^2}\frac{\partial\Delta}{\partial p} \gg 1. \tag{8.14}$$

Let us consider the approximation

$$\left|\frac{1}{E}\frac{\partial\Delta}{\partial E}\right| \ll 1, \tag{8.15}$$

which in the light of the relation (8.32) means that

$$\cos\alpha \gg 2\left|1 - \frac{v_r}{v}\right| - 2, \tag{8.16}$$

which together with the condition $-1 \leqslant \cos\alpha \leqslant 1$ means that

$$v \in \left[\frac{2}{5}v_r, \frac{2}{3}v_r\right] \cup [2v_r, \infty]. \tag{8.17}$$

We shall call (8.15) *ultrahigh speed approximation*. It can be seen straightforwardly that USA satisfies the smallness condition (8.12), *i.e.* is a particular case within the conditions

$$\left|\frac{1}{2pc^2}\frac{\partial\Delta}{\partial p}\right| \leqslant \left|\frac{1}{E}\frac{\partial\Delta}{\partial E}\right| \ll 1, \quad \text{for} \quad \left|\frac{1}{2pc^2}\frac{\partial\Delta}{\partial p}\right| \leqslant 1, \tag{8.18}$$

$$1 \ll \left|\frac{1}{E}\frac{\partial\Delta}{\partial E}\right| \ll \left|1 + \frac{1}{pc^2}\frac{\partial\Delta}{\partial p}\right|, \quad \text{for} \quad \left|\frac{1}{2pc^2}\frac{\partial\Delta}{\partial p}\right| \gg 1. \tag{8.19}$$

In the region of speeds (8.17), which is defined by USA (8.15), the speed correction formula (8.11) becomes

$$\delta v = \frac{v_r\dfrac{1}{E}\dfrac{\partial\Delta}{\partial E} + \dfrac{1}{2E}\dfrac{\partial\Delta}{\partial p}}{1 - \dfrac{1}{E}\dfrac{\partial\Delta}{\partial E}}. \tag{8.20}$$

Let us look on this formula for really small values of $\frac{1}{E}\frac{\partial\Delta}{\partial E}$. Then one can take into account the Taylor power series expansion of (8.20)

$$\delta v = \frac{1}{2E}\frac{\partial\Delta}{\partial p} + v_r\left(1 + \frac{1}{2pc^2}\frac{\partial\Delta}{\partial p}\right)\sum_{n=1}^{\infty}\frac{1}{E^n}\left(\frac{\partial\Delta}{\partial E}\right)^n, \tag{8.21}$$

in which the leading terms are

$$\delta v \approx \frac{1}{2E}\frac{\partial\Delta}{\partial p} + \frac{pc^2}{E^2}\frac{\partial\Delta}{\partial E} + \frac{1}{2E^2}\frac{\partial\Delta}{\partial E}\frac{\partial\Delta}{\partial p} + \frac{pc^2}{E^3}\left(\frac{\partial\Delta}{\partial E}\right)^2 + \frac{1}{2E^3}\left(\frac{\partial\Delta}{\partial E}\right)^2\frac{\partial\Delta}{\partial p} + \ldots \tag{8.22}$$

If, moreover, additionally the condition $\frac{\partial\Delta}{\partial E} \neq 0$ holds, then the speed correction can be presented in the form

$$\delta v = \frac{v_r + \frac{1}{2}\frac{\partial E}{\partial p}}{\left(\frac{1}{E}\frac{\partial\Delta}{\partial E}\right)^{-1} - 1}, \tag{8.23}$$

and in the appropriate Taylor power series expansion

$$\delta v = \left(v_r + \frac{1}{2}\frac{\partial E}{\partial p}\right)\sum_{n=1}^{\infty}\frac{1}{E^n}\left(\frac{\partial\Delta}{\partial E}\right)^n, \tag{8.24}$$

the leading terms are

$$\delta v \approx \frac{pc^2}{E^2}\frac{\partial\Delta}{\partial E} + \frac{1}{2E}\frac{\partial E}{\partial p}\frac{\partial\Delta}{\partial E} + \frac{pc^2}{E^3}\left(\frac{\partial\Delta}{\partial E}\right)^2 + \frac{1}{2E^2}\frac{\partial E}{\partial p}\left(\frac{\partial\Delta}{\partial E}\right)^2 + \ldots \tag{8.25}$$

In the region of speeds opposite to USA

$$\left|\frac{1}{E}\frac{\partial\Delta}{\partial E}\right| \gg 1, \tag{8.26}$$

the expansions (8.22) as well as (8.24) are not still valid. Then, the assumption $\frac{\partial\Delta}{\partial E} \neq 0$ is obligatory because otherwise the region of speed opposite to the ultrahigh speed approximation does not exist. Dividing both sides of the fraction in the speed formula (8.11) by $\frac{\partial\Delta}{\partial E}$ one obtains the speed correction formula

$$\delta v = -\frac{v_r + \frac{1}{2}\frac{\partial E}{\partial p}}{1 - \left(\frac{1}{E}\frac{\partial\Delta}{\partial E}\right)^{-1}}, \tag{8.27}$$

which shows that in such a case the speed correction contributes purely negative constituents, so that the speed is always lower than v_r. In such a situation, the appropriate Taylor power series expansion

$$\delta v = -\left(v_r + \frac{1}{2}\frac{\partial E}{\partial p}\right)\sum_{n=1}^{\infty} E^n \left(\frac{\partial \Delta}{\partial E}\right)^{-n}, \tag{8.28}$$

has the leading terms as follows

$$\delta v \approx v_r + \frac{1}{2}\frac{\partial E}{\partial p} + v_r E\left(\frac{\partial \Delta}{\partial E}\right)^{-1} + v_r E^2\left(\frac{\partial \Delta}{\partial E}\right)^{-2} + \frac{E}{2}\frac{\partial E}{\partial p}\left(\frac{\partial \Delta}{\partial E}\right)^{-1} + \dots \tag{8.29}$$

In this manner, any deformation Δ of Special Relativity can be used for generation of the complement for superluminally moving objects. In other words, within the presented approach for non-trivial Δ's the speed of a moving object can be higher than the speed of light c.

C Non-Parallelism Hypothesis

Let us consider the elementary formula

$$\frac{\partial E}{\partial p} = \frac{\partial E}{\partial p_i}\frac{\partial p_i}{\partial p} = v^i \frac{p_i}{p} = v\cos\alpha, \tag{8.30}$$

where $\alpha = (p^i, v^i)$ is the angle between momentum and velocity vectors of an moving object. In the light of the relation (8.6) this angle is identically equal to 0, *i.e.* $\cos\alpha = 1$. However, such a state of affairs must not be sufficient for description of a motion with superluminal speeds. For this reason we propose

Proposition (The non-parallelism hypothesis). *In a motion with superluminal speed parallelism of the momentum and the velocity vectors of a faster-than-light traveling object is lost.*

Applying *ad hoc* the non-parallelism hypothesis within the formula (8.27) one receives the speed correction in the form

$$\delta v = \frac{v_r + \frac{1}{2} v\cos\alpha}{\left(\frac{1}{E}\frac{\partial \Delta}{\partial E}\right)^{-1} - 1}. \tag{8.31}$$

In such a situation, using definition (8.9) leads to another form of the speed formula

$$v = \frac{v_r}{1 - \left(\frac{1}{E}\frac{\partial \Delta}{\partial E}\right)\left(1 + \frac{1}{2}\cos\alpha\right)}, \tag{8.32}$$

which can be expanded around the non-deformed case, *i.e.* $\cos\alpha = 1$ and $\frac{1}{E}\frac{\partial \Delta}{\partial E} = 0$, as follows

$$\frac{v}{v_r} = \sum_{n=0}^{\infty}\sum_{m=0}^{\infty}\sum_{k=0}^{m} \frac{(-1)^k 2^{-n} 3^{n-m} \Gamma(m-n)\Gamma(n+1)(\cos\alpha)^k \left(\frac{1}{E}\frac{\partial \Delta}{\partial E}\right)^n}{\Gamma(k+1)\Gamma(m+1)\Gamma(m-k+1)\Gamma(-n)\Gamma(n-m+1)}, \tag{8.33}$$

or around $\alpha = 0$

$$\frac{v}{v_r} = \frac{1}{1 - \frac{3}{2E}\frac{\partial \Delta}{\partial E}} - \frac{\frac{1}{E}\frac{\partial \Delta}{\partial E}}{4\left(1 - \frac{3}{2E}\frac{\partial \Delta}{\partial E}\right)^2}\alpha^2 + \frac{\frac{1}{E}\frac{\partial \Delta}{\partial E}\left(1 + \frac{3}{2E}\frac{\partial \Delta}{\partial E}\right)}{48\left(1 - \frac{3}{2E}\frac{\partial \Delta}{\partial E}\right)^3}\alpha^4 + O[\alpha^6]. \tag{8.34}$$

The essential effect is that in the absence of deformation, *i.e.* in the trivial case $\Delta \equiv 0$ which corresponds to Special Relativity, the corrections due to the non-parallelism hypothesis vanish identically. In this manner, the momentum-velocity non-parallelism has any meaning if and only if Special Relativity is superluminally deformed.

Calculating $\cos\alpha$ from the relation (8.32)

$$\cos\alpha = 2\left(\frac{1}{E}\frac{\partial \Delta}{\partial E}\right)^{-1}\left(1 - \frac{v_r}{v}\right) - 2, \tag{8.35}$$

and taking into account the fact that $-1 \leqslant \cos\alpha \leqslant 1$, one obtains the upper and the lower bounds for the speed v

$$\frac{v_r}{1 - \frac{1}{2E}\frac{\partial \Delta}{\partial E}} \leqslant v \leqslant \frac{v_r}{1 - \frac{3}{2E}\frac{\partial \Delta}{\partial E}}. \tag{8.36}$$

D Ab Initio Deformation

Usually, the superluminal deformation Δ of the energy-momentum relation arises from algebraic deformation or non-commutative geometry of phase space. However, in such scenarios deformations are high-order polynomials in the momentum value and energy violating the Poincaré invariance of the Minkowski energy-momentum space, which is the crucial and fundamental feature of Special Relativity. In this section we shall deduce the deformation which, as it will be shown in the next parts of this paper, allows to preserve the Poincaré symmetry.

Let us consider superluminal deformations which are functions of time t hidden within the dynamical parameters E and p, *i.e.*

$$\Delta = \Delta(E(t), p(t)) = \Delta(t). \tag{8.37}$$

In such a situation, the partial derivatives with respect to p and E can be transformed in the following way

$$\frac{\partial \Delta}{\partial p} = \frac{\partial \Delta}{\partial t}\frac{\partial t}{\partial p} = \frac{\dot{\Delta}}{\dot{p}} = \frac{\dot{\Delta}}{F}, \tag{8.38}$$

$$\frac{\partial \Delta}{\partial E} = \frac{\partial \Delta}{\partial t}\frac{\partial t}{\partial E} = \frac{\dot{\Delta}}{\dot{E}} = \frac{\dot{\Delta}}{P}, \tag{8.39}$$

where $F = \dot{p}$ is the value of force acting on a particle, and P is a power. It is easy to see that application of the relations (8.38) and (8.39) within the formula (8.7) leads to the following speed formula

$$v = \frac{P}{F}\frac{pc^2F + \frac{1}{2}\dot{\Delta}}{EP - \dot{\Delta}}, \tag{8.40}$$

which allows to establish the differential equation for the deformation

$$\dot{\Delta} = 2FP\frac{Ev - pc^2}{P + 2Fv}. \tag{8.41}$$

This equation can be easily integrated

$$\begin{aligned}
\Delta &= 2\int_{t_0}^{t} dt' F(t')P(t')\frac{E(t')v(t')-p(t')c^2}{P(t')+2F(t')v(t')} && (8.42)\\
&= 2\int_{t_0}^{t} dt'\frac{dp}{dt'}\frac{dE}{dt'}\frac{E(t')v(t')-p(t')c^2}{\frac{dE}{dt'}+2\frac{dp}{dt'}v(t')}\\
&= 2\int_{t_0}^{t} dt'\frac{dp}{dt'}\frac{dE}{dt'}\frac{E(t')v(t')}{\frac{dE}{dt'}+2\frac{dp}{dt'}v(t')}-2c^2\int_{t_0}^{t} dt'\frac{dp}{dt'}\frac{dE}{dt'}\frac{p(t')}{\frac{dE}{dt'}+2\frac{dp}{dt'}v(t')}\\
&= 2\int_{E(t_0)}^{E(t)} dE(t')\frac{dp}{dt'}\frac{E(t')v(t')}{\frac{dE}{dt'}+2\frac{dp}{dt'}v(t')}-2c^2\int_{p(t_0)}^{p(t)} dp(t')\frac{dE}{dt'}\frac{p(t')}{\frac{dE}{dt'}+2\frac{dp}{dt'}v(t')}\\
&= 2\int_{E(t_0)}^{E(t)} dE(t')\frac{E(t')v(t')}{\frac{dE(t')}{dp(t')}+2v(t')}-2c^2\int_{p(t_0)}^{p(t)} dp(t')\frac{p(t')}{1+2\left(\frac{dE(t')}{dp(t')}\right)^{-1}v(t')}, && (8.43)
\end{aligned}$$

and using the non-parallelism hypothesis

$$\frac{dE(t')}{dp(t')}=\frac{dE(t')}{dp_i(t')}\frac{dp_i(t')}{dp(t')}=v^i(t')\frac{p_i(t')}{p(t')}=v(t')\cos\alpha(t'), \tag{8.44}$$

its solution takes the following form

$$\Delta=2\int_{E(t_0)}^{E(t)} dE(t')\frac{E(t')}{2+\cos\alpha(t')}-2c^2\int_{p(t_0)}^{p(t)} dp(t')\frac{p(t')\cos\alpha(t')}{2+\cos\alpha(t')}. \tag{8.45}$$

Let us evaluate the deformation for the case when the angle α and the energy E, as well as the angle α and the momentum value p, are independent variables. In fact, such a state of affairs is generally true, because angle α is the only free parameter. In such a situation, one obtains the formula

$$\Delta=\frac{2}{2+\cos\alpha}\int_{E(t_0)}^{E(t)} dE(t')E(t')-\frac{2c^2\cos\alpha}{2+\cos\alpha}\int_{p(t_0)}^{p(t)} dp(t')p(t'), \tag{8.46}$$

in which the integrals can be derived straightforwardly and, consequently, the superluminal deformation derived *ab initio* is a second-order polynomial in the momentum value and energy

$$\Delta=\frac{E^2-E_0^2}{2+\cos\alpha}-\frac{(p^2-p_0^2)c^2\cos\alpha}{2+\cos\alpha}, \tag{8.47}$$

where $E_0 = E(t_0)$ and $p_0 = p(t_0)$ are the initial data. We shall call the deformation (8.69) the *ab initio deformation.*

E Ab Initio Energy-Momentum Relation

Applying the *ab initio* deformation (8.69) within the energy-momentum relation (8.1), one receives the ab initio energy-momentum relation

$$E^2 = m^2c^4 + p^2c^2 + \frac{1}{2+\cos\alpha}E^2 - \frac{\cos\alpha}{2+\cos\alpha}p^2c^2 - \Delta_0, \tag{8.48}$$

where Δ_0 is the constant deformation

$$\Delta_0 = \frac{1}{2+\cos\alpha}E_0^2 - \frac{\cos\alpha}{2+\cos\alpha}p_0^2c^2, \tag{8.49}$$

which after minor algebraic manipulations becomes

$$E^2 = \frac{2p^2c^2}{1+\cos\alpha} + m_{eff}^2c^4, \tag{8.50}$$

where m_{eff} is the effective mass arising from the ab initio superluminal deformation

$$m_{eff} = \sqrt{m^2 + \frac{1}{c^4(1+\cos\alpha)}\left(m^2c^4 + \cos\alpha\, p_0^2c^2 - E_0^2\right)}. \tag{8.51}$$

In this manner, the necessary and the sufficient conditions for existence of the new mass in the deformed theory are

$$\cos\alpha \neq 1, \tag{8.52}$$

$$\cos\alpha \neq \frac{E_0^2 - m^2c^4}{p_0^2c^2}. \tag{8.53}$$

The first condition is only the non-parallelism hypothesis, while the second one is exclusion of the specific value of the angle.

The effective mass (8.51) must not be a real number. It is a real number if and only if $m_{eff}^2 \geqslant 0$, *i.e.* the expression under the square root is positive. It leads to the condition

$$E_0^2 \leqslant (2+\cos\alpha)m^2c^4 + p_0^2c^2\cos\alpha, \tag{8.54}$$

which can be understood as the lower bound for the mass parameter

$$m \geqslant \frac{E_0^2 - p_0^2 c^2 \cos\alpha}{(2+\cos\alpha)c^4}, \tag{8.55}$$

or, equivalently, as the lower bound for the angle α

$$\cos\alpha \geqslant \frac{E_0^2 - 2m^2c^4}{p_0^2c^2 + m^2c^4}. \tag{8.56}$$

Let us focus on the case when the effective mass is identically equal to zero, *i.e.* $m_{eff} = 0$. In such a situation, if α is known from experiment then the mass parameter m is established as

$$m = \frac{E_0^2 - p_0^2 c^2 \cos\alpha}{(2+\cos\alpha)c^4}, \tag{8.57}$$

or if m is known from experimental data then

$$\cos\alpha = \frac{E_0^2 - 2m^2c^4}{p_0^2c^2 + m^2c^4}. \tag{8.58}$$

If, moreover, the initial data are constrained by the Special Relativity energy-momentum relation

$$E_0^2 = p_0^2c^2 + m^2c^4, \tag{8.59}$$

then one receives

$$\cos\alpha = 1 - \frac{2m^2c^2}{E_0^2} = \frac{2p_0^2c^2}{E_0^2} - 1. \tag{8.60}$$

Because of the bounds $-1 \leqslant \cos\alpha \leqslant 1$, one obtains the bounds

$$E_0^2 \geqslant m^2c^4 \geqslant 0 \;\text{ and }\; E_0^2 \geqslant p_0^2c^2 \geqslant 0, \tag{8.61}$$

which is consistent with the relation (8.59).

Using the *ab initio* deformed energy-momentum relation (8.50) for calculation of the velocity of a moving object gives the result

$$v^i = \frac{\partial E}{\partial p_i} = \frac{2pc^2}{E(1+\cos\alpha)} \frac{p^i}{p}, \tag{8.62}$$

and, consequently, the speed of such an object is

$$v = \frac{2pc^2}{E(1+\cos\alpha)}. \tag{8.63}$$

It is visible that for the broken non-parallelism hypothesis, *i.e.* when $\cos\alpha = 1$, the speed of a moving object (8.63) becomes

$$v = \frac{pc^2}{E}, \tag{8.64}$$

where energy is the usual relativistic energy

$$E^2 = p^2c^2 + m_{eff}^2c^4,. \tag{8.65}$$

with the appropriate effective mass

$$m_{eff} = \sqrt{m^2 + \frac{1}{c^4}\left(m^2c^4 + p_0^2c^2 - E_0^2\right)}. \tag{8.66}$$

which together with the Special Relativity condition for the initial data

$$E_0^2 = p_0^2c^2 + m^2c^4, \tag{8.67}$$

becomes the non-deformed mass of Special Relativity

$$m_{eff} = m. \tag{8.68}$$

In this manner, the case of $\cos\alpha = 1$ under specific condition (8.67) for the initial data can result in Special Relativity despite the fact that then the ab initio superluminal deformation (8.69) is still present

$$\Delta = \frac{E^2 - p^2c^2}{3} - \frac{E_0^2 - p_0^2c^2}{3}. \tag{8.69}$$

One sees that the most important result of the non-parallelism hypothesis is the deformed speed (8.63). It is easy to check that this formula can be presented in the form

$$v = \frac{2c}{\sqrt{1+\cos\alpha + \frac{1}{2}\left(\frac{m_{eff}c}{p}\right)^2(1+\cos\alpha)^2}}. \tag{8.70}$$

If the effective mass vanishes $m_{eff}=0$ then

$$v=\frac{2c}{\sqrt{1+\cos\alpha}}, \tag{8.71}$$

and, in this manner, in such a situation the speed of a moving object has a superluminal value if and only if

$$\frac{2}{\sqrt{1+\cos\alpha}}\geqslant 1\longrightarrow \cos\alpha\leqslant 3, \tag{8.72}$$

what is always true because the natural bounds of the cosinus function. The case of the non-vanishing effective mass if a little bit difficult. In such a situation, the speed of a moving object is superluminal if and only if

$$\left(\frac{m_{eff}c}{p}\right)^2\cos^4\frac{\alpha}{2}+\cos^2\frac{\alpha}{2}-2\leqslant 0, \tag{8.73}$$

where of course $\cos^2\frac{\alpha}{2}=\frac{1+\cos\alpha}{2}\in[0,1]$. This inequality can be easily solved

$$\cos^2\frac{\alpha}{2}\geqslant\frac{1}{2}\left(\frac{p}{m_{eff}c}\right)^2\left(\sqrt{1+8\left(\frac{m_{eff}c}{p}\right)^2}-1\right), \tag{8.74}$$

and gives the lower bound

$$\cos\alpha\geqslant\left(\frac{p}{m_{eff}c}\right)^2\left(\sqrt{1+8\left(\frac{m_{eff}c}{p}\right)^2}-1\right)-1 \tag{8.75}$$

and because the bounds $\cos^2\frac{\alpha}{2}\in[0,1]$

$$m_{eff}\geqslant\frac{p}{c}. \tag{8.76}$$

In the light of the energy-momentum relation (8.50), the result (8.76) leads to the lower energetic bound

$$E\geqslant\sqrt{\frac{3+\cos\alpha}{1+\cos\alpha}}pc, \tag{8.77}$$

or, equivalently, the lower bound for the cosinus

$$\cos\alpha \geqslant \frac{\left(\dfrac{E}{pc}\right)^2 - 3}{1 - \left(\dfrac{E}{pc}\right)^2}. \tag{8.78}$$

Because the bound $\cos\alpha \leqslant 1$, one receives the upper energetic bound

$$E \leqslant \sqrt{2}pc. \tag{8.79}$$

In this manner, the deformation combined with the non-parallelism hypothesis formally enables exceeding the speed of light c.

F Energy–Momentum Interval

Let us focus our attention on the energy-momentum interval

$$s^2 = \eta_{\mu\nu}p^\mu p^\nu = -E^2 + p^2c^2, \tag{8.80}$$

where $p^\mu = \left[E, p^i c\right]$ is the momentum four-vector and $\eta_{\mu\nu} = \mathrm{diag}(-1,1,1,1)$ is the metric of the Minkowski energy-momentum space. Direct calculation gives

$$s^2 = \frac{\cos\alpha - 1}{\cos\alpha + 1}p^2c^2 - m_{eff}^2c^4. \tag{8.81}$$

Let us perform the sign analysis of the energy-momentum interval (8.81).

1. The case $m_{eff}^2 > 0$ for which

$$\cos\alpha > \frac{E_0^2 - 2m^2c^4}{p_0^2c^2 + m^2c^4}. \tag{8.82}$$

   Then the energy-momentum interval (8.81) is

   (a) light-like if and only if $s^2 = 0$, *i.e.*

$$p = \sqrt{\frac{\cos\alpha + 1}{\cos\alpha - 1}}m_{eff}c, \tag{8.83}$$

   and $E = pc$.

(b) energy-like if and only if $s^2 < 0$, *i.e.*

$$p < \sqrt{\frac{\cos\alpha + 1}{\cos\alpha - 1}} m_{eff} c, \tag{8.84}$$

and $E > pc$.

(c) momentum-like if and only if $s^2 > 0$, *i.e.*

$$p > \sqrt{\frac{\cos\alpha + 1}{\cos\alpha - 1}} m_{eff} c, \tag{8.85}$$

and $E < pc$.

For consistency of all these cases, however, it must be

$$\cos\alpha \leqslant -1 \quad or \quad \cos\alpha > 1, \tag{8.86}$$

what is a manifestly wrong condition.

2. The case $m_{eff}^2 < 0$ for which

$$\cos\alpha < \frac{E_0^2 - 2m^2c^4}{p_0^2c^2 + m^2c^4}. \tag{8.87}$$

Then, the energy-momentum interval (8.81) is

(a) light-like if and only if $s^2 = 0$, *i.e.*

$$p = \sqrt{\frac{1 + \cos\alpha}{1 - \cos\alpha}} |m_{eff}| c, \tag{8.88}$$

and $E = pc$.

(b) energy-like if and only if $s^2 < 0$, *i.e.*

$$p < \sqrt{\frac{1 + \cos\alpha}{1 - \cos\alpha}} |m_{eff}| c, \tag{8.89}$$

and $E > pc$.

(c) momentum-like if and only if $s^2 > 0$, *i.e.*

$$p > \sqrt{\frac{1 + \cos\alpha}{1 - \cos\alpha}} |m_{eff}| c, \tag{8.90}$$

and $E < pc$.

All these cases are consistent if and only if

$$-1 \leqslant \cos\alpha < 1, \tag{8.91}$$

which is equivalent to

$$\frac{1}{3}\left(\frac{E_0^2}{c^4} - \frac{p_0^2}{c^2}\right) < m^2 \leqslant \frac{E_0^2}{c^4} + \frac{p_0^2}{c^2}. \tag{8.92}$$

Otherwise, the momentum value is purely imaginary, *i.e.* non-physical.

3. The case of vanishing effective mass $m_{eff}^2 = 0$, for which

$$\cos\alpha = \frac{E_0^2 - 2m^2c^4}{p_0^2c^2 + m^2c^4}. \tag{8.93}$$

We assume that always $p^2 \geqslant 0$. Then, the energy-momentum interval (8.81) is

(a) light-like if and only if $s^2 = 0$, *i.e.*

$$\cos\alpha = 1, \tag{8.94}$$

and $E = pc$. In this instance

$$m^2 = \frac{1}{3}\left(\frac{E_0^2}{c^4} - \frac{p_0^2}{c^2}\right). \tag{8.95}$$

However, for consistency with the non-parallelism hypothesis it must be $p = 0$ and $E = 0$.

(b) energy-like if and only if $s^2 < 0$, *i.e.*

$$\cos\alpha < -1 \quad or \quad \cos\alpha \geqslant -1, \tag{8.96}$$

and $E > pc$. This case is manifestly non-physical.

(c) momentum-like if and only if $s^2 > 0$, *i.e.*

$$-1 < \cos\alpha \leqslant 1, \tag{8.97}$$

and $E < pc$. In this instance

$$\frac{1}{3}\left(\frac{E_0^2}{c^4} - \frac{p_0^2}{c^2}\right) \leqslant m^2 < \frac{E_0^2}{c^4} + \frac{p_0^2}{c^2}. \tag{8.98}$$

G Violated Poincaré Invariance

The Poincaré invariance is the fundamental feature of Special Relativity and, in general, guarantees the relativistic character of any theory. Let us verify this important feature with respect to our theory.

The problem of the *ab initio* deformed energy-momentum relation (8.50) is the manifest breakdown of the Poincaré invariance. Such a state of affairs can be seen after straightforward performing of the calculation. In the Minkowski energy-momentum space, the Poincaré invariance of any theory means the invariance of the energy-momentum interval

$$s'^2 = s^2, \tag{8.99}$$

which must be preserved under the action of the Poincaré transformation in the Minkowski energy-momentum space of the theory

$$p'^{\mu} = \Lambda^{\mu}_{\nu} p^{\nu} + P^{\mu}, \tag{8.100}$$

where P^{μ} is a constant momentum four-vector, Λ^{μ}_{ν} is a constant matrix, and p'^{μ} is the momentum four-vector transformed under action of this transformation. The detailed tensor formalism of the matrices Λ^{μ}_{ν} can be found *e.g.* in Ref. [356]. The matrix Λ^{μ}_{ν} satisfies the condition

$$\eta_{\mu\nu}\Lambda^{\mu}_{\kappa}\Lambda^{\nu}_{\lambda} = \eta_{\kappa\lambda}, \tag{8.101}$$

which after contraction of its both sides with the combination $\eta^{\mu\nu}\eta^{\kappa\lambda}$ leads to the following result

$$\Lambda^{\alpha\mu}\Lambda^{\nu}_{\lambda} = \eta^{\alpha}_{\lambda}\eta^{\mu\nu}. \tag{8.102}$$

A particular case of this relation is the situation $\alpha = \lambda$, *i.e.*

$$\Lambda^{\alpha\mu}\Lambda^{\nu}_{\alpha} = \eta^{\mu\nu}. \tag{8.103}$$

Contraction of the equation (8.102) with the combination $\eta_{\alpha\beta}\eta_{\mu\kappa}$ allows to establish another important result

$$\Lambda_{\kappa\beta}\Lambda^{\nu}_{\lambda} = \eta_{\beta\lambda}\eta^{\nu}_{\kappa}, \tag{8.104}$$

which in the particular case $\nu = \kappa$ takes the form

$$\Lambda_{\alpha\beta}\Lambda^{\alpha}_{\lambda} = \eta_{\beta\lambda}. \tag{8.105}$$

Application of the Poincaré transformation (8.109) within the energy-momentum interval (8.81) allows to verify the Poincaré invariance. In the present case one obtains the result

$$s'^2 = \frac{\cos\alpha - 1}{\cos\alpha + 1}\left(\eta_{ik}\Lambda^i_j\Lambda^k_i p^j p^i + P^i P_i + P^i \eta_{ik}\Lambda^k_i p^i + P_i\Lambda^i_j p^j\right) c^2 + m^2_{eff}c^4 \quad (8.106)$$

$$= \frac{\cos\alpha - 1}{\cos\alpha + 1}\left(\eta_{ji} p^j p^i + P^i P_i + 2P_i\Lambda^i_j p^j\right) c^2 + m^2_{eff}c^4 \quad (8.107)$$

$$= s^2 + \frac{\cos\alpha - 1}{\cos\alpha + 1} P_i\left(P^i + 2\Lambda^i_j p^j\right) c^2, \quad (8.108)$$

which manifestly breaks the Poincaré invariance condition (8.99).

There are, however, certain situations for which the Poincaré invariance is preserved. The first obvious case is $\cos\alpha = 1$ which does not lead to superluminal speeds. If $\cos\alpha \neq 1$, then there are two special cases defined by the trivial momentum vector $P_i = 0$ or by the following vector

$$P^i = -2\Lambda^i_j p^j, \quad (8.109)$$

for which the Poincaré invariance is preserved. Let us discuss in some detail the case (8.109). Contracting both sides of the relation (8.109) with the combination $\delta_{ki}\Lambda^{jk}$ and taking into account the equation (8.103) for the indexes which are spatial, one receives the result

$$p^j = -\frac{1}{2}\Lambda^j_i P^i, \quad (8.110)$$

which allows to establish the form of spatial part of the Lorentz matrix

$$\Lambda^j_i = -2\frac{p^j P_i}{P^2}. \quad (8.111)$$

The constant momentum four-vector P^μ can be identified with the initial datum momentum four-vector

$$P^\mu = \left[E_0, p^i_0 c\right]. \quad (8.112)$$

In such a situation, $P^i P_i = p_0^2$, and for this reason the matrix Λ^j_i can be presented in following the form

$$\Lambda^j_i = -2\frac{p^j p_{0i}}{p_0^2}. \quad (8.113)$$

The covariant momentum vector can be established by contraction of the contravariant momentum vector (8.110) with the unit matrix

$$p_j = \delta_{jk} p^k = -\frac{1}{2}\Lambda_{ji} P^i, \tag{8.114}$$

and in this manner one can calculate the square of momentum vector

$$p^2 \equiv p^j p_j = \frac{1}{4}\Lambda_{jk}\Lambda^j_i P^i P^k = \frac{1}{4}\delta_{ik} P^i P^k = \frac{1}{4}p_0^2, \tag{8.115}$$

where we have applied the relation (8.105), which gives

$$p = \frac{p_0}{2}. \tag{8.116}$$

In this manner, using the result (8.116), the matrix (8.113) can be presented in the following form

$$\Lambda^j_i = -\frac{p^j}{p}\frac{p_{0i}}{p_0}, \tag{8.117}$$

or after using the constraint (8.116)

$$\Lambda^j_i = -\frac{p^j p_{0i}}{2p^2}. \tag{8.118}$$

In this manner, the matrix Λ^j_i possesses three representations (8.113), (8.117) and (8.118) which are nonequivalent from the point of view of dependence on the momentum vector p^j.

Finally, one can apply the matrix (8.118) to the fundamental relation (8.101) for the Lorentz transformation. The result is non-trivial

$$p_{0i}p_{0j} = p_0^2\delta_{ij} = 4p^2\delta_{ij} = 2pp_0\delta_{ij}, \tag{8.119}$$

and allows to find three nonequivalent forms of the unit matrix. The first one is

$$\delta_{ij} = \frac{p_{0i}}{p_0}\frac{p_{0j}}{p_0}, \tag{8.120}$$

and the other two representations are

$$\delta_{ij} = \frac{1}{2}\frac{p_{0i}}{p}\frac{p_{0j}}{p_0} = \frac{1}{4}\frac{p_{0i}}{p}\frac{p_{0j}}{p}. \tag{8.121}$$

However, since the unit matrix should be a constant matrix the representation involving only the initial datum vector p_0^j is the only justified form. Another two representations of the unit matrix (8.121) cannot be used in the analysis because they depend on the momentum vector value p, i.s. they are not constant matrices. It is the result of the fact that they were obtained by involving the constraint (8.116) which is the relation between values of the momentum vector p^j and the initial datum p_0^j of the momentum vector.

Let us consider the Taylor power series expansion around the point of the initial datum $p^j = p_0^j$ of the matrix Λ_i^j in all the representations (8.113), (8.117), and (8.118). Such a power series can be presented in the form

$$\Lambda_i^j = \sum_{n=0}^{\infty} a_i^j(n) \left(1 - \frac{p^j p_{0j}}{p_0^2}\right)^n = \sum_{n=0}^{\infty} a_i^j(n) \left(1 + \frac{1}{2}\Lambda_j^j\right)^n, \tag{8.122}$$

where the coefficients $a_i^j(n)$ are constant matrices dependent on n and the initial datum p_0^j of the momentum vector

$$a_i^j(n) = \frac{(-1)^n (p_0^j)^n}{n!} \frac{\partial^n \Lambda_i^j}{\partial p^{jn}}(p^j = p_0^j). \tag{8.123}$$

For simplicity, we apply also the notation

$$\Lambda_i^j(p^j = p_0^j) = \Lambda^{(0)}{}_i^j, \tag{8.124}$$

and this quantity depends on the choice of the representation.

The matrix Λ_i^j in the representation (8.113) is linear in p^j

$$\Lambda_i^j = -2\frac{p_{0i}}{p_0^2} p^j. \tag{8.125}$$

so that its analysis is simple. The only non-vanishing coefficients are $n = 0$ and $n = 1$ coefficients of the Taylor series, and in this manner

$$\Lambda_i^j = \Lambda^{(0)}{}_i^j - 2\frac{p_{0i}}{p_0^2}\left(p^j - p_0^j\right), \tag{8.126}$$

where $\Lambda^{(0)}{}_i^j = -2\delta_i^j = -2\dfrac{p_{0i} p_0^j}{p_0^2}$.

The representation (8.118) of the matrix Λ_i^j is more difficult in the analysis. This formula can be presented in a more convenient form

$$\Lambda_i^j = -\frac{p^j p_{0i}}{2\delta_{jk} p^j p^k} = -\frac{p_{0i}}{2\delta_{jk}} \frac{1}{p^k}, \tag{8.127}$$

which shows that now one has to deal with non-linear (reciprocal) dependence on p^j. In such a situation, the nth derivative of the matrix Λ_i^j can be established straightforwardly

$$\frac{\partial^n \Lambda_i^j}{\partial p^{j^n}} = \Lambda_i^j \frac{(-1)^n n!}{(p^j)^n}, \tag{8.128}$$

so that the coefficients of the series are

$$a_i^j(n) = \Lambda^{(0)}{}_i^j, \tag{8.129}$$

where now $\Lambda^{(0)}{}_i^j$ has the form

$$\Lambda^{(0)}{}_i^j = -\frac{1}{2}\delta_i^j. \tag{8.130}$$

In this manner, the corresponding Taylor power series is

$$\Lambda_i^j = \Lambda^{(0)}{}_i^j \sum_{n=0}^{\infty} \left(1 + \frac{1}{2}\Lambda_j^j\right)^n. \tag{8.131}$$

This series, having the form of a geometric series, can be summed up straightforwardly to the result

$$\Lambda_i^j = -2\frac{\Lambda^{(0)}{}_i^j}{\Lambda_j^j}, \tag{8.132}$$

which after using the representation (8.118), leads to the relation (8.119), which confirms its correctness.

A similar procedure can be performed in the case of the representation (8.117) of the matrix Λ_i^j. Then the matrix Λ_i^j

$$\Lambda_i^j = -\frac{p^j}{\sqrt{p^j p_j}} \frac{p_{0i}}{p_0} = -\frac{p^j}{\sqrt{\delta_{jk} p^j p^k}} \frac{p_{0i}}{p_0} \tag{8.133}$$

$$= -\frac{1}{\sqrt{\delta_{jk}}} \sqrt{\frac{p^j}{p^k}} \frac{p_{0i}}{p_0} = -\frac{1}{\sqrt{\delta_{jk}}} \frac{p_{0i}}{p_0} (p^j)^{1/2} (p^k)^{-1/2}, \tag{8.134}$$

is also nonlinear in p^j, and in such a situation one has

$$\Lambda^{(0)}{}_i^j = -\delta_i^j. \tag{8.135}$$

Applying the Leibniz product formula one obtains

$$\frac{\partial^n \Lambda_i^j}{\partial p^{j^n}} = -\frac{1}{\sqrt{\delta_{jk}}}\frac{p_{0i}}{p_0}\sum_{l=0}^{n}\binom{n}{l}\frac{d^{n-l}}{dp^{j^{n-l}}}(p^j)^{1/2}\frac{d^l}{dp^{j^l}}(p^k)^{-1/2}. \tag{8.136}$$

Derivation of the derivatives can be done immediately

$$\frac{d^{n-l}}{dp^{j^{n-l}}}(p^j)^{1/2} = \frac{1}{2^{n-l}}(p^j)^{1/2}\frac{(-1)^{n-l-1}}{(p^j)^n}(p^j)^l(2(n-l)-1)!!, \tag{8.137}$$

$$\frac{d^l}{dp^{j^l}}(p^k)^{-1/2} = \frac{\delta_j^k}{2^l}(p^k)^{-1/2}(-1)^l(p^k)^{-l}(2l+1)!!, \tag{8.138}$$

and, consequently, one obtains

$$\frac{\partial^n \Lambda_i^j}{\partial p^{j^n}} = \Lambda_i^j\frac{(-1)^{n-1}}{2^n}\frac{1}{(p^j)^n}\sum_{l=0}^{n}\binom{n}{l}(2(n-l)-1)!!(2l+1)!!. \tag{8.139}$$

Using the relations for the double factorials

$$(2k+1)!! = \frac{(2k+1)!}{2^k k!} = \frac{2^{k+1}}{\sqrt{\pi}}\Gamma\left(k+\frac{3}{2}\right), \tag{8.140}$$

$$(2k-1)!! = \frac{(2k)!}{2^k k!} = \frac{2^k}{\sqrt{\pi}}\Gamma\left(k+\frac{1}{2}\right), \tag{8.141}$$

and the Newton binomial symbol

$$\binom{n}{l} = \frac{n!}{l!(n-l)!} = \frac{\Gamma(n+1)}{\Gamma(l+1)\Gamma(n-l+1)}, \tag{8.142}$$

one receives

$$\sum_{l=0}^{n}\binom{n}{l}(2(n-l)-1)!!(2l+1)!! = \frac{2^{n+1}n!}{\pi}\sum_{l=0}^{n}\frac{\Gamma\left(l+\frac{3}{2}\right)}{\Gamma(l+1)}\frac{\Gamma\left(n-l+\frac{1}{2}\right)}{\Gamma(n-l+1)}, \tag{8.143}$$

or after summation

$$\sum_{l=0}^{n}\binom{n}{l}(2(n-l)-1)!!(2l+1)!! = \frac{2^{n+1}n!}{\pi}\frac{\pi(n+1)}{2} = 2^n n!(n+1). \quad (8.144)$$

In this manner, one obtains finally

$$\frac{\partial^n \Lambda_i^j}{\partial p^{j^n}} = \Lambda_i^j \frac{1}{(p^j)^n}(-1)^{n-1}n!(n+1), \quad (8.145)$$

and for this reason the series coefficients are

$$a_i^j(n) = -\Lambda^{(0)}{}_i^j(n+1). \quad (8.146)$$

The Taylor power series for this case takes the form

$$\Lambda_i^j = -\Lambda^{(0)}{}_i^j \sum_{n=0}^{\infty}(n+1)\left(1+\frac{1}{2}\Lambda_j^j\right)^n, \quad (8.147)$$

and can be summed up immediately

$$\Lambda_i^j = -\Lambda^{(0)}{}_i^j \frac{1}{\left(1+\frac{1}{2}\Lambda_j^j - 1\right)^2} = -\frac{4\Lambda^{(0)}{}_i^j}{\left(\Lambda_j^j\right)^2}. \quad (8.148)$$

It is easy to check by direct application of the representation (8.117) that the formula (8.148) leads to the result

$$p^j = -\frac{1}{2}p_0^j, \quad (8.149)$$

which is relevant in the light of the constraint (8.116).

H Restoration of Poincaré Invariance

By virtue of the Poincaré invariance the vector P^i must be a constant vector and Λ_j^i must be a constant matrix. In such a situation, the solution (8.110) means that the Poincaré invariance in the case of the ab initio deformed energy-momentum relation is preserved if and only if the momentum vector $p^j = -\frac{1}{2}p_0^j$ is a constant vector. Such a state of affairs is, in general, against the dynamical nature of motion because such a state of affairs means that an object is at rest. On the other hand, the alternative trivial solution $P_i = 0$, which is equivalent to $p_{0i} = 0$, only in the case $P_0 = 0$, which is equivalent to $E_0 = 0$, allows to restore the Lorentz (CP) invariance. Then, however, the Poincaré invariance is not present in the theory. For this reason, in general, the ab initio energy-momentum relation manifestly violates both the Poincaré symmetry and Lorentz symmetry. The only standard case $\cos\alpha = 1$, which does not assure superluminal values of speeds and is against the assumptions of the proposed theory, automatically leads to preserved both the Poincaré invariance and Lorentz invariance.

However, both the Poincaré and Lorentz invariance can be restored straightforwardly by application of the appropriate reinterpretation of the speed of light. Interestingly, it is easy to see that the ab initio energy-momentum relation (8.50) can be transformed into the form of the Einstein equivalence principle

$$E^2 = p^2 c_N^2 + (m_{eff}^N)^2 c_N^4, \tag{8.150}$$

which naturally preserves both the Poincaré and Lorentz invariance with the new speed of light c_N instead of the speed of light c, *i.e.* $c \to c_N$. Such a situation has the place if one introduces *ad hoc* the new (N) speed of light as the effect of motion with superluminal speeds

$$c \to c_N = \sqrt{\frac{2}{1+\cos\alpha}}c = \frac{c}{\cos\frac{\alpha}{2}}, \tag{8.151}$$

which is superluminal, *i.e.* higher than c, if

$$\cos\frac{\alpha}{2} \in (0,1] \longrightarrow \alpha \in [0,\pi). \tag{8.152}$$

For this one must express the effective mass m_{eff} in terms of c_I instead of c and obtain the new effective mass m_{eff}^N in the form

$$m_{eff}^N = \sqrt{m^2 + \frac{1}{1+\cos\alpha}\left(m^2 + \frac{2p_0^2\cos\alpha}{(1+\cos\alpha)c_N^2} - \frac{4E_0^2}{(1+\cos\alpha)^2 c_N^4}\right)}. \tag{8.153}$$

The new energy-momentum interval

$$s_N^2 = -\frac{E^2}{c_N^2} + p^2 = -(m_{eff}^N)^2 c_N^4, \tag{8.154}$$

is independent of energy and momentum values. In this manner, if one considers the Minkowski energy-momentum space of the four-vectors

$$p^{\mu} = \left[E, p^{i} c_N\right], \tag{8.155}$$

then the new energy-momentum interval (8.154) is preserved under the action of both the Poincaré transformation as well as its special case the Lorentz transformation and, consequently, both the Poincaré group and the Lorentz group are symmetries of the theory.

Let us analyze the new energy-momentum interval (8.154) in some detail with respect to its sign. The interval is

1. light-like when $s_N^2 = 0$, *i.e.*

$$m^2 = 2\frac{2E_0^2 - p_0^2 c_N^2 \cos\alpha(1+\cos\alpha)}{c_N^4(2+\cos\alpha)(1+\cos\alpha)^2}. \tag{8.156}$$

This parameter must be a real number. For this reason the following inequality must be satisfied

$$2\cos^4\frac{\alpha}{2} - \cos^2\frac{\alpha}{2} - \left(\frac{E_0}{p_0 c_N}\right)^2 \leqslant 0, \tag{8.157}$$

whose solution

$$\cos^2\frac{\alpha}{2} \in \left[\frac{1}{4}\left(1 - \sqrt{1+8\left(\frac{E_0}{p_0 c_N}\right)^2}\right), \frac{1}{4}\left(1 + \sqrt{1+8\left(\frac{E_0}{p_0 c_N}\right)^2}\right)\right], \tag{8.158}$$

must satisfy the natural bounds $\cos^2\frac{\alpha}{2} \in [0,1]$, *i.e.*

$$\frac{1}{4}\left(1 - \sqrt{1+8\left(\frac{E_0}{p_0 c_N}\right)^2}\right) \leqslant 0, \tag{8.159}$$

$$\frac{1}{4}\left(1 + \sqrt{1+8\left(\frac{E_0}{p_0 c_N}\right)^2}\right) \leqslant 1, \tag{8.160}$$

having the unique solution

$$0 \leqslant E_0 \leqslant p_0 c_N. \tag{8.161}$$

2. energy-like when $s_N^2 < 0$, *i.e.*

$$m^2 > 2\frac{2E_0^2 - p_0^2 c_N^2 \cos\alpha(1+\cos\alpha)}{c_N^4(2+\cos\alpha)(1+\cos\alpha)^2}. \tag{8.162}$$

3. momentum-like when $s_N^2 > 0$, *i.e.*

$$m^2 < 2\frac{2E_0^2 - p_0^2 c_N^2 \cos\alpha(1+\cos\alpha)}{c_N^4(2+\cos\alpha)(1+\cos\alpha)^2}. \tag{8.163}$$

In the cases 2 and 3 the bounds (8.158) and (8.161) are still valid.

I Vanishing Effective Mass

Among many situations within the proposed complement to Special Relativity, the particular and interesting for this paper considerations case is the vanishing effective mass, *i.e.*

$$m_{eff} = 0. \tag{8.164}$$

In such a situation, the energy of a moving object is

$$E = pc_N = \frac{pc}{\cos\frac{\alpha}{2}}, \tag{8.165}$$

so that the value of the cosinus is

$$\cos\frac{\alpha}{2} = \frac{pc}{E} \in (0,1]. \tag{8.166}$$

Then, the energy-momentum interval s_N^2 is light-like, and the mass parameter m is

$$m = \frac{p_0^2}{E_0}\sqrt{\frac{1-\left(\frac{p_0 c}{E_0}\right)^2\left(\frac{pc}{E}\right)^2\left(2\left(\frac{pc}{E}\right)^2-1\right)}{\left(2\left(\frac{pc}{E}\right)^2+1\right)\left(\frac{pc}{E}\right)^4}}. \tag{8.167}$$

When $\cos\alpha = 1$ then also $\cos\frac{\alpha}{2} = 1$, and in this manner it is useful to find the Taylor power series around the point $\frac{pc}{E} = 1$. It is not difficult to see that for $p_0c \neq E_0$ the leading terms are

$$m = \frac{p_0^2}{\sqrt{3}E_0}\left(\sqrt{1-\left(\frac{p_0c}{E_0}\right)^2} - \frac{8+\left(\frac{p_0c}{E_0}\right)^2}{3\sqrt{1-\left(\frac{p_0c}{E_0}\right)^2}}\left(\frac{pc}{E}-1\right)+\right. \tag{8.168}$$

$$\left. + \frac{\left(7+2\left(\frac{p_0c}{E_0}\right)^2\right)\left(7\left(\frac{p_0c}{E_0}\right)^2-4\right)}{6\left(1-\left(\frac{p_0c}{E_0}\right)^2\right)^{3/2}}\left(\frac{pc}{E}-1\right)^2\right) + O\left[\left(\frac{pc}{E}-1\right)^3\right],$$

while for $p_0c = E_0$ the series has purely imaginary coefficients, so that it is better to show the behaviour around the point $\frac{pc}{E} = 0$. In such a situation, the series is

$$m = \frac{E_0}{c^2}\left(\left(\frac{E}{pc}\right)^2 - \frac{1}{2} - \frac{1}{8}\left(\frac{pc}{E}\right)^2 - \frac{1}{16}\left(\frac{pc}{E}\right)^2 + O\left[\left(\frac{pc}{E}\right)^5\right]\right), \tag{8.169}$$

and all the contributions after the first term are negative, so that

$$m < \frac{E_0}{c^2}\left(\frac{E}{pc}\right)^2, \tag{8.170}$$

or in terms of the momentum value

$$p < \sqrt{\frac{p_0}{mc}\frac{E}{c}}. \tag{8.171}$$

In the light of the energy-momentum relation (8.165) one obtains the lower bound for the initial datum p_0

$$p_0 > mc\cos^2\frac{\alpha}{2}, \tag{8.172}$$

or in terms of the new speed of light c_N

$$p_0 > mc_N\cos^3\frac{\alpha}{2}. \tag{8.173}$$

The case of vanishing effective mass has another non-trivial consequence. Namely, in such a situation the speed of a moving object

$$v = c_N = \frac{c}{\cos\dfrac{\alpha}{2}}, \tag{8.174}$$

in the region $\alpha \in [0, \pi)$ is always higher than the speed of light c, but equal to the new speed of light c_N.

J Unit Shift Vector

We have seen that the non-parallelism hypothesis has non-trivial consequences for neutrinos. However, existence of the angle between velocity and momentum vectors of a moving objects means that the difference between these vectors is a non-zero vector. For this reason let us write *ad hoc* the following decomposition

$$v^i = A\frac{p^i}{p} + B\varepsilon^i, \tag{8.175}$$

where ε^i is an unknown unit vector, which we shall call *the unit shift vector*, and A and B are the coefficients which can be straightforwardly and easy established. Multiplying the formula (8.175) by v_i, p_i, and ε_i one obtains respectively

$$v = A\cos\alpha + B\cos\beta, \tag{8.176}$$

$$v\cos\alpha = A + B\cos(\alpha+\beta), \tag{8.177}$$

$$v\cos\beta = A\cos(\alpha+\beta) + B, \tag{8.178}$$

where $\alpha = (p^i, v_i)$, $\beta = (\varepsilon^i, v_i)$, and $\alpha+\beta = (p^i, \varepsilon_i)$ are the angles between the vectors of the decomposition (8.175). The system of equations (8.176)-(8.178) can be easily solved

$$A = v\frac{\sin\beta}{\sin(\alpha+\beta)}, \tag{8.179}$$

$$B = v\frac{\sin\alpha}{\sin(\alpha+\beta)}. \tag{8.180}$$

Let us apply the definition of velocity vector $v^i = \dfrac{\partial E}{\partial p_i}$ for the case $E = E(p)$ to define the energy

$$E = \int v^i dp_i = \int_{p_{0i}}^{p_i} \left(A\frac{p'^i}{p'} + B\varepsilon^i \right) dp'_i, \qquad (8.181)$$

which can be calculated straightforwardly

$$E = \left(\frac{A}{3} + B\cos(\alpha+\beta) \right)(p - p_0), \qquad (8.182)$$

where p_0 is value of the initial datum vector p_{0_i}. One can take *ad hoc* the initial value $p_0 = 0$. In such a situation, one has

$$\frac{E}{p} = \frac{A}{3} + B\cos(\alpha+\beta), \qquad (8.183)$$

and one can establish easily

$$\tan\beta = \frac{\left(\dfrac{E}{pv} - \cos\alpha\right)\sin\alpha}{\cos^2\alpha - \dfrac{E}{pv}\cos\alpha - \dfrac{2}{3}}, \qquad (8.184)$$

and, consequently, the coefficients of the decomposition (8.175) become

$$A = \frac{3}{2}v\left(\cos\alpha - \frac{E}{pv}\right), \qquad (8.185)$$

$$B = -\frac{1}{2}v\sqrt{-3\cos^2\alpha - 6\frac{E}{pv}\cos\alpha + 9\left(\frac{E}{pv}\right)^2 + 4}. \qquad (8.186)$$

Now it is easy to see that the unit shift vector can be easily established from the definition (8.175) as

$$\varepsilon^i = \frac{1}{B}\left(v^i - A\frac{p^i}{p}\right). \qquad (8.187)$$

When $\cos\alpha = 1$, *i.e.* for the Special Relativity case one has

$$A = -\frac{3}{2}v\left(1 - \frac{E}{pv}\right), \qquad (8.188)$$

$$B = -\frac{1}{2}v\sqrt{9\left(\frac{E}{pv}\right)^2 - 6\frac{E}{pv} + 1}, \qquad (8.189)$$

In this manner, the unit shift vector is non-trivial also in this case.

Interestingly, for the case of superluminal neutrinos one has

$$\frac{E}{p} = c_N = v, \tag{8.190}$$

and in this manner in such a situation

$$A = -\frac{3}{2}c_N(1-\cos\alpha), \tag{8.191}$$

$$B = -\frac{1}{2}c_N\sqrt{-3\cos^2\alpha - 6\cos\alpha + 13}, \tag{8.192}$$

Interestingly, if $\cos\alpha = 1$, *i.e.* the non-parallelism hypothesis is neglected, then

$$A = 0, \tag{8.193}$$

$$B = -\frac{\sqrt{13}}{2}c, \tag{8.194}$$

i.e. the coefficient B remains non-zero.

K OPERA Neutrinos

The OPERA Collaboration used in their experiments two beams of neutrinos having the energies 17 GeV and 28 GeV, whose motion was reported as faster-than-light [354]. This result was absent in the previous detections of the neutrino speed, and manifestly contradicts other ones. The OPERA result is that neutrino speed has the following value

$$v = 1.0000248(28)c. \tag{8.195}$$

Let us accept this result as plausible, in spite of the doubts [355]. It is worth stressing that consistent explanation of this result is rather difficult within the standardly accepted and commonly propagated theoretical frameworks offered by modern high energy physics, *i.e.* the Standard Model, its extensions, supersymmetry, string theory, and others. However, it is possible to consistently explain the superluminal neutrinos of the OPERA Collaboration using the deformed Special Relativity with the ab initio deformation, calculated in this chapter, which as the fundamental idea has the non-parallelism hypothesis.

From the point of view of the theory proposed in this chapter we know that the new speed of light is $c_N = c \Big/ \sqrt{\cos\frac{\alpha}{2}}$ and the speed of the moving neutrino is $v = \frac{pc_N^2}{E}$, where α is the angle between the momentum vector and the velocity vector of a moving neutrino, p is the momentum value and $E = \sqrt{p^2 c_N^2 + (m_{eff}^N)^2 c_N^4}$ is the energy of the neutrino. For consistency we need the value of the momentum of neutrino and the value of the effective mass m_{eff}^N.

Let us involve the Standard Model point of view, which says that the mass of the neutrino is exactly zero, *i.e.* that the OPERA neutrinos are the Weyl neutrinos. This is seemingly a manifestly incorrect point of view in the light of the fact that non-zero value of neutrino mass was detected in several experiments. However, our next hypothesis is

Proposition. *The effective mass of superluminal neutrinos vanishes.*

In other words, our proposal is that the Weyl neutrinos are effective massless particles rather than the fundamental massless particles as offered by the Standard Model. The mass parameter m is then given by the formula (8.167), and we propose that this is the free parameter of the theory which is measured by experiments while the effective mass is unmeasurable. In such a situation, the momentum value of the neutrinos is determined as $p = \frac{E}{c_N}$. Such a line of reasoning leads us to the value of the cosinus

$$\cos\frac{\alpha}{2} = 2\frac{c^2}{v^2} - 1 = 0.9999006(92), \tag{8.196}$$

which gives the unique value of the angle between velocity and momentum vectors of a superluminal neutrino

$$\alpha \approx 0.02818651(38)\text{rad} \approx 1.6149682(82)^\circ. \tag{8.197}$$

It should be emphasized that this value is really small and slightly differs from the zero value of the angle in Special Relativity, so the theory is consistent.

Let us establish finally the unit shift vector for the values of the speed c_N and the angle α calculated above. First of all, one can find the following values of the coefficients

$$A = -0.0005958(35)c, \tag{8.198}$$
$$B = -1.0006204(27)c, \tag{8.199}$$

and then taking into account the relation (8.190) one obtains the value of momentum $p = 16.9995779(34)\,\dfrac{\text{GeV}}{c}$ for the case of 17 GeV beam and the value of momentum $p = 27.9993048(33)\,\dfrac{\text{GeV}}{c}$ for the case of 28 GeV beam. In this way, the unit shift vector (8.187) for the OPERA neutrinos becomes

$$\varepsilon^i = -0.9993799(58)\frac{v^i}{c} - 0.0000350(28)\frac{p^i c}{1\text{GeV}}, \tag{8.200}$$

for the 17 GeV beam, and

$$\varepsilon^i = -0.9993799(58)\frac{v^i}{c} - 0.0000212(67)\frac{p^i c}{1\text{GeV}}, \tag{8.201}$$

for the 28 GeV beam. In both these cases the angle β, which is established by the formula (8.184), has the following value

$$\beta \approx -0.0000167(82)\text{rad} \approx -0.0009615(31)^\circ, \tag{8.202}$$

which is much more smaller than value of the angle α established in the previous section.

Let us summarize our approach to explain the nature of OPERA neutrinos. We have build the theory on base of

1. Deformed energy-momentum relation within Special Relativity,
2. Ab initio derivation of the superluminal deformation,
3. The non-parallelism hypothesis,
4. New speed of light restoring the Poincaré invariance,
5. Vanishing effective mass.

Such a collection of assumptions leads to a consistent complement to Special Relativity at superluminal speeds and, possibly, with the last point removed, can be useful for description of any objects moving with superluminal speeds.

Chapter 9

Magnetic Monopoles & Dyons

A Outlook on Yang–Mills Gauge Field Theories

Gauge field theories are the field theories described by the Lagrangian remaining unchanged under action of a continuous group of the local (gauge) transformations (For detailed presentations of this subject Cf. *e.g.* the books in Ref. [349]). One of the fundamental problems of gauge field theories is a consistent quantization procedure for non-Abelian gauge field theories, *i.e.* when a symmetry group is non-commutative. An example of such theories are the Yang–Mills theories, introduced in 1954 by Chen Ning Franklin Yang and Robert Mills [357], which are the non-Abelian gauge field theories in which the gauge transformations are described by the infinitesimal generators of the Lie algebra $\mathfrak{su}(N)$ of the $SU(N)$ gauge group.

In the present-day theoretical physics the Yang–Mills theories are the fundamental paradigm in description of particles and interactions. In particle physics, one can meet a tremendous number of consistent examples of the constructive application of these specific gauge field theories to the fruitful description of observed processes. The well-accepted description of weak interaction is based on the $SU(2)$ Yang–Mills theory, in which the generators are $T_a = \dfrac{\tau_a}{2}$, where τ_a are the Pauli matrices. The $SU(2) \times U(1)$ Yang–Mills connections are identified with massless photons which together with fermions, intermediating massive vector mesons and hypothetical Higgs particle create the Glashow–Weinberg–Salam model, named after Sheldon Lee Glashow, Steven Weinberg, and Mohammad Abdus Salam [358] who received the 1979 Nobel Prize in Physics, unifying weak interactions with electromagnetism and being the core fundament of the Standard Model. An interesting property of the electroweak theory was discovered in 1973 by Makoto Kobayashi and Toshihide Maskawa, co-laureates of the 2008 Nobel Prize in Physics, who observed that in this model CP violation can be accommodated in the presence of at least six quark flavours. Kobayashi and Maskawa introduced the unitary matrix carrying information about the strength of flavour-changing weak decays and specifying the behaviour of quarks during their free propagation as well as participation in the weak interactions. It is worth stressing that many features described by the $SU(3)$ gauge group were considered before the Glashow–Weinberg–Salam model. Over the years 1956-1957 Tsung-Dao Lee and Chen Ning Franklin Yang [359], recipients of the 1957 Nobel Prize in Physics, studied violation of the spatial parity laws, which preserve the CP symmetry in the weak interactions. This feature was first experimentally verified in 1956 by a group led by Chien-Shiung Wu, nicknamed “Chinese Marie Curie”, whose concluded that violation of the parity symmetry by weak interactions.

Quantum chromodynamics, the $SU(3)$ Yang–Mills theory in which the generators are $T_a = \frac{\lambda_a}{2}$, where λ_a are the Gell-Mann matrices named after Murray Gell-Mann who received the 1969 Nobel Prize in Physics, is presently experimentally well-confirmed model of quarks and gluons and the commonly accepted piece of the Standard Model which describes the theory of strong interactions. In 1962, Gell-Mann introduced the $SU(3)$ singlet-octet structure of the known mesons, the octet-decuplet structure for the baryons, and predicted the existence of the Ω^- hyperon [360]. In 1964, Gell-Mann [361] and, independently, George Zweig [362], a graduate student of physics at the California Institute of Technology, postulated the existence on quarks. Gell-Mann introduced the three-quark model of hadrons corresponding to $SU(3)$, which identified the flavor symmetry of the light quarks extending isospin to include strangeness. The pioneering speculations of Zweig since the beginning considered existence of four quarks, and Zweig named them "aces" by analogy with the four playing cards. Implementation of quarks to theory of particles was the milestone for theoretical physics. In 1973, Gell-Mann together with Harald Fritzsch and Heinrich Leutwyler invented the Yang–Mills theory type Lagrangian for quantum chromodynamics [363]. The essential unique feature of this $SU(3)$ Yang–Mills theory is the existence of asymptotic freedom, the phenomenon of vanishing of the forces at short distances discovered in 1973 by Hugh David Politzer, David Jonathan Gross, and Frank Anthony Wilczek [364] who received the 2004 Nobel Prize in Physics. It is worth stressing that the $SU(3)$ group was considered by some theoreticians before Gell-Mann. In 1959 Mineo Ikeda, Shuzo Ogawa, and Yoshio Ohnuki [365] on the basis of the Sakata model, named after Shoichi Sakata [366] who invented the composite model for hadrons based on three fundamental constituents, introduced the $SU(3)$ symmetry for hadrons and predicted existence of the η meson, whereas in 1961 Yuval Ne'eman [367] introduced the $SU(3)$ octet structure of the known mesons and baryons and the classification of hadrons through application of the $SU(3)$ flavour symmetry. In 1963, Nicola Cabibbo [368] considered $SU(3)$ and hadronic weak currents, predicted the leptonic decay rates of hyperons, and introduced the concept of quark mixing angle, known as the Cabibbo angle, to preserve the universality of the weak interaction. In 1964, James Daniel Bjorken and Sheldon Lee Glashow showed in the framework of the $SU(4)$ Yang–Mills theory the existence of a fourth quark which should be a charmed fundamental fermion [369].

As shown in 1976 by Michio Kaku [370], the $SU(4)$ Yang–Mills theories possess solutions constructed out of the Lienard–Wiechert potentials and describing a point-like particle which can execute arbitrary space-time motions. In such a gauge theory, as shown by Jorge Santamarina [371], there exists the solution analogous to the $SU(2)$ Belavin–Polyakov–Schwartz–Tyupkin instanton [372]. The $SU(5)$ Yang–Mills theory is the base of the model proposed in 1974 by Howard Georgi and Sheldon L. Glashow [373] which is the pioneering attempt in construction of Grand Unification Theories. In 1982, Stephen Barr [374] and in 1984 Jean-Pierre Derendinger, Jihn Eui Kim, Dimitrios V. Nanopoulos [375] first contemplated the so-called flipped $SU(5)$ gauge theory as a Grand Unification Theory. Another example of Grand Unification Theory is the $SU(6)$ Yang–Mills theory proposed in 2005 by Andreas Hartanto and Laksana Tri Handoko [376], which is a specific realization of a unification of strong and electroweak interactions and provides a complete form of 35 generators of $SU(6)$. Earlier, the Yang–Mills theory based on the $SU(6)$ gauge group was proposed in 1978 by Suk Koo Yun [377], while the initial applications of $SU(6)$ group to particle physics were considered in 1964 by Feza Gürsey and Luigi Arialdo Radicati di Brozolo [378], and Bunji Sakita [379] whose implemented the $SU(6)$-based classification of hadrons. This theory required eight quarks and eight leptons belonging to 15-plet and singlet representations while gluons and leptoquark and weak vector bosons formed a 35-plet representation. Supersymmetric Grand Unification Theory based on $SU(6)$ and omitting the gauge hierarchy problem, emerging when there is a vast difference between the theoretical and measured fundamental parameters of some, was proposed in 1997 by Zurab Tavartkiladze [380]. In 1981, Jihn Eui Kim [381] proposed a specific flavor unification in the framework of the $SU(7)$ gauge group, which forbids mixings of the b quark with the d and s quarks, and the t quark with the u and c quarks. J.E. Kim constructed the symmetry breaking mechanism which allows the c-t mixing without b-quark mixing. In such a model, a topologically stable magnetic monopole is light and hence there is no the conflict arising from Grand Unification Theories and the Big Bang cosmology.

In 1980, Chung Wook Kim and Claude Roiesnel discussed the Grand Unification Theory based on the $SU(8)$ group, in which the $SU(3)$ group was the horizonal group [382] embedded in the $SU(8)$ grand unified symmetry. Over the years 1981-1982 Jihn Eui Kim and He-Shan Song proposed the so-called “hyperunification $SU(8)$” gauge model [383] of the electroweak, color, and hypercolor interactions, which does not contain the t quark and can be derived from the $SO(8)$ extended supergravity. In this model, some known quarks and leptons are fundamental and the other ones are composite fermions composed of three hypercolored fermions, the total fermion spectrum is under the $SU(5)$ gauge group of the Georgi–Glashow model. Also, over the years 1981-1982, Nilendra G. Deshpande and Philip D. Mannheim [384] proposed the Grand Unification Theory of the strong, electromagnetic, and weak interactions based on the chiral $SU(8)$ gauge group. In this model the chiral symmetry was broken to $SU(3)_C \times SU(2)_L \times U(1)$, which resulted in stability of a proton, massless left-handed neutrinos, and the absence of strong CP violations and axions. Another theory based on the $SU(8)$ group, presenting the flavour unification, was proposed in 1983 by Kwang-Sup Soh, Pong Youl Pac, and Sung Ku Kim [385].

In 1980, Paul Howard Frampton proposed the Grand Unification Theory based on the $SU(5)$ group embedded in $SU(N)$. In this model, imposing the asymptotic freedom, the solution of a diophantine equation leads to a simple $SU(9)$ gauge field theory supporting the existence of the only six flavors of the light quark [386]. Also, the Grand Unification Theory based on the $SU(9)$ group, was proposed in 1981 by Yukihiro Fujimoto and Pasquale Sodano [387]. In this theory, the Frampton $SU(9)$ model is considered in detail as a grand unified theory with the $SU(4)$ horizontal symmetry, and results in a correlation among neutrino, horizontal-gauge-boson, new-fermion masses, and new charged-current processes. In 1982, Savs Dimopoulos, Paul Howard Frampton, Howard Mason Georgi III, and Mark Brian Wise applied $SU(9)$ to construction of the Grand Unification Theory in which there are three families of quarks and leptons [388]. This model automatically eliminated the puzzle of the strong CP violations, with the presence of an invisible axion and the absence of cosmological domain wall problem. Recall that in cosmology topological defects are originally symmetric stable configurations of matter which are generated in the very early universe during phase transitions resulting in the loss of symmetry. The the cosmological domain wall problem arises after a phase transition through the Kibble mechanism in which the various defects, like solitons, cosmic strings, monopoles, textures, and other hybrid creatures, are generated due to spontaneous symmetry breaking of a discrete symmetry. The type of defect depends on the symmetry and the phase transition. Thomas Walter Bannerman Kibble noticed that when two regions of space are causally in contact, *i.e.* they are separated by a distance smaller than the distance which could be traveled by light in the appropriate time, then the fields in these regions can fall into the same vacuum, and when they are not causally in contact then falling of the fields in the same direction in both places is not assured and, consequently, they cannot communicate with each other [389].

In 1983, Paul Howard Frampton and Jihn Eui Kim realized the scenario, originally proposed in 1981 by Edward Witten [390], for a solution to the gauge hierarchy problem in the supersymmetric $SU(7)$ gauge field theory [391]. P.H. Frampton and J.E. Kim regarded the troublesome unwanted particles as superheavy by interlocking interactions between two non-singlet fields, such that the asymptotic freedom of the strong interaction remained up to the grand unification scale. Also in 1983 Paul Howard Frampton together with Takayuki Matsuki and Walter W. Wada re-examined the strong CP violation problem and the resulting idea of invisible axions in the context of supersymmetric $SU(N)$ gauge field theories for the case of the number of colors $N = 5,6,7,8,9$ [392]. 1983 was also a good year for the gauge field theory of gravity based on $SU(9) \supset SU(5) \times O(5)$ which was proposed by Heinz Rudolf Pagels [393]. In 1985, Paul H. Frampton and Sheldon L. Glashow examined the reliability of the simplest predictions due to the $SU(5)$ gauge field theory for the proton lifetime and showed that a specific split of fermion representations improves agreement with experiment [394]. In 1986, Kyungsik Kang, Chung Ku Kim and Jae Kwan Kim [395] discussed symmetry breaking mechanism and the charge operator, *i.e.* the generators of $U(1)$ gauge group, in all possible the symmetry breaking modes reducing $SU(9)$ to $SU(3)_c \times SU(2)_W \times U(1)$. They considered two types of representations that satisfy the requirements of a Grand Unification Theory which were found in $SU(9)$. The first representation trivially embedded the $SU(5)$ group whereas the second one had a non-trivial embedding of $SU(5)$. K. Kang, C.K. Kim, and J.K. Kim calculated the contribution of $U(1)$ eigenvalues to the weak hypercharge in all physically interesting modes of symmetry breaking. They examined evolution of coupling constants to determine the intermediate mass scales responsible for invisible axions and the delayed decay of the proton, and also discussed fermions. In 1980, Jihn Eui Kim discussed the asymptotic properties of the flavor Grand Unification Theory based on the $SU(11)$ group, and in 1981 he considered the idea of an invisible axion and neutrino oscillations in the $SU(11)$ gauge field theory [397].

More complicated models of Grand Unification Theories were also considered. In 1984, Alvaro De Rújula, Howard Georgi, and Sheldon L. Glashow [398] introduced the new idea for Grand Unification Theories known as the trinification model $SU(3) \times SU(3) \times SU(3)$. This idea was seriously taken into account in 1985 by Kaladi Sankaran

Babu, Xiao-Gang He, and Sandip Pakvasa in the context of neutrino masses and proton decay modes [399]. They analyzed among others the spontaneous symmetry breaking to the standard group $SU(3) \times SU(2) \times U(1)$, generation of fermion masses and mixings, stability of the corrections to the neutral leptons masses at the one- and two-loop levels, rather heavy new neutral leptons called neutrettos, and the mass hierarchy of the neutrinos. They assured the stability of the ground-state vacuum and exhibited the Higgs- and gauge-boson mass matrices. In 1988, Haruhiko Nishimura and Akira Okunishi studied trinification in the context of the strong CP violation problem and nucleon stability. In 2003, Scott Willenbrock [401] discussed the trinification model from the viewpoint of the problem of gauge-coupling unification. In 2004 Christopher David Carone and Justin M. Conroy [402] presented the higgsless symmetry breaking mechanism within the model. In 2005, Alison Demaria and Raymond R. Volkas [403] studied the context of symmetry breaking patterns induced by a soliton. In 2006, Joshua Sayre, Soren Wiesenfeldt, and Scott Willenbrock [404] discussed the symmetry breaking mechanism with the Higgs sector. In 2011, Christophe Cauet, Heinrich Pas, and Soren Wiesenfeldt [405] considered the hierarchy problem in the context of the trinification model.

Recently, attention has also been given to the theories based upon product gauge groups, for example $SU(N)_k$, called deconstruction in the context of the Standard Model extensions. The pioneering idea of deconstruction was introduced in 2001 by Nima Arkani-Hamed, Andrew Glen Cohen and Howard Georgi [406] and independently by Christopher Thaddeus Hill, Stefan Pokorski and Jing Wang [407]. The deconstruction due to Arkani-Hamed, Cohen, and Georgi was based on the idea of a new class of four-dimensional perturbative theories for natural electroweak spontaneous symmetry breaking, relying neither on supersymmetry nor on strong dynamics at the TeV scale, and leading to finite radiative corrections to the Higgs mass. Arkani-Hamed, Cohen, and Georgi derived a novel Higgs potential requiring a second light $SU(2)$ doublet scalar. The Hill–Pokorski–Wang deconstruction was based on the gauge invariant Lagrangian in $3+1$ dimensions for N Kaluza–Klein's modes of an $SU(m)$ gauge theory in the higher-dimensional bulk. For the $4+1$-dimensional bulk the effective theory is $\Pi_{i=1}^{N+1} SU(m)_i$ with N chiral $(\bar{m}, m)$ fields connecting the groups sequentially, what can be visualized as a Wilson action for a transverse lattice in x^5 and matches the continuum $4+1$ compactified Lagrangian truncated in the momentum space. Hill, Pokorski, and Wang described the scale dependence of the gauge couplings by the standard renormalization group technique with threshold matching, leading to the effective power law running, and discussed the unitarity constraints and chiral fermions. The idea of the latticized bulk was then developed by Hsin-Chia Cheng, Christopher Thaddeus Hill, Stefan Pokorski, and Jing Wang to construct the manifestly gauge invariant effective Lagrangian in $3+1$ dimensions describing the Standard Model in $4+1$ dimensions, incorporating split generation fermions and exploring naturalness for two Higgs configurations: a universal and a local Higgs vacuum expectation value [408].

In 1974, Gerardus 't Hooft investigated the so-called large-N limit for gauge field theories [409], which is especially useful for considerations of the strong interactions and has been became one of the most popular research trends. Paradoxically, 't Hooft motivated application of this limit to study the properties of the $SU(3)$ quantum chromodynamics. He generalized $SU(3)$ with 3 colors to $SU(N)$ with N colors and showed that this theory simplifies for large N. The expansion in the parameter $1/N$ and the scaling of the coupling constant and gauge fields, leading to the vertices of the Feynman diagrams proportional to N and the propagators proportional to $1/N$, allows to consider the theory of the $U(N)$ group rather than $SU(N)$, and is presently considered as one of the most powerful methods in gauge field theories. It should be also emphasized that the case of massive Yang–Mills theories is still under intensive development. Some proposal, based on the usual path integral quantization, was presented in 1971 by G. 't Hooft [410], and is presently the basis of the Standard Model. The reader interested in more detailed discussion of the basics and applications of the particle physics context of the Lie algebras $SU(N)$ is kindly advised to take into account Refs. [411] and [412].

Massless Yang–Mills theories are characterized by the following Lagrangian density

$$\mathscr{L}_{YM} = -\frac{1}{4\mu_0} F^{\mu\nu a} F_{\mu\nu_a}, \tag{9.1}$$

where $a = 1, \ldots, N^2 - 1$ is a gauge group index, and

$$F^a_{\mu\nu} = \partial_\mu \mathbf{A}^a_\nu - \partial_\nu \mathbf{A}^a_\mu + g f^a_{bc} \mathbf{A}^b_\mu \mathbf{A}^c_\nu, \tag{9.2}$$

is the strength of the gauge field $\mathbf{A}^a_\nu$, known also as the Yang–Mills connection, which is a non-Abelian magnetic potential having dimension N/A. A coupling constant g is assumed to have the physical dimension of the reciprocal of the magnetic flux J/A, and f^a_{bc} are the (antisymmetric) structure constants of the Lie algebra. Elements of the Lie algebra $\mathfrak{su}(N)$, which are the infinitesimal generators T_a of the gauge group $SU(N)$, are traceless hermitian matrices $\mathrm{Tr} T_a = 0$, $T_a^\dagger = T_a$ obeying the commutation relations

$$[T_a, T_b] = i f_{ab}{}^c T_c, \tag{9.3}$$

and the Lie group is produced by formal exponentiation of the Lie algebra. The Euler–Lagrange equations of motion for the Yang–Mills Lagrangian (9.1) can be written in the following form

$$\partial^\mu F^a_{\mu\nu} + g f^a_{bc} \mathbf{A}^{\mu b} F^c_{\mu\nu} = 0, \tag{9.4}$$

and the equations for the dual field strength tensor $*F^{\mu\nu}_a = \frac{1}{2} \varepsilon^{\mu\nu\kappa\lambda} F^a_{\kappa\lambda}$

$$\partial^\mu * F^a_{\mu\nu} + g f^a_{bc} \mathbf{A}^{\mu b} * F^c_{\mu\nu} = 0, \tag{9.5}$$

known as the Bianchi identities for the Yang–Mills connection, similarly to the Abelian situation of the Maxwell electrodynamics are automatically satisfied. It can be seen by straightforward derivation that the explicit form of the Yang–Mills equations (9.4) is

$$\begin{aligned} &\partial^\mu \partial_\mu \mathbf{A}_\nu{}^a - \partial^\nu \partial^\mu \mathbf{A}^a_\mu + g f^a_{bc} \partial^\mu \mathbf{A}^b_\mu \mathbf{A}^c_\nu - g f^a_{bc} \mathbf{A}^{\mu b} \partial_\nu \mathbf{A}^c_\mu \\ &+ 2 g f^a_{bc} \mathbf{A}^b_\mu \partial^\mu \mathbf{A}^c_\nu + g^2 f^a_{bc} f^c_{de} \mathbf{A}^{\mu b} \mathbf{A}^d_\mu \mathbf{A}^e_\nu = 0. \end{aligned} \tag{9.6}$$

Let $\theta^a(x)$ be a gauge vector determining the gauge transformation

$$U(\theta) = \exp\left(i T_a \theta^a(x)\right), \tag{9.7}$$

transforming a gauge field $\mathbf{A}^a_\mu$ in the following way

$$\mathbf{A}^{\theta^a}_\mu T_a = \mathbf{A}^a_\mu U T_a U^{-1} + \frac{1}{ig} \left(\partial_\mu U\right) U^{-1} \tag{9.8}$$

$$= \mathbf{A}^a_\mu \exp\left(i T_b \theta^b\right) T_a \exp\left(-i T_b \theta^b\right) + \frac{1}{g} T_a \partial_\mu \theta^a, \tag{9.9}$$

where g is a coupling constant. In the situation when the gauge transformation is infinitesimal, *i.e.*

$$U(\theta) = 1 + iT_a\theta^a(x) + O[\theta^2], \tag{9.10}$$

then the invariant measure over a gauge group has the product form $[d\theta(x)] = \prod_{j=1}^{N^2-1} d\theta_j$ and, therefore, a gauge field $\mathbf{A}_\mu$ transforms to

$$\mathbf{A}_\mu^{\theta^a} T_a = \mathbf{A}_\mu^a T_a - i[T_a, T_b]\theta^b \mathbf{A}_\mu^a + \frac{1}{g} T_a \partial_\mu \theta^a + O[\theta^2], \tag{9.11}$$

which after using the commutation relations (9.3) and taking into account the fact $T_c = \delta_c^a T_a$ gives

$$\mathbf{A}_\mu^{\theta^a} = \mathbf{A}_\mu^a + \delta_c^a f_{ab}{}^c \theta^b \mathbf{A}_\mu^a + \frac{1}{g}\partial_\mu \theta^a + O[\theta^2]. \tag{9.12}$$

The usual path integral quantization procedure of the massless Yang–Mills theories is based on the method of the Faddeev–Popov determinant named after Ludvig Dmitrievich Faddeev and Victor Nikolaeavich Popov [413]. This method of quantization is based on a restriction of the path integration to a group hypersurface intersecting each gauge group orbit only once. Such a restriction is gauge fixing, *i.e.* a certain intentional selecting of a representative from each gauge group orbit, and in the case of the Yang–Mills theories it often results in global problems. The space of representatives is a submanifold and fits the gauge fixing condition in such a way that every gauge orbit should intersect this hypersurface only once. Because of topological obstructions, in the case of the Yang–Mills theories, it sometimes cannot be performed globally and the only local treatment has a sense. In general, the gauge fixing submanifold cannot have common points with a gauge orbit or have more than one such points, what is called the Gribov ambiguity, named after Vladimir Naumovich Gribov [414], and results in a number of problems on the nonperturbative level. For instance, the Becchi–Rouet–Stora–Tyutin (BRST) symmetry, introduced by Carlo Maria Becchi, Alain Rouet, Raymond Félix Stora [415] and Igor Viktorovich Tyutin [416], for which the gauge fixed Lagrangian is symmetric with respect to the action of global gauge transformations which rotate the gauge fields into the ghost fields, the additional fields which preserve the consistency of the path integral quantization of the Yang–Mills theories. The ghost fields originally arose in Feynman's analysis of the quantization of the Yang–Mills theories, who showed that counter terms should be added to keep unitarity, and are now known as Faddeev–Popov ghosts. Similarly, as in the case of the Gupta–Bleuler quantization, introduced by Suraj Narayan Gupta [417] and Konrad Bleuler [418] for electromagnetic field (See also Ref. [419]), BRST symmetry removes non-physical states from the Fock space. However, from the perturbation theory point of view the Gribov ambiguity in the Coulomb gauge creates the dangerous situation for the formalism.

Strictly speaking, Gribov discovered the impossibility of unambiguous extraction of degrees of freedom of the non-Abelian gauge symmetry due to zero modes if a gauge field is strong with the Coulomb gauge. As the Coulomb gauge does not fix the gauge completely there is an infinite sequence of gauge fields copies, called Gribov copies, which are gauge equivalent that satisfy the Coulomb gauge condition. It should be emphasized that the commonly accepted method for resolving the Gribov ambiguity is restriction of the domain of integration in the path integral to a single Gribov region Ω whose boundary $\partial\Omega$ is called the Gribov horizon. In the original Gribov's analysis to the Landau gauge, *i.e.* the Lorenz gauge with the Feynman asymptotic conditions, the restriction to the region Ω arises from the fact that for a field configuration belonging to Ω and lying near the horizon $\partial\Omega$, there is an equivalent configuration located on the other side of $\partial\Omega$ outside of Ω. Moreover, Gribov expressed the suggestion that confinement of gluons may be due to the restriction of the physical configuration space to Ω. Very soon after Gribov's paper Isadore Manual Singer [420] and Michael Francis Atiyah and John David Stuart Jones [421] showed that the same phenomenon of the Gribov ambiguity appears for any gauge fixing condition. Subsequently, it was pointed out by Daniel Zwanziger [422] that the Gribov region defines the local minimum of the Hilbert norm on each gauge orbit. More specifically, if $U(\theta)$ varies over the group of local gauge transformations and the gauge field $\mathbf{A}_\mu^a$ is fixed, then the Gribov region defines the local minimum of the Hilbert norm

$$\left|\left|\mathbf{A}_\mu^{\theta^a}\right|\right|^2 = \int d^4x \mathbf{A}_\mu^{\theta^a} \mathbf{A}^{\theta^{\mu a}}. \tag{9.13}$$

The existence of the Gribov copies manifestly destroys the path integral quantization of the Yang–Mills theories in the Landau gauge.

Victor N. Gribov and Daniel Zwanziger worked out the treatment of the gauge fields copies in the continuum by introduction of a renormalizable action, which is known as the Gribov–Zwanziger action (For good introduction to the framework Cf. *e.g.* [423]) implementing the restriction of the path integral quantization to the Gribov region, which admittedly removes a subset of the Gribov copies, but not of all copies. In the light of the Dirac canonical quantization scheme [169], however, the degrees of freedom of a symmetry group can be extracted. Certain proposal for partial resolving this unavoidable problem for the Yang–Mills theories was made by Iosif Dmitrievich Manjavidze, Alexei Norairovich Sissakian and Vadim Vladimirovich Voronyuk [424] whose suggested that the Gribov–Atiyah–Jones–Singer result might signify that the gauge fixing problem is unsolvable in the presence of the phase, *i.e.* the Yang–Mills theories are free from the Gribov ambiguity if they are used for the description of observables. The application of this formalism was also deduced earlier by I.D. Manjavidze and A.N. Sissakian [425] who also constructed a suitable perturbation theory [426] based on the factor space which, by its original character, can be called the Manjavidze–Sissakian perturbation theory. The reader interested in more detailed analysis, basics, and advances in the problem of the Gribov ambiguity is advised to familiarize with the wide bibliographical spectrum which can be found in public scientific databases [427].

The equation for a submanifold fitting the gauge fixing condition is given by the system of equations

$$F_a[\mathbf{A}^a_\mu(x)] = 0, \tag{9.14}$$

which fix the gauge, and the equation

$$F_a[\mathbf{A}^{\theta^a}_\mu(x)] = 0, \tag{9.15}$$

is assumed to have an unique solution θ^a for a given gauge field $\mathbf{A}^a_\mu$. In such a situation, one can introduce the matrix

$$\mathbf{M}_{ab} = \frac{\delta F_a[\mathbf{A}^{\theta^a}_\mu]}{\delta\theta^b}, \tag{9.16}$$

whose determinant is called the Faddeev–Popov determinant

$$\det \mathbf{M}_{ab} = \Delta[\mathbf{A}^a_\mu], \tag{9.17}$$

and is given by the identity

$$\Delta^{-1}[\mathbf{A}^a_\mu] = \int [d\theta]\delta\left(F_a[\mathbf{A}^{\theta^a}_\mu(x)]\right). \tag{9.18}$$

In the case of infinitesimal gauge transformations one has also

$$F_a[\mathbf{A}^{\theta^a}_\mu(x)] = F_a[\mathbf{A}^a_\mu(x)] + \int d^{D+1}y \left(\frac{\delta F_a[\mathbf{A}^{\theta^a}_\mu(x)]}{\delta\theta^b(y)}\right)\Bigg|_{\theta=0} \theta^b(y) + O(\theta^2). \tag{9.19}$$

In such a situation, the suitable integrability region, which in the case of $F_a = \partial^\mu \mathbf{A}^{\theta^a}_\mu$ is the Gribov region, is defined as

$$\Omega = \left\{ \mathbf{A}^a_\mu \middle| F_a[\mathbf{A}^{\theta^a}_\mu(x)] = 0, \quad \frac{\delta F_a[\mathbf{A}^{\theta^a}_\mu]}{\delta\theta^b} > 0 \right\}, \tag{9.20}$$

and its boundary $\partial\Omega$ is the horizon analogous to the Gribov horizon.

The geometric interpretation of the gauge fixing can be efficiently understood with application of two concepts: the affine space of all Yang–Mills connections, *i.e.* gauge fields, Sp**A**, and the space $\mathscr{G}$ of all gauge transformations which is known as the gauge group space. The principal bundle $\mathscr{P}(\mathbf{A},\Pi,M,\mathscr{G})$, with the fibre $\mathscr{G}$ and the projection $\Pi : \mathrm{Sp}\mathbf{A} \to \mathrm{Sp}\mathbf{A}/\mathscr{G}$, is infinite dimensional, and the base space $\mathrm{Sp}\mathbf{A}/\mathscr{G}$, which is known as the moduli space, is the physical space of the Yang–Mills theories. An orbit of the gauge group $\mathscr{G}$is the configurations of the transformed gauge field $\mathbf{A}^{\theta^a}_\mu$ for all possible matrices $U(\theta)$. In such a situation, the gauge fixing condition $F^a[\mathbf{A}^\theta_\mu] = 0$ defines gauge section, the surface in the space of all connections crossing each orbit once and only once.

The Gribov ambiguity possesses also unambiguous geometric interpretation. In the quantization method based on the Faddeev–Popov determinant the gauge fixing condition $F^a[\mathbf{A}^\theta_\mu] = 0$ must have only one solution. In fact, the crucial observation due to Gribov, which is beyond perturbation theory and destroys the path integral formulation of the Yang–Mills theories, is that values of the Yang–Mills connections can break this assumption and, in general, the unique solution $U(\theta)$ can be non-existent. The crucial point of the geometric interpretation of the Gribov ambiguity, presented by Isadore Manual Singer, is the non-trivial topology of the principal bundle $\mathscr{P}(\mathbf{A},\Pi,M,\mathscr{G})$ solely arising from $SU(N)$ group and $M = S^4$. Gauge fixing selects only one representative connection in the principal bundle for each element of the base space Sp$\mathbf{A}/\mathscr{G}$, which is possible only locally because the Dirac delta function $\delta\left(F^a[\mathbf{A}^\theta_\mu]\right)$ selects from the path integral over the gauge fields only those connections which lie on the gauge section. Globally, the gauge section does not exist as a result of the non-triviality of the principal bundle, *i.e.* there are configurations where a gauge section is not well defined. More geometrical details of the Yang–Mills theories can be found in Ref. [428].

It is usually accepted in the Standard Model that the electric charge of the massive vector bosons is determined by the unbroken Abelian gauge group $U(1)$, whose is the gauge group of quantum electrodynamics (QED). Also, $U(1)$ is a little group of the rotation in the isospin space about the direction given by the scalar field triplet φ^a whose generator $\dfrac{\varphi_a T^a}{\varphi_0}$ must determine the operator of electric charge

$$Q = \hbar g \frac{\varphi_a}{\varphi_0} T^a, \tag{9.21}$$

and in such a situation the covariant derivative looks like the analogous one in the Maxwell electrodynamics

$$D_\mu = \partial_\mu - i\frac{Q}{\hbar} A_\mu, \tag{9.22}$$

where A_μ is the Abelianization of the gauge field

$$A_\mu = \frac{\varphi_a}{\varphi_0} \mathbf{A}^a_\mu. \tag{9.23}$$

Taking into account the property of infinitesimal generators of the $\mathfrak{su}(N)$ Lie algebra

$$T_a T^b = \frac{1}{2N}\delta^b_a I_N + \frac{1}{2}\left(i f^b_{ac} + d^b_{ac}\right) T^c, \tag{9.24}$$

where d^{ab}_c are symmetric structure constants, using the obvious properties $\mathrm{Tr} I_N = N$ and $\mathrm{Tr} T^c = 0$ one can calculate straightforwardly eigenvalues of the electric charge operator

$$q = \mathrm{Tr} Q = \mathrm{Tr}\left(\hbar g \frac{\varphi}{\varphi_0} T_a T^a\right) = \frac{|\varphi|}{|\varphi_0|} q^{\pm}_{\min}, \tag{9.25}$$

where $q^{\pm}_{\min}$ is the electric charge in the minimum of the Yang–Mills–Higgs configuration, *i.e.* $\varphi = \pm|\varphi_0|$

$$q^{\pm}_{\min} = \pm\frac{\hbar g}{2}, \tag{9.26}$$

so that, in such a situation, the total electric charge identically vanishes $q^{+}_{\min} + q^{-}_{\min} = 0$. After quantization of the scalar field φ^a, one can take into account the vacuum expectation value of Q which, in the light of the fact $\langle\varphi\rangle = |\varphi_0|$, is identical to q. In this manner, in the general case only classically averaged electric charge

$$\langle Q\rangle = \frac{\int q d^{D+1}x}{\int d^{D+1}x} = \frac{q_{\min}}{|\varphi_0|}\frac{\int|\varphi|d^{D+1}x}{\int d^{D+1}x} < \infty, \tag{9.27}$$

which is global, possesses consistent physical significance. Particularly, if

$$\int|\varphi|d^{D+1}x = \langle\varphi\rangle\int d^{D+1}x, \tag{9.28}$$

then one has

$$\langle Q\rangle^{\pm} = q^{\pm}_{\min}, \tag{9.29}$$

and, consequently, the global total electric charge vanishes identically

$$\langle Q\rangle_{TOT} = \langle Q\rangle^{+} + \langle Q\rangle^{-} = 0. \tag{9.30}$$

B Dirac & Wu–Yang Monopole

The first known solution to the Yang–Mills equations, describing a point-like magnetic monopole in the case of the $SU(2)$ gauge group, was found in 1968 by Tai Tsun Wu and Chen Ning Franklin Yang [429]. Recall that for $SU(2)$ the infinitesimal generators are $T_a = \frac{\sigma^a}{2}$ and the coupling constant is $g = \frac{e}{\hbar} = \frac{1}{2\Phi_0}$. This solution, known as the Wu–Yang monopole, has the form $A_\mu = \mathbf{A}^a_\mu T_a$, where the coefficients of this decomposition are

$$\mathbf{A}^a_\mu = \left[0, 2\Phi_0\varepsilon^a_{im}\frac{r^m}{r^2}\right], \tag{9.31}$$

where $f_{abc} = \varepsilon_{abc}$, for which the field strength tensor is

$$F^a_{\mu\nu} = \begin{bmatrix} 0 & \mathbf{0}^T \\ \mathbf{0} & 2\Phi_0 \varepsilon_{ijc} \dfrac{r^a r^c}{r^4} \end{bmatrix}, \tag{9.32}$$

for which the magnetic field induction $B^k = \frac{1}{2}\varepsilon^{kij}F_{ij}$, where $F_{ij} = F^a_{ij}T_a$ can be derived straightforwardly

$$B^k = \frac{1}{2}\varepsilon^{kij}F^a_{ij}T_a = \Phi_0\varepsilon^{kij}\varepsilon_{ijc}\frac{r^c}{r^4}r^aT_a = \Phi_0\varepsilon^{ijk}\varepsilon_{ijc}\frac{r^c}{r^3} = 2\Phi_0\frac{r^k}{r^3} = \frac{1}{gr^2}\frac{r^k}{r}, \tag{9.33}$$

where we have applied the identities $\varepsilon^{ijk}\varepsilon_{ijc} = 2\delta^k_c$ and $r = r^aT_a$. Now one can calculate easily the total flux Φ_B of the magnetic field induction $\mathbf{B} = B^r\frac{\mathbf{r}}{r}$ with the radial component $B^r = \dfrac{1}{gr^2}$

$$\Phi_B = \int_S \mathbf{B}d\mathbf{S} = \int_S B^r dS_r = \int_0^{2\pi}\int_0^{\pi}\frac{1}{gr^2}r^2\sin\theta d\theta d\phi = \frac{4\pi}{g} = 8\pi\Phi_0, \tag{9.34}$$

so that according to the Gauss law for magnetism $\Phi_B = \mu_0 q_M$, the magnetic charge for the Wu–Yang monopole is

$$q_M = \frac{4\pi}{g\mu_0} = \frac{8\pi\Phi_0}{\mu_0}, \tag{9.35}$$

and leads to the following relation

$$\frac{q_M e}{q_P^2 c} = 1, \tag{9.36}$$

which by application of the Dirac quantization condition [430] $\dfrac{q_M e}{q_P^2 c} = \dfrac{n}{2}$ allows to establish the value of the winding number for the non-Abelian Wu–Yang monopole as $n = 2$.

This simplest known non-Abelian solution is radially directed and behaves like $1/r$ in a whole space, which is similar to the behaviour of the Abelian $U(1)$ magnetic monopole of the Maxwell electrodynamics founded in 1931 by Paul Adrien Maurice Dirac [430]

$$\mathbf{A}_\mu = \left[0, 0, 0, \frac{\Phi_0}{r}\tan\frac{\theta}{2}\right]. \tag{9.37}$$

Here $\Phi_0 = \frac{\hbar}{2e} = \frac{E_P t_P}{2\sqrt{\alpha} q_P}$ is the magnetic flux quantum, where $\alpha = \frac{e^2}{4\pi\varepsilon_0 \hbar c}$ is the fine-structure constant, and we have applied the Planck units: the Planck charge $q_P = \frac{e}{\sqrt{\alpha}}$, the Planck length $\ell_P = \sqrt{\frac{\hbar G}{c^3}}$, the Planck time $t_P = \frac{\ell_P}{c}$, and the Planck energy $E_P = \frac{\hbar}{t_P}$. Because the non-Abelian Wu–Yang monopole can be received from something very similar to the Dirac monopole by using a suitable gauge transformation, we shall focus our attention on the Dirac monopole and analyze this case using the Planck units, and discuss certain new aspects of this fundamental solution.

According to Dirac, the condition for an unambiguous physical interpretation for the theory is taking $\beta = 2\pi n$, where β is a change in the phase of a wave function round a closed curve, $\psi = \psi_1 \exp i\beta$. In such a situation, the number $n \in \mathbb{Z}$ is the winding number of a closed curve going around the nodal line, *i.e.* the line on which the zeros of the wave function are placed. The sign of the winding number n is given by the direction of encircling of the nodal line. For the case of Dirac's electron a change in the phase of a wave function is equal to $\beta = \frac{e}{\hbar}\Phi_B$, where Φ_B is the total flux of the magnetic field induction **B**

$$\Phi_B = \int_S \mathbf{B} d\mathbf{S}, \tag{9.38}$$

and S is a closed surface going around the source of the magnetic field induction. According to the Gauss law for magnetism, the total flux Φ_B counts the total magnetic charge q_M

$$\Phi_B = \mu_0 q_M. \tag{9.39}$$

Calculating straightforwardly the magnetic field induction $\mathbf{B} = \nabla \times \mathbf{A}$

$$\mathbf{B} = \frac{\Phi_0}{r^2}\frac{\mathbf{r}}{r} = B^r \frac{\mathbf{r}}{r}, \tag{9.40}$$

the corresponding total flux Φ_B of magnetic field induction can be easily established

$$\Phi_B = \oint_C B^r dS_r = \int_0^{2\pi}\int_0^{\pi} B^r r^2 \sin\theta d\theta d\phi = 4\pi\Phi_0, \tag{9.41}$$

and together with Dirac's argumentation $\beta = 2\pi n$ such a construction leads to the unique relation linking the magnetic charge q_M with the elementary electric charge e

$$\frac{q_M e}{q_P^2 c} = \frac{n}{2}, \tag{9.42}$$

which is known as the Dirac quantization condition, which gives the quantum of the magnetic charge

$$q_M^0 = \frac{q_P^2 c}{2e} = \frac{4\pi\Phi_0}{\mu_0} = \frac{\Phi_B}{\mu_0} = \Phi_H, \tag{9.43}$$

i.e. $q_M = n q_M^0$, where Φ_H is the total flux of the magnetic field $\mathbf{H} = \dfrac{1}{\mu_0}\mathbf{B}$

$$\Phi_H = \int_C \mathbf{H} d\mathbf{S}. \tag{9.44}$$

It should be emphasized that such a construction is fully consistent when there are only free magnetic fields. When magnetic currents are taken into account, this strategy must be somewhat modified, but always the quantum of magnetic charge is determined as the total flux of the magnetic field, *i.e.* $q_M^0 = \Phi_H$, with the magnetic field in the form $\mathbf{H} = \dfrac{1}{\mu_0}\mathbf{B} - \mathbf{M}$. In such a situation, the free magnetic fields are determined the standard way such that the Gauss law for magnetism of the Maxwell electrodynamics remains unchanged

$$\int_S \mathbf{B} d\mathbf{S} = 0, \tag{9.45}$$

and the magnetization field $\mathbf{M}$, which is the only constituent solely responsible for description of the magnetic currents, determines the total magnetic charge q_M with the opposite sign

$$\int_S \mathbf{M} d\mathbf{S} = -q_M. \tag{9.46}$$

It is worth stressing that for the case of linear media the relation $\mathbf{H} = \dfrac{1}{\mu}\mathbf{B}$ holds, and when the magnetic permeability of the material, which is the source of the magnetic currents, is determined through the standard relation $\mu = \mu_r \mu_0$, then the magnetization field is established unambiguously as $\mathbf{M} = \frac{\mu_r - 1}{\mu_r} \frac{1}{\mu_0} \mathbf{B}$. Consequently, in such a situation the magnetic charge identically vanishes $q_M = 0$. However, when the magnetic permeability is not constant, but varies from point to point in space, then the magnetization field has another form

$$\mathbf{M} = \left(1 - \frac{\mu_0}{\mu}\right) \frac{1}{\mu_0} \mathbf{B}, \tag{9.47}$$

and the conditions (9.46) and (9.45) allow to determine the total magnetic charge in the following way

$$\int_S \frac{1}{\mu} \mathbf{B} d\mathbf{S} = q_M. \tag{9.48}$$

Applying the definition of the magnetic field induction $\mathbf{B} = \nabla \times \mathbf{A}$, one can apply the chain rule for differentiation

$$\frac{1}{\mu} \mathbf{B} = \frac{1}{\mu} (\nabla \times \mathbf{A}) = \nabla \times \left(\frac{1}{\mu} \mathbf{A}\right) - \mathbf{A} \times \left(\frac{1}{\mu^2} \nabla \mu\right). \tag{9.49}$$

In such a situation, applying the Stokes theorem

$$\int_S \nabla \times \left(\frac{1}{\mu} \mathbf{A}\right) d\mathbf{S} = \oint_C \frac{1}{\mu} \mathbf{A} d\mathbf{x}, \tag{9.50}$$

the integral in the condition (9.48) becomes

$$\oint_C \frac{1}{\mu} \mathbf{A} d\mathbf{x} - \int_S \mathbf{A} \times \left(\frac{1}{\mu^2} \nabla \mu\right) d\mathbf{S} = -q_M. \tag{9.51}$$

It is now clear that the case of constant magnetic permeability μ, which should identically generate the zero value of the magnetic charge $q_M = 0$, leads to identical vanishing of the first component of the left hand side in the equation (9.51). In this manner, in the general case one can suggest that this part vanishes identically

$$\oint_C \frac{1}{\mu} \mathbf{A} d\mathbf{x} = 0, \tag{9.52}$$

and it can be seen easily that in such a situation the magnetic charge is determined by the following condition

$$\int_S \mathbf{A} \times \left(\frac{1}{\mu^2}\nabla\mu\right) d\mathbf{S} = q_M, \tag{9.53}$$

which can be treated as the suitable Gauss law for magnetism in the case of nonlinear media. The condition (9.52) leads to highly non-trivial ramifications. In the case of constant and non-zero magnetic permeability μ this condition is equivalent to

$$\oint_C \mathbf{A} d\mathbf{x} = 0, \tag{9.54}$$

which removes the Aharonov–Bohm effect. In this manner, it is clear now that the condition (9.52), in the most general situation, for non-linear media establishes the Gauss law for magnetism with the absence of the Aharonov–Bohm effect. Interestingly, in such a situation also $\nabla\mu = 0$, so that the formula (9.53) says that $q_M = 0$, *i.e.* there is no magnetic charge. It is also possible, however, to take into account the Aharonov–Bohm effect. Let us suggest namely the straightforward generalization of the usual effect in the form

$$\oint_C \frac{1}{\mu} \mathbf{A} d\mathbf{x} = n\frac{4\pi\Phi_0}{\mu_c}, \tag{9.55}$$

where μ_c is certain constant magnetic permeability. Now the condition (9.51) can be rewritten in the form of the Dirac quantization condition

$$\frac{Q_M Q_E}{q_P^2 c} = \frac{n}{2}, \tag{9.56}$$

where the effective magnetic charge Q_M and the effective electric charge Q_E are constrained by the relation

$$Q_M Q_E = \frac{\mu_c}{\mu_0}(q_M + \Delta q_M)e, \tag{9.57}$$

where the extra magnetic charge Δq_M is given by the formula

$$\Delta q_M = \int_S \mathbf{A} \times \left(\frac{1}{\mu^2}\nabla\mu\right) d\mathbf{S}. \tag{9.58}$$

The charge constraint (9.57) expresses the ambiguity for the magnetic and the electric charges, which we shall call the charge ambiguity. It can be seen by straightforward analysis that in the most general case the presence of the magnetization field destroys the physical sense of the concept of the Dirac monopole. Namely, in the case of non-zero magnetization field the magnetic monopoles are not free point-like particles, but are confined in the material being the source of the magnetic currents. In such a situation, magnetic monopoles are rather quasi-particles then usual particles similar to electron or proton, and we shall call them *confined magnetic monopoles*. The question is to determine the constant magnetic permeability μ_c. If the charges Q_M, Q_E, q_M, and Δq_M are known from other theoretical argumentation or can be established by experimental data then the constant μ_c can be determined unambiguously by application of the constraint (9.57)

$$\mu_c = \mu_r \mu_0, \tag{9.59}$$

where μ_r is the relative magnetic permeability

$$\mu_r = \frac{Q_M Q_E}{(q_M + \Delta q_M)e}, \tag{9.60}$$

which for non-zero values of the charges can be written as

$$\mu_r = \frac{Q_M}{q_M}\frac{Q_E}{e}\frac{1}{1+\dfrac{\Delta q_M}{q_M}}. \tag{9.61}$$

For $\left|\dfrac{\Delta q_M}{q_M}\right| < 1$ the leading terms of the Taylor series expansion are

$$\mu_r = \frac{Q_M}{q_M}\frac{Q_E}{e}\sum_{n=0}^{\infty}\left(-\frac{\Delta q_M}{q_M}\right)^n = \frac{Q_M}{q_M}\frac{Q_E}{e}\left[1 - \frac{\Delta q_M}{q_M} + \left(\frac{\Delta q_M}{q_M}\right)^2 + O\left[\left(\frac{\Delta q_M}{q_M}\right)^3\right]\right], \tag{9.62}$$

whereas for $\left|\dfrac{\Delta q_M}{q_M}\right| > 1$ the leading terms of the Taylor series expansion are

$$\mu_r = -\frac{Q_M}{q_M}\frac{Q_E}{e}\sum_{n=0}^{\infty}\left(-\frac{q_M}{\Delta q_M}\right)^{n+1} = \frac{Q_M}{q_M}\frac{Q_E}{e}\left[\frac{q_M}{\Delta q_M} - \left(\frac{q_M}{\Delta q_M}\right)^2 + O\left[\left(\frac{\Delta q_M}{q_M}\right)^3\right]\right]. \tag{9.63}$$

Particularly, in the case in which $Q_M = q_M$ and $Q_E = e$ the the relative magnetic permeability is simplified to

$$\mu_r = \frac{1}{1+\dfrac{\Delta q_M}{q_M}}, \tag{9.64}$$

and, consequently, in the case $\left|\dfrac{\Delta q_M}{q_M}\right| < 1$ the Taylor series expansion is

$$\mu_r = \sum_{n=0}^{\infty}(-1)^n\left(\frac{\Delta q_M}{q_M}\right)^n = 1 - \frac{\Delta q_M}{q_M} + \left(\frac{\Delta q_M}{q_M}\right)^2 + O\left[\left(\frac{\Delta q_M}{q_M}\right)^3\right], \tag{9.65}$$

while for $\left|\dfrac{\Delta q_M}{q_M}\right| > 1$ one obtains the Taylor series expansion

$$\mu_r = \sum_{n=0}^{\infty}(-1)^n\left(\frac{q_M}{\Delta q_M}\right)^{n+1} = \frac{q_M}{\Delta q_M} - \left(\frac{q_M}{\Delta q_M}\right)^2 + O\left[\left(\frac{\Delta q_M}{q_M}\right)^3\right]. \tag{9.66}$$

The Dirac monopole is an Abelian monopole arising from the existence of the non-zero magnetic charge implemented *ad hoc* to the Maxwell equations of classical electrodynamics. As the consequence of its manifestly artificial nature, following from purely non-physical reasons for existence of the magnetic charge arising only from the naive straightforward analogy with the Coulomb electric field of an electrically charged point-like particle, such a magnetic monopole formally describes a magnetically charged point-like particle. It can be seen by non-tedious analysis that within the Dirac construction of the magnetic monopole there is present a purely singular magnetic field of a solenoidal line corresponding to a string, which is known as the Dirac string despite that P.A.M. Dirac never suggested and noticed the existence of such a string. Strictly speaking, existence of the Dirac string is a conclusion of the formalism presented by Paul Dirac [430] which in itself is outside this formalism and arises from other arguments. In such a situation, the vector potential possesses singularities and is no longer globally well-defined, so that the existence of a divergence free field is formally justified by existence of the Dirac string.

However, Tai Tsun Wu and Chen Ning Franklin Yang [431] presented also the original approach to the Dirac monopole based on the monopole harmonics, which excludes the necessity of application of the Dirac string. Effectively in the Wu–Yang formalism there are different vector potentials in different regions around the magnetic monopole and, consequently, the Dirac monopole a la Wu–Yang is free of the singularities arising in the approach involving the Dirac string. Let us look on this approach is some detail. We assume that the magnetic monopole is equipped with non-zero magnetic charge e_M. For the northern hemisphere $\theta \in \left[0, \frac{\pi}{2} + \varepsilon\right]$, where $\varepsilon \in \left[0, \frac{\pi}{2}\right]$, the magnetic potential is similar to the Dirac one

$$\mathbf{A}_\mu^N = \left[0, 0, 0, \frac{\mu_0}{4\pi}\frac{e_M}{r}\frac{1-\cos\theta}{\sin\theta}\right] = \left[0, 0, 0, \frac{\mu_0}{4\pi}\frac{e_M}{r}\tan\frac{\theta}{2}\right], \tag{9.67}$$

while for the southern hemisphere $\theta \in \left[\frac{\pi}{2} - \varepsilon, \pi\right]$ one has to deal with a slightly different situation

$$\mathbf{A}_\mu^S = \left[0, 0, 0, \frac{\mu_0}{4\pi}\frac{e_M}{r}\frac{-1-\cos\theta}{\sin\theta}\right] = \left[0, 0, 0, -\frac{\mu_0}{4\pi}\frac{e_M}{r}\left(\tan\frac{\theta}{2}\right)^{-1}\right]. \tag{9.68}$$

Singularities at $\theta = 0$ and $\theta = \pi$ correspond to the Dirac string. On the overlap region $\theta \in \left[\frac{\pi}{2} - \varepsilon, \frac{\pi}{2} + \varepsilon\right]$ the gauge field is the difference which is only a gauge transformation

$$\mathbf{A}_\mu^N - \mathbf{A}_\mu^S = \left[0, 0, 0, \frac{\mu_0}{2\pi}\frac{e_M}{r\sin\theta}\right] = \partial_\mu \chi = \frac{i}{g}\exp\left(ig\chi\right)\partial_\mu \exp\left(-ig\chi\right), \tag{9.69}$$

for which, after taking into account the differentiation in the spherical coordinates $\partial_\mu = \left[\frac{1}{c}\partial_t, \partial_r, \frac{1}{r}\partial_\theta, \frac{1}{r\sin\theta}\partial_\phi\right]$, the parameter χ can be established straightforwardly as

$$\chi(\phi) = \mu_0 e_M \frac{\phi}{2\pi}. \tag{9.70}$$

Let us take into account the wave functions point of view. According to Dirac argumentation, for consistency of the theory on the overlap region the interference effect must be vanishing

$$\psi_O = a_S \psi^S \exp\left[ig \oint dx^\mu \mathbf{A}_\mu^S\right] + a_N \psi^N \exp\left[ig \oint dx^\mu \mathbf{A}_\mu^N\right] = 0, \tag{9.71}$$

where a_S and a_N are non-zero complex constants, which allows to establish the following relation in this region

$$\psi^S = -\frac{a_N a_S^*}{|a_S|^2}\psi^N \exp\left[ig\oint dx^\mu\left(\mathbf{A}_\mu^N - \mathbf{A}_\mu^S\right)\right]. \tag{9.72}$$

When one considers the way of integration as the closed line C, circle say, then the exponent in the expression (9.72) is a special case of the generalized Wilson loop [432]

$$W_C[\mathbf{A}_\mu] = \mathrm{Tr}\mathscr{P}\exp\left[ig\oint_C dx^\mu \mathbf{A}_\mu T^a\right], \tag{9.73}$$

where $\mathscr{P}$ is the path-ordering operator. In the light of the fact (9.69) in the present case using the integral

$$\oint dx^\mu\left(\mathbf{A}_\mu^N - \mathbf{A}_\mu^S\right) = \int_0^{2\pi} d\phi\partial_\phi\chi(\phi) = \chi(2\pi) - \chi(0), \tag{9.74}$$

leads to evaluation of the Wilson loop of the field $\mathbf{A}_\mu^N - \mathbf{A}_\mu^S$

$$W_C[\mathbf{A}_\mu^N - \mathbf{A}_\mu^S] = \exp ig\left(\chi(2\pi) - \chi(0)\right). \tag{9.75}$$

The requirement on the wave function to be a single-valued function, which actually removes the possibility of the Aharonov–Bohm effect and makes the Dirac string undetectable, allows to write out the following quantization condition

$$g\left(\chi(2\pi) - \chi(0)\right) = 2\pi n, \quad n \in \mathbb{Z}, \tag{9.76}$$

which for all the values of the electric charges q and all the magnetic charges q_M constrained by the following relation

$$q_M q = \frac{g\Phi_0}{\alpha} e_M e, \tag{9.77}$$

is equivalent to the Dirac quantization condition

$$\frac{q_M q}{q_P^2 c} = \frac{n}{2}. \tag{9.78}$$

The constraint (9.77) is a syndrome of the ambiguity between the electric charge and the magnetic charge of the Dirac monopole satisfying the Dirac quantization condition. Such a situation is absent in the original analysis due to Dirac as well as when the Dirac string is involved explicitly. It looks like that this important fact has been omitted or neglected by previous authors considering the Dirac monopole. This is the *charge ambiguity* for the Dirac monopole. The simplest situation within the charge ambiguity has a place in the case

$$q_M q = e_M e, \tag{9.79}$$

which corresponds to the following value of the coupling constant

$$g = \frac{\alpha}{\Phi_0} = 2\alpha\frac{e}{\hbar}. \tag{9.80}$$

Interestingly, it can be seen directly that in the specific situation given by the coupling constant of $U(1)$ electrodynamics $g = \frac{e}{\hbar}$, the charge ambiguity condition (9.77) takes the form

$$q_M q = \frac{1}{\alpha} e_M e, \tag{9.81}$$

and can be solved, among others, by the electric charge identical with the Planck charge $q = \frac{e}{\sqrt{\alpha}} = q_P$ and the magnetic charge $q_M = \frac{e_M}{\sqrt{\alpha}} = q_P^M$, where q_P^M is the corresponding Planck magnetic charge. It should be emphasized that the Dirac monopole with the string and the Dirac monopole with no strings are equivalent if and only if the magnetic charge is simply the elementary electric charge multiplied by the speed of light, *i.e.* $e_M = ec$, which can be seen by comparison of the equations (9.37) and (9.67). In such a situation, the constraint (9.77) takes the form $q_M q = g\frac{\hbar}{e}\frac{e^2 c}{\alpha}$, and for the specific situation $g = \frac{e}{\hbar}$ the charge ambiguity becomes $q_M q = \frac{e^2 c}{\alpha}$ and is solved by the Planck charge only, *i.e.* $q = q_P$ and $q_M = q_P c$.

It is worth stressing that the standard formulation of the Big Bang theory, which lays the foundations of modern cosmology of the Early Universe, is rather highly sceptical about the question of the existence and physical significance of magnetic monopoles. There is the so-called monopole problem, also known as the exotic-relics problem, which emerges within Grand Unification Theories (GUT), the term created in 1978 by CERN scholars Andrzej Jerzy Buras, Jonathan Richard Ellis, Mary Katharine Gaillard, and Dimitrios V. Nanopoulos [433] for the $SU(N)$ Yang–Mills theories unifying other gauge theories, and leads to a number of nonstandard ramifications. In the suitable cosmological sector of Grand Unification Theories, during the cooling of the very hot universe there were generously generated very heavy stable magnetic monopoles [434] which should be still existent and be the dominating constituent of the Universe [435]. Grand Unification Theories, however, in themselves are based on the assumption that at high temperatures all fundamental forces are not actually fundamental and arise as the result of the spontaneous symmetry breaking from a single gauge field theory. The monopole problem is usually solved by application of the cosmic inflation which removes them definitely from the theory. *Ad hoc* implementation of the inflation below the temperature of magnetic monopoles production during expansion of the Universe separates monopoles from each other and lowers significantly their observed density.

Factually, Pierre Curie [436] originally proposed existence of magnetic monopoles. He considered the fundamental role of symmetry in physics, and formulated the principle [437]:

> When certain causes produce certain effects, the symmetry elements of the causes must be found in the effects.
>
> When certain effects show some asymmetry, this asymmetry should be reflected in the causes that have given birth.

which is known as the Curie principle. It should be emphasized that Pierre Curie applied this principle to electromagnetism on the empirical level. As the straightforward consequence of his symmetry laws, he recognized the real possibility of the existence of free magnetic charges [438] breaking the symmetry, which he expressed through the statement: *(...) there would be nothing absurd in terms of symmetry, assuming that the asymmetric molecules (...) are naturally loaded with free magnetism.* The problem to find magnetic monopoles in atomic matter, however, is their non-zero magnetic charge and, therefore, magnetic monopoles have presently the status of quasi-particles, whose existence can be verified by the effects which are not similar to the particles like protons, neutrons or electrons which can be directly detected.

The existence of magnetic monopoles has been recently considered seriously in condensed matter physics, where there are verified the models, based on the idea of the Dirac string, which apply infinitely thin connecting flux tubes which are similar to magnetic monopoles. Despite the fact that the existence of magnetic monopoles as particles has never been confirmed unambiguously, the effects making magnetic monopoles respectable physical objects have been observed (Cf. *e.g.* Refs. [439]-[457]). In 2008 Claudio Castelnovo, Roderich Moessner, and Shivaji L. Sondhi [440] proposed description of the spin ice materials, like dysprosium titanate $Dy_2Ti_2O_7$, by application of the Dirac monopoles having the quasi-particle nature. This dipolar spin ice is a model material for geometrically frustrated spin systems, in which the strong magnetic moments of the ion Dy^{3+} occupy a three-dimensional network of corner sharing tetrahedra and form an ideal cubic structure of pyrochlore with the space group being the crystallographic Fedorov group $Fd3m$. Among the four dysprosium spins in a tetrahedron the ferromagnetic nearest neighbour interactions force the geometrical frustration: each spin is aligned parallel to the local [111] axes, however, two spins point inward and two spins outward on each tetrahedron. In the straightforward analogy to the disordered protons in hexagonal water ice such a state of matter is called the spin ice or *2-in & 2-out rule* [458]. Destabilization of one spin of the spin chain in the 2-in & 2-out state breaks the symmetry and results in a pair of the topological defects - neighbouring tetrahedra of the 1-in & 3-out and the 3-in & 1-out states. Continuation of such spin flips over consecutive tetrahedra separates the topological defects and, consequently, leads to creation of the Dirac string of dipoles with a magnetic monopole q_M and a magnetic anti-monopole $-q_M$ at its ends. Diffuse neutron scattering on $Dy_2Ti_2O_7$ and muon spectroscopy on holmium titanate $Ho_2Ti_2O_7$ have provided the plausible evidence for the existence of such magnetic monopoles. The model of the spin ice is based on the magnetic Coulomb interaction, *i.e.* the potential $V(r) = -\frac{\mu_0}{4\pi}\frac{q_M^2}{r}$ where $r = xr_d$, $q_M = 2qc$ is the magnetic charge of an electric dipole of dipole moment $\mu = qr_d$ created by point-like electric charges $+q$ and $-q$ which lie in two vertices on the diamond lattice at the distance r_d. Moreover, the applied lattice structure of the pyrochlore is the kagome lattice, which is an arrangement of the straight lines composed of interlaced triangles in such a way that each point where two straight lines cross has four neighboring points. Detected phenomena among others include: the density and orientation of Dirac strings by diffuse neutron scattering [444], the existence of a magnetic Coulomb phase by polarized neutron scattering [442] and by muon spin rotation [441].

The research trend of seeking for the effects of magnetic monopoles in spin ice has generated more significant results. A particularly essential result for the case of the dysprosium titanate was presented in 2009 by the group led by D. Jonathan P. Morris [444] from the Helmholtz–Zentrum Berlin (HZB). The researchers reported detection of magnetic monopoles understood as quasi-particles. The detection was performed during the low-temperature cooling of a single crystal of $Dy_2Ti_2O_7$, which is a pyrochlore Ising's magnet, in a highly frustrated lattice structure. The researchers applied the method of diffuse neutron scattering to demonstrate that magnetic moments in the spin ice align into interwoven solenoidal tubes, which are the observable versions of the Dirac string. Such a model describes the spin ice through application of the networks of aligned dipoles, and the magnetic field of the ends of the tubes is practically identical to the field of magnetic monopole. The orientation and density of these emergent quasi-particles were intentionally manipulated by the symmetry-breaking magnetic field. The heat capacity of the spin ice was described according to the model of an effective gas of magnetic monopoles, interacting through the magnetic Coulomb interaction, which is based on the Debye–Hückel theory [459] named after Peter Joseph William Debye and Erich Armand Arthur Joseph Hückel whose proposed this model for theoretical explanation of the deformations from ideality in solutions of electrolytes. A similar example of the positive verification of the existence of magnetic monopoles was presented in 2011 by a research team led by Sean Richard Giblin [452]. The researchers created relaxing magnetic monopole currents through application of a magnetic-field pulse to the spin ice crystal $Dy_2Ti_2O_7$ at the temperature of 0.36K, and measured these currents through application induced electromotive force in a solenoid coupled to a sensitive amplifier. They quantitatively described the magnetic monopole currents through application of a chemical kinetic model of point-like charges obeying the Onsager mechanism [460] of the Wien effect of carrier dissociation and recombination [461]. They thus established the values of the microscopic parameters of magnetic monopole motion in the spin ice and identified the distinct roles of free and bound magnetic charges. This research illustrated a basic capacitor effect for magnetic charge and it is believed that 'magnetronic' circuitry will soon be constructed.

C 't Hooft–Polyakov Monopole

Let us return our considerations to the gauge field-theoretical discussion. In 1974, Gerardus 't Hooft [462] considered the $SU(2)$ Yang–Mills–Higgs equations and discovered that one can construct a solution behaving like the Dirac monopole, which is the solution to the Maxwell equations *ad hoc* complemented by the non-zero magnetic charge, which can be non-singular and have a finite energy. G. 't Hooft understood that if the $U(1)$ gauge group, the symmetry of the Maxwell equations, arises from the spontaneous symmetry breaking mechanism then the magnetic monopoles with the $U(1)$ magnetic charge can exist. Gerardus 't Hooft proved that magnetic monopoles are generated when the full gauge group has a compact covering group, which particularly removes the existence of magnetic monopoles from the Standard Model in which the covering group of the electroweak theory is $SU(2)\times\mathbb{R}$ and $\mathbb{R}$ is not compact. On the one hand, topologically non-trivial configurations of the Higgs field emerge if and only if the resulting gauge group consists of non-contractible loops which in the broken gauge symmetry are contractible. On the other hand, a winding number of any loop around the Abelian $U(1)$ gauge group is the same as the loop around the non-Abelian $U(1)$ gauge group and, consequently, this is not contractible in the full gauge group. For this reason, in the electroweak theory the existence of non-singular magnetic monopoles is forbidden by the configurations of the Higgs field. G. 't Hooft suggested the ansatz for the explicit solutions of the Yang–Mills–Higgs equations, *i.e.* the Higgs field and the Yang–Mills connection, in the case of stationary spherical symmetry, which helped to solve numerically these equations and significantly influenced gauge field theories.

It is worth stressing that in 1974 Alexander Markovich Polyakov [463], independently on Gerardus 't Hooft and in the completely different context, focused on the particle spectrum in quantum field theory, suggested the solution to the non-linear wave equation for the φ^4 scalar field theory which formally resembles the ansatz proposed by Gerardus 't Hooft for the magnetic monopole solutions to the Yang–Mills–Higgs equations. However, it should be emphasized that A.M. Polyakov did not presented independently any linkage between his solution and magnetic monopoles. In fact, the original proposal of his analysis of the φ^4 theory is coining the name "hedgehog" for the specific solution which arises from the fact that the internal direction of the Higgs field is determined by the radial direction is space. It should be emphasized also that A.M. Polyakov in his subsequent paper [464], published in 1975 and resulting from unambiguous inspiration by the paper of G. 't Hooft, considered the "hedgehog" solution in the context of magnetic monopoles, but he has not continued this line of thinking in his further research.

Actually, however, in his 1974's paper Alexander M. Polyakov did not considered magnetic monopoles seriously and even neglected their physical role, whereas in his 1974's paper 't Hooft intentionally studied the question of magnetic monopoles in the non-Abelian gauge theories. The state of affairs established by the papers of Gerardus 't Hooft [462] and Alexander M. Polyakov [463, 464] is clear and unambiguous: G. 't Hooft discovered magnetic monopoles in gauge theories, whereas A.M. Polyakov considered them seriously after reading the paper of G. 't Hooft. Despite of such a state of affairs, the commonly accepted names are: the 't Hooft–Polyakov ansatz for the form of the solution to the Yang–Mills–Higgs equations, and the 't Hooft–Polyakov monopole for the magnetic monopole interpretation arising from this ansatz. It is worth stressing that attributing the non-Abelian magnetic monopoles to the name of Alexander M. Polyakov, for the original name "hedgehog" and taking into account the ideas presented by Gerardus 't Hooft, has began since the 1976 papers [465] and today the established name for this solution is the 't Hooft–Polyakov monopole.

From the mathematical point of view the 't Hooft–Polyakov monopole is a smooth non-singular solution to the Yang–Mills–Higgs equations, characterized by a finite energy and, this is also considered, a finite mass. The ramification of straightforward application of the Dirac quantization condition to the 't Hooft–Polyakov monopole is twice the minimum magnetic charge of such a magnetic monopole. This feature is the immediate reflection of the fact that one can apply the $SU(2)$ doublet field having a half-integer electric charge. In the case of the theory with the $SO(3)$ gauge group a magnetic charge of a monopole violates the Dirac quantization condition if there are present in the theory the particles having a half-integer electric charge. However, such a situation could be rather untypical for $SO(3)$ gauge field theories with the minimum magnetic charge, which would also be a non-singular monopole for gauge field theories with the $SU(2)$ gauge group. Fortunately, there is no such solutions for $SO(3)$-based gauge field theories, so all non-singular $SO(3)$ magnetic monopoles are also allowed by the $SU(2)$ magnetic monopoles. The behaviour of the 't Hooft–Polyakov monopole is very similar to the behaviour of a soliton in the case of the four-dimensional non-Abelian gauge field theory. It should be emphasized that in contrast to the Dirac monopole the non-Abelian magnetic monopole is not the result of *ad hoc* implementation to the theory, but is the straightforward ramification of the non-commutative nature of a gauge field theory. In such a theory the effective electromagnetism is obtained through the spontaneous symmetry breaking mechanism.

G. 't Hooft proved that magnetic monopoles, treated as the smooth solutions of the field equations having finite energy, always exist in the gauge field theories in which spontaneous symmetry breaking is present, in the case when the symmetry gauge group permits a compact covering. Particularly, 't Hooft presented the way to construct a magnetic monopole solution to the $SU(2)$ Yang–Mills–Higgs equations, *i.e.* the non-Abelian gauge field theory of the Yang–Mills connection coupled to the Higgs scalar field. For the $SU(N)$ gauge group the Lagrangian of the Yang–Mills–Higgs theory has the form

$$\mathscr{L}_{YMH} = -\frac{1}{4\mu_0}F^a_{\mu\nu}F_a^{\mu\nu} + \frac{E_P\ell_P}{2}D_\mu\varphi^a D^\mu\varphi_a - E_P\ell_P\frac{\lambda}{4}\left(\varphi^a\varphi_a - \varphi_0^2\right)^2, \quad (9.82)$$

where the constant field $\varphi_0^2 = \frac{1}{2M_P^2\ell_P^2}\frac{m^2}{\lambda}$ is the vacuum expectation value of the scalar field φ_a given by the mass m and dimensionless constant λ, and D_μ is the gauge-covariant derivative operator

$$D_\mu = \partial_\mu - igT_b\mathbf{A}^b_\mu, \quad (9.83)$$

The energy density of such a configuration

$$\mathscr{H}_{YMH} = \frac{1}{2}\left(\varepsilon_0 E_k^a E_a^k + \frac{1}{\mu_0} B_k^a B_a^k + E_P \ell_P D_k \varphi^a D^k \varphi_a + E_P \ell_P \frac{\lambda}{2}\left(\varphi^a \varphi_a - \varphi_0^2\right)^2\right) \tag{9.84}$$

where $E_k^a = F_{0k}^a$ is the electric field and $B_k^a = \frac{1}{2}\varepsilon^{kij} F_{ij}^a$ is the magnetic field induction, is minimized $\mathscr{H}_{YMH} = 0$ for the conditions

$$\varphi^a \varphi_a = \varphi_0^2, \tag{9.85}$$

$$D_\mu \varphi^a = 0, \tag{9.86}$$

$$F_{\mu\nu}^a = 0, \tag{9.87}$$

which determine the specific vacuum configuration, known as the Higgs vacuum. Such a vacuum is degenerated, *i.e.* the set of vacuum expectation values of the Higgs field forms a sphere S_∞^2 on which all the points are equivalent because the gauge transformation connects them. For the non-zero value of the constant φ_0 there is spontaneous symmetry breaking which for the $SU(2)$ gauge group has the form $SU(2) \to U(1)$. It should be emphasized that generalization of this type of symmetry breaking mechanism to the case of $SU(N)$ gauge group can be done only in the Cartan–Weyl basis, *i.e.* a basis of a finite-dimensional semi-simple complex Lie algebra consisting of the elements of the Cartan subalgebra of this Lie algebra and root vectors introduced by Hermann Klaus Hugo Weyl [466]. Recall that the Cartan subalgebra, named after Élie Cartan who introduced this concept in his doctoral thesis [467], of finite-dimensional Lie algebra over any field is a nilpotent subalgebra of the Lie algebra which is equal to its normalizer in the Lie algebra. In practical application, this choice corresponds to the situation in which the vacuum expectation value of the scalar field φ is a diagonal matrix having identically vanishing trace. One then must apply the root vectors and $(N-1)$-dimensional Cartan subalgebra of the Lie algebra $\mathfrak{u}(N)$, and the gauge symmetry is spontaneously broken to the maximal torus $SU(N) \to SU(N-1) \times U(1) \to \ldots \to SU(2) \times U(1)^{N-2} \to U(1)^{N-1}$, where the boundary condition is $SU(2) \to U(1)$.

The Euler–Lagrange equations of motion for the case of the Lagrangian of the Yang–Mills–Higgs theory (9.82) can be straightforwardly derived

$$D^\mu D_\mu \varphi_a = -E_P \ell_P \lambda \left(\varphi^c \varphi_c - \varphi_0^2\right) \varphi_a, \tag{9.88}$$
$$\partial_\mu F_a^{\mu\nu} + g f_{abc} \mathbf{A}_\mu^b F_c^{\mu\nu} = \mu_0 \ell_P g f_{abc} \varphi^b D^\nu \varphi^c, \tag{9.89}$$

(For details Cf. *e.g.* Ref. [468]) and the Bianchi identities

$$\partial_\mu * F_a^{\mu\nu} + g f_{abc} \mathbf{A}_\mu^b * F_c^{\mu\nu} = 0. \tag{9.90}$$

Solutions to the Yang–Mills–Higgs equations (9.88)-(9.89) map the two-dimensional sphere, known as the vacuum manifold S_∞^2, onto the boundary of a three-dimensional space, which is also a two-dimensional sphere S^2. These maps are characterized using a winding number $n = 0, \pm 1, \pm 2, \ldots$ which informs how many times the manifold S_∞^2 is covered by a single turn around the spatial boundary S^2. Gerardus 't Hooft showed that a magnetic monopole solution to the Yang–Mills–Higgs equations emerges through the restriction of a specific topological vacuum configuration at the spatial infinity of the Higgs field

$$\varphi^a(r \to \infty) \to \varphi_0 \frac{r^a}{r}, \tag{9.91}$$

called "hedgehog" by Alexander M. Polyakov, which is characterized by the direction in the isospin space in which the vacuum expectation value points are connected with the radial direction in the coordinate space. Such a restriction manifestly mixes the spatial and the group indices and defines a single mapping of the vacuum manifold S_∞^2 onto the spatial asymptotic. The configurations characterized by different values of winding number cannot be continuously deformed into each other. The gauge transformation $U(\theta) = \exp iT^a \theta_a$ rotates the isospin vector to the third axis. If one applies the so-called unitary gauge, which rotates the scalar field everywhere in space to a given direction and is a very specific combing of the hedgehog, then the southern pole singularity of the gauge transformation forbids the procedure from spreading onto the global level. Consequently, the global gauge transformation, which connects the trivial and the hedgehog solutions, is not properly defined and such a situation leads to the existence of the infinite barrier separating these solutions.

The 't Hooft–Polyakov monopole has several non-trivial and remarkable features. Among others, the nonsingular nature in all fields and the magnetic charge identical to the topological charge are the distinguishable properties. Strictly speaking, this is a smooth nonsingular solution to the $SU(2)$ Yang–Mills–Higgs equations arising from the assumptions of spherical symmetry and time independence of the Yang–Mills connection and the Higgs field which, after taking into account the dimensional conventions, can be presented in the following form

$$\mathbf{A}^a_\mu = \left[0, \varepsilon^a_{ik}\frac{1-K(2\Phi_0\varphi_0 gr)}{gr}\frac{r^k}{r}\right], \tag{9.92}$$

$$\varphi^a = \frac{H(2\Phi_0 g\varphi_0 r)}{2\Phi_0 gr}\frac{r^a}{r}, \tag{9.93}$$

known as the 't Hooft–Polyakov ansatz.

In such a situation, the non-Abelian field strength tensor can be calculated immediately

$$F^a_{ij} = \frac{2}{gr^4}\left[-r^2\left(\varepsilon^a_{jk}\partial_i K - \varepsilon^a_{ik}\partial_j K\right)r^k + r^2(1-K)\left(\varepsilon^a_{jk}\delta^k_i - \varepsilon^a_{ik}\delta^k_j\right)\right. \tag{9.94}$$

$$\left. - 2(1-K)\left(\varepsilon^a_{jk}r^k r_i - \varepsilon^a_{ik}r^k r_j\right) + \varepsilon^a_{bc}\varepsilon^b_{ik}\varepsilon^c_{jl}(1-K)^2 r^k r^l\right], \tag{9.95}$$

and, consequently, the corresponding color magnetic field induction can be calculated in straightforward analogy with the Wu–Yang monopole presented above, *i.e.* $B^k = \frac{1}{2}\varepsilon^{kij}F^a_{ij}T_a$. Explicitly, one obtains the result

$$\mathbf{B} = \frac{2}{gr^2}\left[K^2(\xi) - K(\xi) - \xi\frac{dK(\xi)}{d\xi}\right]\frac{\mathbf{r}}{r}, \tag{9.96}$$

where $\xi = 2g\Phi_0\varphi_0 r$ is the dimensionless variable. Consequently, the total flux $\Phi_B = \int \mathbf{B}d\mathbf{S} = \int B^r dS_r = 4\pi r^2 B^r$ of such a magnetic field induction $\mathbf{B} = B^r\frac{\mathbf{r}}{r}$ through the sphere of fixed radius r can be easily established

$$\Phi_B(\xi) = \frac{8\pi}{g}\left[K^2(\xi) - K(\xi) - \xi\frac{dK(\xi)}{d\xi}\right], \tag{9.97}$$

and is determined solely in terms of the arbitrary function $K(\xi)$, which is given by the 't Hooft–Polyakov ansatz (9.92) and its first derivative.

According to the Gauss law for magnetism, one can calculate the magnetic charge contained in the sphere of the fixed radius

$$\Phi_B = \mu_0 q_M, \tag{9.98}$$

and, consequently, one receives the following value of the color magnetic charge

$$q_M = \frac{8\pi}{\mu_0 g}\left[K^2(\xi) - K(\xi) - \xi\frac{dK(\xi)}{d\xi}\right], \tag{9.99}$$

Because this charge should be a constant, this is inadequate expression in general. However there is the specific case

$$K^2(\xi) - K(\xi) - \xi\frac{dK(\xi)}{d\xi} = \frac{\kappa^2 - 1}{4}, \tag{9.100}$$

where κ is a certain dimensionless constant. With such a convention the magnetic charge is a pure constant and has the form

$$q_M = \frac{2\pi}{\mu_0 g}(\kappa^2 - 1). \tag{9.101}$$

Let us consider first the case of trivial magnetic charge, which corresponds to $\kappa = 1$. Then the ordinary differential equation (9.100) can be straightforwardly integrated with the result

$$K(\xi) = \frac{1}{1 - \left|\frac{K_0 + 1}{K_0}\right|\left|\frac{\xi}{\xi_0}\right|}, \tag{9.102}$$

where $K_0 = K(\xi_0)$ and ξ_0 are arbitrary the integration constants. For $\kappa > 0$ and $\kappa \neq 1$ one receives the solution

$$K(\xi) = \frac{1}{2} + \frac{\kappa}{2}\frac{\left|\frac{\kappa + 2K_0 - 1}{\kappa - (2K_0 - 1)}\right|\left|\frac{\xi_0}{\xi}\right|^{\kappa} - 1}{\left|\frac{\kappa + 2K_0 - 1}{\kappa - (2K_0 - 1)}\right|\left|\frac{\xi_0}{\xi}\right|^{\kappa} + 1}, \tag{9.103}$$

whereas for for $\kappa < 0$ one obtains

$$K(\xi) = \frac{1}{2} - \frac{|\kappa|}{2}\frac{\left|\frac{|\kappa| - (2K_0 - 1)}{|\kappa| + (2K_0 - 1)}\right|\left|\frac{\xi}{\xi_0}\right|^{|\kappa|} - 1}{\left|\frac{|\kappa| - (2K_0 - 1)}{|\kappa| + (2K_0 - 1)}\right|\left|\frac{\xi}{\xi_0}\right|^{|\kappa|} + 1}. \tag{9.104}$$

The relation (9.101) can rewritten in the equivalent form

$$\frac{q_M e}{q_P^2 c} = \frac{1}{2\Phi_0 g}\frac{\kappa^2 - 1}{2}, \tag{9.105}$$

so that in the light of the Dirac quantization condition $\frac{q_M e}{q_P^2 c} = \frac{n}{2}$ one receives the quantization condition

$$\frac{1}{2\Phi_0 g}(\kappa^2 - 1) = n, \tag{9.106}$$

which can be interpreted in two different ways. The first interpretation is quantization of the coupling constant

$$g_n = \frac{1}{2n\Phi_0}(\kappa^2 - 1), \tag{9.107}$$

while the constant κ remains non quantized, free parameter and, consequently, also the function K is still continuous. The second interpretation is quantization of the constant κ

$$\kappa_n = \pm\sqrt{1 + 2g\Phi_0 n}, \tag{9.108}$$

whereas the coupling constant g is a free parameter. In such a situation, the function $K(\xi)$ possesses two independent kinds of discrete values

$$K_n^+(\xi) = \frac{1}{2} + \frac{\sqrt{1+2g\Phi_0 n}}{2}\frac{\left|\frac{\sqrt{1+2g\Phi_0 n}+2K_0-1}{\sqrt{1+2g\Phi_0 n}-(2K_0-1)}\right|\left|\frac{\xi_0}{\xi}\right|^{\sqrt{1+2g\Phi_0 n}}-1}{\left|\frac{\sqrt{1+2g\Phi_0 n}+2K_0-1}{\sqrt{1+2g\Phi_0 n}-(2K_0-1)}\right|\left|\frac{\xi_0}{\xi}\right|^{\sqrt{1+2g\Phi_0 n}}+1}, \tag{9.109}$$

and

$$K_n^-(\xi) = \frac{1}{2} - \frac{\sqrt{1+2g\Phi_0 n}}{2}\frac{\left|\frac{\sqrt{1+2g\Phi_0 n}-(2K_0-1)}{\sqrt{1+2g\Phi_0 n}+(2K_0-1)}\right|\left|\frac{\xi}{\xi_0}\right|^{\sqrt{1+2g\Phi_0 n}}-1}{\left|\frac{\sqrt{1+2g\Phi_0 n}-(2K_0-1)}{\sqrt{1+2g\Phi_0 n}+(2K_0-1)}\right|\left|\frac{\xi}{\xi_0}\right|^{\sqrt{1+2g\Phi_0 n}}+1}, \tag{9.110}$$

and both these situations are consistent if and only if

$$n \geqslant -\frac{1}{2g\Phi_0}. \tag{9.111}$$

It can be seen by straightforward calculation that

$$K_0^{\pm}(\xi_0) = \frac{1}{1+\left|\dfrac{1-K_0}{K_0}\right|^{\pm 1}}, \tag{9.112}$$

so if one wishes to satisfy the relation

$$K_0^{+}(\xi_0) = K_0^{-}(\xi_0), \tag{9.113}$$

then one receives the following value of the the integration constant K_0

$$K_0 = \frac{1}{2}, \tag{9.114}$$

and, consequently, $K_0^{\pm}(\xi_0) = K_0$. In this manner, one obtains the following simplified discretization

$$K_n^{\pm}(\xi) = \frac{1}{2} \pm \frac{\sqrt{1+2g\Phi_0 n}}{2} \frac{\left|\dfrac{\xi_0}{\xi}\right|^{\pm\sqrt{1+2g\Phi_0 n}} - 1}{\left|\dfrac{\xi_0}{\xi}\right|^{\pm\sqrt{1+2g\Phi_0 n}} + 1} \tag{9.115}$$

$$= \frac{1 \pm \sqrt{1+2g\Phi_0 n}}{2} - \frac{\pm\sqrt{1+2g\Phi_0 n}}{\left|\dfrac{\xi_0}{\xi}\right|^{\pm\sqrt{1+2g\Phi_0 n}} + 1}. \tag{9.116}$$

which corresponds to the non-discretized function

$$K(\xi) = \frac{1+\kappa}{2} - \frac{\kappa}{\left|\dfrac{\xi_0}{\xi}\right|^{\kappa} + 1}. \tag{9.117}$$

This function in itself is interesting. Its Taylor power series expansion for $\xi > \xi_0$ has the form

$$K(\xi) = \frac{1+\kappa}{2} - \kappa \sum_{i=0}^{\infty} (-1)^i \left|\frac{\xi_0}{\xi}\right|^{\kappa i}, \tag{9.118}$$

while for $\xi < \xi_0$ one receives

$$K(\xi) = \frac{1+\kappa}{2} + \kappa \sum_{i=1}^{\infty} (-1)^i \left| \frac{\xi}{\xi_0} \right|^{\kappa i}. \tag{9.119}$$

For the gauge-invariant description in the presence of the scalar field, ’t Hooft [462] constructed the Abelian field strength tensor which for the $SU(2)$ gauge theory can be presented in the form

$$F_{\mu\nu} = F^a_{\mu\nu} \frac{\varphi_a}{|\varphi|} + \frac{1}{g} f_{abc} \frac{\varphi^a}{|\varphi|} D_\mu \frac{\varphi^b}{|\varphi|} D_\mu \frac{\varphi^c}{|\varphi|}, \tag{9.120}$$

which is not the only choice considered in the literature discussing non-Abelian gauge fields. In fact, the definition of the corresponding electromagnetic (which must be Abelian!) field strength tensor in a non-Abelian gauge theory is always dependent on a choice. For example, Peter Goddard and David Ian Olive [469] one can find another kind of Abelianization $F_{\mu\nu} = F^a_{\mu\nu} \frac{\varphi_a}{|\varphi|}$ which, strictly speaking, is the straightforward projection of the non-Abelian strength tensor onto the unit vector field created from the scalar field in the isospin space. The question arises: whether in the presence of scalar field it is necessary, in general, to Abelianize the Yang–Mills gauge field strength tensor by mixing the scalar field with the Yang–Mills connections? This question is, in fact, an unsolved problem of modern theoretical physics and for this reason we have presented above possibly the simplest Abelianization of the Yang–Mills field strength tensor for the theory of the non-Abelian field coupled to a scalar field $F_{\mu\nu} = F^a_{\mu\nu} T_a$, which does not mix the scalar field with gauge fields. Similarly, to the Abelianization based on the ’t Hooft tensor (9.120), this way also leads to gauge-invariant description, but we shall not focus on its results in the further part of this book.

Let us return to considerations of the well-known facts about the ’t Hooft–Polyakov monopole. By application of the ’t Hooft–Polyakov ansatz (9.92)-(9.93) one can establish

$$B^a_k = \frac{1}{gr^2} \left[\left(1 - K^2 + \xi \frac{dK}{d\xi} \right) \frac{r^a r_n}{r^2} - \xi \frac{dK}{d\xi} \delta^a_n \right], \tag{9.121}$$

$$D_i \varphi^a = \frac{1}{2\Phi_0 g r^2} \left[\left(\xi \frac{dH}{d\xi} - KH - H \right) \frac{r^a r_i}{r^2} + \delta^a_i KH \right], \tag{9.122}$$

so that after rather tedious algebraic manipulations the total energy calculated from the energy density (9.84) is

$$E = (2\Phi_0 g)\frac{4\pi\varphi_0}{\mu_0 g^2}\int_0^\infty \frac{d\xi}{\xi^2}\left\{\xi^2\left(\frac{dK}{d\xi}\right)^2 + \frac{1}{2}(K^2-1)^2 + \right. \tag{9.123}$$

$$\left. + \ 4\pi\alpha\left[K^2H^2 + \frac{1}{2}\left(\xi\frac{dH}{d\xi} - H\right)^2 + \frac{\lambda}{(2\Phi_0 g)^2}\frac{1}{4}\left(H^2-\xi^2\right)^2\right]\right\}.$$

Applying the relations

$$M_H = \sqrt{2\lambda}\,\varphi_0 M_P \ell_P, \tag{9.124}$$

$$M_W = 2\Phi_0 g \varphi_0 M_P \ell_P, \tag{9.125}$$

arising from the Higgs mechanism of spontaneous symmetry breaking for the non-Abelian gauge field theories of coupled Yang–Mills connections and the scalar field, where M_H is the mass of Higgs particle and M_W is the mass of a vector boson, one receives

$$(2\Phi_0 g)\frac{4\pi\varphi_0}{\mu_0 g^2} = \frac{1}{2\Phi_0 g}\frac{M_W c^2}{\alpha} = \frac{M_H c^2}{2\sqrt{\lambda}\alpha}, \tag{9.126}$$

$$\frac{\lambda}{(2\Phi_0 g)^2} = \frac{M_H^2}{4M_W^2}, \tag{9.127}$$

In the standard case of $SU(2)$ Yang–Mills–Higgs theory, for which identically $g = e/\hbar$ or equivalently $2\Phi_0 g = 1$, this formula is simplified to

$$E = \frac{M_W c^2}{\alpha}\int_0^\infty \frac{d\xi}{\xi^2}\left\{\xi^2\left(\frac{dK}{d\xi}\right)^2 + \frac{1}{2}(K^2-1)^2 + \right. \tag{9.128}$$

$$\left. + \ 4\pi\alpha\left[K^2H^2 + \frac{1}{2}\left(\xi\frac{dH}{d\xi} - H\right)^2 + \frac{\lambda}{4}\left(H^2-\xi^2\right)^2\right]\right\},$$

and simply $\lambda = \frac{M_H^2}{4M_W^2}$. In this manner, the mass of 't Hooft–Polyakov monopole can be established by $E = M_M c^2$ as

$$M_M = \frac{M_W}{\alpha} C\left(\alpha, \frac{M_H^2}{M_W^2}\right), \tag{9.129}$$

where $C\left(\alpha, \frac{M_H^2}{M_W^2}\right)$ is a constant.

Interestingly, in the case of the Higgs–Hubble inflaton discussed in the cosmological part of this book, one has

$$\varphi_0 = \frac{2\sqrt{\pi}}{3\ell_P}, \tag{9.130}$$

and if this inflaton is the Higgs particle then its mass is

$$M_H = \frac{4}{3}\sqrt{\pi\lambda}M_P, \tag{9.131}$$

and in the light of the Markov hypothesis [337] $M_H < M_P$ one has

$$\lambda \leqslant \frac{9}{16\pi} \approx 0.17905. \tag{9.132}$$

It is easy to see that the mass of the corresponding vector boson breaks the Markov hypothesis

$$M_W = \frac{2}{3}\sqrt{\pi}M_P > M_P. \tag{9.133}$$

Applying the Higgs–Hubble inflaton one obtains the mass of the corresponding $SU(2)$ 't Hooft–Polyakov monopole

$$M_M = \frac{2\sqrt{\pi}}{3}\frac{M_P}{\alpha}C(\alpha,\lambda), \tag{9.134}$$

where $\lambda = \frac{M_H^2}{4M_W^2}$. If, moreover, one takes into account the upper bound $\lambda = \frac{9}{16\pi}$, which corresponds to the mass $M_W = \frac{2\sqrt{\pi}}{3}M_H$, then the mass of such a magnetic monopole is

$$M_M = \frac{2\sqrt{\pi}}{3}\frac{M_P}{\alpha}C(\alpha). \tag{9.135}$$

The Euler–Lagrange equations of motion arising from the energy (9.128)

$$\frac{\delta E}{\delta K} - \frac{d}{d\xi}\frac{\delta E}{\delta\frac{dK}{d\xi}} = 0, \tag{9.136}$$

$$\frac{\delta E}{\delta H} - \frac{d}{d\xi}\frac{\delta E}{\delta\frac{dH}{d\xi}} = 0, \tag{9.137}$$

can be derived straightforwardly

$$\xi^2 \frac{d^2K}{d\xi^2} = KH^2 + 4\pi\alpha K(K^2-1), \tag{9.138}$$

$$\xi^2 \frac{d^2H}{d\xi^2} = 2K^2H + \lambda H(H^2-\xi^2), \tag{9.139}$$

and the boundary conditions are

$$K(\xi \to 0) \to 1 + O(\xi), \tag{9.140}$$

$$H(\xi \to 0) < O(\xi), \tag{9.141}$$

$$K(\xi \to \infty) \to 0, \tag{9.142}$$

$$H(\xi \to \infty) \sim \xi. \tag{9.143}$$

For $\frac{\lambda}{(2\Phi_0 g)^2} \gg 1$ one can suggest $H(\xi) = \xi(1-h(\xi))$, where $h(\xi)$ is a small correction. Then one obtains the equation

$$\frac{d^2h}{h\xi^2} + \frac{2}{\xi}\frac{dh}{d\xi} = 2\frac{\lambda}{(2\Phi_0 g)^2}h, \tag{9.144}$$

having the following solution

$$h(\xi) = K_0\left(\frac{\xi}{\xi_0}\right) \sim \exp\left(-\frac{\xi}{\xi_0}\right), \tag{9.145}$$

where $\xi_0 = \sqrt{\frac{\lambda}{(2\Phi_0 g)^2}} = \frac{M_H}{2M_W}$, and K_0 is a modified Bessel function.

The total magnetic flux for diverse charge states of the magnetic monopole is, in fact, the integration of the topological charge for different homotopic classes in mapping the triplet Higgs field on the unit sphere in the field space onto the infinitive sphere in the three dimensional configuration space. In this manner, considerations of the magnetic charge for the 't Hooft–Polyakov monopole are usually based on identification of the magnetic charge with the so-called topological charge which can be defined as follows

$$q_M = \frac{1}{\mu_0 g}\int d^2\xi \sqrt{\det \partial_\mu \frac{\varphi^a}{|\varphi|}\partial_\nu \frac{\varphi_a}{|\varphi|}}, \tag{9.146}$$

and it should be emphasized that such a definition is valid only in the Higgs vacuum. The calculation can be made straightforwardly with the result

$$q_M = \frac{4\pi n}{\mu_0 g}, \tag{9.147}$$

where n is an integer, which can be rewritten as the non-Abelian version of the Dirac quantization condition

$$\frac{q_M q}{q_P^2 c} = \frac{n}{2}, \tag{9.148}$$

where the electric charge q has the form

$$q = (2\Phi_0 g)e, \tag{9.149}$$

which for the $SU(2)$ coupling constant $2\Phi_0 g = 1$ gives identically $q = e$.

The magnetic charge can be also established in another way, directly from the usual Gauss law for magnetism

$$q_M = \frac{1}{\mu_0}\int \mathbf{B}d\mathbf{S}, \tag{9.150}$$

in which the magnetic field induction is Abelianized in the Higgs vacuum

$$\mathbf{B} = \mathbf{B}^a \frac{\varphi_a}{|\varphi|}. \tag{9.151}$$

The calculation is easy to do

$$q_M = \frac{1}{\mu_0 \varphi_0}\int dV B_k^a D^k \varphi_a = \tag{9.152}$$

$$= -\frac{4\pi}{\mu_0 g}\int_0^\infty \frac{d\xi}{\xi^2}\left[(K^2-1)\left(\xi\frac{dH}{d\xi} - H\right) + 2KH\xi\frac{dK}{d\xi}\right] \tag{9.153}$$

$$= \frac{4\pi}{\mu_0 g}\int_0^\infty d\xi \frac{d}{d\xi}\left(\frac{1-K^2}{\xi}\right) = \frac{4\pi}{\mu_0 g} = \frac{1}{2\Phi_0 g}\frac{ec}{\alpha}. \tag{9.154}$$

The 't Hooft–Polyakov monopole is a nonsingular topological soliton, similar to the Dirac monopole, arising from the theory of Yang–Mills connection coupled to a Higgs field which as a result of the Higgs mechanism spontaneously breakdowns to a subgroup $\mathscr{H}$ of a gauge group $\mathscr{G}$. The 't Hooft–Polyakov monopole is a finite-energy smooth solution which is localized around $r=0$ and for $r \gg 0$ and, as a result of the Higgs mechanism, its radial behaviour is reduced to the behaviour of the Dirac monopole. It should be emphasized that the 't Hooft monopole itself is non-singular at and near the origin.

D Julia–Zee Dyons

Magnetic monopoles have an identically vanishing electric charge. However, there is straightforward generalization of magnetic monopoles to the case of non-zero both electric and magnetic charges, known as dyons, which was first proposed in the late 1960s by Julian Seymour Schwinger as the phenomenological alternative to quarks [472] and then in 1968 by Daniel Zwanziger [473]. J.S. Schwinger extended the Dirac quantization condition and applied the dyon model to describe a particle, whose features were analogous to the flavor-neutral J/ψ meson consisting of a charm quark and a charm antiquark, called generally charmonium, discovered in 1974 by a Brookhaven National Laboratory (BNL) team led by Samuel Chao Chung Ting at the Massachusetts Institute of Technology (MIT) [474] and independently by a Stanford Linear Accelerator Center group headed by Burton Richter [475]. In 1971, Iwo Bialynicki-Birula and Zofia Bialynicka-Birula [476], by application of the hydrodynamic formulation of quantum mechanics without using wave-functions and electromagnetic potentials, described the non-relativistic quantum theory of dyons moving in an arbitrary external electromagnetic field, in which the Dirac quantization condition guarantees a consistency of motion of the probability fluid. In the same year Iwo Bialynicki-Birula [477] described relativistic quantum mechanics of dyons and found the exact solutions and the energy levels to the case of one dyon moving in the electromagnetic field of another dyon. In 1979, Edward Witten [478] introduced the vacuum angle θ to describe a possible source of CP violation by effectively arising the electric charge of the 't Hooft–Polyakov monopole.

In the context of the Yang–Mills equations dyons were introduced in 1975 by Bernard Julia and Anthony Zee [479] who also generalized the 't Hooft–Polyakov ansatz for the case of dyons by

$$\mathbf{A}_0^a = \frac{J(\xi)}{gr}\frac{r^a}{r}, \tag{9.155}$$

with the suitable boundary conditions

$$J(r \to 0) \to 0, \tag{9.156}$$

$$J(r \to \infty) \to Cr, \tag{9.157}$$

where C is a constant. Then the electric charge of the dyon is simply given by the Gauss law for electricity

$$\frac{q}{\varepsilon_0} = \int \mathbf{E} d\mathbf{S}, \tag{9.158}$$

calculated for the Abelianized electric field

$$\mathbf{E} = \mathbf{E}^a \frac{\varphi_a}{|\varphi|}, \tag{9.159}$$

with the color electric field $E_k^a = \frac{1}{c} F_{0k}^a$. The elementary calculation gives

$$q = \frac{\varepsilon_0}{\varphi_0} \int dV E_k^a D^k \varphi_a = \frac{4\pi\varepsilon_0 c}{g} C = \frac{C}{2\Phi_0 g} \frac{e}{\alpha} = \frac{e}{\alpha} \frac{C}{2\Phi_0 g} = \frac{q_M}{c} C. \tag{9.160}$$

There is the so-called Julia–Zee correspondence between the scalar field φ^a and the time component of the vector potential, which in the isospin space is parallel to the direction of the Higgs field, which can be considered as an additional triplet of the scalar fields, *i.e.* $\varphi^a \leftrightarrows \mathbf{A}_0^a$.

The only known analytical solution to the problem is the limit of vanishing potential, *i.e.* $\lambda \to 0$, known as the Bogomol'nyi–Prasad–Sommerfield limit named after Manoj K. Prasad, Charles Michael Sommerfield [470], and Evgeny Borisovich Bogomoln'yi [471], which generates the specific instance of the 't Hooft–Polyakov monopole. In this limit, the Higgs field is massless but the full gauge invariance remains broken by the non-zero vacuum expectation value of the Higgs field. It is easy to see that the energy of the Yang–Mills–Higgs system calculated using the energy density (9.84)

$$\mathscr{E}_{YMH} = \frac{1}{2} \int d\mathrm{V} \left(\varepsilon_0 E_k^a E_a^k + \frac{1}{\mu_0} B_k^a B_a^k + E_P \ell_P D_k \varphi^a D^k \varphi_a + E_P \ell_P \frac{\lambda}{2} \left(\varphi^a \varphi_a - \varphi_0^2 \right)^2 \right) \tag{9.161}$$

where $d\mathrm{V}$ is a volume element, can be presented in the equivalent form

$$\mathscr{E}_{YMH} = \frac{\varepsilon_0}{2}\int d\mathrm{V}\left(E_a^k - c\sqrt{\mu_0 E_P \ell_P}\sin\alpha_c D^k\varphi_a\right)\left(E_k^a - c\sqrt{\mu_0 E_P \ell_P}\sin\alpha_c D_k\varphi^a\right)$$
$$+\frac{1}{2\mu_0}\int d\mathrm{V}\left(B_a^k - \sqrt{\mu_0 E_P \ell_P}\cos\alpha_c D^k\varphi_a\right)\left(B_k^a - \sqrt{\mu_0 E_P \ell_P}\cos\alpha_c D_k\varphi^a\right)$$
$$+\varepsilon_0 c\sqrt{\mu_0 E_P \ell_P}\sin\alpha_c\int d\mathrm{V}E_a^k D_k\varphi^a + \frac{1}{\mu_0}\sqrt{\mu_0 E_P \ell_P}\cos\alpha_c\int d\mathrm{V}B_a^k D_k\varphi^a$$
$$+E_P\ell_P\frac{\lambda}{4}\int d\mathrm{V}\left(\varphi^a\varphi_a - \varphi_0^2\right)^2, \tag{9.162}$$

where α_c is any real angular parameter, which in the case of a dyon relates the angle relating the electric and the magnetic charges

$$\tan\alpha_c = \frac{qc}{q_M}. \tag{9.163}$$

Let us consider now the minimum of the energy (9.162) for any scalar field in the limit $\lambda \to 0$. Such a situation occurs if the Bogomol'nyi equations are satisfied

$$E_a^k = 2\Phi_0 c\sqrt{4\pi\alpha}\sin\alpha_c D^k\varphi_a, \tag{9.164}$$
$$B_a^k = 2\Phi_0\sqrt{4\pi\alpha}\cos\alpha_c D^k\varphi_a, \tag{9.165}$$

where we have applied the elementary identity $\mu_0 E_P\ell_P = 4\pi\alpha(2\Phi_0)^2$. Applying the Gauss law for electricity in the case of the Abelianized color electric field $\mathbf{E} = \mathbf{E}^a\dfrac{\varphi_a}{\varphi_0}$

$$\frac{1}{\varphi_0}\int d\mathrm{V}E_a^k D_k\varphi^a = \int \mathbf{E}d\mathbf{S} = \frac{q}{\varepsilon_0}, \tag{9.166}$$

and the Gauss law for magnetism in the case of the Abelianized color magnetic field induction $\mathbf{B} = \mathbf{B}^a\dfrac{\varphi_a}{\varphi_0}$

$$\frac{1}{\varphi_0}\int d\mathrm{V}B_a^k D_k\varphi^a = \int \mathbf{B}d\mathbf{S} = \mu_0 q_M, \tag{9.167}$$

together with the simple identity $2\Phi_0 ec = E_P\ell_P$, one receives

$$\mathscr{E}_{YMH} = E_P\ell_P\varphi_0\sqrt{4\pi\alpha}\left(\frac{q}{e}\sin\alpha_c + \frac{q_M}{ec}\cos\alpha_c\right) + E_P\ell_P\frac{\lambda}{4}\int dV\left(\varphi^a\varphi_a - \varphi_0^2\right)^2. \tag{9.168}$$

In this manner, taking into account the Einstein mass-energy relation $\mathcal{E}_{YMH} = Mc^2$, where M is the mass of a charged object, in general one obtains the Bogomol'nyi bound

$$M \geqslant M_P \ell_P \varphi_0 \sqrt{4\pi\alpha} \left(\frac{q}{e} \sin\alpha_c + \frac{q_M}{ec} \cos\alpha_c \right), \tag{9.169}$$

where the equality holds in the Bogomol'nyi–Prasad–Sommerfield limit. In the case of 't Hooft–Polyakov monopole in the Bogomol'nyi–Prasad–Sommerfield limit one has

$$M = M_P \ell_P \varphi_0 \sqrt{4\pi\alpha} \frac{q_M}{ec}, \tag{9.170}$$

and the total energy density of such a solution is

$$\mathcal{H} = \frac{4\pi\alpha}{\mu_0} (2\Phi_0)^2 D_k \varphi^a D^k \varphi_a = \frac{1}{\cos^2\alpha_c} \frac{\mathbf{B}^2}{\mu_0} = \frac{\varepsilon_0}{\sin^2\alpha_c} \mathbf{E}^2. \tag{9.171}$$

Note that in such a specific situation the Bogomol'nyi equations (9.164) and (9.165) become

$$E_k^a = 0, \tag{9.172}$$

$$B_a^k = 2\Phi_0 \sqrt{4\pi\alpha} D^k \varphi_a, \tag{9.173}$$

and the second equation after taking into account the relations (9.121)-(9.122) leads to the pair of the first order ordinary differential equations

$$\xi \frac{dH_\alpha}{d\xi} = H_\alpha + 1 - K^2, \tag{9.174}$$

$$\xi \frac{dK}{d\xi} = -K H_\alpha, \tag{9.175}$$

where $H_\alpha = \sqrt{4\pi\alpha} H$, which can be solved straightforwardly

$$K = \frac{\xi}{\sinh\xi}, \tag{9.176}$$

$$H = \frac{1}{\sqrt{4\pi\alpha}} (\xi \coth\xi - 1). \tag{9.177}$$

Let us consider the non-Abelianized magnetic charge (9.99) for the solution (9.176). It can be seen by straightforward easy derivation that in such a situation

$$q_M = \frac{8\pi}{\mu_0 g}\left[\frac{\xi^2(1+\cosh\xi)}{\sinh^2\xi} - 2\frac{\xi}{\sinh\xi}\right]. \tag{9.178}$$

Such a magnetic charge has several interesting features. First of all it is quickly vanishing in the limit $\xi \to 0$ and is identically equal to zero for $\xi = 0$. The maximum of the magnetic charge (9.178) is obtained for $\xi = 0.397432$ and has the value

$$q_M^{max} = \frac{8\pi}{\mu_0 g}2.77207. \tag{9.179}$$

Moreover, the magnetic charge has also the stationary point, obtained at the point $\xi = 0.235545$, which has the value

$$q_M^{st} = \frac{8\pi}{\mu_0 g}1.15827. \tag{9.180}$$

D1 Generalization of Bogomol'nyi Equations

Let us generalize the Bogomol'nyi–Prasad–Sommerfield solution to the case of the non-zero value of λ and, in fact, to more general forms of the scalar field potential. Let us consider the purely electromagnetic energy

$$\begin{aligned}\mathscr{E}_{EM} = \int d\mathrm{V}\Bigg\{ &\frac{\varepsilon_0}{2}\left[E_k^a - c\sqrt{\mu_0 E_P\ell_P}\alpha_1 D_k\varphi^a - c\sqrt{\mu_0 V}\beta_1\left(\delta_1\delta_k^a + \delta_2\frac{\varphi^a\varphi_k}{|\varphi|^2}\right)\right] \\ &\times\left[E_a^k - c\sqrt{\mu_0 E_P\ell_P}\alpha_1 D^k\varphi_a - c\sqrt{\mu_0 V}\beta_1\left(\delta_1\delta_a^k + \delta_2\frac{\varphi_a\varphi^k}{|\varphi|^2}\right)\right] \\ &+\frac{1}{2\mu_0}\left[B_k^a - \sqrt{\mu_0 E_P\ell_P}\alpha_2 D_k\varphi^a - \sqrt{\mu_0 V}\beta_2\left(\delta_1'\delta_k^a + \delta_2'\frac{\varphi^a\varphi_k}{|\varphi|^2}\right)\right] \\ &\times\left[B_a^k - \sqrt{\mu_0 E_P\ell_P}\alpha_2 D^k\varphi_a - \sqrt{\mu_0 V}\beta_2\left(\delta_1'\delta_a^k + \delta_2'\frac{\varphi_a\varphi^k}{|\varphi|^2}\right)\right]\Bigg\},\end{aligned} \tag{9.181}$$

where $V = E_P\ell_P\frac{\lambda}{4}\left(\varphi^a\varphi_a - \varphi_0^2\right)^2$ is the Higgs potential, and α_1, α_2, β_1, β_2, δ_1, δ_2, δ_1', δ_2' are dimensionless constant coefficients. It can be seen by straightforward analysis that the energy (9.161) of the Yang–Mills–Higgs system can be rewritten in the form

$$\mathscr{E}_{YMH} = \mathscr{E}_{EM} + \sqrt{\frac{E_P\ell_P}{\mu_0}} \int d\mathrm{V} \left(\alpha_1 \frac{E^a_k}{c} + \alpha_2 B^a_k\right) D^k \varphi_a$$
$$+ \sqrt{\frac{1}{\mu_0}} \int d\mathrm{V} \left[\beta_1 \left(\delta_1 \delta^k_a + \delta_2 \frac{\varphi_a \varphi^k}{|\varphi|^2}\right) \frac{E^a_k}{c} + \beta_2 \left(\delta'_1 \delta^k_a + \delta'_2 \frac{\varphi_a \varphi^k}{|\varphi|^2}\right) B^a_k\right] \sqrt{V}$$
$$- \sqrt{E_P\ell_P} \int d\mathrm{V} \left[\alpha_1 \beta_1 \left(\delta_1 \delta^k_a + \delta_2 \frac{\varphi_a \varphi^k}{|\varphi|^2}\right) + \alpha_2 \beta_2 \left(\delta'_1 \delta^k_a + \delta'_2 \frac{\varphi_a \varphi^k}{|\varphi|^2}\right)\right] \sqrt{V} D_k \varphi^a, \tag{9.182}$$

where the following consistency conditions hold

$$\alpha_1^2 + \alpha_2^2 = 1, \tag{9.183}$$

$$\beta_1^2 \left(\delta_1 \delta^k_a + \delta_2 \frac{\varphi_a \varphi^k}{|\varphi|^2}\right) \left(\delta'_1 \delta^a_k + \delta'_2 \frac{\varphi^a \varphi_k}{|\varphi|^2}\right)$$
$$+ \beta_2^2 \left(\delta'_1 \delta^k_a + \delta'_2 \frac{\varphi_a \varphi^k}{|\varphi|^2}\right) \left(\delta_1 \delta^a_k + \delta_2 \frac{\varphi^a \varphi_k}{|\varphi|^2}\right) = 1. \tag{9.184}$$

Applying the Gauss laws for electricity (9.166) and magnetism (9.167), one receives

$$\sqrt{\frac{E_P\ell_P}{\mu_0}} \int d\mathrm{V} \left(\alpha_1 \frac{E^a_k}{c} + \alpha_2 B^a_k\right) D^k \varphi_a = E_P\ell_P \varphi_0 \sqrt{4\pi\alpha} \left(\alpha_1 \frac{q}{e} + \alpha_2 \frac{q_M}{ec}\right). \tag{9.185}$$

The problem is an appropriate understanding of the last two integrals in the energy (9.182). This part of the energy can be rewritten in the following form

$$\sqrt{\frac{1}{\mu_0}} \int d\mathrm{V} \left[\frac{1}{c} \left(E^a_k - c\sqrt{\mu_0 E_P\ell_P} \alpha_1 D_k \varphi^a\right) \beta_1 \left(\delta_1 \delta^k_a + \delta_2 \frac{\varphi_a \varphi^k}{|\varphi|^2}\right)\right.$$
$$\left. + \left(B^a_k - \sqrt{\mu_0 E_P\ell_P} \alpha_2 D_k \varphi^a\right) \beta_2 \left(\delta'_1 \delta^k_a + \delta'_2 \frac{\varphi_a \varphi^k}{|\varphi|^2}\right)\right] \sqrt{V}, \tag{9.186}$$

or, equivalently,

$$\sqrt{\frac{1}{\mu_0}}\int dV\Bigg[\beta_1\left(\frac{E_k^a}{c}-\sqrt{\mu_0 E_P\ell_P}\alpha_1 D_k\varphi^a-\sqrt{\mu_0 V}\beta_1\left(\delta_1\delta_k^a+\delta_2\frac{\varphi^a\varphi_k}{|\varphi|^2}\right)\right)$$
$$\times\left(\delta_1\delta_a^k+\delta_2\frac{\varphi_a\varphi^k}{|\varphi|^2}\right)$$
$$+\beta_2\left(B_k^a-\sqrt{\mu_0 E_P\ell_P}\alpha_2 D_k\varphi^a-\sqrt{\mu_0 V}\beta_2\left(\delta_1'\delta_k^a+\delta_2'\frac{\varphi^a\varphi_k}{|\varphi|^2}\right)\right)$$
$$\times\left(\delta_1'\delta_a^k+\delta_2'\frac{\varphi_a\varphi^k}{|\varphi|^2}\right)\Bigg]\sqrt{V}+$$
$$+\int dV\Bigg[\beta_1^2\left(\delta_1\delta_k^a+\delta_2\frac{\varphi^a\varphi_k}{|\varphi|^2}\right)\left(\delta_1\delta_a^k+\delta_2\frac{\varphi_a\varphi^k}{|\varphi|^2}\right)+$$
$$+\beta_2^2\left(\delta_1'\delta_k^a+\delta_2'\frac{\varphi^a\varphi_k}{|\varphi|^2}\right)\left(\delta_1'\delta_a^k+\delta_2'\frac{\varphi_a\varphi^k}{|\varphi|^2}\right)\Bigg]V. \tag{9.187}$$

In the light of the consistency condition (9.184), one has

$$\int dV\Bigg[\beta_1^2\left(\delta_1\delta_k^a+\delta_2\frac{\varphi^a\varphi_k}{|\varphi|^2}\right)\left(\delta_1\delta_a^k+\delta_2\frac{\varphi_a\varphi^k}{|\varphi|^2}\right)+$$
$$+\beta_2^2\left(\delta_1'\delta_k^a+\delta_2'\frac{\varphi^a\varphi_k}{|\varphi|^2}\right)\left(\delta_1'\delta_a^k+\delta_2'\frac{\varphi_a\varphi^k}{|\varphi|^2}\right)\Bigg]V=\int dVV, \tag{9.188}$$

and, in this manner, the energy (9.182) takes the form

$$\mathscr{E}_{YMH}=\mathscr{E}_{EM}+E_P\ell_P\varphi_0\sqrt{4\pi\alpha}\left(\alpha_1\frac{q}{e}+\alpha_2\frac{q_M}{ec}\right)$$
$$+\sqrt{\frac{1}{\mu_0}}\int dV\Bigg[\beta_1\left(\frac{E_k^a}{c}-\sqrt{\mu_0 E_P\ell_P}\alpha_1 D_k\varphi^a-\sqrt{\mu_0 V}\beta_1\left(\delta_1\delta_k^a+\delta_2\frac{\varphi^a\varphi_k}{|\varphi|^2}\right)\right)$$
$$\times\left(\delta_1\delta_a^k+\delta_2\frac{\varphi_a\varphi^k}{|\varphi|^2}\right)$$
$$+\beta_2\left(B_k^a-\sqrt{\mu_0 E_P\ell_P}\alpha_2 D_k\varphi^a-\sqrt{\mu_0 V}\beta_2\left(\delta_1'\delta_k^a+\delta_2'\frac{\varphi^a\varphi_k}{|\varphi|^2}\right)\right)$$
$$\times\left(\delta_1'\delta_a^k+\delta_2'\frac{\varphi_a\varphi^k}{|\varphi|^2}\right)\Bigg]\sqrt{V}+\int dVV. \tag{9.189}$$

Let us consider now the minimum of this energy which outside the Higgs vacuum case $\varphi = \pm\varphi_0$, which corresponds also to the Bogomol'nyi–Prasad–Sommerfield limit, is satisfied for

$$E_k^a = 2c\Phi_0\sqrt{4\pi\alpha}\left[\alpha_1 D_k\varphi^a + \sqrt{\frac{V}{E_P\ell_P}}\beta_1\left(\delta_1\delta_k^a + \delta_2\frac{\varphi^a\varphi_k}{|\varphi|^2}\right)\right], \quad (9.190)$$

$$B_k^a = 2\Phi_0\sqrt{4\pi\alpha}\left[\alpha_2 D_k\varphi^a + \sqrt{\frac{V}{E_P\ell_P}}\beta_2\left(\delta_1'\delta_k^a + \delta_2'\frac{\varphi^a\varphi_k}{|\varphi|^2}\right)\right], \quad (9.191)$$

and because in such a situation $\mathscr{E}_{EM} = 0$ one receives

$$\mathscr{E}_{YMH}^0 = E_P\ell_P\varphi_0\sqrt{4\pi\alpha}\left(\alpha_1\frac{q}{e} + \alpha_2\frac{q_M}{ec}\right) + \int d\mathrm{V}V. \quad (9.192)$$

In such a situation, the mass $M = \frac{1}{c^2}\mathscr{E}_{YMH}^0$ of the electromagnetic configuration is

$$M = M_P\ell_P\varphi_0\sqrt{4\pi\alpha}\left(\alpha_1\frac{q}{e} + \alpha_2\frac{q_M}{ec}\right) + \frac{1}{c^2}\int d\mathrm{V}V. \quad (9.193)$$

It is visible now that, in general, such mass can be identically vanishing. Such a situation exists in the case

$$\int d\mathrm{V}V = -E_P\ell_P\varphi_0\sqrt{4\pi\alpha}\left(\alpha_1\frac{q}{e} + \alpha_2\frac{q_M}{ec}\right). \quad (9.194)$$

Interestingly, when the relation holds

$$\frac{\alpha_1}{\alpha_2} = -\frac{q_M}{qc}, \quad (9.195)$$

then one has the condition

$$\int d\mathrm{V}V = 0, \quad (9.196)$$

which includes, as the particular situation, the Bogomol'nyi–Prasad–Sommerfield limit. In general, however, in such a situation the scalar field potential V must not be identically vanishing – only its volume integral vanishes identically. Similarly, in the general situation (9.193) the particular case

$$M = M_P\ell_P\varphi_0\sqrt{4\pi\alpha}\left(\alpha_1\frac{q}{e} + \alpha_2\frac{q_M}{ec}\right), \quad (9.197)$$

is true for all scalar fields potentials satisfying the condition (9.196), which includes the Bogomol'nyi–Prasad–Sommerfield limit. In a general situation, the bound

$$M \geqslant M_P \ell_P \varphi_0 \sqrt{4\pi\alpha}\left(\alpha_1 \frac{q}{e} + \alpha_2 \frac{q_M}{ec}\right), \tag{9.198}$$

which in the case $\alpha_1 = \sin\alpha_c$, $\alpha_2 = \cos\alpha_c$ corresponds to the Bogomol'nyi bound, is true if and only if

$$\int d\mathrm{V}V \geqslant 0. \tag{9.199}$$

In the case of the negative value of the volume integral

$$\int d\mathrm{V}V < 0. \tag{9.200}$$

one has another bound

$$M \leqslant M_P \ell_P \varphi_0 \sqrt{4\pi\alpha}\left(\alpha_1 \frac{q}{e} + \alpha_2 \frac{q_M}{ec}\right). \tag{9.201}$$

Let us focus our attention on the consistency conditions (9.183) and (9.184). The condition (9.183) can be solved straightforwardly. There are eight inequivalent types of solutions which can be parameterized by the trigonometric form

$$\alpha_1 = \pm\sin\alpha_c \qquad \alpha_2 = \pm\cos\alpha_c, \tag{9.202}$$

$$\alpha_1 = \pm\cos\alpha_c \qquad \alpha_2 = \pm\sin\alpha_c. \tag{9.203}$$

The second condition (9.184) is much more difficult to analyze. This relation can be presented in the form

$$\beta_1^2\left(D\delta_1^2 + \delta_2^2 + 2\delta_1\delta_2\right) + \beta_2^2\left(D\delta_1'^2 + \delta_2'^2 + 2\delta_1'\delta_2'\right) = 1, \tag{9.204}$$

where $D = \delta_k^a \delta_a^k$ is the dimension of space. It is easy to see that this relation can be parameterized as the equation

$$A^2 + B^2 = 1, \tag{9.205}$$

which has also eight trigonometric type solutions

$$A = \pm\sin\beta_c \qquad B = \pm\cos\beta_c, \tag{9.206}$$

$$A = \pm\cos\beta_c \qquad B = \pm\sin\beta_c, \tag{9.207}$$

where β_c is a certain new angle variable which must not be related to α_c in general. In such instance one obtains the relations

$$\delta_2 = \delta_1\left(-1\pm\sqrt{\frac{A^2}{\beta_1^2\delta_1^2}+1-D}\right), \tag{9.208}$$

$$\delta_2' = \delta_1'\left(-1\pm\sqrt{\frac{B^2}{\beta_2^2\delta_1'^2}+1-D}\right), \tag{9.209}$$

or, equivalently,

$$\delta_1 = \frac{\delta_2}{D}\left(-1\pm\sqrt{\frac{A^2}{\beta_1^2\delta_2^2}+1-D}\right), \tag{9.210}$$

$$\delta_1' = \frac{\delta_2'}{D}\left(-1\pm\sqrt{\frac{B^2}{\beta_2^2\delta_2'^2}+1-D}\right). \tag{9.211}$$

These solutions are real if and only if

$$\beta_1^2\delta_1^2 \geqslant \frac{A^2}{D-1}, \tag{9.212}$$

$$\beta_2^2\delta_1'^2 \geqslant \frac{B^2}{D-1}, \tag{9.213}$$

$$\beta_1^2\delta_2^2 \geqslant \frac{A^2}{D-1}, \tag{9.214}$$

$$\beta_2^2\delta_2'^2 \geqslant \frac{B^2}{D-1}. \tag{9.215}$$

The 't Hooft–Polyakov ansatz supplemented by the Julia–Zee ansatz

$$\mathbf{A}_\mu^a = \left[2\Phi_0\varphi_0\frac{J(\xi)}{\xi}\frac{r^a}{r}, 2\Phi_0\varphi_0\varepsilon_{ik}^a\frac{1-K(\xi)}{\xi}\frac{r^k}{r}\right], \tag{9.216}$$

leads to the relations

$$E_k^a = c\frac{(2\Phi_0\varphi_0)^2 g}{\xi^2}\left[\left(\xi\frac{dJ}{d\xi}-J(K+1)\right)\frac{r^a r_n}{r^2}+JK\delta_n^a\right], \tag{9.217}$$

$$B_k^a = \frac{(2\Phi_0\varphi_0)^2 g}{\xi^2}\left[\left(1-K^2+\xi\frac{dK}{d\xi}\right)\frac{r^a r_n}{r^2}-\xi\frac{dK}{d\xi}\delta_n^a\right], \tag{9.218}$$

$$D_i\varphi^a = \varphi_0\frac{(2\Phi_0\varphi_0)g}{\xi^2}\left[\left(\xi\frac{dH}{d\xi}-KH-H\right)\frac{r^a r_i}{r^2}+\delta_i^a KH\right], \tag{9.219}$$

which, together with the equations (9.190) and (9.191), lead to the following system of equations

$$\xi\frac{dJ}{d\xi}-J(K+1) = \alpha_1\sqrt{4\pi\alpha}\left(\xi\frac{dH}{d\xi}-H(K+1)\right)+\frac{\sqrt{4\pi\alpha}}{(2\Phi_0 g)\varphi_0^2}\sqrt{\frac{\mathrm{V}}{E_P\ell_P}}\beta_1\delta_2\xi^2, \tag{9.220}$$

$$-JK = \alpha_1\sqrt{4\pi\alpha}KH+\frac{\sqrt{4\pi\alpha}}{(2\Phi_0 g)\varphi_0^2}\sqrt{\frac{\mathrm{V}}{E_P\ell_P}}\beta_1\delta_1\xi^2, \tag{9.221}$$

$$\xi\frac{dK}{d\xi}-K^2+1 = \alpha_2\sqrt{4\pi\alpha}\left(\xi\frac{dH}{d\xi}-H(K+1)\right)+\frac{\sqrt{4\pi\alpha}}{(2\Phi_0 g)\varphi_0^2}\sqrt{\frac{\mathrm{V}}{E_P\ell_P}}\beta_2\delta_2'\xi^2, \tag{9.222}$$

$$-\xi\frac{dK}{d\xi} = \alpha_2\sqrt{4\pi\alpha}KH+\frac{\sqrt{4\pi\alpha}}{(2\Phi_0 g)\varphi_0^2}\sqrt{\frac{\mathrm{V}}{E_P\ell_P}}\beta_2\delta_1'\xi^2. \tag{9.223}$$

which describes a dyon in general. The case of $J=0$ corresponds to a magnetic monopole. In such a situation, both the equations (9.222) and (9.223) remain unchanged, whereas the equations (9.220) and (9.221) are simplified to

$$\alpha_2\sqrt{4\pi\alpha}\left(\xi\frac{dH}{d\xi}-H(K+1)\right) = -\frac{\sqrt{4\pi\alpha}}{(2\Phi_0 g)\varphi_0^2}\sqrt{\frac{\mathrm{V}}{E_P\ell_P}}\frac{\alpha_2}{\alpha_1}\beta_1\delta_2\xi^2, \tag{9.224}$$

$$\alpha_2\sqrt{4\pi\alpha}KH = -\frac{\sqrt{4\pi\alpha}}{(2\Phi_0 g)\varphi_0^2}\sqrt{\frac{\mathrm{V}}{E_P\ell_P}}\frac{\alpha_2}{\alpha_1}\beta_1\delta_1\xi^2. \tag{9.225}$$

Taking into account the equations (9.224) and (9.225), one can transform the equations (9.222) and (9.223) to the system

$$\xi\frac{dK}{d\xi}-K^2+1 = \frac{\sqrt{4\pi\alpha}}{(2\Phi_0 g)\varphi_0^2}\sqrt{\frac{\mathrm{V}}{E_P\ell_P}}\left(\beta_2\delta_2'-\frac{\alpha_2}{\alpha_1}\beta_1\delta_2\right)\xi^2, \tag{9.226}$$

$$\xi\frac{dK}{d\xi} = \frac{\sqrt{4\pi\alpha}}{(2\Phi_0 g)\varphi_0^2}\sqrt{\frac{\mathrm{V}}{E_P\ell_P}}\left(\frac{\alpha_2}{\alpha_1}\beta_1\delta_1-\beta_2\delta_1'\right)\xi^2, \tag{9.227}$$

which after minor algebraic manipulations can be solved

$$K=\pm\sqrt{1+\frac{\sqrt{4\pi\alpha}}{(2\Phi_0 g)\varphi_0^2}\sqrt{\frac{\mathrm{V}}{E_P\ell_P}}\left(\frac{\alpha_2}{\alpha_1}\beta_1(\delta_2+\delta_1)-\beta_2(\delta_1'+\delta_2')\right)\xi^2}. \tag{9.228}$$

On the one side, one can use the equation (9.225) to reduce the equation (9.224)

$$\xi\frac{dH}{d\xi}-H=-\frac{1}{(2\Phi_0 g)\varphi_0^2}\sqrt{\frac{\mathrm{V}}{E_P\ell_P}}\frac{\beta_1}{\alpha_1}(\delta_2-\delta_1)\xi^2. \tag{9.229}$$

On the other hand, taking into account the equation (9.225) together with the solution (9.228), one receives the explicit solution for H

$$H=\frac{\mp\dfrac{1}{\alpha_2(2\Phi_0 g)\varphi_0^2}\sqrt{\dfrac{\mathrm{V}}{E_P\ell_P}}\dfrac{\alpha_2}{\alpha_1}\beta_1\delta_1\xi^2}{\sqrt{1+\dfrac{\sqrt{4\pi\alpha}}{(2\Phi_0 g)\varphi_0^2}\sqrt{\dfrac{\mathrm{V}}{E_P\ell_P}}\left(\dfrac{\alpha_2}{\alpha_1}\beta_1(\delta_2+\delta_1)-\beta_2(\delta_1'+\delta_2')\right)\xi^2}}. \tag{9.230}$$

In this manner, the solution can be presented as follows: the Higgs field is

$$\varphi^a=\frac{\mp\dfrac{1}{\alpha_2(2\Phi_0 g)\varphi_0}\sqrt{\dfrac{\mathrm{V}}{E_P\ell_P}}\dfrac{\alpha_2}{\alpha_1}\beta_1\delta_1\xi}{\sqrt{1+\dfrac{\sqrt{4\pi\alpha}}{(2\Phi_0 g)\varphi_0^2}\sqrt{\dfrac{\mathrm{V}}{E_P\ell_P}}\left(\dfrac{\alpha_2}{\alpha_1}\beta_1(\delta_2+\delta_1)-\beta_2(\delta_1'+\delta_2')\right)\xi^2}}\frac{r^a}{r}, \tag{9.231}$$

while the gauge field is

$$\mathbf{A}_\mu^a=\left[0,\frac{2\Phi_0\varphi_0}{\xi}\left(1\mp\sqrt{1+\frac{G}{\varphi_0^2}\sqrt{\frac{\mathrm{V}}{E_P\ell_P}}\xi^2}\right)\varepsilon_{ik}^a\frac{r^k}{r}\right], \tag{9.232}$$

where we have introduced the constant

$$G=\frac{\sqrt{4\pi\alpha}}{2\Phi_0 g}\left(\frac{\alpha_2}{\alpha_1}\beta_1(\delta_2+\delta_1)-\beta_2(\delta_1'+\delta_2')\right). \tag{9.233}$$

Now one can apply the solution (9.230) within the equation (9.229) and, consequently, one receives the differential equation for the scalar field potential V

$$\frac{d\mathrm{V}}{d\xi}=\frac{-4\pm\dfrac{\delta_2-\delta_1}{\delta_1}\left(1+\dfrac{G}{\varphi_0^2}\sqrt{\dfrac{\mathrm{V}}{E_P\ell_P}}\xi^2\right)^{3/2}}{2+\dfrac{G}{\varphi_0^2}\sqrt{\dfrac{\mathrm{V}}{E_P\ell_P}}\xi^2}\frac{\mathrm{V}}{\xi}, \tag{9.234}$$

Application of the Bogomol'nyi–Prasad–Sommerfield limit, which is equivalent to the limit $\mathrm{V} \to 0$, to the case of the magnetic monopole gives the solutions $H=0$ and $K=\pm 1$. Consequently, in such a situation the Higgs field identically vanishes $\varphi^a=0$ and for the case $K=1$ one obtains trivial gauge field $\mathbf{A}^a_\mu=0$, while for the case $K=-1$ the solution is non-trivial and similar to the Dirac monopole

$$\mathbf{A}^a_\mu = \left[0, \frac{4\Phi_0\varphi_0}{\xi}\varepsilon^a_{ik}\frac{r^k}{r}\right], \tag{9.235}$$

In such a situation, the equation (9.234) gives the identity $0=0$ and, consequently, the scalar field potential V is arbitrary.

Let us consider some particular case of the equation for the scalar field potential (9.234), which is given by identically vanishing constant G. Such a situation corresponds to the specific constraint between the constants

$$\frac{\beta_1}{\alpha_1}(\delta_2+\delta_1) = \frac{\beta_2}{\alpha_2}(\delta_1'+\delta_2'). \tag{9.236}$$

Then the equation (9.234) is simplified to

$$\frac{d\mathrm{V}}{\mathrm{V}} = \left(2 \pm \frac{\delta_2-\delta_1}{2\delta_1}\right)\frac{d\xi}{\xi}, \tag{9.237}$$

and can be solved by straightforward integration

$$\mathrm{V}_\pm(\xi) = \mathrm{V}(\xi_0)\left(2 \pm \frac{\delta_2-\delta_1}{2\delta_1}\right)\frac{\xi}{\xi_0}, \tag{9.238}$$

where ξ_0 and $\mathrm{V}(\xi_0)$ are the integration constants. For simplicity, one can consider the following choice

$$\xi_{0\pm} = 2 \pm \frac{\delta_2-\delta_1}{2\delta_1}, \tag{9.239}$$

and in such a situation the scalar field potential is $\mathrm{V}_\pm(\xi) = \mathrm{V}(\xi_{0\pm})\xi$. The reader interested in more detailed analysis and advances in magnetic monopoles is kindly advised to look through the books in Ref. [480], the presentation in the papers [481].

Part III

Quantum Gravity

Chapter 10

Review of Quantum General Relativity

In this chapter we shall present certain standard strategy having the fundamental status for quantization of General Relativity. Namely, these are the $3+1$ Arnowitt–Deser–Misner Hamiltonian formulation of General Relativity and the Dirac method of canonical primary quantization which lead to the Wheeler–DeWitt equation and the concept of the Wheeler superspace.

A Arnowitt–Deser–Misner Decomposition

Let us take into account the four-dimensional pseudo-Riemannian manifold (M,g) (For detailed differential geometric discussion Cf. *e.g.* Refs. [161, 162, 548, 549]) equipped with metric tensor $g_{\mu\nu}$ of signature $(1,3)$, the Christoffel symbols $\Gamma^{\rho}_{\mu\nu}$, the Riemann–Christoffel curvature tensor $R^{\lambda}_{\mu\alpha\nu}$, the Ricci curvature tensor $R_{\mu\nu}$, and the Ricci scalar curvature ${}^{(4)}R$

$$\Gamma^{\rho}_{\mu\nu} = \frac{1}{2}g^{\rho\sigma}\left(g_{\mu\sigma,\nu}+g_{\sigma\nu,\mu}-g_{\mu\nu,\sigma}\right), \tag{10.1}$$

$$R^{\lambda}_{\mu\alpha\nu} = \Gamma^{\lambda}_{\mu\nu,\alpha}-\Gamma^{\lambda}_{\mu\alpha,\nu}+\Gamma^{\lambda}_{\sigma\alpha}\Gamma^{\sigma}_{\mu\nu}-\Gamma^{\lambda}_{\sigma\nu}\Gamma^{\sigma}_{\mu\alpha}, \tag{10.2}$$

$$R_{\mu\nu} = R^{\lambda}_{\mu\lambda\nu}=\Gamma^{\lambda}_{\mu\nu,\lambda}-\Gamma^{\lambda}_{\mu\lambda,\nu}+\Gamma^{\lambda}_{\sigma\lambda}\Gamma^{\sigma}_{\mu\nu}-\Gamma^{\lambda}_{\sigma\nu}\Gamma^{\sigma}_{\mu\lambda}, \tag{10.3}$$

$${}^{(4)}R = g^{\mu\nu}R_{\mu\nu}, \tag{10.4}$$

where a holonomic basis [32] was chosen. In General Relativity (For much more detailed books in its basics and applications Cf. *e.g.* Refs. [32, 171, 173]) the manifold M is identified with space-time, and presence of Matter fields reflected by non-zero stress-energy tensor[1] $T_{\mu\nu}$ is then studied. In such a situation, the Einstein tensor

$$G_{\mu\nu} = R_{\mu\nu} - \frac{1}{2}g_{\mu\nu}{}^{(4)}R, \tag{10.5}$$

allows to construct the Einstein field equations

$$G_{\mu\nu} + \Lambda g_{\mu\nu} = \kappa T_{\mu\nu}, \tag{10.6}$$

where $\kappa = \dfrac{8\pi G}{c^4} \approx 2.076 \cdot 10^{-43}\ \mathrm{N}^{-1}$ is the Einstein constant, and Λ is the cosmological constant.

To construct the Hamilton formulation of General Relativity, it is necessary to foliate a space-time manifold M with a family of space-like hypersurfaces, called also *slices*. In general, such procedure is possible when M is a globally hyperbolic, *i.e.* pseudo-Riemannian, manifold which is the usual situation in General Relativity. Let $t(x^\mu)$ be a scalar field, an arbitrary single-valued function of coordinates x^μ, such that the foliation $t = const$ corresponds with a family of nonintersecting space-like hypersurfaces $\Sigma(t)$, also known as constant-time hypersurfaces. Let us denote by y^i the coordinates on all hypersurfaces $\Sigma(t)$. Let us choose a concrete hypersurface Σ defined by a parametric equations $x^\mu = x^\mu(y^i)$, where the superscript $i = 1, \ldots, 3$ labels coordinates intrinsic to Σ. Equivalently, hypersurface Σ can be selected by any restriction in the form $f(x^\mu) = 0$. In such a situation, $\partial_\mu f(x^\mu)$ is a normal to Σ which, if is not null, allows to define the unit normal vector field to Σ as $n^\mu n_\mu = -1$. Then the normal vector field is given by the formula

$$n_\mu = -\frac{\partial_\mu f}{\sqrt{|\partial_\mu f \partial^\mu f|}}, \quad n^\mu \partial_\mu f > 0. \tag{10.7}$$

In other words, $\Sigma(t)$ are such that the unit normal to the hypersurfaces can be chosen to be future-directed time-like vector field $n_\mu \sim \partial_\mu t$ satisfying the condition $n^\mu n_\mu = -1$.

[1]Some authors call $T_{\mu\nu}$ the energy-momentum tensor.

Let γ be a congruence of curves intersecting the space-like hypersurfaces $\Sigma(t)$ which, in general, are not geodesics nor orthogonal to $\Sigma(t)$. Let t be an affine parameter along the congruence γ, and let us denote by t^μ the tangent vector to γ. Then the following relation is satisfied

$$t^\mu \partial_\mu t = 1. \tag{10.8}$$

An arbitrary fixed curve γ_F is a mapping between points on all hypersurfaces $\Sigma(t)$

$$\gamma_F : P \in \Sigma(t) \mapsto P' \in \Sigma(t') \mapsto P'' \in \Sigma(t'') \mapsto \ldots P^{(n)} \in \Sigma(t^{(n)}), \tag{10.9}$$

where the index n is an integer, and fixing the coordinates on arbitrary two hypersurfaces leads to constant coordinates y^i for an arbitrary value of n. In this manner, the coordinate system (t, y^i) in M is established. Assuming a transformation between this coordinate system and another coordinate system x^μ: $x^\mu = x^\mu(t, y^i)$ one can determine the tangent vector to the congruence γ

$$t^\mu = (\partial_t x^\mu)_{y^i} = \delta_t^\mu, \text{ in } (t, y^i) \tag{10.10}$$

as well as the tangent vectors on hypersurfaces $\Sigma(t)$

$$e_i^\mu = (\partial_{y^i} x^\mu)_t = \delta_i^\mu, \text{ in } (t, y^i). \tag{10.11}$$

In any coordinates the relation is satisfied

$$\mathcal{L}_t e_i^\mu = 0. \tag{10.12}$$

Let us use the unit normal vector field to the hypersurfaces in the form

$$n_\mu = -N\partial_\mu t, \tag{10.13}$$
$$n_\mu e_i^\mu = 0, \tag{10.14}$$

where N is called the lapse scalar, which is a function normalizing the vector field n_μ. In general, $t^\mu \nparallel n^\mu$ and, therefore, the tangent vector t^μ can be decomposed in the basis (n^μ, e_i^μ)

$$t^\mu = N n^\mu + N^i e_i^\mu, \tag{10.15}$$

where N^i is a three-vector valued function called the shift vector. The coordinate transformation $x^\mu = x^\mu(t, y^i)$ allows to write in (t, y^i)

$$\begin{aligned} dx^\mu &= t^\mu dt + e_i^\mu dy^i = (Nn^\mu + N^i e_i^\mu)dt + e_i^\mu dy^i \\ &= (Ndt)n^\mu + (dy^i + N^i dt)e_i^\mu, \end{aligned} \tag{10.16}$$

and hence one can evaluate the space-time interval

$$\begin{aligned} ds^2 &= g_{\mu\nu}dx^\mu dx^\nu = dx^\mu dx_\mu \\ &= [(Ndt)n^\mu + (dy^i + N^i dt)e_i^\mu][(Ndt)n_\mu + (dy^j + N^j dt)e_{\mu j}] \\ &= (n^\mu n_\mu)N^2 dt^2 + (dy^i + N^i dt)(dy^j + N^j dt)e_i^\mu e_{\mu j} \\ &= -N^2 dt^2 + h_{ij}(dy^i + N^i dt)(dy^j + N^j dt) \\ &= -\left(N^2 - N_i N^i\right) dt^2 + N_i dx^i dt + N_j dx^j dt + h_{ij} dx^i dx^j, \end{aligned} \tag{10.17}$$

where h_{ij} is induced metric on $\Sigma(t)$

$$h_{ij} = g_{\mu\nu} e_i^\mu e_j^\nu, \tag{10.18}$$

which actually expresses the Pythagoras theorem between two points lying on two distinguishable constant time hypersurfaces and was investigated in pioneering works of Richard Lewis Arnowitt, Stanley Deser, and Charles William Misner [170]. For this reason, space-time metric tensor $g_{\mu\nu}$ of an arbitrary Lorentzian manifold satisfying the Einstein field equations (10.6) decomposes into space and time as follows

$$g_{\mu\nu} = \begin{bmatrix} -N^2 + N_i N^i & N_j \\ N_i & h_{ij} \end{bmatrix}, \tag{10.19}$$

where $N^j = h^{ij} N_i$ is the contravariant shift vector, and the spatial metric satisfies the orthogonality condition

$$h_{ik} h^{kj} = \delta_i^j. \tag{10.20}$$

Completeness relations for the metric are

$$g_{\mu\nu} = -n_\mu n_\nu + h_{ij} e_\mu^i e_\nu^j. \tag{10.21}$$

It can be verified straightforwardly that

$$\sqrt{-g} = N\sqrt{h}, \tag{10.22}$$

where $g = \det g_{\mu\nu}$ and $h \det h_{ij}$, while the inverted metric has the form

$$g^{\mu\nu} = \begin{bmatrix} -\frac{1}{N^2} & \frac{N^j}{N^2} \\ \frac{N^i}{N^2} & h^{ij} - \frac{N^i N^j}{N^2} \end{bmatrix}. \tag{10.23}$$

Completeness relations for the inverse metric are

$$g^{\mu\nu} = -n^\mu n^\nu + h^{ij} e_i^\mu e_j^\nu. \tag{10.24}$$

The second fundamental form of a slice is called the extrinsic curvature tensor or induced curvature and has the form

$$K_{ij} = n_{\mu;\nu} e_i^\mu e_j^\nu = -\nabla_{(i} n_{j)} - n_{(i} a_{j)}, \tag{10.25}$$

where a_j is called the acceleration of the unit normal vector field

$$a_j = n^i n_{j|i}, \tag{10.26}$$

and its trace, called the intrinsic curvature, has the form

$$K = K_i^i = h^{ij} K_{ij} = n^\mu_{;\mu}. \tag{10.27}$$

The hypersurface is called convex when the congruence is diverging, *i.e.* $K > 0$, and concave when the congruence is converging, *i.e.* $K < 0$. The tangent vector satisfies the Gauss–Weingarten equations

$$e^\alpha_{i;\beta} e_j^\beta = \Gamma^k_{ij} e_k^\alpha + K_{ij} n^\mu, \tag{10.28}$$

and the Gauss–Codazzi equations [550, 551, 552] can be derived by straightforward computation

$$R_{\mu\nu\kappa\lambda} e_i^\mu e_j^\nu e_k^\kappa e_l^\lambda = R_{ijkl} - K_{il} K_{jk} + K_{ik} K_{jl}, \tag{10.29}$$

$$R_{\mu\nu\kappa\lambda} n^\mu e_i^\nu e_j^\kappa e_k^\lambda = K_{ij|k} - K_{ik|j}, \tag{10.30}$$

which by using the decomposition of the Ricci curvature tensor and Ricci scalar curvature

$$R_{\mu\nu} = -R_{\kappa\mu\lambda\nu} n^\kappa n^\lambda + h^{ij} R_{\kappa\mu\lambda\nu} e_i^\kappa e_j^\lambda, \tag{10.31}$$

$$R = -2h^{kl} R_{\kappa\mu\lambda\nu} n^\kappa n^\lambda e_k^\mu e_l^\nu + h^{kl} h^{ij} R_{\kappa\mu\lambda\nu} e_i^\kappa e_j^\lambda e_k^\mu e_l^\nu, \tag{10.32}$$

can be presented in terms of the Einstein tensor (10.5)

$$\begin{aligned} 2G_{\mu\nu}n^{\mu}n^{\nu} &= {}^{(3)}R + K^{ij}K_{ij} + K^2, &(10.33)\\ G_{\mu\nu}e^{\mu}_{j}n^{\nu} &= K^{i}_{j|i} - K_{,j}. &(10.34)\end{aligned}$$

Another identity, called the Ricci equations

$$\mathcal{L}_n K_{ij} = n^{\mu}n^{\nu}e^{\kappa}_{i}e^{\lambda}_{j}R_{\mu\nu\kappa\lambda} - \frac{1}{N}N_{|ij} - K_{ik}K^{k}_{j}, \tag{10.35}$$

can be also derived by straightforward computation, which we omit here. By using the completeness relations (10.24) and the fact

$$R_{\kappa\mu\lambda\nu}n^{\kappa}n^{\mu}n^{\lambda}n^{\nu} = 0, \tag{10.36}$$

one can see that the first term in the relation (10.32) reduces to $-2R_{\mu\nu}n^{\mu}n^{\nu}$. By using the relations

$$\begin{aligned} R_{\mu\nu}n^{\mu}n^{\nu} &= 2\left(n^{\mu}_{;[\nu}n^{\nu}\right)_{;\mu]} + 2n^{\mu}_{;[\mu}n^{\nu}_{;\nu]}, &(10.37)\\ n^{\mu}_{;\nu}n^{\nu}_{;\mu} &= K^{ij}K_{ij}, &(10.38)\end{aligned}$$

to the reduced first term in (10.32), and the Gauss–Codazzi equations (10.29) to the second term in (10.32)

$$\begin{aligned} h^{kl}h^{ij}R_{\kappa\mu\lambda\nu}e^{\kappa}_{i}e^{\lambda}_{j}e^{\mu}_{k}e^{\nu}_{l} &= h^{kl}h^{ij}\left(R_{ijkl} - K_{il}K_{jk} + K_{ik}K_{jl}\right) = \\ &= {}^{(3)}R + K^2 - K^{ij}K_{ij}, &(10.39)\end{aligned}$$

one obtains the three-dimensional evaluation of the four-dimensional Ricci scalar curvature

$${}^{(4)}R = {}^{(3)}R + K^2 - K^{ij}K_{ij} - 2\left(n^{\mu}_{;\nu}n^{\nu} - n^{\mu}n^{\nu}_{;\nu}\right)_{;\mu}. \tag{10.40}$$

We have denoted by the stroke on the left of an index the intrinsic covariant differentiation with respect to the coordinate labeled by this index. Two indices before the stroke means taking two times the intrinsic covariant derivative with respect to each of the indices. For instance, for vector V_i and tensor T_{ij} the intrinsic covariant differentiation is defined in the following way

$$\begin{aligned} V_{i|j} &= \nabla_j V_i = \partial_j V_i - \Gamma^{k}_{ji}V_k, &(10.41)\\ T_{ij|k} &= \nabla_k T_{ij} = \partial_k T_{ij} - T_{lj}\Gamma^{l}_{ik} - T_{il}\Gamma^{l}_{jk}, &(10.42)\end{aligned}$$

where Γ^k_{ij} are the spatial Christoffel symbols

$$\Gamma^k_{ij} = \frac{1}{2} h^{kl} \left(h_{il,j} + h_{lj,i} - h_{ij,l} \right). \tag{10.43}$$

The induced metric h_{ij} and the extrinsic curvature tensor K_{ij} are the dynamical variables which describe the geometry of a submanifold ∂M by the $3+1$ decomposed Einstein field equations. The pair (h_{ij}, K_{ij}) describes the local geometry of a single space-like (constant time) hypersurface ∂M, and then the evolution of the global four-dimensional geometry can be formulated in terms of the one-parameter family of the dynamical variables $(h_{ij}(t), K_{ij}(t))$ describing evolution of the local three-dimensional geometry of the space-like hypersurfaces ∂M_t. For consistency, one must also specify the relation between the time evolution operator ∂_t and the vector field n normal to ∂M_t

$$\partial_t = Nn + N^i \partial_i, \tag{10.44}$$

where N and N_i are called the lapse function and the shift vector, respectively. Albeit, in general stationarity of Matter fields, *i.e.* $T_{\mu\nu} \equiv 0$, results in the existence of a global time-like Killing vector field $\mathscr{K}_\mu$ for metric tensor $g_{\mu\nu}$.

One can choose a coordinate system in such a way that the Killing vector field equals $\dfrac{\partial}{\partial t}$ and the foliation $t = const$ is space-like. In such a situation, the metric tensor depends at most on a spatial coordinates x^i and, therefore, the time t can be treated as a global coordinate [553]. Let us introduce such a coordinate system chosen by this gauge condition in such a way that an induced three-dimensional boundary space is a constant time t hypersurface. Then the space-time boundary ∂M becomes an embedded space and satisfies the Nash embedding theorem (For detailed discussion of the theorem, its consequences and advanced development Cf. *e.g.* Refs. [554, 555, 556, 557]).

B Classical and Quantum Geometrodynamics

Let us consider the enveloping space-time manifold M which is compact and possesses a space-like boundary $(\partial M, h)$ equipped with the induced space metric h_{ij} and the second fundamental form K_{ij}. Let the topology of such four-dimensional space-time manifold be $\Sigma \times \mathbb{R}$, where Σ is an unrestricted topology of the three-dimensional space. In such a situation, the Einstein field equations (10.6) can be generated as the Euler–Lagrange equations of motion by using the Hilbert–Palatini action principle [163, 164] with respect to the fundamental field which for General Relativity is metric tensor $g_{\mu\nu}$

$$\frac{\delta S[g]}{\delta g_{\mu\nu}} = 0, \tag{10.45}$$

which must be complemented by the suitable boundary condition

$$\delta g_{\mu\nu}|_{\partial M} = 0, \tag{10.46}$$

and applied to the Einstein–Hilbert action complemented by the York–Gibbons–Hawking boundary action [533, 558], *i.e.* the action of a four-geometry with fixed an induced three-geometry of a boundary

$$S[g] = \frac{1}{c}\int_M d^4x\sqrt{-g}\left(-\frac{1}{2\kappa}{}^{(4)}R + \frac{\Lambda}{\kappa}\right) + S_\phi[g] - \frac{1}{\kappa c}\int_{\partial M} d^3x\sqrt{h}K, \tag{10.47}$$

where $S_\phi[g]$ is the action of Matter fields

$$S_\phi[g] = \frac{1}{c}\int_M d^4x\sqrt{-g}L_\phi. \tag{10.48}$$

Einstein field equations (10.6) can be obtained by straightforward computation of the variation $\delta S = \delta S_{EH} + \delta S_{YGH} + \delta S_\phi = 0$ on ∂M, where

$$\delta S_G = \frac{1}{2\kappa c}\int_M d^4x\sqrt{-g}\left(G_{\mu\nu} + \Lambda g_{\mu\nu}\right)\delta g^{\mu\nu}, \tag{10.49}$$

$$\delta S_{YGH} = \frac{1}{2\kappa c}\int_{\partial M} d^3y\sqrt{|h|}h^{\mu\nu}n^\rho\delta g_{\mu\nu,\rho}, \tag{10.50}$$

$$\delta S_\phi = -\frac{1}{2c}\int_M d^4x\sqrt{-g}T_{\mu\nu}\delta g^{\mu\nu}, \tag{10.51}$$

where by $S_G = S_{EH} + S_{YGH}$ we have denoted the geometric part of the total action functional (10.47). Moreover, the variational principle allows to establish the relation between the stress-energy tensor and the Lagrangian of Matter fields in the following form

$$T_{\mu\nu} = -\frac{2}{\sqrt{-g}}\frac{\delta\left(\sqrt{-g}L_\phi\right)}{\delta g^{\mu\nu}}. \tag{10.52}$$

When the cosmological constant vanishes identically $\Lambda = 0$, then the global time-like Killing vector field $\mathscr{K}_\mu$ on space-time manifold M exists. Recall (For more advanced and abstractive approach we suggest *e.g.* the books in Ref. [559]) that such a field arises from vanishing of the Lie derivative with respect to this field of the metric tensor $g_{\mu\nu}$

$$\mathscr{L}_K g_{\mu\nu} = \lim_{\varepsilon\to 0}\frac{g_{\mu\nu}(\tilde{x}) - \tilde{g}_{\mu\nu}(\tilde{x})}{\varepsilon} = 0, \tag{10.53}$$

where $\tilde{g}_{\mu\nu}(\tilde{x})$ is the metric tensor $g_{\mu\nu}(x)$ transformed under the infinitesimal transformation - an isometric mapping

$$\tilde{x}^\mu = x^\mu + \varepsilon\mathscr{K}^\mu, \tag{10.54}$$

which is equivalent to the Killing equation

$$\nabla_{(\mu}\mathscr{K}_{\nu)}(x) = 0. \tag{10.55}$$

In other words, the Killing vector fields are the infinitesimal generators of isometries.

For the case of a positive value of the cosmological constant $\Lambda > 0$ the Killing vector field $\mathscr{K}_\mu$ does not exist, and space-like boundary ∂M only foliates an exterior to the horizons on geodesic lines. Therefore, in such a situation, the Arnowitt–Deser–Misner decomposition (10.19) is a gauge of the field of metric. In all these situations, however, the total action (10.47) evaluated for the $3+1$ decomposed metric tensor (10.19) takes the form of the Hamilton action functional

$$S[g] = \int dtL, \tag{10.56}$$

where L is the total Lagrangian expressed through the $3+1$ splitting. The most important contribution to the theory is the geometric part of the Einstein–Hilbert Lagrangian which is

$$\begin{aligned}\sqrt{g}\left(-{}^{(4)}R+2\Lambda\right) &= N\sqrt{h}\left(-K_{ij}K^{ij}+K^2-{}^{(3)}R+2\Lambda\right)\\ &+ 2\partial_0\left(\sqrt{h}K\right)-2\partial_i\left(\sqrt{h}(KN^i-h^{ij}N_{|j})\right), \qquad (10.57)\end{aligned}$$

and because the last two terms are total derivatives they can be dropped when performing canonical formulation. The Lagrangian related to the York–Gibbons–Hawking boundary action in itself is a total derivative and, therefore, this term does not play any role here. Analysis of both the Lagrangian of Matter fields and the cosmological constant term can be done easily, and as a result one obtains the following Lagrangian of the total theory (10.56)

$$L=\int_{\partial M}d^3xN\sqrt{h}\left(K^2-K_{ij}K^{ij}-{}^{(3)}R+2\Lambda+2\kappa\rho\right). \qquad (10.58)$$

The Einstein field equations can be decomposed in the $3+1$ splitting. As a result one obtains the evolutionary equations for the induced metric h_{ij} and the intrinsic curvature K_{ij}

$$\partial_t h_{ij} = N_{i|j}+N_{j|i}-2NK_{ij}, \qquad (10.59)$$

$$\begin{aligned}\partial_t K_{ij} &= -N_{|ij}+N(R_{ij}+KK_{ij}-2K_{ik}K^k_j)+N^kK_{ij|k}+K_{ik}N^k_{|j}+K_{jk}N^k_{|i}-\\ &- \kappa N\left[S_{ij}-\frac{1}{2}h_{ij}(S-\rho)\right], \qquad (10.60)\end{aligned}$$

where the dot means differentiation with respect to the time coordinate. It can be seen by straightforward computation that the determinant of spatial metric $h=\det h_{ij}$ and the extrinsic curvature tensor $K=K^i_i$ satisfy the equations

$$\partial_t \ln\sqrt{h} = -NK+N^i_{|i}, \qquad (10.61)$$

$$\partial_t K = -h^{ij}N_{|ij}+N\left(K^{ij}K_{ij}+\frac{\kappa}{2}(S+\rho)\right)+N^iK_{|i}. \qquad (10.62)$$

Here ρ, called the energy density, is double projection of the stress-energy tensor onto the normal vector field

$$\rho=T(n,n)=T_{\mu\nu}n^\mu n^\nu, \qquad (10.63)$$

and n^μ is the normal vector field following from the $3+1$ splitting

$$n^\mu = \left[\frac{1}{N}, -\frac{N^i}{N}\right], \tag{10.64}$$

$$n_\mu = [-N, 0_i]^T, \tag{10.65}$$

where $0_i = [0,0,0]^T$ is the null three-vector. The tensor S_{ij}, called the spatial stress tensor, is double projection of the stress-energy tensor onto the spatial metric, and S is its trace which we will call the spatial stress density

$$S_{ij} = T(h,h) = T_{\mu\nu} h^\mu_i h^\nu_j, \tag{10.66}$$

$$S = h^{ij} S_{ij}, \tag{10.67}$$

where $h^\mu_\nu = \delta^\mu_\nu + n^\mu n_\nu$. Straightforward calculation gives

$$S - \rho = T, \tag{10.68}$$

where $T = g^{\mu\nu} T_{\mu\nu}$ is the trace of the stress-energy tensor. In the light of the Einstein field equations (10.6) one obtains

$$T = \frac{G}{\kappa} + \frac{4\Lambda}{\kappa}, \tag{10.69}$$

where G is the trace of the Einstein tensor which can be computed straightforwardly

$$G = g^{\mu\nu} G_{\mu\nu} = g^{\mu\nu}\left(R_{\mu\nu} - \frac{1}{2} g_{\mu\nu} {}^{(4)}R\right) = {}^{(4)}R - 2{}^{(4)}R = -{}^{(4)}R. \tag{10.70}$$

In this manner, one obtains the constraint between the spatial stress tensor, the energy density, the cosmological constant and the Ricci scalar curvature of an enveloping space-time manifold

$$S - \rho = \frac{4\Lambda}{\kappa} - \frac{{}^{(4)}R}{\kappa}. \tag{10.71}$$

It can be seen by straightforward computation that the total Lagrangian (10.58) leads to the Euler–Lagrange equations of motion. The first equation is

$$2\kappa h^{-1}\left(h_{ik} h_{jl} - \frac{1}{2} h_{ij} h_{kl}\right) \frac{\delta S[g]}{\delta h_{ij}} \frac{\delta S[g]}{\delta h_{kl}} - \frac{1}{2\kappa}\left({}^{(3)}R - 2\Lambda - 2\kappa\rho\right) = 0, \tag{10.72}$$

while the second one has the form

$$\pi^{ij}_{|j} + J^i = 0, \tag{10.73}$$

where π^{ij} is the momentum conjugated to the induced metric

$$\pi^{ij} = \frac{\delta S[g]}{\delta h_{ij}} = \frac{\delta L}{\delta\left(\partial_t h_{ij}\right)} = -\frac{\sqrt{h}}{2\kappa}\left(K^{ij} - h^{ij}K\right), \tag{10.74}$$

and J^i, called the momentum density, is the stress-energy tensor projected onto the normal vector field and the spatial metric

$$J^i = T(n,h) = T_{\mu\nu}n^{\mu}h^{\nu i}. \tag{10.75}$$

The resulting dynamical evolution (10.72) is the Hamilton–Jacobi equation (For details of classical mechanics Cf. *e.g.* Ref. [560]) to the case of General Relativity. Originally, the equation (10.72) divided by $\sqrt{h}$ with $\Lambda = 0$ and $\rho = 0$ was derived by Asher Peres [561], and for this reason we call it the Peres equation. Interestingly, John A. Wheeler [175] called the Hamilton–Jacobi equation of General Relativity (10.72) the Einstein–Hamilton–Jacobi equation. The Peres equation defines the classical geometrodynamics.

The total Lagrangian (10.58) can be analyzed by the Hamiltonian approach. Let us determine the canonical momenta

$$\pi_{\phi} = \frac{\delta L}{\delta\left(\partial_t \phi\right)}, \tag{10.76}$$

$$\pi = \frac{\delta L}{\delta\left(\partial_t N\right)} = 0, \tag{10.77}$$

$$\pi^i = \frac{\delta L}{\delta\left(\partial_t N_i\right)} = 0, \tag{10.78}$$

conjugated to Matter fields, lapse function, shift vector, respectively. Then the Legendre transformation [560] allows to rewrite the total Lagrangian in the form

$$L = \int_{\partial M} d^3x \left[\frac{1}{2\kappa}\left(\pi_{\phi}\partial_t\phi + \pi\partial_t N + \pi^i\partial_t N_i + \pi^{ij}\partial_t h_{ij}\right) - NH - N_i H^i\right], \tag{10.79}$$

where the quantities H and H^i are defined as

$$H = \frac{\sqrt{h}}{2\kappa}\left(K^2 - K_{ij}K^{ij} - {}^{(3)}R + 2\Lambda + 2\kappa\rho\right), \quad (10.80)$$
$$H^i = -2\pi^{ij}_{\ \ |j} - 2J^i = -2\partial_j\pi^{ij} - h^{il}\left(2\partial_k h_{jl} - \partial_l h_{jk}\right)\pi^{jk} - 2J^i, \quad (10.81)$$

where ${}^{(3)}R = h^{ij}R_{ij}$ is the Ricci scalar curvature of a three-dimensional embedded space. Application of the time-preservation [169] to the primary constraints

$$\pi \approx 0, \quad (10.82)$$
$$\pi^i \approx 0, \quad (10.83)$$

leads to the secondary constraints

$$H \approx 0, \quad (10.84)$$
$$H^i \approx 0, \quad (10.85)$$

called the Hamiltonian (scalar) constraint which yields the dynamics, and the diffeomorphism (vector) constraint which merely reflects the spatial diffeoinvariance. Bryce S. DeWitt [176] showed that the quantities H^i are generators of the spatial diffeomorphisms $\widetilde{x}^i = x^i + \xi^i$

$$i\left[h_{ij}, \int_{\partial M} H_a\xi^a d^3x\right] = -h_{ij,k}\xi^k - h_{kj}\xi^k_{\ ,i} - h_{ik}\xi^k_{\ ,j}, \quad (10.86)$$
$$i\left[\pi^{ij}, \int_{\partial M} H_a\xi^a d^3x\right] = -\left(\pi^{ij}\xi^k\right)_{,k} + \pi^{kj}\xi^i_{\ ,k} + \pi^{ik}\xi^j_{\ ,k}, \quad (10.87)$$

where we have denoted the intrinsic covariant components $H_i = h_{ij}H^j$. Application of the structure constants of the diffeomorphism group, which can be presented in the compact form

$$c^a_{ij} = \delta^a_i\delta^b_j\delta^{(3)}_{,b}(x,z)\delta^{(3)}(y,z) - (x \to y). \quad (10.88)$$

Applying the relations (10.86) and (10.87) one can derive the first-class constraints algebra

$$i\left[H_i(x), H_j(y)\right] = \int_{\partial M} H_a c^a_{ij} d^3z, \quad (10.89)$$
$$i\left[H(x), H_i(y)\right] = H\delta^{(3)}_{,i}(x,y), \quad (10.90)$$

while involving the elementary relation

$$\delta\left(\sqrt{h}{}^{(3)}R\right) = \sqrt{h}h^{ij}h^{kl}\left(\delta h_{ik,jl} - \delta h_{ij,kl}\right) - \sqrt{h}\left[R^{ij} - \frac{{}^{(3)}R}{2}h^{ij}\right]\delta h_{ij}, \quad (10.91)$$

allows to establish the third bracket

$$i\left[\int_{\partial M} H\xi_1 d^3x, \int_{\partial M} H\xi_2 d^3x\right] = \int_{\partial M} H^a\left(\xi_{1,a}\xi_2 - \xi_1\xi_{2,a}\right)d^3x. \quad (10.92)$$

The constraints algebra (10.89)-(10.92) was derived first by B.S. DeWitt, and for this reason we shall call it *the DeWitt algebra.*

The method of canonical primary quantization appropriate for constrained systems was investigated by Dirac [169] and developed for needs of quantum geometrodynamics by Ludvig Dmitrievich Faddeev [562] (For general analysis and discussion see also Refs. [499, 552]). In the light of the general method applied to the present situation ,the canonical commutation relations are

$$i\left[\pi^{ij}(x), h_{kl}(y)\right] = \frac{1}{2}\left(\delta^i_k\delta^j_l + \delta^i_l\delta^j_k\right)\delta^{(3)}(x,y), \quad (10.93)$$

$$i\left[\pi^i(x), N_j(y)\right] = \delta^i_j\delta^{(3)}(x,y), \quad (10.94)$$

$$i\left[\pi(x), N(y)\right] = \delta^{(3)}(x,y). \quad (10.95)$$

The solutions or rather representations of the momenta operators satisfying the commutators (10.93)-(10.95) is the question of choice. In quantum geometrodynamics, the Wheeler metric representation is usually taken into account. In such a representation the momenta operators are analogous to the momentum operator in quantum mechanics

$$\pi = -i\frac{\delta}{\delta N}, \quad (10.96)$$

$$\pi^i = -i\frac{\delta}{\delta N_i}, \quad (10.97)$$

$$\pi^{ij} = -i\frac{\delta}{\delta h_{ij}}, \quad (10.98)$$

and applied to the Hamiltonian constraint (10.80) yield the Wheeler–DeWitt equation [176, 175]

$$\left\{2\kappa G_{ijkl}\frac{\delta^2}{\delta h_{ij}\delta h_{kl}} + \frac{1}{2\kappa}\sqrt{h}\left({}^{(3)}R - 2\Lambda - 2\kappa\rho\right)\right\}\Psi[h_{ij},\phi] = 0, \quad (10.99)$$

where G_{ijkl} is the DeWitt supermetric on the configurational space of General Relativity called the Wheeler superspace $S(\partial M)$ (For more detailed analysis and discussion Cf. *e.g.* Refs. [175, 176, 563, 564, 565, 566])

$$G_{ijkl} = \frac{1}{2\sqrt{h}}\left(h_{ik}h_{jl} + h_{il}h_{jk} - h_{ij}h_{kl}\right). \tag{10.100}$$

Other first-class constraints satisfy the canonical commutation relations

$$\begin{aligned}
\left[\pi(x), \pi^i(y)\right] &= 0, && (10.101)\\
\left[\pi(x), H^i(y)\right] &= 0, && (10.102)\\
\left[\pi^i(x), H^j(y)\right] &= 0, && (10.103)\\
\left[\pi^i(x), H(y)\right] &= 0, && (10.104)
\end{aligned}$$

and after the canonical primary quantization are the supplementary conditions on wave functional $\Psi[h_{ij},\phi]$. The primary constraints lead to the equations

$$\begin{aligned}
-i\frac{\delta\Psi[h_{ij},\phi]}{\delta N} &= 0, && (10.105)\\
-i\frac{\delta\Psi[h_{ij},\phi]}{\delta N_i} &= 0. && (10.106)
\end{aligned}$$

The diffeomorphism constraint also leads to such a condition

$$i\left(\frac{\delta\Psi[h_{ij},\phi]}{\delta h_{ij}}\right)_{|j} = J^i\Psi[h_{ij},\phi], \tag{10.107}$$

which can be rewritten in explicit form

$$\left[i\frac{\partial}{\partial x^j}\frac{\delta}{\delta h_{ij}} + ih^{il}\left(h_{jl,k} - \frac{1}{2}h_{jk,l}\right)\frac{\delta}{\delta h_{jk}} - J^i\right]\Psi[h_{ij},\phi] = 0. \tag{10.108}$$

The diffeomorphism constraint can be simply reduced

$$\begin{aligned}
-2\partial_j\pi^{ij} - h^{il}\left(2h_{jl,k} - h_{jk,l}\right)\pi^{jk} - 2J^i &=\\
-2\partial_j\pi^{ij} - h^{il}\left(2h_{jl,k} - h_{jk,l}\right)h^j_i h^k_j\pi^{ij} - 2J^i &=\\
-2\left[\partial_j + h^{il}h^j_i h^k_j\left(h_{jl,k} - \frac{1}{2}h_{jk,l}\right)\right]\pi^{ij} - 2J^i &=\\
-2\left[\partial_j + h^{lk}\left(h_{jl,k} - \frac{1}{2}h_{jk,l}\right)\right]\pi^{ij} - 2J^i. & && (10.109)
\end{aligned}$$

Application of the relations

$$h^{lk}h_{jl,k} = \left(h^{lk}h_{jl}\right)_{,k} - h^{lk}_{,k}h_{jl} = \partial_k\delta^k_j - h^{lk}_{,k}h_{jl} = -h^{lk}_{,k}h_{jl} \quad (10.110)$$

$$h^{lk}h_{jk,l} = \left(h^{lk}h_{jk}\right)_{,l} - h^{lk}_{,l}h_{jk} = \partial_l\delta^l_j - h^{lk}_{,l}h_{jk} = -h^{lk}_{,l}h_{jk}, \quad (10.111)$$

and manipulations in the indices

$$-h^{lk}_{,l}h_{jk} = -h^{kl}_{,l}h_{jk} = -h^{lk}_{,k}h_{jl}, \quad (10.112)$$

allows to rewrite the diffeomorphism constraint in the form

$$H^i = -2\left(\partial_j - \frac{1}{2}h_{jl}h^{lk}_{,k}\right)\pi^{ij} - 2J^i \approx 0, \quad (10.113)$$

which by using the identity $h_{jl}h^{lk}_{,k} = \partial_k\delta^k_j - h_{jl,k}h^{kl} = -h_{jl,k}h^{kl}$ becomes

$$H^i = -2\left(\partial_j + \frac{1}{2}h_{jl,k}h^{kl}\right)\pi^{ij} - 2J^i \approx 0. \quad (10.114)$$

For this reason, the canonical primary quantization of the diffeomorphism constraint leads to the following equation

$$\left[i\left(\partial_j + \frac{1}{2}h_{jl,k}h^{kl}\right)\frac{\delta}{\delta h_{ij}} - J^i\right]\Psi[h_{ij},\phi] = 0. \quad (10.115)$$

The DeWitt supermetric (10.100) and the analogous metric following from the Peres equation (10.72) by multiplication of both sides by $2\sqrt{h}$ are not the same despite the fact that the same procedure is applied. In fact, the relation between both the metrics arises from the change

$$h_{ik}h_{jl} \longleftrightarrow \frac{h_{ik}h_{jl} + h_{il}h_{jk}}{2} = h_{i(k}h_{jl)}, \quad (10.116)$$

which means that B.S. DeWitt applied symmetrization $h_{i(k}h_{jl)}$ instead of $h_{ik}h_{jl}$, and for this reason included in quantum geometrodynamics one more stratum than A. Peres in derivation of classical geometrodynamics. The equation (10.116) in itself is non-trivial, because it generates

$$h_{ik}h_{jl} = h_{il}h_{jk}, \quad (10.117)$$

which after multiplication of both sides by $h^{ik}h^{jl}$ leads to

$$D^2 = h^k_l h^l_k = h^k_k = D, \tag{10.118}$$

what is true if and only if the dimensionality of embedded space is $D=1$ or $D=0$. Naturally, it is not true in general. However, in the case when one uses the DeWitt method then one obtains the Peres equation with the DeWitt supermetric, *i.e.*

$$2\kappa G_{ijkl}\frac{\delta S[g]}{\delta h_{ij}}\frac{\delta S[g]}{\delta h_{kl}} + \frac{1}{2\kappa}\sqrt{h}\left({}^{(3)}R - 2\Lambda - 2\kappa\rho\right) = 0, \tag{10.119}$$

and canonical primary quantization of such classical geometrodynamics leads to the Wheeler–DeWitt equation. The classical geometrodynamics is argued by the fact that only such a procedure establishes straightforward equivalence between quantization of the Hamiltonian constraint and quantization of the Euler–Lagrange equations of motion for geometrodynamics of an embedded space of arbitrary dimensionality D, including the situation $D=3$ which we are studying in this book. In fact, J.A. Wheeler [175] did not argue using the symmetrization $h_{i(k}h_{jl)}$ and in this way he studied the quantized Peres equation rather than the Wheeler–DeWitt equation. DeWitt [176] established the supermetric (10.100) and derived the Wheeler–DeWitt equation. The difference is crucial because in modern quantum geometrodynamics the Wheeler superspace is defined by the DeWitt supermetric which possesses non-trivial properties. The diffeomorphism constraint (10.85) with (10.81) and the equation of motion (10.73) differs only by the constant factor -2, and for this reason lead to the same quantum and classical conditions independently on the DeWitt supermetric. This is because this condition merely reflects diffeoinvariance, while the Wheeler–DeWitt equation or the Peres equation define quantum and classical dynamics, respectively. In other words, in quantum geometrodynamics studying diffeomorphism constraint is worthless from the dynamical point of view.

B.S. DeWitt [176] argued that application of the Wheeler metric representation for closed finite worlds results in the feature: the wave functional $\Psi[h_{ij},\phi]$ depends only on components of three-metric, *i.e.* is $\Psi[h_{ij}]$. In such a situation, the equations (10.108) express the necessary and sufficient conditions for diffeoinvariance of the wave functional

$\Psi[h_{ij}]$. For finite worlds it means that the wave functional depends only on the geometry of an embedded space. He proposed to construct the related structure of $\Psi[h_{ij}]$ by the hypersurface integrals which can be constructed out of products of the Riemann–Christoffel curvature tensor and its covariant derivatives, with the topology of three-space itself being separately specified. DeWitt discussed differences between infinite and finite worlds. He proposed that in the finite case one can replace the wave functional $\Psi[h_{ij}]$ by $\Psi[{}^{(3)}\mathfrak{G}]$, where ${}^{(3)}\mathfrak{G}$ is the three-geometry. For description of the quantum gravity he introduced $\mathfrak{M}$ - the set of all possible three-geometries which a finite world may possess, called today *midisuperspace*, and asked for topological issues related to $\mathfrak{M}$.

The Wheeler–DeWitt equation has never been solved in general. The only known solution is the Hartle–Hawking wave function [534], called the wave function of the Universe, which is the path integral

$$\Psi[h_{ij}] = N\int_C \delta g(x)\exp(iS_E[g]), \tag{10.120}$$

$$\Psi[h_{ij},\phi] = N\int_C \delta g\delta\phi\exp(iS[g,\phi]), \tag{10.121}$$

where N is the normalization factor, and $S_E[g]$ and $S[g,\phi]$ are the geometric part of the total action and the total action, respectively. The path integral is over all four-geometries with a space-like boundary on which induced metric h_{ij} is fixed. Albeit, the Feynman path integrals (10.120) and (10.121) are only a kind of tautology following from the fact that the Wheeler-DeWitt equation is usually understood as the analogue of the Schrödinger equation, *i.e.* non-relativistic quantum mechanics. Despite the fact that this strategy is familiar, this has never led to evaluation of the Hartle–Hawking wave function for the general case. Strictly speaking, the functional integrals of quantum geometrodynamics (10.120)-(10.121) can be established straightforwardly and easily for very few particular cases. Such a situation is the straightforward legacy of the fact that, in general, the integration strategy which is based on the Feynman path integral technique carries difficult computational level and, above all, is an ill-defined mathematical procedure. Such wave functionals formally lead to the Wheeler–DeWitt equation and, for this reason, are its solutions. Factually, the wave function of the Universe defines only one class of wave functionals.

C Wheeler Superspace

The mathematical structure of geometrodynamics is determined by the Wheeler superspace, *i.e.* the configurational space of General Relativity, which is the space of all equivalence classes of metric fields of General Relativity related by the action of the diffeomorphism group of a compact, connected, orientable, Hausdorff, C^∞ three-dimensional space-like manifold without boundary ∂M. Superspace is the factor space

$$S(\partial M) = \frac{Riem(\partial M)}{Diff(\partial M)}, \tag{10.122}$$

where $Riem(\partial M)$ is the space of all C^∞ Riemannian metrics on the boundary ∂M, while $Diff(\partial M)$ is the group of all C^∞ diffeomorphisms of ∂M that preserve orientation.

Superspace as a space of all orbits of the diffeomorphism group in itself is a connected, second-countable, metrizeable space. From a topological point of view, the space $Riem(\partial M)$ is an open positive convex cone in the infinite dimensional vector space of all smooth C^∞ symmetric second-rank tensor fields over ∂M having the point-set topology, *i.e.*

$$\forall \lambda \in \mathbb{R}_+ \cap \forall h_{ij} \in Riem(\partial M) : \lambda h_{ij} \in Riem(\partial M). \tag{10.123}$$

This vector space is a locally convex topological vector space possessing a translation-invariant metric $\bar{d}$ which induces its topology and defines completeness of the space, *i.e.* is a Fréchet space. The metric can be chosen in such a way that $Diff(\partial M)$ preserves distances, and the space $Riem(\partial M)$ inherits the metric and for this reason is a metrizable topological space that is also paracompact and second countable. Factoring out $Diff(\partial M)$ transits the topological information concerning ∂M to the quotient space $S(\partial M)$. There are two problems. The first is the case of closed ∂M equipped with metrics with a non-trivial isometry group, for which $S(\partial M)$ is not manifold. As showed by Arthur Elliot Fischer [564], in such a situation $Diff(\partial M)$ does not act freely and, consequently, $S(\partial M)$ is a stratified manifold with nested sets of strata ordered according to the dimension of the isometry groups. However, then there is a consistent way to resolving the singularities [567] which involves the frame bundle $F(\partial M)$ (For details of the theory of bundles Cf. *e.g.* Ref. [568]) such that the quotient space

$$\frac{Riem(\partial M)\times F(\partial M)}{Diff(\partial M)}, \tag{10.124}$$

is the refinement of superspace.

The action of the diffeomorphism group $Diff(\partial M)$ is now free because non-trivial isometries fixing a frame are removed. If ϕ is such an isometry, one can apply the exponential map and the relation which is valid for any isometry $\phi\circ\exp=\exp\circ\phi_\star$ to show that the subset of points in ∂M fixed by an isometry ϕ is open. Since this set is also closed and ∂M is connected, ϕ must be the identity. The definition (10.122) can be more refined if one restricts the group of diffeomorphisms to the proper subgroup of those diffeomorphisms that fix a preferred point, called $\infty\in\partial M$ and the tangent space at this point

$$Diff_F(\partial M)=\left\{\phi\in Diff(\partial M)|\phi(\infty)=\infty,\phi_\star(\infty)=id|_{T_\infty\partial M}\right\}. \tag{10.125}$$

Then, the quotient $Riem(\partial M)\times F(\partial M)/Diff(\partial M)$ is isomorphic to

$$S_F(\partial M)=\frac{Riem(\partial M)}{Diff_F(\partial M)}, \tag{10.126}$$

but a preferred point ∞ must be chosen arbitrarily. The space $S_F(\partial M)$ is known as *the extended superspace*.

A problematic situation is created also by asymptotic flatness. If one treats ∂M as one-point compactification of a manifold with one end then diffeomorphisms have to respect the asymptotic geometry and for this reason the extended superspace is the appropriate concept. The extended superspace would have been unnecessary in the closed case if one restricted attention to those manifolds ∂M which do not allow for metrics with continuous symmetries, *i.e.* whose degree of symmetry is zero. Recall that the degree of symmetry of a manifold ∂M is defined as

$$\deg(\partial M)=\sup_{h_{ij}\in Riem(\partial M)}\dim J(\partial M,h_{ij}), \tag{10.127}$$

where $J(\partial M,h_{ij})$ is the isometry group of $(\partial M,h_{ij})$

$$J(\partial M,h_{ij})=\left\{\phi\in Diff(\partial M)|\phi^\star h_{ij}=h_{ij}\right\}, \tag{10.128}$$

and when the dimension of a manifold is $D = \dim \partial M$ then

$$\dim J(\partial M, h_{ij}) \leqslant \frac{D(D+1)}{2}. \tag{10.129}$$

$J(\partial M, h_{ij})$ is compact if ∂M is compact. If ∂M allows for an effective action of a compact group G then it clearly allows for a metric h_{ij} on which G acts as isometries just average any Riemannian metric over G. For compact ∂M the degree of symmetry is zero if and only if ∂M cannot support action of the circle group $SO(2)$.

A list of three-manifolds with $deg > 0$ was compiled by Arthur Elliot Fischer [569], and with $\deg = 0$ by Arthur Elliot Fischer and Vincent Edward Moncrief [567]. Because projection

$$Riem(\partial M) \rightarrow S_F(\partial M) \tag{10.130}$$

is continuous, and $Diff_F(\partial M)$ acts continuously on the space $Riem(\partial M)$, the topology of extended superspace is quotient and open. For arbitrary two geometries $x, y \in S_F(\partial M)$ a metric on $S_F(\partial M)$

$$d(x,y) = \sup_{\phi_x,\phi_y \in Diff_F(\partial M)} \bar{d}(\phi_x^\star x, \phi_y^\star y), \tag{10.131}$$

where $\bar{d}$ is mentioned above a translation-invariant metric on $Riem(\partial M)$, turns $S_F(\partial M)$ into a connected, metrizable and second countable topological space. Hence $Riem(\partial M)$ and $S_F(\partial M)$ are perfectly connected topological spaces satisfying the axioms of strongest separability and countability. The basic geometric idea is to regard $Riem(\partial M)$ as a principal fibre bundle with structure group $Diff_F(\partial M)$ and the extended superspace $S_F(\partial M)$

$$Diff_F(\partial M) \xrightarrow{\ i\ } Riem(\partial M) \xrightarrow{\ p\ } S_F(\partial M)$$

where the i, p are inclusion and projection maps, respectively. This is made possible by the so-called slice theorems and the fact that the group acts freely and properly. This bundle structure has two significant consequences regarding the geometry and topology of $S_F(\partial M)$.

The topologically trivial space $Riem(\partial M)$ can be visualized as the box fibred by the action of the diffeomorphism group $Diff_F(\partial M) \in Diff(\partial M)$ generated by the diffeomorphism constraint, where orbits of the diffeomorphism group $Diff(\partial M)$ are represented by straight-lines in the box.

The quotient space $S(\partial M)$ obtains non-trivial topology from $Diff(\partial M)$ with an orbit represented by one point only. The subgroup $Diff_F(\partial M)$ acts as isometries on the DeWitt supermetric on a space of all C^∞ Riemannian metrics on the boundary ∂M – $Riem(\partial M)$. Superspace $S(\partial M)$ is the set of geometries of an embedding ∂M, which are equivalence classes of isometric Riemannian metrics. By the Metrization Theorem for Superspace, $S(\partial M)$ is a connected, second-countable, metrizeable space. In other words, a countable basis of open sets exists for its topology, and there also exists a metric on $S(\partial M)$ inducing such a topology. It should be emphasized that the partially-ordered set of conjugacy classes of compact subgroups of $Diff(\partial M)$ indexes a partition of $S(\partial M)$, *i.e.* is a set of nonempty subspaces $\{\Sigma_\alpha\}$ such that

$$S(\partial M) = \bigcup_\alpha \Sigma_\alpha, \tag{10.132}$$

$$\Sigma_\alpha \cap \Sigma_\beta \neq \emptyset \Rightarrow \alpha = \beta. \tag{10.133}$$

A partition is a manifold partition if an each subspace Σ_α is a manifold. All geometries with the same kind of symmetry have homeomorphic neighbourhoods and, therefore, create a manifold in the Wheeler superspace $S(\partial M)$. The neighbourhoods of all symmetric geometries are not homeomorphic to the neighbourhoods of all nonsymmetric geometries and, therefore, Superspace $S(\partial M)$ is not a manifold.

According to the Decomposition Theorem of Superspace, the space $S(\partial M)$ can be decomposed by its subspaces $S_G(\partial M)$ onto a countable, partially-ordered, C^∞-Fréchet manifold partition. According to the Stratification Theorem for Superspace, any subspace $S_G(\partial M)$ is an inverted stratification indexed by the type of symmetry, *i.e.* geometries with a given symmetry are completely contained within the boundary of less symmetric geometries. There is also the theorem established by Domenico J.W. Giulini which states that in a neighbourhood of the round three-sphere in the superspace $S(\partial M)$ the DeWitt supermetric is an infinite-dimensional Lorentzian metric, *i.e.* is of signature $(-1,\infty)$. However, at each point of the space ∂M the DeWitt supermetric defines a Lorentzian metric on the $1+5$ dimensional space of symmetric second-rank tensors at that point, which can be identified with the homogeneous quotient space

$$\frac{GL(3,\mathbb{R})}{SO(3)} \cong \frac{SL(3,\mathbb{R})}{SO(3)\times\mathbb{R}_+}. \tag{10.134}$$

Domenico Giulini together with Claus Kiefer [570] showed that the Lorentzian signature of the DeWitt supermetric has nothing to do with the Lorentzian signature of the space-time metric, *i.e.* it persists in the Euclidean gravity. But this is related to the attractivity of gravity. Arbitrary two points within the Wheeler superspace which differ by an action of the diffeomorphism constraint are gauge equivalent and hence physically indistinguishable. However, the question of whether and when the diffeomorphism constraint actually generates all diffeomorphisms of ∂M is still unsolved.

In fact, General Relativity is a dynamical system on the cotangent bundle, *i.e.* phase space, built over $S(\partial M)$. The topology of superspace is inherited from the topology of ∂M. The Hamiltonian evolution is varying embedding of space ∂M into space-time M. Hence the images of an embedded space have the same topological type, which reflects the fact that the classical geometrodynamics transitions of topology are impossible. This is not implied by the Einstein field equations, but is a consequence of the restriction to space-times admitting a global space-like foliation. There is a number of solutions to the Einstein field equations which do not satisfy such a requirement, *i.e.* such space-times cannot be constructed by integrating the Gauss–Codazzi equations with some reasonable initial data. From the Hartle–Hawking wave function point of view topology changing classical solutions should not be removed as possible contributors in the Feynman path integral. In the evolutionary formulation of the Einstein field equations, there is no space-time to start with. Only solutions of the dynamical equations construct the space-time. Then one can interpret the time dependence of the induced metric as being brought about by 'wafting' three-space through space-time by an embedding. Initially, there is only a space-like submanifold of unrestricted topology.

The deformations of the space-like hypersurfaces, *i.e.* infinitesimal changes of embeddings $\mathscr{E} : \partial M \mapsto M$, possess non-trivial kinematics. The generators of normal and tangential deformations are

$$\mathscr{N}_N = \int_{\partial M} d^3x N(x) n^\mu [y(x)] \frac{\delta}{\delta y^\mu(x)}, \tag{10.135}$$

$$\mathscr{T}_{N^i} = \int_{\partial M} d^3x N^i(x) y^\mu_{,i}(x) \frac{\delta}{\delta y^\mu(x)}, \tag{10.136}$$

where $y^\mu_{,i} = \partial_i y^\mu$, and y^μ and x^i are the local coordinates on M and ∂M, respectively. The generators (10.135) and (10.136) can be understood as the tangent vectors to the space of embeddings of ∂M into the manifold M. An embedding is locally given by four functions $y^\mu(x)$, such that the 3×4 matrix $y^\mu_{,i}$ has its maximum rank equal to 3. We have denoted by n^μ the components of the normal to the image $\mathscr{E}(\partial M)\in M$, which are functionals of $y^\mu(x)$, *i.e.* $n^\mu = n^\mu[y(x)]$. It can be seen by straightforward calculation that the generators (10.135) and (10.136) satisfy the following collection of commutation relations

$$\left[\mathscr{T}_{N^i},\mathscr{T}_{N'^i}\right] = -\mathscr{T}_{\left[N^i,N'^i\right]}, \tag{10.137}$$

$$\left[\mathscr{T}_{N^i},\mathscr{N}_N\right] = -\mathscr{N}_{N^i(N)}, \tag{10.138}$$

$$\left[\mathscr{N}_N,\mathscr{N}_{N'}\right] = -\mathscr{T}_{N\nabla_h N' - N'\nabla_h N}, \tag{10.139}$$

where $\nabla_h N = \left(h^{ab}\partial_b N\right)\partial_a$, and the Lie brackets (10.138) and (10.139) were obtained by taking variations of the basic identities $g_{\mu\nu}n^\mu n^\nu = -1$ and $g_{\mu\nu}y^\mu_{,i}n^\nu = 0$. The space-time vector field

$$V = Nn^\mu\partial_\mu + N^i\partial_i, \tag{10.140}$$

induces the foliation-dependent decomposition of the tangent vector $\mathrm{T}(V)$ at $Y\in \mathrm{Emb}(\partial M,M)$

$$\mathrm{T}(V) = \int_{\partial M} d^3x V^\mu(y(x))\frac{\delta}{\delta y^\mu(x)}, \tag{10.141}$$

obeying the Lie algebra

$$\left[\mathrm{T}(V),\mathrm{T}\left(V'\right)\right] = \mathrm{T}\left(\left[V,V'\right]\right), \tag{10.142}$$

which means that $V\mapsto \mathrm{T}(V)$ is a Lie homomorphism from the tangent-vector fields on M to the tangent-vector fields on $\mathrm{Emb}(\partial M,M)$. In this sense, the Lie algebra of the four-dimensional diffeomorphism group is implemented on the phase space of arbitrary generally covariant theory whose phase space includes the embedding variables, *i.e.* is so-called parameterized theory. Decomposing the vector (10.141) into normal and tangential components with respect to the leaves of the embedding at which the tangent-vector field to $Emb(\partial M,M)$ is evaluated, yields embedding-dependent parametrization of $\mathrm{T}(V)$ by (N,N^i)

$$\mathrm{T}(N,N^i) = \int_{\partial M} d^3x\left[Nn^\mu[y(x)] + N^i(x)y^\mu_{,i}(x)\right]\frac{\delta}{\delta y^\mu(x)}. \tag{10.143}$$

Computing the functional derivatives of n with respect to y, one can establish the commutator of deformation generators

$$\left[\mathrm{T}\left(N,N^i\right),\mathrm{T}\left(N',N'^i\right)\right]=-\mathrm{T}\left(N'',N'''^i\right), \tag{10.144}$$

where we have denoted

$$N'' = N^i(N)-N'^i(N'), \tag{10.145}$$
$$N'''^i = [N^i,N'^i]+N\nabla_h N'-N'\nabla_h N. \tag{10.146}$$

The situation can be easily visualized. Composition of two infinitesimal hypersurface deformations with parameters (N,N^i) and (N',N'^i) that maps $\partial M\mapsto\partial M_1$ and $\partial M_1\mapsto\partial M_{12}$ respectively, differs by the hypersurface deformation with parameters (N'',N'''^i) given by the composition that maps with the same parameters but in the opposite order.

$$\begin{array}{ccccc}
\partial M_1 & \xleftarrow{(N,N^i)} & \partial M & \xrightarrow{(N',N'^i)} & \partial M_2\\
{\scriptstyle (N',N'^i)}\downarrow & & & & \downarrow{\scriptstyle (N,N^i)}\\
\partial M_{12} & & \xleftarrow[(N'',N'''^i)]{} & & \partial M_{21}
\end{array} \tag{10.147}$$

Hamiltonian General Relativity is a particular Lie-anti representation of the algebra (10.137)-(10.138) as a Hamilton system on the phase space of physical fields. When the latter merely depends on the induced metric h_{ij} on ∂M, then the unconstrained phase space is the cotangent bundle $T^\star Riem(\partial M)$ over the space $Riem(\partial M)$, parameterized by the pair (h_{ij},π_{ij}). In this simplest situation one can search for

$$(N,N^i)\mapsto\left(\mathscr{H}(N,N^i):T^\star Riem(\partial M)\to\mathbb{R}\right) \tag{10.148}$$

where $\mathscr{H}$ is a distribution, the test functions are N and N^i, with values in real-valued functions on $T^\star Riem(\partial M)$

$$\mathscr{H}(N,N^i)[h_{ij},\pi_{ij}]=\int_{\partial M}d^3x\left(N(x)H(x)+N_i(x)H^i(x)\right). \tag{10.149}$$

The fundamental condition is that the Poisson brackets between two values of $\mathscr{H}(N,N^i)$ are

$$\left\{\mathscr{H}\left(N,N^i\right),\mathscr{H}\left(N',N'^i\right)\right\}=\mathscr{H}\left(N'',N''^i\right) \tag{10.150}$$

The essential question is recovering the action of the Lie algebra of four-dimensional diffeomorphism on the extended phase space including embedding variables.

The Hamiltonian Lie-anti representation of the diffeomorphism algebra (10.137)-(10.138) can be constructed by the identification

$$\mathscr{N}_N \mapsto \mathscr{H}(N) = \int_{\partial M} d^3xN(x)H(x), \tag{10.151}$$

$$\mathscr{T}_{N^i} \mapsto \mathscr{D}(N^i) = \int_{\partial M} d^3xN_i(x)H^i(x), \tag{10.152}$$

and the resulting algebra is the Lie algebra of the diffeomorphism group $Diff(\partial M)$

$$\left\{\mathscr{D}\left(N^i\right),\mathscr{D}\left(N'^i\right)\right\} = \mathscr{D}\left(\left[N^i,N'^i\right]\right), \tag{10.153}$$

$$\left\{\mathscr{D}\left(N^i\right),\mathscr{H}\left(N\right)\right\} = \mathscr{H}\left(N^i\left(N\right)\right), \tag{10.154}$$

$$\left\{\mathscr{H}\left(N\right),\mathscr{H}\left(N'\right)\right\} = \mathscr{D}\left(N\nabla_hN'-N'\nabla_hN\right), \tag{10.155}$$

which means that geometrodynamics does not define the extraordinary situation. Any four dimensional $Diff(\partial M)$-invariant theory will gives rise to this same algebra. The Lie algebra of $Diff(\partial M)$, namely, is universally satisfied for a theory considered in an arbitrary $space+time$ decomposition, which is formed by hypersurface foliations and is spacetime covariant. The universality of the Lie algebra of $Diff(\partial M)$ suggests searching for its another, possibly more general, representations on a given phase space.

The theorem due to Karel Vaclav Kuchař, Claudio Teitelboim, and Sergio Andrés Hojman [571], which states that for the the unique two-parameter family, given by κ and Λ, of realizations for the algebra (10.151)-(10.152) equipped with the constraints H and H^i, in which the conjugated momentum can be expressed through the extrinsic curvature tensor and the kinetic term is multiplied by an overall $n^\mu n_\mu$, is the most general Hamiltonian representation of the universal diffeomorphism Lie algebra (10.153)-(10.155) on $T^*Riem(\partial M)$, *i.e.* the space tangent to all C^∞ Riemannian metrics on the boundary ∂M, up to the residual canonical transformations having the following form

$$\pi^{ij} \mapsto \pi^{ij} + \frac{\delta F[h_{ij}]}{\delta h_{ij}}, \tag{10.156}$$

where F is some function invariant with respect to action of the diffeomorphism group $Diff(\partial M)$ determined on the space of all C^∞ Riemannian metrics on the boundary ∂M $Riem(\partial M)$.

Superspace possesses analogous ambiguity to the Aharonov–Bohm effect [312] following from the lack of simple connectedness of topology. This depends on the topology of ∂M, *i.e.* contractibility of the space $Riem(\partial M)$ results in the relation for n-th homotopy group

$$\pi_n\left(\frac{Riem(\partial M)}{Diff_F(\partial M)}\right) \cong \pi_{n-1}\left(Diff_F(\partial M)\right), \tag{10.157}$$

where $n \geqslant 1$. For $n=1$ one obtains the fundamental group of the extended superspace

$$\pi_1\left(S_F(\partial M)\right) \cong \pi_0\left(Diff_F(\partial M)\right) = \frac{Diff_F(\partial M)}{Diff_F^0(\partial M)} = MCG_F(\partial M), \tag{10.158}$$

where $Diff_F^0(\partial M)$ is the identity component of $Diff_F(\partial M)$, and $MCG_F(\partial M)$ denotes abbreviation of the name mapping-class group for frame fixing diffeomorphisms [566].

Uniqueness of representations of the Poisson brackets (10.150) has much more serious ambiguity. It is namely labeled by an additional $\mathbb{C}$-valued parameter, called the Barbero–Immirzi parameter, obtained due to connection variables (See Ref. [572] and the books [322, 526]). In such a situation, one does not represent the Poisson brackets (10.150) but a semi-direct product of these brackets with the Lie algebra of $SU(2)$ gauge transformations, and by taking the quotient with respect to this algebra the Poisson brackets (10.150) are represented non-locally. The Poisson brackets (10.150) do not seem to apply in the case of connection variables, because the connection variable does not permit an interpretation as a space-time gauge field restricted to space-like hypersurfaces and the dynamics generated by the constraints does not permit the interpretation of being induced by an appropriately moving hypersurface through a space-time with fixed geometric structures on it. The homotopy groups of the extended superspace, given by the formula (10.157), encode much of its global topology. They were investigated by Domenico J.W. Giulini [566] and Donald Michael Witt [573], and its quantum gravity context was investigated in the contributions due to John Friedman and Rafael Sorkin [574], Christopher John Isham [575], and Rafael Sorkin [576].

Topological invariants of $S_F(\partial M)$ are also topological invariants of ∂M, while in general homotopy invariants of ∂M are not homotopy invariants of $S_F(\partial M)$. Such a situation means that one can distinguish homotopy equivalent but non-homeomorphic three-manifolds by looking at homotopy invariants of their associated superspaces. When ∂M is a connected closed orientable three-manifold then the homology and cohomology groups are determined by the fundamental group. Namely, the first four (zeroth to third, the only non-trivial ones) homology and cohomology groups are

$$H_\star = (\mathbb{Z}, A\pi_1, FA\pi_1, \mathbb{Z}), \tag{10.159}$$
$$H^\star = (\mathbb{Z}, FA\pi_1, A\pi_1, \mathbb{Z}), \tag{10.160}$$

where A and F are the operation of Abelianization of a group and the operation of taking into account the free part of a finitely generated Abelian group, respectively. Particularly interesting is the fundamental group of the extended superspace. The analogy with quantum mechanics already suggests that its classes of inequivalent irreducible unitary representations correspond to a superselection structure which here might serve as a fingerprint of the topology of ∂M in the quantum theory. The sectors might, for instance, correspond to various statistics that preserve or violate a naively expected spin-statistics correlation (Cf. *e.g.* papers in Ref. [577]).

General three-manifolds can be understood by surgery (For theory Cf. *e.g.* books in Ref. [578], for applications Cf. *e.g.* the paper by Grigori Yakovlevich Perelman [579]). Particularly cutting along certain embedded two manifolds, such that the remaining pieces are simpler, is frequently applied technique. As the crucial and essential example, let us consider those simplifications which are achieved by cutting along embedded two-spheres. An essential two-sphere is one which does not bound a three-ball and a splitting two-sphere is one whose complement has two connected components. Let us consider closed three-manifold ∂M, which is cutting along an essential splitting two-sphere, capping off the two-sphere boundary of each remaining component by a three-disk, and this process is repeated for each of the remaining closed three-manifolds. This process stops after a finite number of steps where the resulting components are uniquely determined up to diffeomorphisms, orientation preserving if oriented manifolds are considered, and permutation [580]. The process stops at the stage at which none of the remaining components, $\Pi_1, \ldots, \Pi_n$, allows for essential splitting two-spheres, *i.e.* at which each Π_i is a prime manifold, *i.e.* such a manifold for which each embedded two-sphere either bounds a three-disc or does not split; it is called irreducible if each embedded two-sphere bounds a three-disc.

In the latter case the second homotopy group π_2 must be trivial. Otherwise the sphere theorem is violated. The theorem says that if $\pi_2 M = 0$ for connected three-manifold ∂M, then there is either an embedded two-sphere S^2 in M representing a non-trivial element in $\pi_2 M$, or an embedded two-sided real projective plane $\mathbb{R}\mathrm{P}^2$ in ∂M such that the composition of the cover $S^2 \to \mathbb{R}\mathrm{P}^2$ with the inclusion $\mathbb{R}\mathrm{P}^2 \hookrightarrow \partial M$ represents a non-trivial element of $\pi_2 M$. This theorem ensures existence of a non-trivial element of π_2 which could be represented by an embedded S^2. Conversely, it arises from the validity of the Poincaré conjecture that for trivial π_2 implies irreducibility. There is precisely one non-irreducible prime three-manifold, and that is the handle $S^1 \times S^2$. Hence three-manifold is prime when it is either handle or has trivial π_2. For general three-manifold ∂M being the connected sum of primes $\Pi_1, \ldots, \Pi_n$, there is the general method to establish $MCG_F(\partial M)$ in terms of $MCG_F(\Pi_i)$ which looks on the effect of elements in $MCG_F(\partial M)$ on the fundamental group of ∂M. As ∂M is the connected sum of primes, and as the connected sums in D dimensions are taken along $D-1$ spheres which are simply-connected for $D \geqslant 3$, the fundamental group of the connected sum is the free product of the fundamental groups of the primes for $D \geqslant 3$.

The group $MCG_F(\partial M)$ now naturally behaves like automorphisms of $\pi(\partial M)$ by simply taking the image of a based loop that represents an element in $\pi(\partial M)$ by a based (same basepoint) diffeomorphism which represents the class in $MCG_F(\partial M)$. Consequently, there is a natural map of $MCG_F(\partial M)$ into the group of automorphisms of the fundamental group of the D-dimensional manifold ∂M

$$d_F : MCG_F(\partial M) \to \mathrm{Aut}(\pi_1(\partial M)). \tag{10.161}$$

The known finite presentations of the automorphism group, *i.e.* its characterization by making use of a finite number of generators and finite number of relations, of free products in terms of presentations of the automorphisms of the individual factors and additional generators, basically exchanging isomorphic factors and conjugating whole factors by individual elements of others, can now be applied in order to establish finite presentations of the group $MCG_F(\partial M)$, provided finite presentations for all prime factors are known. Such a presentation of the automorphism group of free products was first constructed by David Izrailevich Fouxe–Rabinovitch [581], and its modern realizations with essential improvements were performed by Darryl John McCullough and Andrew Gardner Miller [582], and Nicholas David Gilbert [583]. This situation would be much more complicated if the diffeomorphism group $Diff(\partial M)$ rather than $Diff_F(\partial M)$, at least the diffeomorphisms fixing a preferred point, would be taken into account. Only for the case of the subgroup $Diff_F(\partial M)$, or the slightly larger group of diffeomorphisms fixing the point, it is generally true that the mapping-class group of a prime factor injects into the mapping-class group of the connected sum in which it emerges. For more detailed discussion, compare the discussion presented by Domenico J.W. Giulini in Ref. [584].

Strictly speaking, it is also necessary know which elements are contained in the kernel of the map (10.161). The cotangent bundle over the Wheeler superspace is not the fully reduced phase space for the case when the Matter fields are absent in General Relativity. This only takes into account the diffeomorphism constraints and leaves the scalar constraint unreduced. However, under certain specific requirements, the scalar constraint can be solved by application of the "conformal method" which leaves only the conformal equivalence class of three-dimensional geometries as the physical configurations. In all such instances, the fully reduced phase space is the cotangent bundle over the conformal superspace, which is given by replacement of the diffeomorphism group $Diff(\partial M)$ with the following semi-direct product

$$Diff^C(\partial M) = C(\partial M) \rtimes Diff(\partial M), \tag{10.162}$$

where $C(\partial M)$ denotes the Abelian group of the conformal re-scalings that acts on a space of all C^∞ Riemannian metrics on the boundary ∂M – $Riem(\partial M)$ by the pointwise multiplication $(f, h_{ij}) \mapsto f h_{ij}$, where $f : \partial M \to \mathbb{R}_+$. The right action of the map $(f, \phi) \in Diff_C(\partial M)$ on an arbitrary D-dimensional metric $h_{ij} \in Riem(\partial M)$ is then determined by the relation

$$R_{(f,\phi)}(h_{ij}) = f\phi^\star h_{ij}, \tag{10.163}$$

and, consequently, application of the property

$$R_{(f,\phi)} R_{(f',\phi')} = R_{(f,\phi)(f',\phi')} \tag{10.164}$$

allows to establish the semi-direct product structure

$$(f, \phi)(f', \phi') = (f'(f \circ \phi), \phi \circ \phi'). \tag{10.165}$$

In the light of the fact

$$(ff') \circ \phi = (f \circ \phi)(f' \circ \phi), \tag{10.166}$$

the diffeomorphism group $Diff(\partial M)$ indeed acts as the automorphisms of the Abelian group of the conformal re-scalings $C(\partial M)$.

The conformal superspace and the extended conformal superspace would then be defined as follows

$$CS(\partial M) = \frac{Riem(\partial M)}{Diff^C(\partial M)}, \tag{10.167}$$

$$CS_F(\partial M) = \frac{Riem(\partial M)}{Diff_F^C(\partial M)}, \tag{10.168}$$

where we have used the notation

$$Diff_F^C(\partial M) = C(\partial M) \rtimes Diff_F(\partial M). \tag{10.169}$$

Since the group $C(\partial M)$ is contractible, the topologies of the manifolds $Diff^C(\partial M)$ and $Diff_F^C(\partial M)$ are those of the manifolds $Diff(\partial M)$ and $Diff_F(\partial M)$ which also transcend to the quotient spaces whenever the groups act freely.

In the first situation this is essentially achieved by restricting to manifolds of the vanishing degree of symmetry, whereas in the second situation this follows almost as before, with the sole exception being (S^3, h_{ij}) with h_{ij} conformal to the round metric. Let

$$CJ(\partial M, h_{ij} = \{\phi \in Diff(\partial M) | \phi^\star h_{ij} = f h_{ij}, f : \partial M \to \mathbb{R}_+\} \qquad (10.170)$$

be the group of conformal isometries. For compact ∂M it is known to be compact except if and only if $\partial M = S^3$ and h_{ij} is conformal to the round metric [585]. Hence, for $\partial M \neq S^3$, we can average h_{ij} over the compact group $CJ(\partial M, h_{ij})$ and obtain a new Riemannian metric h'_{ij} in the conformal equivalence class of h_{ij} for which $CJ(\partial M, h_{ij})$ acts as proper isometries. Therefore, it cannot contain non-trivial elements fixing a frame. Hence in contrast, the geometry for conformal superspace differs insofar from that discussed above as the conformal modes that formed the negative directions of the DeWitt supermetric.

The horizontal subspaces, orthogonal to the orbits of $Diff_F^C(\partial M)$, are now given by the transverse and traceless symmetric two-tensors. In that sense the geometry of conformal superspace, if defined as before by some ultralocal bilinear form on $Riem(\partial M)$, is manifestly positive due to the absence of trace terms and hence less pathological than the superspace metric discussed above. It might seem that its physical significance is less clear, as there is now no constraint left that may be said to induce this particular geometry. It is hard to say whether it is a realistic hope to understand superspace and conformal superspace, its cotangent bundle being the space of solutions to the Einstein field equations, well enough to actually gain a sufficiently complete understanding of its automorphism group. An interesting strategy lies in the attempt to understand the solution space directly in a group-, or Lie algebra-, theoretic fashion in terms of a quotient G^∞/H^∞, where G^∞ is an infinite dimensional group (Lie algebra) that (locally) acts transitively on the space of solutions and H^∞ is a suitable subgroup (algebra), usually the fixed-point set of an involutive automorphism of G.

The basis for the hope that this might work, in general, is the fact that it works for the subset of stationary axially symmetric vacuum solutions of the Einstein field equations for which the infinite dimensional group G^∞ coincides with the Geroch group introduced by Robert Geroch [586]. The Geroch group is generated by two non-commuting subgroups: the Matzner–Misner group, implemented by Richard Alfred Matzner and Charles William Misner [587], which contains linear combinations of the two Killing vector fields associated with the axial symmetry and stationarity of the space-time, and the Ehlers group, named after Jürgen Ehlers [588], which is a finite-dimensional transformation group of stationary vacuum solutions of the Einstein field equations, which maps space-times to other space-times. The Ehlers group has plenty of applications: from generation of new solutions to a proof of integrable system character of stationary axially symmetric space-times.

One can classify the homotopy equivalent but non-homeomorphic three-dimensional ∂M by looking at the homotopy invariants of their associated superspaces $S_F(\partial M)$. Adequate examples are certain types of the lens spaces $L(p,q)$, for detailed discussion Cf. *e.g.* Ref. [589], which are prime manifolds. The lens spaces were introduced to mathematics by Heinrich Franz Friedrich Tietze [590] as the simplest possible examples of 3-manifolds obtained by identifying faces of a polyhedron. They were both first known three-manifolds not determined only by homology and fundamental group, as well as the most simple closed manifolds for which the homotopy type does not determine the homeomorphism type. James Waddell Alexander [591] proved that the lens spaces $L(5;1)$ and $L(5;2)$ are not homeomorphic despite the fact that their fundamental groups are isomorphic and their homology groups are identical, despite the fact that their homotopy type is not the same. The homotopy type of the lens spaces is the same and, therefore, other lens spaces have isomorphic fundamental groups and homology. However, in general the homeomorphism type is not the same for the lens spaces and for this reason the lens spaces can be formed from the geometric topology of manifolds as distinct from algebraic topology.

In the case of dimension 3 the lens spaces are defined as the quotient space of a three-sphere S^3

$$S^3 = \left\{(z_1, z_2) \in \mathbb{C} \times \mathbb{C} \middle| |z_1|^2 + |z_2|^2 = 1\right\}, \tag{10.171}$$

presented as the union of two solid tori

$$A_+ = \left\{|z| = 1 - |w|^2, |w| \leqslant \frac{\sqrt{2}}{2}\right\}, \tag{10.172}$$

$$A_- = \left\{|z| \leqslant \frac{\sqrt{2}}{2}, |w| = 1 - |z|^2\right\}, \tag{10.173}$$

whose common boundary torus is given as the zero level of the function $f(z,w) = |z|^2 - |w|^2$. These solid tori are invariant under the isometric free action of the cyclic group $\mathbb{Z}_p$ of order p

$$\mathbb{Z}_p = \left\{1, \varepsilon, \varepsilon^2, \ldots, \varepsilon^{p-1}\right\}, \tag{10.174}$$

where ε is a primitive p-th root of unity

$$\varepsilon = \exp\frac{2\pi i}{p}. \tag{10.175}$$

In other words, the three-dimensional lens space is the orbit space

$$L(p,q) = \frac{S^3}{\mathbb{Z}_p}, \tag{10.176}$$

where (p,q) is the pair of relatively prime (coprime) integers with $p > 1$. Moreover, after taking the quotient of the sphere, these solid tori are again transformed into solid tori into which the lens space $L(p,q)$ splits. The function $f(z,w)$ defined initially on the sphere generates a smooth function on $L(p,q)$. The levels of this function define some foliation on the lens space. The three-manifold $L(p,q)$ can be visualized easily. Namely, the equator of a 3-ball, *i.e.* solid ball in $\mathbb{R}^3$, is divided into p equal segments, so that the upper and lower hemispheres become p-sided polygons. These hemispherical faces are then identified by rotation through $2\pi\frac{q}{p}$ about the vertical symmetry axis, where $0 \leqslant q < p$ and $(p,q) = 1$. If a corner is introduced along the equator of the 3-ball it assumes the lens-shaped appearance that gave these manifolds their name, the term lens space being introduced by Herbert Karl Johannes Seifert and William Richard Maximilian Hugo Threlfall [592]. Heinrich Tietze noted that the lens spaces $L(p,q)$ may also be described as the manifold with a genus one Heegaard diagram consisting of a curve on the boundary of a solid torus which winds around p times latitudinally and q times meridianally.

The action of the cyclic group $\mathbb{Z}_p$ on the three-sphere S^3 is given by

$$z_1' = \varepsilon z_1, \tag{10.177}$$
$$z_2' = \varepsilon^q z_2. \tag{10.178}$$

In this way each set of p equidistant points on the equator is identified to a single point. The fundamental group of $L(p,q)$ is independent of q

$$\pi_1\left(L(p,q)\right) = \mathbb{Z}_p, \tag{10.179}$$

and the higher homotopy groups are those whose universal covering space is $L(1,0) = S^3$ with p sheets, *i.e.* p-fold 3-sphere. Therefore, also the standard invariants (10.159) and (10.160) taken for $L(p,q)$ are sensitive only to p. Two lens spaces $L(p,q)$ and $L(p,q')$ are

1. homotopy equivalent if and only if $qq' = \pm n^2 \mod p$ where $n \in \mathbb{Z}$ [593]

2. homeomorphic if and only if $q' = \pm q^{\pm 1} \mod p$ [594]

3. orientation-preserving homeomorphic if and only if $q' = q^{\pm 1} \mod p$ [594].

The torsion linking form is the invariant allowing to perform the homotopy classification of three-dimensional lens spaces and the Reidemeister torsion, named after Kurt Werner Friedrich Reidemeister [594], allows to make the homeomorphism classification. This invariant was formalized and generalized to the higher dimensional situations by Reidemeister's student Wolfgang Franz [595]. It is well known that Georges De Rham worked on the Reidemeister torsion [596], and his work began the rapid developments in this topic. De Rham together with Michel André Kervaire and Serge Maumary wrote a book devoted to the concept of invariant torsion [597]. The latter was performed also by Kurt Reidemeister [598] as a classification up to piecewise linear homeomorphism (For PL topology Cf. *e.g.* Ref. [599]), and E.J. Brody [600] showed that the Reidemeister construction is a classification up to homeomorphism. The generalization of the Reidemeister torsion to the case of Riemannian manifolds, known as the analytic torsion, was made by Daniel Burrill Ray and Isadore Manual Singer [601]. Jeff Cheeger [602] and Werner Müller [603] proved the Ray–Singer conjecture saying that for the case of compact Riemannian manifolds the concepts of analytic torsion and the Reidemeister torsion coincide. The Reidemeister torsion is the invariant which distinguish between spaces which are homotopy equivalent but non-homeomorphic and, consequently, can be the source of geometric topology as a distinct research field. This concept can be used for classification of the lens spaces. The lens spaces are determined by a simple homotopy type, and there are no normal invariants, like *e.g.* the characteristic classes, or the surgery obstruction. For example, Józef Henryk Przytycki and Akira Yasuhara [604] formulated the knot-theoretic classification: for a closed curve C in the universal cover of the lens space which lifts to a knot having a trivial Alexander polynomial, computation of the torsion linking form on the pair (C,C) gives the homeomorphism classification. Paolo Salvatore and Riccardo Longoni [605] showed that homotopy equivalent but not homeomorphic lens spaces may have configuration spaces with different homotopy types, which can be detected by different Massey products. The theory of the Reidemeister torsion and the analytic torsion, however, is not the main theme of this book, and we kindly advice the reader interested in more advanced discourse to take into account the professional literature [606].

The lens spaces were also natural subjects for investigations of a more algebraic topological nature. In this case, Marcel Rueff [607] showed that there exists a degree 1 map $L(p,q) \to L(p,q')$ if and only if $qq' \equiv r^2(\mod p)$, for some r. As showed by Ulrich Brehm and Jacek Świątkowski [608], lens spaces can be triangularized with few simplices. Lens space in the dimension three is the Seifert fiber space [609], *i.e.* a S^1-bundle (circle bundle) over a two-dimensional orbifold. In the dimension three the equivalence of the combinatorial and topological classifications, called *Hauptvermutung* and proved by Edwin Evariste Moise [610], is validated (For detailed discussion Cf. *e.g.* Ref. [611]). For given p, we aim to attach a homotopy invariant to $L(p,q)$ that depends on q. For this, one needs the Bockstein homomorphism. The short exact sequence of coefficient groups

$$0 \longrightarrow \mathbb{Z} \xrightarrow{p} \mathbb{Z} \longrightarrow \mathbb{Z}_p \longrightarrow 0 \tag{10.180}$$

induces the long exact sequence

$$\ldots \longrightarrow H^n(X;\mathbb{Z}) \xrightarrow{p} H^n(X;\mathbb{Z}) \xrightarrow{\rho} H^n(X;\mathbb{Z}_p) \xrightarrow{\beta_0} H^{n+1}(X;\mathbb{Z}) \longrightarrow \ldots \tag{10.181}$$

Let us take into account the homomorphism

$$\beta = \rho \circ \beta_0 : H^n(X;\mathbb{Z}_p) \to H^{n+1}(X;\mathbb{Z}_p), \tag{10.182}$$

which, as well as β^0 itself, is called the Bockstein homomorphism. It is natural in a space X, and it is called a cohomology operation. The Hurewicz theorem, the Universal Coefficient Theorem, and the Poincaré duality allow to compute the cohomologies ($L = L(p,q)$)

$$H_1(L;\mathbb{Z}) \approx \mathbb{Z}_p, \quad H^1(L;\mathbb{Z}) \approx 0, \quad H_1(L;\mathbb{Z}_p) \approx H^1(L;\mathbb{Z}_p) \approx \mathbb{Z}_p, \tag{10.183}$$

$$H_2(L;\mathbb{Z}) \approx 0, \quad H^2(L;\mathbb{Z}) \approx \mathbb{Z}_p, \quad H_2(L;\mathbb{Z}_p) \approx H^2(L;\mathbb{Z}_p) \approx \mathbb{Z}_p, \tag{10.184}$$

$$H_3(L;\mathbb{Z}) \approx \mathbb{Z}, \quad H^3(L;\mathbb{Z}) \approx \mathbb{Z}, \quad H_3(L;\mathbb{Z}_p) \approx H^3(L;\mathbb{Z}_p) \approx \mathbb{Z}_p. \tag{10.185}$$

and the exact sequence

$$H^1(L;\mathbb{Z}) \longrightarrow H^1(L;\mathbb{Z}_p) \xrightarrow{\beta_0} H^2(L;\mathbb{Z}) \xrightarrow{0} H^2(L;\mathbb{Z}) \xrightarrow{\rho} H^2(L;\mathbb{Z}_p) \tag{10.186}$$

shows that β_0 and ρ are isomorphisms, and therefore

$$\beta : H^1(L;\mathbb{Z}_p) \to H^2(L;\mathbb{Z}), \tag{10.187}$$

is an isomorphism.

D Problems of Geometrodynamics

The key problem of quantum geometrodynamics are observables. The important notion is the space of solutions of all constraints $\mathscr{F}_0$ which is a subspace of a functional space $\mathscr{F}$ of quantum states represented in the Schrödinger picture by wave functionals of the three-metric $\Psi[h_{ij}]$. Physical states must belong to $\mathscr{F}_0$ in order to be invariant under the symmetries encoded in constraints. The Dirac observables must commute with all the first-class constraints generating gauge transformations, $[O,H]\Psi = 0$, so the action of an observable on a physical state does not project the state out of the space of physical states $\mathscr{F}_0$. An inner product must be defined on $\mathscr{F}_0$ in order to obtain an Hilbert space of physical normalized state vectors.

Karel Vaclav Kuchař [612] considered the problem of observables, and concluded that the observables are defined by non-vanishing Poisson brackets with all of the constraints, but claimed that the rightness of such a treatment is justified for the diffeomorphism constraint, while is manifestly wrong for the Hamiltonian constraint because H generates the dynamics between the constant-time hypersurfaces. Both a hypersurface as well as its points in themselves are not directly observable. Albeit, the values of the canonical pair (q_{ij}, π^{ij}) are evidently distinguishable on an initial hypersurface and on an evolved hypersurface and, moreover, the effects of the evolution are not possible to observe when there is a lack of difference between these values. For consistency, Kuchař introduced two types of variable, *i.e.* observables and perennials. In his approach observables are the diffeoinvariant dynamical variables which do not commute with the Hamiltonian constraint, meanwhile perennials are the observables commuting with the Hamiltonian constraint and cannot be observed because the Hamiltonian constraint should not be seen as a generator of the gauge transformations. For this reason the observables do not act on the space of physical solutions $\mathscr{F}_0$.

In spite of a number of details, in general the strategy proposed by Kuchař can be summarized concisely. First of all, one must find the four kinematic variables $X^A : \partial M \mapsto M$, where $A = 0, 1, 2, 3$, which represent a space-like embedding of a space-like hypersurface ∂M into the space-time manifold M. These scalar fields represent the space-time positions and observables evolving along M, which are the true gravitational degrees of freedom and the dynamical variables separated out from these fields on the level of phase space. The second point is to interpret the constraints as conditions and identify the momenta P_A conjugated to X_A, which determine the evolution of the degrees of freedom between the hypersurfaces, with the energy-momenta of the degrees of freedom. Such a procedure involves solving the constraints on the classical level, necessity of the internal time, and quantization formulated in terms of the Tomonaga–Schwinger equation [613, 614]

$$i\frac{\delta\Psi[\phi^r(x)]}{\delta X^A(x)} = h_A\left(x; X^B, \phi^r, p^s\right)\Psi[\phi^r(x)], \tag{10.188}$$

where $r, s = 1, 2$ and the variables X^A are treated as classical, like time in quantum mechanics. There arise problems which include multiple-choice, no global time, problem in definition of the Hamiltonian h_A, and many others. J. David Brown and Karel V. Kuchař [615] introduced matter variables, which label space-time points and are coupled to space-time geometry, instead of functionals of the gravitational variables. They proposed to take into account a dust field filling all space and playing the role of time, which includes an internal time variable against which systems can evolve, and which can play a role of the fixed background for the construction of quantum gravity. In the Brown–Kuchař formalism the Schrödinger equation can be written out and the emerging Hamiltonian does not depend on the dust variables.

Other solution of the problem of time, the so-called unimodular gravity, was proposed by William George Unruh [616], who modified General Relativity such that the cosmological constant is a dynamical variable for which the conjugate is taken to be cosmic time. The result is that the Hamiltonian constraint is augmented by a cosmological constant term giving the modified Hamiltonian constraint $\Lambda + \sqrt{h}H = 0$. The presence of this extra term and cosmic time τ unfreezes the dynamics and leads to the τ-dependent Schrödinger equation.

Bryce Seligman DeWitt [176] also tried to solve the problem of time in the framework of quantum geometrodynamics. It is the problem of extracting a notion of time from timeless dynamics described by the Wheeler–DeWitt equation. A consequence of the timeless nature of this equation is the problematic implementation of an inner product for state vectors. In analogy to the inner product obtained from the Klein–Gordon equation, B.S. DeWitt proposed the definition the inner product of two solutions of the Wheeler–DeWitt equation

$$(\Psi_a, \Psi_b) = Z \int \Psi_a^*[{}^{(3)}\mathfrak{G}] \prod_x \left(d\Sigma^{ij} G_{ijkl} \frac{\overrightarrow{\delta}}{i\delta h_{kl}} - \frac{\overleftarrow{\delta}}{i\delta h_{kl}} G_{ijkl} d\Sigma^{ij} \right) \Psi_b[{}^{(3)}\mathfrak{G}], \tag{10.189}$$

where the product is taken over all the points of a three-dimensional embedded space ∂M, integration is over a $5 \times \infty^3$-dimensional surface in $S(\partial M)$, $d\Sigma^{ij}(x)$ is the surface element of the topological product of a set of 5-dimensional hypersurfaces $\Sigma(x)$ chosen at each point of ∂M, and Z is the normalization constant. The Klein–Gordon inner product (10.189) is invariant under the deformation of the $5 \times \infty^3$ surface, but is not positively defined and vanishes for real solutions of the Wheeler–DeWitt equation. Moreover, by such a treatment, all problems related to the Klein–Gordon equation, like *e.g.* no separation into positive and negative frequencies and the negative probability, in general are available for the Wheeler-DeWitt equation.

There are also other problems within quantum geometrodynamics following from the Wheeler–DeWitt equation. First of all, it is the initial data problem. Namely, in quantum geometrodynamics the classical space-time is the history of space geometry governed by the deterministic evolution. There arises an uncertainty relation between intrinsic and extrinsic geometry due to the Lie bracket of induced metric and its conjugated momentum. For this reason interpretation of the properties of quantum space-time is unclear. The second important problem is that, for consistency, the standardly applied quantization of the constraints needs a regularization method to be selected. However, factor orderings lead to a non-unique result and quantum anomalies, *i.e.* the most terrible theoretical ambiguities. The problem of indefiniteness of measure in the Wheeler superspace arises from the definition of the inner product. The canonical variables are present in the non-polynomial way in the Wheeler–DeWitt equation, which in itself gives rise to problematic analysis. As we have mentioned earlier, the Wheeler–DeWitt equation has never been solved in general, the only simplest Feynman's path integral solutions are discussed and the general integrability problem seems to be omitted. Another question is the interpretation of both wave functionals solving the Wheeler–DeWitt equation as well as their normalization and superpositions. In fact, no Dirac observable of the quantum geometrodynamics is known. In fact, both classical and quantum geometrodynamics are time-independent evolutions, but both the problem of time and, therefore, also the quantum evolution are still unsolved.

The model of quantum cosmology presented in the previous chapter has the structure of the Wheeler superspace which, however, for simplicity was not stated explicitly. In the situation given by space-time equipped with the Friedmann–Lemaître–Robertson–Walker metric, the configurational space is the stratum of superspace known under the name *minisuperspace*. We shall continue studying metric tensors considered according to the Arnowitt–Deser–Misner decomposition. Such space-times define all isotropic solutions of the Einstein field equations and in itself create another stratum of superspace known as *midisuperspace*. The midisuperspace models of quantum cosmology and quantum gravity are not presently the most popular theme in modern theoretical gravitational physics. In general, quantum geometrodynamics by its functional nature has the status of not a very well-defined mathematical theory rather than a theory of quantum gravity possessing physical significance. However, this is only ostensible problem, and we shall see in the next chapters that this problem can be omitted. We shall present the new constructive approach to such theories based on well established methods of quantum field theory, which leads to plausible sounding quantum gravity phenomenology. The resulting plausibility is not a coincidence, but is the straightforward ramification of the constructive application of the methods of quantum field theory which possess the established significance for physics. It should be emphasized that such a strategy is fully justified for one-dimensional quantum gravities. Albeit, both its applicability to and usefulness for other possible situations are the good questions for further research. On the one hand, one can suppose *ad hoc* that the one-dimensionality of quantum gravity is its universal *physical* feature, and other situations are non-physical. On other hand, one can suppose that the universal nature of one-dimensional quantum gravity arises as the result of some projection of higher-dimensional quantum gravities. However, such a reasoning in itself is the attempt to preserve *ad hoc* applicability of quantum field theory for the consistent and constructive formulation of quantum gravity, while in the light of the recent trends the legitimateness of quantum field theoretic methods applied to rather non usual situations in itself is a moot point. It should be emphasized that everything what is widely applied and developed in modern theoretical physics is primarily rooted in the methodology offered by quantum field theory. The best possible example of such a state of affairs is the string theory which in itself is a quantum field theory, and is one of the most popular and intriguing research trends. For this reason, the consistency of the necessity of making the construction of the adequate quantum field theoretic formalism for quantum gravity, *i.e.* quantum theory of gravitational field or quantum field theory of gravity, is fully argued. It should be emphasized that any logical argument must not be satisfied by Nature, and for this reason the results received by the adequate formalism should receive an empirical verification. Otherwise, the physical meaning of the theory will remain unclear.

Factually, the quantum geometrodynamics based on the Wheeler–DeWitt equation is historically the first consistent and constructive attempt to formulate the quantization of General Relativity and the resulting model of quantum gravity. Actually, however, quantum geometrodynamics has became the most influential motivation for development of both alternative theories of quantum gravity based on quantum geometrodynamics as well as searching for completely different standpoints. In the next chapters of this book we shall present the model of quantum gravity which is strictly based on the Wheeler-DeWitt equation (10.99) briefly discussed above. The basics and the applications of the Arnowitt–Deser–Misner formulation of General Relativity, detailed considerations of the classical and the quantum geometrodynamics, and ramifications of the Wheeler–DeWitt equation have been studied intensively in the scientific and research literature for more than fifty years (Cf. *e.g.* Refs. [617]-[888]). In the absence of other constructive competitors, the quantum geometrodynamics remains the most consistent and constructive theory of quantum gravitational fields and, factually, the only one having real chances for predictions of the constructive quantum gravity phenomenology.

E Alternative Approaches to Quantum Gravity

There are several alternative approaches to quantum gravity. One such point of view arises from application of the Ashtekar Hamiltonian formulation of General Relativity [889], which applies the Einstein–Cartan theory with a complex connection, and is based on the so-called Ashtekar's new variables. Carlo Rovelli and Lee Smolin [890] used Ashtekar's new variables to investigation of the loop representation of quantum General Relativity. This research trend has been popularized by the research by Abhay Ashtekar, Carlo Rovelli, Lee Smolin, Theodore A. Jacobson, and Jerzy Krzysztof Lewandowski [891], in the quantum gravitational context, and by Martin Bojowald [892] in the quantum cosmological considerations. The resulting theories are known under the names loop quantum gravity/cosmology and take into account the fundamental role of diffeomorphisms, including the diffeomorphism constraint which in the quantum geometrodynamics, formulated in terms of the Wheeler–DeWitt equation, does not play a crucial role for dynamics and merely reflects diffeoinvariance. In loop quantum gravity an important role in played by the theorem due to A. Ashtekar and J. Lewandowski [893] on the bijection between unitary representations of a certain group, which is known as the Ashtekar–Lewandowski group, on a fixed finite-dimensional Hilbert space and the spectrum points of a particular $C^\star$ algebra generated by the Wilson loop functions, the well known approach to Yang–Mills theories. Its main aim is to describe the equivalence classes of gauge equivalent connections and their ideal limits. For the $SU(2)$ the proof was given by Ashtekar and Lewandowski, and is based on the previous results of Abhay Ashtekar and Christopher John Isham [894] and Roscoe C. Giles [895]. For a general case of $SU(N)$ and $U(N)$ gauge groups the proof using the Mandelstam identities was suggested in the paper by Ashtekar and Lewandowski [896], where another approach to their theorem is presented. This involves arbitrary compact connects structure groups by using the technique of projective limits of projective families of the Hausdorff spaces, and there is no counterpart of the the so-called hoop group. Recently, the research trend of loop quantization received the well-established research status and is still under intensive development (See, *e.g.* the papers collected in Ref. [897]).

Loop quantization program offered by Ashtekar and Lewandowski is devoted to constructing the consistent and constructive model of the Hamiltonian non-perturbative quantization of pure and constrained gauge theories. Particularly, the General Relativity written in terms of Ashtekar's new variables and the three-dimensional Chern–Simons theory is a constrained gauge theory. The non-perturbative level of the quantization procedure is, according to Ashtekar's views, essential to study quantum aspects of gravity. The perturbation models are background dependent, *i.e.* take into account existence of a base geometry – the Minkowski space-time – and perturbate it by implementation of space-time distortions which violate the Einstein equivalence principle. Loop quantization does not require new structures, like strings or supersymmetry, applies new techniques and for this reason is very close to the conventional canonical quantum field theory. The program is based on steps. The first is the individuation of a collection of gauge-invariant functions, which happens to be the Wilson loop functions, on the configuration space of the classical theory (which is the Wheeler superspace in the case of General Relativity) which is complete because contains all gauge-inequivalent degrees of freedom of the theory. The second step is construction of the $C^{\star}$ algebra generated by such functions and leads to the quantization by using the Gelfand–Naimark–Segal representation, named after Izrail Moiseevich Gelfand, Mark Aronovich Naimark [898], and Irving Ezra Segal [899], of this $C^{\star}$-algebra carried on a certain Hilbert space, which becomes the kinematical Hilbert space of the quantum theory. The $C^{\star}$ algebra is then constructed. The last step is promotion of the configuration space observables and their conjugate momenta to self-adjoint operators on the kinematical Hilbert space, which makes the canonical commutation relations essential. If there are another constraints, like the diffeomorphism and the Hamiltonian constraints of General Relativity, then they are forced by averages of the operator equations satisfied by the vector states belonging to the kinematical Hilbert space. The physical Hilbert space is a subspace of the kinematical Hilbert space, which contains the states solving the constraints. Presently, the kinematical part of the program is complete, while the dynamical part is an open problem of the loop quantization program.

The Arnowitt–Deser–Misner and the Ashtekar Hamiltonian formulations of General Relativity present manifestly different strategies for understanding and searching for solutions to the Einstein field equations. This heritage possesses straightforward reflection in the evident differences between quantum geometrodynamics and loop quantum gravity/cosmology following from the Arnowitt–Deser–Misner and the Ashtekar Hamiltonian formulations of General Relativity, respectively. Quantum gravity, formulated in terms of the Wheeler–DeWitt equation, is commonly understood as the established model, while loop quantum gravity, similarly to string theory, despite the fact that is presently under intensive development due to a number of theoreticians and mathematicians, still presents a highly speculative level of discussion. The attempts of quantum geometrodynamics were practically obscured by the alternative loop quantum gravity approach, while the Wheeler–DeWitt equation in itself still needs development and is the most probable source of a consistent and constructive quantum gravity phenomenology.

There are a number of alternative evolution schemes, known under the overall term *numerical relativity* (For modern analysis and discussion Cf. *e.g.* Ref. [900]), which presently in general is not taken into account in construction of the models of quantum gravity. The privileged strong position of the Arnowitt–Deser–Misner and Ashtekar Hamiltonian formulations of General Relativity arises from their straightforward roots in the Hamiltonian analysis, because the primary canonical quantization procedure is one of the ramifications of the Hamiltonian analysis and is naturally applicable in such a formulation. Standardly, alternative evolution schemes are strictly based on these two canonical formulations, are their extensions, modifications, or their particular cases. The crucial issue, however, which connects all the numerical schemes, is the consistent formulation of the Cauchy problem for the Einstein field equations.

The pioneering approach to the Cauchy problem for General Relativity in the case of the analytic initial data was proposed by Georges Darmois [901] in 1927 and André Lichnerowicz [902] in 1939. In 1944, Lichnerowicz [903] proposed the first known $3+1$ formalism based on the conformal decomposition of a spatial metric. In 1952, Yvonne Fourès-Bruhat (Choquet-Bruhat) [904] presented the formulation of the Cauchy problem for C^5 initial data by application of the local existence and uniqueness in harmonic coordinates, which in 1956 resulted in the $3+1$ formalism in moving frame [905]. In 1962, Richard Lewis Arnowitt, Stanley Deser, and Charles William Misner [170] proposed the formalism of the $3+1$ splitting which is based on the Hamiltonian analysis of General Relativity and lays the foundations for the present-day models of quantum gravity. Soon after, in 1972, James Wesley York, Jr. [906] considered the problem of gravitational dynamical degrees of freedom related to the conformal spatial metric, and in 1974 together with Niall Ó Murchadha [907] implemented the conformal transverse-traceless (CTT) method for solving the resulting constraint equations. In 1977, Larry Smarr [908] considered a two-dimensional axially symmetric head-on collision of two black holes and produced the first numerical solution beyond the spherical symmetry of the Cauchy problem for asymptotically flat space-times. In 1978, Larry Smarr and James W. York, Jr. [909] proposed using the so-called radiation gauge for numerical relativity which resulted in an elliptic-hyperbolic system with an asymptotic transverse-traceless (TT) behaviour. In 1983, James Maxwell Bardeen and Tsvi Piran [910] considered two-dimensional computations of partially constrained schemes. Takashi Nakamura [911] in 1983, and Richard F. Stark and Tsvi Piran [912] in 1985 applied a two-dimensional axially symmetric gravitational collapse to a black hole. In 1986, Abhay Ashtekar [889] proposed the new variables mentioned above, which presently are the most serious alternative to the ADM formalism and one of their ramifications is loop quantum gravity. In 1987, Takashi Nakamura, Ken-ichi Oohara, and Yasufumi Kojima [913] tested evolution of pure gravitational wave space-times in spherical coordinates. In 1989, Carles Bona García and Joan Massó [914], in 1995 Yvonne Choquet-Bruhat and James Wesley York, Jr. [915], and in 2001 Lawrence E. Kidder, Mark A. Scheel and Saul Arno Teukolsky [916] considered the first-order symmetric hyperbolic formulations of the Einstein field equations within the formalism of the $3+1$ splitting. Masaru Shibata and Takashi Nakamura [917] in 1995, and independently Thomas W. Baumgarte and Stuart L. Shapiro [918] in 1999 investigated the so-called BSSN formulation, which results from the conformal decomposition of the $3+1$ Einstein field equations and promotion of some connection function as an independent variable. In 1999, James Wesley York, Jr. [919] implemented the conformal thin-sandwich (CTS) method for solving the constraint equations. In 2000, Masaru Shibata [920] performed three-dimensional full computation of binary neutron star merger, which was the first full general relativistic three-dimensional solution of the Cauchy problem in the astrophysical context. In 2000, Sean A. Hayward [921] proposed a new scheme involving a dual-null decomposition of space-time and removing second-order terms from the Einstein field equations, which would vanish in the case of spherically symmetric space-time. In 2004, Silvano Bonazzola, Eric Gourgoulhon, Philippe Grandclément, and Jérôme Novak [922] proposed the constrained scheme based on maximal slicing and the Dirac gauge.

All these approaches of numerical relativity, in fact, construct the modern approach to General Relativity which is based on approximation methods. The spectrum of the applications of these models is too wide for discussion in this book. We shall focus here on the most fundamental one - the Arnowitt–Deser–Misner decomposition.

Chapter 11

Global One-Dimensionality Conjecture

A Introduction

This chapter is devoted to our proposition for the consistent formulation of the quantum gravity theory. The physical role of quantum aspects of gravity is the fundamental problem of modern theoretical physics. For instance, due to the absence of a consistent theory of quantum gravity we are not able to understand the physics of our Universe at the Planck scale level. Factually, despite of the fact that there has been a number of significant efforts for such a theorization (For the various approaches Cf. *e.g.* Refs. [322], [348] and [482]-[530]), the present-day point of view is very far from understanding the significance of quantized gravitational fields for the physical phenomena at high and ultra-high energies. In this chapter, we propose a very simple but eventually effective model of quantum gravity which can be helpful for the clarification of many quantum aspects of gravity and leading to constructive quantum gravity phenomenology. However, the simplicity of the theory of quantum gravity presented here in itself is very far from triviality and the arguments leading to this model are not obvious and have not formulated in the hitherto existing literature. On the one hand, the model is manifestly *ad hoc* but, on the other hand, is a straightforward ramification of the quantum geometrodynamics based on the Wheeler–DeWitt equation. Moreover, our model in itself is a certain particular realization of the general theory of the Wheeler superspace. Our model is significantly simpler than the usual quantum geometrodynamics and for this reason is able to generate new ramifications in the light of the straightforward analogy with the established phenomenological models of theoretical physics.

The formalism of classical and quantum field theory, so celebrated in modern physics, yields a plausible phenomenology for a number of experimental data coming from a rich spectrum of observations. In this chapter this point of view is considered as the foundation for the construction of a simple theory of quantum gravity. We shall perform such a construction by application of the standard strategy which results in the Wheeler–DeWitt equation. The difference between our model and the standard quantum geometrodynamics is fundamental and is contained in the suitable approach to Matter fields resulting in the specific behaviour of the wave functionals solving the equivalent of the Wheeler–DeWitt equation. The $3+1$ splitting of a general relativistic metric tensor and the canonical primary quantization of the appropriate Hamiltonian and diffeomorphism constraints are employed in the way well-grounded in all the approaches to quantization of gravitation. The modification of the standard quantum geometrodynamics is done on the basis of the *global one-dimensionality conjecture*, which in itself is not beyond the quantum geometrodynamics and arises from the straightforward and strict analogy with the generic cosmological model [531] presented in the earlier chapters of this book. The crucial proposal for this model of quantum gravity is the ansatz which can be summarized by four brief statements

1. Investigation of the global one-dimensionality conjecture, *i.e.* taking into account a certain specific one-dimensional nature of Matter fields and the wave functional,

2. Reduction of quantum geometrodynamics, which results in the one-dimensional theory characteristic for bosonic fields,

3. Application of the Hamilton equations of motion, yielding the corresponding one-dimensional Dirac equation,

4. Expression of the supposition that the quantum gravity is a one-dimensional field theory, and performing its secondary quantization.

Application of the Hamilton equations of motion allows to establish the appropriate one-dimensional Dirac equation and the corresponding Clifford algebra. The secondary quantization method, based on the static Fock space and the diagonalization procedure consisting of the Bogoliubov transformation and the Heisenberg equations of motion, yields the correctly defined quantum field theory formulated in terms of the static Fock repère associated with the initial data. We derive the general form of the one-dimensional wave functional and discuss the corresponding three-dimensional manifolds. Quantum correlations of the field are associated with physical scales. Mathematically, we employ the one-dimensional functional integrals and, therefore, despite the fact that the model of quantum gravity corresponds to the trend initiated by Stephen William Hawking and his collaborators [532]-[547] derivation of its solutions is significantly simplified.

B Γ-Scalar-Flat Space-Times

Let us consider first the relation (10.71), *i.e.*

$$S-\rho=\frac{4\Lambda}{\kappa}-\frac{^{(4)}R}{\kappa}. \tag{11.1}$$

In the light of the rule (10.40) one has

$$^{(4)}R={}^{(3)}R+K^2-K^{ij}K_{ij}, \tag{11.2}$$

where we have omitted the total derivative, because its vanishing in Hamiltonian analysis. Applying the Hamiltonian constraint

$$^{(3)}R+K^2-K^{ij}K_{ij}-2\Lambda-2\kappa\rho\approx 0, \tag{11.3}$$

one obtains another important relation between the energy density and the spatial stress density of Matter fields

$$\rho=\frac{2\Lambda}{\kappa}-S, \tag{11.4}$$

which can be presented as the equation for the cosmological constant

$$\Lambda=\frac{\kappa}{2}(S+\rho), \tag{11.5}$$

and gives the insight into the nature of the cosmological constant. In other words, the cosmological constant is an arithmetic mean of the spatial stress density and the energy density multiplied by the Einstein constant κ. One can, however, also apply the difference (10.68) $S-\rho=T$ together with the equation (11.5) and establish

$$\rho = \frac{\Lambda}{\kappa}-\frac{T}{2}, \tag{11.6}$$

$$S = \frac{\Lambda}{\kappa}+\frac{T}{2}. \tag{11.7}$$

Application of the definition $\rho=T_{\mu\nu}n^\mu n^\nu$ within the equation (11.6) allows to rewrite this equation in the following form

$$T_{\mu\nu}\left(g^{\mu\nu}+2n^\mu n^\nu\right)=\frac{2\Lambda}{\kappa}, \tag{11.8}$$

and solve it immediately with respect to the stress-energy tensor

$$T_{\mu\nu} = \frac{2\Lambda}{\kappa} \frac{1}{g^{\mu\nu} + 2n^\mu n^\nu}. \tag{11.9}$$

Let us consider the right hand side of this expression. Taking into account the following identities

$$g^{\mu\nu} n_\mu n_\nu = g_{\mu\nu} n^\mu n^\nu = n^\mu n_\mu = n_\mu n^\mu = -1, \tag{11.10}$$
$$g^{\mu\nu} g_{\mu\nu} = 4, \tag{11.11}$$

one receives

$$\begin{aligned} \frac{1}{g^{\mu\nu} + 2n^\mu n^\nu} &= \frac{(g_{\mu\nu} + 2n_\mu n_\nu)}{(g^{\mu\nu} + 2n^\mu n^\nu)(g_{\mu\nu} + 2n_\mu n_\nu)} = \\ &= \frac{(g_{\mu\nu} + 2n_\mu n_\nu)}{g^{\mu\nu} g_{\mu\nu} + 4(n^\mu n_\mu)^2 + 4n^\mu n_\mu} = \frac{1}{4}(g_{\mu\nu} + 2n_\mu n_\nu), \end{aligned} \tag{11.12}$$

and for this reason the stress-energy tensor (11.9) becomes

$$T_{\mu\nu} = \frac{\Lambda}{2\kappa} g_{\mu\nu} + \frac{\Lambda}{\kappa} n_\mu n_\nu, \tag{11.13}$$

and its covariant form also can be also derived straightforwardly

$$T^{\mu\nu} = \frac{\Lambda}{2\kappa} g^{\mu\nu} + \frac{\Lambda}{\kappa} n^\mu n^\nu. \tag{11.14}$$

Now it can be unambiguously recognized that the resultant stress-energy tensor (11.14) coincides with the stress-energy tensor of the perfect fluid (Cf. *e.g.* Ref. [923])

$$T^{\mu\nu} = p g^{\mu\nu} + \left(\frac{p}{c^2} + \mu\right) u^\mu u^\nu, \tag{11.15}$$

for which the four-velocity u^μ equals the unit normal vector field multiplied by the speed of light c, *i.e.*

$$u^\mu = c n^\mu, \tag{11.16}$$
$$u^\mu u_\mu = -c^2, \tag{11.17}$$

and the isotropic pressure p and the mass density μ are as follows

$$p = \frac{\Lambda}{2\kappa}, \tag{11.18}$$

$$\mu = \frac{\Lambda}{2c^2\kappa}. \tag{11.19}$$

Now the trace of the stress-energy tensor can be established by straightforward easy computation

$$\begin{aligned} T &= g^{\mu\nu}T_{\mu\nu} = \frac{\Lambda}{2\kappa}g^{\mu\nu}g_{\mu\nu} + \frac{\Lambda}{\kappa}g^{\mu\nu}n_\mu n_\nu = \frac{\Lambda}{2\kappa}4 + \frac{\Lambda}{\kappa}n^\mu n_\mu = \\ &= \frac{2\Lambda}{\kappa} - \frac{\Lambda}{\kappa} = \frac{\Lambda}{\kappa}, \end{aligned} \tag{11.20}$$

and, consequently, the energy density (11.6) and the spatial stress density (11.7) of Matter fields have the values

$$\rho = \frac{\Lambda}{2\kappa}, \tag{11.21}$$

$$S = \frac{3\Lambda}{2\kappa}. \tag{11.22}$$

The momentum density related to such a situation can be established straightforwardly

$$J^i = T_{\mu\nu}n^\mu h^{\nu i} = \left(\frac{\Lambda}{2\kappa}g_{\mu\nu} + \frac{\Lambda}{\kappa}n_\mu n_\nu\right)n^\mu h^{\nu i} \tag{11.23}$$

$$= \frac{\Lambda}{2\kappa}n_\nu h^{\nu i} - \frac{\Lambda}{\kappa}n_\nu h^{\nu i} = -\frac{\Lambda}{2\kappa}n_\nu h^{\nu i} \tag{11.24}$$

$$= -\frac{\Lambda}{2\kappa}n_\nu\left(g^{\nu i} + n^\nu n^i\right) = -\frac{\Lambda}{2\kappa}\left(n^i - n^i\right) = 0. \tag{11.25}$$

For this reason, the classical geometrodynamics becomes

$$2\kappa G_{ijkl}\frac{\delta S[g]}{\delta h_{ij}}\frac{\delta S[g]}{\delta h_{kl}} + \frac{\sqrt{h}}{2\kappa}\left({}^{(3)}R - 3\Lambda\right) = 0, \tag{11.26}$$

while the Wheeler–DeWitt equation is

$$\left\{2\kappa G_{ijkl}\frac{\delta^2}{\delta h_{ij}\delta h_{kl}} + \frac{\sqrt{h}}{2\kappa}\left({}^{(3)}R - 3\Lambda\right)\right\}\Psi[h_{ij},\phi] = 0, \tag{11.27}$$

and the quantized diffeomorphism constraint is

$$i\left(\partial_j+\frac{1}{2}h_{jl,k}h^{kl}\right)\frac{\delta\Psi[h_{ij},\phi]}{\delta h_{ij}}=0. \tag{11.28}$$

In other words, in such a situation both the classical and quantum geometrodynamics become purely geometrical.

In such a situation, a good question is what is the Lagrangian of Matter fields describing such a situation. Rewriting the stress-energy tensor (11.13) in the form

$$T_{\mu\nu}=\frac{1}{2}\left(3p+\mu c^2\right)g_{\mu\nu}, \tag{11.29}$$

and using the definition (10.52) of $T_{\mu\nu}$ following from the Hilbert–Palatini action principle one obtains the equation

$$-\frac{2}{\sqrt{-g}}\frac{\delta}{\delta g^{\mu\nu}}\left(\sqrt{-g}L_\phi\right)=\frac{1}{2}\left(3p+\mu c^2\right)g_{\mu\nu}, \tag{11.30}$$

which can be presented in the equivalent form

$$\delta\left(\sqrt{-g}L_\phi\right)=-\frac{1}{4}\left(3p+\mu c^2\right)\sqrt{-g}g_{\mu\nu}\delta g^{\mu\nu}. \tag{11.31}$$

Using the Jacobi formula for differentiating a determinant

$$\delta g=gg^{\mu\nu}\delta g_{\mu\nu}=-gg_{\mu\nu}\delta g^{\mu\nu}, \tag{11.32}$$

allows to write the equation (11.31) as

$$L_\phi\delta\sqrt{-g}+\sqrt{-g}\delta L_\phi=\frac{1}{2}\left(3p+\mu c^2\right)\delta\sqrt{-g}. \tag{11.33}$$

However, because the parameters p and μ are constant, one has uniquely

$$L_\phi = \frac{1}{2}\left(3p+\mu c^2\right), \tag{11.34}$$

$$\delta L_\phi = 0. \tag{11.35}$$

Applying the relations (11.18) and (11.19) to the formula (11.34) one receives finally

$$L_\phi=\frac{\Lambda}{\kappa}=-\rho. \tag{11.36}$$

The spatial stress density $S = h^{ij}S_{ij}$ together with the formula (11.22) can be used for derivation of the spatial stress tensor

$$S_{ij} = \frac{S}{3}h_{ij} = \frac{\Lambda}{2\kappa}h_{ij}, \tag{11.37}$$

and together with the difference

$$S - \rho = \frac{\Lambda}{\kappa}, \tag{11.38}$$

allow to establish the tensor

$$-\kappa\left[S_{ij} - \frac{1}{2}h_{ij}(S-\rho)\right] = 0. \tag{11.39}$$

Hence, in such a situation, the evolutionary equations for the extrinsic curvature tensor and the intrinsic curvature are given by

$$\partial_t K_{ij} = -N_{|ij} + N\left(R_{ij} + KK_{ij} - 2K_{ik}K^k_j\right) + N^k K_{ij|k} + K_{ik}N^k_{|j} + K_{jk}N^k_{|i}, \tag{11.40}$$

$$\partial_t K = -h^{ij}N_{|ij} + N\left(K^{ij}K_{ij} + \Lambda\right) + N^i K_{|i}, \tag{11.41}$$

while the evolutionary equations for induced metrics and its determinant remain unchanged, *i.e.*

$$\partial_t h_{ij} = N_{i|j} + N_{j|i} - 2NK_{ij}, \tag{11.42}$$

$$\partial_t \ln\sqrt{h} = -NK + N^i_{|i}. \tag{11.43}$$

However, the spatial stress tensor (11.37) is not unique. Because $h^{ij}n_i n_j = n^i n_i = 1$, there are another inequivalent choices

$$S_{ij} = Sn_i n_j, \tag{11.44}$$

$$S_{ij} = \frac{S}{4}\left(h_{ij} + n_i n_j\right), \tag{11.45}$$

$$S_{ij} = \frac{S}{2}\left(h_{ij} - n_i n_j\right), \tag{11.46}$$

and also a much more general form

$$S_{ij} = \frac{S}{3\alpha + \beta}\left(\alpha h_{ij} + \beta n_i n_j\right) = S_{ij}(\alpha, \beta), \tag{11.47}$$

where α and β are any numbers. All these forms of the spatial stress tensor have the same value of trace, but in the basis (h_{ij}, n_i) the two-parameter family (11.47) is the most general solution. In other words, all relations between the spatial stress density S and the energy density ρ are validated when the spatial stress tensor has the form (11.47). It is easy to see that the following Poisson algebra is satisfied

$$\{S_{ij}(\alpha,\beta), S_{kl}(\alpha',\beta')\} = S\frac{\alpha\alpha'\chi_{ijkl} + \beta\beta'\lambda_{ijkl} + \alpha\beta'\zeta_{ijkl} + \alpha'\beta\zeta_{klij}}{9\alpha\alpha' + \beta\beta' + 3\alpha\beta' + 3\alpha'\beta}, \quad (11.48)$$

where we have introduced the tensors

$$\chi_{ijkl} = \{h_{ij}, h_{kl}\}, \quad (11.49)$$
$$\lambda_{ijkl} = \{n_i n_j, n_k n_l\}, \quad (11.50)$$
$$\zeta_{ijkl} = \{h_{ij}, n_k n_l\}. \quad (11.51)$$

Therefore, (11.48) will be called *the stress algebra*. Let us consider the right hand side of the equation (11.48). When the bracketed quantities are classical (C), *i.e.* are not operators corresponding to the classical quantities, the Poisson brackets are easy to establish. Let us denote such classical Poisson brackets as

$$\sigma_{ijkl}(\alpha,\beta,\alpha',\beta') = \{S_{ij}(\alpha,\beta), S_{kl}(\alpha',\beta')\}_C, \quad (11.52)$$

or in the explicit form

$$\sigma_{ijkl}(\alpha,\beta,\alpha',\beta') = 2S\frac{\alpha\alpha' h_{ij}h_{kl} + \beta\beta' n_i n_j n_k n_l + \alpha\beta' h_{ij} n_k n_l + \alpha'\beta n_i n_j h_{kl}}{9\alpha\alpha' + \beta\beta' + 3\alpha\beta' + 3\alpha'\beta}. \quad (11.53)$$

It can be seen by straightforward computation that

$$\sigma_{jl}(\alpha,\beta,\alpha',\beta') = h^{ik}\sigma_{ijkl} = 2S\frac{\alpha\alpha' h_{jl} + (\beta\beta' + \alpha\beta' + \alpha'\beta) n_j n_l}{9\alpha\alpha' + \beta\beta' + 3\alpha\beta' + 3\alpha'\beta}, \quad (11.54)$$

$$\sigma(\alpha,\beta,\alpha',\beta') = h^{jl}\sigma_{jl} = 2S\frac{3\alpha\alpha' + \beta\beta' + \alpha\beta' + \alpha'\beta}{9\alpha\alpha' + \beta\beta' + 3\alpha\beta' + 3\alpha'\beta}. \quad (11.55)$$

For this reason, for generality, let us consider the spatial stress tensor (11.47). Then one has

$$-\kappa\left[S_{ij} - \frac{1}{2}h_{ij}(S-\rho)\right] = \frac{\Lambda}{2}\frac{\beta}{3\alpha+\beta}(h_{ij} - 3n_i n_j). \quad (11.56)$$

In this manner, the evolutionary equations (11.41), (11.42), and (11.43) remain unchanged, but time evolution of the extrinsic curvature tensor is

$$\begin{aligned}\partial_t K_{ij} &= -N_{|ij} + N\left[R_{ij} + KK_{ij} - 2K_{ik}K^k_j + \frac{\Lambda}{2}\frac{\beta}{3\alpha+\beta}\left(h_{ij} - 3n_i n_j\right)\right] + \\ &+ N^k K_{ij|k} + K_{ik}N^k_{|j} + K_{jk}N^k_{|i}\end{aligned} \tag{11.57}$$

However, in the light of the Einstein field equations one can express the trace of stress-energy tensor by the Ricci scalar curvature and the cosmological constant, *i.e.*

$$T = -\frac{^{(4)}R}{\kappa} + \frac{4\Lambda}{\kappa} \tag{11.58}$$

Applying this fact to the equation (11.20), one obtains

$$^{(4)}R \equiv 3\Lambda, \tag{11.59}$$

which means that the space-time manifold is a four-dimensional pseudo-Riemannian manifold of the constant Ricci scalar curvature given by the cosmological constant up to a constant multiplier equal to the dimensionality of an embedded space $D = 3$. In the light of the relation (11.13), the right hand side of the Einstein field equations (10.6) is

$$\kappa T_{\mu\nu} = \frac{1}{2}\Lambda g_{\mu\nu} + \Lambda n_\mu n_\nu, \tag{11.60}$$

while the left hand side of the Einstein field equations is

$$R_{\mu\nu} - \frac{1}{2}{}^{(4)}R g_{\mu\nu} + \Lambda g_{\mu\nu} = R_{\mu\nu} - \frac{1}{2}\Lambda g_{\mu\nu}, \tag{11.61}$$

and, for this reason, the Einstein field equations for such a situation are

$$R_{\mu\nu} = \Lambda\left(g_{\mu\nu} + n_\mu n_\nu\right) = \Lambda h_{\mu\nu}, \tag{11.62}$$

where we have applied the completeness relations (10.21), in which

$$h_{\mu\nu} = h_{ij}e^i_\mu e^j_\nu. \tag{11.63}$$

Let us consider the contravariant form of $h_{\mu\nu}$

$$h^{\mu\nu} = g^{\mu\kappa}g^{\nu\lambda}h_{\kappa\lambda} = h_{ij}g^{\mu\kappa}e^i_\kappa g^{\nu\lambda}e^j_\lambda = h^{kl}(h_{ik}g^{\mu\kappa}e^i_\kappa)(h_{jl}g^{\nu\lambda}e^j_\lambda) = h^{kl}e^\mu_k e^\nu_l, \tag{11.64}$$

where we have applied the notation

$$e_k^{\mu} = h_{ik} g^{\mu\kappa} e_{\mu}^{i}. \tag{11.65}$$

Applying the inverted way one obtains

$$h_{\mu\nu} = g_{\mu\kappa} g_{\nu\lambda} h^{\kappa\lambda} = g_{\mu\kappa} g_{\nu\lambda} h^{kl} e_k^{\kappa} e_l^{\lambda} = h^{kl} (g_{\mu\kappa} e_k^{\kappa})(g_{\nu\lambda} e_l^{\lambda}) = h^{kl} e_{\mu k} e_{\nu l}, \tag{11.66}$$

where we have applied the notation

$$e_{\nu k} = g_{\mu\nu} e_k^{\mu}, \tag{11.67}$$

following from transition between the completeness relations for metric (10.21) and the completeness relations for inverse metric (10.24). In this manner, the equation (11.62) bacomes

$$R_{\mu\nu} = \left(\Lambda g_{\mu\kappa} g_{\nu\lambda}\right) h^{ij} e_i^{\kappa} e_j^{\lambda}, \tag{11.68}$$

and application of the Ricci curvature tensor evaluated on the three-boundary (10.31)

$$R_{\mu\nu} = -R_{\kappa\mu\lambda\nu} n^{\kappa} n^{\lambda} + R_{\kappa\mu\lambda\nu} h^{ij} e_i^{\kappa} e_j^{\lambda}, \tag{11.69}$$

to the equation (11.68) leads to the system of equations

$$\left\{ \begin{array}{c} R_{\kappa\mu\lambda\nu} n^{\kappa} n^{\lambda} = 0 \\ \left(R_{\kappa\mu\lambda\nu} - \Lambda g_{\kappa\mu} g_{\lambda\nu}\right) h^{\kappa\lambda} = 0 \end{array} \right. . \tag{11.70}$$

The first equation in (11.70) expresses the property that the double projection of the Riemann–Christoffel curvature tensor onto the unit normal vector field vanishes, while the second one expresses the fact that the projection onto the metric $h^{\kappa\lambda}$ of the tensor

$$A_{\kappa\mu\lambda\nu} := R_{\kappa\mu\lambda\nu} - \Lambda g_{\kappa\mu} g_{\lambda\nu}, \tag{11.71}$$

vanishes. In other words, the space-time manifold is characterized by the Riemann–Christoffel curvature tensor

$$R_{\kappa\mu\lambda\nu} = \Lambda g_{\kappa\mu} g_{\lambda\nu} + A_{\kappa\mu\lambda\nu}, \tag{11.72}$$

where $A_{\kappa\mu\lambda\nu}$ is the tensor satisfying the equations

$$\Lambda n_\mu n_\nu + A_{\kappa\mu\lambda\nu} n^\kappa n^\lambda = 0, \quad (11.73)$$

$$A_{\kappa\mu\lambda\nu} h^{\kappa\lambda} = 0, \quad (11.74)$$

where the first equation arises from the first equation of the system (11.70). The equation (11.73) projected onto $n_\kappa n_\lambda$ leads to

$$A_{\kappa\mu\lambda\nu} = -\Lambda n_\mu n_\nu n_\kappa n_\lambda + \Gamma_{\kappa\mu\lambda\nu}, \quad (11.75)$$

where the tensor $\Gamma_{\kappa\mu\lambda\nu}$ satisfies the equations

$$\Gamma_{\kappa\mu\lambda\nu} n^\kappa n^\lambda = 0, \quad (11.76)$$

$$\Gamma_{\kappa\mu\lambda\nu} h^{\kappa\lambda} = 0, \quad (11.77)$$

and the second equation was deduced from application of the tensor (11.75) to the equation (11.74). In this manner, by application of the completeness relations for the metric $g_{\mu\nu}$, the Riemann–Christoffel curvature tensor describing the considered space-time has the form

$$R_{\kappa\mu\lambda\nu} = \Lambda\left(g_{\kappa\mu} g_{\lambda\nu} - n_\mu n_\nu n_\kappa n_\lambda\right) + \Gamma_{\kappa\mu\lambda\nu}, \quad (11.78)$$

with the tensor $\Gamma_{\kappa\mu\lambda\nu}$ being a solution of the equations (11.76) and (11.77). The problem is to solve the system of equations (11.76)–(11.77) in general, but we shall not perform such procedure in this book. In general, the Riemann–Christoffel curvature tensor (11.78) describes such four-dimensional space-times which are characterized by the Ricci scalar curvature ${}^{(4)}R = 3\Lambda$ and the Ricci curvature tensor $R_{\mu\nu} = \Lambda(g_{\mu\nu} + n_\mu n_\nu)$. Interestingly, one can compute these curvatures immediately using the relation (11.78)

$$R_{\mu\nu} = g^{\kappa\lambda} R_{\kappa\mu\lambda\nu} = \Lambda\left(g_{\mu\nu} + n_\mu n_\nu\right) + \Gamma_{\mu\nu}, \quad (11.79)$$

$${}^{(4)}R = g^{\mu\nu} R_{\mu\nu} = 3\Lambda + {}^{(4)}\Gamma, \quad (11.80)$$

which gives equations for the contractions of the tensor $\Gamma_{\kappa\mu\lambda\nu}$

$$\Gamma_{\mu\nu} := g^{\kappa\lambda} \Gamma_{\kappa\mu\lambda\nu} = 0, \quad (11.81)$$

$${}^{(4)}\Gamma := g^{\kappa\lambda} \Gamma_{\mu\nu} = 0. \quad (11.82)$$

Moreover, double projection of the Ricci curvature tensor (11.79) onto the unit normal vector field leads to

$$R_{\mu\nu}n^{\mu}n^{\nu} = \Gamma_{\mu\nu}n^{\mu}n^{\nu}, \tag{11.83}$$

i.e. in this projective sense the $\Gamma_{\mu\nu}$ curvature tensor carries the same information as the Ricci curvature tensor.

Interestingly, one can compute the Weyl curvature tensor

$$\begin{aligned} W_{\mu\kappa\nu\lambda} &= R_{\mu\kappa\nu\lambda} + \frac{1}{2}\left(g_{\mu\lambda}R_{\nu\kappa} + g_{\kappa\nu}R_{\lambda\mu} - g_{\mu\nu}R_{\lambda\kappa} - g_{\kappa\lambda}R_{\nu\mu}\right) + \\ &+ \frac{1}{3}{}^{(4)}R\left(g_{\mu\nu}g_{\lambda\kappa} - g_{\mu\lambda}g_{\nu\kappa}\right), \end{aligned} \tag{11.84}$$

which for the considered situation takes the form

$$W_{\mu\kappa\nu\lambda} = \Gamma_{\kappa\mu\lambda\nu} + \Lambda\left(g_{\kappa\mu}g_{\lambda\nu} + 2g_{\mu[\lambda}g_{\nu]\kappa} + 4g_{[\mu(\lambda}n_{\nu)}n_{\kappa]} - n_{\mu}n_{\nu}n_{\kappa}n_{\lambda}\right). \tag{11.85}$$

It means that the curvature tensor $\Gamma_{\kappa\mu\lambda\nu}$ is not the Weyl tensor. The contractions of the Weyl curvature tensor are

$$W_{\kappa\lambda} = g^{\mu\nu}W_{\mu\kappa\nu\lambda} = \Gamma_{\kappa\lambda} - \Lambda\left(g_{\kappa\lambda} + n_{\kappa}n_{\lambda}\right) = 2\Gamma_{\kappa\lambda} - R_{\kappa\lambda}, \tag{11.86}$$

$${}^{(4)}W = g^{\kappa\lambda}W_{\kappa\lambda} = 2{}^{(4)}\Gamma - {}^{(4)}R = {}^{(4)}\Gamma - 3\Lambda, \tag{11.87}$$

which means that in the particular case considered in this section

$$\Gamma_{\kappa\lambda} = \frac{R_{\kappa\lambda} + W_{\kappa\lambda}}{2}, \tag{11.88}$$

$${}^{(4)}\Gamma = \frac{{}^{(4)}R + {}^{(4)}W}{2}. \tag{11.89}$$

We shall call $\Gamma_{\mu\nu}$ *the* Γ *curvature tensor*, and ${}^{(4)}\Gamma$ *the* Γ *scalar curvature*. Then the space-time considered above is the $\Gamma_{\mu\nu}$-flat manifold of the zero Γ scalar curvature, which we shall call *the* Γ*-scalar-flat manifolds*. The equations (11.79) and (11.80) can be used for construction of the left hand side of the Einstein field equations

$$G_{\mu\nu} + \Lambda g_{\mu\nu} = \frac{1}{2}\Lambda g_{\mu\nu} + \Lambda n_{\mu}n_{\nu} + \Gamma_{\mu\nu} + \frac{1}{2}{}^{(4)}\Gamma g_{\mu\nu}, \tag{11.90}$$

and, in the light of the fact that the stress-energy tensor is given by (11.13), the right hand side of the Einstein field equations

$$\kappa T_{\mu\nu} = \frac{1}{2}\Lambda g_{\mu\nu} + \Lambda n_{\mu}n_{\nu}, \tag{11.91}$$

and the Einstein field equations expressed in terms of the Γ curvatures take the form of the vacuum field equations

$$\Gamma_{\mu\nu}+\frac{1}{2}{}^{(4)}\Gamma g_{\mu\nu}=0. \tag{11.92}$$

In other words, the Γ curvatures are the curvatures which for blatantly non stationary space-time given by the Ricci scalar curvature ${}^{(4)}R=3\Lambda$, the Ricci curvature tensor $R_{\mu\nu}=\Lambda(g_{\mu\nu}+n_{\mu}n_{\nu})$, and the stress-energy tensor (11.91) makes the non stationary solution of the Einstein field equations the space-time obeying vacuum field equations (11.92). The constructive hypothesis is

Hypothesis (The Γ Curvatures Hypothesis). *In general, the Γ curvatures, transforming non stationary four-dimensional Einstein field equations to the vacuum field equations (11.92), can be constructed only by using the Riemann–Christoffel curvature tensor, its contractions with space-time metric, and combinations of all these quantities.*

This hypothesis can be easily justified in a general case. If there is some fixed space-time metric $\tilde{g}_{\mu\nu}$ described by the curvatures

$$\tilde{R}_{\mu\nu\kappa\lambda} = R_{\mu\nu\kappa\lambda}+\Gamma_{\mu\nu\kappa\lambda}, \tag{11.93}$$

$$\tilde{R}_{\mu\nu} = R_{\mu\nu}+\Gamma_{\mu\nu}, \tag{11.94}$$

$${}^{(4)}\tilde{R} = {}^{(4)}R+{}^{(4)}\Gamma, \tag{11.95}$$

obeying the Einstein field equations

$$\tilde{R}_{\mu\nu}-\frac{1}{2}\tilde{g}_{\mu\nu}\tilde{R}+\Lambda\tilde{g}_{\mu\nu}=\kappa\tilde{T}_{\mu\nu}, \tag{11.96}$$

and the space-time metric $g_{\mu\nu}$ described by the curvatures $R_{\mu\nu\kappa\lambda}$, $R_{\mu\nu}$, R obeying the Einstein field equations

$$R_{\mu\nu}-\frac{1}{2}g_{\mu\nu}R+\Lambda g_{\mu\nu}=\kappa T_{\mu\nu}, \tag{11.97}$$

then the Γ curvatures hypothesis for the metric $\tilde{g}_{\mu\nu}$ is true

$$\Gamma_{\mu\nu}+\frac{1}{2}{}^{(4)}\Gamma\tilde{g}_{\mu\nu}=0, \tag{11.98}$$

if and only if the condition holds

$$\left(\frac{1}{2}{}^{(4)}R-\Lambda\right)\left(g_{\mu\nu}-\tilde{g}_{\mu\nu}\right)=\kappa\left(\tilde{T}_{\mu\nu}-T_{\mu\nu}\right). \qquad (11.99)$$

For the case

$${}^{(4)}R=2\Lambda, \qquad (11.100)$$

the condition (11.99) gives identically

$$\tilde{T}_{\mu\nu}=T_{\mu\nu}, \qquad (11.101)$$

and the metrics $g_{\mu\nu}$ and $\tilde{g}_{\mu\nu}$ are arbitrary. Otherwise, however, the condition (11.99) can be understood as the solution for the metric $\tilde{g}_{\mu\nu}$

$$\tilde{g}_{\mu\nu}=g_{\mu\nu}-\frac{2\kappa}{{}^{(4)}R-2\Lambda}\left(\tilde{T}_{\mu\nu}-T_{\mu\nu}\right), \qquad (11.102)$$

which only in the case $\tilde{T}_{\mu\nu}=T_{\mu\nu}$, which in general is not satisfied, gives trivially $\tilde{g}_{\mu\nu}=g_{\mu\nu}$. In this manner, in such a situation, the Γ curvatures remains undetermined.

The situation changes in the case when the Γ curvatures hypothesis is true for the metric $g_{\mu\nu}$

$$\Gamma_{\mu\nu}+\frac{1}{2}{}^{(4)}\Gamma g_{\mu\nu}=0. \qquad (11.103)$$

Then the relation holds

$$\left(\frac{1}{2}{}^{(4)}R-\Lambda+{}^{(4)}\Gamma\right)\left(g_{\mu\nu}-\tilde{g}_{\mu\nu}\right)=\kappa\left(\tilde{T}_{\mu\nu}-T_{\mu\nu}\right). \qquad (11.104)$$

When

$${}^{(4)}\Gamma=\Lambda-\frac{1}{2}{}^{(4)}R, \qquad (11.105)$$

then also

$$\Gamma_{\mu\nu}=\frac{1}{4}{}^{(4)}g_{\mu\nu}\left({}^{(4)}R-2\Lambda\right). \qquad (11.106)$$

Otherwise, one receives

$${}^{(4)}\Gamma=\Lambda-\frac{1}{2}{}^{(4)}R-\frac{\kappa}{\gamma}\left(\tilde{T}_{\mu\nu}-T_{\mu\nu}\right)\left(\tilde{g}^{\mu\nu}+g^{\mu\nu}\right), \qquad (11.107)$$

where $\gamma = g_{\mu\nu}\tilde{g}^{\mu\nu} - \tilde{g}_{\mu\nu}g^{\mu\nu}$ and, consequently,

$$\Gamma_{\mu\nu} = -\frac{1}{2}\left({}^{(4)}R - 2\Lambda\right)g_{\mu\nu} - \kappa\frac{\beta}{\gamma}\left(\tilde{T}_{\mu\nu} - T_{\mu\nu}\right), \qquad (11.108)$$

where $\beta = \tilde{g}^{\mu\nu}g_{\mu\nu} + 4$. In this manner, the hypothesis is proven.

The Ricci curvature tensor (11.62) can be presented in the form

$$R_{\mu\nu} = \Lambda g_{\mu\nu} + \Lambda n_{\mu}n_{\nu}, \qquad (11.109)$$

and, for this reason, the second term on the right hand side of the equation (11.62) can be interpreted as the correction to the four-dimensional Einstein manifold, *i.e.* the four-dimensional Riemannian manifold for which the Ricci curvature tensor is proportional to metric $R_{\mu\nu} = \lambda g_{\mu\nu}$ and, therefore, the scalar curvature is constant ${}^{(4)}R = 4\lambda$ (For advanced discussion of general Einstein manifolds *e.g.* the well-known Arthur L. Besse's[1] book [924]), defined by the sign identical to the cosmological constant $\lambda = \Lambda$. In other words, the situation of the perfect fluid solution presented in this section corresponds to the deformation of the four-dimensional Einstein manifolds of sign $\lambda = \Lambda$

$$R_{\mu\nu} = \Lambda g_{\mu\nu} + \Delta_{\mu\nu}, \qquad (11.110)$$

where $\Delta_{\mu\nu} = \Lambda n_{\mu}n_{\nu}$ is *the deformation curvature tensor*, for which

$${}^{(4)}R = 4\Lambda + \Delta, \qquad (11.111)$$

where $\Delta = g^{\mu\nu}\Delta_{\mu\nu} = -\Lambda$ is *the deformation scalar curvature*. Comparison of the result (11.111) with the equation (11.80) leads to expression of the Γ scalar curvature by the deformation scalar curvature

$${}^{(4)}\Gamma = \Lambda + \Delta, \qquad (11.112)$$

[1]Arthur L. Besse is a fictional author of mathematics, a pseudonym of a group of French differential geometers: Geneviève Avérous, Lionel Bérard-Bergery, Marcel Berger, Jean-Pierre Bourguignon, Andrzej Derdziński, Dennis Michael De Turck, Paul Gauduchon, Nigel James Hitchin, Josette Houillot, Hermann Karcher, Jerry Lawrence Kazdan, Norihito Koiso, Jacques Lafontaine, Pierre Pansu, Albert Polombo, John Alden Thorpe, and Liane Valere, headed by Marcel Berger, following the model of Nicolas Bourbaki.

so that in the light of the result (11.107) one receives for such a case

$$\Delta = -\frac{1}{2}{}^{(4)}R - \frac{\kappa}{\gamma}\left(\tilde{T}_{\mu\nu} - T_{\mu\nu}\right)\left(\tilde{g}^{\mu\nu} + g^{\mu\nu}\right), \tag{11.113}$$

and, consequently,

$$g_{\mu\nu} = \tilde{g}_{\mu\nu} - \frac{2\kappa}{2\Lambda - {}^{(4)}R}\left(\tilde{T}_{\mu\nu} - T_{\mu\nu}\right), \tag{11.114}$$

where $\tilde{g}_{\mu\nu}$ is any fixed metric.

Let us analyze the presented situation concisely. For the case of the identically vanishing cosmological constant $\Lambda \equiv 0$ the four-dimensional space-time manifold is the Ricci-flat manifold. This is, however, the straightforward ramification of the fact that we have calculated the value of the cosmological constant by application of the relation (10.40), *i.e.* the Ricci scalar curvature evaluated on the space-time boundary ∂M. Such a situation means that, in this particular case, the enveloping space-time is the Ricci-flat manifold from the point of view of embedded space. It does not mean, however, that the space-time is flat in general, because of the Riemann–Christoffel curvature tensor must not be vanishing identically when both the Ricci curvature tensor and the Ricci scalar curvature are identically vanishing. This in itself is a highly non-trivial consequence in the light of the fact that in such a situation both the cosmological constant and the stress-energy tensor are, in general, non-vanishing and arbitrary. This suggests that from the space point of view space-time looks like vacuum space-time. Such a situation, however, should be understood like a local property, *i.e.* related to the quantum gravity given by the Wheeler–DeWitt equation, rather than the property of the classical space-time. Moreover, it should be emphasized that the topology of such a space-time is still unrestricted, because there are many possible topologies of the four-dimensional Ricci-flat manifold. The Ricci-flat manifolds are the particular case of the Einstein manifolds, for which the sign is trivial $\lambda = 0$. Such a class of the Einstein manifolds includes, for example, the Calabi–Yau manifolds [925] and the hyper-Kähler manifolds [926] which recently have attracted intensive interest by a number of theoreticians and mathematicians (For some particular applications Cf. *e.g.* the papers in Ref. [927]), especially in the context of the string theory. Recall that in the case of the four-dimensional space-time every Calabi–Yau manifolds is a hyper-Kähler manifold. There are many simpler vacuum solutions of the Einstein field equations. For example, the flat Minkowski space-time is the simplest vacuum solution, and the various non-trivial situations include the Schwarzschild space-time and the Kerr space-time describing the geometry of space-time around a non-rotating spherical mass and a rotating massive body, respectively.

C Ansatz for Wave Functionals

According to the evaluation (11.2), the Ricci scalar curvature ${}^{(4)}R$ of the enveloping space-time manifold is expressed by the Ricci scalar curvature ${}^{(3)}R$ of the embedded space, and its extrinsic curvature tensor K_{ij} and intrinsic curvature K. All these embedding characterizations, in general, are functionals of induced metric h_{ij}. Moreover, the trace of the spatial stress $S \equiv T(h,h)$ as a double projection of the stress-energy tensor onto induced metric is also a functional of the metric h_{ij}. The cosmological constant can be treated as a constant functional of h_{ij}. In this manner, the energy density of Matter fields ρ is at most a functional of the metric h_{ij} and, consequently, one has the general functional dependence

$$\rho[h_{ij}] = -\frac{4\Lambda}{\kappa} + S[h_{ij}] + \frac{1}{\kappa}\left({}^{(3)}R + K^2 - K^{ij}K_{ij}\right) = \frac{2\Lambda}{\kappa} - S[h_{ij}], \qquad (11.115)$$

which allows to establish

$$S[h_{ij}] = \frac{3\Lambda}{\kappa} - \frac{1}{2\kappa}\left({}^{(3)}R + K^2 - K^{ij}K_{ij}\right), \qquad (11.116)$$

$$\rho[h_{ij}] = -\frac{\Lambda}{\kappa} + \frac{1}{2\kappa}\left({}^{(3)}R + K^2 - K^{ij}K_{ij}\right). \qquad (11.117)$$

Actually, both the relations (11.116) and (11.117) are straightforward ramifications of application of the Hamiltonian constraint, *i.e.* strictly speaking, they have a sense only within the geometrodynamics. The functional nature of their left hand sides is a straightforward conclusion of the functional character of their right hand sides. Such a situation implies the non-trivial physical significance. Namely, in the light of the fact that both the spatial stress density S and the energy density ρ are the appropriate projections of the stress-energy tensor of Matter fields, they depend on Matter fields and their derivatives. Consequently, in the light of the functional nature of the relations (11.116) and (11.117) one can conclude that such a situation is equivalent to the statement that, in general, Matter fields are functionals of the metric h_{ij},

$$\phi = \phi[h_{ij}]. \tag{11.118}$$

In this manner, the DeWitt wave functional

$$\Psi[h_{ij},\phi] = \Psi[h_{ij},\phi[h_{ij}]] \equiv \Psi[h_{ij}], \tag{11.119}$$

is fully justified, and the Wheeler–DeWitt equation becomes

$$\left\{2\kappa G_{ijkl}\frac{\delta^2}{\delta h_{ij}\delta h_{kl}} + \frac{\sqrt{h}}{2\kappa}\left({}^{(3)}R[h_{ij}] - 2\Lambda - 2\kappa\rho[h_{ij}]\right)\right\}\Psi[h_{ij}] = 0. \tag{11.120}$$

However, the crucial general problem is solving the Wheeler–DeWitt equation in general. As we have mentioned earlier, the Wheeler–DeWitt equation has never been solved in general, and even taking into account the DeWitt wave functional this not simplify this general problem because $\Psi[h_{ij}]$ is still a functional but not a function. It means that it is not clear how to treat $\Psi[h_{ij}]$ mathematically. We shall present here the strategy for solution of the Wheeler–DeWitt equation which is based on the DeWitt wave functional but reduces the functional $\Psi[h_{ij}]$ to a function. In itself, such a reduction defines a new model of quantum gravity within the quantum geometrodynamics formulated in terms of the Wheeler–DeWitt equation which, however, will not be governed by the Wheeler–DeWitt equation.

To start deductions, we should rethink the quantum geometrodynamics (11.120), particularly the structure of the DeWitt wave functional $\Psi[h_{ij}]$. The fundamental interpretation of the Wheeler–DeWitt equation, as the result of the primary canonical quantization, is the Schrödinger equation or Klein–Gordon equation. In both these situations, however, a wave function is always a scalar field. The operator acting on the wave functional in the quantum geometrodynamics (11.120) is always a scalar and is a functional determined on the configurational space which, in the present case, is the Wheeler superspace

$$2\kappa G_{ijkl}\frac{\delta^2}{\delta h_{ij}\delta h_{kl}}+\frac{\sqrt{h}}{2\kappa}\left({}^{(3)}R[h_{ij}]-2\Lambda-2\kappa\rho[h_{ij}]\right)=\hat{\mathscr{O}}[h_{ij}], \qquad (11.121)$$

which is similar to the case of the Schrödinger equation or Klein–Gordon equation, in which the operator acting on the wave function is a functional determined on the configurational space, *i.e.* the product space $\mathbb{R}^4$. Moreover, the differential operator of the Wheeler–DeWitt equation is the $Diff(\partial M)$-invariant. This suggests clearly that the DeWitt wave functional $\Psi[h_{ij}]$ must be a function invariant with respect to action of the diffeomorphism group, *i.e.* must be a function of other $Diff(\partial M)$-invariant quantities. Furthermore, for full consistency these diffeoinvariant quantities must be constructed using the induced metric h_{ij}, $f=f(h_{ij})=inv$, for example. Then, however, according to the Kuchař formalism, the wave functional $\Psi(f)$ inevitably will become an observable or a perennial and, above all, if one expresses the differential operator (11.121) by these invariant quantities then one can treat these invariants as solution of the problem of time in quantum geometrodynamics by identification of the time t with the invariant of induced metric, *i.e.* $t\equiv f$. If a wave functional is an usual function, then one can also perform straightforwardly and in an extraordinary simple way the formalism of secondary quantization and construct the theory of quantum gravity which is the quantum field theory of gravity. Let us apply such a strategy to the case of quantum geometrodynamics.

Step 1: Global One-Dimensionality Conjecture

According to the DeWitt construction, based on the Wheeler metric representation, $\Psi[h_{ij}]$ is a functional of the 3×3 symmetric matrix of induced metric. It suggests that the wave functional is a single functional

$$\Psi[h_{ij}]=\Psi\left[\left[\begin{array}{ccc} h_{11} & h_{12} & h_{13}\\ h_{12} & h_{22} & h_{23}\\ h_{13} & h_{23} & h_{33}\end{array}\right]\right]. \qquad (11.122)$$

However, such a reasoning is not unique. The wave functional must not be a single functional but rather is a 3×3 symmetric matrix whose elements are dependent on a single element of induced metric

$$\Psi[h_{ij}] = \begin{bmatrix} \Psi[h_{11}] & \Psi[h_{12}] & \Psi[h_{13}] \\ \Psi[h_{12}] & \Psi[h_{22}] & \Psi[h_{23}] \\ \Psi[h_{13}] & \Psi[h_{23}] & \Psi[h_{33}] \end{bmatrix}. \tag{11.123}$$

The still unsolved problem of quantum gravity is the reduction procedure

$$\Psi\left[\begin{bmatrix} h_{11} & h_{12} & h_{13} \\ h_{12} & h_{22} & h_{23} \\ h_{13} & h_{23} & h_{33} \end{bmatrix}\right] \to \begin{bmatrix} \Psi[h_{11}] & \Psi[h_{12}] & \Psi[h_{13}] \\ \Psi[h_{12}] & \Psi[h_{22}] & \Psi[h_{23}] \\ \Psi[h_{13}] & \Psi[h_{23}] & \Psi[h_{33}] \end{bmatrix}. \tag{11.124}$$

This is evidently a perfectionist situation because, in general, the wave functional can be considered as a 3×3 symmetric matrix whose elements functionals of several elements of induced metric.

Albeit, the straightforward analogy with quantum mechanics suggests that the wave functional $\Psi[h_{ij}]$ is a classical scalar field like usual wave function in quantum mechanics based on the Schrödinger equation, *i.e.* in such a situation $\Psi[h_{ij}]$ is a single functional. Let us accept such a state of things. For realization of this idea the wave functional should be dependent on a scalar function of induced metric h_{ij}, which must be an invariant of the induced matrix as well as invariant with respect to action of the diffeomorphism group. The Cayley–Hamilton theorem for any 3×3 square matrix $\mathbf{h}$ states that the matrix obeys its characteristic equation

$$\mathbf{h}^3 - c_1\mathbf{h}^2 - c_2\mathbf{h} + c_3\mathbf{I}_{3\times 3} = 0, \tag{11.125}$$

where the coefficients of the polynomial

$$c_1 = \mathrm{Tr}\mathbf{h}, \tag{11.126}$$

$$c_2 = \frac{(\mathrm{Tr}\mathbf{h})^2 - \mathrm{Tr}\mathbf{h}^2}{2}, \tag{11.127}$$

$$c_3 = -\det\mathbf{h}, \tag{11.128}$$

are the invariants of the matrix $\mathbf{h}$. A scalar valued matrix function $\Psi(h_{ij})$ that depends merely on the three invariants of a symmetric 3×3 matrix

$$\Psi(h_{ij}) = \Psi(c_1, c_2, c_3), \tag{11.129}$$

is independent of rotations of the coordinate system and is called *objective function*. The invariants c_1 and c_2, however, are irrelevant because they do not carry full information about the matrix h_{ij}. The third invariant c_3 as a function of all elements of a matrix carries full information about the matrix. In $3+1$ decomposition the determinant is the diffeoinvariant function of a 3×3 induced metric h_{ij}. It suggests that the invariant dimension is $\det h_{ij}$. Then the wave functional reduces to

$$\Psi(h_{ij}) = \Psi(c_3) = \Psi(h). \tag{11.130}$$

and the quantum geometrodynamics becomes a one-dimensional quantum mechanics. We shall call $\det h_{ij}$ *the global dimension*, because it is a function of local dimensions (coordinates) and some free parameters, and (11.130) *the global one-dimensionality conjecture*. We shall call *generalized dimensions* other, possibly more convenient, invariants constructed as $f(h)$. We shall call *objective quantum gravity* a theory of quantum gravity related to wave functionals (11.129), and *global one-dimensional quantum gravity* the theory of quantum gravity related to the wave functionals (11.130).

Such a global one-dimensional wave function can be constructed in the following way. Suppose that Matter fields are, in general, functionals dependent on the one global variable

$$\phi = \phi[h], \tag{11.131}$$

which is the determinant $h = \det h_{ij}$ of induced metric on ∂M. Recall that in the dimension 3 one has

$$h = \frac{1}{3}\varepsilon^{ijk}\varepsilon^{lmn}h_{il}h_{jm}h_{kn}, \tag{11.132}$$

where ε^{abc} is the three-dimensional Levi–Civita tensor

$$\varepsilon^{abc} = \frac{(a-b)(b-c)(c-a)}{2}. \tag{11.133}$$

As the crucial point of the model let us assume that quantum gravity is globally one-dimensional. As a result the DeWitt wave functional becomes one-dimensional wave function

$$\Psi[h_{ij}] \to \Psi(h), \tag{11.134}$$

and the Wheeler–DeWitt equation is

$$\left\{-2\kappa G_{ijkl}\frac{\delta^2}{\delta h_{ij}\delta h_{kl}}-\frac{h^{1/2}}{2\kappa}\left({}^{(3)}R-2\Lambda-2\kappa\rho[h]\right)\right\}\Psi(h)=0. \qquad (11.135)$$

By straightforward analogy with the generic cosmology [531] the global one-dimensionality conjecture (11.134) describes isotropic space-times and is related to the strata of the Wheeler superspace, the so-called midisuperspace, in which wave functionals are usual functions of one variable.

Step 2: Reduction of Quantum Geometrodynamics

Let us consider the Jacobi formula for the determinant of the space-time metric

$$\delta g = g g^{\mu\nu}\delta g_{\mu\nu}, \qquad (11.136)$$

which can be rewritten in components

$$\delta g = g\left(g^{00}\delta g_{00}+g^{ij}\delta g_{ij}+g^{0j}\delta g_{0j}+g^{i0}\delta g_{i0}\right). \qquad (11.137)$$

The $3+1$ splitting (10.19) allows to determine the partial variations

$$\begin{aligned}
\delta g_{00} &= -\delta N^2+N^iN^j\delta h_{ij}+h_{ij}N^i\delta N^j+h_{ij}N^j\delta N^i, &(11.138)\\
\delta g_{ij} &= \delta h_{ij}, &(11.139)\\
\delta g_{0j} &= h_{ij}\delta N^i+N^i\delta h_{ij}, &(11.140)\\
\delta g_{i0} &= h_{ij}\delta N^j+N^j\delta h_{ij}, &(11.141)
\end{aligned}$$

as well as the total variation

$$\delta g = N^2\delta h + h\delta N^2. \qquad (11.142)$$

Collecting all, one obtains the result relevant for the induced metric

$$N^2\delta h = N^2 h h^{ij}\delta h_{ij}, \qquad (11.143)$$

which allows to establish the Jacobian matrix for transformation of variables $h_{ij}\to h$

$$\mathscr{J}(h_{ij},h)=\frac{\delta(h)}{\delta(h_{ij})}=\frac{\delta h}{\delta h_{ij}}\equiv h h^{ij}. \qquad (11.144)$$

Because the approximation (11.134) the functional derivative $\frac{\delta}{\delta h_{ij}}$ acts on a wave functional depending only on h. It allows us to express the functional derivative with respect h_{ij} through the functional derivative $\frac{\delta}{\delta h}$. Therefore, one has

$$\frac{\delta\Psi[h]}{\delta h_{ij}} = hh^{ij}\frac{\delta\Psi[h]}{\delta h}. \tag{11.145}$$

Consequently, application of (11.145) within the differential operator of the Wheeler–DeWitt equation (11.135) leads to

$$G_{ijkl}\frac{\delta^2}{\delta h_{ij}\delta h_{kl}} = G_{ijkl}h^{ij}h^{kl}h^2\frac{\delta^2}{\delta h^2}. \tag{11.146}$$

In this manner, the reduction is given by the double projection of the DeWitt supermetric onto induced metric

$$\begin{aligned}
G_{ijkl}h^{ij}h^{kl} &= \frac{1}{2\sqrt{h}}\left(h_{ik}h_{jl}+h_{il}h_{jk}-h_{ij}h_{kl}\right)h^{ij}h^{kl} = \\
&= \frac{1}{2\sqrt{h}}\left(h_{ik}h^{kl}h^{ij}h_{jl}+h_{il}h^{ij}h_{jk}h^{kl}-h_{ij}h^{ij}h_{kl}h^{kl}\right) = \\
&= \frac{1}{2\sqrt{h}}\left(\delta_i^l\delta_l^i+\delta_l^j\delta_j^l-\delta_i^i\delta_k^k\right) = \\
&= \frac{1}{2\sqrt{h}}\left(\delta_i^i+\delta_j^j-(\delta_i^i)^2\right) = \\
&= \frac{1}{2\sqrt{h}}\left(2\delta_i^i-(\delta_i^i)^2\right) = \\
&= \frac{1}{2\sqrt{h}}\left(2\cdot 3-(3)^2\right) = -\frac{3}{2}h^{-1/2},
\end{aligned} \tag{11.147}$$

where we have used the relations for the three-dimensional embedded space $h^{ab}h_{bc} = h^a_c$, $h^a_a = \delta^a_a = \mathrm{Tr}h_{ab} = 3$. Joining the formulas (11.146) and (11.147), one obtains finally the transformation

$$G_{ijkl}\frac{\delta^2}{\delta h_{ij}\delta h_{kl}} = -\frac{3}{2}h^{3/2}\frac{\delta^2}{\delta h^2}, \tag{11.148}$$

which leads to the quantum geometrodynamics

$$\left[\frac{\delta^2}{\delta h^2} - \frac{1}{6\kappa^2 h}\left({}^{(3)}R - 2\Lambda - 2\kappa\rho[h]\right)\right]\Psi(h) = 0. \tag{11.149}$$

Because the relation (11.144) arises due to $3+1$ approximation, the formula (11.147) is an approximation within the ansatz.

Step 3: Dimensional Reduction

The quantum geometrodynamics (11.149) can be rewritten in the form of the one-dimensional Klein–Gordon equation

$$\left(\frac{\delta^2}{\delta h^2}+\omega^2\right)\Psi=0, \tag{11.150}$$

where ω^2 is the parameter given by the formula

$$\begin{aligned}\omega^2 &= -\frac{1}{6\kappa^2}\frac{1}{h}\left({}^{(3)}R-2\Lambda-2\kappa\rho\right)= &(11.151)\\ &= -\frac{1}{6\kappa^2}\frac{1}{h}(K_{ij}K^{ij}-K^2), &(11.152)\end{aligned}$$

where the Hamiltonian constraint was involved in the second line. In general, the squared mass can be positive, negative or even vanishing identically. The equation (11.150) can be treated as the classical-field-theoretical Euler–Lagrange equations of motion arising from stationarity of the action functional

$$S[\Psi]=\int\delta hL\left(\Psi,\frac{\delta\Psi}{\delta h}\right), \tag{11.153}$$

where $L=L\left(\Psi,\dfrac{\delta\Psi}{\delta h}\right)$ is the field-theoretic Lagrange function

$$\begin{aligned}L &= \frac{1}{2}\left(\frac{\delta\Psi}{\delta h}\right)^2-\frac{\omega^2}{2}\Psi^2= &(11.154)\\ &= \frac{1}{2}\Pi_\Psi^2-\frac{\omega^2}{2}\Psi^2, &(11.155)\end{aligned}$$

where Π_Ψ is the momentum conjugated to the classical scalar field Ψ

$$\Pi_\Psi=\frac{\partial L}{\partial\left(\dfrac{\delta\Psi}{\delta h}\right)}=\frac{\delta\Psi}{\delta h}. \tag{11.156}$$

The action $S[\Psi]$ is a field-theoretic action functional in the classical field Ψ and, therefore, the arbitrary dependence on the variable h of the mass $m = m[h]$ does not play any role, *i.e.* it behaves as a coefficient, in derivation of the Euler–Lagrange equations of motion

$$\delta S[\Psi] = \int \delta h \left[\frac{\partial L}{\partial \Psi} - \frac{\delta}{\delta h} \frac{\partial L}{\partial \left(\frac{\delta \Psi}{\delta h} \right)} \right] \delta \Psi + \int \delta h \frac{\delta}{\delta h} \left(\frac{\partial L}{\partial \Psi} \delta \Psi \right) = 0, \quad (11.157)$$

which gives the result

$$\frac{\partial L}{\partial \Psi} - \frac{\delta}{\delta h} \frac{\partial L}{\partial \left(\frac{\delta \Psi}{\delta h} \right)} = 0, \quad (11.158)$$

where we have considered *ad hoc* the field theoretical condition of vanishing of the boundary term

$$\int \delta h \frac{\delta}{\delta h} \left(\frac{\partial L}{\partial \Psi} \delta \Psi \right) = \int \delta \left(\frac{\partial L}{\partial \Psi} \delta \Psi \right) = \frac{\partial L}{\partial \Psi} \delta \Psi \bigg|_0 = 0. \quad (11.159)$$

It can be seen by straightforward computation that the equation (11.158) coincides with (11.150).

By application of the conjugate momentum Π_Ψ one rewrites the equation (11.150) in the following form

$$\frac{\delta \Pi_\Psi}{\delta h} + \omega^2 \Psi = 0, \quad (11.160)$$

and, therefore, the equations (11.156) and (11.160) are the system of canonical Hamilton equations of motion

$$\frac{\delta}{\delta h} \Psi = \frac{\delta}{\delta \Pi_\Psi} H(\Psi, \Pi_\Psi), \quad (11.161)$$

$$\frac{\delta}{\delta h} \Pi_\Psi = -\frac{\delta}{\delta \Psi} H(\Psi, \Pi_\Psi), \quad (11.162)$$

where the Hamilton function $H(\Psi, \Pi_\Psi)$ is obtained from the Lagrange function (11.155) by the Legendre transformation

$$H(\Psi, \Pi_\Psi) = \Pi_\Psi \frac{\delta \Psi}{\delta h} - L\left(\Psi, \frac{\delta \Psi}{\delta h} \right) = \quad (11.163)$$

$$= \frac{1}{2} \Pi_\Psi^2 - \frac{\omega^2}{2} \Psi^2. \quad (11.164)$$

If one recognizes the kinetic T and potential V energies as

$$T = \frac{1}{2}\Pi_{\Psi}^{2}, \tag{11.165}$$

$$V = \frac{1}{2}\omega^{2}\Psi^{2}, \tag{11.166}$$

then the Hamilton function (11.164) is $H = T - V$ and the Lagrange function (11.155) is $L = T + V$, which means that the field theory presented above is the Euclidean field theory of a simple harmonic oscillator characterized by the parameter ω which plays the role of frequency. In this context, the classical scalar field - the wave function Ψ - becomes the generalized coordinate.

Let us introduce the two-component field

$$\Phi = \begin{bmatrix} \Pi_{\Psi} \\ \Psi \end{bmatrix}, \tag{11.167}$$

whose components satisfy the system of canonical Hamilton equations (11.156) and (11.160) , which can be easily rewritten in the form of the vector equation

$$\left(-i \begin{bmatrix} 0 & -i \\ i & 0 \end{bmatrix} \frac{\delta}{\delta h} - \begin{bmatrix} -\frac{1}{\Pi_{\Psi}}\frac{\delta}{\delta \Pi_{\Psi}} & 0 \\ 0 & -\frac{1}{\Psi}\frac{\delta}{\delta \Psi} \end{bmatrix} H\left(\Psi, \Pi_{\Psi}\right) \right) \Phi = 0, \tag{11.168}$$

which for the particular situation determined by the Hamilton function (11.164) leads to the appropriate one-dimensional Dirac equation for the classical two-component field Φ

$$\left(-i\gamma \frac{\delta}{\delta h} - M \right) \Phi = 0, \tag{11.169}$$

where M is the mass matrix of the field Φ

$$M = \begin{bmatrix} -1 & 0 \\ 0 & -\omega^{2} \end{bmatrix}, \tag{11.170}$$

and the γ matrix is the Pauli matrix σ_y

$$\gamma = \sigma_y = \begin{bmatrix} 0 & -i \\ i & 0 \end{bmatrix}, \tag{11.171}$$

obeying the following algebra

$$\gamma^2 = \mathbf{I}_2, \quad \{\gamma, \gamma\} = 2\mathbf{I}_2, \quad \mathbf{I}_2 = \begin{bmatrix} 1 & 0 \\ 0 & 1 \end{bmatrix}. \tag{11.172}$$

The algebra (11.172) is the four-dimensional Clifford algebra over the complex vector space $\mathbb{C}^2$ (For basics and advances in Clifford algebras Cf. *e.g.* Ref. [928])

$$\mathscr{C}\ell_2(\mathbb{C}) = \mathscr{C}\ell_0(\mathbb{C}) \otimes \mathrm{M}_2(\mathbb{C}) \cong \mathrm{M}_2(\mathbb{C}) = \mathbb{C} \oplus \mathbb{C}, \tag{11.173}$$

where $\mathscr{C}\ell_n \equiv \mathscr{C}\ell_{n,0}$, and $\mathrm{M}_2(\mathbb{C})$ denotes algebra of all 2×2 matrices over $\mathbb{C}$. The Clifford algebra $\mathscr{C}\ell_{2,0}(\mathbb{C})$ possesses a two-dimensional complex representation. Restriction to the pinor group $\mathrm{Pin}_{2,0}(\mathbb{R})$ yields a complex representation of the two-dimensional pinor group, *i.e.* the two-dimensional spinor representation, whereas restriction to the spinor group $\mathrm{Spin}_{2,0}(\mathbb{R})$ splits $\mathscr{C}\ell_{1,1}(\mathbb{R})$ into a sum of two half spin representations of dimension 1, *i.e.* the one dimensional Weyl representations. There is the isomorphism

$$\mathrm{Spin}_{2,0}(\mathbb{R}) \cong \mathrm{U}(1) \cong \mathrm{SO}(2), \tag{11.174}$$

and the spinor group $\mathrm{Spin}_{2,0}(\mathbb{R})$ acts on a 1-sphere S^1 in such a way that one has a fibre bundle with fibre $\mathrm{Spin}_{1,0}(\mathbb{R})$

$$\mathrm{Spin}_{1,0}(\mathbb{R}) \longrightarrow \mathrm{Spin}_{2,0}(\mathbb{R}) \longrightarrow S^1, \tag{11.175}$$

and the homotopy sequence is

$$\pi_1\left(\mathrm{Spin}_{1,0}(\mathbb{R})\right) \longrightarrow \pi_1\left(\mathrm{Spin}_{2,0}(\mathbb{R})\right) \longrightarrow \pi_1\left(S^1\right). \tag{11.176}$$

The Clifford algebra $\mathscr{C}\ell_2(\mathbb{C})$ can be generated by complexification

$$\mathscr{C}\ell_2(\mathbb{C}) \cong \mathscr{C}\ell_{1,1}(\mathbb{R}) \otimes \mathscr{C}\ell_0(\mathbb{C}), \tag{11.177}$$

where $\mathscr{C}\ell_{1,1}(\mathbb{R})$ is the four-dimensional Clifford algebra over the real vector space $\mathbb{R}^{2,0}$

$$\mathscr{C}\ell_{1,1}(\mathbb{R}) \cong \mathrm{M}_2(\mathbb{R}) \otimes \mathscr{C}\ell_0(\mathbb{R}) \cong \mathrm{M}_2(\mathbb{R}), \tag{11.178}$$

with $M_2(\mathbb{R})$ being algebra of 2×2 matrices over $\mathbb{R}$, and

$$\mathscr{C}\ell_0(\mathbb{R}) = \mathbb{R}, \tag{11.179}$$

$$\mathscr{C}\ell_0(\mathbb{C}) = \mathbb{C}. \tag{11.180}$$

The Clifford algebra (11.178) can be decomposed into a direct sum of central simple algebras isomorphic to matrix algebra over $\mathbb{R}$

$$\mathscr{C}\ell_{1,1}(\mathbb{R}) = \mathscr{C}\ell^{+}_{1,1}(\mathbb{R}) \oplus \mathscr{C}\ell^{-}_{1,1}(\mathbb{R}), \tag{11.181}$$

$$\mathscr{C}\ell^{\pm}_{1,1}(\mathbb{R}) = \frac{1\pm\gamma}{2}\mathscr{C}\ell_{1,1}(\mathbb{R}) \cong \mathbb{R}, \tag{11.182}$$

as well as into a tensor product

$$\mathscr{C}\ell_{1,1}(\mathbb{R}) = \mathscr{C}\ell_{2,0}(\mathbb{R}) \otimes \mathscr{C}\ell_{0,0}(\mathbb{R}), \tag{11.183}$$

$$\mathscr{C}\ell_{2,0}(\mathbb{R}) = M_2(\mathbb{R}) \otimes \mathscr{C}\ell_{0,0}(\mathbb{R}) \cong M_2(\mathbb{R}). \tag{11.184}$$

D Field Quantization in Static Fock Space

The one-dimensional Dirac equation (11.169) can be quantized

$$\left(-i\gamma\frac{\delta}{\delta h} - M\right)\hat{\Phi} = 0, \tag{11.185}$$

according to the bosonic canonical commutation relations

$$i\left[\hat{\Pi}_\Psi[h'], \hat{\Psi}[h]\right] = \delta(h'-h), \tag{11.186}$$

$$i\left[\hat{\Pi}_\Psi[h'], \hat{\Pi}_\Psi[h]\right] = 0, \tag{11.187}$$

$$i\left[\hat{\Psi}[h'], \hat{\Psi}[h]\right] = 0, \tag{11.188}$$

following from the fact that one has the one-dimensional situation in which there is no difference between bosons and fermions. Particles governed by one-dimensional quantum evolutions are *axions* and, consequently, the secondary quantized one-dimensional Dirac equation (11.185) describes axions obeying the Bose–Einstein statistics, which are *gravitons* in our understanding.

Let us apply the Fock space formalism, which allows to write out explicitly the decomposition of the solution

$$\hat{\Phi} = \mathbf{Q}\mathfrak{B}, \tag{11.189}$$

where $\mathbf{Q}$ is the matrix of secondary quantization

$$\mathbf{Q} = \begin{bmatrix} \sqrt{\dfrac{1}{2\omega}} & \sqrt{\dfrac{1}{2\omega}} \\ -i\sqrt{\dfrac{\omega}{2}} & i\sqrt{\dfrac{\omega}{2}} \end{bmatrix}, \tag{11.190}$$

and $\mathfrak{B} = \mathfrak{B}[h]$ is a dynamical repère

$$\mathfrak{B} = \left\{ \begin{bmatrix} \mathsf{G}[h] \\ \mathsf{G}^\dagger[h] \end{bmatrix} : \left[\mathsf{G}[h'], \mathsf{G}^\dagger[h]\right] = \delta\left(h' - h\right), \left[\mathsf{G}[h'], \mathsf{G}[h]\right] = 0 \right\}, \tag{11.191}$$

on the Fock space of creation and annihilation operators

$$\mathscr{F} = \left(\mathsf{G}, \mathsf{G}^\dagger\right). \tag{11.192}$$

Application of the decomposition (11.189) yields the Heisenberg equations of motion modified by the non-diagonal components

$$\frac{\delta \mathfrak{B}}{\delta h} = \mathbf{X}\mathfrak{B}, \tag{11.193}$$

where $\mathbf{X}$ is the matrix

$$\mathbf{X} = \begin{bmatrix} -i\omega & \dfrac{1}{2\omega}\dfrac{\delta\omega}{\delta h} \\ \dfrac{1}{2\omega}\dfrac{\delta\omega}{\delta h} & i\omega \end{bmatrix}. \tag{11.194}$$

Let us suppose that there is another repère $\mathfrak{F}$ in the Fock space, which is determined by the Bogoliubov transformation

$$\mathfrak{F} = \begin{bmatrix} u & v \\ v^* & u^* \end{bmatrix} \mathfrak{B}, \tag{11.195}$$

where the Bogoliubov coefficients u and v form the Gauss–Lobachevsky–Bolyai hyperbolic space and obey the constraint

$$|u|^2 - |v|^2 = 1, \tag{11.196}$$

and together with the parameter Ω are functionals of h. As the second requirement, let us suppose also that dynamics of the repère is governed by the Heisenberg equations of motion

$$\frac{\delta \mathfrak{F}}{\delta h} = \begin{bmatrix} -i\Omega & 0 \\ 0 & i\Omega \end{bmatrix} \mathfrak{F}. \tag{11.197}$$

Application of the system of equations (11.195)- (11.197) to the equations (11.193) leads to the equation for the vector of the Bogoliubov coefficients

$$\mathbf{b} = \begin{bmatrix} u \\ v \end{bmatrix}, \tag{11.198}$$

which is given by the following vector equation

$$\frac{\delta \mathbf{b}}{\delta h} = \mathbf{X}\mathbf{b}, \tag{11.199}$$

and gives trivial value of the unknown

$$\Omega \equiv 0. \tag{11.200}$$

Therefore, the conjectured repère $\mathfrak{F}$ becomes the static Fock repère with respect to the initial data (I)

$$\mathfrak{F} = \left\{ \begin{bmatrix} \mathsf{G}_I \\ \mathsf{G}_I^\dagger \end{bmatrix} : \left[\mathsf{G}_I, \mathsf{G}_I^\dagger\right] = 1, [\mathsf{G}_I, \mathsf{G}_I] = 0 \right\}, \tag{11.201}$$

and the vacuum state $|0\rangle$ is correctly defined

$$\mathsf{G}_I |0\rangle = 0, \tag{11.202}$$

$$\langle 0| \mathsf{G}_I^\dagger = 0. \tag{11.203}$$

Integrability of the system of equations (11.199) is the crucial element of the scheme presented above. The Bogoliubov transformation (11.195), however, suggests application of the superfluid parametrization

$$u = e^{i\theta} \cosh \phi, \tag{11.204}$$

$$v = e^{i\theta} \sinh \phi, \tag{11.205}$$

where θ and ϕ are the angles which for the present situation are

$$\theta = \pm i \int_{h_I}^{h} \omega' \delta h', \tag{11.206}$$

$$\phi = \ln \sqrt{\left| \frac{\omega_I}{\omega} \right|}, \tag{11.207}$$

where $\omega' = \omega(h')$ and ω_I is the initial datum of the parameter ω

$$\omega_I = -\frac{1}{\sqrt{6}\kappa}, \tag{11.208}$$

which yield the Bogoliubov coefficients in the following form

$$u = \frac{\mu+1}{2\sqrt{\mu}}\exp\left\{i\int_{h_I}^{h}\omega'\delta h'\right\}, \tag{11.209}$$

$$v = \frac{\mu-1}{2\sqrt{\mu}}\exp\left\{-i\int_{h_I}^{h}\omega'\delta h'\right\}, \tag{11.210}$$

where $\mu = \dfrac{\omega}{\omega_I}$. For convenience, one can apply also the reciprocal of μ, *i.e.* the parameter $\lambda = \dfrac{\omega_I}{\omega} = \dfrac{1}{\mu}$

$$\lambda = \sqrt{\left|\frac{h}{{}^{(3)}R - 2\Lambda - 2\kappa\rho}\right|} = \sqrt{\left|\frac{h}{K_{ij}K^{ij} - K^2}\right|}, \tag{11.211}$$

and we understand $\lambda \equiv \lambda[h]$, $\lambda' = \lambda[h']$.

Consequently, the integrability problem is solved by the equation

$$\hat{\Phi} = \mathbf{QG}\mathfrak{F}, \tag{11.212}$$

where $\mathbf{G}$ is the monodromy matrix

$$\mathbf{G} = \begin{bmatrix} \dfrac{1+\mu}{2\sqrt{\mu}}\exp\left\{-i\int_{h_I}^{h}\omega'\delta h'\right\} & \dfrac{1-\mu}{2\sqrt{\mu}}\exp\left\{i\int_{h_I}^{h}\omega'\delta h'\right\} \\ \dfrac{1-\mu}{2\sqrt{\mu}}\exp\left\{-i\int_{h_I}^{h}\omega'\delta h'\right\} & \dfrac{1+\mu}{2\sqrt{\mu}}\exp\left\{i\int_{h_I}^{h}\omega'\delta h'\right\} \end{bmatrix}. \tag{11.213}$$

Now it can be seen straightforwardly that the presented version of quantum geometrodynamics formulates quantum gravity as a quantum field theory of gravity, where the quantum gravitational field is associated with configuration of embedded space and given by the decomposition (11.212) in the static Fock space. In this manner, one can write out straightforwardly conclusions following from the global one-dimensional model of quantum gravity.

It must be noticed that the functional measure δh in any integral of the form $\int \delta h' f[h']$ for the case of a fixed configuration of space, *i.e.* $h = const$, becomes the Riemann–Lebesgue measure dh. However, because h in general is a smooth function of space-time coordinates and free parameters, the measure δh as a total variation over space-time coordinates is the Lebesgue–Stieltjes measure which can be rewritten as the Riemann–Lebesgue measure on space-time. In the most general case $h = h(x_0, x_1, x_2, x_3)$, one can use the transformation

$$\delta h = \frac{\partial^4 h(x_0, x_1, x_2, x_3)}{\partial x_0 \partial x_1 \partial x_2 \partial x_3} d^4 x, \tag{11.214}$$

where $d^4 x = dx_0 dx_1 dx_2 dx_3$, and compute the integral

$$\int \delta h' f[h'] = \int d^4 x' \frac{\partial^4 h(x'_0, x'_1, x'_2, x'_3)}{\partial x'_0 \partial x'_1 \partial x'_2 \partial x'_3} f(x'_0, x'_1, x'_2, x'_3). \tag{11.215}$$

In this manner, the transformation (11.214) establishes the relation between the Wheeler superspace and space-time.

The initial datum condition $\omega = \omega_I$ generates the equation for the initial manifold

$$^{(3)}R^{(I)} - 2\Lambda - 2\kappa \rho_I = h_I, \tag{11.216}$$

or equivalently

$$K^{(I)}_{ij} K^{(I)ij} - K^{(I)2} = h^I, \tag{11.217}$$

where the superscript I means the initial value of given quantity. The quantum evolution (11.150) in such a situation takes the form

$$\left(\frac{\delta^2}{\delta h_I^2} - \frac{1}{6\kappa^2} \right) \Psi(h_I) = 0, \tag{11.218}$$

and after taking into account the suitable boundary conditions

$$\Psi(h_I = h_0) = \Psi_0, \tag{11.219}$$

$$\left. \frac{\delta \Psi(h_I)}{\delta h_I} \right|_{h_I = h_0} = \Pi^0_\Psi, \tag{11.220}$$

can be solved straightforwardly

$$\Psi(h_I) = \Psi_0 \cosh \left\{ \frac{h_I - h_0}{8\pi\sqrt{6}} \right\} + \sqrt{6}\kappa \Pi^0_\Psi \sinh \left\{ \frac{h_I - h_0}{8\pi\sqrt{6}} \right\}. \tag{11.221}$$

E Several Implications

The quantum field-theoretic geometrodynamics was formulated. However, still we do not know what is the role of the one-dimensional wave function which solves the equation (11.135). The same problem is to define any geometric quantities related to the midisuperspace quantum geometrodynamics. The quantum field theory of gravity (11.212) has also unprecise significance. Let us present now several conclusions arising from the previous section, which shall clarify our doubts in some detail.

E1 Global 1D Wave Function

The one-dimensional Dirac equation (12.20) can be rewritten in the form of Schrödinger equation

$$\frac{\delta\Phi}{\delta h} = H\Phi, \tag{11.222}$$

where H is the hermitian Hamiltonian

$$H = i\gamma M = \begin{bmatrix} 0 & -\omega^2 \\ 1 & 0 \end{bmatrix} = \begin{bmatrix} 0 & -\dfrac{\omega_I^2}{\lambda^2} \\ 1 & 0 \end{bmatrix}, \tag{11.223}$$

yielding the evolution operator

$$U[h,h_I] = \exp\int_{h_I}^{h} \delta h' H[h'] = \exp\begin{bmatrix} 0 & -\int_{h_I}^{h}\delta h'\omega^2[h'] \\ h-h_I & 0 \end{bmatrix}, \tag{11.224}$$

where $h \geqslant h_I$, which is explicitly

$$U[h,h_I] = \begin{bmatrix} \cos f[h,h_I] & -\left(\int_{h_I}^{h}\delta h'\omega'^2\right)\dfrac{\sin f[h,h_I]}{f[h,h_I]} \\ (h-h_I)\dfrac{\sin f[h,h_I]}{f[h,h_I]} & \cos f[h,h_I] \end{bmatrix}, \tag{11.225}$$

where $f[h,h_I]$ is the functional

$$f[h,h_I] = \sqrt{(h-h_I)\int_{h_I}^{h}\omega'^2\delta h'}, \tag{11.226}$$

so that the solution of the equation (12.26) is

$$\Phi[h,h_I] = U[h,h_I]\Phi[h_I]. \tag{11.227}$$

Straightforward elementary algebraic manipulations allow to determine the global one-dimensional wave function as

$$\Psi[h,h_I]] = \Psi^I \cos f[h,h_I] + \Pi^I_\Psi (h-h_I)\frac{\sin f[h,h_I]}{f[h,h_I]}, \tag{11.228}$$

and, similarly, the canonical conjugate momentum is

$$\Pi_\Psi[h,h_I] = \Pi^I_\Psi \cos f[h,h_I] - \Psi^I \left(\int_{h_I}^{h} \delta h' \omega'^2\right)\frac{\sin f[h,h_I]}{f[h,h_I]}, \tag{11.229}$$

where Ψ^I and Π^I_Ψ are the initial data

$$\Psi^I = \Psi[h_I], \tag{11.230}$$

$$\Pi^I_\Psi = \Pi_\Psi[h_I] = \left.\frac{\delta\Psi}{\delta h}\right|_{h=h_I}. \tag{11.231}$$

Because $U^\dagger[h,h_I] = U^T[h,h_I]$, one sees that

$$\Psi^\star[h,h_I] = \Psi[h,h_I], \tag{11.232}$$

$$\Pi^\star_\Psi[h,h_I] = \Pi_\Psi[h,h_I], \tag{11.233}$$

and, for consistency, it also must be $(\Psi^I)^\star = \Psi^I$, $(\Pi^I_\Psi)^\star = \Pi^I_\Psi$. The probability density in the quantum mechanics is

$$\Omega[h,h_I] = \Phi^\dagger[h,h_I]\Phi[h,h_I] = \Phi^\dagger[h_I]U^\dagger[h,h_I]U[h,h_I]\Phi[h_I]. \tag{11.234}$$

Computing the matrix $U^\dagger[h,h_I]U[h,h_I]$

$$U^\dagger[h,h_I]U[h,h_I] = \begin{bmatrix} \cos^2 f[h,h_I] + \left[(h-h_I)\frac{\sin f[h,h_I]}{f[h,h_I]}\right]^2 & \left(h-h_I-\int_{h_I}^{h}\delta h'\omega'^2\right)\frac{\sin 2f[h,h_I]}{2f[h,h_I]} \\ \left(h-h_I-\int_{h_I}^{h}\delta h'\omega'^2\right)\frac{\sin 2f[h,h_I]}{2f[h,h_I]} & \cos^2 f[h,h_I] + \left[\int_{h_I}^{h}\delta h'\omega'^2\frac{\sin f[h,h_I]}{f[h,h_I]}\right]^2 \end{bmatrix}, \tag{11.235}$$

and taking into account that

$$\Phi^{\dagger}[h_I] = \begin{bmatrix} \Pi_{\Psi}^I \\ \Psi^I \end{bmatrix}^{\dagger} = \begin{bmatrix} (\Pi_{\Psi}^I)^{\star} \\ (\Psi^I)^{\star} \end{bmatrix}^T = \begin{bmatrix} \Pi_{\Psi}^I \\ \Psi^I \end{bmatrix}^T = \left[\Pi_{\Psi}^I, \Psi^I\right], \tag{11.236}$$

one obtains finally

$$\Omega[h,h_I] = A[h,h_I]\left(\Psi^I\right)^2 + 2B[h,h_I]\Psi^I\Pi_{\Psi}^I + C[h,h_I]\left(\Pi_{\Psi}^I\right)^2, \tag{11.237}$$

where the functional coefficients $A[h,h_I]$, $B[h,h_I]$, $C[h,h_I]$ in (11.237) are

$$A[h,h_I] = \cos^2 f[h,h_I] + \left(\int_{h_I}^{h} \delta h' \omega'^2\right)^2 \left(\frac{\sin f[h,h_I]}{f[h,h_I]}\right)^2, \tag{11.238}$$

$$B[h,h_I] = \left(h - h_I - \int_{h_I}^{h} \delta h' \omega'^2\right) \frac{\sin 2f[h,h_I]}{2f[h,h_I]}, \tag{11.239}$$

$$C[h,h_I] = \cos^2 f[h,h_I] + (h-h_I)^2 \left(\frac{\sin f[h,h_I]}{f[h,h_I]}\right)^2, \tag{11.240}$$

The initial data Ψ^I and Π_{Ψ}^I are not arbitrary, but constrained by the normalization condition for the probability density

$$\int_{h_I}^{h_F} \Omega[h']\delta h' = 1, \tag{11.241}$$

where h_F is some maximal value of h, which in the explicit form leads to the algebraic equation

$$C[h_I](\Pi_{\Psi}^I)^2 + 2B[h_I]\Psi^I\Pi_{\Psi}^I + A[h_I](\Psi^I)^2 - 1 = 0, \tag{11.242}$$

with the coefficients $A[h_I]$, $B[h_I]$, $C[h_I]$ given by the integrals

$$A[h_I] = \int_{h_I}^{h_F} \delta h' A[h',h_I], \tag{11.243}$$

$$B[h_I] = \int_{h_I}^{h_F} \delta h' B[h',h_I], \tag{11.244}$$

$$C[h_I] = \int_{h_I}^{h_F} \delta h' C[h',h_I]. \tag{11.245}$$

The equation (11.242) can be solved straightforwardly. As a result one obtains

$$\Pi_{\Psi}^I = -\frac{B[h_I]}{C[h_I]}\Psi^I \pm \sqrt{\frac{B^2[h_I] - A[h_I]C[h_I]}{C^2[h_I]}(\Psi^I)^2 + \frac{1}{C[h_I]}}. \tag{11.246}$$

Application of the explicit form of initial datum of the conjugate momentum $\Pi^I_\Psi = \dfrac{\delta\Psi^I}{\delta h_I}$ to the equation (11.242) yields the differential equation for the initial datum of the classical scalar field Ψ^I

$$C[h_I]\left(\frac{\delta\Psi^I}{\delta h_I}\right)^2 + 2B[h_I]\Psi^I\frac{\delta\Psi^I}{\delta h_I} + A[h_I](\Psi^I)^2 - 1 = 0, \tag{11.247}$$

which is very hard to solve in general. However, there is the case defined by the values of the coefficients $A[h_I] = A$, $B[h_I] = B$, $C[h_I] = C$ which are independent of h_I. In such a situation, the equation (11.247) possesses solutions which are easy to extract

$$\Psi^I_\mp = f_\pm^{(-1)}\left(\mp\frac{h_I}{C} + C_1\right), \tag{11.248}$$

where C_1 is the integration constant, and $f_\pm(x)$ are the functions

$$f_\pm(x) = \pm\frac{B}{AC}\left\{\frac{1}{2}\ln\left|Ax^2 - 1\right| \pm \operatorname{artanh}\left[\frac{Bx}{\sqrt{C + (B^2 - AC)x^2}}\right] \mp \right.$$
$$\left. \mp\frac{\sqrt{B^2 - AC}}{B}\ln\left|2\left(B^2 - AC\right)x + 2\sqrt{B^2 - AC}\sqrt{C + (B^2 - AC)x^2}\right|\right\}, \tag{11.249}$$

The solution (11.248) was determined computationally, and using it one can construct the analytical solutions straightforwardly. Let us introduce the following parameter

$$x = \pm\frac{h_I - h_0}{C}, \tag{11.250}$$

where for consistency we have taken $C_1 = \mp\dfrac{h_0}{C}$. Differentiating the solutions (11.248) with respect to x one obtains

$$\frac{\delta\Psi^I_\mp}{\delta\left(\pm\dfrac{h_I}{C} + C_1\right)} = \pm C\frac{\delta\Psi^I_\mp}{\delta h_I} = -\frac{1}{f'_\mp(x)}, \tag{11.251}$$

where $f'(x) = \dfrac{df(x)}{dx}$, and for this reason

$$\frac{\delta\Psi^I_\mp}{\delta h_I} = \mp\frac{1}{Cf'_\mp(x)}. \tag{11.252}$$

Substitution of the derivative (11.252) to the equation (11.247) gives

$$\frac{1}{C(f'_{\mp}(x))^2} \mp \frac{2B}{Cf'_{\mp}(x)}\Psi^I_{\mp} + A(\Psi^I_{\mp})^2 - 1 = 0, \tag{11.253}$$

or equivalently one obtains the quadratic equation for Ψ^I

$$AC(f'_{\mp}(x))^2(\Psi^I_{\mp})^2 \mp 2Bf'_{\mp}(x)\Psi^I_{\mp} + 1 - C(f'(x))^2 = 0. \tag{11.254}$$

The equation (11.254) has two solutions

$$\Psi^I_{\mp} = \pm\frac{B}{ACf'_{\mp}(x)}\left(1 + \frac{1}{ABC}\sqrt{B^2 - AC + 4AC^2(f'_{\mp}(x))^2}\right), \tag{11.255}$$

where the derivative $f'_{\mp}(x)$ can be established straightforwardly

$$f'_{\mp}(x) = \pm\frac{B}{C}\left[\frac{x}{Ax^2-1} \mp \frac{B}{\sqrt{C+(B^2-AC)x^2}}\left(\frac{x^2}{Ax^2-1} - \frac{C}{B^2}\right)\right]. \tag{11.256}$$

Substitution of the derivative (11.256) to the formula (11.255) gives

$$\Psi^I_{\mp} = \left(\frac{Ax}{Ax^2-1} \mp \frac{AB}{\sqrt{C+(B^2-AC)x^2}}\left(\frac{x^2}{Ax^2-1} - \frac{C}{B^2}\right)\right)^{-1/2} \times$$
$$\left\{1 + \frac{1}{AC}\left[1 - \frac{AC}{B^2} + \frac{4Ax^2}{(Ax^2-1)^2} + \frac{4AB}{C+(B^2-AC)x^2}\left(\frac{x^2}{Ax^2-1} - \frac{C}{B^2}\right)^2 \pm \right.\right.$$
$$\left.\left.\frac{ABx}{(Ax^2-1)\sqrt{C+(B^2-AC)x^2}}\left(\frac{x^2}{Ax^2-1} - \frac{C}{B^2}\right)\right]^{1/2}\right\}. \tag{11.257}$$

E2 Unitary Three-Manifolds

In general, the evolution operator (11.225) is non-unitary, *i.e.* it can be shown by straightforward calculation that the condition

$$U^\dagger U = UU^\dagger = \mathbf{1}_2, \tag{11.258}$$

is broken. However, there are situations in the theory for which the evolution operator (11.225) is unitary. It can be proved easily that for the unitarity of U the necessary and sufficient conditions are

$$\left(h-h_I\int_{h_I}^{h}\delta h'\omega'^2\right)\frac{\sin 2f[h,h_I]}{2f[h,h_I]} = 0, \tag{11.259}$$

$$\cos^2 f[h,h_I]+\left(\int_{h_I}^{h}\delta h'\omega'^2\right)^2\left(\frac{\sin f[h,h_I]}{f[h,h_I]}\right)^2 = 1, \tag{11.260}$$

$$\left[\left(\int_{h_I}^{h}\delta h'\omega'^2\right)^2-(h-h_I)^2\right]\left(\frac{\sin f[h,h_I]}{f[h,h_I]}\right)^2 = 0. \tag{11.261}$$

These equations have two solutions. The first solution is trivial

$$f[h,h_I]=0, \tag{11.262}$$

and corresponds to the initial datum

$$h=h_I. \tag{11.263}$$

In such a situation, the evolution operator is trivially equal to the unit 2×2 matrix, which in fact means that there is no evolution. The second solution, however, is non-trivial

$$f^2[h,h_I]=-(h-h_I)^2, \tag{11.264}$$

and corresponds to the equation

$$\omega^2=-1, \tag{11.265}$$

which is associated with the purely imaginary value of ω. Using the definition (11.151), the condition (11.265) generates the equation for the embedded three-dimensional space

$$^{(3)}R=6\kappa^2h+2\Lambda+2\kappa\rho, \tag{11.266}$$

or, equivalently, using the second definition (11.152)

$$K_{ij}K^{ij}-K^2=6\kappa^2h. \tag{11.267}$$

In this instance, the unitary evolution operator has the form

$$U[h,h_I]=\begin{bmatrix}\cosh(h-h_I) & \sinh(h-h_I)\\ -\sinh(h-h_I) & \cosh(h-h_I)\end{bmatrix}, \tag{11.268}$$

and is the rotation matrix of the unitary Lie group $U(1) \cong SO(2)$, where the angle of the rotation is $i(h-h_I)$. Hence, in such a situation, the classical scalar field Ψ and its conjugate momentum field Π_Ψ are

$$\Psi[h,h_I] = \Psi^I \cosh(h-h_I) - \Pi^I_\Psi \sinh(h-h_I), \quad (11.269)$$
$$\Pi_\Psi[h,h_I] = \Psi^I \sinh(h-h_I) + \Pi^I_\Psi \cosh(h-h_I), \quad (11.270)$$

and the corresponding probability density equals

$$\Omega[h,h_I] = \left(\Psi^I\right)^2 \cosh^2(h-h_I) + \left(\Pi^I_\Psi\right)^2 \sinh^2(h-h_I) - \quad (11.271)$$
$$- \quad 2\Psi^I\Pi^I \sinh(h-h_I)\cosh(h-h_I). \quad (11.272)$$

In this manner, the normalization condition (11.241) generates the equation for the initial datum

$$\alpha(h_F-h_I)\left(\Pi^I_\Psi\right)^2 - 2\beta(h_F-h_I)\Psi^I\Pi^I + \gamma(h_F-h_I)\left(\Psi^I\right)^2 = \frac{1}{h_F-h_I}, \quad (11.273)$$

where the coefficients-functions of h_I are

$$\alpha(h_F-h_I) = \frac{\sinh(2(h_F-h_I)) - 2(h_F-h_I)}{4}, \quad (11.274)$$
$$\beta(h_F-h_I) = \frac{1}{2}\sinh^2(2(h_F-h_I)), \quad (11.275)$$
$$\gamma(h_F-h_I) = \frac{\sinh(2(h_F-h_I)) + 2(h_F-h_I)}{4}. \quad (11.276)$$

Application of the basic definition of initial datum of the conjugate momentum $\Pi^I_\Psi = \dfrac{\delta\Psi_I}{\delta h_I}$ to the equation (11.277) gives the differential equation for the initial datum of the scalar field Ψ_I

$$\alpha(h_F-h_I)\left(\frac{\delta\Psi_I}{\delta h_I}\right)^2 - 2\beta(h_F-h_I)\Psi^I\frac{\delta\Psi_I}{\delta h_I} + \gamma(h_F-h_I)\left(\Psi^I\right)^2 - \frac{1}{h_F-h_I} = 0. \quad (11.277)$$

for finite values of h_F this equation is difficult to solve straightforwardly. When the upper limit is infinite, *i.e.* $h_F \to \infty$, then the third term in (11.277) vanishes identically whereas the coefficients α, β, and γ tends to infinity. Then however, one can divide both sides of the equation (11.277) by $\frac{1}{2}\sinh(2(h_F-h_I))$ and obtain the finite limit

$$\left(\frac{\delta\Psi_I}{\delta h_I}\right)^2 - 2\Psi^I\frac{\delta\Psi_I}{\delta h_I} + \left(\Psi^I\right)^2 = \left(\frac{\delta\Psi_I}{\delta h_I} - \Psi^I\right)^2 = 0. \quad (11.278)$$

Using the boundary condition $\Psi_I(h_0) = \Psi_0$, this equation can be solved immediately

$$\Psi_I = \Psi_0 \exp\{h_I - h_0\}. \qquad (11.279)$$

Because the equations (11.266) and (11.267) are related to the unitary operator of evolution (11.268) we shall call their solutions *the unitary three-manifolds*. Interestingly, the equation (11.266) suggests that, in general, the unitary three-manifolds are deformations of the three-manifolds defined by the Ricci scalar-curvature proportional to determinant of induced metric, *i.e.* to the global dimension

$$^{(3)}R = 6\kappa^2 h, \qquad (11.280)$$

and deformation is due to the cosmological constant Λ and the energy density ρ of Matter fields. Because the three-dimensional manifolds defined by (11.280) are also the unitary three-manifolds, we shall call them *the global unitary three-manifolds*.

E3 Fourier Analysis

The quantum gravity given by the one-dimensional Klein–Gordon equation (11.150)

$$\left(\frac{\delta^2}{\delta h^2} + \omega^2[h]\right)\Psi(h) = 0, \qquad (11.281)$$

$$\omega^2[h] = -\frac{1}{6\kappa^2}\frac{1}{h}\left({}^{(3)}R - 2\Lambda - 2\kappa\rho\right), \qquad (11.282)$$

can be considered as the equation for the three-dimensional scalar curvature $^{(3)}R$

$$^{(3)}R = 2(\Lambda + \kappa\rho) + 6\kappa^2\varphi(h)h, \qquad (11.283)$$

where we have introduced the function

$$\varphi(h) = \frac{1}{\Psi(h)}\frac{\delta^2\Psi(h)}{\delta h^2}. \qquad (11.284)$$

In the stationary case

$$\rho \equiv 0 \cap \Lambda \equiv 0 \quad \text{or} \quad \rho = -\frac{\Lambda}{\kappa}, \qquad (11.285)$$

one obtains from (11.283) that

$$^{(3)}R = 6\kappa^2\varphi_n h, \tag{11.286}$$

where φ_n are eigenvalues determined by the equation

$$\frac{\delta^2\Psi}{\delta h^2} = \varphi_n\Psi. \tag{11.287}$$

If one wishes to consider the non-stationary situation, then it is easy to see that

$$\varphi(h) = \varphi_n - \frac{2}{6\kappa^2}\frac{\Lambda+\kappa\rho}{h}. \tag{11.288}$$

One can apply the analytical form of the classical scalar field Ψ

$$\Psi = \sum_{n=0}^{\infty} a_n(h-h_I)^n, \tag{11.289}$$

to the eigenequation (11.287)

$$\sum_{n=-\infty}^{\infty}(n+1)(n+2)a_{n+2}(h-h_I)^n = \varphi_n\sum_{n=-\infty}^{\infty}a_n(h-h_I)^n, \tag{11.290}$$

and express eigenvalues φ_n by the series coefficients a_n

$$\varphi_n = (n+1)(n+2)\frac{a_{n+2}}{a_n}. \tag{11.291}$$

However, because the series coefficients are defined as

$$a_n = \frac{1}{n!}\left.\frac{\delta^n\Psi}{\delta h^n}\right|_{h=h_I}, \tag{11.292}$$

one obtains from (11.291)

$$\varphi_n = \left.\frac{\dfrac{\delta^{n+2}}{\delta h^{n+2}}\Psi(h)}{\dfrac{\delta^n}{\delta h^n}\Psi(h)}\right|_{h=h_I}. \tag{11.293}$$

In the light of the fact

$$\frac{\delta^{n+2}\Psi}{\delta h^{n+2}} = \frac{\delta^n}{\delta h^n}\left(\frac{\delta^2\Psi}{\delta h^2}\right) = -\frac{\delta^n}{\delta h^n}\left(\omega^2[h]\Psi(h)\right), \tag{11.294}$$

where we have applied the equation (11.281), one has

$$\varphi_n = - \left. \frac{\frac{\delta^n}{\delta h^n}\left(\omega^2[h]\Psi(h)\right)}{\frac{\delta^n}{\delta h^n}\Psi(h)} \right|_{h=h_I} . \tag{11.295}$$

Let us assume that there are generalized Fourier transforms

$$\widetilde{\Psi}(s) = \int \delta h e^{-2i\pi sh}\Psi(h), \tag{11.296}$$

$$\widetilde{\omega^2}(s) = \int \delta h e^{-2i\pi sh}\omega^2[h], \tag{11.297}$$

as well as the inverted Fourier transforms

$$\Psi(h) = \int \delta s e^{2i\pi sh}\widetilde{\Psi}(s), \tag{11.298}$$

$$\omega^2[h] = \int \delta s e^{2i\pi sh}\widetilde{\omega^2}(s). \tag{11.299}$$

Applying the generalized Leibniz product formula

$$\frac{\delta^n}{\delta h^n}\left(\omega^2[h]\Psi(h)\right) = \sum_{r=0}^{n}\binom{n}{r}\left(\frac{\delta^r}{\delta h^r}\omega^2[h]\right)\left(\frac{\delta^{n-r}}{\delta h^{n-r}}\Psi(h)\right), \tag{11.300}$$

one obtains

$$\begin{aligned}\frac{\frac{\delta^n}{\delta h^n}\left(\omega^2[h]\Psi(h)\right)}{\frac{\delta^n}{\delta h^n}\Psi(h)} &= \frac{1}{\frac{\delta^n}{\delta h^n}\Psi(h)}\sum_{r=0}^{n}\binom{n}{r}\left(\frac{\delta^r}{\delta h^r}\omega^2[h]\right)\left(\frac{\delta^{n-r}}{\delta h^{n-r}}\Psi(h)\right) = \\ &= \frac{\frac{\delta^n}{\delta h^n}\Psi(h)}{\frac{\delta^n}{\delta h^n}\Psi(h)}\sum_{r=0}^{n}\binom{n}{r}\left(\frac{\delta^r}{\delta h^r}\omega^2[h]\right)\left(\frac{\delta^{-r}}{\delta h^{-r}}\Psi(h)\right) = \\ &= \sum_{r=0}^{n}\binom{n}{r}\left(\frac{\delta^r}{\delta h^r}\omega^2[h]\right)\left(\frac{\delta^{-r}}{\delta h^{-r}}\Psi(h)\right). \end{aligned} \tag{11.301}$$

In this manner, using the inverted Fourier transforms (11.298) and (11.299) gives

$$\frac{\delta^r}{\delta h^r}\omega^2[h] = \int \delta s e^{2i\pi sh}(2i\pi s)^r\widetilde{\omega^2}(s), \tag{11.302}$$

$$\frac{\delta^{-r}}{\delta h^{-r}}\Psi(h) = \int \delta s e^{2i\pi sh}(2i\pi s)^{-r}\widetilde{\Psi}(s), \tag{11.303}$$

and for this reason the formula (11.301) becomes

$$\frac{\frac{\delta^n}{\delta h^n}\left(\omega^2[h]\Psi(h)\right)}{\frac{\delta^n}{\delta h^n}\Psi(h)} = \int\int \delta s \delta s' e^{2i\pi('+s')h}\left[\sum_{r=0}^{n}\binom{n}{r}\left(\frac{s}{s'}\right)^r\right]\widetilde{\omega^2}(s)\widetilde{\Psi}(s'). \tag{11.304}$$

Applying the standard summation procedure

$$\sum_{r=0}^{n}\binom{n}{r}x^r = (1+x)^n, \tag{11.305}$$

one obtains

$$\frac{\frac{\delta^n}{\delta h^n}\left(\omega^2[h]\Psi(h)\right)}{\frac{\delta^n}{\delta h^n}\Psi(h)} = \int\int \delta s \delta s' e^{2i\pi(s'+s)h}\left(1+\frac{s}{s'}\right)^n\widetilde{\omega^2}(s)\widetilde{\Psi}(s'). \tag{11.306}$$

For this reason, the eigenvalues (11.295) are

$$\varphi_n = -(n+1)(n+2)\int\int \delta s \delta s' e^{2i\pi(s+s')h}\left(1+\frac{s}{s'}\right)^n\widetilde{\omega^2}(s)\widetilde{\Psi}(s')\bigg|_{h=h_I}. \tag{11.307}$$

Applying the Fourier transforms (11.296) and (11.297), one receives

$$\widetilde{\omega^2}(s)\widetilde{\Psi}(s') = \int\int \delta h \delta h' e^{-2i\pi(sh+s'h')}\omega^2[h]\Psi(h'), \tag{11.308}$$

and for this reason one obtains finally

$$\varphi_n = -\int\int \delta h \delta h' \mathscr{G}_n(h-h')\omega^2[h]\Psi(h')\bigg|_{h=h_I,h'=h_I} = \tag{11.309}$$

$$= -\int\int \delta h \delta h' \mathscr{G}_n(h-h')\frac{\delta^2\Psi(h)}{\delta h^2}\frac{\Psi(h')}{\Psi(h)}\bigg|_{h=h_I,h'=h_I} = \tag{11.310}$$

$$= -\int\int \delta h \delta h' \mathscr{G}_n(h-h')\frac{\delta\Pi_\Psi(h)}{\delta h}\frac{\Psi(h')}{\Psi(h)}\bigg|_{h=h_I,h'=h_I}, \tag{11.311}$$

where in the second line we have used equations of motion and in the third line we have applied definition of the conjugate momentum field. The kernel $\mathscr{G}(h-h')$ is given by the relation

$$\mathscr{G}_n(h-h') = \int\int \delta s \delta s' e^{2i\pi s'(h-h')}\left(1+\frac{s'}{s}\right)^n. \tag{11.312}$$

Estimation of the kernel (11.312) is the crucial element of the proposed analysis. To make it consistently, let us determine the range of s and s'. In quantum mechanics, the Fourier analysis transforms theory to the momentum space representation. Therefore, s and s' are the momenta conjugated to h and h', respectively.

The double integral in the kernel (11.312) can be transformed as

$$\int \delta s' e^{2i\pi s'(h-h')} \left[\int \delta s \left(1+\frac{s}{s'}\right)^n\right], \tag{11.313}$$

and, therefore, one sees that the variable s is *internal*, while the variable s' is *external*. Let us take *ad hoc* the finite range of the internal variable $s \in [0,S]$ and the infinite range of the external variable $s' \in [0,\infty]$. The internal integral can be easily computed

$$\int \delta s \left(1+\frac{s}{s'}\right)^n = s' \int \delta\left(1+\frac{s}{s'}\right)\left(1+\frac{s}{s'}\right)^n = \tag{11.314}$$

$$= s' \int_{t=1+\frac{s}{s'}} \delta t t^n = \frac{s'}{n+1}\left(1+\frac{S}{s'}\right)^{n+1}, \tag{11.315}$$

and for this reason the double integral (11.313) becomes

$$\frac{1}{n+1}\int \delta s' e^{2i\pi s'(h-h')} s' \left(1+\frac{S}{s'}\right)^{n+1}. \tag{11.316}$$

In the range $s' \in [0,\infty]$ the integral in (11.316) converges for $n < 1$

$$\int \delta s' e^{2i\pi s'(h-h')} s' \left(1+\frac{S}{s'}\right)^{n+1} = -\frac{\Gamma(1-n)}{4\pi^2(h-h')^2} U\left(-1-n,-1,-2i\pi S\left(h-h'\right)\right), \tag{11.317}$$

where $\Gamma(z)$ is the Euler gamma-function

$$\Gamma(z) = \int_0^\infty t^{z-1} e^{-t} dt, \tag{11.318}$$

and $U(a,b,z)$ is the Tricomi confluent hypergeometric function (Cf. *e.g.* Ref. [929] for basic knowledge about special functions)

$$U(a,b,z) = \frac{\Gamma(1-b)}{\Gamma(a-b+1)} M(a,b,z) + \frac{\Gamma(b-1)}{\Gamma(a)} z^{1-b} M(a-b+1,2-b,z), \tag{11.319}$$

where $M(a,b,z)$ is the Kummer confluent hypergeometric function

$$M(a,b,z) = \sum_{n=0}^{\infty} \frac{(a)_n}{(b)_n} \frac{z^n}{n!} = 1 + \frac{\Gamma(b)}{\Gamma(a)} \sum_{n=1}^{\infty} \frac{\Gamma(a+n)}{\Gamma(b+n)} \frac{z^n}{n!}, \tag{11.320}$$

where $(\alpha)_n$ are the Pochhammer symbols

$$(\alpha)_n = \frac{\Gamma(\alpha+n)}{\Gamma(\alpha)}, \quad (\alpha)_0 = 1, \tag{11.321}$$

and for $\Re a > 0$ there is the integral relation

$$\frac{(a)_n}{(b)_n} = \frac{\Gamma(b)}{\Gamma(a)\Gamma(b-a)} \int_0^1 t^{a-1+n}(1-t)^{b-a-1} dt. \tag{11.322}$$

Applying the values $a = 1-n > 0$, $b = 3$, $z = -2i\pi S(h-h')$ to the Kummer transformation

$$U(a,b,z) = z^{1-b} U(1+a-b, 2-b, z), \tag{11.323}$$

one receives

$$U\left(-1-n,-1,-2i\pi S\left(h-h'\right)\right) = -4\pi^2 S^2 (h-h')^2 U\left(1-n,3,-2i\pi S\left(h-h'\right)\right), \tag{11.324}$$

and for this reason the integral (11.317) becomes

$$\int \delta s' e^{2i\pi s'(h-h')} s' \left(1+\frac{S}{s'}\right)^{n+1} = S^2 \Gamma(1-n) U\left(1-n,3,-2i\pi S\left(h-h'\right)\right). \tag{11.325}$$

Because now $1-n > 0$ the Tricomi confluent hypergeometric functions can be computed in the integral representation

$$U(a,b,z) = \frac{1}{\Gamma(a)} \int_0^{\infty} e^{-zt} t^{a-1} (1+t)^{b-a-1} dt. \tag{11.326}$$

In this manner, the kernel (11.312) (for $n < 1$) is given by the formula

$$\mathscr{G}_n(h-h') = S^2 \frac{\Gamma(1-n)}{n+1} U\left(1-n,3,-2i\pi S\left(h-h'\right)\right), \tag{11.327}$$

and, consequently, the eigenvalue (11.291) can be evaluated as

$$\varphi_n = -S^2 \frac{\Gamma(1-n)}{n+1} \int\int \delta h \delta h' U\left(1-n,3,-2i\pi S\left(h-h'\right)\right) \Upsilon(h,h')\Big|_{h=h_I, h'=h_I}, \tag{11.328}$$

where we have introduced the symbol $\Upsilon(h,h')$

$$\Upsilon(h,h') = \frac{\Psi(h')}{\Psi(h)}\frac{\delta^2\Psi(h)}{\delta h^2} = \frac{\Psi(h')}{\Psi(h)}\frac{\delta\Pi_\Psi(h)}{\delta h}, \tag{11.329}$$

which can be straightforwardly established when both wave function $\Psi(h)$ and its second derivative $\frac{\delta^2\Psi(h)}{\delta h^2}$ or, equivalently, the derivative of its conjugate momentum field $\frac{\delta\Pi_\Psi(h)}{\delta h}$, are given explicitly.

E4 Quantum Correlations

Using matrices (11.213) and (11.189), and the relation (11.212) one derives the quantum field

$$\hat{\Psi}(h) = \sqrt{\frac{\omega_I}{8}}\frac{1}{\omega}\left(\exp\left\{-i\int_{h_I}^{h}\omega'\delta h'\right\}\mathsf{G}_I + \exp\left\{i\int_{h_I}^{h}\omega'\delta h'\right\}\mathsf{G}_I^\dagger\right). \tag{11.330}$$

Let us take into account the n-particle one-point quantum states determined as

$$|h,n\rangle \equiv \hat{\Psi}^n|0\rangle = \left(\sqrt{\frac{\omega_I}{8}}\frac{1}{\omega}\exp\left\{\int_{h_I}^{h}\omega'\delta h'\right\}\right)^n \mathsf{G}_I^{\dagger n}|0\rangle, \tag{11.331}$$

which yields two-point correlators $\mathrm{Cor}_{n'n}(h',h) \equiv \langle n',h'|h,n\rangle$ or explicitly

$$\begin{aligned} \mathrm{Cor}_{n'n}(h',h) &= \left(\frac{\omega_I}{8}\right)^{(n'+n)/2}\exp\left\{i\left(n'\int_{h'}^{h_I} + n\int_{h_I}^{h}\right)\omega''\delta h''\right\}\times \\ &\times \frac{\langle 0|\mathsf{G}_I^{n'}\mathsf{G}_I^{\dagger n}|0\rangle}{\omega'^{n'}\omega^n}. \end{aligned} \tag{11.332}$$

Basically, one obtains

$$\mathrm{Cor}_{00}(h,h) = \mathrm{Cor}_{00}(h',h) = \mathrm{Cor}_{00}(h_I,h_I) = \langle 0|0\rangle, \tag{11.333}$$

$$\mathrm{Cor}_{11}(h_I,h_I) = \frac{1}{8\omega_I}, \tag{11.334}$$

$$\frac{\mathrm{Cor}_{n'n}(h_I,h_I)}{[\mathrm{Cor}_{11}(h_I,h_I)]^{(n'+n)/2}} = \langle 0|\mathsf{G}_I^{n'}\mathsf{G}_I^{\dagger n}|0\rangle, \tag{11.335}$$

and by elementary algebraic manipulations one receives

$$\mathrm{Cor}_{11}(h',h) = \frac{\sqrt{\mathrm{Cor}_{11}(h',h')\mathrm{Cor}_{11}(h,h)}}{\mathrm{Cor}_{11}(h_I,h_I)} \exp\left\{i\int_{h'}^{h} \omega''\delta h''\right\}, \tag{11.336}$$

$$\frac{\mathrm{Cor}_{nn}(h',h)}{\mathrm{Cor}_{00}(h_I,h_I)} = \left[\frac{\mathrm{Cor}_{11}(h',h)}{\mathrm{Cor}_{00}(h_I,h_I)}\right]^n, \tag{11.337}$$

$$\frac{\mathrm{Cor}_{11}(h,h)}{\mathrm{Cor}_{00}(h_I,h_I)} = \left(\frac{\omega_I}{\omega}\right)^2 \mathrm{Cor}_{11}(h_I,h_I). \tag{11.338}$$

Straightforwardly from (11.338) one can relate the size scale with quantum correlations

$$\lambda = \frac{\omega_I}{\omega} = \sqrt{\frac{\mathrm{Cor}_{11}(h,h)}{\mathrm{Cor}_{11}(h_I,h_I)\mathrm{Cor}_{00}(h_I,h_I)}}, \tag{11.339}$$

and, consequently, one receives the formulas

$$\frac{\mathrm{Cor}_{n'n}(h,h)}{\mathrm{Cor}_{n'n}(h_I,h_I)} = \lambda^{n'+n} \exp\left\{-i(n'-n)\int_{h_I}^{h} \omega''\delta h''\right\}, \tag{11.340}$$

$$\frac{\mathrm{Cor}_{11}(h',h)}{\mathrm{Cor}_{00}(h_I,h_I)\mathrm{Cor}_{11}(h_I,h_I)} = \lambda'\lambda \exp\left\{i\int_{h'}^{h} \omega''\delta h''\right\}, \tag{11.341}$$

$$\frac{\mathrm{Cor}_{nn}(h',h)}{\mathrm{Cor}_{00}(h_I,h_I)} = \lambda'^n\lambda^n[\mathrm{Cor}_{11}(h_I,h_I)]^n \exp\left\{in\int_{h'}^{h} \omega''\delta h''\right\} \tag{11.342}$$

Finally, let us note that making use of the definition (11.208) and the result (11.334) gives the physical interpretation of the Einstein constant

$$\kappa = -\frac{4\sqrt{6}}{3}\mathrm{Cor}_{11}(h_I,h_I). \tag{11.343}$$

In this manner, in the presented model the whole physical information about quantum gravity is solely contained in the free parameters of the theory, *i.e.* the parameter ω and its initial datum ω_I or, equivalently, size λ or mass μ scales. It is evident that the quantum correlations are strictly determined by these fundamental quantities only. In other words, measurement of quantum correlations can be used for deduction of values of the fundamental parameters of the theory.

Despite the fact that the presented ramifications have purely formal character, they have shown manifestly general features of the proposed theory of quantum gravity. Let us examine its further consequences.

Chapter 12

Invariant Global Dimension

A Invariant Global Quantum Gravity

Let us include, to the model presented in the previous chapter, the explicit presence of Matter fields, and extend the global one-dimensional wave function $\Psi(h)$ to the suitable functional $\Psi[h,\phi]$, which we shall call *the extended global wave functional*. We preserve the concept of the global dimension as the determinant of the spatial metric $h = \det h_{ij}$.

The theory of quantum gravity proposed in the previous chapter can be rewritten in the generalized form

$$\left(\frac{\delta^2}{\delta h^2} + V_{eff}[h,\phi]\right)\Psi[h,\phi] = 0. \tag{12.1}$$

where V_{eff} is *the (effective) gravitational potential*

$$V_{eff} \equiv \frac{1}{6\kappa^2}\left(-\frac{{}^{(3)}R}{h} + \frac{2\Lambda}{h} + \frac{2\kappa}{h}\rho[\phi]\right). \tag{12.2}$$

The manifestly singular behaviour $\sim 1/h$ of the potential (12.2), however, can be regularized by the suitable change of variables

$$h \rightarrow \xi = \xi[h], \tag{12.3}$$

which generates the adequate Jacobi formula

$$\delta\xi = \left(\frac{\delta\xi}{\delta h}\right) hh^{ij}\delta h_{ij}. \tag{12.4}$$

The dimension $\xi[h]$, as a functional of the global dimension h, is also diffeoinvariant. Applying the change of variables (12.3) within the equation (12.1) one obtains

$$\left\{\left(\frac{\delta\xi}{\delta h}\right)^2 \frac{\delta^2}{\delta\xi^2} + V_{eff}[\xi,\phi]\right\}\Psi[\xi,\phi] = 0, \tag{12.5}$$

and, therefore, for all nonsingular situations $\frac{\delta\xi}{\delta h} \neq 0$ it can be rewritten in a more convenient form

$$\left\{\frac{\delta^2}{\delta\xi^2} + V[\xi,\phi]\right\}\Psi[\xi,\phi] = 0, \tag{12.6}$$

where the potential $V[\xi,\phi]$ is

$$V[\xi,\phi] = \left(\frac{\delta\xi}{\delta h}\right)^{-2} V_{eff}[\xi,\phi]. \tag{12.7}$$

We shall call it *the generalized gravitational potential*. In fact, the choice of the invariant dimension ξ is a kind of the choice of a gauge for the theory of quantum gravity.

Naturally, the generic gauge is the global dimension $\xi[h] \equiv h$. Other situations can be generated straightforwardly from this fundamental case. There are many possible choices of gauge for quantum gravity. However, note that the particular choice

$$\xi = \frac{2}{\kappa}\sqrt{\frac{h}{6}}, \tag{12.8}$$

which is associated to the measure

$$\delta\xi = \frac{1}{\kappa}\sqrt{\frac{h}{6}}h^{ij}\delta h_{ij}, \tag{12.9}$$

removes the singularity $1/h$ from the gravitational potential $V_{eff}[h,\phi]$ (12.2) and, consequently, the equation (12.6) reads

$$\left\{\frac{\delta^2}{\delta\xi^2} - \left({}^{(3)}R[\xi] - 2\Lambda - 2\kappa\rho[\phi]\right)\right\}\Psi[\xi,\phi] = 0. \tag{12.10}$$

The appropriate normalization condition should be chosen as

$$\int |\Psi[\xi,\phi]|^2 \delta\mu(\xi,\phi) = 1, \tag{12.11}$$

where $\delta\mu(\xi,\phi) = \delta\xi\delta\phi$ is the invariant product functional measure. Let us notice that similarly to the measure δh, the measure $\delta\sqrt{h}$ is also the Lebesgue–Stieltjes (Radon) type integral measure, which can be transformed to the Riemann–Lebesgue measure over a space-time. In the general case $h = h(x_0,x_1,x_2,x_3)$ one obtains

$$\delta\sqrt{h} = \frac{\partial^4\sqrt{h}}{\partial x_0\partial x_1\partial x_2\partial x_3} d^4x. \tag{12.12}$$

By the special role of the change of variables (12.8) we shall call this dimension *invariant global dimension*, and the theory of quantum gravity (12.10) will be called *the invariant global quantum gravity*.

B One-Dimensional Dirac Equation

The equation (12.6) can be derived as the Euler–Lagrange equations of motion by the field theoretical variational principle $\delta S[\Psi] = 0$ applied to the action functional

$$\begin{aligned} S[\Psi] &= -\frac{1}{2}\int \delta\xi\delta\phi\Psi[\xi,\phi]\left(\frac{\delta^2}{\delta\xi^2} + V[\xi,\phi]\right)\Psi[\xi,\phi] = \\ &= -\frac{1}{2}\int \delta\phi\Psi[\xi,\phi]\frac{\delta\Psi[\xi,\phi]}{\delta\xi} + \\ &+ \frac{1}{2}\int \delta\xi\delta\phi\left\{\left(\frac{\delta\Psi[\xi,\phi]}{\delta\xi}\right)^2 + V[\xi,\phi]\Psi^2[\xi,\phi]\right\}, \end{aligned} \tag{12.13}$$

where we have applied integration by parts. One can choose the coordinate system in the Wheeler superspace by the condition of vanishing of the material surface term

$$-\frac{1}{2}\int \delta\phi\Psi[\xi,\phi]\frac{\delta\Psi[\xi,\phi]}{\delta\xi} = 0. \tag{12.14}$$

In the material coordinates the action functional (12.13) is reduced to the action of the Euclidean field theory

$$S[\Psi] \equiv \int \delta\xi\delta\phi L\left[\Psi[\xi,\phi], \frac{\delta\Psi[\xi,\phi]}{\delta\xi}\right], \tag{12.15}$$

where the Euclidean Lagrangian has the form

$$L\left[\Psi[\xi,\phi],\frac{\delta\Psi[\xi,\phi]}{\delta\xi}\right]=\frac{1}{2}\left(\frac{\delta\Psi[\xi,\phi]}{\delta\xi}\right)^2+\frac{V[\xi,\phi]}{2}\Psi^2[\xi,\phi], \tag{12.16}$$

and the corresponding canonical momentum conjugated to the classical scalar-valued field $\Psi[\xi,\phi]$ is

$$\Pi_\Psi[\xi,\phi]=\frac{\partial L}{\partial\left(\dfrac{\delta\Psi[\xi,\phi]}{\delta\xi}\right)}=\frac{\delta\Psi[\xi,\phi]}{\delta\xi}. \tag{12.17}$$

Therefore, the choice of the coordinate system (12.14) actually means the choice of the orthogonal coordinates in the space field-theoretic phase space $(\Psi[\xi,\phi],\Pi_\Psi[\xi,\phi])$

$$\forall\xi,\phi\in\Sigma(\xi,\phi):\int\delta\phi\Psi[\xi,\phi]\Pi_\Psi[\xi,\phi]=0, \tag{12.18}$$

where by $\Sigma(\xi,\phi)$ we denote the midisuperspace strata of the Wheeler superspace. Using the conjugate momentum (12.17), the equation (12.6) can be rewritten in the form

$$\frac{\delta\Pi_\Psi[\xi,\phi]}{\delta\xi}+V[\xi,\phi]\Psi[\xi,\phi]=0, \tag{12.19}$$

and together with the equation (12.17) creates the canonical Hamilton equations of motion, yielding the appropriate one-dimensional Dirac equation

$$\left(-i\sigma_y\frac{\delta}{\delta\xi}-M[\xi,\phi]\right)\Phi[\xi,\phi]=0, \tag{12.20}$$

where $\Phi[\xi,\phi]$ is the two-component classical field

$$\Phi[\xi,\phi]=\begin{bmatrix}\Pi_\Psi[\xi,\phi]\\ \Psi[\xi,\phi]\end{bmatrix}, \tag{12.21}$$

$M[\xi,\phi]$ is the mass matrix of this field

$$M[\xi,\phi]=\begin{bmatrix}1 & 0\\ 0 & V[\xi,\phi]\end{bmatrix}, \tag{12.22}$$

and the σ_y is the Pauli matrix

$$\sigma_y = \begin{bmatrix} 0 & -i \\ i & 0 \end{bmatrix}, \tag{12.23}$$

$$\sigma_y^2 = \begin{bmatrix} 1 & 0 \\ 0 & 1 \end{bmatrix} \equiv \mathbf{I}_2, \tag{12.24}$$

which in itself creates the Clifford algebra $\mathscr{C}\ell_2(\mathbb{C})$

$$\{\sigma_y, \sigma_y\} = 2\mathbf{I}_2. \tag{12.25}$$

C Cauchy-like Wave Functionals

The one-dimensional Dirac equation (12.20) can be rewritten in the form of the Schrödinger equation

$$i\frac{\delta\Phi[\xi,\phi]}{\delta\xi} = H[\xi,\phi]\Phi[\xi,\phi], \tag{12.26}$$

where the Hamiltonian is

$$H[\xi,\phi] = i\begin{bmatrix} 0 & -V[\xi,\phi] \\ 1 & 0 \end{bmatrix}. \tag{12.27}$$

Solution of the ξ-evolution (12.26) can be written out straightforwardly

$$\Phi[\xi,\phi] = U[\xi,\phi]\Phi[\xi^I,\phi], \tag{12.28}$$

where $\Phi[\xi^I,\phi]$ is the initial datum vector with respect to ξ only, and $U[\xi,\phi]$ is the operator of ξ-evolution

$$U = \exp\left\{-i\int_{\Sigma(\xi)} \delta\xi' H[\xi',\phi]\right\} = \exp\{-i\Omega(\xi,\phi)\langle H\rangle(\xi,\phi)\}, \tag{12.29}$$

where $\Sigma(\xi)$ is the finite integration region in the subset of midisuperspace, which we shall call ξ-*space*, $\Omega = V(\Sigma(\phi,\xi))$ is the volume of the full configuration space, and $\langle H\rangle(\phi)$ is the Hamiltonian averaged over midisuperspace

$$\Omega(\xi,\phi) = \int_{\Sigma(\xi,\phi)} \delta\xi'\delta\phi', \tag{12.30}$$

$$\langle H\rangle(\xi,\phi) = \frac{1}{\Omega(\xi,\phi)}\int_{\Sigma(\xi)} \delta\xi' H[\xi',\phi], \tag{12.31}$$

where $\Sigma(\xi,\phi)$ is the midisuperspace

$$\Sigma(\xi,\phi) = \Sigma(\xi)\times\Sigma(\phi), \tag{12.32}$$

which is assumed to be a finite integration region. Explicitly one obtains

$$U[\xi,\phi] = \mathbf{I}_2\cos\left[\Omega(\xi,\phi)\sqrt{\langle V\rangle(\xi,\phi)}\right] + \\ + \begin{bmatrix} 0 & -\sqrt{\langle V\rangle(\xi,\phi)} \\ \dfrac{1}{\sqrt{\langle V\rangle(\xi,\phi)}} & 0 \end{bmatrix} \sin\left[\Omega(\xi,\phi)\sqrt{\langle V\rangle(\xi,\phi)}\right], \tag{12.33}$$

where we have introduced the averaged generalized gravitational potential

$$\langle V\rangle(\xi,\phi) = \frac{1}{\Omega(\xi,\phi)}\int_{\Sigma(\xi)} \delta\xi' V[\xi',\phi]. \tag{12.34}$$

Elementary algebraic manipulations yield the extended global wave functional

$$\begin{aligned} \Psi[\xi,\phi] &= \Psi[\xi^I,\phi]\cos\left[\Omega(\xi,\phi)\sqrt{\langle V\rangle(\xi,\phi)}\right] + \\ &+ \Pi_\Psi[\xi^I,\phi]\frac{\sin\left[\Omega(\xi,\phi)\sqrt{\langle V\rangle(\xi,\phi)}\right]}{\sqrt{\langle V\rangle(\xi,\phi)}}, \end{aligned} \tag{12.35}$$

and the canonical conjugate momentum as the solution is

$$\begin{aligned} \Pi_\Psi[\xi,\phi] &= \Pi_\Psi[\xi^I,\phi]\cosh\left[\Omega(\xi,\phi)\sqrt{\langle V\rangle(\xi,\phi)}\right] + \\ &- \Psi[\xi^I,\phi]\sqrt{\langle V\rangle(\xi,\phi)}\sinh\left[\Omega(\xi,\phi)\sqrt{\langle V\rangle(\xi,\phi)}\right], \end{aligned} \tag{12.36}$$

where $\Psi[\xi^I,\phi]$ and $\Pi_\Psi[\xi^I,\phi]$ are initial data with respect to ξ only. Applying, however, the equation (12.17) to (12.36), one obtains the relation

$$\Pi_\Psi[\xi,\phi] = \frac{\Pi_\Psi[\xi^I,\phi]}{\sqrt{\langle V\rangle(\xi,\phi)}}\frac{\delta}{\delta\xi}\left[\Omega(\xi,\phi)\sqrt{\langle V\rangle(\xi,\phi)}\right]\cos\left[\Omega(\xi,\phi)\sqrt{\langle V\rangle(\xi,\phi)}\right]$$
$$-\left\{\Psi[\xi^I,\phi]\frac{\delta}{\delta\xi}\left[\Omega(\xi,\phi)\sqrt{\langle V\rangle(\xi,\phi)}\right] - \Pi_\Psi[\xi^I,\phi]\frac{\delta}{\delta\xi}\left[\frac{1}{\sqrt{\langle V\rangle(\xi,\phi)}}\right]\right\}$$
$$\times\sin\left[\Omega(\xi,\phi)\sqrt{\langle V\rangle(\xi,\phi)}\right], \tag{12.37}$$

which after calculation of the functional derivatives

$$\frac{\delta}{\delta\xi}\left[\Omega(\xi,\phi)\sqrt{\langle V\rangle(\xi,\phi)}\right] = \frac{\sqrt{\langle V\rangle(\xi,\phi)}}{2}\left(\frac{\delta\Omega(\xi,\phi)}{\delta\xi}+1\right), \tag{12.38}$$

and

$$\frac{\delta}{\delta\xi}\left[\frac{1}{\sqrt{\langle V\rangle(\xi,\phi)}}\right] = \frac{1}{2}\frac{1}{\Omega(\xi,\phi)\sqrt{\langle V\rangle(\xi,\phi)}}\left(\frac{\delta\Omega(\xi,\phi)}{\delta\xi}-1\right) \tag{12.39}$$

and using them within the formula (12.37) yields

$$\Pi_\Psi[\xi,\phi] = \Pi_\Psi[\xi^I,\phi]\frac{1}{2}\left(\frac{\delta\Omega(\xi,\phi)}{\delta\xi}+1\right)\cos\left[\Omega(\xi,\phi)\sqrt{\langle V\rangle(\xi,\phi)}\right]$$
$$-\left[\Psi[\xi^I,\phi]\frac{\sqrt{\langle V\rangle(\xi,\phi)}}{2}\left(\frac{\delta\Omega(\xi,\phi)}{\delta\xi}+1\right)\right.$$
$$\left.-\frac{\Pi_\Psi[\xi^I,\phi]}{2\Omega(\xi,\phi)\sqrt{\langle V\rangle(\xi,\phi)}}\left(\frac{\delta\Omega(\xi,\phi)}{\delta\xi}-1\right)\right]\sin\left[\Omega(\xi,\phi)\sqrt{\langle V\rangle(\xi,\phi)}\right]. \tag{12.40}$$

After comparison with (12.36) one obtains the equations

$$\frac{1}{2}\left(\frac{\delta\Omega(\xi,\phi)}{\delta\xi}+1\right) = 1, \tag{12.41}$$

and

$$\Psi[\xi^I,\phi]\frac{1}{2}\left(\frac{\delta\Omega(\xi,\phi)}{\delta\xi}+1\right) - \frac{\Pi_\Psi[\xi^I,\phi]}{\Omega(\xi,\phi)\langle V\rangle(\xi,\phi)}\frac{1}{2}\left(\frac{\delta\Omega(\xi,\phi)}{\delta\xi}-1\right) = \Psi[\xi^I,\phi], \tag{12.42}$$

The equation (12.41) yields the relation

$$\frac{\delta\Omega}{\delta\xi} = 1 = \int_{\Sigma(\phi)} \delta\phi', \tag{12.43}$$

where the last integral arises by the formula (12.30). Application of the result (12.43) to the equation (12.42) leads to the self-consistent identity $\Psi[\xi^I,\phi] = \Psi[\xi^I,\phi]$. Such a situation means that the volume of midisuperspace $\Omega(\xi,\phi)$ is ϕ-invariant

$$\Omega(\xi,\phi) = \int_{\Sigma(\xi,\phi)} \delta\xi'\delta\phi' = \int_{\Sigma(\phi)} \delta\phi' \int_{\Sigma(\xi)} \delta\xi' = \int_{\Sigma(\xi)} \delta\xi' = \Omega(\xi), \tag{12.44}$$

i.e. does not depend on Matter fields.

Directly from (12.35) the probability density can be deduced easily as

$$\begin{aligned} |\Psi[\xi,\phi]|^2 &= (\Psi[\xi^I,\phi])^2 \cos^2\left[\Omega(\xi)\sqrt{\langle V\rangle(\xi,\phi)}\right] \\ &+ (\Pi_\Psi[\xi^I,\phi])^2 \left(\frac{\sin\left[\Omega(\xi)\sqrt{\langle V\rangle(\xi,\phi)}\right]}{\sqrt{\langle V\rangle(\xi,\phi)}}\right)^2 \\ &+ \Psi[\xi^I,\phi]\Pi_\Psi[\xi^I,\phi]\frac{\sin\left[2\Omega(\xi)\sqrt{\langle V\rangle(\xi,\phi)}\right]}{\sqrt{\langle V\rangle(\xi,\phi)}}, \end{aligned} \tag{12.45}$$

and in the light of the relation (12.18) it simplifies to

$$\begin{aligned} |\Psi[\xi,\phi]|^2 &= (\Psi[\xi^I,\phi])^2 \cos^2\left[\Omega(\xi)\sqrt{\langle V\rangle(\xi,\phi)}\right] \\ &+ (\Pi_\Psi[\xi^I,\phi])^2 \left(\frac{\sin\left[\Omega(\xi)\sqrt{\langle V\rangle(\xi,\phi)}\right]}{\sqrt{\langle V\rangle(\xi,\phi)}}\right)^2. \end{aligned} \tag{12.46}$$

Let us take into account *ad hoc* the following separation conditions

$$\Psi[\xi^I,\phi] = \Psi[\xi^I]\Gamma_\Psi[\phi], \tag{12.47}$$

$$\Pi_\Psi[\xi^I,\phi] = \Pi_\Psi[\xi^I]\Gamma_\Pi[\phi], \tag{12.48}$$

where Γ_Ψ and Γ_Π are functionals of ϕ only and $\Psi[\xi^I]$, and $\Pi_\Psi[\xi^I]$ are constant functionals. Applying the usual normalization condition

$$\int_{\Sigma(\xi,\phi)} |\Psi[\xi',\phi']|^2 \delta\xi'\delta\phi' = 1, \tag{12.49}$$

one obtains the simple constraint for the initial datum

$$A(\Pi_\Psi[\xi^I])^2 + B(\Psi[\xi^I])^2 - 1 = 0, \tag{12.50}$$

where the constants A and B are given by the integrals

$$A = \int_{\Sigma(\xi,\phi)} \Gamma_\Pi[\phi'] \left(\frac{\sin\left[\Omega(\xi')\sqrt{\langle V\rangle(\xi',\phi')}\right]}{\sqrt{\langle V\rangle(\xi',\phi')}} \right)^2 \delta\xi'\delta\phi', \tag{12.51}$$

$$B = \int_{\Sigma(\xi,\phi)} \Gamma_\Psi[\phi'] \cos^2\left[\Omega(\xi')\sqrt{\langle V(\xi',\phi')\rangle}\right] \delta\xi'\delta\phi', \tag{12.52}$$

which according to our assumption are convergent, finite, and independent on the initial datum ξ^I. The equation (12.50), however, can be solved straightforwardly. As a result one obtains

$$\Pi_\Psi[\xi^I] = \pm\sqrt{\frac{1}{A} - \frac{B}{A}(\Psi[\xi^I])^2}, \tag{12.53}$$

which together with the definition

$$\Pi_\Psi[\xi^I, \phi] = \frac{\delta\Psi[\xi^I, \phi]}{\delta\xi^I}, \tag{12.54}$$

and the separability conditions (12.47)-(12.48) yields the differential equation for the initial datum of the field $\Psi[\xi]$

$$\frac{1}{\Gamma[\phi]}\frac{\delta\Psi[\xi^I]}{\delta\xi^I} = \pm\sqrt{\frac{1}{A} - \frac{B}{A}(\Psi[\xi^I])^2}, \tag{12.55}$$

where $\Gamma[\phi]$ is the coefficient dependent on Matter fields only

$$\Gamma[\phi] \equiv \frac{\Gamma_\Pi[\phi]}{\Gamma_\Psi[\phi]}, \tag{12.56}$$

which can be integrated straightforwardly

$$\sqrt{A}\int \frac{\delta\Psi[\xi^I]}{\sqrt{1 - B(\Psi[\xi^I])^2}} = \pm\Gamma[\phi]\xi^I + C, \tag{12.57}$$

where C is the integration constant, with the result

$$\sqrt{\frac{A}{B}}\arcsin\left\{\sqrt{\frac{B}{A}}\Psi[\xi^I]\right\} = \pm\Gamma[\phi]\xi^I + C, \tag{12.58}$$

so that after elementary algebraic manipulations one obtains

$$\Psi[\xi^I] = \sqrt{\frac{A}{B}}\sin\theta(\xi^I,\phi), \tag{12.59}$$

where $\theta(\xi^I,\phi)$ is the phase

$$\theta(\xi^I,\phi) = \sqrt{\frac{B}{A}}\left(\pm\Gamma[\phi]\xi^I + C\right), \tag{12.60}$$

Albeit, because $\Psi[\xi^I]$ must be a functional of ξ^I only, it must hold $\Gamma[\phi] = \Gamma_0$, where Γ_0 is a constant independent of ϕ and ξ^I, for which the phase $\theta(\xi^I,\phi)$ is reduced to

$$\theta(\xi^I) = \sqrt{\frac{B}{A}}\left(\pm\Gamma_0\xi^I + C\right). \tag{12.61}$$

Taking into account the relation (12.53), one obtains finally

$$\Psi[\xi^I] = \sqrt{\frac{A}{B}}\sin\theta(\xi^I), \tag{12.62}$$

$$\Pi_\Psi[\xi^I] = \pm\sqrt{\frac{1}{A} - \sin^2\theta(\xi^I)}. \tag{12.63}$$

In the light of the equation (12.18), however, one of the relations

$$\sin\theta(\xi^I) = 0, \tag{12.64}$$

$$\sin\theta(\xi^I) = \pm\sqrt{\frac{1}{A}}, \tag{12.65}$$

is always true. One sees that both these conditions define discrete values of the initial datum of the invariant global dimension ξ_I. Namely, the first relation (12.64) leads to the solution

$$\sqrt{\frac{B}{A}}\left(\pm\Gamma_0\xi^I + C\right) = k\pi, \tag{12.66}$$

where $k \in \mathbb{Z}$ is an integer, which leads to

$$\xi^I = \pm\frac{1}{\Gamma_0}\left(\sqrt{\frac{A}{B}}k\pi - C\right). \tag{12.67}$$

Similarly, the second relation (12.65) can be solved immediately

$$\xi^I = \pm\frac{1}{\Gamma_0}\left(\pm\sqrt{\frac{A}{B}}\arcsin\sqrt{\frac{1}{A}} - C\right). \tag{12.68}$$

For the first case one has

$$\Psi[\xi^I] = 0, \tag{12.69}$$

$$\Pi_\Psi[\xi^I] = \pm\sqrt{\frac{1}{A}}, \tag{12.70}$$

and for the second one

$$\Psi[\xi^I] = \pm\sqrt{\frac{1}{B}}, \tag{12.71}$$

$$\Pi_\Psi[\xi^I] = 0. \tag{12.72}$$

Finally, one sees that the invariant global wave functional (12.35) is

$$\Psi[\xi,\phi] = \pm\Gamma_\Psi[\phi]\Gamma_0\sqrt{\frac{1}{A}}\frac{\sin\left[\Omega(\xi)\sqrt{\langle V\rangle(\xi,\phi)}\right]}{\sqrt{\langle V\rangle(\xi,\phi)}}, \tag{12.73}$$

in the first case (12.64), and

$$\Psi[\xi,\phi] = \pm\Gamma_\Psi[\phi]\sqrt{\frac{1}{B}}\cos\left[\Omega(\xi)\sqrt{\langle V\rangle(\xi,\phi)}\right], \tag{12.74}$$

for the second one (12.65).

Applying the normalization condition (12.49) to the solutions (12.73) and (12.74), one receives the equations

$$|\Gamma_\Psi[\phi]\Gamma_0|^2 = 1, \tag{12.75}$$

$$\Gamma_\Psi[\phi]\Gamma_0 = 1, \tag{12.76}$$

which can be solved easy and lead to the relation

$$\Gamma_\Psi[\phi] = \frac{1}{\Gamma_0}. \tag{12.77}$$

Therefore, one obtains finally

$$\Psi_1[\xi,\phi] = \pm\sqrt{\frac{1}{|A|}}\frac{\sin\left[\Omega(\xi)\sqrt{\langle V\rangle(\xi,\phi)}\right]}{\sqrt{\langle V\rangle(\xi,\phi)}}, \tag{12.78}$$

$$\Psi_2[\xi,\phi] = \pm\sqrt{\frac{1}{|B|}}\cos\left[\Omega(\xi)\sqrt{\langle V\rangle(\xi,\phi)}\right], \tag{12.79}$$

where now the constants A and B are equal to

$$A = \int_{\Sigma(\xi,\phi)}\left(\frac{\sin\left[\Omega(\xi')\sqrt{\langle V\rangle(\xi',\phi')}\right]}{\sqrt{\langle V\rangle(\xi',\phi')}}\right)^2 \delta\xi'\delta\phi', \tag{12.80}$$

$$B = \frac{1}{\Gamma_0}\int_{\Sigma(\xi,\phi)}\cos^2\left[\Omega(\xi')\sqrt{\langle V\rangle(\xi',\phi')}\right]\delta\xi'\delta\phi'. \tag{12.81}$$

In this manner, the general solutions of the theory of quantum gravity can be now constructed straightforwardly by using the solutions (12.73) and (12.74), in which the integrals must be implemented

$$\Omega(\xi) = \int_{\Sigma(\xi)}\delta\xi' = \xi, \tag{12.82}$$

$$\langle V\rangle(\xi,\phi) = \frac{1}{\Omega(\xi)}\int_{\Sigma(\xi)}\delta\xi' V[\xi',\phi], \tag{12.83}$$

where the ξ-measure is

$$\delta\xi = \frac{1}{\kappa}\frac{\delta h}{\sqrt{6h}}. \tag{12.84}$$

However, because the generalized gravitational potential $V[\xi,\phi]$ has the form of an algebraic sum

$$V[\xi,\phi] = -{}^{(3)}R + 2\Lambda + 2\kappa\rho, \tag{12.85}$$

one has very convenient separability

$$\langle V\rangle(\xi,\phi) = -\frac{1}{\Omega(\xi)}\int_{\Sigma(\xi)}\delta\xi'\ {}^{(3)}R + 2\Lambda + \frac{2\kappa}{\Omega(\xi)}\int_{\Sigma(\phi)}\delta\phi'\int_{\Sigma(\xi)}\delta\xi'\rho. \quad (12.86)$$

Therefore, for a concretely given geometry of a three-dimensional space-like embedded space, one should estimate the functionally averaged three-dimensional Ricci scalar

$$\langle {}^{(3)}R\rangle = \frac{1}{\Omega(\xi)}\int_{\Sigma(\xi)}\delta\xi'\ {}^{(3)}R, \quad (12.87)$$

and the functionally averaged energy density of Matter fields

$$\langle\rho\rangle = \frac{1}{\Omega(\xi)}\int_{\Sigma(\phi)}\delta\phi'\int_{\Sigma(\xi)}\delta\xi'\rho, \quad (12.88)$$

and using these quantities construct the functionally averaged generalized gravitational potential

$$\langle V\rangle(\xi,\phi) = -\langle {}^{(3)}R\rangle + 2\Lambda + 2\kappa\langle\rho\rangle. \quad (12.89)$$

Applying this averaging method, one can construct the solutions (12.73) and (12.74) straightforwardly. If the Ricci scalar curvature ${}^{(3)}R$ of an embedded space and/or the energy density of Matter fields ρ are functions of a space-time point $x^\mu = (x^0,x^1,x^2,x^3)$ or any one space-time coordinate, the integration procedure is understood as the procedure of calculating the Lebesgue–Stieltjes integral. For instance, in the case when the integrand is a function of x^μ the Lebesgue–Stieltjes measure takes the following form

$$\delta\xi = \frac{\partial^4\xi}{\partial x_0\partial x_1\partial x_2\partial x_3}d^4x, \quad (12.90)$$

or in terms of the determinant of induced metric h

$$\delta\xi = \frac{1}{\kappa}\frac{\partial^4 h}{\partial x_0\partial x_1\partial x_2\partial x_3}\frac{d^4x}{\sqrt{6h}}. \quad (12.91)$$

D Inverted Transformation

There is the problem of inverted transformation

$$\xi \to h_{ij}, \tag{12.92}$$

within the obtained solutions. In fact, this problem is strictly related to the problem of finding of induced metric if one knows its determinant. Such a procedure cannot be performed analytically, but there are functional methods which enable performing of this step.

The solutions (12.73) and (12.74) expressed through h_{ij} are

$$\Psi_1[h_{ij},\phi] = \pm\sqrt{\frac{1}{|A|}}\frac{\sin\left[\Omega(h_{ij})\sqrt{\langle V\rangle(h_{ij},\phi)}\right]}{\sqrt{\langle V\rangle(h_{ij},\phi)}}, \tag{12.93}$$

$$\Psi_2[h_{ij},\phi] = \pm\sqrt{\frac{1}{|B|}}\cos\left[\Omega(h_{ij})\sqrt{\langle V\rangle(h_{ij},\phi)}\right], \tag{12.94}$$

where now the constants A and B are equal to

$$A = \frac{1}{\kappa}\int_{\Sigma(h_{ij},\phi)}\left(\frac{\sin\left[\Omega(h'_{ij})\sqrt{\langle V\rangle(h'_{ij},\phi')}\right]}{\sqrt{\langle V\rangle(h'_{ij},\phi')}}\right)^2\sqrt{\frac{h'}{6}}h^{ij\prime}\delta h'_{ij}\delta\phi', \tag{12.95}$$

$$B = \frac{1}{\kappa\Gamma_0}\int_{\Sigma(h_{ij},\phi)}\cos^2\left[\Omega(h'_{ij})\sqrt{\langle V\rangle(h'_{ij},\phi')}\right]\sqrt{\frac{h'}{6}}h^{ij\prime}\delta h'_{ij}\delta\phi', \tag{12.96}$$

and assumed to be convergent and finite. The solutions (12.93) and (12.94) are two independent states transformed to the standard quantum geometrodynamics, but in general they are not solutions of the Wheeler–DeWitt equation.

Interestingly, in such a situation the normalization condition (12.49) becomes

$$\frac{1}{\kappa}\int_{\Sigma(h_{ij},\phi)}|\Psi[h'_{ij},\phi']|^2\sqrt{\frac{h'}{6}}h^{ij\prime}\delta h'_{ij}\delta\phi' = 1, \tag{12.97}$$

and can be used to construct another physical interpretation of the Einstein constant

$$\kappa = \int_{\Sigma(h_{ij},\phi)}|\Psi[h'_{ij},\phi']|^2\sqrt{\frac{h'}{6}}h^{ij\prime}\delta h'_{ij}\delta\phi'. \tag{12.98}$$

E Hilbert Space and Superposition

However, because both equations of quantum geometrodynamics (10.99) and (12.10) are linear in the field Ψ, in general the superposition

$$\Psi = \sum_{i=1,2} \alpha_i \Psi_i, \tag{12.99}$$

where α_i are arbitrary constants, and Ψ_i are (12.93) and (12.94), should be also a solution. The normalization (12.97) of (12.99) gives

$$|\alpha_1|^2 + |\alpha_2|^2 + (\alpha_1^\star \alpha_2 + \alpha_1 \alpha_2^\star) I = 1, \tag{12.100}$$

where I is the integral

$$I = \frac{1}{\sqrt{|A||B|}} \frac{1}{\kappa} \int_{\Sigma(h_{ij},\phi)} \frac{\sin\left[2\Omega(h'_{ij})\sqrt{\langle V\rangle(h'_{ij},\phi')}\right]}{2\sqrt{\langle V\rangle(h'_{ij},\phi')}} \sqrt{\frac{h'}{6}} h^{ij\prime} \delta h'_{ij} \delta\phi'. \tag{12.101}$$

For vanishing $I = 0$ one obtains form (12.100) simply

$$|\alpha_2| = \sqrt{1 - |\alpha_1|^2}, \quad |\alpha_1| \geqslant 1. \tag{12.102}$$

The case of $I \neq 0$ is much more complicated. Note that the equation (12.100) can be rewritten in form

$$(\alpha_1 + \alpha_2 I)\alpha_1^\star + (\alpha_2 + \alpha_1 I)\alpha_2^\star = 0, \tag{12.103}$$

which leads to the result

$$\frac{\alpha_1^\star}{\alpha_2^\star} = \frac{-\alpha_1 I + \alpha_2}{\alpha_1 + \alpha_2 I}, \tag{12.104}$$

or, equivalently,

$$C\alpha_1^\star = -\alpha_1 I + \alpha_2, \tag{12.105}$$
$$C\alpha_2^\star = \alpha_1 + \alpha_2 I, \tag{12.106}$$

where $0 \neq C \in \mathbb{R}$ is a constant. The relations (12.105)-(12.106) lead to

$$C|\alpha_1|^2 = -\alpha_1^2 I + \alpha_2 \alpha_1, \tag{12.107}$$
$$C|\alpha_2|^2 = \alpha_1 \alpha_2 + \alpha_2^2 I. \tag{12.108}$$

Mutual addition and application of (12.100) yields

$$CI[(\alpha_1^\star-\alpha_2)\alpha_2+(\alpha_2^\star+\alpha_1)\alpha_1]=\alpha_1\alpha_2+\alpha_2\alpha_1, \tag{12.109}$$

which generates the equations

$$\begin{aligned} CI(\alpha_1^\star-\alpha_2) &= \alpha_1, & (12.110)\\ CI(\alpha_2^\star+\alpha_1) &= \alpha_2. & (12.111)\end{aligned}$$

Complex decomposition of (12.110)-(12.111) leads to

$$\begin{aligned} \Re\alpha_2 &= (CI-1)\Re\alpha_1, & (12.112)\\ \Im\alpha_2 &= (CI-1)\Im\alpha_1, & (12.113)\end{aligned}$$

or equivalently

$$\begin{aligned} \alpha_2 &= (CI-1)\alpha_1, & (12.114)\\ |\alpha_2|^2 &= (CI-1)^2|\alpha_1|^2. & (12.115)\end{aligned}$$

Employing (12.114)-(12.115) within the constraint (12.100) yields

$$\frac{1}{|\alpha_1|^2}=IC^2+(I^2-2I)C-I+2. \tag{12.116}$$

Because the natural condition $\dfrac{1}{|\alpha_1|^2}>0$ one has the inequality for C

$$IC^2+I(I-2)C-(I-2)>0, \tag{12.117}$$

where $I\geqslant 0$. One sees that the case $I=0$ gives $2>0$ what is true. The inequality (12.117) has two different solutions which are dependent on $\mathrm{sgn}(I-2)$. When $0\leqslant I<2$ then the solution is

$$C\in(C_-,C_+). \tag{12.118}$$

When $I>2$ the solution is somewhat different

$$C\in(-\infty,C_-)\cup(C_+,\infty). \tag{12.119}$$

Here are the constants

$$C_\mp=\frac{2-I}{2}\mp\frac{1}{2}\sqrt{\frac{I^3-4I^2+8I-8}{I}}. \tag{12.120}$$

For consistency, the constants $C_{\mp}$ must be real numbers. For this it is necessary and sufficient to satisfy the condition

$$I^3 - 4I^2 + 8I - 8 \geqslant 0, \tag{12.121}$$

which is satisfied for $I \geqslant 2$ or $I < 0$. Because $I > 0$, the first case is true and, therefore, (12.119) is the true region of validity of C. Applying the definition of I, A and B, one receives the following inequality

$$\begin{aligned}
&\int_{\Sigma(h_{ij},\phi)} \frac{\sin\left[2\Omega(h'_{ij})\sqrt{\langle V\rangle(h'_{ij},\phi')}\right]}{2\sqrt{\langle V\rangle(h'_{ij},\phi')}} \sqrt{h'}h^{ij'}\delta h'_{ij}\delta\phi' \geqslant \\
&\frac{2}{\sqrt{|\Gamma_0|}}\left(\int_{\Sigma(h_{ij},\phi)}\left(\frac{\sin\left[\Omega(h'_{ij})\sqrt{\langle V\rangle(h'_{ij},\phi')}\right]}{\sqrt{\langle V\rangle(h'_{ij},\phi')}}\right)^2 \sqrt{h'}h^{ij'}\delta h'_{ij}\delta\phi'\right)^{1/2} \times \\
&\left(\int_{\Sigma(h_{ij},\phi)} \cos^2\left[\Omega(h'_{ij})\sqrt{\langle V\rangle(h'_{ij},\phi')}\right] \sqrt{h'}h^{ij'}\delta h'_{ij}\delta\phi'\right)^{1/2},
\end{aligned} \tag{12.122}$$

or in terms of the invariant global dimension

$$\begin{aligned}
&\int_{\Sigma(\xi,\phi)} \frac{\sin\left[2\Omega(\xi')\sqrt{\langle V\rangle(\xi',\phi')}\right]}{2\sqrt{\langle V\rangle(\xi',\phi')}} \delta\xi'\delta\phi' \geqslant \\
&\frac{2}{\sqrt{|\Gamma_0|}}\left(\int_{\Sigma(\xi,\phi)}\left(\frac{\sin\left[\Omega(\xi')\sqrt{\langle V\rangle(\xi',\phi')}\right]}{\sqrt{\langle V\rangle(\xi',\phi')}}\right)^2 \delta\xi'\delta\phi'\right)^{1/2} \times \\
&\left(\int_{\Sigma(\xi,\phi)} \cos^2\left[\Omega(\xi')\sqrt{\langle V\rangle(\xi',\phi')}\right] \delta\xi'\delta\phi'\right)^{1/2}.
\end{aligned} \tag{12.123}$$

Let us consider the concept of an inner product space, called also the pre-Hilbert space. Such a space is a vector space equipped with an additional structure called an inner product

$$\langle\bullet,\bullet\rangle : V \times V \to \mathbb{F}, \tag{12.124}$$

where V is a vector space over the field of scalars $\mathbb{F}$. The pre-Hilbert space is the Hilbert space if is complete as a normed space under the induced norm $||\bullet||$. Every Hilbert space is the Banach space. The structure of an inner product allows to build intuitive geometrical notions such as the length of a vector or the angle between two vectors, and deduce orthogonality between vectors by vanishing of the inner product. Pre-Hilbert spaces generalize Euclidean spaces to vector spaces of any, including infinite, dimension and are the theme of functional analysis (For more detailed discussion of functional analysis Cf. *e.g.* Ref. [930]). The Lebesgue space $L^2(\Sigma,\mu)$, where Σ is the configurational space equipped with the Lebesgue measure μ, is also called the space of square-integrable functions, which is the special case of L^p-spaces (For more details Cf. *e.g.* Ref. [931]) and, moreover, is a Hilbert space. This particular Hilbert space is the core fundament of quantum mechanics, in which wave functions belong to L^2.

Let us construct the Lebesgue space L^2 for the theory of quantum gravity *in accordance* with the superposition principle. In such a situation, the configurational space is the midisuperspace $\Sigma(\xi,\phi)$ with the measure $\mu=\delta\xi\delta\phi$. The Lebesgue space $L^2(\Sigma(\xi,\phi),\delta\xi\delta\phi)$ is the Hilbert space with the inner product

$$\langle f(\xi,\phi),g(\xi,\phi)\rangle=\int_{\Sigma(\xi,\phi)}f^\star(\xi',\phi')g(\xi',\phi')\delta\xi'\delta\phi', \tag{12.125}$$

satisfying the conditions

$$\langle f(\xi,\phi),g(\xi,\phi)\rangle = \overline{\langle f(\xi,\phi),g(\xi,\phi)\rangle}, \tag{12.126}$$

$$\langle af(\xi,\phi),g(\xi,\phi)\rangle = a\langle f(\xi,\phi),g(\xi,\phi)\rangle, \tag{12.127}$$

$$\langle f(\xi,\phi)+h(\xi,\phi),g(\xi,\phi)\rangle = \langle f(\xi,\phi),g(\xi,\phi)\rangle+\langle h(\xi,\phi),g(\xi,\phi)\rangle. \tag{12.128}$$

The induced norm

$$||f(\xi,\phi)||=\sqrt{\langle f(\xi,\phi),f(\xi,\phi)\rangle}=\left(\int_{\Sigma(\xi,\phi)}|f(\xi',\phi')|^2\delta\xi'\delta\phi'\right)^{1/2}<\infty, \tag{12.129}$$

is homogeneous

$$||af(\xi,\phi)||=|a|\,||f(\xi,\phi)||, \tag{12.130}$$

and satisfies the triangle inequality

$$||f(\xi,\phi)+g(\xi,\phi)||\leqslant||f(\xi,\phi)||+||g(\xi,\phi)||. \tag{12.131}$$

For the orthogonal situation

$$\langle f(\xi,\phi), g(\xi,\phi)\rangle = 0, \tag{12.132}$$

the Pythagoras theorem holds

$$||f(\xi,\phi)+g(\xi,\phi)||^2 = ||f(\xi,\phi)||^2 + ||g(\xi,\phi)||^2. \tag{12.133}$$

The parallelogram law

$$||f(\xi,\phi)+g(\xi,\phi)||^2 + ||f(\xi,\phi)-g(\xi,\phi)||^2 = 2\,||f(\xi,\phi)||^2 + 2\,||g(\xi,\phi)||^2, \tag{12.134}$$

is a necessary and sufficient condition for the existence of a scalar product corresponding to a given norm. This scalar product arises from the polarization identity

$$||f(\xi,\phi)+g(\xi,\phi)||^2 = ||f(\xi,\phi)||^2 + ||g(\xi,\phi)||^2 + \Re\,\langle f(\xi,\phi), g(\xi,\phi)\rangle, \tag{12.135}$$

called also the law of cosinuses, and equals

$$\begin{aligned}(f(\xi,\phi), g(\xi,\phi)) &= \frac{||f(\xi,\phi)+g(\xi,\phi)||^2 - ||f(\xi,\phi)-g(\xi,\phi)||^2}{2} = \\ &= \Re\,\langle f(\xi,\phi), g(\xi,\phi)\rangle.\end{aligned} \tag{12.136}$$

When $f_i(\xi,\phi)$, $i=1,\ldots,N$, are orthogonal vectors then the following relation holds

$$\sum_{i=1}^{N} ||f_i(\xi,\phi)|| = \left|\left|\sum_{i=1}^{N} f_i(\xi,\phi)\right|\right|. \tag{12.137}$$

When V is a complete pre-Hilbert space, *i.e.* is the Hilbert space, then the equation (12.137) becomes the Parseval identity

$$\sum_{i=1}^{\infty} ||f_i(\xi,\phi)|| = \left|\left|\sum_{i=1}^{\infty} f_i(\xi,\phi)\right|\right|, \tag{12.138}$$

provided that the infinite series on the left hand side is convergent. Recall that completeness of the space is necessary for convergence of the sequence of partial sums on the space

$$s_k = \sum_{i=1}^{k} ||f_i(\xi,\phi)||, \tag{12.139}$$

which is a Cauchy sequence.

The inner product and the norm can be used for definition of the angle α between two vectors $f(\xi,\phi)$ and $g(\xi,\phi)$

$$\alpha\left(f(\xi,\phi),g(\xi,\phi)\right)=\arccos\frac{\langle f(\xi,\phi),g(\xi,\phi)\rangle}{\|f(\xi,\phi)\|\,\|g(\xi,\phi)\|}, \qquad (12.140)$$

and it can be shown by straightforward computation that the Cauchy–Bunyakovsky–Schwarz inequality holds

$$|\langle f(\xi,\phi),g(\xi,\phi)\rangle|\leqslant\|f(\xi,\phi)\|\,\|g(\xi,\phi)\|. \qquad (12.141)$$

Taking into account

$$f(\xi,\phi) = \frac{\sin\left[\Omega(\xi)\sqrt{\langle V\rangle(\xi,\phi)}\right]}{\sqrt{\langle V\rangle(\xi,\phi)}}, \qquad (12.142)$$

$$g(\xi,\phi) = \cos\left[\Omega(\xi)\sqrt{\langle V\rangle(\xi,\phi)}\right], \qquad (12.143)$$

one obtains

$$\langle f(\xi,\phi),g(\xi,\phi)\rangle = \int_{\Sigma(\xi,\phi)}\frac{\sin\left[2\Omega(\xi')\sqrt{\langle V\rangle(\xi',\phi')}\right]}{2\sqrt{\langle V\rangle(\xi',\phi')}}\delta\xi'\delta\phi', \qquad (12.144)$$

$$\|f(\xi,\phi)\|^2 = \int_{\Sigma(\xi,\phi)}\left(\frac{\sin\left[\Omega(\xi')\sqrt{\langle V\rangle(\xi',\phi')}\right]}{\sqrt{\langle V\rangle(\xi',\phi')}}\right)^2\delta\xi'\delta\phi', \qquad (12.145)$$

$$\|g(\xi,\phi)\|^2 = \int_{\Sigma(\xi,\phi)}\cos^2\left[\Omega(\xi)\sqrt{\langle V\rangle(\xi,\phi)}\right]\delta\xi'\delta\phi'. \qquad (12.146)$$

For this reason, one can rewrite the integral (12.101) as

$$I=\frac{\langle f(\xi,\phi),g(\xi,\phi)\rangle}{\|f(\xi,\phi)\|\,\|g(\xi,\phi)\|}, \qquad (12.147)$$

or using the definition (12.140)

$$I=\cos\alpha\left(f(\xi,\phi),g(\xi,\phi)\right). \qquad (12.148)$$

In this manner, the inequality (12.123) can be rewritten as

$$\cos\alpha\left(f(\xi,\phi),g(\xi,\phi)\right)\geqslant\frac{2}{\sqrt{|\Gamma_0|}} \qquad (12.149)$$

whereas the Cauchy–Bunyakovsky–Schwarz inequality (12.141) gives natural bound for cosinus

$$|\cos\alpha\left(f(\xi,\phi),g(\xi,\phi)\right)| \leqslant 1. \tag{12.150}$$

Hence the superposition principle requires

$$\frac{2}{\sqrt{|\Gamma_0|}} \leqslant |\cos\alpha\left(f(\xi,\phi),g(\xi,\phi)\right)| \leqslant 1, \tag{12.151}$$

or, equivalently,

$$\alpha\left(f(\xi,\phi),g(\xi,\phi)\right) \in \left[\arccos\frac{2}{\sqrt{|\Gamma_0|}} + 2k\pi, 2\pi + 2k\pi\right], \quad k \in \mathbb{Z}, \tag{12.152}$$

which is consistent if and only if

$$\sqrt{|\Gamma_0|} \geqslant 2. \tag{12.153}$$

Let us consider the wave functions (12.93) and (12.94) expressed through the invariant global dimension

$$\Psi_1[\xi,\phi] = \frac{1}{\sqrt{|A|}}\frac{\sin\left[\Omega(\xi)\sqrt{\langle V\rangle(\xi,\phi)}\right]}{\sqrt{\langle V\rangle(\xi,\phi)}} = \frac{1}{\sqrt{|A|}}f(\xi,\phi), \tag{12.154}$$

$$\Psi_2[\xi,\phi] = \frac{1}{\sqrt{|B|}}\cos\left[\Omega(\xi)\sqrt{\langle V\rangle(\xi,\phi)}\right] = \frac{1}{\sqrt{|B|}}g(\xi,\phi), \tag{12.155}$$

where the constant A is

$$A = \int_{\Sigma(\xi,\phi)}\left(\frac{\sin\left[\Omega(\xi')\sqrt{\langle V\rangle(\xi',\phi')}\right]}{\sqrt{\langle V\rangle(\xi',\phi')}}\right)^2 \delta\xi'\delta\phi' \equiv ||f(\xi,\phi)||^2, \tag{12.156}$$

while the constant B is

$$B = \frac{1}{|\Gamma_0|}\int_{\Sigma(\xi,\phi)}\cos^2\left[\Omega(\xi')\sqrt{\langle V\rangle(\xi',\phi')}\right]\delta\xi'\delta\phi' \equiv \frac{1}{|\Gamma_0|}||g(\xi,\phi)||^2. \tag{12.157}$$

In other words, one receives

$$\Psi_1[\xi,\phi] = \frac{f(\xi,\phi)}{||f(\xi,\phi)||}, \tag{12.158}$$

$$\Psi_2[\xi,\phi] = \sqrt{|\Gamma_0|}\frac{g(\xi,\phi)}{||g(\xi,\phi)||}, \tag{12.159}$$

$$I = \cos\alpha\left(\Psi_1[\xi,\phi],\Psi_2[\xi,\phi]\right). \tag{12.160}$$

The wave functions (12.158) and (12.159) are elements of the Lebesgue space $L^2\left(\Sigma(\xi,\phi),\delta\xi\delta\phi\right)$ with the scalar product

$$(\Psi_1(\xi,\phi),\Psi_2(\xi,\phi)) = \Re\left\langle\Psi_1(\xi,\phi),\Psi_2(\xi,\phi)\right\rangle, \tag{12.161}$$

where the inner product is

$$\langle\Psi_1(\xi,\phi),\Psi_2(\xi,\phi)\rangle = \int_{\Sigma(\xi,\phi)} \Psi_1^\star(\xi',\phi')\Psi_2(\xi',\phi')\delta\xi'\delta\phi', \tag{12.162}$$

and the induced norm

$$||\Psi_i(\xi,\phi)|| = \left(\int_{\Sigma(\xi,\phi)} |\Psi_i(\xi',\phi')|^2\delta\xi'\delta\phi'\right)^{1/2}, \tag{12.163}$$

where $i=1,2$. It is easy to see that

$$(\Psi_1(\xi,\phi),\Psi_2(\xi,\phi)) = \sqrt{|\Gamma_0|}I, \tag{12.164}$$

and because $I \geqslant 2$, the scalar product satisfies the inequality

$$(\Psi_1(\xi,\phi),\Psi_2(\xi,\phi)) \geqslant 2\sqrt{|\Gamma_0|}. \tag{12.165}$$

In the light of the inequality (12.153) one has

$$(\Psi_1(\xi,\phi),\Psi_2(\xi,\phi)) \geqslant 4. \tag{12.166}$$

It can be seen also that

$$||\Psi_1|| = 1, \tag{12.167}$$

$$||\Psi_2|| = \sqrt{|\Gamma_0|} \geqslant 2, \tag{12.168}$$

i.e. the state Ψ_1 is a ray whereas the state Ψ_2 is not a ray in the Hilbert space. However, because the superposition is given by

$$\Psi = \alpha_1\Psi_1 + \alpha_2\Psi_2, \tag{12.169}$$

one can selected the constants α_1 and α_2 in such a way that the wave function Ψ_2 will be normalized to unity. The choice is easy to deduce. Because Ψ_1 is normalized to unity the coefficient α_1 is an arbitrary constant, while the coefficient α_2 should be exchanged on

$$\alpha_2' = \frac{1}{\sqrt{|\Gamma_0|}}\alpha_2, \tag{12.170}$$

where α_2 is an arbitrary constant. Still, however, in general the problem of normalization of the state Ψ_2 is unsolved. Redefinition of the scalar product

$$(\Psi_1(\xi,\phi),\Psi_2(\xi,\phi)) = \frac{\langle\Psi_1(\xi,\phi),\Psi_2(\xi,\phi)\rangle}{||\Psi_1(\xi,\phi)||\,||\Psi_2(\xi,\phi)||} = I \geqslant 2, \tag{12.171}$$

does not work consistently in the light of (12.148). The only solution is to apply *ad hoc* the scaling

$$\Psi_2' = \frac{1}{\sqrt{|\Gamma_0|}}\Psi_2 \leqslant \frac{1}{\sqrt{2}}\Psi_2, \tag{12.172}$$

which, however, is consistent with the choice (12.170).

Another alternative can be constructed as follows. Let us assume that the superposition state

$$\Psi' = \alpha_1'\Psi_1' + \alpha_2'\Psi_2', \tag{12.173}$$

is such that the states Ψ_1' and Ψ_2' are orthonormal, *i.e.* are orthogonal rays in the Hilbert space

$$\begin{aligned} ||\Psi_1'|| &= 1, && (12.174)\\ ||\Psi_2'|| &= 1, && (12.175)\\ (\Psi_1',\Psi_2') &= 0. && (12.176) \end{aligned}$$

The problem is to construct such constants α_1', α_2', and states Ψ_1' and Ψ_2' which satisfy these requirements. This problem can be solved with help of the Gram–Schmidt algorithm, in which the orthonormal states Ψ_i' (here $i = 1, 2$) are expressed through the non-orthonormal states Ψ_i by using the Gram determinants

$$\begin{aligned} G_0 &= 1, && (12.177)\\ G_1 &= |(\Psi_1,\Psi_1)|, && (12.178)\\ G_2 &= \det\begin{bmatrix} (\Psi_1,\Psi_1) & (\Psi_1,\Psi_2) \\ (\Psi_1,\Psi_2) & (\Psi_2,\Psi_2) \end{bmatrix}, && (12.179) \end{aligned}$$

as follows

$$\Psi_1' = \frac{1}{\sqrt{G_0G_1}}\Psi_1, \tag{12.180}$$

$$\Psi_2' = \frac{1}{\sqrt{G_1G_2}}\det\begin{bmatrix} (\Psi_1,\Psi_1) & (\Psi_1,\Psi_2) \\ \Psi_1 & \Psi_2 \end{bmatrix}. \tag{12.181}$$

It can be seen by straightforward computation that

$$G_1 = 1, \tag{12.182}$$

$$G_2 = |\Gamma_0|\left(\frac{1}{\sqrt{|\Gamma_0|}} - I^2\right), \tag{12.183}$$

and, consequently,

$$\Psi_1' = \Psi_1, \tag{12.184}$$

$$\Psi_2' = \left(\frac{1}{\sqrt{|\Gamma_0|}} - I^2\right)^{-1/2}\left(-I\Psi_1 + \frac{1}{\sqrt{|\Gamma_0|}}\Psi_2\right). \tag{12.185}$$

One sees now that the second term in (12.185) converges with the proposed scaled state (12.172). In this manner, the superposition state (12.173) can be rewritten as

$$\Psi' = \alpha_1'\Psi_1 + \alpha_2'\Psi_2', \tag{12.186}$$

where now $\Psi_2' = \frac{1}{\sqrt{|\Gamma_0|}}\Psi_2$ is the scaled state (12.172) and

$$\alpha_1' = \alpha_1 - \alpha_2 I\left(\frac{1}{\sqrt{|\Gamma_0|}} - I^2\right)^{-1/2}, \tag{12.187}$$

$$\alpha_2' = \alpha_2\left(\frac{1}{\sqrt{|\Gamma_0|}} - I^2\right)^{-1/2}. \tag{12.188}$$

The states Ψ_1 and Ψ_2 can be expressed through application of the orthonormal states Ψ_1' and Ψ_2'

$$\Psi_1 = \Psi_1', \tag{12.189}$$

$$\Psi_2 = \sqrt{|\Gamma_0|}\left(\Psi_2' + I\left(\frac{1}{\sqrt{|\Gamma_0|}} - I^2\right)^{1/2}\Psi_1'\right), \tag{12.190}$$

and for this reason

$$
\begin{aligned}
(\Psi_1, \Psi_2) &= \left(\Psi_1', \sqrt{|\Gamma_0|}\left(\Psi_2' + I\left(\frac{1}{\sqrt{|\Gamma_0|}} - I^2\right)^{1/2}\Psi_1'\right)\right) = \\
&= \sqrt{|\Gamma_0|}\left[(\Psi_1', \Psi_2') + I\left(\frac{1}{\sqrt{|\Gamma_0|}} - I^2\right)^{1/2}(\Psi_1', \Psi_1')\right] = \\
&= \sqrt{|\Gamma_0|}I\left(\frac{1}{\sqrt{|\Gamma_0|}} - I^2\right)^{1/2}.
\end{aligned}
\tag{12.191}
$$

In the light of the relation (12.164) one obtains the equation

$$
\sqrt{|\Gamma_0|}I\left(\frac{1}{\sqrt{|\Gamma_0|}} - I^2\right)^{1/2} = \sqrt{|\Gamma_0|}I, \tag{12.192}
$$

which gives uniquely and unambiguously the result

$$
\left(\frac{1}{\sqrt{|\Gamma_0|}} - I^2\right)^{1/2} = 1, \tag{12.193}
$$

which can be presented in the equivalent form

$$
I^2 = \frac{1}{\sqrt{|\Gamma_0|}} - 1. \tag{12.194}
$$

In the light of the fact $I \geqslant 2$ one has

$$
\sqrt{|\Gamma_0|} \leqslant \frac{1}{5}, \tag{12.195}
$$

which, however, is inconsistent in the light of the fact (12.153). In this manner, the problem of superposition in the theory of quantum gravity cannot be solved by application of the Gram–Schmidt algorithm of orthonormalization to the Hilbert space constructed in this subsection.

F Problem of Time

It is easy to see that within the invariant global one-dimensional quantum gravity the problem of time is naturally solved. Namely, the role of physical time is played by the invariant global dimension

$$\xi = \frac{2}{\kappa}\sqrt{\frac{h}{6}} \sim t_{\text{physical}}, \tag{12.196}$$

where τ is a reference constant, which can be taken *ad hoc* as the Planck time, *i.e.* $\tau = t_P$. Such an identification is justified by the fact that ξ is invariant with respect to action of the spatial diffeomorphisms group, because the quantity $\sqrt{h}$ is such a diffeoinvariant. In this manner, in the sense of Kuchař ξ is an observable.

Such a situation is analogous to the model of quantum cosmology presented in the previous chapters of this book, in which the role of affine time parameter is played by the cosmic scale factor parameter a. This parameter is assumed to be a function of the conformal or cosmic time. For this reason, the proposed model of quantum gravity can be rewritten in a more fashionable form

$$\left(\frac{d^2}{dt^2} + V[t]\right)\Psi(t,\phi) = 0. \tag{12.197}$$

This global one-dimensional evolutionary equation in itself defines nontrivial dynamics. Namely, this is the time-dependent $0+1$-dimensional Schrödinger wave equation.

Chapter 13

Examples of Invariant 1D Wave Functions

Let us present with no detailed computations the wave functionals associated to the three classical solutions of the Einstein field equations.

A Minkowski Space-Time

Let us consider first an empty space with no cosmological constant, *i.e.* the Minkowski space-time. In such an instance, in the Cartesian coordinates, the spatial metric coincides with the metric of the Euclidean space

$$h_{ij} = \begin{bmatrix} 1 & 0 & 0 \\ 0 & 1 & 0 \\ 0 & 0 & 1 \end{bmatrix}, \tag{13.1}$$

and, therefore, $h = 1$ so that the invariant global dimension is

$$\xi = \frac{2}{\kappa}\frac{1}{\sqrt{6}}. \tag{13.2}$$

The Minkowski space-time is characterized by

$$^{(3)}R = 0, \quad \rho = 0, \quad \Lambda = 0, \tag{13.3}$$

and for this reason the averaged generalized gravitational potential is

$$\langle V\rangle(\xi,\phi) = 0. \tag{13.4}$$

In this manner, the wave functions for this case are

$$\Psi_1 = \frac{1}{\sqrt{|A|}}\xi, \tag{13.5}$$

$$\Psi_2 = \frac{1}{\sqrt{|B|}}, \tag{13.6}$$

where the the integration constants A and B are

$$A = \frac{4}{9\kappa^3\sqrt{6}}, \tag{13.7}$$

$$B = \frac{2}{\Gamma_0\kappa\sqrt{6}}. \tag{13.8}$$

In other words, the wave functions

$$\Psi_1 = \sqrt{\frac{3}{2}\kappa\sqrt{6}}, \tag{13.9}$$

$$\Psi_2 = \sqrt{\frac{1}{2}|\Gamma_0|\kappa\sqrt{6}}, \tag{13.10}$$

are constant on the midisuperspace. This case is singular because the constant wave functions cannot be normalized to unity separately. However, taking the superposed state

$$\Psi = \alpha_1\Psi_1 + \alpha_2\Psi_2, \tag{13.11}$$

one obtains the normalization condition

$$3|\alpha_1|^2 + |\Gamma_0||\alpha_2|^2 + (\alpha_1^\star\alpha_2 + \alpha_1\alpha_2^\star)\sqrt{3|\Gamma_0|} = 1, \tag{13.12}$$

having the following solutions

$$\alpha_1 = \frac{\beta_1}{\sqrt{3}}\exp(i\alpha), \tag{13.13}$$

$$\alpha_2 = \frac{\beta_2}{\sqrt{|\Gamma_0|}}\exp(i\alpha), \tag{13.14}$$

where α is an arbitrary real phase, and the constants β_1 and β_2 are constrained by

$$\beta_2 = \pm 1 - \beta_1. \tag{13.15}$$

In other words, the superposed states

$$\Psi_{\pm}^{M} = \exp(i\alpha)\left(\frac{\beta}{\sqrt{3}}\Psi_1 + \frac{\pm 1-\beta}{\sqrt{|\Gamma_0|}}\Psi_2\right) = \pm\exp(i\alpha)\sqrt{\frac{\kappa}{2}\sqrt{6}}, \tag{13.16}$$

are consistent wave functions of the Minkowski space-time. Because the Minkowski vacuum is nonlinearly stable, the wave functions (13.16) are the reference states for another interpretation of the Einstein constant

$$\kappa = \frac{2\left|\Psi_{\pm}^{M}\right|^2}{\sqrt{6}}. \tag{13.17}$$

B Kasner Space-Time

Let us consider the simple solution of the Einstein field equations describing an anisotropic universe without matter, called the Kasner metric. The spatial part of this metric is

$$h_{ij} = \begin{bmatrix} \left(\frac{t}{t_0}\right)^{2p_1} & 0 & 0 \\ 0 & \left(\frac{t}{t_0}\right)^{2p_2} & 0 \\ 0 & 0 & \left(\frac{t}{t_0}\right)^{2p_3} \end{bmatrix}, \tag{13.18}$$

where ct is the time coordinate and t_0 is a time constant, and

$$\sum_i p_i = 1, \tag{13.19}$$

$$\sum_i p_i^2 = 1. \tag{13.20}$$

In this instance, the global dimension is

$$h = \left(\frac{t}{t_0}\right)^2, \tag{13.21}$$

while the invariant global dimension is

$$\xi = \frac{2}{\kappa\sqrt{6}}\frac{t}{t_0}. \tag{13.22}$$

Because the cosmological constant and the Matter fields are absent here, and the three-dimensional Ricci scalar curvature vanishes, one has $\langle V\rangle = 0$. In this manner, the wave functions of the Kasner space-time are

$$\Psi_1^K = \frac{1}{\sqrt{|A|}}\frac{2}{\kappa\sqrt{6}}\frac{t}{t_0}, \tag{13.23}$$

$$\Psi_2^K = \frac{1}{\sqrt{|B|}}, \tag{13.24}$$

where the integration constants are

$$A = \frac{4}{9\kappa^3\sqrt{6}}\left(\frac{T}{t_0}\right)^3, \tag{13.25}$$

$$B = \frac{1}{\Gamma_0}\frac{2}{\kappa\sqrt{6}}\left(\frac{T}{t_0}\right), \tag{13.26}$$

where T is some reference value of time t.

C Schwarzschild Space-Time

The second situation which we shall present here is the Schwarzschild space-time. This metric is a spherically symmetric vacuum solution of the Einstein field equations with no cosmological constant, *i.e.* a situation ${}^{(3)}R[h]=0$, $\Lambda=0$, $\rho=0$. It means that in this case also $\langle V\rangle(h_{ij},\phi)=0$. The spatial part of the metric has the form

$$h_{ij} = \begin{bmatrix} \left(1-\frac{r_S}{r}\right)^{-1} & 0 & 0 \\ 0 & r^2 & 0 \\ 0 & 0 & r^2\sin^2\theta \end{bmatrix}, \tag{13.27}$$

where r_S is the Schwarzschild radius of the spherically symmetric non-rotating object of mass M

$$r_S = \frac{2GM}{c^2} = \frac{\kappa}{4\pi}Mc^2. \tag{13.28}$$

Therefore, in this case the global dimension is

$$h = \frac{r^4\sin^2\theta}{1-\frac{r_S}{r}}, \tag{13.29}$$

so that the invariant global dimension is

$$\xi = \frac{2}{\kappa}\frac{r^2\sin\theta}{\sqrt{6\left(1-\frac{r_S}{r}\right)}}, \tag{13.30}$$

and the total volume of the Schwarzschild midisuperspace is $\Omega(\xi) = \xi$. The the integration constants A and B can be established easy

$$A = \int \Omega^2(\xi)\delta\xi = \frac{\Xi^3}{3} = \frac{8}{\kappa^3}\frac{R^6\sin^3\Theta}{\left(6\left(1-\frac{r_S}{R}\right)\right)^{3/2}}, \tag{13.31}$$

$$B = \frac{1}{\Gamma_0}\int \delta\xi = \frac{\Xi}{\Gamma_0} = \frac{2}{\Gamma_0\kappa}\frac{R^2\sin\Theta}{\sqrt{6\left(1-\frac{r_S}{R}\right)}}, \tag{13.32}$$

where $\Xi = \xi(R,\Theta)$ is the constant parameter

$$\Xi = \frac{2}{\kappa}\frac{R^2\sin\Theta}{\sqrt{6\left(1-\frac{r_S}{R}\right)}}. \tag{13.33}$$

In this manner, the wave functions of the Schwarzschild space-time are

$$\Psi_1^S = \sqrt{\frac{\kappa}{2}\sqrt{6}}\sqrt{\frac{1}{R^2\sin\Theta}\sqrt{1-\frac{r_S}{R}}}\sqrt{\frac{1-\frac{r_S}{R}}{1-\frac{r_S}{r}}\frac{r^2\sin\theta}{R^2\sin\Theta}}, \tag{13.34}$$

$$\Psi_2^S = \sqrt{\frac{\kappa}{2}|\Gamma_0|\sqrt{6}}\sqrt{\frac{1}{R^2\sin\Theta}\sqrt{1-\frac{r_S}{R}}}. \tag{13.35}$$

D (Anti-) De Sitter Space-Time

Let us consider now the spherically symmetric solution of the vacuum Einstein field equations in presence of the cosmological constant, *i.e.* the lambdavacuum Einstein field equations

$$R_{\mu\nu} = \Lambda g_{\mu\nu}, \tag{13.36}$$

called the (Anti-) De Sitter space-time. Such a space-time is a submanifold of the Minkowski space-time of one higher dimension described by the hyperboloid of one sheet

$$-x_0^2 + x_1^2 + x_2^2 + x_3^2 = \pm\alpha^2, \tag{13.37}$$

where the sign plus describes De Sitter space-time, whereas the sign minus is related to the Anti-De Sitter space. The (Anti-) De Sitter space-time is an Einstein manifold

$$R_{\mu\nu} = \pm\frac{3}{\alpha^2} g_{\mu\nu}, \tag{13.38}$$

which in the light of the lambdavacuum Einstein field equations (13.36) the cosmological constant is

$$\Lambda = \pm\frac{3}{\alpha^2}. \tag{13.39}$$

The three-dimensional Ricci scalar curvature of the (Anti-) De Sitter space-time can be established easily

$${}^{(3)}R = h^{ij} R_{ij} = 3\Lambda, \tag{13.40}$$

and because there is no Matter fields $\rho = 0$, the averaged generalized gravitational potential has the value

$$\langle V \rangle(\xi, \phi) = -3\Lambda + 2\Lambda = -\Lambda. \tag{13.41}$$

In other words, for the De Sitter space-time

$$\langle V \rangle(\xi, \phi) = -\frac{3}{\alpha^2}, \tag{13.42}$$

while for the Anti-De Sitter space-time

$$\langle V\rangle(\xi,\phi) = \frac{3}{\alpha^2}. \tag{13.43}$$

The spatial part of the (Anti-) De Sitter metric is

$$h_{ij} = \begin{bmatrix} \left(1-\frac{\Lambda}{3}r^2\right)^{-1} & 0 & 0 \\ 0 & r^2 & 0 \\ 0 & 0 & r^2\sin^2\theta \end{bmatrix}, \tag{13.44}$$

and, for this reason, the global dimension is

$$h = \frac{r^4\sin^2\theta}{1-\frac{\Lambda}{3}r^2}, \tag{13.45}$$

whereas the invariant global dimension is

$$\xi = \frac{2}{\kappa\sqrt{6}}\frac{r^2\sin\theta}{\sqrt{1-\frac{\Lambda}{3}r^2}}, \tag{13.46}$$

and of course the volume of the related midisuperspace is $\Omega(\xi) = \xi$.

Let us consider first the case of the De Sitter space-time, *i.e.* $\Lambda = \frac{3}{\alpha^2}$. The integration constants can be evaluated as follows

$$A = \int\left(\frac{\sinh\left(\sqrt{\Lambda}\xi\right)}{\sqrt{\Lambda}}\right)^2\delta\xi = \frac{1}{\Lambda}\left(-\frac{\Xi}{2}+\frac{\sinh\left(2\sqrt{\Lambda}\Xi\right)}{4\sqrt{\Lambda}}\right), \tag{13.47}$$

$$B = \frac{1}{\Gamma_0}\int\cosh^2\left(\sqrt{\Lambda}\xi\right)\delta\xi = \frac{1}{\Gamma_0}\left(\frac{\Xi}{2}+\frac{\cosh\left(2\sqrt{\Lambda}\Xi\right)}{4\sqrt{\Lambda}}\right), \tag{13.48}$$

where Ξ is the reference constant

$$\Xi = \frac{2}{\kappa\sqrt{6}}\frac{R^2\sin\Theta}{\sqrt{1-\frac{\Lambda}{3}R^2}}. \tag{13.49}$$

For this reason, the wave functions of the De Sitter space-time are

$$\Psi_1^{DS} = \frac{1}{\sqrt{|A|}}\frac{1}{\sqrt{\Lambda}}\sinh\left(\frac{2\sqrt{\Lambda}}{\kappa\sqrt{6}}\frac{r^2\sin\theta}{\sqrt{1-\frac{\Lambda}{3}r^2}}\right), \tag{13.50}$$

$$\Psi_2^{DS} = \frac{1}{\sqrt{|B|}}\cosh\left(\frac{2\sqrt{\Lambda}}{\kappa\sqrt{6}}\frac{r^2\sin\theta}{\sqrt{1-\frac{\Lambda}{3}r^2}}\right). \tag{13.51}$$

Similarly, for the case of Anti-De Sitter space-time one obtains the following wave functions

$$\Psi_1^{ADS} = \frac{1}{\sqrt{|A|}}\frac{1}{\sqrt{|\Lambda|}}\sin\left(\frac{2\sqrt{|\Lambda|}}{\kappa\sqrt{6}}\frac{r^2\sin\theta}{\sqrt{1+\frac{|\Lambda|}{3}r^2}}\right), \tag{13.52}$$

$$\Psi_2^{ADS} = \frac{1}{\sqrt{|B|}}\cos\left(\frac{2\sqrt{|\Lambda|}}{\kappa\sqrt{6}}\frac{r^2\sin\theta}{\sqrt{1+\frac{|\Lambda|}{3}r^2}}\right), \tag{13.53}$$

where $|\Lambda| = \frac{3}{\alpha^2}$, and the integration constants A and B are

$$A = \int\left(\frac{\sin\left(\sqrt{|\Lambda|}\xi\right)}{\sqrt{|\Lambda|}}\right)^2\delta\xi = \frac{1}{|\Lambda|}\left(\frac{\Xi}{2}-\frac{\sin\left(2\sqrt{|\Lambda|}\Xi\right)}{4\sqrt{|\Lambda|}}\right), \tag{13.54}$$

$$B = \frac{1}{\Gamma_0}\int\cos^2\left(\sqrt{|\Lambda|}\xi\right)\delta\xi = \frac{1}{\Gamma_0}\left(\frac{\Xi}{2}+\frac{\cos\left(2\sqrt{|\Lambda|}\Xi\right)}{4\sqrt{|\Lambda|}}\right), \tag{13.55}$$

where Ξ is the reference constant

$$\Xi = \frac{2}{\kappa\sqrt{6}}\frac{R^2\sin\Theta}{\sqrt{1+\frac{|\Lambda|}{3}R^2}}. \tag{13.56}$$

E (Anti-) De Sitter–Schwarzschild Space-Time

The (Anti-) De Sitter–Schwarzschild space-time is the case joining the (Anti-) De Sitter and Schwarzschild space-times. This is the spherically symmetric solution of lambdavacuum Einstein's field equations with suitable boundary conditions. The spatial part of space-time metric has the form

$$h_{ij} = \begin{bmatrix} \left(1 - \frac{r_S}{r} - \frac{\Lambda}{3} r^2\right)^{-1} & 0 & 0 \\ 0 & r^2 & 0 \\ 0 & 0 & r^2 \sin^2 \theta \end{bmatrix}. \tag{13.57}$$

In this manner, the global dimension is

$$h = \frac{r^4 \sin^2 \theta}{1 - \frac{r_S}{r} - \frac{\Lambda}{3} r^2}, \tag{13.58}$$

while the invariant global dimension is

$$\xi = \frac{2}{\kappa\sqrt{6}} \frac{r^2 \sin\theta}{\sqrt{1 - \frac{r_S}{r} - \frac{\Lambda}{3} r^2}}, \tag{13.59}$$

and of course the volume of the related midisuperspace is $\Omega(\xi) = \xi$. The wave functions of the De Sitter–Schwarzschild space-time, called also the Kottler space-time, can be written in the form

$$\Psi_1^{DS-S} = \frac{1}{\sqrt{|A|}} \frac{1}{\sqrt{\Lambda}} \sinh\left(\frac{2\sqrt{\Lambda}}{\kappa\sqrt{6}} \frac{r^2 \sin\theta}{\sqrt{1 - \frac{r_S}{r} - \frac{\Lambda}{3} r^2}} \right), \tag{13.60}$$

$$\Psi_2^{DS-S} = \frac{1}{\sqrt{|B|}} \cosh\left(\frac{2\sqrt{\Lambda}}{\kappa\sqrt{6}} \frac{r^2 \sin\theta}{\sqrt{1 - \frac{r_S}{r} - \frac{\Lambda}{3} r^2}} \right). \tag{13.61}$$

where the integration constants A and B are

$$A = \int \left(\frac{\sinh\left(\sqrt{\Lambda}\xi\right)}{\sqrt{\Lambda}} \right)^2 \delta\xi = \frac{1}{\Lambda}\left(-\frac{\Xi}{2} + \frac{\sinh\left(2\sqrt{\Lambda}\Xi\right)}{4\sqrt{\Lambda}} \right), \quad (13.62)$$

$$B = \frac{1}{\Gamma_0}\int \cosh^2\left(\sqrt{\Lambda}\xi\right)\delta\xi = \frac{1}{\Gamma_0}\left(\frac{\Xi}{2} + \frac{\cosh\left(2\sqrt{\Lambda}\Xi\right)}{4\sqrt{\Lambda}} \right), \quad (13.63)$$

where Ξ is the reference constant

$$\Xi = \frac{1}{\kappa\sqrt{6}} \frac{R^2 \sin\Theta}{\sqrt{1 - \frac{r_S}{R} - \frac{\Lambda}{3}R^2}}. \quad (13.64)$$

Similarly, the wave functions of the Anti-De Sitter–Schwarzschild space-time are

$$\Psi_1^{ADS-S} = \frac{1}{\sqrt{|A|}}\frac{1}{\sqrt{|\Lambda|}} \sin\left(\frac{2\sqrt{|\Lambda|}}{\kappa\sqrt{6}} \frac{r^2 \sin\theta}{\sqrt{1 - \frac{r_S}{r} + \frac{|\Lambda|}{3}r^2}} \right), \quad (13.65)$$

$$\Psi_2^{ADS-S} = \frac{1}{\sqrt{|B|}} \cos\left(\frac{2\sqrt{|\Lambda|}}{\kappa\sqrt{6}} \frac{r^2 \sin\theta}{\sqrt{1 - \frac{r_S}{r} + \frac{|\Lambda|}{3}r^2}} \right), \quad (13.66)$$

where $|\Lambda| = \frac{3}{\alpha^2}$, and the integration constants A and B are

$$A = \int \left(\frac{\sin\left(\sqrt{|\Lambda|}\xi\right)}{\sqrt{|\Lambda|}} \right)^2 \delta\xi = \frac{1}{|\Lambda|}\left(\frac{\Xi}{2} - \frac{\sin\left(2\sqrt{|\Lambda|}\Xi\right)}{4\sqrt{|\Lambda|}} \right), \quad (13.67)$$

$$B = \frac{1}{\Gamma_0}\int \cos^2\left(\sqrt{|\Lambda|}\xi\right)\delta\xi = \frac{1}{\Gamma_0}\left(\frac{\Xi}{2} + \frac{\cos\left(2\sqrt{|\Lambda|}\Xi\right)}{4\sqrt{|\Lambda|}} \right), \quad (13.68)$$

where Ξ is the reference constant

$$\Xi = \frac{2}{\kappa\sqrt{6}} \frac{R^2 \sin\Theta}{\sqrt{1 - \frac{r_S}{R} + \frac{|\Lambda|}{3}R^2}}. \quad (13.69)$$

F Kerr Space-Time

Kerr's space-time is a solution of the vacuum Einstein field equations for a spherical body rotating with the angular momentum J. Because the spatial Ricci scalar curvature is zero, the cosmological constant is not present, and Matter fields are absent, in such a situation the averaged generalized gravitational potential identically vanishes. The spatial part of the Kerr metric has the form

$$h_{ij} = \begin{bmatrix} \dfrac{r^2+\alpha^2\cos^2\theta}{r^2-r_s r+\alpha^2} & 0 & 0 \\ 0 & r^2+\alpha^2\cos^2\theta & 0 \\ 0 & 0 & r^2+\alpha^2+\dfrac{\alpha^2 r_S r\sin^2\theta}{r^2+\alpha^2\cos^2\theta} \end{bmatrix}, \quad (13.70)$$

where α is the coefficient related to the angular momentum and the mass of the rotating object

$$\alpha = \frac{J}{Mc}. \quad (13.71)$$

In this manner, the global dimension is

$$h = \frac{\left(r^2+\alpha^2\cos^2\theta\right)^2}{r^2-r_s r+\alpha^2}\left(r^2+\alpha^2+\frac{\alpha^2 r_S r\sin^2\theta}{r^2+\alpha^2\cos^2\theta}\right), \quad (13.72)$$

while the invariant global dimension is

$$\xi = \frac{2}{\kappa\sqrt{6}}\left(r^2+\alpha^2\cos^2\theta\right)\sqrt{\frac{\left(r^2+\alpha^2\right)^2-\alpha^2\left(r^2-r_S r+\alpha^2\right)\sin^2\theta}{\left(r^2-r_s r+\alpha^2\right)\left(r^2+\alpha^2\cos^2\theta\right)}}, \quad (13.73)$$

and, of course, the volume of the Kerr midisuperspace is $\Omega(\xi)=\xi$. Introducing the reference parameter

$$\Xi = \frac{2}{\kappa\sqrt{6}}\left(R^2+\alpha^2\cos^2\Theta\right)\sqrt{\frac{\left(R^2+\alpha^2\right)^2-\alpha^2\left(R^2-r_S R+\alpha^2\right)\sin^2\Theta}{\left(R^2-r_s R+\alpha^2\right)\left(R^2+\alpha^2\cos^2\Theta\right)}}, \quad (13.74)$$

one can write out the integration constants A and B

$$A = \frac{\Xi^3}{3}, \quad (13.75)$$

$$B = \frac{1}{\Gamma_0}\Xi, \quad (13.76)$$

so that the wave functions of the Kerr space-time have the form

$$\Psi_1^{Kerr} = 2\frac{r^2+\alpha^2\cos^2\theta}{\kappa\sqrt{|A|}\sqrt{6}}\sqrt{\frac{\left(r^2+\alpha^2\right)^2-\alpha^2\left(r^2-r_Sr+\alpha^2\right)\sin^2\theta}{\left(r^2-r_sr+\alpha^2\right)\left(r^2+\alpha^2\cos^2\theta\right)}}, \quad (13.77)$$

$$\Psi_2^{Kerr} = \frac{1}{\sqrt{|B|}}. \quad (13.78)$$

G Kerr–Newman Space-Time

The Kerr–Newman space-time is a solution of the electrovacuum Einstein field equations called also Einstein–Maxwell equations, which are a combination of the Einstein field equations with the stress-energy tensor of the electromagnetic field and no cosmological constant, and the source-free Maxwell equations

$$R_{\mu\nu} = \kappa T_{\mu\nu}^{em}, \quad (13.79)$$

$$F_{\mu\nu;\kappa} + F_{\nu\kappa;\mu} + F_{\kappa\mu;\nu} = 0, \quad (13.80)$$

$$F^{\mu\nu}_{;\nu} = 0, \quad (13.81)$$

where $T_{\mu\nu}^{em}$ is the stress-energy tensor of the electromagnetic field, which can be given *ad hoc* in analogy with the case of flat space-time

$$T_{\mu\nu}^{em} = \frac{1}{\mu_0}\left(F_\mu^\alpha F_{\alpha\nu} - \frac{1}{4}g_{\mu\nu}F_{\alpha\beta}F^{\alpha\beta}\right), \quad (13.82)$$

and $F_{\mu\nu}$ is the Faraday tensor

$$F_{\mu\nu} = \begin{bmatrix} 0 & \frac{E_x}{c} & \frac{E_y}{c} & \frac{E_z}{c} \\ -\frac{E_x}{c} & 0 & -B_z & -B_y \\ -\frac{E_y}{c} & B_z & 0 & -B_x \\ -\frac{E_z}{c} & -B_y & B_x & 0 \end{bmatrix}, \quad (13.83)$$

$$F^{\mu\nu} = \begin{bmatrix} 0 & -\frac{E_x}{c} & -\frac{E_y}{c} & -\frac{E_z}{c} \\ \frac{E_x}{c} & 0 & -B_z & -B_y \\ \frac{E_y}{c} & B_z & 0 & -B_x \\ \frac{E_z}{c} & -B_y & B_x & 0 \end{bmatrix}, \quad (13.84)$$

where $\vec{E} = [E_x, E_y, E_z]$ and $\vec{B} = [B_x, B_y, B_z]$ are three-vectors of the electric and the magnetic field, respectively. The speed of light c can be expressed in terms of the vacuum permeability μ_0 and the vacuum permittivity ε_0 by the formula $\varepsilon_0 \mu_0 = 1/c^2$.

There is the question how to derive the stress-energy tensor of the electromagnetic field (13.82) by using the definition

$$T_{\mu\nu} = -\frac{2}{\sqrt{-g}} \frac{\delta(\sqrt{-g}\mathscr{L})}{\delta g^{\mu\nu}}, \tag{13.85}$$

following from the Hilbert–Palatini action principle, and describing Matter fields characterized by the Lagrangian $\mathscr{L}$ in a four-dimensional Riemannian space-time with metric $g^{\mu\nu}$. The Lagrangian of the Maxwell electromagnetic field has the form (For more general approach Cf. *e.g.* Ref. [932])

$$\mathscr{L} = -\frac{1}{4\mu_0} F_{\alpha\beta} F^{\alpha\beta}. \tag{13.86}$$

Using the Jacobi formula $\delta\sqrt{-g} = -\frac{1}{2}\sqrt{-g} g_{\mu\nu} \delta g^{\mu\nu}$, the stress-energy tensor (13.85) can be rewritten in a more convenient form

$$T_{\mu\nu} = -2\frac{\delta \mathscr{L}}{\delta g^{\mu\nu}} + g_{\mu\nu}\mathscr{L}, \tag{13.87}$$

so that the problem is to establish the functional derivative

$$\frac{\delta \mathscr{L}}{\delta g^{\mu\nu}} = -\frac{1}{4\mu_0}\frac{\delta}{\delta g^{\mu\nu}}\left(F_{\alpha\beta}F^{\alpha\beta}\right) = -\frac{1}{4\mu_0}\left(\frac{\delta F_{\alpha\beta}}{\delta g^{\mu\nu}} F^{\alpha\beta} + F_{\alpha\beta}\frac{\delta F^{\alpha\beta}}{\delta g^{\mu\nu}}\right). \tag{13.88}$$

Let us notice that the tensor $F_{\alpha\beta}$ can be presented in a more complex form

$$\begin{aligned} F_{\alpha\beta} &= g^{\gamma}_{\alpha} g^{\delta}_{\beta} F_{\gamma\delta} = \left(g^{\gamma}_{\mu} g^{\mu\nu} g_{\nu\alpha}\right)\left(g^{\delta}_{\nu} g^{\nu\mu} g_{\mu\beta}\right) F_{\gamma\delta} = \\ &= \left(g^{\gamma}_{\mu} g^{\mu\nu} g_{\nu\alpha}\right)\left(g^{\delta}_{\nu} g^{\mu\nu} g_{\mu\beta}\right) F_{\gamma\delta} = g^{\mu\nu} g^{\mu\nu}\left(g^{\gamma}_{\mu} g_{\nu\alpha} g^{\delta}_{\nu} g_{\mu\beta}\right) F_{\gamma\delta}. \end{aligned} \tag{13.89}$$

In this manner, one has

$$
\begin{aligned}
\frac{\delta F_{\alpha\beta}}{\delta g^{\mu\nu}} &= 2g^{\mu\nu}\left(g^{\gamma}_{\mu}g_{\nu\alpha}g^{\delta}_{\nu}g_{\mu\beta}\right)F_{\gamma\delta} = 2g^{\gamma}_{\mu}g^{\delta}_{\nu}\left(g^{\mu\nu}g_{\nu\alpha}\right)g_{\mu\beta}F_{\gamma\delta} = \\
&= 2\left(g^{\gamma}_{\mu}g^{\delta}_{\nu}g^{\mu}_{\alpha}g_{\mu\beta}\right)F_{\gamma\delta} = 2\left(g^{\gamma}_{\mu}g^{\mu}_{\alpha}\right)g^{\delta}_{\nu}g_{\mu\beta}F_{\gamma\delta} = \\
&= 2\left(g^{\gamma}_{\alpha}g^{\delta}_{\nu}g_{\mu\beta}\right)F_{\gamma\delta} = 2\left(g^{\delta}_{\nu}g_{\mu\beta}\right)g^{\gamma}_{\alpha}F_{\gamma\delta} = 2\left(g^{\delta}_{\nu}g_{\mu\beta}\right)F_{\alpha\delta} = \\
&= -2g_{\mu\beta}\left(g^{\delta}_{\nu}F_{\delta\alpha}\right) = -2g_{\mu\beta}F_{\nu\alpha},
\end{aligned}
\tag{13.90}
$$

and, consequently,

$$
\frac{\delta F_{\alpha\beta}}{\delta g^{\mu\nu}}F^{\alpha\beta} = -2g_{\mu\beta}F_{\nu\alpha}F^{\alpha\beta} = 2F_{\nu\alpha}\left(g_{\mu\beta}F^{\beta\alpha}\right) = 2F_{\nu\alpha}F^{\alpha}_{\mu} = 2F^{\alpha}_{\mu}F_{\alpha\nu}. \tag{13.91}
$$

Similarly, one can rewrite $F^{\alpha\beta}$ in the form

$$
F^{\alpha\beta} = g^{\alpha\kappa}g^{\beta\lambda}F_{\kappa\lambda}, \tag{13.92}
$$

and, consequently, one has

$$
\begin{aligned}
\frac{\delta F^{\alpha\beta}}{\delta g^{\mu\nu}} &= \frac{\delta g^{\alpha\kappa}}{\delta g^{\mu\nu}}g^{\beta\lambda}F_{\kappa\lambda} + g^{\alpha\kappa}\frac{\delta g^{\beta\lambda}}{\delta g^{\mu\nu}}F_{\kappa\lambda} + g^{\alpha\kappa}g^{\beta\lambda}\frac{\delta F_{\kappa\lambda}}{\delta g^{\mu\nu}} = \\
&= g^{\alpha}_{\mu}g^{\kappa}_{\nu}g^{\beta\lambda}F_{\kappa\lambda} + g^{\alpha\kappa}g^{\beta}_{\mu}g^{\lambda}_{\nu}F_{\kappa\lambda} + g^{\alpha\kappa}g^{\beta\lambda}\frac{\delta F_{\kappa\lambda}}{\delta g^{\mu\nu}} = \\
&= -g^{\alpha}_{\mu}g^{\kappa}_{\nu}g^{\beta\lambda}F_{\lambda\kappa} - g^{\alpha\kappa}g^{\beta}_{\mu}g^{\lambda}_{\nu}F_{\lambda\kappa} + g^{\alpha\kappa}g^{\beta\lambda}\frac{\delta F_{\kappa\lambda}}{\delta g^{\mu\nu}} = \\
&= -g^{\alpha}_{\mu}g^{\kappa}_{\nu}F^{\beta}_{\kappa} - g^{\alpha\kappa}g^{\beta}_{\mu}F_{\nu\kappa} + g^{\alpha\kappa}g^{\beta\lambda}\frac{\delta F_{\kappa\lambda}}{\delta g^{\mu\nu}} = \\
&= -g^{\alpha}_{\mu}F^{\beta}_{\nu} - g^{\beta}_{\mu}g^{\alpha\kappa}F_{\nu\kappa} + g^{\alpha\kappa}g^{\beta\lambda}\frac{\delta F_{\kappa\lambda}}{\delta g^{\mu\nu}} = \\
&= g^{\alpha}_{\mu}F^{\beta}_{\nu} + g^{\beta}_{\mu}g^{\alpha\kappa}F_{\kappa\nu} + g^{\alpha\kappa}g^{\beta\lambda}\frac{\delta F_{\kappa\lambda}}{\delta g^{\mu\nu}} = \\
&= g^{\alpha}_{\mu}F^{\beta}_{\nu} + g^{\beta}_{\mu}F^{\alpha}_{\nu} + g^{\alpha\kappa}g^{\beta\lambda}\frac{\delta F_{\kappa\lambda}}{\delta g^{\mu\nu}}.
\end{aligned}
\tag{13.93}
$$

For this reason, one obtains

$$
\begin{aligned}
F_{\alpha\beta}\frac{\delta F^{\alpha\beta}}{\delta g^{\mu\nu}} &= F_{\alpha\beta}\left[g^{\alpha}_{\mu}F^{\beta}_{\nu}+g^{\beta}_{\mu}F^{\alpha}_{\nu}+g^{\alpha\kappa}g^{\beta\lambda}\frac{\delta F_{\kappa\lambda}}{\delta g^{\mu\nu}}\right]= \\
&= \left(F_{\alpha\beta}g^{\alpha}_{\mu}\right)F^{\beta}_{\nu}+\left(F_{\alpha\beta}g^{\beta}_{\mu}\right)F^{\alpha}_{\nu}+\left(F_{\alpha\beta}g^{\alpha\kappa}g^{\beta\lambda}\right)\frac{\delta F_{\kappa\lambda}}{\delta g^{\mu\nu}}= \\
&= \left(-F_{\beta\alpha}g^{\alpha}_{\mu}\right)F^{\beta}_{\nu}-F_{\alpha\mu}F^{\alpha}_{\nu}-\left(F_{\beta\alpha}g^{\alpha\kappa}g^{\beta\lambda}\right)\frac{\delta F_{\kappa\lambda}}{\delta g^{\mu\nu}}= \\
&= -F_{\beta\mu}F^{\beta}_{\nu}+F^{\alpha}_{\nu}F_{\alpha\mu}-\left(F^{\kappa}_{\beta}g^{\beta\lambda}\right)\frac{\delta F_{\kappa\lambda}}{\delta g^{\mu\nu}}= \\
&= F^{\beta}_{\nu}F_{\beta\mu}+F^{\alpha}_{\nu}F_{\alpha\mu}-F^{\kappa\lambda}\frac{\delta F_{\kappa\lambda}}{\delta g^{\mu\nu}}= \\
&= 2F^{\alpha}_{\nu}F_{\alpha\mu}-2F^{\alpha}_{\mu}F_{\alpha\nu}=0,
\end{aligned}
\tag{13.94}
$$

where we have applied the identity

$$
F^{\alpha}_{\nu}F_{\alpha\mu}=g^{\alpha\beta}F_{\beta\nu}F_{\alpha\mu}=g^{\beta\alpha}F_{\alpha\mu}F_{\beta\nu}=F^{\beta}_{\mu}F_{\beta\nu}=F^{\alpha}_{\mu}F_{\alpha\nu}. \tag{13.95}
$$

Taking into account (13.91) and (13.94), one receives

$$
\frac{\delta F_{\alpha\beta}}{\delta g^{\mu\nu}}F^{\alpha\beta}+F_{\alpha\beta}\frac{\delta F^{\alpha\beta}}{\delta g^{\mu\nu}}=-2F^{\alpha}_{\mu}F_{\alpha\nu}. \tag{13.96}
$$

For this reason, one obtains

$$
\frac{\delta\mathscr{L}}{\delta g^{\mu\nu}}=-\frac{1}{4\mu_0}\left(-2F^{\alpha}_{\mu}F_{\alpha\nu}\right)=\frac{1}{2\mu_0}F^{\alpha}_{\mu}F_{\alpha\nu}=-\frac{1}{2\mu_0}F^{\alpha}_{\mu}F_{\alpha\nu}, \tag{13.97}
$$

where in the last step we have changed the order of indexes in F^{α}_{μ}, which is invisible by only the inconvenient notation. In this manner, one receives finally

$$
T^{em}_{\mu\nu}=-2\frac{\delta\mathscr{L}}{\delta g^{\mu\nu}}+g_{\mu\nu}\mathscr{L}=\frac{1}{\mu_0}\left(F^{\alpha}_{\mu}F_{\alpha\nu}-\frac{1}{4}g_{\mu\nu}F_{\alpha\beta}F^{\alpha\beta}\right), \tag{13.98}
$$

which coincides with the stress-energy tensor of the electromagnetic field (13.82) given *ad hoc* in analogy with the case of the flat space-time.

For the case of the Kerr–Newman space-time, in general, the spatial Ricci scalar curvature and energy density of Matter fields are manifestly non-zero. Let us calculate them straightforwardly. The spatial Ricci curvature tensor can be taken from (13.79) as $R_{ij} = \kappa T^{em}_{ij}$, where

$$T^{em}_{ij} = \frac{1}{\mu_0}\left(F_i^{\alpha}F_{\alpha j} - \frac{1}{4}h_{ij}F_{\alpha\beta}F^{\alpha\beta}\right), \tag{13.99}$$

is the spatial part of the stress-energy tensor for the electromagnetic field. If one takes into account the metric of the Minkowski space-time $\eta_{\mu\nu}$, then the spatial part of the stress-energy tensor of the electromagnetic field is

$$T^{em}_{ij}\Big|_{g_{\mu\nu}=\eta_{\mu\nu}} = -\sigma^{flat}_{ij}, \tag{13.100}$$

where σ^{flat}_{ij} is the Maxwell stress tensor of the electromagnetic field

$$\sigma^{flat}_{ij} = \varepsilon_0 E_i E_j + \frac{1}{\mu_0}B_i B_j - \frac{1}{2}\left(\varepsilon_0 \vec{E}^2 + \frac{1}{\mu_0}\vec{B}^2\right)\delta_{ij}. \tag{13.101}$$

Consequently, the straightforward generalization of the Maxwell stress tensor of the electromagnetic field to the case of non-flat space-time is the tensor

$$\sigma_{ij} = -\frac{1}{\mu_0}\left(F_i^{\alpha}F_{\alpha j} - \frac{1}{4}h_{ij}F_{\alpha\beta}F^{\alpha\beta}\right), \tag{13.102}$$

which we shall call the the curved-space Maxwell stress tensor of the electromagnetic field. Therefore, the three-dimensional Ricci scalar curvature ${}^{(3)}R = h^{ij}R_{ij}$ up to the minus sign becomes the curved-space Maxwell stress tensor of the electromagnetic field projected onto the induced metric

$$\begin{aligned} {}^{(3)}R &= -\kappa h^{ij}\sigma_{ij} = \kappa h^{ij}\frac{1}{\mu_0}\left(F_i^{\alpha}F_{\alpha j} - \frac{1}{4}h_{ij}F_{\alpha\beta}F^{\alpha\beta}\right) = \\ &= \frac{\kappa}{\mu_0}\left(h^{ij}F_i^{\alpha}F_{\alpha j} - \frac{3}{4}F_{\alpha\beta}F^{\alpha\beta}\right). \end{aligned} \tag{13.103}$$

where we have used the identity $h^{ij}h_{ij} = 3$. Using the transformation

$$h^{ij}F_i^{\alpha}F_{\alpha j} = h^{ij}h_{ik}F^{k\alpha}F_{\alpha j} = F^{j\alpha}F_{\alpha j} = F^{\alpha j}F_{j\alpha}, \tag{13.104}$$

and the definition

$$F_{\alpha\beta}F^{\alpha\beta} = F^{\alpha j}F_{\alpha j} + F^{\alpha 0}F_{\alpha 0} = -F^{\alpha j}F_{j\alpha} - F^{\alpha 0}F_{0\alpha}, \tag{13.105}$$

together with the properties of the Faraday tensor

$$F_{\alpha\beta}F^{\alpha\beta} = 2\left(\vec{B}^2 - \frac{\vec{E}^2}{c^2}\right), \tag{13.106}$$

$$F_{\alpha 0}F^{0\alpha} = \frac{\vec{E}^2}{c^2}, \tag{13.107}$$

one obtains

$$h^{ij}F_i^{\alpha}F_{\alpha j} = -F_{\alpha\beta}F^{\alpha\beta} - F^{\alpha 0}F_{0\alpha} = -2\left(\vec{B}^2 - \frac{\vec{E}^2}{c^2}\right) - \frac{\vec{E}^2}{c^2} = 2\vec{B}^2 + \frac{\vec{E}^2}{c^2}. \tag{13.108}$$

Collecting all together, one receives finally

$${}^{(3)}R = \frac{\kappa}{\mu_0}\left[2\vec{B}^2 + \frac{\vec{E}^2}{c^2} - \frac{3}{2}\left(\vec{B}^2 - \frac{\vec{E}^2}{c^2}\right)\right] = \frac{\kappa}{2}\left(\frac{1}{\mu_0}\vec{B}^2 + 5\varepsilon_0\vec{E}^2\right). \tag{13.109}$$

Similarly, one can calculate the energy density of the electromagnetic field. Application of the definition $n^\mu = n^\beta g^\mu_\beta$ leads to the result

$$\begin{aligned} \rho &= T_{\mu\nu}n^\mu n^\nu = \frac{1}{\mu_0}\left(F_{\mu\alpha}g^{\alpha\beta}F_{\nu\beta} - \frac{1}{4}g_{\mu\nu}F_{\alpha\beta}F^{\alpha\beta}\right)n^\mu n^\nu = \\ &= \frac{1}{\mu_0}\left(F_{\mu\alpha}(g^{\alpha\beta}n^\mu n^\nu)F_{\nu\beta} - \frac{1}{4}g_{\mu\nu}n^\mu n^\nu F_{\alpha\beta}F^{\alpha\beta}\right). \end{aligned} \tag{13.110}$$

The second term in the formula (13.110) can be easily transformed using the identity $g_{\mu\nu}n^\mu n^\nu = n^\mu n_\mu = -1$. The first term, however, is not so easy to transform. Let us notice that

$$g^{\alpha\beta}n^\mu n^\nu = g^{\alpha\mu}g^{\nu\beta}(g_{\mu\nu}n^\mu n^\nu) = -g^{\alpha\mu}g^{\nu\beta}, \tag{13.111}$$

where we have applied $g_{\mu\nu}n^\mu n^\nu = n_\nu n^\nu = -1$. In this manner,

$$\begin{aligned} F_{\mu\alpha}(g^{\alpha\beta}n^\mu n^\nu)F_{\nu\beta} &= -F_{\mu\alpha}(g^{\alpha\mu}g^{\nu\beta})F_{\nu\beta} = F_{\mu\alpha}(g^{\alpha\mu}g^{\nu\beta})F_{\beta\nu} = \\ &= F^\mu_\mu F^\nu_\nu = F_{\mu\nu}(g^{\nu\mu}g_{\nu\mu})F^{\mu\nu} = \\ &= F_{\mu\nu}(g^{\nu\mu}g_{\mu\nu})F^{\mu\nu} = 4F_{\mu\nu}F^{\mu\nu}, \end{aligned} \tag{13.112}$$

where we have applied the identities $F^{\mu}_{\mu} = F_{\mu\nu}g^{\nu\mu}$, $F^{\nu}_{\nu} = g_{\nu\mu}F^{\mu\nu}$, and $g^{\nu\mu}g_{\mu\nu} = 4$. In this manner, one receives finally the energy density

$$\begin{aligned} \rho &= \frac{1}{\mu_0}\left(4F_{\alpha\beta}F^{\alpha\beta} + \frac{1}{4}F_{\alpha\beta}F^{\alpha\beta}\right) = \frac{17}{4\mu_0}F_{\alpha\beta}F^{\alpha\beta} = \\ &= \frac{17}{2\mu_0}\left(\vec{B}^2 - \frac{\vec{E}^2}{c^2}\right) = \frac{17}{2}\left(\frac{1}{\mu_0}\vec{B}^2 - \varepsilon_0\vec{E}^2\right), \end{aligned} \tag{13.113}$$

where we have applied the property (13.106) of the Faraday tensor. Consequently, the generalized gravitational potential is

$$\begin{aligned} V = -{}^{(3)}R + 2\kappa\rho = -\frac{\kappa}{2}\left(\frac{1}{\mu_0}\vec{B}^2 + 5\varepsilon_0\vec{E}^2\right) + 17\kappa\left(\frac{1}{\mu_0}\vec{B}^2 - \varepsilon_0\vec{E}^2\right) = \\ = \frac{33\kappa}{2}\left(\frac{1}{\mu_0}\vec{B}^2 - \frac{13}{11}\varepsilon_0\vec{E}^2\right). \end{aligned} \tag{13.114}$$

Now it is easy to see that V averaged over midisuperspace is

$$\langle V\rangle = \frac{33\kappa}{2}\left(\frac{1}{\mu_0}\langle\vec{B}^2\rangle - \frac{13}{11}\varepsilon_0\langle\vec{E}^2\rangle\right), \tag{13.115}$$

where $\langle\vec{B}^2\rangle$ and $\langle\vec{E}^2\rangle$ are the midisuperspace means of the squared fields $\vec{B}$ and $\vec{E}$, respectively. One has explicitly

$$\langle\vec{B}^2\rangle = \frac{1}{\Omega(\xi)}\int \vec{B}^2\delta\xi, \tag{13.116}$$

$$\langle\vec{E}^2\rangle = \frac{1}{\Omega(\xi)}\int \vec{E}^2\delta\xi. \tag{13.117}$$

Because, in general, the fields $\vec{B}$ and $\vec{E}$ are functions of space-time coordinates, it is evident that both averages (13.116) and (13.117) must be treated as functions of space-time coordinates, of course after performing in all the functional integrals the suitable transformation from midisuperspace to space-time $\xi \to \xi(x)$. In other words, one has

$$\langle V\rangle(\xi,\phi) = \langle V\rangle(x). \tag{13.118}$$

Because the considerations presented above are independent of the concrete form of the electrovacuum solution, they can be applied to any solution of the Einstein–Maxwell equations. Interestingly, one can *ad hoc* add the cosmological constant

$$\langle V\rangle = \frac{33\kappa}{2}\left(\frac{1}{\mu_0}\langle\vec{B}^2\rangle - \frac{13}{11}\varepsilon_0\langle\vec{E}^2\rangle\right) + 2\Lambda, \tag{13.119}$$

so that the averaged generalized gravitational potential can be presented by two equivalent ways

$$\begin{aligned}\langle V\rangle &= \frac{33\kappa}{2}\left(\frac{1}{\mu_0}\langle\vec{B}^2\rangle - \varepsilon_0\langle\vec{E}^2\rangle\right) + 2\Lambda - 3\kappa\varepsilon_0\langle\vec{E}^2\rangle = &(13.120)\\ &= \frac{39\kappa}{2}\left(\frac{1}{\mu_0}\langle\vec{B}^2\rangle - \varepsilon_0\langle\vec{E}^2\rangle\right) + 2\Lambda - \frac{3\kappa}{\mu_0}\langle\vec{B}^2\rangle. &(13.121)\end{aligned}$$

If the averages $\langle\vec{B}^2\rangle = B_\Lambda^2$ and $\langle\vec{E}^2\rangle = E_\Lambda^2$ are constant then one can take

$$2\Lambda = 3\kappa\varepsilon_0 E_\Lambda^2 = \frac{3\kappa}{\mu_0}B_\Lambda^2, \tag{13.122}$$

and the averaged generalized gravitational potential vanishes

$$\langle V\rangle = 0. \tag{13.123}$$

The condition (13.122, however, leads to the relation

$$E_\Lambda = cB_\Lambda, \tag{13.124}$$

which can be used to define the speed of light

$$c := \frac{E_\Lambda}{B_\Lambda}, \tag{13.125}$$

if one knows E_Λ and B_Λ.

Let us return to the Kerr–Newman space-time. The spatial part of the Kerr–Newman metric has the form

$$h_{ij} = \begin{bmatrix} \frac{r^2+\alpha^2\cos^2\theta}{\Delta} & 0 & 0 \\ 0 & r^2+\alpha^2\cos^2\theta & 0 \\ 0 & 0 & \frac{(r^2+\alpha^2)^2-\alpha^2\Delta\sin^2\theta}{r^2+\alpha^2\cos^2\theta}\sin^2\theta \end{bmatrix}, \tag{13.126}$$

where $\Delta = r^2\alpha^2 - r_S r + r_Q^2$ and, for this reason, the global dimension can be deduced easy

$$h = \left(r^2+\alpha^2\cos^2\theta\right)\left[r^2+\alpha^2\cos^2\theta+\frac{(r^2+\alpha^2)(r_S r - r_Q^2)}{r^2+\alpha^2-r_S r+r_Q^2}\right]\sin^2\theta, \quad (13.127)$$

so that the invariant global dimension is

$$\xi = \frac{2}{\kappa\sqrt{6}}\sqrt{\left(r^2+\alpha^2\cos^2\theta\right)\left[r^2+\alpha^2\cos^2\theta+\frac{(r^2+\alpha^2)(r_S r - r_Q^2)}{r^2+\alpha^2-r_S r+r_Q^2}\right]}\sin\theta. \quad (13.128)$$

The volume of the Kerr-Newman midisuperspace is $\Omega(\xi) = \xi$. Here r_S is the Schwarzschild radius and r_Q is a length-scale corresponding to the electric charge Q of the mass

$$r_Q^2 = \frac{Q^2 G}{4\pi\varepsilon_0 c^4} = \frac{Q^2}{4\pi\varepsilon_0}\frac{\kappa}{8\pi}. \quad (13.129)$$

There are three possible situations. Namely, when the averaged generalized gravitational potential (13.115) is 1° vanishing, 2° positive, 3° negative. The first case is rather trivial. $\langle V\rangle$ vanishes if and only if

$$\langle\vec{B}^2\rangle = \frac{13}{11}\frac{\langle\vec{E}^2\rangle}{c^2}, \quad (13.130)$$

which in fact means that

$$|\vec{B}| = \sqrt{\frac{13}{11}}\frac{|\vec{E}|}{c} \approx 1.087\frac{|\vec{E}|}{c}. \quad (13.131)$$

In such a situation, one can construct straightforwardly and easily the wave functions of the Kerr–Newman space-time

$$\Psi_1^{KN} = \frac{2\sin\theta}{\kappa\sqrt{|A|}\sqrt{6}}\sqrt{\left(r^2+\alpha^2\cos^2\theta\right)\left[r^2+\alpha^2\cos^2\theta+\frac{(r^2+\alpha^2)(r_S r - r_Q^2)}{r^2+\alpha^2-r_s r+r_Q^2}\right]}, \quad (13.132)$$

$$\Psi_2^{KN} = \frac{1}{\sqrt{|B|}}, \quad (13.133)$$

where the integration constants A and B are

$$A = \frac{\Xi^3}{3}, \tag{13.134}$$

$$B = \frac{1}{\Gamma_0}\Xi, \tag{13.135}$$

where Ξ is the reference constant

$$\Xi = \frac{2}{\kappa\sqrt{6}}\sqrt{(R^2+\alpha^2\cos^2\Theta)\left[R^2+\alpha^2\cos^2\Theta+\frac{(R^2+\alpha^2)(r_S R - r_Q^2)}{R^2+\alpha^2-r_s R+r_Q^2}\right]}\sin\Theta. \tag{13.136}$$

The second situation, *i.e.* positive $\langle V\rangle$, is defined for

$$\langle\vec{B}^2\rangle > \frac{13}{11}\frac{\langle\vec{E}^2\rangle}{c^2}, \tag{13.137}$$

which can be presented in the form

$$|\vec{B}| > \sqrt{\frac{13}{11}}\frac{|\vec{E}|}{c}. \tag{13.138}$$

In such a situation, the integration constants A and B are not easy to calculate for general fields $\vec{B}$ and $\vec{E}$, but can be presented in the compact form

$$A = \int\left[\frac{\sin\left(\xi\sqrt{\frac{33\kappa}{2}\left(\frac{1}{\mu_0}\langle\vec{B}^2\rangle-\frac{13}{11}\varepsilon_0\langle\vec{E}^2\rangle\right)}\right)}{\sqrt{\frac{33\kappa}{2}\left(\frac{1}{\mu_0}\langle\vec{B}^2\rangle-\frac{13}{11}\varepsilon_0\langle\vec{E}^2\rangle\right)}}\right]^2\delta\xi, \tag{13.139}$$

$$B = \int\cos^2\left(\xi\sqrt{\frac{33\kappa}{2}\left(\frac{1}{\mu_0}\langle\vec{B}^2\rangle-\frac{13}{11}\varepsilon_0\langle\vec{E}^2\rangle\right)}\right)\delta\xi, \tag{13.140}$$

and the wave functions of the Kerr–Newman space-time are

$$\Psi_1^{KN} = \frac{1}{\sqrt{|A|}} \frac{\sin\left(\xi\sqrt{\frac{33\kappa}{2}\left(\frac{1}{\mu_0}\langle\vec{B}^2\rangle - \frac{13}{11}\varepsilon_0\langle\vec{E}^2\rangle\right)}\right)}{\sqrt{\frac{33\kappa}{2}\left(\frac{1}{\mu_0}\langle\vec{B}^2\rangle - \frac{13}{11}\varepsilon_0\langle\vec{E}^2\rangle\right)}}, \quad (13.141)$$

$$\Psi_2^{KN} = \frac{1}{\sqrt{|B|}} \cos\left(\xi\sqrt{\frac{33\kappa}{2}\left(\frac{1}{\mu_0}\langle\vec{B}^2\rangle - \frac{13}{11}\varepsilon_0\langle\vec{E}^2\rangle\right)}\right). \quad (13.142)$$

The case of the negative averaged generalized gravitational potential can be considered analogously. In this instance

$$|\vec{B}| < \sqrt{\frac{13}{11}}\frac{|\vec{E}|}{c}. \quad (13.143)$$

So that the wave functions of the Kerr–Newman space-time are

$$\Psi_1^{KN} = \frac{1}{\sqrt{|A|}} \frac{\sinh\left(\xi\sqrt{\frac{33\kappa}{2}\left|\frac{1}{\mu_0}\langle\vec{B}^2\rangle - \frac{13}{11}\varepsilon_0\langle\vec{E}^2\rangle\right|}\right)}{\sqrt{\frac{33\kappa}{2}\left|\frac{1}{\mu_0}\langle\vec{B}^2\rangle - \frac{13}{11}\varepsilon_0\langle\vec{E}^2\rangle\right|}}, \quad (13.144)$$

$$\Psi_2^{KN} = \frac{1}{\sqrt{|B|}} \cosh\left(\xi\sqrt{\frac{33\kappa}{2}\left|\frac{1}{\mu_0}\langle\vec{B}^2\rangle - \frac{13}{11}\varepsilon_0\langle\vec{E}^2\rangle\right|}\right). \quad (13.145)$$

where the the integration constants A and B are

$$A = \int \left[\frac{\sinh\left(\xi\sqrt{\frac{33\kappa}{2}\left|\frac{1}{\mu_0}\langle\vec{B}^2\rangle - \frac{13}{11}\varepsilon_0\langle\vec{E}^2\rangle\right|}\right)}{\sqrt{\frac{33\kappa}{2}\left|\frac{1}{\mu_0}\langle\vec{B}^2\rangle - \frac{13}{11}\varepsilon_0\langle\vec{E}^2\rangle\right|}}\right]^2 \delta\xi, \quad (13.146)$$

$$B = \int \cosh^2\left(\xi\sqrt{\frac{33\kappa}{2}\left|\frac{1}{\mu_0}\langle\vec{B}^2\rangle - \frac{13}{11}\varepsilon_0\langle\vec{E}^2\rangle\right|}\right)\delta\xi, \quad (13.147)$$

H Reissner–Nordström Space-Time

Another solution of the electrovacuum Einstein field equations is the Reissner–Nordström metric describing a spherically symmetric static massive charged object. Therefore, in this case one has also

$$\langle V\rangle = \frac{33\kappa}{2}\left(\frac{1}{\mu_0}\langle\vec{B}^2\rangle - \frac{13}{11}\varepsilon_0\langle\vec{E}^2\rangle\right), \tag{13.148}$$

and the metric is changed. The spatial part of Reissner–Nordström space-time metric is

$$h_{ij} = \begin{bmatrix} \left(1-\frac{r_S}{r}+\frac{r_Q^2}{r^2}\right)^{-1} & 0 & 0 \\ 0 & r^2 & 0 \\ 0 & 0 & r^2\sin^2\theta \end{bmatrix}, \tag{13.149}$$

so that the global dimension is

$$h = \frac{r^4\sin^2\theta}{1-\frac{r_S}{r}+\frac{r_Q^2}{r^2}} \tag{13.150}$$

while the invariant global dimension has the form

$$\xi = \frac{2}{\kappa\sqrt{6}}\frac{r^2\sin\theta}{\sqrt{1-\frac{r_S}{r}+\frac{r_Q^2}{r^2}}} \tag{13.151}$$

and the volume of the Reissner–Nordström midisuperspace is $\Omega(\xi) = \xi$. There are three possible situations defined by absolute values of the electric and magnetic fields. For the case of negative $\langle V\rangle$ the wave functions of the Reissner–Nordström space-time are

$$\Psi_1^{RN} = \frac{1}{\sqrt{|A|}}\frac{\sinh\left(\xi\sqrt{\frac{33\kappa}{2}\left|\frac{1}{\mu_0}\langle\vec{B}^2\rangle - \frac{13}{11}\varepsilon_0\langle\vec{E}^2\rangle\right|}\right)}{\sqrt{\frac{33\kappa}{2}\left|\frac{1}{\mu_0}\langle\vec{B}^2\rangle - \frac{13}{11}\varepsilon_0\langle\vec{E}^2\rangle\right|}}, \tag{13.152}$$

$$\Psi_2^{RN} = \frac{1}{\sqrt{|B|}}\cosh\left(\xi\sqrt{\frac{33\kappa}{2}\left|\frac{1}{\mu_0}\langle\vec{B}^2\rangle - \frac{13}{11}\varepsilon_0\langle\vec{E}^2\rangle\right|}\right). \tag{13.153}$$

where the the integration constants A and B are

$$A = \int \left[\frac{\sinh\left(\xi\sqrt{\frac{33\kappa}{2}\left|\frac{1}{\mu_0}\langle\vec{B}^2\rangle - \frac{13}{11}\varepsilon_0\langle\vec{E}^2\rangle\right|}\right)}{\sqrt{\frac{33\kappa}{2}\left|\frac{1}{\mu_0}\langle\vec{B}^2\rangle - \frac{13}{11}\varepsilon_0\langle\vec{E}^2\rangle\right|}} \right]^2 \delta\xi, \tag{13.154}$$

$$B = \int \cosh^2\left(\xi\sqrt{\frac{33\kappa}{2}\left|\frac{1}{\mu_0}\langle\vec{B}^2\rangle - \frac{13}{11}\varepsilon_0\langle\vec{E}^2\rangle\right|}\right)\delta\xi. \tag{13.155}$$

When $\langle V\rangle$ is positive then

$$A = \int \left[\frac{\sin\left(\xi\sqrt{\frac{33\kappa}{2}\left(\frac{1}{\mu_0}\langle\vec{B}^2\rangle - \frac{13}{11}\varepsilon_0\langle\vec{E}^2\rangle\right)}\right)}{\sqrt{\frac{33\kappa}{2}\left(\frac{1}{\mu_0}\langle\vec{B}^2\rangle - \frac{13}{11}\varepsilon_0\langle\vec{E}^2\rangle\right)}} \right]^2 \delta\xi, \tag{13.156}$$

$$B = \int \cos^2\left(\xi\sqrt{\frac{33\kappa}{2}\left(\frac{1}{\mu_0}\langle\vec{B}^2\rangle - \frac{13}{11}\varepsilon_0\langle\vec{E}^2\rangle\right)}\right)\delta\xi, \tag{13.157}$$

and the wave functions of the Reissner–Nordström space-time are

$$\Psi_1^{RN} = \frac{1}{\sqrt{|A|}} \frac{\sin\left(\xi\sqrt{\frac{33\kappa}{2}\left(\frac{1}{\mu_0}\langle\vec{B}^2\rangle - \frac{13}{11}\varepsilon_0\langle\vec{E}^2\rangle\right)}\right)}{\sqrt{\frac{33\kappa}{2}\left(\frac{1}{\mu_0}\langle\vec{B}^2\rangle - \frac{13}{11}\varepsilon_0\langle\vec{E}^2\rangle\right)}}, \tag{13.158}$$

$$\Psi_2^{RN} = \frac{1}{\sqrt{|B|}} \cos\left(\xi\sqrt{\frac{33\kappa}{2}\left(\frac{1}{\mu_0}\langle\vec{B}^2\rangle - \frac{13}{11}\varepsilon_0\langle\vec{E}^2\rangle\right)}\right). \tag{13.159}$$

For vanishing $\langle V\rangle$ one has

$$\Psi_1^{RN} = \frac{2}{\kappa\sqrt{|A|}\sqrt{6}} \frac{r^2\sin\theta}{\sqrt{1 - \frac{r_S}{r} + \frac{r_Q^2}{r^2}}}, \tag{13.160}$$

$$\Psi_2^{RN} = \frac{1}{\sqrt{|B|}}, \tag{13.161}$$

where the integration constants A and B are

$$A = \frac{\Xi^3}{3}, \tag{13.162}$$

$$B = \frac{1}{\Gamma_0}\Xi, \tag{13.163}$$

where Ξ is the reference constant

$$\Xi = \frac{2}{\kappa\sqrt{6}} \frac{R^2 \sin\Theta}{\sqrt{1 - \frac{r_S}{R} + \frac{r_Q^2}{R^2}}}. \tag{13.164}$$

I Gödel Space-Time

Let us consider the solution of the Einstein field equations with the presence of the cosmological constant and the stress-energy tensor of dust

$$R_{\mu\nu} - \frac{1}{2} g_{\mu\nu} {}^{(4)}R + \Lambda g_{\mu\nu} = \kappa T_{\mu\nu}, \tag{13.165}$$

$$T_{\mu\nu} = \varepsilon u_\mu u_\nu, \tag{13.166}$$

which is given by the Gödel space-time

$$g_{\mu\nu} = \begin{bmatrix} -a^2 & 0 & 0 & -a^2 e^x \\ 0 & a^2 & 0 & 0 \\ 0 & 0 & a^2 & 0 \\ -a^2 e^x & 0 & 0 & -\frac{1}{2} a^2 e^{2x} \end{bmatrix}, \tag{13.167}$$

$$g^{\mu\nu} = \begin{bmatrix} \frac{1}{a^2} & 0 & 0 & -\frac{2e^{-x}}{a^2} \\ 0 & \frac{1}{a^2} & 0 & 0 \\ 0 & 0 & \frac{1}{a^2} & 0 \\ -\frac{2e^{-x}}{a^2} & 0 & 0 & \frac{2e^{-2x}}{a^2} \end{bmatrix}. \tag{13.168}$$

having the following spatial part

$$h_{ij} = \begin{bmatrix} a^2 & 0 & 0 \\ 0 & a^2 & 0 \\ 0 & 0 & -\frac{1}{2}a^2 e^{2x} \end{bmatrix}, \quad h^{ij} = \begin{bmatrix} \frac{1}{a^2} & 0 & 0 \\ 0 & \frac{1}{a^2} & 0 \\ 0 & 0 & \frac{2e^{-2x}}{a^2} \end{bmatrix}, \tag{13.169}$$

where $a = a(x,y,z)$. In this instance, the global dimension is

$$|h| = \frac{a^6}{2} e^{2x}, \tag{13.170}$$

so the invariant global dimension has the form

$$\xi = \frac{1}{\kappa\sqrt{3}} a^3 e^x, \tag{13.171}$$

and, therefore, the volume of the Gödel midisuperspace is $\Omega(\xi) = \xi$. The Gödel metric satisfies the Einstein field equations for which the cosmological constant is related to the parameter a by the relation

$$\Lambda = -\frac{1}{2a^2}. \tag{13.172}$$

The energy density ε of the dust is

$$\varepsilon = \frac{1}{\kappa a^2}. \tag{13.173}$$

The Gödel metric can be decomposed in the ADM $3+1$ form. It is easy to see that such an *ad hoc* decomposition generates the equations

$$-N^2 + N_i N^i = -a^2, \tag{13.174}$$

$$N_i = \left[0, 0, -a^2 e^x\right]. \tag{13.175}$$

Taking into account the spatial metric (13.169) one receives

$$N^i = h^{ij} N_j = \left[0, 0, -2e^{-x}\right], \tag{13.176}$$

which gives $N_i N^i = 2a^2$ and for this reason

$$N = \sqrt{3}a. \tag{13.177}$$

In this manner, one can establish the normal unit vector field

$$
\begin{aligned}
n^{\mu} &= \left[\frac{1}{N}, -\frac{N^i}{N}\right] = \frac{1}{\sqrt{3}}\left[\frac{1}{a}, 0, 0, \frac{2e^{-x}}{a}\right], && (13.178)\\
n_{\mu} &= [-N, 0_i] = \sqrt{3}\left[-a, 0, 0, 0\right]. && (13.179)
\end{aligned}
$$

The problem is to choose the velocity vector u_μ according to the general rules

$$
\begin{aligned}
u_\mu u^\mu &= -1, && (13.180)\\
u_\mu n^\mu &= -1, && (13.181)\\
u_\mu n^\mu u_\nu n^\nu &= 1. && (13.182)
\end{aligned}
$$

However, it can be seen by straightforward calculation that the following choice

$$
\begin{aligned}
u^{\mu} &= \frac{1}{\sqrt{3}}\left[\frac{1}{a}, 0, 0, 0\right], && (13.183)\\
u_{\mu} &= g_{\mu\nu}u^{\nu} = \sqrt{3}\left[-a, 0, 0, -ae^{x}\right], && (13.184)
\end{aligned}
$$

satisfies the conditions (13.180)-(13.182). The contraction of the Einstein field equations (13.165)-(13.166) with metric $g^{\mu\nu}$ leads to

$$
{}^{(4)}R = 4\Lambda + \kappa\varepsilon, \tag{13.185}
$$

so that the equations (13.165)-(13.166) can be presented in the form

$$
R_{\mu\nu} = \Lambda g_{\mu\nu} + \kappa\varepsilon\left(u_\mu u_\nu + \frac{1}{2}g_{\mu\nu}\right), \tag{13.186}
$$

which after including the fact that for the Gödel Universe one has

$$
\kappa\varepsilon = -2\Lambda, \tag{13.187}
$$

one receives the four-dimensional Ricci curvature tensor

$$
R_{\mu\nu} = \kappa\varepsilon u_\mu u_\nu = -2\Lambda u_\mu u_\nu. \tag{13.188}
$$

In this manner, the three-dimensional Ricci curvature tensor has the form

$$
R_{ij} = -2\Lambda u_i u_j, \tag{13.189}
$$

and, consequently, the three-dimensional Ricci scalar curvature is

$$^{(3)}R = h^{ij}R_{ij} = -2\Lambda u_i u^i = 0. \tag{13.190}$$

Similarly, one can establish the energy density of Matter fields

$$\rho = T_{\mu\nu}n^{\mu}n^{\nu} = \varepsilon u_{\mu}n^{\mu}u_{\nu}n^{\nu} = \varepsilon = -2\frac{\Lambda}{\kappa}. \tag{13.191}$$

In this manner, the generalized gravitational potential for the Gödel Universe has the following form

$$V = -^{(3)}R + 2\Lambda + 2\kappa\rho = -2\Lambda = \kappa\varepsilon = \frac{1}{a^2} > 0, \tag{13.192}$$

so that

$$\langle V\rangle = \frac{1}{a^3e^x}\int\frac{d(a^3e^x)}{a^2} = \frac{1}{a^3e^x}\int\frac{3a^2dae^x + a^3e^xdx}{a^2} = \tag{13.193}$$

$$= \frac{1}{a^3e^x}\int(3e^xda + ae^xdx), \tag{13.194}$$

and, for this reason,

$$\sqrt{\langle V\rangle}(\xi,\phi) = \frac{1}{a^{3/2}e^{x/2}}\sqrt{\int(3e^xda + ae^xdx)}. \tag{13.195}$$

In this manner, the wave functions of the Gödel dust Universe are

$$\Psi_1^G = \frac{1}{\sqrt{|A|}}\frac{\sin\left(\frac{a^{3/2}e^{x/2}}{\kappa\sqrt{3}}\sqrt{\int(3e^xda + ae^xdx)}\right)}{\frac{1}{a^{3/2}e^{x/2}}\sqrt{\int(3e^xda + ae^xdx)}}, \tag{13.196}$$

$$\Psi_2^G = \frac{1}{\sqrt{|B|}}\cos\left(\frac{a^{3/2}e^{x/2}}{\kappa\sqrt{3}}\sqrt{\int(3e^xda + ae^xdx)}\right), \tag{13.197}$$

where $\mathrm{sgn}(a) \neq 0$, and the integration constants A and B are

$$A = \frac{1}{2\sqrt{3}S_P}\int\left(\frac{\sin\left(\frac{a^{3/2}e^{x/2}}{\kappa\sqrt{3}}\sqrt{\int(3e^xda + ae^xdx)}\right)}{\frac{1}{a^{3/2}e^{x/2}}\sqrt{\int(3e^xda + ae^xdx)}}\right)^2 d(a^3e^x), \tag{13.198}$$

$$B = \frac{1}{\Gamma_0}\frac{1}{2\sqrt{3}S_P}\int\cos^2\left(\frac{a^{3/2}e^{x/2}}{\kappa\sqrt{3}}\sqrt{\int(3e^xda + ae^xdx)}\right)d(a^3e^x). \tag{13.199}$$

Of course, when $a = const$ the situation is much more simpler, but we will not derive these particular results.

J Einstein–Rosen Space-Time

The example of vacuum solution is also the Einstein–Rosen space-time describing the cylindrical gravitational waves having the following form

$$h_{ij} = \begin{bmatrix} e^{2\gamma-2\psi} & 0 & 0 \\ 0 & e^{2\psi} & 0 \\ 0 & 0 & r^2 e^{-2\psi} \end{bmatrix}, \tag{13.200}$$

where r is the radial distance from the z axis, and $\psi = \psi(t,r)$ and $\gamma = \gamma(t,r)$ are functions satisfying the equations

$$\ddot{\psi} = \frac{\psi'}{r} + \psi'', \tag{13.201}$$

$$\frac{\gamma'}{r} = \psi'^2 + \dot{\psi}^2, \tag{13.202}$$

$$\frac{\dot{\gamma}}{r} = 2\dot{\psi}\psi'. \tag{13.203}$$

In other words, when one solves the equation (13.201) then the function γ can be established by

$$\gamma = \int \left[r\left(\psi'^2 + \dot{\psi}^2\right) dr + 2r\dot{\psi}\psi' dt \right]. \tag{13.204}$$

In this manner, the global dimension for the Einstein–Rosen space-time is

$$h = r^2 e^{2\gamma-2\psi}, \tag{13.205}$$

while the invariant global dimension has the form

$$\xi = \frac{1}{\sqrt{6}S_P} r e^{\gamma-\psi}. \tag{13.206}$$

The generalized gravitational potential vanishes in this case, so that the wave functions are

$$\Psi_1^{ER} = \frac{1}{\sqrt{|A|}} \frac{2}{\kappa\sqrt{6}} r e^{\gamma-\psi}, \tag{13.207}$$

$$\Psi_2^{ER} = \frac{1}{\sqrt{|B|}}, \tag{13.208}$$

where the integration constants A and B are determined by the relations

$$A = \frac{\Xi^3}{3}, \tag{13.209}$$

$$B = \frac{1}{\Gamma_0}\Xi, \tag{13.210}$$

where Ξ is the reference constant

$$\Xi = \frac{2}{\kappa\sqrt{6}} R e^{\gamma(T,R)-\psi(T,R)}, \tag{13.211}$$

where T and R are reference values of t and R, respectively.

K Taub–Newman–Unti–Tamburino Space-Time

Let us consider the generalized axisymmetric solution of the vacuum Einstein field equations presented in the Weyl canonical coordinates

$$h_{ij} = \begin{bmatrix} e^{2\gamma-2\psi} & 0 & 0 \\ 0 & e^{2\gamma-2\psi} & 0 \\ 0 & 0 & r^2e^{-2\psi} - a^2e^{2\gamma} \end{bmatrix}, \tag{13.212}$$

for which the Einstein field equations are

$$\triangle\psi = 0, \tag{13.213}$$

$$\frac{\gamma'}{r} = \psi'^2 + \dot{\psi}^2, \tag{13.214}$$

$$\frac{\gamma_z'}{r} = 2\dot{\psi}\psi'. \tag{13.215}$$

Here r is the radial distance from the axis of symmetry z. In such a situation, the global dimension is

$$h = e^{4\gamma-4\psi}\left(r^2e^{-2\psi} - a^2e^{2\gamma}\right), \tag{13.216}$$

so that the invariant global dimension is

$$\xi = \frac{2}{\kappa\sqrt{6}} e^{2\gamma-2\psi}\sqrt{r^2e^{-2\psi} - a^2e^{2\gamma}}. \tag{13.217}$$

For this case the cosmological constant as well as Matter fields are absent. Therefore, the three-dimensional Ricci scalar curvature and the generalized gravitational potential are trivial. The wave functions of any static axisymmetric vacuum solution can be established as

$$\Psi_1^{SAVS} = \frac{1}{\sqrt{|A|}}\frac{2}{\kappa\sqrt{6}}e^{2\gamma-2\psi}\sqrt{r^2e^{-2\psi}-a^2e^{2\gamma}}, \quad (13.218)$$

$$\Psi_2^{SAVS} = \frac{1}{\sqrt{|B|}}, \quad (13.219)$$

where the integration constants A and B are

$$A = \frac{\Xi^3}{3}, \quad (13.220)$$

$$B = \frac{1}{\Gamma_0}\Xi, \quad (13.221)$$

where Ξ is the reference constant

$$\Xi = \frac{2}{\kappa\sqrt{6}}e^{2\gamma(R,Z)-2\psi(R,Z)}\sqrt{R^2e^{-2\psi(R,Z)}-a^2(R,Z)e^{2\gamma(R,Z)}}, \quad (13.222)$$

where R and Z are reference values of r and z. Let us see the wave functions of two particular cases.

The Taub–Newman–Unti–Tamburino space-time is the particular case of the static axisymmetric vacuum space-time characterized by

$$e^{2\psi} = \frac{(r_+ + r_-)^2 - (r_S^2 + 4l^2)}{(r_+ + r_- + r_S)^2 + 4l^2}, \quad (13.223)$$

$$e^{2\gamma} = \frac{(r_+ + r_-)^2 - (r_S^2 + 4l^2)}{4r_+r_-}, \quad (13.224)$$

$$a = \frac{2l(r_+ - r_-)}{\sqrt{r_S^2 + 4l^2}}, \quad (13.225)$$

$$r_\pm^2 = r^2 + \left(z \pm \frac{1}{2}\sqrt{r_S^2 + 4l^2}\right)^2, \quad (13.226)$$

and, in this manner, the results for this case can be obtained straightforwardly

$$\begin{aligned}
\Psi_1^{Taub-NUT} &= \frac{1}{\sqrt{|A|}}\frac{2}{\kappa\sqrt{6}}\frac{(r_+ + r_- + r_S)^2 + 4l^2}{4r_+r_-}\left[r^2\frac{(r_+ + r_- + r_S)^2 + 4l^2}{(r_+ + r_-)^2 - (r_S^2 + 4l^2)} + \right. \\
&+ \left.\frac{l^2(r_+ - r_-)^2}{r_+r_-}\left(1 - \frac{(r_+ + r_-)^2}{r_S^2 + 4l^2}\right)\right]^{1/2}, \qquad (13.227) \\
\Psi_2^{Taub-NUT} &= \frac{1}{\sqrt{|B|}}, \qquad (13.228)
\end{aligned}$$

where the integration constants A and B are

$$\begin{aligned}
A &= \frac{1}{9216\kappa^3\sqrt{6}}\frac{((R_+ + R_- + r_S)^2 + 4l^2)^3}{R_+^3R_-^3}\left[r^2\frac{(R_+ + R_- + r_S)^2 + 4l^2}{(R_+ + R_-)^2 - (r_S^2 + 4l^2)} + \right. \\
&+ \left.\frac{l^2(R_+ - R_-)^2}{R_+R_-}\left(1 - \frac{(R_+ + R_-)^2}{r_S^2 + 4l^2}\right)\right]^{3/2}, \qquad (13.229) \\
B &= \frac{2}{\Gamma_0\kappa\sqrt{6}}\frac{(R_+ + R_- + r_S)^2 + 4l^2}{4R_+R_-}\left[r^2\frac{(R_+ + R_- + r_S)^2 + 4l^2}{(R_+ + R_-)^2 - (r_S^2 + 4l^2)} + \right. \\
&+ \left.\frac{l^2(R_+ - R_-)^2}{R_+R_-}\left(1 - \frac{(R_+ + R_-)^2}{r_S^2 + 4l^2}\right)\right]^{1/2}, \qquad (13.230)
\end{aligned}$$

where

$$R_\pm^2 = R^2 + \left(Z \pm \frac{1}{2}\sqrt{r_S^2 + 4l^2}\right)^2. \qquad (13.231)$$

Chapter 14

Functional Objective Geometry

A Effective Scalar Curvature

Let us assume that the concrete form of the effective gravitational potential V_{eff} (12.2) is fixed *ad hoc* as a functional or function of the global dimension h. In such a situation, one can express the Ricci scalar of a three-dimensional embedded space as follows

$$^{(3)}R = 2\Lambda + 2\kappa\rho - 6\kappa^2 h V_{eff}, \tag{14.1}$$

whereas the global one-dimensional quantum gravity is given by the following evolutionary equation

$$\left(\frac{\delta^2}{\delta h^2} + V_{eff}\right)\Psi[h] = 0. \tag{14.2}$$

In this manner, the quantum gravity is the system describing geometry (14.1), and quantum mechanics (14.2) of an embedded space.

The quantum gravity (14.1)-(14.2) in itself is non-trivial. In fact, this can be expressed by a more general notation

$$^{(3)}R = f[h_{ij}], \tag{14.3}$$

$$\frac{\delta^2\Psi[h]}{\delta h^2} = -V_{eff}[h_{ij}]\Psi[h], \tag{14.4}$$

where both $f[h_{ij}]$ as well as $V_{eff}[h_{ij}]$ are scalar-valued functionals of an induced three-dimensional metric h_{ij}, *i.e.* are *objective functionals*

$$f[h_{ij}] = f[c_1, c_2, c_3], \tag{14.5}$$

$$\Psi[h_{ij}] = \Psi[c_1, c_2, c_3], \tag{14.6}$$

where c_1, c_2, and c_3 are the 3×3 matrix invariants of induced metric h_{ij}

$$c_1 = \mathrm{Tr}\mathbf{h}, \tag{14.7}$$

$$c_2 = \frac{(\mathrm{Tr}\mathbf{h})^2 - \mathrm{Tr}\mathbf{h}^2}{2}, \tag{14.8}$$

$$c_3 = -\det\mathbf{h}, \tag{14.9}$$

which according to the Cayley–Hamilton theorem are the coefficients of the characteristic polynomial of the matrix h_{ij}

$$\mathbf{h}^3 - c_1\mathbf{h}^2 - c_2\mathbf{h} + c_3\mathbf{I}_{3\times3} = 0. \tag{14.10}$$

In this manner, we shall call *the functional objective geometry* the quantum gravity given by system of equations (14.3)-(14.4).

The Ricci scalar curvature (14.1) in itself is highly non-trivial. This quantity is responsible for description of the intrinsic geometry of a three-dimensional embedded space in terms of some kind of effective matter. For this reason, we shall call this curvature by the notion *the effective scalar curvature* and study its physical meaning in this section. For convenience let us present the effective gravitational potential V_{eff} as an algebraic sum of three elementary energetic constituents

$$V_{eff} = V_G + V_C + V_M, \tag{14.11}$$

where the geometric (G), the cosmological (C), and the material (M) contributions are given by the following formulas

$$V_G = -\frac{1}{6\kappa^2}\frac{^{(3)}R}{h}, \tag{14.12}$$

$$V_C = \frac{1}{6\kappa^2}\frac{2\Lambda}{h}, \tag{14.13}$$

$$V_M = \frac{1}{6\kappa^2}\frac{2\kappa}{h}\rho. \tag{14.14}$$

One can list several typical examples of the physical scenarios within the global one-dimensional model of quantum gravity, with respect to the concretely fixed form of the effective gravitational potential V_{eff}.

1. The case of the constant non-vanishing effective gravitational potential $V_{eff} = V_c \neq 0$. In such a situation, the Ricci scalar curvature of the embedded space and the global one-dimensional quantum gravity are given by the equations

$$^{(3)}R = 2\Lambda + 2\kappa\rho - 6\kappa^2 h V_c, \tag{14.15}$$

$$\left(\frac{\delta^2}{\delta h^2} + V_c\right)\Psi_c[h] = 0, \tag{14.16}$$

where $\Psi_c[h]$ is a wave functional related to $V_{eff} = V_c$.

2. The case of the trivial effective gravitational potential $V_{eff} = 0$. In such a situation, the three-dimensional Ricci scalar curvature and the global one-dimensional quantum gravity are

$$^{(3)}R = 2\Lambda + 2\kappa\rho, \tag{14.17}$$

$$\frac{\delta^2}{\delta h^2}\Psi_0[h] = 0, \tag{14.18}$$

where Ψ_0 is a "free" wave functional related to $V_{eff} = 0$.

3. The case when the sum of the geometric and cosmological contributions is identically vanishing $V_G + V_C = 0$, but the material contribution does not vanish identically $V_M \neq 0$. In such a situation, the three-dimensional Ricci scalar curvature and the global one-dimensional quantum gravity are

$$^{(3)}R = 2\Lambda, \tag{14.19}$$

$$\left(\frac{\delta^2}{\delta h^2} - \frac{1}{6\kappa^2}\frac{2\kappa}{h}\rho[h]\right)\Psi_M[h] = 0, \tag{14.20}$$

where Ψ_M is a "material" wave functional related to the material contribution $V_M \neq 0$.

4. The case when the sum of the geometric and material contributions is identically vanishing $V_G + V_M = 0$, while in general the cosmological contribution is non-trivial $V_C \neq 0$. In such a situation, the Ricci scalar curvature of the embedded space and the global one-dimensional quantum gravity are respectively

$$^{(3)}R = 2\kappa\rho, \tag{14.21}$$

$$\left(\frac{\delta^2}{\delta h^2} + \frac{1}{6\kappa^2}\frac{2\Lambda}{h}\right)\Psi_C[h] = 0. \tag{14.22}$$

Here Ψ_C is the "cosmological" wave functional related to the cosmological contribution $V_C \neq 0$.

5. The case when the sum of the cosmological and material contributions is identically vanishing $V_C + V_M = 0$, whereas the geometric contribution is non-zero $V_G \neq 0$. In such a situation, the energy density of Matter fields and the global one-dimensional quantum gravity are given by the equations

$$\rho = -\frac{\Lambda}{\kappa}, \tag{14.23}$$

$$\left(\frac{\delta^2}{\delta h^2} - \frac{1}{6\kappa^2}\frac{{}^{(3)}R}{h}\right)\Psi_G[h] = 0, \tag{14.24}$$

where Ψ_G is the "geometric" wave functional related to the geometric contribution $V_G \neq 0$.

6. A more general explicit form of the effective gravitational potential can be constructed in the spirit of complex analysis (For basics and advances Cf. *e.g* Ref. [933]). Let us consider *ad hoc* functional generalization of the Laurent series expansion in the global dimension h of the effective gravitational potential $V_{eff}[h]$

$$V_{eff}[h] = \sum_{-\infty}^{\infty} a_n (h - h_0)^n, \tag{14.25}$$

in an infinitesimal neighborhood, *i.e.* in a one-sphere (circle) of radius h_ε, of any *ad hoc* fixed initial value h_0 of the global dimension

$$C(h_\varepsilon) = \{h : |h - h_0| < h_\varepsilon\}. \tag{14.26}$$

The numbers a_n are the series coefficients determined by the classical functional integral

$$a_n = \frac{1}{2\pi i}\int_{C(h_\varepsilon)} \frac{V_{eff}[h]}{(h - h_0)^{n+1}}\delta h, \tag{14.27}$$

which is straightforward functional generalization of the Cauchy integral formula with the Lebesgue–Stieltjes measure – the classical functional Radon measure δh.

Let us take into considerations the most general choice of h_0 which similarly to h is assumed to be a complex number. In such a situation, the Ricci scalar curvature of a three-dimensional embedded space takes the following form

$$^{(3)}R = 2\Lambda + 2\kappa\rho - 6\kappa^2 \sum_{-\infty}^{\infty} b_n (h - h_0)^n, \tag{14.28}$$

where b_n is the combined series coefficient

$$b_n = a_{n-1} + h_0 a_n = \frac{1}{2\pi i}\int_{C(h_\varepsilon)} \frac{h}{(h-h_0)^{n+1}} V_{eff}[h]\delta h, \tag{14.29}$$

and the global one-dimensional model of quantum gravity is then defined by the equation

$$\left(\frac{\delta^2}{\delta h^2} + \sum_{-\infty}^{\infty} a_n (h - h_0)^n\right)\Psi[h] = 0. \tag{14.30}$$

Making use of the triangle inequality one can write

$$|b_n| \leqslant |a_{n-1}| + |h_0||a_n| \tag{14.31}$$

so that it can be deduced straightforwardly that

$$\frac{|b_n|}{|a_n|} \leqslant \frac{|a_{n-1}|}{|a_n|} + |h_0|. \tag{14.32}$$

Applying the well-known inequality for any Riemann integral

$$\left|\int f\right| \leqslant \int |f|, \tag{14.33}$$

where f is considered as the Riemann-integrable function and the integral is considered as defined, to the coefficients a_n and b_n one obtains the inequality

$$|a_n| \leqslant \frac{1}{h_\varepsilon^{n+1}}\frac{1}{2\pi}\int_{C(h_\varepsilon)} |V_{eff}|\,\delta h \leqslant \frac{1}{h_\varepsilon^{n+1}}|a_{-1}|, \tag{14.34}$$

where a_{-1} is the residue of the effective gravitational potential at the fixed point $h = h_0$. This residue is determined by the straightforward functional generalization of the Cauchy integral formula

$$a_{-1} = \mathrm{Res}(V_{eff}, h_0) = \frac{1}{2\pi i}\int_{C(h_\varepsilon)} V_{eff}\,\delta h, \tag{14.35}$$

where $C(h_\varepsilon)$ traces out circle around the fixed point h_0 in counterclockwise manner on the punctured disk $D = \{z : 0 < |h - h_0| < R\}$. If the center of the circle $h = h_0$ is a pole of order n, then the residue is defined by the simple limiting procedure

$$\mathrm{Res}(V_{eff}, h_0) = \frac{1}{\Gamma(n)}\lim_{h\to h_0}\frac{\delta^{n-1}}{\delta h^{n-1}}\left((h - h_0)V_{eff}\right). \tag{14.36}$$

Making use of the following reasoning

$$|a_n| = \left|\frac{1}{2\pi i}\int_{C(h_\varepsilon)}\frac{V_{eff}}{(h-h_0)^{n+1}}\right| \leqslant \frac{1}{h_\varepsilon}\left|\frac{1}{2\pi i}\int_{C(h_\varepsilon)}\frac{V_{eff}}{(h-h_0)^{n}}\right| = \frac{1}{h_\varepsilon}|a_{n-1}|, \tag{14.37}$$

one sees that for any n the inequality holds

$$\frac{|a_{n-1}|}{|a_n|} \geqslant h_\varepsilon, \tag{14.38}$$

and hence the inequality (14.32) can be rewritten as

$$\frac{|b_n|}{|a_n|} \geqslant h_\varepsilon + |h_0|. \tag{14.39}$$

In the light of the triangle inequality one can write

$$|b_{n+1}| = |a_n + h_0 a_{n+1}| \leqslant |a_n| + |h_0||a_{n+1}|, \tag{14.40}$$

and because of the relation

$$\frac{|a_{n+1}|}{|a_n|} \leqslant \frac{1}{h_\varepsilon}, \tag{14.41}$$

one receives the following relation

$$\frac{|b_{n+1}|}{|a_n|} \leqslant 1 + \frac{|h_0|}{h_\varepsilon}. \tag{14.42}$$

Finally, application of the inequality (14.39) in the equivalent form

$$\frac{|a_n|}{|b_n|} \leqslant \frac{1}{h_\varepsilon + |h_0|}. \tag{14.43}$$

lead us to the following upper bound

$$\frac{|b_{n+1}|}{|a_n|}\frac{|a_n|}{|b_n|} = \frac{|b_{n+1}|}{|b_n|} \leqslant \frac{1}{h_\varepsilon + |h_0|}\left(1 + \frac{|h_0|}{h_\varepsilon}\right). \tag{14.44}$$

Another bound for $\dfrac{|b_{n+1}|}{|b_n|}$ can be obtained as follows. In the light of the definition (14.32) one can write

$$a_n = \frac{b_n - a_{n-1}}{h_0}, \tag{14.45}$$

and, consequently, one can deduce the recursive relation for the coefficients b_n

$$b_{n+1} = \frac{b_n - a_{n-1}}{h_0} + h_0 a_{n+1}, \tag{14.46}$$

which after simple algebraic manipulations leads to the relation

$$h_0 b_{n+1} + a_{n-1} = h_0 a_{n+1} + b_n. \tag{14.47}$$

This equation can be rewritten as

$$1 = \left|\frac{h_0 b_{n+1}}{h_0 a_{n+1} + b_n} + \frac{a_{n-1}}{h_0 a_{n+1} + b_n}\right|, \tag{14.48}$$

which after taking into account the triangle inequality gives the relation

$$\left|\frac{h_0 b_{n+1}}{h_0 a_{n+1} + b_n} + \frac{a_{n-1}}{h_0 a_{n+1} + b_n}\right| \leqslant \left|\frac{h_0 b_{n+1}}{h_0 a_{n+1} + b_n}\right| + \left|\frac{a_{n-1}}{h_0 a_{n+1} + b_n}\right|, \tag{14.49}$$

which leads to the conclusion

$$|h_0 a_{n+1} + b_n| \leqslant |h_0||b_{n+1}| + |a_{n-1}|. \tag{14.50}$$

Once again, making use of the triangle inequality one has

$$|h_0 a_{n+1} + b_n| \leqslant |h_0||a_{n+1}| + |b_n|, \tag{14.51}$$

and, consequently,

$$|h_0||b_{n+1}| - |b_n| \leqslant |h_0||a_{n+1}| - |a_{n-1}|. \tag{14.52}$$

This inequality can be rewritten in the following form

$$|h_0|\frac{|b_{n+1}|}{|a_n|} - \frac{|b_n|}{|a_n|} \leqslant |h_0|\frac{|a_{n+1}|}{|a_n|} - \frac{|a_{n-1}|}{|a_n|}, \tag{14.53}$$

or more conveniently

$$|h_0|\frac{|b_{n+1}|}{|b_n|} - 1 \leqslant \frac{|a_n|}{|b_n|}\left(|h_0|\frac{|a_{n+1}|}{|a_n|} - \frac{|a_{n-1}|}{|a_n|}\right). \tag{14.54}$$

In the light of the inequality (14.43) and the relation

$$|h_0|\frac{|a_{n+1}|}{|a_n|} - \frac{|a_{n-1}|}{|a_n|} \leqslant \frac{|h_0|}{h_\varepsilon} - h_\varepsilon, \tag{14.55}$$

one obtains finally the following bound

$$\frac{|b_{n+1}|}{|b_n|} \leqslant \frac{1}{|h_0|}\left(1 + \frac{1}{h_\varepsilon + |h_0|}\left(\frac{|h_0|}{h_\varepsilon} - h_\varepsilon\right)\right), \tag{14.56}$$

which taken together with the previous result (14.44) allows to deduce the inequality for the fixed point

$$|h_0|(|h_0| - 1) \geqslant 0, \tag{14.57}$$

which can be resolved immediately and gives the consistency condition for the fixed point h_0

$$|h_0| \in \{0\} \cup [1, \infty). \tag{14.58}$$

Naturally, there is plenty of other possibilities for the choice of a concrete form of the effective gravitational potential $V_{eff}[h]$. However, in the next section we shall discuss only a particular situation.

B Newton–Coulomb Potential

B1 Residual Approximation

Let us consider the situation in which the series coefficients of the effective gravitational potential are

$$a_n = \begin{cases} a_{-1} = const & \text{for } n = -1 \\ 0 & \text{for } n \neq -1 \end{cases}. \tag{14.59}$$

We shall call such a case *the residual approximation*. In this approximation the effective gravitational potential (12.2) takes the form

$$V_{eff} = \frac{a_{-1}}{h - h_0}, \tag{14.60}$$

which is formally the Newton–Coulomb potential, *i.e.* has the behaviour like $1/h$ where h is interpreted as a kind of radial quantity. The value of the coefficient a_{-1} is unknown, but it is assumed that this coefficient exists. It can be verified by straightforward easy calculation that in such a situation the combined series coefficients b_n are

$$b_n = \begin{cases} b_{-1} = h_0 a_{-1} & \text{for } n = -1 \\ b_0 = a_{-1} & \text{for } n = 0 \\ 0 & \text{for } n \neq -1, 0 \end{cases}, \tag{14.61}$$

and, consequently, the Ricci scalar curvature of the three-dimensional space becomes

$${}^{(3)}R = 2\Lambda + 2\kappa\rho - 6\kappa^2 a_{-1}\left(1 + \frac{h_0}{h - h_0}\right), \tag{14.62}$$

whereas the equation (12.1), defining the global one-dimensional model of quantum gravity. takes the form

$$\left(\frac{\delta^2}{\delta h^2} + \frac{a_{-1}}{h - h_0}\right)\Psi = 0. \tag{14.63}$$

The Ricci scalar curvature (14.68) defines certain states of the geometry of the three-dimensional embedded space. However, even when one considers the case of vacuum, *i.e.* when both the energy density of Matter fields and cosmological constant are identically vanishing $\rho = 0$,

$\Lambda = 0$, it is rather difficult to establish a three dimensional metric tensor for which the Ricci scalar curvature behaves like

$$^{(3)}R \sim 1 + \frac{h_0}{h - h_0}. \tag{14.64}$$

Interestingly, in the most general situation the residue of the three-dimensional Ricci scalar curvature calculated at a fixed point h_0 is

$$\mathrm{Res}(^{(3)}R, h_0) = 2\kappa \mathrm{Res}(\rho, h_0) - 6\kappa^2 a_{-1} h_0, \tag{14.65}$$

i.e. it can be taken *ad hoc* identical to zero if and only if the residue of the energy density of Matter fields is

$$\mathrm{Res}(\rho, h_0) = 3\kappa a_{-1} h_0. \tag{14.66}$$

If one takes *ad hoc* the following relation

$$a_{-1} = \frac{\Lambda}{3\kappa^2}, \tag{14.67}$$

then the Ricci scalar curvature of an induced three-dimensional geometry of embedded space takes the form

$$^{(3)}R = 2\kappa\rho - \frac{2\Lambda h_0}{h - h_0}, \tag{14.68}$$

and its residue at a fixed point h_0

$$\mathrm{Res}(^{(3)}R, h_0) = 2\kappa \mathrm{Res}(\rho, h_0) - 2\Lambda h_0, \tag{14.69}$$

identically vanishes if and only if the residue of the energy density of Matter fields at a fixed point h_0 has the value

$$\mathrm{Res}(\rho, h_0) = \frac{\Lambda}{\kappa} h_0. \tag{14.70}$$

In such a situation, also the geometry of an embedded three-dimensional space manifold is automatically Ricci-flat if and only if the energy density of Matter fields takes the following form

$$\rho = \frac{\Lambda}{\kappa} \frac{h_0}{h - h_0}. \tag{14.71}$$

It is easy to see that another possible Ricci-flat three-dimensional manifold is obtained for the identically vanishing cosmological constant $\Lambda = 0$ and the following value of the energy density of Matter fields

$$\rho = 3\kappa a_{-1} \left(1 + \frac{h_0}{h - h_0}\right). \tag{14.72}$$

B2 Newton–Coulomb Quantum Gravity

In the most general situation three-dimensional spaces having induced metrics characterized by the Ricci scalar curvature of the form (14.64) are not yet known explicitly in literature. However, it is evident that in the particular situation $h_0 = 0$, which is fully consistent with the general condition (14.58), the state of affairs is determined by much more simplified equations

$$^{(3)}R = 2\Lambda + 2\kappa\rho - 6\kappa^2 a_{-1}, \tag{14.73}$$

$$\left(\frac{\delta^2}{\delta h^2} + \frac{a_{-1}}{h}\right)\Psi = 0. \tag{14.74}$$

Let us consider this particular case as the basic situation. We shall call the global one-dimensional quantum gravity described by the system of equations (14.73)-(14.74) *the Newton–Coulomb quantum gravity*.

As an example, we shall consider first the case of constant Ricci curvature. It is not difficult to see that in the most general view such a situation corresponds to the identically vanishing energy density of Matter fields

$$\rho \equiv 0. \tag{14.75}$$

We shall call such a case *the Newton–Coulomb stationary quantum gravity*. In such particular situation, the Ricci scalar curvature of three-dimensional embedded space becomes

$$^{(3)}R = 2\Lambda - 6\kappa^2 a_{-1} = const, \tag{14.76}$$

and, consequently, the Ricci curvature tensor, which characterizes the intrinsic geometry of the manifolds, describes the three-dimensional Einstein manifolds [924]

$$R_{ij} = \lambda h_{ij}, \tag{14.77}$$

where the sign λ of the Einstein manifolds, in these specific conditions, is completely defined by parameters of the Newton–Coulomb stationary quantum gravity – the cosmological constant and the residue of the effective gravitational potential – as follows

$$\frac{2}{3}\Lambda - 2\kappa^2 a_{-1} = \lambda. \tag{14.78}$$

Interestingly, the crucial consequence of identical vanishing of the energy density of Matter fields is the property of *maximal symmetry* of the Einstein manifolds described by the sign (14.78). For this reason, the Newton–Coulomb stationary quantum gravity possesses highly nontrivial geometrical interpretation: such a situation describes embedded three-dimensional manifolds which are *maximally symmetric Einstein manifolds*.

Consequently, one can deduce straightforwardly the classification of the three-dimensional embedded spaces, which are maximally symmetric three-dimensional Einstein manifolds (14.77), with respect to the value of the sign λ (14.78) of a manifold in dependence on the cosmological constant and the residue a_{-1}:

Conclusion (Classification of maximally symmetric three-dimensional Einstein manifolds). *The Newton–Coulomb stationary quantum gravity, defined by the effective gravitational potential* $V_{eff}[h] = \dfrac{a_{-1}}{h}$, *determines the three-dimensional embedded spaces which are the maximally symmetric three-dimensional Einstein manifolds, characterized by the sign of the form (14.78). There is the classification of such manifolds with respect to the cosmological constant* Λ *and the value* a_{-1} *of the residue of the effective gravitational potential*

1. *If the sign of manifold is non-zero* $\lambda \neq 0$ *and the residue of the effective gravitational potential is a negative real* $a_{-1} = -|\alpha|$, *then the effective gravitational potential* $V_{eff}[h]$ *corresponds to the Newtonian attractive potential energy*

$$V_{eff} = -\frac{|\alpha|}{h} \sim -\frac{m_1 m_2}{h}. \tag{14.79}$$

   (a) *If the cosmological constant is a positive real* $\Lambda = +|\Lambda|$ *then the maximally symmetric Einstein three-manifolds are characterized by the positive Ricci scalar curvature*

$${}^{(3)}R = \frac{2}{3}|\Lambda| + 2\kappa^2|\alpha|. \tag{14.80}$$

   (b) *If the cosmological constant is a negative real* $\Lambda = -|\Lambda|$ *then the maximally symmetric Einstein three-manifolds are characterized by the Ricci scalar curvature*

$${}^{(3)}R = -\frac{2}{3}|\Lambda| + 2\kappa^2|\alpha|, \tag{14.81}$$

which is

$$ {}^{(3)}R \begin{cases} <0, & \textit{iff} \quad |\Lambda| > 3\kappa^2|\alpha| \\ =0, & \textit{iff} \quad |\Lambda| = \kappa^2|\alpha| \\ >0, & \textit{iff} \quad |\Lambda| < 3\kappa^2|\alpha| \end{cases} \tag{14.82} $$

2. *If the sign of manifold is non-zero* $\lambda \neq 0$ *and the residue of the effective gravitational potential is a positive real* $a_{-1} = +|\alpha|$, *then the effective potential* $V_{eff}[h]$ *becomes the Coulomb repulsive potential energy*

$$ V_{eff} = \frac{|\alpha|}{h} \sim \frac{q_1 q_2}{h}. \tag{14.83} $$

   (a) *If the cosmological constant is a negative real* $\Lambda = -|\Lambda|$ *then the maximally symmetric Einstein three-manifolds are characterized by negative Ricci scalar curvature*

$$ {}^{(3)}R = -\frac{2}{3}|\Lambda| - 2\kappa^2|\alpha|. \tag{14.84} $$

   (b) *If the cosmological constant is a positive real* $\Lambda = +|\Lambda|$ *then the maximally symmetric Einstein three-manifolds are characterized by the Ricci scalar curvature*

$$ {}^{(3)}R = \frac{2}{3}|\Lambda| - 2\kappa^2|\alpha|, \tag{14.85} $$

   which is

$$ {}^{(3)}R \begin{cases} <0, & \textit{iff} \quad |\Lambda| < 3\kappa^2|\alpha| \\ =0, & \textit{iff} \quad |\Lambda| = \kappa^2|\alpha| \\ >0, & \textit{iff} \quad |\Lambda| > 3\kappa^2|\alpha| \end{cases} \tag{14.86} $$

3. *If the sign of manifold is identically vanishing* $\lambda = 0$, i.e. *the maximally symmetric Einstein three-manifolds are Ricci-flat manifolds, then one can determine uniquely the value of the residue of the effective gravitational potential as follows*

$$ a_{-1} = \pm\frac{|\Lambda|}{3\kappa^2}. \tag{14.87} $$

   In such a situation, one obtains the values of the cosmological constant

$$ |\Lambda| \sim \begin{cases} r_g(m_1)r_g(m_2) & \text{for the Newton law} \\ r_e(q_1)r_e(q_2) & \text{for the Coulomb law} \end{cases} \tag{14.88} $$

where m is the mass of a body generating Newtonian gravitational field in vacuum and $r_g(m) = \frac{2Gm}{c^2}$ is its gravitational radius, q is the charge generating Coulombic electrical field in vacuum and $r_e(q) = q\sqrt{\frac{G}{4\pi\varepsilon_0 c^4}}$ is its electrical radius.

Note that, in fact, by taking into account *ad hoc* the relation for the series coefficients (14.27), the residue a_{-1} of the effective gravitational potential is the Cauchy integral of V_{eff} at a fixed point $h_0 = 0$

$$a_{-1} = \text{Res}\left[\frac{1}{6\kappa^2}\frac{1}{h}\left(-{}^{(3)}R + 2\Lambda + 2\kappa\rho\right), h = 0\right], \tag{14.89}$$

and its value can be straightforwardly established as

$$\begin{aligned} a_{-1} &= \frac{1}{6\kappa^2}\left(-{}^{(3)}R + 2\Lambda + 2\kappa\rho\right)\Big|_{h=0} \\ &= -\frac{{}^{(3)}R_0}{6\kappa^2} + \frac{\Lambda}{3\kappa^2} + \frac{\kappa\rho_0}{3\kappa^2}, \end{aligned} \tag{14.90}$$

where subscript "0" means the value of a quantity calculated in $h = 0$.

B3 Quantum Mechanical Interpretation

When one associates the effective gravitational potential $V_{eff} = \frac{a_{-1}}{h}$ with the Newton or Coulomb potential energy, then the global dimension becomes a spatial distance $r = \sqrt{x^2 + y^2 + z^2}$

$$h \equiv r \quad , \tag{14.91}$$

and for this reason the evolution (14.63) obtains the interpretation of a stationary radial Schrödinger-type wave equation

$$\left(\frac{d^2}{dr^2} + \frac{\mp|\alpha|}{r}\right)\Psi(r) = 0, \tag{14.92}$$

which corresponds with a quantum Kepler problem. There is a lot of metrics h_{ij} possessing the same determinant, for instance the following choice

$$h_{ij} = r^{1/3} r_{ij}, \tag{14.93}$$

where r_{ij} is the $SO(3)$ group rotation matrix, which can be expressed through the Euler angles (θ, φ, ϕ)

$$r_{ij}(\theta, \varphi, \phi) \equiv r^{(3)}_{il}(\theta) r^{(2)}_{lk}(\varphi) r^{(3)}_{kj}(\phi) \quad , \tag{14.94}$$

where $r^{(p)}_{ij}(\vartheta)$ are rotation matrices around a p-axis

$$r^{(3)}_{ij}(\vartheta) = \begin{bmatrix} \cos\vartheta & -\sin\vartheta & 0 \\ \sin\vartheta & \cos\vartheta & 0 \\ 0 & 0 & 1 \end{bmatrix}, \tag{14.95}$$

$$r^{(2)}_{ij}(\vartheta) = \begin{bmatrix} \cos\vartheta & 0 & \sin\vartheta \\ 0 & 1 & 0 \\ -\sin\vartheta & 0 & \cos\vartheta \end{bmatrix}, \tag{14.96}$$

defines the quantum gravity related to this problem.

The interpretation of the radial wave equation (14.92) as a kind of the stationary radial Schrödinger equation for a particle equipped with energy E, mass m, and potential energy V (For more detailed discussion of the radial Schrödinger equation Cf. *e.g.* Ref. [934]) is most natural and means that the quantum gravity is a kind of non-relativistic quantum mechanics. Such a situation possesses non-trivial consequences and establishes the linkage between the model of quantum gravity and atomic and nuclear physics.

Let us recall that applying the substitution $\Psi(r) = rR(r)$ where $R(r)$ is the radial part of the wave function of a particle, one can present the radial Schrödinger equation in the following form

$$\left[\frac{d^2}{dr^2} - \left(\frac{2m}{\hbar^2}V + \frac{l(l+1)}{r^2}\right)\right]\Psi(r) = \frac{2mE}{\hbar^2}\psi(r), \tag{14.97}$$

where l is one of subscripts of the spherical harmonics $Y_l^m(\theta, \phi)$ of degree l and order m

$$Y_l^m(\theta, \phi) = \sqrt{\frac{2l+1}{4\pi}\frac{(l-m)!}{(l+m)!}} P_{lm}(\cos\theta) e^{im\phi}, \tag{14.98}$$

which is the angular part of the wave function of a particle. Here $P_{lm}(x)$ are associated Legendre polynomials

$$P_{lm}(x) = \frac{1}{2^l l!}(1-x^2)^{m/2} \frac{d^{l+m}}{dx^{l+m}}(x^2-1)^l. \tag{14.99}$$

Taking into account *ad hoc* the values $E=0$, $l=0$ and $-\frac{2m}{\hbar^2}V=V_{eff}$, one receives excellent coincidence between the Schrödinger equation and the quantum gravity (14.92). This nice property allows to propose straightforwardly the clear and unambiguous physical interpretation of the quantum gravity's wave functional $\Psi(r)$. Namely, it is strictly related to the wave function $\psi(r,\theta,\phi)=R(r)Y_0^m(\theta,\phi)$ of a particle equipped with mass m and having the total energy $E=0$ and the potential energy $V=-\frac{\hbar^2}{2m}\frac{\pm|\alpha|}{r}$. The necessary identification can be performed immediately

$$\Psi(r)=r\frac{\psi(r,\theta,\phi)}{Y_0^0(\theta,\phi)}, \tag{14.100}$$

where $Y_0^0(\theta,\phi)=\frac{1}{2\sqrt{\pi}}$ is the only non-trivial value of $Y_0^m(\theta,\phi)$.

There is, however, an alternative linkage between the quantum gravity and the radial Schrödinger equation . Let the quantum gravity's wave functional $\Psi(r)$ be the radial part $R(r)$ of the wave function of a particle

$$\psi(r,\theta,\phi)=\Psi(r)Y_l^m(\theta,\phi), \tag{14.101}$$

equipped with mass m and total energy E, and moving in *arbitrary* potential $V(r)$. In such a situation, $\Psi(r)$ satisfies the radial Schrödinger equation

$$-\frac{1}{r^2}\frac{d}{dr}\left(r\frac{d\Psi}{dr}\right)+\left(\frac{2m}{\hbar^2}V+\frac{l(l+1)}{r^2}\right)\Psi=\frac{2mE}{\hbar^2}\Psi, \tag{14.102}$$

or, equivalently,

$$\frac{d^2\Psi}{dr^2}+\frac{1}{r}\frac{d\Psi}{dr}-\left(\frac{2m}{\hbar^2}rV+\frac{l(l+1)}{r}\right)\Psi=-\frac{2mE}{\hbar^2}r\Psi. \tag{14.103}$$

These equations can be separated on the system of two equations

$$\frac{d^2\Psi}{dr^2}-\frac{2m}{\hbar^2}rV\Psi = 0, \tag{14.104}$$

$$\frac{d\Psi}{dr}-l(l+1)\Psi+\frac{2mE}{\hbar^2}r^2\Psi = 0. \tag{14.105}$$

The equation (14.104) is exactly the wave equation (14.92) if and only if the following relationship is satisfied

$$V(r) = -\frac{\hbar^2}{2mr}V_{eff}. \tag{14.106}$$

The equation (14.105) must be solved. It is easy to see that this equation can be integrated straightforwardly

$$\int_{\Psi(r_0)}^{\Psi(r)} \frac{d\Psi}{\Psi} = \int_{r_0}^{r} \left(l(l+1) - \frac{2mE}{\hbar^2}r^2 \right) dr, \tag{14.107}$$

where r_0 is some reference value of r, which gives

$$\ln\left|\frac{\Psi(r)}{\Psi(r_0)}\right| = l(l+1)(r-r_0) - \frac{2mE}{\hbar^2}\frac{r^3-r_0^3}{3}, \tag{14.108}$$

and results in the solution

$$\Psi(r) = \Psi(r_0)\exp\left\{-l(l+1)r_0 + \frac{2mE}{3\hbar^2}r_0^3\right\}\exp\left\{l(l+1)r - \frac{2mE}{3\hbar^2}r^3\right\}. \tag{14.109}$$

Now one can differentiate the equation (14.105) with respect to r and obtains

$$\begin{aligned} \frac{d^2\Psi}{dr^2} &= l(l+1)\frac{d\Psi}{dr} - \frac{2mE}{\hbar^2}\left(2r\Psi + r^2\frac{d\Psi}{dr}\right) = \\ &= \left[l(l+1) - \frac{2mE}{\hbar^2}r^2\right]\frac{d\Psi}{dr} - \frac{4mE}{\hbar^2}r\Psi = \\ &= \left\{\left[l(l+1) - \frac{2mE}{\hbar^2}r^2\right]^2 - \frac{4mE}{\hbar^2}r\right\}\Psi. \end{aligned} \tag{14.110}$$

Application of the equation (14.104) to the result (14.110) leads to

$$\frac{2m}{\hbar^2}rV = \left\{\left[l(l+1) - \frac{2mE}{\hbar^2}r^2\right]^2 - \frac{4mE}{\hbar^2}r\right\}, \tag{14.111}$$

which allows to establish a highly non-trivial form of the potential V

$$V(r) = \frac{\hbar^2}{2m}\frac{l^2(l+1)^2}{r} - 2E - 2El(l+1)r + \frac{2mE^2}{\hbar^2}r^3. \tag{14.112}$$

Using the identification (14.106) and the explicit form of the effective gravitational potential $V_{eff} = \frac{\pm|\alpha|}{r}$, one receives the non-trivial equation for r

$$\left(\frac{2mE}{\hbar^2}\right)^2 r^5 - \frac{4mE}{\hbar^2} l(l+1) r^3 - \frac{4mE}{\hbar^2} r^2 + l^2(l+1)^2 r \pm |\alpha| = 0, \quad (14.113)$$

which is very difficult to solve in general, but significantly simplifies for the suggested value of the quantum number $l = 0$

$$\left(\frac{2mE}{\hbar^2}\right)^2 r^5 - \frac{4mE}{\hbar^2} r^2 \pm |\alpha| = 0. \quad (14.114)$$

The equation (14.114) can be solved easily. Let us focus on the real solution which for the Coulombic case is

$$\begin{aligned} r &= \left[\frac{\hbar^2}{2mE}\left(\frac{mE}{\hbar^2}|\alpha|\right)^{1/3}\left(\sqrt{1-\frac{32}{27}\frac{2mE}{\hbar^2|\alpha|^2}}-1\right)^{1/3} + \right. \\ &+ \left. \frac{2}{3}\left(\frac{mE}{\hbar^2}|\alpha|\right)^{-1/3}\left(\sqrt{1-\frac{32}{27}\frac{2mE}{\hbar^2|\alpha|^2}}-1\right)^{-1/3}\right]^{1/2}, \end{aligned} \quad (14.115)$$

while for the Newtonian case one obtains

$$\begin{aligned} r &= \left[\frac{\hbar^2}{2mE}\left(\frac{mE}{\hbar^2}|\alpha|\right)^{1/3}\left(\sqrt{1-\frac{32}{27}\frac{2mE}{\hbar^2|\alpha|^2}}+1\right)^{1/3} + \right. \\ &+ \left. \frac{2}{3}\left(\frac{mE}{\hbar^2}|\alpha|\right)^{-1/3}\left(\sqrt{1-\frac{32}{27}\frac{2mE}{\hbar^2|\alpha|^2}}+1\right)^{-1/3}\right]^{1/2}, \end{aligned} \quad (14.116)$$

Recall that in our theory $h = r$, *i.e.* in fact the results (14.115)-(14.116) fix the value of the determinant of the space metric h and, consequently, define the geometry of the three-dimensional embedded space. The distance, however, must be a real. The necessary condition to satisfy such a requirement defines the upper bound for the energy E

$$E \leqslant \frac{27}{32}\frac{\hbar^2}{2m}|\alpha|^2. \quad (14.117)$$

The algebraic equation (14.113) is of 5th order. Solutions to such equations, in the most general situation, should be found by the methods of the Galois group. Despite such state of affairs and its inexorable consequence, which causes great difficulties in explicit finding out of the real and positive solutions, a certain simpler model solution can be constructed by a straightforward and much more elementary method. Let us make the substitution for the unknown $r = x + iy$ within the algebraic equation (14.113), where the imaginary part y will be taken as identically vanishing $y \to 0$ at a certain stage of the procedure. In such a situation, the algebraic equation (14.113) is equivalent to the statement that both its real and imaginary part identically vanish. Such a method generates the following system of equations

$$\left(\frac{2mE}{\hbar^2}\right)^2 y^4 + 2\left(\frac{2mE}{\hbar^2}\right)\left(l(l+1) - 5\left(\frac{2mE}{\hbar^2}\right)x^2\right)y^2 -$$
$$-\left(\frac{2mE}{\hbar^2}\right)x\left(4 + 6l(l+1)x - 5\left(\frac{2mE}{\hbar^2}\right)x^3\right) + l^2(l+1)^2 = 0, \quad (14.118)$$

$$5\left(\frac{2mE}{\hbar^2}\right)^2 xy^4 + 2\left(\frac{2mE}{\hbar^2}\right)\left(1 + 3l(l+1)x - 5\left(\frac{2mE}{\hbar^2}\right)x^3\right)y^2$$
$$+x\left(-2\left(\frac{2mE}{\hbar^2}\right)x + \left(l(l+1) - \left(\frac{2mE}{\hbar^2}\right)^2\right)^2\right) \pm |\alpha| = 0. \quad (14.119)$$

Now one can implement the value $y = 0$, what implies that $x = r$, and obtain a significantly simpler system of equations for the radial variable r

$$-\left(\frac{2mE}{\hbar^2}\right)r\left(4 + 6l(l+1)r - 5\left(\frac{2mE}{\hbar^2}\right)r^3\right) + l^2(l+1)^2 = 0, \quad (14.120)$$

$$l^2(l+1)^2 r - \left(\frac{2mE}{\hbar^2}\right)r^2\left(2 + 2l(l+1)r - \left(\frac{2mE}{\hbar^2}\right)r^3\right) \pm |\alpha| = 0, \quad (14.121)$$

which can be presented in the more convenient form

$$\left(\frac{2mE}{\hbar^2}\right)r^2 = \frac{l^2(l+1)^2 r}{\left(4 + 6l(l+1)r - 5\left(\frac{2mE}{\hbar^2}\right)r^3\right)}, \quad (14.122)$$

$$\left(\frac{2mE}{\hbar^2}\right)r^2 = \frac{l^2(l+1)^2 r \pm |\alpha|}{\left(2 + 2l(l+1)r - \left(\frac{2mE}{\hbar^2}\right)r^3\right)}, \quad (14.123)$$

and lead to the relation

$$\frac{l^2(l+1)^2 r}{\left(4+6l(l+1)r-5\left(\frac{2mE}{\hbar^2}\right)r^3\right)}=\frac{l^2(l+1)^2 r\pm|\alpha|}{\left(2+2l^2(l+1)^2 r-\left(\frac{2mE}{\hbar^2}\right)r^3\right)}. \tag{14.124}$$

The relation (14.124), however, can be rewritten in a much more convenient form of the 4th order algebraic equation

$$4\left(\frac{2mE}{\hbar^2}\right)l^2(l+1)^2r^4\pm 5\left(\frac{2mE}{\hbar^2}\right)|\alpha|r^3-4l^3(l+1)^3r^2-$$
$$2l(l+1)\left(l(l+1)\pm 3|\alpha|\right)r-(\pm 4|\alpha|)=0. \tag{14.125}$$

In other words, the 5th order algebraic equation, which is very difficult to solve in general, has been reduced to the 4th order algebraic equation which in the most general situation can be solved straightforwardly by elementary algebra methods. In our interest is to extract the real and positive solutions of the algebraic equation (14.125) which is given by the following formula

$$r=-\frac{\pm 5|\alpha|}{16l^2(l+1)^2}+\frac{1}{2}\left(\frac{1}{2^{2/3}3A_l}\left(45\frac{|\alpha|^2}{l(l+1)}-81(\pm|\alpha|)+\frac{4\hbar^2l^4(l+1)^4}{mE}\right)+\right.$$
$$\left.\frac{25|\alpha|^2}{64l^4(l+1)^4}+\frac{\hbar^2 l(l+1)}{3mE}+\frac{\hbar^2A_l}{2^{10/3}3mEl^2(l+1)^2}\right)^{1/2}+\frac{1}{2}\left(\frac{25|\alpha|^2}{32l^4(l+1)^4}+\right.$$
$$\frac{2\hbar^2 l(l+1)}{3mE}-\frac{1}{2^{2/3}3A_l}\left(45\frac{|\alpha|^2}{l(l+1)}-81(\pm|\alpha|)+\frac{4\hbar^2l^4(l+1)^4}{mE}\right)-$$
$$\frac{\hbar^2A_l}{2^{10/3}3mEl^2(l+1)^2}+\frac{1}{4l(l+1)}\left(\frac{2\hbar^2l(l+1)}{mE}-\frac{\pm 125|\alpha|^3}{64l^5(l+1)^5}+\frac{\pm 7\hbar^2|\alpha|}{2mE}\right)\times$$
$$\left(\frac{25|\alpha|^2}{64l^4(l+1)^4}+\frac{\hbar^2l(l+1)}{3mE}+\frac{\hbar^2A_l}{2^{10/3}3mEl^2(l+1)^2}+\right.$$
$$\left.\left.\frac{1}{2^{2/3}3A_l}\left(45\frac{|\alpha|^2}{l(l+1)}-81(\pm|\alpha|)+\frac{4\hbar^2l^4(l+1)^4}{mE}\right)\right)^{-1/2}\right)^{1/2}, \tag{14.126}$$

where

$$A_l = \frac{2mE}{\hbar^2}\left(-\frac{\pm 675\hbar^2|\alpha|^3}{2mE} + \frac{\hbar^4 l^6(l+1)^6}{2(mE)^2}\left(\frac{351|\alpha|^2}{l(l+1)} - \frac{\pm 297|\alpha|}{l(l+1)} - \frac{8\hbar^2 l^3(l+1)^3}{mE} + 54\right) + \frac{\hbar^2}{2mE}\left(-\frac{\hbar^2 l^6(l+1)^6}{mE}\left(\frac{45|\alpha|^2}{l(l+1)} + \frac{4\hbar^2 l^4(l+1)^4}{mE} - 81(\pm|\alpha|)\right)^3 + \left(\pm 675|\alpha|^3 - \frac{\hbar^2 l^8(l+1)^8}{mE}\left(\frac{351|\alpha|^2}{l(l+1)} - \frac{\pm 297|\alpha|}{l(l+1)} - \frac{8\hbar^2 l^3(l+1)^3}{mE}\right)\right)^2 + 54\right)^{1/2}\right)^{1/3}. \tag{14.127}$$

The solution for $l = 0$ can not be deduced from these formulas. However, the result can be obtained from the equation by straightforward implementation of $l = 0$. The solution is

$$r = \left(\frac{2}{5}\frac{\hbar^2}{mE}\right)^{1/3}. \tag{14.128}$$

Another separation of the radial Schrödinger equation (14.103) is possible. Namely, one can rewrite this equation as the system

$$\frac{d^2\Psi}{dr^2} - \left(\frac{2m}{\hbar^2}rV + \frac{l(l+1)}{r}\right)\Psi = 0, \tag{14.129}$$

$$\frac{d\Psi}{dr} + \frac{2mE}{\hbar^2}r^2\Psi = 0. \tag{14.130}$$

The procedure analogous to the previous separation gives the solution

$$\Psi(r) = \Psi(r_0)\exp\left\{\frac{2mE}{3\hbar^2}r_0^3\right\}\exp\left\{-\frac{2mE}{3\hbar^2}r^3\right\}, \tag{14.131}$$

the potential

$$V(r) = -\frac{\hbar^2}{2m}\frac{l(l+1)}{r^2} - 2E + \frac{2mE^2}{\hbar^2}r^3, \tag{14.132}$$

and the equation for r

$$\left(\frac{2mE}{\hbar^2}\right)^2 r^4 - \frac{4mE}{\hbar^2}r - \frac{\hbar^2}{2m}l^2(l+1)^2 \pm |\alpha| = 0. \tag{14.133}$$

This equation can be solved for arbitrary l with no problems. There are real and positive solutions if and only if

$$\pm|\alpha| \leqslant \frac{\hbar^2}{2m} l^2(l+1)^2. \tag{14.134}$$

Then there is only one real and positive solution given by

$$\begin{aligned} r = &\frac{1}{2}\left\{\frac{2\hbar^2}{mE}\left[\frac{2^{-1/3}}{(mE/\hbar^2)^{2/3}}\beta_l^{1/3} + \frac{2^{2/3}\alpha_l}{3(mE/\hbar^2)^{4/3}}\beta_l^{-1/3}\right]^{-1/2}\right.\\ &\left.-\frac{2^{-1/3}}{(mE/\hbar^2)^{2/3}}\beta_l^{1/3} - \frac{2^{2/3}\alpha_l}{3(mE/\hbar^2)^{4/3}}\beta_l^{-1/3}\right\}^{1/2}\\ &+\frac{1}{2}\left(\frac{2^{-1/3}}{(mE/\hbar^2)^{2/3}}\beta_l^{1/3} + \frac{2^{2/3}\alpha_l}{3(mE/\hbar^2)^{4/3}}\beta_l^{-1/3}\right)^{1/2}, \end{aligned} \tag{14.135}$$

where we have introduced the shortened notation

$$\alpha_l = -\frac{\hbar^2}{2m} l^2(l+1)^2 \pm |\alpha|, \tag{14.136}$$

$$\beta_l = 1 + \sqrt{1 - \frac{4}{27}\frac{\hbar^4}{m^2E^2}\alpha_l^3}. \tag{14.137}$$

In such a situation, it is easy to deduce the solution for $l = 0$. With

$$\beta_0 = 1 + \sqrt{1 - (\pm 1)^3 \frac{4}{27}\frac{\hbar^4|\alpha|^3}{m^2E^2}}, \tag{14.138}$$

one obtains the following solution

$$\begin{aligned} r = &\frac{1}{2}\left\{\frac{2\hbar^2}{mE}\left[\frac{2^{-1/3}}{(mE/\hbar^2)^{2/3}}\beta_0^{1/3} + \frac{\pm 2^{2/3}|\alpha|}{3(mE/\hbar^2)^{4/3}}\beta_0^{-1/3}\right]^{-1/2}\right.\\ &\left.-\frac{2^{-1/3}}{(mE/\hbar^2)^{2/3}}\beta_0^{1/3} - \frac{\pm 2^{2/3}|\alpha|}{3(mE/\hbar^2)^{4/3}}\beta_0^{-1/3}\right\}^{1/2}\\ &+\frac{1}{2}\left(\frac{2^{-1/3}}{(mE/\hbar^2)^{2/3}}\beta_0^{1/3} + \frac{\pm 2^{2/3}|\alpha|}{3(mE/\hbar^2)^{4/3}}\beta_0^{-1/3}\right)^{1/2}. \end{aligned} \tag{14.139}$$

C Boundary Conditions for Wave Functionals

In this section we shall consider certain solutions of the global one-dimensional model of quantum gravity (12.1) for the case of the residual approximation of the effective gravitational potential V_{eff} implemented in the previous section. In the most general situation, the considered quantum mechanical evolution

$$\left(\frac{\delta^2}{\delta h^2} \mp \frac{|\alpha|}{h}\right)\Psi^{\mp}[h] = 0, \tag{14.140}$$

is solved by two types of wave functions $\Psi^{\mp}$ where the attractive wave functions $\Psi^{-}_{G}[h]$ are associated with the Newton-like effective gravitational potential, and the repulsive ones $\Psi^{+}[h]$ are associated with the Coulomb-like effective gravitational potential. Because of the manifest one-dimensionality of the functional evolutionary equation (14.140), one can solve this equation in the framework of the theory of ordinary differential equations by interpretation of the functional derivative as the ordinary one, *i.e.* $\frac{\delta}{\delta h} = \frac{d}{dh}$, and the wave functional as a wave function $\Psi[h] = \Psi(h)$ with no loss of generality.

In this manner, the problem to solve is given by the second order ordinary differential equation

$$\left(\frac{d^2}{dh^2} \mp \frac{|\alpha|}{h}\right)\Psi^{\mp}(h) = 0, \tag{14.141}$$

which is well-known in the mathematical physics literature. The general solution of such a differential equation can be constructed straightforwardly by making use of the Bessel functions J_n and Y_n for the case of the attractive potential

$$\Psi^{-}[h] = \sqrt{|\alpha|h}\left[C_1^{-} J_1\left(2\sqrt{|\alpha|h}\right) + 2iC_2^{-} Y_1\left(2\sqrt{|\alpha|h}\right)\right], \tag{14.142}$$

and by making use of the modified Bessel I_n and K_n for the case of the repulsive potential

$$\Psi^{+}[h] = -\sqrt{|\alpha|h}\left[C_1^{+} I_1\left(2\sqrt{|\alpha|h}\right) + 2C_2^{+} K_1\left(2\sqrt{|\alpha|h}\right)\right], \tag{14.143}$$

where $C_1^{\pm}$ and $C_2^{\pm}$ are the integration constants. In concrete calculations one can take the standard definitions of the Bessel functions of the first and second kind [929], $J_\alpha(x)$ and $Y_\alpha(x)$, which are

$$J_\alpha(x) = \frac{1}{\pi}\int_0^\pi dt \cos(x\cos t - \alpha t), \quad (14.144)$$

$$Y_\alpha(x) = \frac{J_\alpha(x)\cos(\alpha\pi) - J_{-\alpha}(x)}{\sin(\alpha\pi)}, \quad (14.145)$$

as well as the modified Bessel functions of the first and the second kind, $I_\alpha(x)$ and $K_\alpha(x)$, which are

$$I_\alpha(x) = \frac{1}{\pi}\int_0^\pi dt \exp(x\cos t)\cos(\alpha t), \quad (14.146)$$

$$K_\alpha(x) = \frac{\pi}{2}\frac{I_{-\alpha}(x) - I_\alpha(x)}{\sin(\alpha\pi)}. \quad (14.147)$$

Recall that standardly the values of the Bessel functions of second kind and the modified Bessel functions of second kind for any integers n can be obtained by application of the limiting procedure

$$Y_n(x) = \lim_{\alpha\to n} Y_\alpha(x), \quad (14.148)$$

$$K_n(x) = \lim_{\alpha\to n} K_\alpha(x). \quad (14.149)$$

In further part of this section we shall present solutions to the one-dimensional quantum mechanics (14.141) with respect to three general types of boundary conditions for the general solutions (14.142) and (14.143).

C1 Boundary Conditions I

Let us consider the global one-dimensional quantum gravity (12.1) with the boundary conditions for some selected initial value of the dimension $h = h_I$:

$$\Psi[h_I] = \Psi_I, \quad (14.150)$$

$$\frac{\delta\Psi}{\delta h}[h_I] = \Psi'_I. \quad (14.151)$$

For construction of the solution one can use the regularized hypergeometric functions ${}_p\tilde{F}_q$

$$ {}_p\tilde{F}_q\left(\begin{array}{c} a_1,\ldots,a_p \\ b_1,\ldots,b_q \end{array};x\right) = \frac{{}_pF_q\left(\begin{array}{c} a_1,\ldots,a_p \\ b_1,\ldots,b_q \end{array};x\right)}{\Gamma(b_1)\ldots\Gamma(b_q)}, \tag{14.152} $$

where ${}_pF_q$ is the confluent hypergeometric function

$$ {}_pF_q\left(\begin{array}{c} a_1,\ldots,a_p \\ b_1,\ldots,b_q \end{array};x\right) = \sum_{r=0}^{\infty} \frac{(a_1)_r\ldots(a_p)_r}{(b_1)_r\ldots(b_q)_r}\frac{x^r}{r!}, \tag{14.153} $$

where $(a)_r = \dfrac{\Gamma(a+r)}{\Gamma(a)}$ are the Pochhammer symbols. Taking into account the boundary conditions (14.150)-(14.151), the general solutions (14.142) and (14.143) can be presented in the following form

$$ \Psi^- = C_1^-\left(2\sqrt{|\alpha|h}\right)K_1\left(2\sqrt{|\alpha|h}\right) + C_2^-\left(2\sqrt{|\alpha|h}\right)^2 {}_0\tilde{F}_1\left(\begin{array}{c} - \\ 2 \end{array};|\alpha|h\right), \tag{14.154} $$

where the integration constants are

$$ \begin{aligned} C_1^- &= \Psi_I\,{}_0\tilde{F}_1\left(\begin{array}{c} - \\ 1 \end{array};|\alpha|h_I\right) - \Psi_I' h_I\,{}_0\tilde{F}_1\left(\begin{array}{c} - \\ 2 \end{array};|\alpha|h_I\right), & (14.155) \\ C_2^- &= \frac{1}{2}\left(\Psi_I K_0\left(2\sqrt{|\alpha|h_I}\right) + \Psi_I'\sqrt{\frac{h_I}{|\alpha|}}K_1\left(2\sqrt{|\alpha|h_I}\right)\right), & (14.156) \end{aligned} $$

for the Newton-like case, and

$$ \Psi^+ = C_1^+\left(2\sqrt{|\alpha|h}\right)Y_1\left(2\sqrt{|\alpha|h}\right) + C_2^+\left(2\sqrt{|\alpha|h}\right)^2 {}_0\tilde{F}_1\left(\begin{array}{c} - \\ 2 \end{array};-|\alpha|h\right), \tag{14.157} $$

where the integration constants are

$$ \begin{aligned} C_1^+ &= \frac{\pi}{2}\left(\Psi_I' h_I\,{}_0\tilde{F}_1\left(\begin{array}{c} - \\ 2 \end{array};-|\alpha|h_I\right) - \Psi_I\,{}_0\tilde{F}_1\left(\begin{array}{c} - \\ 1 \end{array};-|\alpha|h_I\right)\right), & (14.158) \\ C_2^+ &= \frac{\pi}{2}\left(\Psi_I Y_0\left(2\sqrt{|\alpha|h_I}\right) - \Psi_I'\sqrt{\frac{h_I}{|\alpha|}}Y_1\left(2\sqrt{|\alpha|h_I}\right)\right), & (14.159) \end{aligned} $$

for the Coulomb-like case.

C2 Boundary Conditions II

The second case which we want to present in this paper, are the boundary conditions for 1st and 2nd functional derivatives

$$\frac{\delta\Psi}{\delta h}[h_I] = \Psi'_I, \tag{14.160}$$

$$\frac{\delta^2\Psi}{\delta h^2}[h_I] = \Psi''_I. \tag{14.161}$$

Using the hypergeometric functions, one can express the solution for the attractive case as follows

$$\Psi^- = C_1^-\left(2\sqrt{|\alpha|h}\right)K_1\left(2\sqrt{|\alpha|h}\right)+C_2^-\left(2\sqrt{|\alpha|h}\right)^2{}_0\tilde{F}_1\left(\ \overline{2}\ ;|\alpha|h\right), \tag{14.162}$$

where C_1^- and C_2^- are constants defined as

$$C_1^- = -h_I\left(\Psi'_I{}_0\tilde{F}_1\left(\ \overline{2}\ ;|\alpha|h_I\right)-\frac{\Psi''_I}{|\alpha|}{}_0\tilde{F}_1\left(\ \overline{1}\ ;|\alpha|h_I\right)\right), \tag{14.163}$$

$$C_2^- = \frac{1}{2}\sqrt{\frac{h_I}{|\alpha|}}\left(\Psi''_I\sqrt{\frac{h_I}{|\alpha|}}K_0\left(2\sqrt{|\alpha|h_I}\right)+\Psi'_I K_1\left(2\sqrt{|\alpha|h_I}\right)\right) \tag{14.164}$$

Similarly, for the repulsive case one obtains easily

$$\Psi^+ = C_1^+\left(2\sqrt{|\alpha|h}\right)Y_1\left(2\sqrt{|\alpha|h}\right)+C_2^+\left(2\sqrt{|\alpha|h}\right)^2{}_0\tilde{F}_1\left(\ \overline{2}\ ;-|\alpha|h\right), \tag{14.165}$$

where the integration constants are

$$C_1^+ = \frac{\pi h_I}{2}\left(\Psi'_I{}_0\tilde{F}_1\left(\ \overline{2}\ ;-|\alpha|h_I\right)+\frac{\Psi''_I}{|\alpha|}{}_0\tilde{F}_1\left(\ \overline{1}\ ;-|\alpha|h_I\right)\right), \tag{14.166}$$

$$C_2^+ = \frac{\pi}{4}\sqrt{\frac{h_I}{|\alpha|}}\left(\Psi''_I\sqrt{\frac{h_I}{|\alpha|}}Y_0\left(2\sqrt{|\alpha|h_I}\right)+\Psi'_I Y_1\left(2\sqrt{|\alpha|h_I}\right)\right). \tag{14.167}$$

C3 Boundary Conditions III

The third possible choice of the boundary conditions for the considered problem has the following form

$$\Psi[h_I] = \Psi_I, \tag{14.168}$$

$$\frac{\delta^2\Psi}{\delta h^2}[h_I] = \Psi_I''. \tag{14.169}$$

These conditions are formally improper for the problem, because they lead to manifestly singular solutions. In such a situation, however, one can present the solutions in the form in which the integration constants are formally singular. For the case of the attractive Newton-like potential one has

$$\Psi_G^- = C_1^-\left(2\sqrt{|\alpha|h}\right)K_1\left(2\sqrt{|\alpha|h}\right) + C_2^-\left(2\sqrt{|\alpha|h}\right)^2{}_0\tilde{F}_1\left(\,\overline{}\,;\,2\,;|\alpha|h\right), \tag{14.170}$$

where the integration constants are ($\varepsilon \to 0$)

$$C_1^- = \frac{2}{\varepsilon}\sqrt{|\alpha|h_I}\left(\Psi_I - \frac{h_I}{|\alpha|}\Psi_I''\right){}_0\tilde{F}_1\left(\,\overline{}\,;\,2\,;|\alpha|h_I\right), \tag{14.171}$$

$$C_2^- = \frac{1}{\varepsilon}\left(\Psi_I - \frac{h_I}{|\alpha|}\Psi_I''\right)K_1\left(2\sqrt{|\alpha|h_I}\right). \tag{14.172}$$

Similarly, for case of the repulsive Coulomb-like potential one obtains

$$\Psi_G^+ = C_1^+\left(2\sqrt{|\alpha|h}\right)Y_1\left(2\sqrt{|\alpha|h}\right) + C_2^+\left(2\sqrt{|\alpha|h}\right)^2{}_0\tilde{F}_1\left(\,\overline{}\,;\,2\,;-|\alpha|h\right), \tag{14.173}$$

where the integration constants are ($\varepsilon \to 0$)

$$C_1^+ = \frac{2}{\varepsilon}\sqrt{|\alpha|h_I}\left(\Psi_I + \frac{h_I}{|\alpha|}\Psi_I''\right){}_0\tilde{F}_1\left(\,\overline{}\,;\,2\,;-|\alpha|h_I\right), \tag{14.174}$$

$$C_2^+ = \frac{1}{\varepsilon}\left(\Psi_I + \frac{h_I}{|\alpha|}\Psi_I''\right)Y_1\left(2\sqrt{|\alpha|h_I}\right). \tag{14.175}$$

There is the question of how to regularize these solutions to obtain the non-singular integration constants. There is no general method for such a productive procedure. Let us propose some constructive regularization. For regularization of the solution let us take into account the following *ansatz for boundary conditions*

$$\pm\frac{h_I}{|\alpha|}\Psi_I^{\pm\prime\prime}+\Psi_I^{\pm}\equiv\varepsilon f_{\pm}[h_I,|\alpha|], \tag{14.176}$$

where $f_{\pm}[h_I,|\alpha|]\neq 0$ are some nonsingular functionals of h_I and $|\alpha|$, which are presently unknown and arbitrary. For formal correctness and consistency of the method, we shall implement the limiting procedure $\varepsilon\to 0$ at a certain level of the regularization process. The sign $+$ is related to the Newton-like case, and the sign $-$ to the Coulomb-like case. It can be seen by straightforward calculation that in such a situation the singularity of the solutions (14.170) and (14.173) can be removed. We are going to show now that the ansatz (14.176) allows to express the problem of choice of the initial data by integral equations for the functionals $f_{\pm}[h_I,|\alpha|]$. For basics and applications of the theory of integral equations we suggest to Cf. *e.g.* the books in Ref. [935].

For the attractive Newton-like case the initial value Ψ_I can be obtained as follows

$$\begin{aligned}\Psi_I^{-}=&-|\alpha|h_I\,{}_0F_1\left(\begin{matrix}-\\2\end{matrix};|\alpha|h_I\right)\left[c_1^{-}+2\varepsilon\sqrt{|\alpha|}\int_1^{h_I}\frac{dt}{\sqrt{t}}f_{-}[t,|\alpha|]K_1\left(2\sqrt{|\alpha|t}\right)\right]\\&+2\sqrt{|\alpha|h_I}K_1\left(2\sqrt{|\alpha|h_I}\right)\left[c_2^{-}+\varepsilon|\alpha|\int_1^{h_I}dt\,f_{-}[t,|\alpha|]\,{}_0F_1\left(\begin{matrix}-\\2\end{matrix};|\alpha|t\right)\right],\end{aligned} \tag{14.177}$$

and, similarly, for the repulsive Coulomb-like case one receives

$$\begin{aligned}\Psi_I^{+}=&|\alpha|h_I\,{}_0F_1\left(\begin{matrix}-\\2\end{matrix};-|\alpha|h_I\right)\left[c_1^{+}-\varepsilon\pi\sqrt{|\alpha|}\int_1^{h_I}\frac{dt}{\sqrt{t}}f_{+}[t,|\alpha|]Y_1\left(2\sqrt{|\alpha|t}\right)\right]\\&+2i\sqrt{|\alpha|h_I}Y_1\left(2\sqrt{|\alpha|h_I}\right)\left[c_2^{+}-\varepsilon\frac{i\pi}{2}|\alpha|\int_1^{h_I}dt\,f_{+}[t,|\alpha|]\,{}_0F_1\left(\begin{matrix}-\\2\end{matrix};-|\alpha|t\right)\right],\end{aligned} \tag{14.178}$$

where $c_{1,2}^{\pm}$ are now the non-singular integration constants. The functionals $f_{\pm}[h_I,|\alpha|]$ can be established by straightforward application of the ansatz (14.176) within the general solutions (14.170) and (14.173).

Such an application yields the following results

$$\Psi_I^- = 8|\alpha|h_I K_1\left(2\sqrt{|\alpha|h_I}\right) {}_0\tilde{F}_1\left(\frac{-}{2};|\alpha|h_I\right) f_-[h_I,|\alpha|], \quad (14.179)$$

$$\Psi_I^+ = 8|\alpha|h_I Y_1\left(2\sqrt{|\alpha|h_I}\right) {}_0\tilde{F}_1\left(\frac{-}{2};-|\alpha|h_I\right) f_+[h_I,|\alpha|]. \quad (14.180)$$

Employing straightforwardly these results in the equations (14.177) and (14.178) one obtains the integral equations for the functionals $f_\pm$. For the Coulomb-like situation one receives the following Volterra integral equation of the second kind

$$f_-[h_I,|\alpha|] = g^-[h_I,|\alpha|] + \varepsilon\int_1^{h_I} dt\mathscr{K}^-(t,|\alpha|,h_I)f_-[t,|\alpha|], \quad (14.181)$$

where the function $g^-[h_I,|\alpha|]$ causing the non-homogeneity is

$$g^-[h_I,|\alpha|] \doteq -\frac{c_1^-}{8K_1\left(2\sqrt{|\alpha|h_I}\right)} + \frac{c_2^-}{2\left(2\sqrt{|\alpha|h_I}\right) {}_0\tilde{F}_1\left(\frac{-}{2};|\alpha|h_I\right)}, \quad (14.182)$$

and the kernel $\mathscr{K}^-(t,|\alpha|,h_I)$ has the form

$$\mathscr{K}^-(t,|\alpha|,h_I) = \frac{|\alpha|}{\sqrt{t}}\left[\frac{{}_0F_1\left(\frac{-}{2};|\alpha|t\right)}{2\left(2\sqrt{|\alpha|h_I}\right) {}_0\tilde{F}_1\left(\frac{-}{2};|\alpha|h_I\right)} - \right.$$
$$\left. - \frac{\sqrt{h_I}K_1\left(2\sqrt{|\alpha|t}\right)}{2\left(2\sqrt{|\alpha|h_I}\right)K_1\left(2\sqrt{|\alpha|h_I}\right)}\right]. \quad (14.183)$$

A similar procedure can be performed for the Newton-like situation. In this instance, the Volterra integral equation of the second kind is

$$f_+[h_I,|\alpha|] = g^+[h_I,|\alpha|] + \varepsilon\int_1^{h_I} dt\mathscr{K}^+(t,|\alpha|,h_I)f_+[t,|\alpha|] \quad (14.184)$$

where the function $g^-[h_I,|\alpha|]$ causing the non-homogeneity is

$$g^+[h_I,|\alpha|] = \frac{c_1^+}{8Y_1\left(2\sqrt{|\alpha|h_I}\right)} + \frac{ic_2^+}{2\left(2\sqrt{|\alpha|h_I}\right) {}_0\tilde{F}_1\left(\frac{-}{2};-|\alpha|h_I\right)}, \quad (14.185)$$

and the kernel $\mathscr{K}^+(t,|\alpha|,h_I)$ has the form

$$\mathscr{K}^+(t,|\alpha|,h_I) = \frac{\pi|\alpha|}{2\sqrt{t}}\left[-\frac{\sqrt{h_I}Y_1\left(2\sqrt{|\alpha|t}\right)}{2\left(2\sqrt{|\alpha|h_I}\right)Y_1\left(2\sqrt{|\alpha|h_I}\right)} - \right.$$

$$\left. - \frac{i\,{}_0F_1\left(\frac{-}{2};-|\alpha|t\right)}{2\left(2\sqrt{|\alpha|h_I}\right){}_0\tilde{F}_1\left(\frac{-}{2};-|\alpha|h_I\right)}\right]. \quad (14.186)$$

The number ε plays the role analogous to the eigenvalue in linear algebra. It is evidently seen that the integral operators acting on the functionals $f_\pm$ are non-singular when the limiting procedure $\varepsilon \to 0$ is performed. Such a property guarantees stability and effectiveness of the regularization method given by the ansatz (14.176). For this reason, one can perform the limit $\varepsilon \to 0$ straightforwardly within the Volterra integral equations of the second kind (14.181) and (14.184), and extract the unknown functionals

$$f_\pm[h_I,|\alpha|] = g^\pm[h_I,|\alpha|]. \quad (14.187)$$

Taking into account elementary properties of the hypergeometric function, the final results can be presented in the following form

$$f_-[h_I,|\alpha|] = \frac{-c_1^-}{8K_1\left(2\sqrt{|\alpha|h_I}\right)} + \frac{c_2^-}{4I_1\left(2\sqrt{|\alpha|h_I}\right)}, \quad (14.188)$$

$$f_+[h_I,|\alpha|] = \frac{c_1^+}{8Y_1\left(2\sqrt{|\alpha|h_I}\right)} + \frac{ic_2^+}{4J_1\left(2\sqrt{|\alpha|h_I}\right)}. \quad (14.189)$$

In this manner, the conditions for the initial data in the considered situation given by the improper boundary conditions (14.168)-(14.169) and in the light to the ansatz (14.176) cannot be chosen arbitrarily, but according to the following selection rules

$$\Psi_I^- = \sqrt{|\alpha|h_I}\left[-c_1^- I_1\left(2\sqrt{|\alpha|h_I}\right) + 2c_2^- K_1\left(2\sqrt{|\alpha|h_I}\right)\right], \quad (14.190)$$

$$\Psi_I^+ = \sqrt{|\alpha|h_I}\left[c_1^+ J_1\left(2\sqrt{|\alpha|h_I}\right) + 2ic_2^+ Y_1\left(2\sqrt{|\alpha|h_I}\right)\right]. \quad (14.191)$$

It should be emphasized that the proposed ansatz for the boundary conditions (14.176) is not unique, and can be replaced by another proposal. However, as we have mentioned earlier, this type of the regularization method assures stability and effectiveness of the limiting procedure $\varepsilon \to 0$ and, therefore, in general guarantees consistency of the method.

The proposed ansatz (14.176) made formulation of the regularization procedure in terms of the Volterra integral equation of the second kind consistent. However, the resulting Volterra integral equation was not solved because the limit $\varepsilon \to 0$ was performed. Albeit, it should be emphasized also that in the light of the theory of integral equations, the obtained Volterra integral equations of the second kind should be solved for arbitrary ε and then the limit $\varepsilon \to 0$ should be performed. Such a solution can be constructed straightforwardly by application of the Born–Neumann series, named after Max Born and Carl Gottfried Neumann. If one rewrites the Volterra integral equations (14.181) and (14.184) in a more convenient symbolic form

$$f_{\pm}[h_I, |\alpha|] = g^{\pm}[h_I, |\alpha|] + \varepsilon \mathsf{K}^{\pm} f_{\pm}[t, |\alpha|], \tag{14.192}$$

where the integral operators $\mathsf{K}^{\pm}$ are defined as

$$\mathsf{K}^{\pm} = \int_{1}^{h_I} dt \mathscr{K}^{\pm}(t, |\alpha|, h_I), \tag{14.193}$$

then the solution to the equation (14.192) can be written out straightforwardly by making use of the Born–Neumann series

$$f_{\pm}[h_I, |\alpha|] = g^{\pm}[h_I, |\alpha|] + \sum_{n=1}^{\infty} \varepsilon^n \mathsf{K}^{\pm n} g^{\pm}[h_I, |\alpha|]. \tag{14.194}$$

It can be seen straightforwardly now that by application of the limiting procedure $\varepsilon \to 0$ one obtains the result $f_{\pm}[h_I, |\alpha|] = g^{\pm}[h_I, |\alpha|]$ which has been received in our earlier analysis. The unsolved problem is, however, the convergence of the Born–Neumann series as well as definiteness of the integral operators $\mathsf{K}^{\pm}$ because in the presented situation the integral kernels $\mathsf{K}^{\pm}(t, |\alpha|, h_I)$ are special functions. Anyway, in the most general situation the presented method reflects the typical problems arising from application of the improper boundary conditions.

Chapter 15

Ab Initio Thermodynamics of Space Quanta Æther

The quantum field theory formulated in terms of the static Fock repère is the most natural approach to consistent formulation of the state of thermodynamic equilibrium. The practical efficiency of such a method is emphasized by the fact that in spite of the system being manifestly in non-equilibrium (for details Cf. *e.g.* Ref. [936]) application of the Fock space formalism leads to the properly defined state of equilibrium. In the particular situation defined by the global one-dimensional quantum gravity, the equilibrium is strictly related to the ensemble of the quanta of space which in itself creates the model of Æther. Such a situation creates the most natural conditions for the straightforward application of the first principles of statistical mechanics [937], which will result in the *ab initio* formulation of the thermodynamics of such an Æther of space quanta.

In this chapter, we shall present the constructive approach in which the simplest possible approximation is applied. The approach, however, in itself is the most fundamental contribution to the thermodynamics of the Æther. Namely, as the example of the effective computational strategy for the thermodynamics, we shall apply the so-called *one-particle approximation*, based on the corresponding method of *the one-particle density matrix*. The simplicity of the one-particle approximation is emphasized by the fact that in such a situation the density operator D is equivalent to the occupation number operator. Hence, the thermodynamic equilibrium will be determined with respect to the static Fock repère in the stable Bogoliubov vacuum and, therefore, the one-particle density matrix in equilibrium $\mathbb{D}$ will be established according to the von Neumann–Heisenberg picture.

A Entropy I: Analytic Approach

The density operator D in the state of non-equilibrium, which possesses manifestly dynamical nature by its existence in a dynamical Fock repère, has been expressed in the static Fock repère $\mathfrak{F}$ by application of the equilibrium density matrix $\mathbb{D}$ as follows

$$\mathsf{D} = \mathsf{G}^{\dagger}\mathsf{G} = \mathfrak{F}^{\dagger}\mathbb{D}\mathfrak{F}, \tag{15.1}$$

where in the case of global one-dimensional quantum gravity one the density matrix has the form

$$\mathbb{D} = \begin{bmatrix} \dfrac{(\mu+1)^2}{4\mu} & \dfrac{1-\mu^2}{4\mu} \\ \dfrac{1-\mu^2}{4\mu} & \dfrac{(\mu-1)^2}{4\mu} \end{bmatrix}, \tag{15.2}$$

here $\mu = \dfrac{\omega}{\omega_I}$ is the mass scale of the system of space quanta. Note that in the present situation

$$\mathbb{D}^2 = (\mathrm{Tr}\mathbb{D})\mathbb{D}, \tag{15.3}$$

$$\det\mathbb{D} = 0. \tag{15.4}$$

Vanishing of the determinant means that in the one-particle approximation the corresponding thermodynamics is *irreversible*, or in other words that the thermodynamic processes in the Æther of space quanta are irreversible.

Employing the density matrix (15.90), one can establish the value of the occupation number

$$N = \frac{\mathrm{Tr}\left(\mathbb{D}^2\right)}{\mathrm{Tr}\mathbb{D}} = \frac{\mathrm{Tr}\left((\mathrm{Tr}\mathbb{D})\mathbb{D}\right)}{\mathrm{Tr}\mathbb{D}} = \frac{(\mathrm{Tr}\mathbb{D})^2}{\mathrm{Tr}\mathbb{D}} = \mathrm{Tr}\mathbb{D} = \frac{\mu^2+1}{2\mu}, \tag{15.5}$$

and the entropy can be derived from its basic definition

$$S = \frac{\mathrm{Tr}(\mathbb{D}\ln\mathbb{D})}{\mathrm{Tr}\mathbb{D}}. \tag{15.6}$$

The problem is to derive $\ln\mathbb{D}$. It can be performed effectively by application of the matrix Taylor series

$$\ln\mathbb{D} = \ln[(\mathbb{D}-\mathbb{I})+\mathbb{I}] = \sum_{n=1}^{\infty}\frac{(-1)^{n+1}}{n}(\mathbb{D}-\mathbb{I})^n, \tag{15.7}$$

and the matrix Newton binomial series

$$(\mathbb{D}-\mathbb{I})^n = \sum_{k=1}^{n}\binom{n}{k}\mathbb{D}^k(-\mathbb{I})^{n-k} = \sum_{k=1}^{n}(-1)^{n-k}\binom{n}{k}\mathbb{D}^k = \tag{15.8}$$

$$= \sum_{k=1}^{n}(-1)^{n-k}\binom{n}{k}(\mathrm{Tr}\mathbb{D})^{k-1}\mathbb{D}, \tag{15.9}$$

where $\binom{n}{m}$ is the Newton binomial symbol

$$\binom{n}{m} = \frac{n!}{m!(n-m)!}, \tag{15.10}$$

which allows to write out the following entropy formula

$$S = \sum_{n=1}^{\infty}\sum_{k=1}^{n}\frac{(-1)^{k-1}}{n}\binom{n}{k-1}S_k, \tag{15.11}$$

where S_k are *cluster entropies*

$$S_k = \frac{\mathrm{Tr}(\mathbb{D}^k)}{\mathrm{Tr}\mathbb{D}} = \frac{\mathrm{Tr}\left((\mathrm{Tr}\mathbb{D})^{k-1}\mathbb{D}\right)}{\mathrm{Tr}\mathbb{D}} = \frac{(\mathrm{Tr}\mathbb{D})^{k-1}\mathrm{Tr}\mathbb{D}}{\mathrm{Tr}\mathbb{D}} = N^{k-1}. \tag{15.12}$$

We shall call the series expansion (15.11) *the cluster series*. Its summation is the method for obtaining entropy in the analytical way.

According to a certain mathematical tradition, the series expansion (15.11) converges if and only if the spectral radius ρ of the difference $\mathbb{D}-\mathbb{I}$, where $\mathbb{I}$ is 2×2 unit matrix, is

$$\rho(\mathbb{D}-\mathbb{I}) < 1. \tag{15.13}$$

Straightforward calculation leads to the condition for the mass scale

$$\mu \in [1;2+\sqrt{3}), \tag{15.14}$$

and the result of the summation procedure is

$$S = \frac{\zeta(1)}{2}\left(\frac{\mu^2-1}{\mu^2+1}\right)^2 + \frac{\mu^4+6\mu^2+1}{(\mu^2+1)^2}\ln\frac{(\mu-1)^2}{2\mu}, \tag{15.15}$$

where $\zeta(s)$ is the Riemann zeta function

$$\zeta(s) = \sum_{n=1}^{\infty}\frac{1}{n^s}, \tag{15.16}$$

and $\zeta(1)$ is manifestly divergent.

Note that by straightforward application of the Hagedorn hadronization formula $\omega \sim T_H$ [938], where m is the mass of the system, one can establish the hadronized temperature as

$$\frac{T_H}{T_I} = \mu. \tag{15.17}$$

By the relation $\Delta\omega \sim \Delta T_H$ one obtains the hadronized temperature normalized to T_I value

$$\frac{\Delta T_H}{T_I} = \mu_{\max} - \mu_{\min} = 1+\sqrt{3} \approx 2.732, \tag{15.18}$$

so that one can establish the ratio

$$\frac{T_H}{\Delta T_H} \in \left[\frac{\sqrt{3}-1}{2}, \frac{\sqrt{3}+1}{2}\right), \tag{15.19}$$

which allows to determine the reciprocal

$$\frac{\Delta T_H}{T_H} \in \left(-(\sqrt{3}+1), \sqrt{3}+1\right]. \tag{15.20}$$

Defining anisotropy as $\delta T_H = \Delta T_H - T_H$, one derives

$$\frac{\delta T_H}{T_H} \in \left(-(\sqrt{3}+2), \sqrt{3}\right], \tag{15.21}$$

so that half of the difference anisotropy is exactly

$$\Delta\left(\frac{\delta T_H}{T_H}\right) = \left(\frac{\Delta T_H}{T_H}\right)_{\max} - \left(\frac{\Delta T_H}{T_H}\right)_{\min} = 2(\sqrt{3}+1) = 2\frac{\Delta T_H}{T_I}. \tag{15.22}$$

In this manner, one obtains

$$\Delta T_H = \frac{T_I}{2}\Delta\left(\frac{\delta T_H}{T_H}\right) \approx 2.732 T_I. \tag{15.23}$$

The difference (15.23) can be identified with a background temperature, *i.e.* $\Delta T_H \equiv T_B$. For the initial datum $T_I \sim 1\text{K}$ it is very close to the averaged Cosmic Microwave Background Radiation temperature, $T_{CMB} \approx 2.725\text{K}$. Because the quantity $|T_{CMB} - T_B|$ is small $7 \cdot 10^{-3}\text{K}$, one can deduce that next approximations will result in successive contributions to the temperature T_{CMB}. In other words,

$$T_{CMB} = T_B + \ldots. \tag{15.24}$$

In the dynamical Fock repère the Hamiltonian operator is

$$\mathsf{H} = \frac{m}{2}\left(\mathsf{G}^\dagger\mathsf{G} + \mathsf{G}\mathsf{G}^\dagger\right) = \mathfrak{F}^\dagger \mathbb{H} \mathfrak{F}, \tag{15.25}$$

where $\mathbb{H}$ is the Hamiltonian matrix in the static Fock repère

$$\mathbb{H} = \begin{bmatrix} \frac{\omega_I}{4}\left(1+\mu^2\right) & \frac{\omega_I}{4}\left(1-\mu^2\right) \\ \frac{\omega_I}{4}\left(1-\mu^2\right) & \frac{\omega_I}{4}\left(1+\mu^2\right) \end{bmatrix}, \tag{15.26}$$

which for a fixed mass scale has a discrete spectrum

$$\text{Spec}\ \mathbb{H} = \left\{\frac{\omega_I}{2}\mu^2, \frac{\omega_I}{2}\right\}. \tag{15.27}$$

The internal energy, calculated from the Hamiltonian matrix (15.26), is

$$U = \frac{\text{Tr}(\mathbb{D}\mathbb{H})}{\text{Tr}\mathbb{D}} = \frac{\omega_I}{4}(\mu^2+1). \tag{15.28}$$

The Hamiltonian matrix $\mathbb{H}$, however, consists the constant term $\mathbb{H}_I$

$$\mathbb{H}_I = \begin{bmatrix} \frac{\omega_I}{4} & \frac{\omega_I}{4} \\ \frac{\omega_I}{4} & \frac{\omega_I}{4} \end{bmatrix} \tag{15.29}$$

which can be eliminated by simple renormalization

$$\mathbb{H} \to \mathbb{H}' = \mathbb{H} - \mathbb{H}_I = \begin{bmatrix} \frac{\omega_I}{4}\mu^2 & -\frac{\omega_I}{4}\mu^2 \\ -\frac{\omega_I}{4}\mu^2 & \frac{\omega_I}{4}\mu^2 \end{bmatrix}. \tag{15.30}$$

The spectrum of the renormalized Hamiltonian matrix is

$$\text{Spec } \mathbb{H}' = \left\{ \frac{\omega_I}{2}\mu^2, 0 \right\}, \tag{15.31}$$

and straightforward computation of the renormalized internal energy yields the following result

$$U' = \frac{\text{Tr}(\mathbb{D}\mathbb{H}')}{\text{Tr}\mathbb{D}} = \frac{\omega_I}{4}\mu^2 \equiv U - U_I, \tag{15.32}$$

where $U_I = \frac{\omega_I}{4}$ is the constant term, which possesses the property of the Eulerian homogeneity of degree 2, *i.e.*

$$U'[\alpha\mu] = \alpha^2 U'[\mu]. \tag{15.33}$$

In this manner, the thermodynamics describing space quanta Æther can be formulated in the standard way of the Eulerian systems.

The three elementary and fundamental physical characterizations, *i.e.* the occupation number N, the internal energy U, and the entropy S, have been derived and, therefore, one can conclude the appropriate thermodynamics by straightforward application of the first principles. Actually, the entropy (15.91) is manifestly divergent due to the presence of formal infinity $\zeta(1)$. Straightforward calculation shows that the temperature $T = \frac{\delta U}{\delta S}$ arising from the entropy (15.91) contains the term with the singularity $\zeta(1)$. However, it is also visible that such a temperature possesses a finite limit if and only if one re-scales the initial datum of the mass to formal infinity, *i.e.* $\omega_I \to \omega_I \zeta(1)$. In the light of the fact that the mass m is related to length l like $m \sim 1/l$, performing of the limiting procedure $\omega_I \to \infty$ corresponds with introduction *purely point object* $l_I \to 0$ to the theory.

Scaling of the initial datum is not a good procedure, however, because it has no well-defined physical meaning. There is another possibility for reorganization of the troublesome divergence. Namely, it can be seen by straightforward calculation that the entropy renormalization $S \to \dfrac{S}{\zeta(1)}$ with performing the formal limit $\zeta(1) \to \infty$ leads to the equivalent result for the thermodynamics with no need t apply unclear scaling of the initial datum. Such entropy renormalization corresponds to the initial quantum state of the embedded three-dimensional space being a purely point object, and yields perfect accordance with the second law of thermodynamics

$$S \longrightarrow S' = \lim_{\zeta(1)\to\infty} \frac{S}{\zeta(1)} = \frac{1}{2}\left(\frac{\mu^2-1}{\mu^2+1}\right)^2 \geqslant 0. \tag{15.34}$$

Calculating the temperature T' of space quanta, one obtains the formula

$$T' = \frac{\delta U'}{\delta S'} = \omega_I \frac{(\mu^2+1)^3}{8(\mu^2-1)}, \tag{15.35}$$

and one sees that initially, *i.e.* for $\mu = 1$, temperature is infinite. Such a situation describes the *Hot Big Bang (HBB) phenomenon*. It can be seen that after the HBB point the system is cooled right up until the mass scale reaches the value

$$\mu_{PT} = \sqrt{2} \approx 1.414, \tag{15.36}$$

and then is warmed, which means that the value μ_{PT} is the phase transition point.

This phenomenon is better visible when one computes the energetic heat capacity C_U

$$C_U = T\frac{\delta S'}{\delta T} = \frac{\delta U'}{\delta T} = \frac{(\mu^2-1)^2}{(\mu^2-2)(\mu^2+1)^2}, \tag{15.37}$$

which possesses the singularity at the point μ_{PT}. Application of the generalized law of equipartition

$$\frac{\delta U'}{\delta T'} = \frac{f}{2}, \tag{15.38}$$

allows to establish the variability of the number of degrees of freedom

$$f = 2C_U. \tag{15.39}$$

The Helmholtz free energy $F = U' - T'S'$

$$F = -\frac{\omega_I}{16}(\mu^4 - 4\mu^2 - 1), \tag{15.40}$$

is finite for the finite value of the initial datum ω_I, increases since the initial point $\mu = 1$ until the phase transition μ_{PT}, and then decreases. Therefore, the *thermal equilibrium point* is the HBB point related to the initial point $\mu_{eq} = 1$. In the region of mass scales for which $1 \leqslant \mu < \mu_{PT}$ the mechanical isolation is absent, but it is present after the phase transition, *i.e.* in the region $\mu > \mu_{PT}$.

Calculating the chemical potential

$$\varpi = \frac{\delta F}{\delta N} = -\omega_I \frac{\mu^3(\mu^2 - 2)}{2(\mu^2 - 1)}, \tag{15.41}$$

one can see straightforwardly that at the HBB point $\mu_{eq} = 1$ this thermodynamic potential diverges and at the phase transition point μ_{PT} identically vanishes. Applying the chemical potential (15.41) together with the occupation number N and the Helmholtz free energy F yields the appropriate free energy defined by the Landau grand potential Ω

$$\Omega = F - \varpi N = \omega_I \frac{3\mu^6 + \mu^4 - 11\mu^2 - 1}{16(\mu^2 - 1)}, \tag{15.42}$$

and, therefore, the corresponding Massieu–Planck free entropy Ξ can be also derived straightforwardly

$$\Xi = -\frac{\Omega}{T} = -\frac{3\mu^6 + \mu^4 - 11\mu^2 - 1}{2(\mu^2 + 1)^3}. \tag{15.43}$$

Consequently, the grand partition function Z is established as

$$Z = e^{\Xi} = \exp\left\{-\frac{3\mu^6 + \mu^4 - 11\mu^2 - 1}{2(\mu^2 + 1)^3}\right\}. \tag{15.44}$$

The feature of the second order Eulerian homogeneity of the system of space quanta yields the equation of state $\frac{PV}{T} = \ln Z = \Xi$ and determines the product of pressure P and volume V as

$$PV = -\Omega, \tag{15.45}$$

and together with the appropriate Gibbs–Duhem equation

$$V\delta P = S'\delta T + N\delta\varpi \tag{15.46}$$

allows to establish the value of the pressure

$$|P| = \exp\left\{-\int\left(\frac{S}{\Omega}\delta T + \frac{N}{\Omega}\delta\varpi\right)\right\}. \tag{15.47}$$

Similarly, the first law of thermodynamics

$$-\delta\Omega = S'\delta T + P\delta V + N\delta\varpi, \tag{15.48}$$

together with the equation of state (15.45) determine the volume

$$V = \frac{|\Omega|}{|P|}, \tag{15.49}$$

which is positive by definition. Regarding the equation of state (15.45) for $\Omega = -|\Omega| < 0$ the pressure is $P = |P|$, whereas for $\Omega = |\Omega| > 0$ the pressure has the value $P = -|P|$. In this manner, one receives

$$P = P(\mu_0)\exp\left[\int_{\mu_0}^{\mu}\frac{8t\left(t^6 - 3t^2 + 4\right)}{3t^8 - 2t^6 - 12t^4 + 10t^2 + 1}dt\right], \tag{15.50}$$

where μ_0 is some reference value of μ. Regarding the relation (15.49), V is a fixed parameter and its value can be established as follows

$$V = \frac{\omega_I}{P(\mu_0)}\frac{3\mu^6 + \mu^4 - 11\mu^2 - 1}{16(\mu^2 - 1)}\exp\left[-\int_{\mu_0}^{\mu}\frac{8t\left(t^6 - 3t^2 + 4\right)}{3t^8 - 2t^6 - 12t^4 + 10t^2 + 1}dt\right]. \tag{15.51}$$

One can also easily determine two other thermodynamical potentials: the Gibbs free energy $G = U - TS + PV$ and the enthalpy $H = U + PV$. The results are as follows

$$G = -\frac{\omega_I\mu^2\left(\mu^4 - \mu^2 - 2\right)}{4\left(\mu^2 - 1\right)}, \tag{15.52}$$

$$H = -\frac{\omega_I\left(3\mu^6 - 3\mu^4 - 7\mu^2 - 1\right)}{16\left(\mu^2 - 1\right)}. \tag{15.53}$$

Equivalently, the thermodynamics of space quanta Æther can be expressed through the size scale $\lambda = \frac{1}{\mu}$. There are the relations relating both scales with an occupation number

$$\lambda = N\left(1 \mp \sqrt{1-\frac{1}{N^2}}\right), \tag{15.54}$$

$$\mu = N\left(1 \pm \sqrt{1-\frac{1}{N^2}}\right), \tag{15.55}$$

which in the limit of infinite N are equal

$$\lambda(N=\infty) = \{0,\infty\}, \tag{15.56}$$

$$\mu(N=\infty) = \{\infty,0\}. \tag{15.57}$$

Therefore, there are two possible asymptotic behaviours. The first situation, $\lambda = 0$ and $\mu = \infty$, can be interpreted as a *black hole* as well as with the Hot Big Bang. The second situation, given by $\lambda = \infty$ and $\mu = 0$, defines a stable classical physical object - the classical space-time.

One can establish the number n of space quanta generated from the stable Bogoliubov vacuum determined by the static Fock space related to the initial data. We shall call such class of states *vacuum space quanta*. By definition this is the vacuum expectation value of the one-particle density operator. In other words,

$$n = \frac{\langle 0|\mathsf{D}|0\rangle}{\langle 0|0\rangle} = \frac{\langle 0|\mathsf{G}^\dagger\mathsf{G}|0\rangle}{\langle 0|0\rangle}. \tag{15.58}$$

Straightforward application of the Bogoliubov transformation and the canonical commutation relations of the static Fock space leads to

$$\begin{aligned} \mathsf{G}^\dagger\mathsf{G} &= \left(u\mathsf{G}_I^\dagger - v\mathsf{G}_I\right)\left(-v^*\mathsf{G}_I^\dagger + u^*\mathsf{G}_I\right) = \\ &= |v|^2\mathsf{G}_I\mathsf{G}_I^\dagger + |u|^2\mathsf{G}_I^\dagger\mathsf{G}_I - v^*u\mathsf{G}_I^\dagger\mathsf{G}_I^\dagger - vu^*\mathsf{G}_I\mathsf{G}_I = \\ &= |v|^2 + \left(|u|^2+|v|^2\right)\mathsf{G}_I^\dagger\mathsf{G}_I - v^*u\mathsf{G}_I^\dagger\mathsf{G}_I^\dagger - vu^*\mathsf{G}_I\mathsf{G}_I. \end{aligned} \tag{15.59}$$

and using the properties of the stable Bogoliubov vacuum gives

$$\langle 0|\mathsf{G}^\dagger\mathsf{G}|0\rangle = |v|^2\langle 0|0\rangle + \left(|u|^2+|v|^2\right)\langle 0|\mathsf{G}_I^\dagger\mathsf{G}_I|0\rangle - \tag{15.60}$$

$$- \; v^*u\langle 0|\mathsf{G}_I^\dagger\mathsf{G}_I^\dagger|0\rangle - vu^*\langle 0|\mathsf{G}_I\mathsf{G}_I|0\rangle = \tag{15.61}$$

$$= |v|^2\langle 0|0\rangle. \tag{15.62}$$

Therefore, the number of vacuum space quanta is

$$n = |v|^2 = \frac{(\mu - 1)^2}{4\mu}. \tag{15.63}$$

In this manner, the mass scale has two values

$$\mu_{\pm}(n) = \left(\sqrt{n} \pm \sqrt{n+1}\right)^2, \tag{15.64}$$

corresponding to two independent phases of the Æther of space quanta. For convenience we shall call the phase described by the plus sign *the positive phase of the Æther*, and the phase described by the minus sign *the negative phase of the Æther*. The phases have completely different asymptotic behaviours. Namely,

$$\lim_{n\to\infty} \mu_{\pm}(n) = \begin{cases} \infty & \text{for the positive phase} \\ 0 & \text{for the negative phase} \end{cases}. \tag{15.65}$$

Let us consider the asymptotic thermodynamics $n \to \infty$ of the Æther. The case of the positive phase is characterized by

$$T \to \infty, \tag{15.66}$$
$$C_U \to 0, \tag{15.67}$$
$$f \to 0, \tag{15.68}$$
$$F \to \infty, \tag{15.69}$$
$$\varpi \to \infty, \tag{15.70}$$
$$\Omega \to \infty, \tag{15.71}$$
$$\Xi \to -\frac{3}{2}, \tag{15.72}$$
$$Z \to e^{-3/2}, \tag{15.73}$$
$$G \to \infty, \tag{15.74}$$
$$H \to \infty, \tag{15.75}$$
$$P \to P_\infty, \tag{15.76}$$
$$V \to \infty, \tag{15.77}$$

where P_∞ is the (constant) value of the pressure in the asymptotic value $\mu = \infty$ which can be assessed numerically. For instance, in the trivial case $\mu_0 = 0$ one obtains $P_\infty \approx P(0)\exp(23.9527)$, whereas for $\mu_0 = 1$ one receives $P_\infty \approx P(1)\exp(510959)$.

Similarly, one can analyse the negative phase

$$T \rightarrow -\frac{\omega_I}{8}, \tag{15.78}$$

$$C_U \rightarrow -\frac{1}{2}, \tag{15.79}$$

$$f \rightarrow -1, \tag{15.80}$$

$$F \rightarrow \frac{\omega_I}{16}, \tag{15.81}$$

$$\varpi \rightarrow 0, \tag{15.82}$$

$$\Omega \rightarrow \frac{\omega_I}{16}, \tag{15.83}$$

$$\Xi \rightarrow \frac{1}{2}, \tag{15.84}$$

$$Z \rightarrow e^{1/2}, \tag{15.85}$$

$$G \rightarrow 0, \tag{15.86}$$

$$H \rightarrow -\frac{\omega_I}{16}, \tag{15.87}$$

$$P \rightarrow P_0, \tag{15.88}$$

$$V \rightarrow \frac{\omega_I}{16P_0}, \tag{15.89}$$

where P_0 is the (constant) value of the pressure in $\mu = 0$ which can be assessed numerically. For instance when $\mu_0 = 0$ one has $P_0 = P(0)$. For $\mu_0 = 1$ one obtains $P_0 \approx P(1)\exp(-151.536)$.

In this section we have presented the next implication of the global one-dimensional quantum gravity. It was shown that this algorithm yields a constructive, consistent, and plausible quantum gravity phenomenology, that is thermodynamics of Æther, in the discussed situation describing the behaviour of the space quanta. The theory of quantum gravity as well as the thermodynamics can be applied to any general relativistic space-times whose metrics can be presented in the form of the $3+1$ splitting. Such space-time satisfy the Mach principle, *i.e.* are defined by isotropic metric tensors. Their importance for elementary particle physics, cosmology and high energy astrophysics is experimentally confirmed; one can say that these are the phenomenological space-times.

As the example of *ab initio* formulation of thermodynamics we have employed the one-particle approximation of the density matrix. Application of the renormalization method to the entropy and to the Hamiltonian matrix resulted in the second order Eulerian homogeneity property. The Landau grand potential Ω and the Massieu–Planck free entropy Ξ were employed to ensure consistent physical description. The grand partition function Z and thermodynamic volume V were determined constructively. Other thermodynamical potentials were derived in the framework of the *entropic formalism*, which agrees with the first and the second principles of thermodynamics. Physical information following from the thermodynamics of space quanta Æther is the crucial point of the construction presented in this section. Actually, the proposed approach differs from another ones (Cf. *e.g.* [939, 940, 941, 942]) by *ab initio* treatment of the quantum gravity phenomenology.

Studying of particular physical situations in the framework of the proposed approach seems to be the most important prospective arising from the thermodynamics of the Æther of space quanta. From the experimental point of view the presented considerations are evidently useful, because bosonic systems are common in high energy physics.

B Entropy II: Algebraic Approach

Let us consider the space quanta Æther according to the method of the grand canonical ensemble. First of all, let us express the static one-particle density matrix by the Bogoliubov coefficients

$$\mathbb{D} = \left[\begin{array}{cc} |u|^2 & -uv \\ -u^*v^* & |v|^2 \end{array}\right]. \tag{15.90}$$

The basic quantity is an entropy, which for an arbitrary quantum system is defined by the standard Boltzmann–von Neumann formula

$$S = \frac{\mathrm{Tr}\,(\mathbb{D}\ln\mathbb{D})}{\mathrm{Tr}\mathbb{D}}, \tag{15.91}$$

and in the present situation can be immediately computed from the density matrix (15.90).

The problem is to calculate logarithm of the one-particle density matrix $\ln\mathbb{D}$. This can be performed by application of the linear algebra methods, particularly the polynomial long division algorithm, the characteristic polynomial, and the Cayley–Hamilton theorem (For some details of basic and advanced algebra Cf. *e.g.* books in Ref. [943]). Let us present the method in detail for any analytical function of matrix $\mathbb{D}$. Let us consider the characteristic polynomial $ch_{\mathbb{D}}(\lambda)$ of the matrix $\mathbb{D}$, where λ is any eigenvalue of $\mathbb{D}$. If $p(\lambda)$ is an analytical function, *i.e.* possesses power series expansion, then by using the division transformation one can present $p(\lambda)$ as the dividend whose divisor is the characteristic polynomial

$$p(\lambda) = q(\lambda) ch_{\mathbb{D}}(\lambda) + r(\lambda). \tag{15.92}$$

In other words, the problem is to establish the remainder polynomial $r(\lambda)$ and the quotient polynomial $q(\lambda)$. Recall that according to the Cayley–Hamilton theorem, named after Arthur Cayley and William Rowan Hamilton, the characteristic polynomial of square matrix $\mathbb{D}$ over a commutative ring, like *e.g.* the field of real or complex numbers, evaluated on this matrix vanishes identically

$$ch_{\mathbb{D}}(\mathbb{D}) = 0, \tag{15.93}$$

so that, consequently, the evaluation $p(\mathbb{D})$ is exactly equal to the reminder polynomial evaluated on the matrix $\mathbb{D}$

$$p(\mathbb{D}) = r(\mathbb{D}). \tag{15.94}$$

If matrix $\mathbb{D}$ is a matrix of dimension $n \times n$, then its characteristic polynomial $ch_{\mathbb{D}}(\lambda)$ is a polynomial of degree n whose coefficients are invariants of matrix $\mathbb{D}$. Therefore, the remainder polynomial $r(\lambda)$ must be of the order $n-1$ at most. There is some problem when there are eigenvalues of $\mathbb{D}$ for which the function $p(\lambda)$ has a singularity. Then, however, we shall not include such eigenvalues, and for determination of the quotient $q(\lambda)$ and the remainder polynomial $r(\lambda)$ we shall differentiate the relation (15.92) n times and evaluate all $n+1$ relations on the non-singular eigenvalues of $\mathbb{D}$. Such a procedure generates the system of $n+1$ equations which allows to establish the coefficients of the remainder polynomial $r(\lambda)$ as well as leads to the quotient $q(\lambda)$ as the result of solving an appropriate differential equation of the degree

n at most. It should be emphasized that the quotient $q(\lambda)$, as a solution of the ordinary differential equation, is assured to be an analytical function.

Let us apply such a method to the situation given by the 2×2 matrix $\mathbb{D}$ having the zero determinant $\det\mathbb{D} = 0$, and the function $p(x) = \ln x$. The characteristic polynomial of the matrix $\mathbb{D}$ is

$$ch_{\mathbb{D}}(\lambda) = \det(\mathbb{D} - \lambda\mathbb{I}) = \lambda^2 - (\mathrm{Tr}\mathbb{D})\lambda, \tag{15.95}$$

and its eigenvalues are $\lambda = 0$ and $\lambda = \mathrm{Tr}\mathbb{D}$. Application of the Cayley–Hamilton theorem gives the following constraint

$$\mathbb{D}^2 - (\mathrm{Tr}\mathbb{D})\mathbb{D} = 0, \tag{15.96}$$

i.e. determines the square of the matrix as $\mathbb{D}^2 = (\mathrm{Tr}\mathbb{D})\mathbb{D}$. Let us write out the basic relation (15.92) for the present case

$$\ln\lambda = q(\lambda)\left(\lambda^2 - (\mathrm{Tr}\mathbb{D})\lambda\right) + a_0 + a_1\lambda, \tag{15.97}$$

where a_0 and a_1 are the coefficients of the reminder polynomial $r(\lambda)$. The problem is to establish both the quotient function $q(\lambda)$ as well as coefficients a_0 and a_1. First of all, let us note that the eigenvalue $\lambda = 0$ is the singularity of the function $\ln\lambda$, which means that we shall not consider this eigenvalue. For determination of the three unknown quantities let us differentiate the relation (15.97) two times. The results are as follows

$$\begin{aligned}
\frac{1}{\lambda} &= q'(\lambda)\left(\lambda^2 - (\mathrm{Tr}\mathbb{D})\lambda\right) + q(\lambda)(2\lambda - \mathrm{Tr}\mathbb{D}) + a_1, & (15.98)\\
-\frac{1}{\lambda^2} &= q''(\lambda)\left(\lambda^2 - (\mathrm{Tr}\mathbb{D})\lambda\right) + 2q'(\lambda)(2\lambda - \mathrm{Tr}\mathbb{D}) + 2q(\lambda). & (15.99)
\end{aligned}$$

We take into account the fact that evaluation of the characteristic polynomial on an eigenvalue is zero. Application of this fundamental property in the case of the eigenvalue $\lambda = \mathrm{Tr}\mathbb{D}$ leads to a significant simplification of the system of equations (15.97), (15.98) and (15.99)

$$\begin{aligned}
\ln\lambda &= a_0 + a_1\lambda, & (15.100)\\
\frac{1}{\lambda} &= q(\lambda)\lambda + a_1, & (15.101)\\
-\frac{1}{\lambda^2} &= 2q'(\lambda)\lambda + 2q(\lambda), & (15.102)
\end{aligned}$$

which can be presented in the following form

$$a_0 = \ln\lambda - a_1\lambda, \tag{15.103}$$

$$a_1 = \frac{1}{\lambda} - q(\lambda)\lambda, \tag{15.104}$$

$$\lambda q'(\lambda) + q(\lambda) + \frac{1}{2\lambda^2} = 0. \tag{15.105}$$

The equation (15.105) is the differential equation for the quotient function $q(\lambda)$ and can be solved straightforwardly

$$q(\lambda) = \frac{1}{2\lambda^2} + \frac{C}{\lambda}, \tag{15.106}$$

where C is the integration constant. Because we are still interested in the concrete eigenvalue $\lambda = \mathrm{Tr}\mathbb{D}$, one receives

$$q(\mathrm{Tr}\mathbb{D}) = \frac{1}{2(\mathrm{Tr}\mathbb{D})^2} + \frac{C}{\mathrm{Tr}\mathbb{D}}, \tag{15.107}$$

and, for this reason,

$$a_1 = \frac{1}{\mathrm{Tr}\mathbb{D}} - q(\mathrm{Tr}\mathbb{D})\mathrm{Tr}\mathbb{D} = \frac{1}{2\mathrm{Tr}\mathbb{D}} - C, \tag{15.108}$$

$$a_0 = \ln\mathrm{Tr}\mathbb{D} - a_1\mathrm{Tr}\mathbb{D} = \ln\mathrm{Tr}\mathbb{D} - \frac{1}{2} + C\mathrm{Tr}\mathbb{D}. \tag{15.109}$$

In this manner, one can compute the function $\ln\mathbb{D}$ as follows

$$\ln\mathbb{D} = a_0\mathbb{I} + a_1\mathbb{D} = \left(\ln\mathrm{Tr}\mathbb{D} - \frac{1}{2} + C\mathrm{Tr}\mathbb{D}\right)\mathbb{I} + \left(\frac{1}{2\mathrm{Tr}\mathbb{D}} - C\right)\mathbb{D}, \tag{15.110}$$

and, consequently, one obtains

$$\mathbb{D}\ln\mathbb{D} = \left(\ln\mathrm{Tr}\mathbb{D} - \frac{1}{2} + C\mathrm{Tr}\mathbb{D}\right)\mathbb{D} + \left(\frac{1}{2\mathrm{Tr}\mathbb{D}} - C\right)\mathbb{D}^2 = \tag{15.111}$$

$$= \left(\ln\mathrm{Tr}\mathbb{D} - \frac{1}{2} + C\mathrm{Tr}\mathbb{D}\right)\mathbb{D} + \left(\frac{1}{2\mathrm{Tr}\mathbb{D}} - C\right)(\mathrm{Tr}\mathbb{D})\mathbb{D} = \tag{15.112}$$

$$= \left(\ln\mathrm{Tr}\mathbb{D} - \frac{1}{2} + C\mathrm{Tr}\mathbb{D}\right)\mathbb{D} + \left(\frac{1}{2} - C\mathrm{Tr}\mathbb{D}\right)\mathbb{D} = \tag{15.113}$$

$$= \left(\ln\mathrm{Tr}\mathbb{D} - \frac{1}{2} + C\mathrm{Tr}\mathbb{D} + \frac{1}{2} - C\mathrm{Tr}\mathbb{D}\right)\mathbb{D} = \tag{15.114}$$

$$= (\ln\mathrm{Tr}\mathbb{D})\,\mathbb{D}. \tag{15.115}$$

Now it is easy to establish the entropy

$$S = \frac{\mathrm{Tr}(\mathbb{D}\ln\mathbb{D})}{\mathrm{Tr}\mathbb{D}} = \frac{\mathrm{Tr}\left((\ln\mathrm{Tr}\mathbb{D})\mathbb{D}\right)}{\mathrm{Tr}\mathbb{D}} = \frac{(\ln\mathrm{Tr}\mathbb{D})\mathrm{Tr}\mathbb{D}}{\mathrm{Tr}\mathbb{D}} = \ln\mathrm{Tr}\mathbb{D}. \tag{15.116}$$

In our situation, the trace of the density matrix is

$$\mathrm{Tr}\mathbb{D} = |u|^2 + |v|^2 = 2|v|^2 + 1 \tag{15.117}$$

and, for this reason, one obtains the following entropy formula

$$S = \ln\left(2|v|^2 + 1\right) = -\ln\Sigma, \tag{15.118}$$

where Σ is the quantum statistics of the system of space quanta

$$\Sigma = \frac{1}{2|v|^2 + 1} = \frac{1}{2n+1}, \tag{15.119}$$

where n is the number of vacuum space quanta (15.62).

Let us compute the thermodynamical potentials for the received entropy (15.118) and the same values of the internal energy U, and the occupation number N established in the previous section, *i.e.*

$$S = \ln\left[\frac{\mu^2+1}{2\mu}\right], \tag{15.120}$$

$$U = \frac{\omega_I}{4}\mu^2, \tag{15.121}$$

$$N = \frac{\mu^2+1}{2\mu}. \tag{15.122}$$

The most important is, of course, the temperature of the system

$$T = \frac{\omega_I \mu^2}{2}\frac{\mu^2+1}{\mu^2-1}, \tag{15.123}$$

which for $\mu > 1$ is manifestly negative. The heat capacity has the form

$$C_U = \frac{1}{2} + \frac{1}{\mu^4 - 2\mu^2 - 1}, \tag{15.124}$$

and the number of degrees of freedom is

$$f = 1 + \frac{2}{\mu^4 - 2\mu^2 - 1}. \tag{15.125}$$

The Helmholtz free energy can be also derived straightforwardly

$$F = \frac{\omega_I \mu^2}{4}\left(1 - 2\frac{\mu^2+1}{\mu^2-1}\ln\left[\frac{\mu^2+1}{2\mu}\right]\right), \tag{15.126}$$

and the chemical potential has the form

$$\varpi = -\frac{2\omega_I \mu^3 \left(\mu^4 - 2\mu^2 - 1\right)}{\left(\mu^2-1\right)^3}\ln\left[\frac{\mu^2+1}{2\mu}\right]. \tag{15.127}$$

The Landau grand potential and the Massieu–Planck free entropy are respectively

$$\Omega = \frac{\omega_I \mu^2}{4}\left(1 + 2\frac{\left(\mu^2-3\right)\left(\mu^2+1\right)^2}{\left(\mu^2-1\right)^3}\ln\left[\frac{\mu^2+1}{2\mu}\right]\right), \tag{15.128}$$

$$\Xi = -\frac{\mu^2-1}{2\left(\mu^2+1\right)} - \frac{\left(\mu^2-3\right)\left(\mu^2+1\right)}{\left(\mu^2-1\right)^2}\ln\left[\frac{\mu^2+1}{2\mu}\right]. \tag{15.129}$$

The grand partition function can be established as

$$Z = \exp\left[-\frac{\mu^2-1}{2\left(\mu^2+1\right)}\right]\left(\frac{\mu^2+1}{2\mu}\right)^{-\frac{\left(\mu^2-3\right)\left(\mu^2+1\right)}{\left(\mu^2-1\right)^2}}. \tag{15.130}$$

The pressure is

$$P = P(\mu_0)\exp\left[-4\int_{\mu_0}^{\mu}\frac{\left(t^2-1\right)^2\left(t^4-2t^2-1\right) + 4\left(t^4+4t^2+1\right)\ln\left[\frac{t^2+1}{2t}\right]}{t\left(t^2-1\right)\left(\left(t^2-1\right)^3 + 2\left(t^2-3\right)\left(t^2+1\right)^2\ln\left[\frac{t^2+1}{2t}\right]\right)}dt\right] \tag{15.131}$$

where μ_0 is some reference value of μ, and the thermodynamical volume has the form

$$V = \frac{\omega_I \mu^2}{4P(\mu_0)}\left|1 + 2\frac{\left(\mu^2-3\right)\left(\mu^2+1\right)^2}{\left(\mu^2-1\right)^3}\ln\left[\frac{\mu^2+1}{2\mu}\right]\right| \times \tag{15.132}$$

$$\times \exp\left[4\int_{\mu_0}^{\mu}\frac{\left(t^2-1\right)^2\left(t^4-2t^2-1\right) + 4\left(t^4+4t^2+1\right)\ln\left[\frac{t^2+1}{2t}\right]}{t\left(t^2-1\right)\left(\left(t^2-1\right)^3 + 2\left(t^2-3\right)\left(t^2+1\right)^2\ln\left[\frac{t^2+1}{2t}\right]\right)}dt\right].$$

The Gibbs free energy and enthalpy are respectively

$$G = \frac{\omega_I \mu^2 (\mu^2+1)(\mu^4-2\mu^2-1)}{(\mu^2-1)^3} \ln\left[\frac{\mu^2+1}{2\mu}\right], \quad (15.133)$$

$$H = -\frac{\omega_I \mu^2 (\mu^2-3)(\mu^2+1)^2}{2(\mu^2-1)^3} \ln\left[\frac{\mu^2+1}{2\mu}\right]. \quad (15.134)$$

There is the question about asymptotic $n \to \infty$ thermodynamics of the Æther. The positive phase is then

$$T \to \infty, \quad (15.135)$$

$$C_U \to \frac{1}{2}, \quad (15.136)$$

$$f \to 1, \quad (15.137)$$

$$F \to -\infty, \quad (15.138)$$

$$\varpi \to -\infty, \quad (15.139)$$

$$\Omega \to \infty, \quad (15.140)$$

$$\Xi \to -\infty, \quad (15.141)$$

$$Z \to 0, \quad (15.142)$$

$$G \to -\infty, \quad (15.143)$$

$$H \to -\infty, \quad (15.144)$$

$$P \to P_\infty, \quad (15.145)$$

$$V \to \infty, \quad (15.146)$$

where P_∞ is the numerical value of the pressure at the asymptotic value $\mu = \infty$. For example in the trivial situation $\mu_0 = 0$ one obtains the value $P_\infty \approx P(0)\exp(473.822)$, whereas when $\mu_0 = 1$ one receives another result $P_\infty \approx P(1)\exp(0.208615) \approx 1.23197 P(1)$.

Similarly, the asymptotic thermodynamics of the negative phase of the space quanta Æther can be analyzed. The results are as follows

$$T \rightarrow 0, \tag{15.147}$$
$$C_U \rightarrow -\frac{1}{2}, \tag{15.148}$$
$$f \rightarrow -1, \tag{15.149}$$
$$F \rightarrow 0, \tag{15.150}$$
$$\varpi \rightarrow 0, \tag{15.151}$$
$$\Omega \rightarrow 0, \tag{15.152}$$
$$\Xi \rightarrow -\infty, \tag{15.153}$$
$$Z \rightarrow -\infty, \tag{15.154}$$
$$G \rightarrow 0, \tag{15.155}$$
$$H \rightarrow 0, \tag{15.156}$$
$$P \rightarrow P_0, \tag{15.157}$$
$$V \rightarrow 0, \tag{15.158}$$

where P_0 is the numerical value of pressure for the value $\mu = 0$. For example, in the trivial situation $\mu_0 = 0$ one has $P_0 = P(0)$, while when $\mu_0 = 1$ one obtains $P_0 \approx P(1)\exp(-510956)$.

C Summary

In this chapter we have presented two strategies for calculation of the thermodynamics of the Æther of space quanta. Both the strategies were based on the one-particle approximation of the density matrix, while the difference was the choice of the method for calculation of the entropy of the system. We have obtained two different points of view on the thermodynamics, and the physical consistency of both the strategies should be verified through confrontation with suitable experimental data.

Chapter 16

Higher Dimensional Quantum Gravity

We have seen that the global one-dimensionality conjecture in the case of the four-dimensional space-time, gives the natural possibility to formulate quantum gravity as a quantum field theory. This non-trivial feature of this approach leads to natural question about the generalization of the model to the case of the higher-dimensional Lorentzian space-time.

In this chapter we shall carry out the straightforward generalization of our approach to quantum gravity to the case of any space-time satisfying the higher-dimensional equivalent of the Einstein field equations of General Relativity. There is evident formal similarity between the generalized model and the four-dimensional global one-dimensional quantum gravity, but it should be emphasized that the higher-dimensional model of quantum gravity contains a certain new contribution in comparison to the four-dimensional model. This modification removes the specific approximation within the proposed ansatz, and makes the approach the most general possible. This modification, however, does not change the general idea of the one-dimensional quantum gravity leading to an extraordinary simple formulation of quantum gravity in terms of the most fundamental notions of quantum field theory. The essential change, carried by the higher-dimensional quantum gravity, is straightforward possibility to study the linkages between the quantum gravity and the string theory.

A Higher Dimensional Geometrodynamics

Let us consider the space-time which is a $D+1$-dimensional pseudo-Riemannian manifold (M,g) equipped with metric tensor $g_{\mu\nu}$ of the Lorentzian signature $(1,D)$. In such a situation, the Christoffel symbols $\Gamma^{\rho}_{\mu\nu}$, the Riemann–Christoffel curvature tensor $R^{\lambda}_{\mu\alpha\nu}$, the Ricci curvature tensor $R_{\mu\nu}$, and the Ricci scalar curvature ${}^{(D+1)}R$ are

$$\Gamma^{\rho}_{\mu\nu} = \frac{1}{2}g^{\rho\sigma}\left(g_{\mu\sigma,\nu}+g_{\sigma\nu,\mu}-g_{\mu\nu,\sigma}\right), \tag{16.1}$$

$$R^{\lambda}_{\mu\alpha\nu} = \Gamma^{\lambda}_{\mu\nu,\alpha}-\Gamma^{\lambda}_{\mu\alpha,\nu}+\Gamma^{\lambda}_{\sigma\alpha}\Gamma^{\sigma}_{\mu\nu}-\Gamma^{\lambda}_{\sigma\nu}\Gamma^{\sigma}_{\mu\alpha}, \tag{16.2}$$

$$R_{\mu\nu} = R^{\lambda}_{\mu\lambda\nu}=\Gamma^{\lambda}_{\mu\nu,\lambda}-\Gamma^{\lambda}_{\mu\lambda,\nu}+\Gamma^{\lambda}_{\sigma\lambda}\Gamma^{\sigma}_{\mu\nu}-\Gamma^{\lambda}_{\sigma\nu}\Gamma^{\sigma}_{\mu\lambda}, \tag{16.3}$$

$${}^{(D+1)}R = g^{\mu\nu}R_{\mu\nu}, \tag{16.4}$$

and the Einstein field equations hold

$$R_{\mu\nu}-\frac{1}{2}g_{\mu\nu}{}^{(D+1)}R+\Lambda g_{\mu\nu}=\kappa T_{\mu\nu}, \tag{16.5}$$

where $\kappa=\dfrac{8\pi G}{c^4}\approx 2.08\cdot 10^{-43}\ \mathrm{N}^{-1}$ is the Einstein constant, Λ is the cosmological constant, and $T_{\mu\nu}$ is the stress-energy tensor of Matter fields.

The Hamiltonian formulation of General Relativity adequate for an arbitrary four-dimensional pseudo-Riemannian space-time, which was presented in the previous chapters of this book, can be easily generalized to the case when the space-time is characterized by the metric of Lorentzian signature $(1,D)$ and has the dimension $D+1$. Such space-times are known in literature under the name of higher-dimensional space-times, and in such a situation the Arnowitt–Deser–Misner decomposition can be straightforwardly generalized. For this, of course, it is necessary to foliate the space-time manifold M with slices, *i.e.* an adequate family of constant-time hypersurfaces. Such a space-like slicing can be preformed if and only if M is globally hyperbolic manifold and, fortunately, the space-times with the Lorentzian signature naturally belong to the class of such manifolds.

Let $t(x^{\mu})$ be a scalar field on M such that the foliation $t=const$ defines a family of nonintersecting space-like hypersurfaces $\Sigma(t)$. Let y^i be the coordinates on all Σs, and let us fix a concrete hypersurface Σ restricted by parametric equations $x^{\mu}=x^{\mu}(y^i)$, where $i=1,\ldots,D$ labels coordinates intrinsic to Σ, or equivalently by equation $f(x^\mu)=0$. Then, the $D+1$-vector field $\partial_\mu f(x^\mu)$ is a normal to Σ which, if it is not null, defines the future-directed, time-like, unit normal vector field to Σ

$$n_\mu = -\frac{\partial_\mu f}{\sqrt{|\partial_\mu f \partial^\mu f|}} \sim \partial_\mu t, \quad n^\mu \partial_\mu f > 0, \tag{16.6}$$

which satisfies the condition $n^\mu n_\mu = -1$.

Let a congruence of curves γ, which in general are not geodesics nor orthogonal to hypersurfaces $\Sigma(t)$, be intersecting $\Sigma(t)$, t be an affine parameter along γ, and t^μ be a tangent vector to γ. Then

$$t^\mu \partial_\mu t = 1. \tag{16.7}$$

An arbitrary fixed curve γ_F is a mapping between points on all $\Sigma(t)$s

$$\gamma_F : P \in \Sigma(t) \mapsto P' \in \Sigma(t') \mapsto P'' \in \Sigma(t'') \mapsto \ldots P^{(n)} \in \Sigma(t^{(n)}), \quad n \in \mathbb{Z}_+ \tag{16.8}$$

and fixation of coordinates on arbitrarily selected two hypersurfaces Σ in the recurrence way establishes constant values of y^i for each n. Such construction determines the coordinate system (t,y^i) in the space-time manifold M. For other coordinate system x^μ: $x^\mu = x^\mu(t,y^i)$ the tangent vectors t^μ and e_i^μ in (t,y^i) are

$$t^\mu = (\partial_t x^\mu)_{y^i} = \delta_t^\mu, \tag{16.9}$$
$$e_i^\mu = (\partial_{y^i} x^\mu)_t = \delta_i^\mu. \tag{16.10}$$

In any coordinates the Lie derivative of the tangent vectors e_i^μ on a hypersurface vanishes identically, *i.e.* $\mathcal{L}_t e_i^\mu = 0$. Let us consider the unit normal vector field n_μ to a hypersurface $\Sigma(t)$

$$n_\mu = -N\partial_\mu t, \tag{16.11}$$

such that $n_\mu e_i^\mu = 0$, where N is the lapse scalar. In general $t^\mu \nparallel n^\mu$, so the tangent vector t^μ can be decomposed in the basis (n^μ, e_i^μ)

$$t^\mu = Nn^\mu + N^i e_i^\mu, \tag{16.12}$$

where N^i is the shift D-vector. Using transformation of coordinates $x^\mu = x^\mu(t,y^i)$, one can express infinitesimal change of coordinate $D+1$-vector in the coordinates (t,y^i)

$$dx^\mu = t^\mu dt + e_i^\mu dy^i = (Nn^\mu + N^i e_i^\mu)dt + e_i^\mu dy^i = (Ndt)n^\mu + (dy^i + N^i dt)e_i^\mu, \tag{16.13}$$

and hence the space-time interval $ds^2 = g_{\mu\nu}dx^\mu dx^\nu = dx^\mu dx_\mu$ is

$$ds^2 = -\left(N^2 - N_i N^i\right) dt^2 + N_i dx^i dt + N_j dx^j dt + h_{ij} dx^i dx^j, \tag{16.14}$$

where h_{ij} is induced metric on a hypersurface $\Sigma(t)$

$$h_{ij} = g_{\mu\nu} e_i^\mu e_j^\nu. \tag{16.15}$$

A space-time metric $g_{\mu\nu}$, obeying the Einstein field equations (16.5), is

$$g_{\mu\nu} = \begin{bmatrix} -N^2 + N_i N^i & N_j \\ N_i & h_{ij} \end{bmatrix}, \tag{16.16}$$

$$g^{\mu\nu} = \frac{1}{N^2} \begin{bmatrix} -1 & N^j \\ N^i & N^2 h^{ij} - N^i N^j \end{bmatrix}, \tag{16.17}$$

where $N^j = h^{ij} N_i$, and the condition holds

$$h_{ik} h^{kj} = \delta_i^j. \tag{16.18}$$

For $D = 3$ (16.16) is the Arnowitt–Deser–Misner decomposition and, in spite of straightforward generalization, the general case has not been derived in literature. The appropriate completeness relations are

$$g_{\mu\nu} = -n_\mu n_\nu + h_{ij} e^i_\mu e^j_\nu, \tag{16.19}$$

$$g^{\mu\nu} = -n^\mu n^\nu + h^{ij} e_i^\mu e_j^\nu. \tag{16.20}$$

It can be verified straightforwardly that

$$\sqrt{-g} = N\sqrt{h}, \tag{16.21}$$

where $g = \det g_{\mu\nu}$ and $h = \det h_{ij}$. The second fundamental form of a slice, called the extrinsic curvature tensor or induced curvature, is

$$K_{ij} = n_{\mu;\nu} e_i^\mu e_j^\nu = -\nabla_{(i} n_{j)} - n_{(i} n^k n_{j)|k}, \tag{16.22}$$

and its trace, called the intrinsic curvature, has the form

$$K = K_i^i = h^{ij} K_{ij} = n^\mu_{;\mu}. \tag{16.23}$$

Σ is convex when the congruence is diverging ($K > 0$) and concave when the congruence is converging ($K < 0$). The tangent vectors to $\Sigma(t)$ satisfy the Gauss–Weingarten equations

$$e^\alpha_{i;\beta} e^\beta_j = \Gamma^k_{ij} e^\alpha_k + K_{ij} n^\mu, \tag{16.24}$$

and the Gauss–Codazzi equations can be derived straightforwardly

$$R_{\mu\nu\kappa\lambda}e^{\mu}_{i}e^{\nu}_{j}e^{\kappa}_{k}e^{\lambda}_{l} = R_{ijkl} - K_{il}K_{jk} + K_{ik}K_{jl}, \tag{16.25}$$

$$R_{\mu\nu\kappa\lambda}n^{\mu}e^{\nu}_{i}e^{\kappa}_{j}e^{\lambda}_{k} = K_{ij|k} - K_{ik|j}. \tag{16.26}$$

Applying the decomposition of the Ricci curvature tensor and Ricci scalar curvature

$$R_{\mu\nu} = -R_{\kappa\mu\lambda\nu}n^{\kappa}n^{\lambda} + h^{ij}R_{\kappa\mu\lambda\nu}e^{\kappa}_{i}e^{\lambda}_{j}, \tag{16.27}$$

$${}^{(D)}R = -2h^{kl}R_{\kappa\mu\lambda\nu}n^{\kappa}n^{\lambda}e^{\mu}_{k}e^{\nu}_{l} + h^{kl}h^{ij}R_{\kappa\mu\lambda\nu}e^{\kappa}_{i}e^{\lambda}_{j}e^{\mu}_{k}e^{\nu}_{l}, \tag{16.28}$$

one obtains projections of the Einstein tensor $G_{\mu\nu} = R_{\mu\nu} - \frac{1}{2}g_{\mu\nu}{}^{(D+1)}R$

$$2G_{\mu\nu}n^{\mu}n^{\nu} = {}^{(D)}R + K^{ij}K_{ij} + K^{2}, \tag{16.29}$$

$$G_{\mu\nu}e^{\mu}_{j}n^{\nu} = K^{i}_{j|i} - K_{,j}. \tag{16.30}$$

Also, the Ricci equations

$$\mathscr{L}_{n}K_{ij} = n^{\mu}n^{\nu}e^{\kappa}_{i}e^{\lambda}_{j}R_{\mu\nu\kappa\lambda} - \frac{1}{N}N_{|ij} - K_{ik}K^{k}_{j}, \tag{16.31}$$

are easy derivable. Similarly, to the $3+1$-dimensional case, the stroke on the left of an index means the intrinsic covariant differentiation with respect to the coordinate labeled by this index. Two indices before the stroke means taking twice the intrinsic covariant derivative with respect to each of the indices. For any (r,s) tensor one has

$$T^{a_1\dots a_r}_{b_1\dots b_s|k} = \nabla_k T^{a_1\dots a_r}_{b_1\dots b_s} = \partial_k T^{a_1\dots a_r}_{b_1\dots b_s} + \sum_{i=1}^{r-1}\Gamma^{a_i}_{ck}T^{a_1\dots a_{i-1}ca_{i+1}\dots a_r}_{b_1\dots b_s} - \sum_{i=1}^{s-1}\Gamma^{c}_{b_ik}T^{a_1\dots a_r}_{b_1\dots b_{i-1}cb_{i+1}\dots b_s}, \tag{16.32}$$

where Γ^{k}_{ij} are the spatial Christoffel symbols

$$\Gamma^{k}_{ij} = \frac{1}{2}h^{kl}\left(h_{il,j} + h_{lj,i} - h_{ij,l}\right). \tag{16.33}$$

Applying the completeness relations (16.20) for the inverted metric and the following elementary relation

$$R_{\kappa\mu\lambda\nu}n^{\kappa}n^{\mu}n^{\lambda}n^{\nu} = 0, \tag{16.34}$$

one sees that the first term in the formula (16.28) reduces to $-2R_{\mu\nu}n^\mu n^\nu$. Application of the relations

$$R_{\mu\nu}n^\mu n^\nu = 2\left(n^\mu_{;[\nu}n^\nu\right)_{;\mu]} + 2n^\mu_{;[\mu}n^\nu_{;\nu]}, \tag{16.35}$$

$$n^\mu_{;\nu}n^\nu_{;\mu} = K^{ij}K_{ij}, \tag{16.36}$$

to the reduced first term in the formula (16.28), and the Gauss–Codazzi equations (16.25) to the second terms of this formula, *i.e.*

$$h^{kl}h^{ij}R_{\kappa\mu\lambda\nu}e^\kappa_i e^\lambda_j e^\mu_k e^\nu_l = h^{kl}h^{ij}\left(R_{ijkl} - K_{il}K_{jk} + K_{ik}K_{jl}\right) = {}^{(D)}R + K^2 - K^{ij}K_{ij}, \tag{16.37}$$

leads to the D-dimensional evaluation of the $D+1$-dimensional Ricci scalar curvature

$${}^{(D+1)}R = {}^{(D)}R + K^2 - K^{ij}K_{ij} - 2\left(n^\mu_{;\nu}n^\nu - n^\mu n^\nu_{;\nu}\right)_{;\mu}. \tag{16.38}$$

In the $D+1$ approach an induced spatial metric h_{ij} and the extrinsic curvature tensor K_{ij} are the appropriate dynamical variables describing the evolution of geometry of submanifold ∂M. In such a situation, the geometry evolves according to the Einstein field equations. The pair (h_{ij}, K_{ij}) describes the local geometry of a single space-like hypersurface ∂M, and then the evolution of the global space-time geometry can be formulated in terms of the one-parameter family $(h_{ij}(t), K_{ij}(t))$ describing the evolution of the local space geometry of the constant-time hypersurfaces ∂M_t. For consistency, the relation between the time evolution operator ∂_t and the vector field n normal to ∂M_t must be specified

$$\partial_t = Nn + N^i\partial_i. \tag{16.39}$$

Let an enveloping space-time manifold M be compact and possess a space-like boundary $(\partial M, h)$ equipped with the pair (h_{ij}, K_{ij}). Let topology of the space-time manifold M have the product form $\Sigma \times \mathbb{R}$, where Σ is an unrestricted topology of the space-like boundary ∂M. Then, the Einstein field equations (16.5) are the Euler–Lagrange equations of motion, following from the Hilbert–Palatini action principle, supplemented by the suitable boundary conditions, with respect to the metric tensor $g_{\mu\nu}$

$$\frac{\delta S[g]}{\delta g_{\mu\nu}} = 0, \tag{16.40}$$

$$\delta g_{\mu\nu}|_{\partial M} = 0. \tag{16.41}$$

This variational principle must be applied to the Einstein–Hilbert action complemented by the appropriate York–Gibbons–Hawking boundary action, *i.e.* the action of a $D+1$-geometry with fixed an induced D-geometry of a space-like boundary

$$S[g] = \frac{1}{c}\int_M d^{D+1}x\sqrt{-g}\left(-\frac{1}{2\kappa}{}^{(D+1)}R + \frac{\Lambda}{\kappa}\right) - \frac{1}{\kappa c}\int_{\partial M} d^D x\sqrt{h}K + S_\phi[g], \tag{16.42}$$

where $S_\phi[g]$ is the action of Matter fields

$$S_\phi[g] = \frac{1}{c}\int_M d^{D+1}x\sqrt{-g}L_\phi. \tag{16.43}$$

The Einstein field equations (16.5) result from vanishing of the variation $\delta S = \delta S_G + \delta S_\phi = 0$ on ∂M, where $S_G = S_{EH} + S_{YGH}$ and

$$\delta S_G = \frac{1}{2\kappa c}\int_M d^{D+1}x\sqrt{-g}\left(G_{\mu\nu} + \Lambda g_{\mu\nu}\right)\delta g^{\mu\nu}, \tag{16.44}$$

$$\delta S_{YGH} = \frac{1}{2\kappa c}\int_{\partial M} d^D y\sqrt{|h|}h^{\mu\nu}n^\rho\,\delta g_{\mu\nu,\rho}, \tag{16.45}$$

$$\delta S_\phi = -\frac{1}{2c}\int_M d^{D+1}x\sqrt{-g}T_{\mu\nu}\delta g^{\mu\nu}. \tag{16.46}$$

Moreover, the variational principle establishes the relationship

$$T_{\mu\nu} = -\frac{2}{\sqrt{-g}}\frac{\delta}{\delta g^{\mu\nu}}\left(\sqrt{-g}L_\phi\right). \tag{16.47}$$

In general, the stationarity of Matter fields, which simply means that the stress-energy tensor is identically vanishing $T_{\mu\nu} \equiv 0$, guarantees existence of a global time-like Killing vector field $\mathscr{K}_\mu$ for a metric tensor $g_{\mu\nu}$. One can specify a coordinate system such that the Killing vector field is time-like and the foliation $t = const$ is space-like. In such a situation, a metric tensor $g_{\mu\nu}$ depends at most on spatial coordinates x^i, and time coordinate ct becomes a global coordinate. Let us specify a coordinate system in such a way that an induced space is constant-time hypersurface. If the cosmological constant vanishes $\Lambda = 0$, then a global time-like Killing vector field on the space-time manifold M always exists. When the cosmological constant is positive $\Lambda > 0$ then the Killing vector field $\mathscr{K}_\mu$ does not exist, and the space-like boundary ∂M

only foliates an exterior to the horizons on geodesic lines. The deduced $D+1$ decomposition (16.16) of the space-time metric is then a gauge of the fundamental field $g_{\mu\nu}$.

The action functional (16.42) evaluated for the space-time metric decomposed according to the $D+1$ splitting (16.16) has the Hamiltonian form, *i.e.* this is the time integral of the Lagrangian L

$$S[g] = \int dt L. \tag{16.48}$$

According to a certain traditional point of view, L is interpreted as the total Lagrangian of the higher-dimensional General Relativity in the $D+1$ splitting. The Lagrangian related to the York–Gibbons–Hawking boundary action is a total derivative and, therefore, is not essential in the Hamiltonian analysis. Contributions due to Matter fields and cosmological constant are extraordinarily simple to analysis. The most complicated piece of the problem is the geometric part of the Einstein–Hilbert Lagrangian, *i.e.*

$$\sqrt{g}\left(-{}^{(D+1)}R+2\Lambda\right) = N\sqrt{h}\left(-K_{ij}K^{ij}+K^2-{}^{(D)}R+2\Lambda\right) \tag{16.49}$$

$$+ \; 2\partial_0\left(\sqrt{h}K\right)-2\partial_i\left(\sqrt{h}(KN^i-h^{ij}N_{|j})\right), \tag{16.50}$$

and because the last two terms are total derivatives they can be dropped when a canonical formulation is performed. As a result one obtains the following form of the Lagrangian

$$L = \frac{1}{2\kappa}\int_{\partial M} d^D x N\sqrt{h}\left(K^2-K_{ij}K^{ij}-{}^{(D)}R+2\Lambda+2\kappa\rho\right). \tag{16.51}$$

The Einstein field equations (16.5) can also be decomposed using the $D+1$ splitting. As a result one obtains the system of the differential equations describing the higher-dimensional evolution for the induced metric h_{ij} and the intrinsic curvature K_{ij}

$$\partial_t h_{ij} = N_{i|j}+N_{j|i}-2NK_{ij}, \tag{16.52}$$

$$\partial_t K_{ij} = -N_{|ij}+N(R_{ij}+KK_{ij}-2K_{ik}K^k_j)+N^kK_{ij|k}+K_{ik}N^k_{|j}+K_{jk}N^k_{|i} - \kappa N\left[S_{ij}-\frac{S-\rho}{D-2}h_{ij}\right], \tag{16.53}$$

where the energy density ρ is the double projection of the stress-energy tensor onto the normal vector field

$$\rho = T(n,n) = T_{\mu\nu}n^{\mu}n^{\nu}, \tag{16.54}$$

the spatial stress tensor S_{ij} is the double projection of the stress-energy tensor onto the spatial metric, and the spatial stress density S is its trace

$$S_{ij} = T(h,h) = T_{\mu\nu}h_i^{\mu}h_j^{\nu}, \tag{16.55}$$

$$S = h^{ij}S_{ij}, \tag{16.56}$$

$h_{\nu}^{\mu} = \delta_{\nu}^{\mu} + n^{\mu}n_{\nu}$, and n^{μ} is the normal vector field following from the $D+1$ splitting

$$n^{\mu} = \left[\frac{1}{N}, -\frac{N^i}{N}\right], \quad n_{\mu} = \left[-N, \mathbf{0}_i\right]^T, \tag{16.57}$$

where $\mathbf{0}_i = \underbrace{[0,\dots,0]}_{D \text{ times}}$ is the D-dimensional zero vector.

The Lagrangian (16.51) leads to the Euler–Lagrange equations of motion. The first is the scalar equation

$$2\kappa h^{-1}\left(h_{ik}h_{jl} - \frac{1}{2}h_{ij}h_{kl}\right)\frac{\delta S[g]}{\delta h_{ij}}\frac{\delta S[g]}{\delta h_{kl}} - \frac{1}{2\kappa}\left({}^{(D)}R - 2\Lambda - 2\kappa\rho\right) = 0, \tag{16.58}$$

and the second is the D-vector equation

$$\pi^{ij}_{\ |j} + J^i = 0, \tag{16.59}$$

where the momentum conjugated to the induced metric is

$$\pi^{ij} = \frac{\delta L}{\delta\left(\partial_t h_{ij}\right)} = -\frac{\sqrt{h}}{2\kappa}\left(K^{ij} - h^{ij}K\right), \tag{16.60}$$

and the momentum density J^i is the stress-energy tensor projected onto the normal vector field and the spatial metric

$$J^i = T(n,h) = T_{\mu\nu}n^{\mu}h^{\nu i}. \tag{16.61}$$

The equation (16.58) is the Hamilton–Jacobi equation of General Relativity and defines the classical D-dimensional geometrodynamics.

Let us analyze in some detail the higher-dimensional Lagrangian (16.51). The canonical momenta of the theory are

$$\pi_\phi = \frac{\delta L}{\delta(\partial_t \phi)}, \tag{16.62}$$

$$\pi = \frac{\delta L}{\delta(\partial_t N)} = 0, \tag{16.63}$$

$$\pi^i = \frac{\delta L}{\delta(\partial_t N_i)} = 0. \tag{16.64}$$

Then, using the Legendre transformation allows to rewrite the total Lagrangian in the form

$$L = \int_{\partial M} d^D x \left[\frac{1}{2\kappa} \left(\pi_\phi \partial_t \phi + \pi \partial_t N + \pi^i \partial_t N_i + \pi^{ij} \partial_t h_{ij} \right) - NH - N_i H^i \right], \tag{16.65}$$

where the quantities

$$H = \frac{\sqrt{h}}{2\kappa} \left(K^2 - K_{ij} K^{ij} - {}^{(D)}R + 2\Lambda + 2\kappa\rho \right), \tag{16.66}$$

$$H^i = -2\pi^{ij}_{\ \ |j} - 2J^i = -2\partial_j \pi^{ij} - h^{il} \left(2h_{jl,k} - h_{jk,l} \right) \pi^{jk} - 2J^i, \tag{16.67}$$

satisfy the first-class constraints algebra

$$i\left[H_i(x), H_j(y)\right] = \int_{\partial M} H_a c^a_{ij} d^D z, \tag{16.68}$$

$$i\left[H(x), H_i(y)\right] = H \delta^{(D)}_{,i}(x,y), \tag{16.69}$$

while involving of the elementary relation

$$\delta\left(\sqrt{h}{}^{(D)}R\right) = \sqrt{h} h^{ij} h^{kl} \left(\delta h_{ik,jl} - \delta h_{ij,kl} \right) - \sqrt{h} \left[R^{ij} - \frac{1}{2} h^{ij(D)} R \right] \delta h_{ij}, \tag{16.70}$$

allows to establish the third bracket

$$i\left[\int_{\partial M} H \xi_1 d^D x, \int_{\partial M} H \xi_2 d^D x \right] = \int_{\partial M} H^a \left(\xi_{1,a} \xi_2 - \xi_1 \xi_{2,a} \right) d^D x. \tag{16.71}$$

Moreover, the quantities H^i are generators of the spatial diffeomorphisms $\widetilde{x}^i = x^i + \xi^i$

$$i\left[h_{ij}, \int_{\partial M} H_a \xi^a d^D x \right] = -h_{ij,k}\xi^k - h_{kj}\xi^k_{\ ,i} - h_{ik}\xi^k_{\ ,j}, \tag{16.72}$$

$$i\left[\pi^{ij}, \int_{\partial M} H_a \xi^a d^D x \right] = -\left(\pi^{ij} \xi^k \right)_{,k} + \pi^{kj} \xi^i_{\ ,k} + \pi^{ik} \xi^j_{\ ,k}, \tag{16.73}$$

where $H_i = h_{ij}H^j$ and

$$c^a_{ij} = \delta^a_i \delta^b_j \delta^{(D)}_{,b}(x,z)\delta^{(D)}(y,z) - (x \rightarrow y), \tag{16.74}$$

are the structure constants of the D-dimensional spatial diffeomorphism group.

Application of the time-preservation to the primary constraints

$$\pi \approx 0, \tag{16.75}$$
$$\pi^i \approx 0, \tag{16.76}$$

leads to the secondary constraints

$$H \approx 0, \tag{16.77}$$
$$H^i \approx 0. \tag{16.78}$$

The Hamiltonian (scalar) constraint (16.77) yields the dynamics, and the diffeomorphism (vector) constraint (16.78) merely reflects the spatial diffeoinvariance.

The canonical primary quantization method allows to establish the following commutation relations

$$i\left[\pi^{ij}(x), h_{kl}(y)\right] = \frac{1}{2}\left(\delta^i_k\delta^j_l + \delta^i_l\delta^j_k\right)\delta^{(D)}(x,y), \tag{16.79}$$
$$i\left[\pi^i(x), N_j(y)\right] = \delta^i_j\delta^{(D)}(x,y), \tag{16.80}$$
$$i\left[\pi(x), N(y)\right] = \delta^{(D)}(x,y). \tag{16.81}$$

A representation of the momenta operators obeying the canonical commutation relations (16.79)-(16.81) is dependent on choice. Using the Wheeler metric representation, one has

$$\pi = -i\frac{\delta}{\delta N}, \tag{16.82}$$
$$\pi^i = -i\frac{\delta}{\delta N_i}, \tag{16.83}$$
$$\pi^{ij} = -i\frac{\delta}{\delta h_{ij}}. \tag{16.84}$$

Application of these relations to the Hamiltonian constraint (16.77) yields the Wheeler–DeWitt equation

$$\left\{2\kappa G_{ijkl}\frac{\delta^2}{\delta h_{ij}\delta h_{kl}}+\frac{\sqrt{h}}{2\kappa}\left({}^{(D)}R-2\Lambda-2\kappa\rho\right)\right\}\Psi[h_{ij},\phi]=0, \tag{16.85}$$

where G_{ijkl} is the DeWitt supermetric

$$G_{ijkl}=\frac{1}{2\sqrt{h}}\left(h_{ik}h_{jl}+h_{il}h_{jk}-h_{ij}h_{kl}\right), \tag{16.86}$$

on the Wheeler superspace, which in the present case is the configuration space of the geometries of a D-dimensional space embedded in an enveloping $D+1$-dimensional space-time.

Other first-class constraints satisfy the following canonical commutation relations

$$\left[\pi(x),\pi^i(y)\right] = 0, \tag{16.87}$$
$$\left[\pi(x),H^i(y)\right] = 0, \tag{16.88}$$
$$\left[\pi^i(x),H^j(y)\right] = 0, \tag{16.89}$$
$$\left[\pi^i(x),H(y)\right] = 0, \tag{16.90}$$

and after application of the canonical primary quantization method are the supplementary conditions on wave functional $\Psi[h_{ij},\phi]$. The primary constraints lead to the equations

$$-i\frac{\delta\Psi[h_{ij},\phi]}{\delta N} = 0, \tag{16.91}$$
$$-i\frac{\delta\Psi[h_{ij},\phi]}{\delta N_i} = 0. \tag{16.92}$$

The diffeomorphism constraint also leads to such a condition

$$i\left(\frac{\delta\Psi[h_{ij},\phi]}{\delta h_{ij}}\right)_{|j}=J^i\Psi[h_{ij},\phi], \tag{16.93}$$

which merely reflects diffeoinvariance.

B Higher Dimensional Global Optimization

The geometry of an enveloping $D+1$-dimensional space-time can be evaluated through the geometry of the D-dimensional space whose characteristics are, in general, functionals of induced space metric h_{ij}. The spatial stress density S and the cosmological constant Λ are also functionals of the D-metric h_{ij} and, consequently, the energy density of Matter fields ρ must be also a functional of the space metric h_{ij}. Such a situation implies the non-trivial ramification – Matter fields are, in the most general case, functionals of the induced D-metric h_{ij}. In this manner, in general the DeWitt wave function

$$\Psi[h_{ij},\phi] \equiv \Psi[\mathfrak{G}^D], \tag{16.94}$$

is fully justified choice for the wave functional. The problem is, however, to solve the Wheeler–DeWitt equation, what has never been made in general, even in the case of a four-dimensional space-time. Even taking *ad hoc* the ansatz $\Psi[\mathfrak{G}^D] = \Psi[h_{ij}]$ does not simplify the situation – there is still a functional dependence on $D \times D$ matrix h_{ij}, and it is not clear how to treat $\Psi[h_{ij}]$. In this manner, the structure of a wave functional should be thoroughly re-considered. From quantum mechanics point of view $\Psi[h_{ij}]$ is a scalar-valued functional in the metric h_{ij}. The Wheeler–DeWitt operator is $Diff(\partial M)$-invariant and, consequently, full diffeoinvariance is guaranteed if and only if $\Psi[h_{ij}]$ is a diffeoinvariant, *i.e.* is a function of other $Diff(\partial M)$-invariants which must be constructed from h_{ij} only.

B1 D-Dimensional Global One Dimensionality

By straightforward generalization of the DeWitt construction $\Psi[h_{ij}]$ is a functional of the $D \times D$ symmetric matrix h_{ij}. The analogy with quantum mechanics suggests that $\Psi[h_{ij}]$, as the Schrödinger wave function, is a classical scalar-valued field, *i.e.* $\Psi[h_{ij}]$ is a single functional. It means that we have to deal with the *global optimization of quantum geometrodynamics*. For realization of such a point of view $\Psi[h_{ij}]$ should be dependent on diffeoinvariant matrix invariants of the D-metric h_{ij}.

In the light of the Cayley–Hamilton theorem the matrix invariants c_n of $D \times D$-dimensional matrix $h_{ij} = \mathbf{h}$ are the coefficients of its characteristic polynomial and satisfy the relation

$$\mathbf{h}^D - c_1\mathbf{h}^{D-1} - c_2\mathbf{h}^{D-2} - \ldots + c_D\mathbf{I}_{D\times D} = 0, \tag{16.95}$$

where $\mathbf{I}_{D\times D} = \mathrm{diag}[\underbrace{1,\ldots,1}_{D\text{ times}}]$ is the $D \times D$-dimensional unit matrix. For the case $k = 1,\ldots,D-1$ the matrix invariants c_k can be established in the recursive way by application of the Newton identities

$$kc_k = p_k - \sum_{i=1}^{k-1} c_i p_{k-i}, \tag{16.96}$$

where $p_k = \mathrm{Tr}\mathbf{h}^k$ and, consequently,

$$c_1 = p_1, \tag{16.97}$$

$$c_2 = \frac{1}{2}p_2 - \frac{1}{2!}p_1^2, \tag{16.98}$$

$$c_3 = \frac{1}{3}p_3 - \frac{1}{2}p_1p_2 + \frac{1}{3!}p_1^3, \tag{16.99}$$

$$c_4 = \frac{1}{4}p_4 - \frac{1}{3}p_1p_3 + \frac{1}{4}p_1^2p_2 - \frac{1}{2\cdot 4}p_2^2 - \frac{1}{4!}p_1^4, \tag{16.100}$$

$$c_5 = \frac{1}{5}p_5 - \frac{1}{4}p_1p_4 - \frac{1}{6}p_2p_3 - \frac{1}{12}p_1^3p_2 + \frac{1}{6}p_1^2p_3 + \frac{1}{8}p_1p_2^2 + \frac{1}{5!}p_1^5, \tag{16.101}$$

and so on. For $k = D$ the coefficient has the form

$$c_D = (-1)^D \det\mathbf{h}. \tag{16.102}$$

In the light of the quantum-mechanical arguments presented above, the wave functional $\Psi(h_{ij})$ should be a scalar-valued function and diffeoinvariant. Such restrictions are satisfied by the wave functional which depends only on the matrix invariants of induced space metric h_{ij}. In other words, from the mathematical point of view, it is fully justified to limit generally ill-defined mathematical considerations of functional-differential nature to the objective functions

$$\Psi[h_{ij}] = \Psi(c_1, c_2, \ldots, c_D), \tag{16.103}$$

i.e. remain unchanged under rotations of the coordinate system. Similarly, as in the case of the $3+1$ dimensional space-time, we shall call

$h = \det h_{ij}$ *the global dimension* and *higher dimensional objective quantum gravity* the quantum gravity model described by the evolution equation (16.103). In this manner, we shall consider the following model of higher-dimensional quantum gravity

$$\left\{-2\kappa G_{ijkl}\frac{\delta^2}{\delta h_{ij}\delta h_{kl}} - \frac{h^{1/2}}{2\kappa}\left({}^{(D)}R - 2\Lambda - 2\kappa\rho[h]\right)\right\}\Psi(c_1,\ldots,c_D) = 0. \tag{16.104}$$

Such a conjecture can be used to describe the isotropic space-times, and is related to the strata of the Wheeler superspace which is known as the midisuperspace.

B2 Reduction of Wheeler–DeWitt Equation

The crucial problem is evaluation of the operator

$$G_{ijkl}\frac{\delta^2}{\delta h_{ij}\delta h_{kl}}\Psi(c_1,c_2,\ldots,c_D), \tag{16.105}$$

which can be performed as follows. Let us consider the obvious identity

$$\frac{\delta}{\delta h_{ij}}\Psi(c_1,c_2,\ldots,c_D) = \sum_{n=1}^{D}\frac{\delta c_n}{\delta h_{ij}}\frac{\delta}{\delta c_n}\Psi(c_1,c_2,\ldots,c_D), \tag{16.106}$$

together with the elementary relation

$$\frac{\delta}{\delta c_n}\Psi(c_1,c_2,\ldots,c_D) = \frac{\delta h}{\delta c_n}\frac{\delta}{\delta h}\Psi(h), \tag{16.107}$$

arising from the assumption that any matrix invariant c_n is a functional in the global dimension h. Applying the Jacobi formula for determinant of the space-time metric $\delta g = g g^{\mu\nu}\delta g_{\mu\nu}$ in the $D+1$ decomposition, one obtains also analogous Jacobi's formula for the space metric determinant $\delta h = h h^{ij}\delta h_{ij}$, and for this reason

$$\frac{\delta h}{\delta c_n}\frac{\delta}{\delta h}\Psi(h) = h h^{ij}\frac{\delta h_{ij}}{\delta c_n}\frac{\delta}{\delta h}\Psi(h). \tag{16.108}$$

Applying this fact within the formula (16.106), one receives

$$\frac{\delta}{\delta h_{ij}}\Psi(c_1,c_2,\ldots,c_D) = h h^{ij}\left(\sum_{n=1}^{D}\frac{\delta c_n}{\delta h_{ij}}\frac{\delta h_{ij}}{\delta c_n}\right)\frac{\delta}{\delta h}\Psi(h), \tag{16.109}$$

and after calculation of the sum

$$\sum_{n=1}^{D}\frac{\delta c_n}{\delta h_{ij}}\frac{\delta h_{ij}}{\delta c_n}=\sum_{n=1}^{D}\frac{\delta c_n}{\delta c_n}=\sum_{n=1}^{D}1=\underbrace{1+\ldots+1}_{D\text{ times}}=D, \tag{16.110}$$

one can establish the final result as

$$\frac{\delta}{\delta h_{ij}}\Psi(c_1,c_2,\ldots,c_D)=Dhh^{ij}\frac{\delta}{\delta h}\Psi(h). \tag{16.111}$$

Consequently, one receives the following evaluation of the differential operator (16.105)

$$G_{ijkl}\frac{\delta^2}{\delta h_{ij}\delta h_{kl}}\Psi(c_1,c_2,\ldots,c_D)=DG_{ijkl}h^{ij}h^{kl}h^2\frac{\delta^2}{\delta h^2}\Psi(h). \tag{16.112}$$

In this manner, the reduction of the quantum geometrodynamics is solely determined through the double projection of the DeWitt supermetric onto induced metric. It can be seen by straightforward calculation that this projection has the form

$$G_{ijkl}h^{ij}h^{kl} \quad = \quad -\frac{D(D-2)}{2}h^{-1/2}, \tag{16.113}$$

where we have applied the basic relationships for D-dimensional embedded space: $h^{ab}h_{bc}=\delta^a_c$, $\delta^a_a=D$. Application of the relation (16.113) within the formula (16.112) leads us to the following transformation

$$G_{ijkl}\frac{\delta^2}{\delta h_{ij}\delta h_{kl}}\Psi(h_{ij})=-\frac{D^2(D-2)}{2}h^{3/2}\frac{\delta^2}{\delta h^2}\Psi(h), \tag{16.114}$$

which allows to write out the higher dimensional quantum geometrodynamics (16.104) in the form

$$\left[2\kappa\frac{D^2(D-2)}{2}h^{3/2}\frac{\delta^2}{\delta h^2}-\frac{h^{1/2}}{2\kappa}\left({}^{(D)}R-2\Lambda-2\kappa\rho[h]\right)\right]\Psi(h)=0. \tag{16.115}$$

Making elementary algebraic manipulations, one receives finally the equation

$$\left[\frac{\delta^2}{\delta h^2}-\frac{{}^{(D)}R-2\Lambda-2\kappa\rho[h]}{2\kappa^2D^2(D-2)}\frac{1}{h}\right]\Psi(h)=0, \tag{16.116}$$

which is consistent for the dimensions of space different from $D\neq 0,2$.

B3 One-Dimensional Dirac Equation

Higher dimensional quantum geometrodynamics (16.115) can be rewritten in the form of the one-dimensional Klein–Gordon equation

$$\left(\frac{\delta^2}{\delta h^2}+\omega^2\right)\Psi=0, \tag{16.117}$$

where ω^2 is the parameter

$$\omega^2=-\frac{{}^{(D)}R-2\Lambda-2\kappa\rho}{2\kappa^2D^2(D-2)}\frac{1}{h}=\frac{K^2-K_{ij}K^{ij}}{2\kappa^2D^2(D-2)}\frac{1}{h}, \tag{16.118}$$

which can be positive, negative or vanishing identically. The equation (16.117) can be treated as the Euler–Lagrange equation of motion arising from the stationarity of the action

$$S[\Psi]=\int\delta hL\left(\Psi,\frac{\delta\Psi}{\delta h}\right), \tag{16.119}$$

where L is the Lagrange function

$$L=\frac{1}{2}\Pi_\Psi^2-\frac{\omega^2}{2}\Psi^2, \tag{16.120}$$

and Π_Ψ is the momentum conjugated to the classical scalar field Ψ

$$\Pi_\Psi=\frac{\partial L}{\partial\left(\dfrac{\delta\Psi}{\delta h}\right)}=\frac{\delta\Psi}{\delta h}. \tag{16.121}$$

$S[\Psi]$ is the action in the classical field Ψ and, therefore, h behaves as a parameter in ω, *i.e.* is not differentiated in the derivation of the Euler–Lagrange equations of motion

$$\frac{\partial L}{\partial\Psi}-\frac{\delta}{\delta h}\frac{\partial L}{\partial\left(\dfrac{\delta\Psi}{\delta h}\right)}=0, \tag{16.122}$$

with the consistency condition of the vanishing boundary term

$$\int\delta\left(\frac{\partial L}{\partial\Psi}\delta\Psi\right)=\frac{\partial L}{\partial\Psi}\delta\Psi\bigg|_0=0. \tag{16.123}$$

It can be seen by straightforward calculation that the equation (16.122) coincides with the equation (16.117). Through application of the conjugate momentum Π_Ψ within the equation (16.117), one can make the order reduction of this equation

$$\frac{\delta \Pi_\Psi}{\delta h} + \omega^2 \Psi = 0, \tag{16.124}$$

and, consequently, the equations (16.121) and (16.124) take the form of the canonical Hamilton equations of motion

$$\frac{\delta}{\delta h}\Psi = \frac{\delta}{\delta \Pi_\Psi} H(\Psi, \Pi_\Psi), \tag{16.125}$$

$$\frac{\delta}{\delta h}\Pi_\Psi = -\frac{\delta}{\delta \Psi} H(\Psi, \Pi_\Psi), \tag{16.126}$$

where the Hamilton function $H(\Psi, \Pi_\Psi)$ is obtained from the Lagrange function (16.120) by the Legendre transformation

$$H(\Psi, \Pi_\Psi) = \Pi_\Psi \frac{\delta \Psi}{\delta h} - L\left(\Psi, \frac{\delta \Psi}{\delta h}\right) = \frac{1}{2}\Pi_\Psi^2 - \frac{\omega^2}{2}\Psi^2. \tag{16.127}$$

If one introduces the classical two-component field $\Phi = [\Pi_\Psi \ \ \Psi]^T$ then the canonical Hamilton equations of motion (16.125) and (16.126) can be presented as the one-dimensional Dirac equation

$$\left(-i\sigma_y \frac{\delta}{\delta h} - M\right)\Phi = 0, \tag{16.128}$$

where M is the mass matrix of the field Φ

$$M = \begin{bmatrix} -1 & 0 \\ 0 & -\omega^2 \end{bmatrix}, \tag{16.129}$$

and σ_y is the Pauli matrix

$$\sigma_y = \begin{bmatrix} 0 & -i \\ i & 0 \end{bmatrix}, \tag{16.130}$$

which in itself satisfies the self-created algebra

$$\sigma_y^2 = \mathbf{I}_2, \quad \{\sigma_y, \sigma_y\} = 2\mathbf{I}_2, \quad \mathbf{I}_2 = \begin{bmatrix} 1 & 0 \\ 0 & 1 \end{bmatrix}, \tag{16.131}$$

which is the Clifford algebra over the complex vector space $\mathbb{C}^2$

$$\mathscr{C}\ell_2(\mathbb{C}) = \mathscr{C}\ell_0(\mathbb{C}) \otimes \mathrm{M}_2(\mathbb{C}) \cong \mathrm{M}_2(\mathbb{C}) = \mathbb{C} \oplus \mathbb{C}, \tag{16.132}$$

where $\mathscr{C}\ell_n \equiv \mathscr{C}\ell_{n,0}$, and $\mathrm{M}_2(\mathbb{C})$ denotes the algebra of all 2×2 matrices over the field of complex numbers $\mathbb{C}$. The Clifford algebra $\mathscr{C}\ell_{2,0}(\mathbb{C})$ possesses a two-dimensional complex representation. Restriction to the pinor group $\mathrm{Pin}_{2,0}(\mathbb{R})$ yields a complex representation of the two-dimensional pinor group, *i.e.* the two-dimensional spinor representation, whereas restriction to the spinor group $\mathrm{Spin}_{2,0}(\mathbb{R})$ splits $\mathscr{C}\ell_{1,1}(\mathbb{R})$ into the sum of two half spin representations of dimension one, *i.e.* the one dimensional Weyl representations. There is the isomorphism

$$\mathrm{Spin}_{2,0}(\mathbb{R}) \cong \mathrm{U}(1) \cong \mathrm{SO}(2), \tag{16.133}$$

and the spinor group $\mathrm{Spin}_{2,0}(\mathbb{R})$ acts on a 1-sphere S^1 in such a way that one has a fibre bundle with fibre $\mathrm{Spin}_{1,0}(\mathbb{R})$

$$\mathrm{Spin}_{1,0}(\mathbb{R}) \longrightarrow \mathrm{Spin}_{2,0}(\mathbb{R}) \longrightarrow S^1, \tag{16.134}$$

and the homotopy sequence is

$$\pi_1\left(\mathrm{Spin}_{1,0}(\mathbb{R})\right) \longrightarrow \pi_1\left(\mathrm{Spin}_{2,0}(\mathbb{R})\right) \longrightarrow \pi_1\left(S^1\right). \tag{16.135}$$

The Clifford algebra $\mathscr{C}\ell_2(\mathbb{C})$ can be generated by complexification

$$\mathscr{C}\ell_2(\mathbb{C}) \cong \mathscr{C}\ell_{1,1}(\mathbb{R}) \otimes \mathscr{C}\ell_0(\mathbb{C}), \tag{16.136}$$

where $\mathscr{C}\ell_{1,1}(\mathbb{R})$ denotes the four-dimensional Clifford algebra over the real vector space $\mathbb{R}^{2,0}$

$$\mathscr{C}\ell_{1,1}(\mathbb{R}) \cong \mathrm{M}_2(\mathbb{R}) \otimes \mathscr{C}\ell_0(\mathbb{R}) \cong \mathrm{M}_2(\mathbb{R}), \tag{16.137}$$

with $\mathrm{M}_2(\mathbb{R})$ being algebra of 2×2 matrices over $\mathbb{R}$, and

$$\mathscr{C}\ell_0(\mathbb{R}) = \mathbb{R}, \quad \mathscr{C}\ell_0(\mathbb{C}) = \mathbb{C}. \tag{16.138}$$

The Clifford algebra (16.137) can be decomposed into a direct sum of central simple algebras isomorphic to matrix algebra over $\mathbb{R}$

$$\mathscr{C}\ell_{1,1}(\mathbb{R}) = \mathscr{C}\ell^+_{1,1}(\mathbb{R}) \oplus \mathscr{C}\ell^-_{1,1}(\mathbb{R}), \tag{16.139}$$

$$\mathscr{C}\ell^\pm_{1,1}(\mathbb{R}) = \frac{1 \pm \gamma}{2}\mathscr{C}\ell_{1,1}(\mathbb{R}) \cong \mathbb{R}, \tag{16.140}$$

as well as into a tensor product

$$\mathscr{C}\ell_{1,1}(\mathbb{R}) = \mathscr{C}\ell_{2,0}(\mathbb{R}) \otimes \mathscr{C}\ell_{0,0}(\mathbb{R}), \tag{16.141}$$

$$\mathscr{C}\ell_{2,0}(\mathbb{R}) = \mathrm{M}_2(\mathbb{R}) \otimes \mathscr{C}\ell_{0,0}(\mathbb{R}) \cong \mathrm{M}_2(\mathbb{R}). \tag{16.142}$$

B4 Field Quantization in Static Fock Repère

The one-dimensional Dirac equation (16.128) can be straightforwardly quantized by the secondary canonical quantization method

$$\left(-i\sigma_y\frac{\delta}{\delta h}-M\right)\hat{\Phi}=0, \tag{16.143}$$

according to the bosonic canonical commutation relations

$$i\left[\hat{\Pi}_\Psi[h'],\hat{\Psi}[h]\right] = \delta(h'-h), \tag{16.144}$$
$$i\left[\hat{\Pi}_\Psi[h'],\hat{\Pi}_\Psi[h]\right] = 0, \tag{16.145}$$
$$i\left[\hat{\Psi}[h'],\hat{\Psi}[h]\right] = 0. \tag{16.146}$$

The choice of the bosonic relations is argued by the one-dimensionality of the quantized theory. In such a situation, there is no difference between bosons and fermions. Particles obeying one-dimensional statistics are called *axions*. Therefore, in the quantum field theory (16.143) axions are the quanta of gravity, *i.e. gravitons*.

Application of the Fock space formalism suggests using the explicit decomposition of the field operator

$$\hat{\Phi}=\mathbf{Q}\mathfrak{B} \tag{16.147}$$

where $\mathbf{Q}$ is the matrix of the second quantization

$$\mathbf{Q}=\frac{1}{\sqrt{2}}\begin{bmatrix}\frac{1}{\sqrt{\omega}} & \frac{1}{\sqrt{\omega}}\\ -i\sqrt{\omega} & i\sqrt{\omega}\end{bmatrix}, \tag{16.148}$$

and $\mathfrak{B}=\mathfrak{B}[h]$ is a dynamical repère

$$\mathfrak{B}=\left\{\begin{bmatrix}\mathsf{G}[h]\\ \mathsf{G}^\dagger[h]\end{bmatrix}:\left[\mathsf{G}[h'],\mathsf{G}^\dagger[h]\right]=\delta\left(h'-h\right),\left[\mathsf{G}[h'],\mathsf{G}[h]\right]=0\right\}, \tag{16.149}$$

in the Fock space of creation and annihilation operators. Application of the decomposition (16.147) to the one-dimensional Dirac equation (16.143) yields the modified Heisenberg equations of motion

$$\frac{\delta\mathfrak{B}}{\delta h}=\mathbf{X}\mathfrak{B}, \tag{16.150}$$

where **X** is the evolution matrix

$$\mathbf{X} = \begin{bmatrix} -i\omega & \dfrac{1}{2\omega}\dfrac{\delta\omega}{\delta h} \\ \dfrac{1}{2\omega}\dfrac{\delta\omega}{\delta h} & i\omega \end{bmatrix}. \tag{16.151}$$

Let us assume *ad hoc* existence of another repère $\mathfrak{F}$ determined by the Bogoliubov transformation

$$\mathfrak{F} = \begin{bmatrix} u & v \\ v^* & u^* \end{bmatrix} \mathfrak{B}, \tag{16.152}$$

where the hyperbolic constraint $|u|^2 - |v|^2 = 1$ holds, and its dynamics is *ad hoc* governed by the Heisenberg equations of motion

$$\frac{\delta \mathfrak{F}}{\delta h} = \begin{bmatrix} -i\Omega & 0 \\ 0 & i\Omega \end{bmatrix} \mathfrak{F}. \tag{16.153}$$

Application of such a strategy to the modified Heisenberg equations of motion (16.150) leads to the vector equation

$$\frac{\delta \mathbf{b}}{\delta h} = \mathbf{X}\mathbf{b}, \quad \mathbf{b} = \begin{bmatrix} u \\ v \end{bmatrix}, \tag{16.154}$$

and gives trivially $\Omega \equiv 0$. Consequently, the new repère $\mathfrak{F}$ is the static Fock repère which can be related to initial data (I)

$$\mathfrak{F} = \left\{ \begin{bmatrix} \mathsf{G}_I \\ \mathsf{G}_I^\dagger \end{bmatrix} : \left[\mathsf{G}_I, \mathsf{G}_I^\dagger\right] = 1, [\mathsf{G}_I, \mathsf{G}_I] = 0 \right\}, \tag{16.155}$$

with the correctly defined vacuum state $|0\rangle$

$$\mathsf{G}_I |0\rangle = 0, \tag{16.156}$$

$$\langle 0| \mathsf{G}_I^\dagger = 0. \tag{16.157}$$

The equations (16.154) are most important in the above scheme. The Bogoliubov transformation (16.152) suggests application of the superfluid parametrization

$$u = e^{i\theta} \cosh \phi, \tag{16.158}$$

$$v = e^{i\theta} \sinh \phi, \tag{16.159}$$

with the angles

$$\theta = \pm i\int_{h_I}^{h}\omega'\delta h', \tag{16.160}$$

$$\phi = \ln\sqrt{\left|\frac{\omega_I}{\omega}\right|}, \tag{16.161}$$

where $\omega' = \omega(h')$, and

$$\omega_I = -\frac{1}{\kappa\sqrt{2}\pi D\sqrt{D-2}}, \tag{16.162}$$

is the initial datum of the parameter ω. The angles (16.160) and (16.161) determine the Bogoliubov coefficients

$$u = \frac{\mu+1}{2\sqrt{\mu}}\exp\left\{i\int_{h_I}^{h}\omega'\delta h'\right\}, \tag{16.163}$$

$$v = \frac{\mu-1}{2\sqrt{\mu}}\exp\left\{-i\int_{h_I}^{h}\omega'\delta h'\right\}, \tag{16.164}$$

where $\mu = \dfrac{\omega}{\omega_I}$. Consequently, the theory is given by the equation

$$\hat{\Phi} = \mathbf{Q}\mathbf{G}\mathfrak{F}, \tag{16.165}$$

where $\mathbf{G}$ is the monodromy matrix

$$\mathbf{G} = \begin{bmatrix} \frac{1+\mu}{2\sqrt{\mu}}\exp\left\{-i\int_{h_I}^{h}\omega'\delta h'\right\} & \frac{1-\mu}{2\sqrt{\mu}}\exp\left\{i\int_{h_I}^{h}\omega'\delta h'\right\} \\ \frac{1-\mu}{2\sqrt{\mu}}\exp\left\{-i\int_{h_I}^{h}\omega'\delta h'\right\} & \frac{1+\mu}{2\sqrt{\mu}}\exp\left\{i\int_{h_I}^{h}\omega'\delta h'\right\} \end{bmatrix}. \tag{16.166}$$

Now it is clear that the presented higher-dimensional quantum geometrodynamics formulates the model of quantum gravity as a quantum field theory, where the graviton is associated with the configuration of an embedded space and given by the decomposition (16.165) in the static Fock space.

The initial value $\omega = \omega_I$ generates the equation for the set of initial configurations of space

$${}^{(3)}R^{(I)} - 2\Lambda - 2\kappa\rho_I = h_I, \tag{16.167}$$

or equivalently in terms of extrinsic curvatures

$$K_{ij}^{(I)}K^{(I)ij} - K^{(I)2} = h_I. \tag{16.168}$$

In such a situation, the quantum evolution (16.117) becomes

$$\left(\frac{\delta^2}{\delta h_I^2} - \frac{1}{2\kappa^2 D^2(D-2)}\right)\Psi(h_I) = 0, \tag{16.169}$$

and after taking into account the suitable boundary conditions

$$\Psi(h_I = h_0) = \Psi_0, \tag{16.170}$$

$$\left.\frac{\delta\Psi(h_I)}{\delta h_I}\right|_{h_I=h_0} = \Pi_\Psi^0, \tag{16.171}$$

its solution can be constructed straightforwardly

$$\Psi(h_I) = \Psi_0 \cosh\left\{\frac{h_I - h_0}{\kappa\sqrt{2}D\sqrt{D-2}}\right\} + \sqrt{6}\kappa\Pi_\Psi^0 \sinh\left\{\frac{h_I - h_0}{\kappa\sqrt{2}D\sqrt{D-2}}\right\}. \tag{16.172}$$

C Gauge Quantum Gravity

We have seen by straightforward analysis that the Newton identities for the matrix invariants c_k of induced space metric $\mathbf{h}$ for the indexes $k = 1, \ldots, D-1$ allow to express them in terms of the quantities $p_k = \mathrm{Tr}\mathbf{h}^k$ in the recursive way

$$kc_k = p_k - \sum_{i=1}^{k-1} c_i p_{k-i}, \tag{16.173}$$

while for the closing index $k = D$ the matrix invariant is $c_D = (-1)^D h$, where $h = \det\mathbf{h}$ is the determinant of induced space metric $\mathbf{h}$, *i.e.* the global dimension of the quantum gravity model. In our analysis we have assumed implicitly that the functional derivative with respect to the global dimension of all matrix invariants is non-zero

$$\frac{\delta c_k}{\delta h} \neq 0, \tag{16.174}$$

i.e. all matrix invariants are functionals of the global dimension h. In this manner, an essential problem which, in fact, lays the foundations of the higher-dimensional quantum gravity model presented above, is to establish the functional relation between the trace of any power of induced space metric $\mathbf{h}$ and the global dimension h. It is worth stressing that such a relation is not known in literature and, therefore, the deductions presented in this section present an essentially new content.

In this section we shall base our point of view on quantum gravity on certain arguments of the general framework of gauge field theories. We shall focus on making use of the functional methods, which are the modern methodology of gauge field theories. In our approach, we shall work in the spirit of the method of the Faddeev–Popov determinant, which in the present context shall apply to any square matrix $\mathbf{M}_{ab}$ of dimension $D \times D$ where $D+1$ is the dimension of the enveloping Lorentzian space-time. The determinant of any such matrix $\mathbf{M}_{ab}$ can be established by the standard definition

$$\frac{1}{\det \mathbf{M}_{ab}} = \int [d\theta(x)]\delta\left(\int \mathbf{M}_{ab}\delta\theta^b\right). \tag{16.175}$$

Interestingly, in such a situation the dimension N of the gauge group $SU(N)$ is the square root of the dimension $D+1$ of the space-time

$$N = \sqrt{D+1}, \tag{16.176}$$

i.e. we shall consider the gauge group $SU(\sqrt{D+1})$ for which the space-time dimension $D+1$ must be the square of natural number N. It is easy to see that for large values of D

$$N = \sqrt{D}\left(1 + \frac{1}{2D} + O\left(\frac{1}{D^3}\right)\right) \tag{16.177}$$

and, consequently, in such a situation the gauge group is

$$G = SU\left(\sqrt{D}\right). \tag{16.178}$$

When D is small then the suitable approximation is

$$N = 1 + \frac{D}{2} + O(D^2), \tag{16.179}$$

and for this reason in such a case the gauge group is

$$G = SU\left(1+\frac{D}{2}\right). \tag{16.180}$$

Applying the well-known relation valid for any square matrix $\mathbf{M}_{ab}$

$$\det \mathbf{M}_{ab} = \exp \operatorname{Tr} \ln \mathbf{M}_{ab}, \tag{16.181}$$

one receives the following result

$$\operatorname{Tr} \ln \mathbf{M}_{ab} = -\ln \int [d\theta(x)]\delta\left(\int \mathbf{M}_{ab}\delta\theta^b\right), \tag{16.182}$$

and, in this manner, one obtains finally

$$\operatorname{Tr} \mathbf{N}_{ab} = -\ln \int [d\theta(x)]\delta\left(\int \exp \mathbf{N}_{ab}\delta\theta^b\right), \tag{16.183}$$

where formally $\mathbf{N}_{ab} = \ln \mathbf{M}_{ab}$.

The relation (16.183) is crucial for our deduction because its particular case is the trace of any power of induced space metric $\mathbf{h}_{ab}$

$$\operatorname{Tr} \mathbf{h}_{ab}{}^k = -\ln\left|\int [d\theta(x)]\delta\left(\int \exp \mathbf{h}_{ab}{}^k\delta\theta^b\right)\right|. \tag{16.184}$$

The whole problem, therefore, is contained in making the evaluation of the delta function $\delta\left(\int \exp \mathbf{h}_{ab}{}^k\delta\theta^b\right)$ which, however, can be performed by application of the elementary property of the Dirac delta function generalized to a matrix case

$$\delta(\mathbf{g}_a(\mathbf{X}_{ab})) = \sum_{\chi} \frac{\delta(\mathbf{X}_{ab} - \chi_{ab})}{\det\left(\left.\dfrac{\delta \mathbf{g}_a(\mathbf{X}_{ab})}{\delta \mathbf{X}_{ab}}\right|_{\mathbf{X}_{ab}=\chi_{ab}}\right)}, \tag{16.185}$$

where the sum symbol $\sum_{\chi}$ means the summation over the collection $\{\chi_{ab}\}$ of the matrices χ_{ab} which are the roots of the vector-matrix equation $\mathbf{g}_a(\mathbf{X}_{ab}) = 0$. Application of such a generalization to the situation (16.184) allows to obtain the following result

$$\delta\left(\int \exp \mathbf{h}_{ab}{}^k\delta\theta^b\right) =$$

$$= \sum_{\eta}\left[\det\left(\left.\frac{\delta \int \exp \mathbf{h}_{ab}{}^k\delta\theta^b}{\delta \int \mathbf{h}_{ab}\delta\theta^b}\right|_{\mathbf{h}_{ab}=\eta_{ab}}\right)\right]^{-1} \delta\left(\int (\mathbf{h}_{ab} - \eta_{ab})\delta\theta^b\right), \tag{16.186}$$

where the matrices η_{ab} are the roots of the functional integral equation $\int \exp \mathbf{h}_{ab}{}^k \delta\theta^b = 0$. Calculating the functional derivative

$$\frac{\delta \int \exp \mathbf{h}_{ab}{}^k \delta\theta^b}{\delta \int \mathbf{h}_{ab} \delta\theta^b} = \frac{\frac{\delta}{\delta\theta^b} \int \exp \mathbf{h}_{ab}{}^k \delta\theta^b}{\frac{\delta}{\delta\theta^b} \int \mathbf{h}_{ab} \delta\theta^b} = \frac{\exp \mathbf{h}_{ab}{}^k}{\mathbf{h}_{ab}}, \tag{16.187}$$

one can easily establish the determinant

$$\det\left(\left. \frac{d \int \exp \mathbf{h}_{ab}{}^k \delta\theta^b}{d \int \mathbf{h}_{ab} \delta\theta^b} \right|_{\mathbf{h}_{ab}=\eta_{ab}} \right) = \det\left(\frac{\exp \eta_{ab}{}^k}{\eta_{ab}} \right) = \frac{\det \exp \eta_{ab}{}^k}{\det \eta_{ab}}, \tag{16.188}$$

where we have made use of the well-known Cauchy determinant formula $\det \frac{\mathbf{P}}{\mathbf{Q}} = \det \mathbf{P}\mathbf{Q}^{-1} = \det \mathbf{P} \det \mathbf{Q}^{-1} = \frac{\det \mathbf{P}}{\det \mathbf{Q}}$. Application of the elementary identity connecting a trace and a determinant of an arbitrary square matrix $\mathbf{P}$ $\det \exp \mathbf{P} = \exp \mathrm{Tr} \mathbf{P}$ allows to derive the following simple relationship

$$\left[\det\left(\left. \frac{d \int \exp \mathbf{h}_{ab}{}^k \delta\theta^b}{d \int \mathbf{h}_{ab} \delta\theta^b} \right|_{\mathbf{h}_{ab}=\eta_{ab}} \right) \right]^{-1} = \frac{\det \eta_{ab}}{\exp \mathrm{Tr} \eta_{ab}{}^k}, \tag{16.189}$$

which can be straightforwardly applied to the evaluation of the Dirac delta function

$$\delta\left(\int \exp \mathbf{h}_{ab}{}^k \delta\theta^b \right) = \sum_\eta \frac{\det \eta_{ab}}{\exp \mathrm{Tr} \eta_{ab}{}^k} \delta\left(\int (\mathbf{h}_{ab} - \eta_{ab}) \delta\theta^b \right), \tag{16.190}$$

which applied to the formula (16.184) gives

$$\mathrm{Tr} \mathbf{h}_{ab}{}^k = -\ln \left| \int [d\theta(x)] \sum_\eta \frac{\det \eta_{ab}}{\exp \mathrm{Tr} \eta_{ab}{}^k} \delta\left(\int (\mathbf{h}_{ab} - \eta_{ab}) \delta\theta^b \right) \right|. \tag{16.191}$$

Because the matrices η_{ab} are obviously independent on the gauge vector θ^a, one can rewrite the trace-determinant formula (16.191) in the following form

$$\mathrm{Tr} \mathbf{h}_{ab}{}^k = -\ln \left| \sum_\eta \frac{\det \eta_{ab}}{\exp \mathrm{Tr} \eta_{ab}{}^k} \int [d\theta(x)] \delta\left(\int (\mathbf{h}_{ab} - \eta_{ab}) \delta\theta^b \right) \right|, \tag{16.192}$$

which in the light of the elementary relation

$$\int [d\theta(x)]\delta\left(\int (\mathbf{h}_{ab} - \eta_{ab})\delta\theta^b\right) = \frac{1}{\det(\mathbf{h}_{ab} - \eta_{ab})}, \tag{16.193}$$

leads to the following trace-determinant formula

$$\mathrm{Tr}\mathbf{h}_{ab}{}^k = -\ln\left|\sum_\eta \exp\left(-\mathrm{Tr}\eta_{ab}{}^k\right)\frac{\det\eta_{ab}}{\det(\mathbf{h}_{ab} - \eta_{ab})}\right|. \tag{16.194}$$

Now it can be seen straightforwardly that in the most general situation all $\mathrm{Tr}\mathbf{h}_{ab}{}^k$ depend on $\det\mathbf{h}_{ab}$. For this one should calculate the derivative

$$\frac{\delta\mathrm{Tr}\mathbf{h}_{ab}{}^k}{\delta\det\mathbf{h}_{ab}} = \frac{\sum_\eta \exp\left(-\mathrm{Tr}\eta_{ab}{}^k\right)\det\eta_{ab}\frac{\delta}{\delta\det\mathbf{h}_{ab}}\left(\frac{1}{\det(\mathbf{h}_{ab} - \eta_{ab})}\right)}{\sum_\eta \exp\left(-\mathrm{Tr}\eta_{ab}{}^k\right)\frac{\det\eta_{ab}}{\det(\mathbf{h}_{ab} - \eta_{ab})}}. \tag{16.195}$$

Using the elementary derivatives

$$\frac{\delta}{\delta\det\mathbf{h}_{ab}}\left(\frac{1}{\det(\mathbf{h}_{ab} - \eta_{ab})}\right) = -\frac{1}{(\det(\mathbf{h}_{ab} - \eta_{ab}))^2}\frac{\delta\det(\mathbf{h}_{ab} - \eta_{ab})}{\delta\det\mathbf{h}_{ab}}, \tag{16.196}$$

and

$$\frac{\delta\det(\mathbf{h}_{ab} - \eta_{ab})}{\delta\det\mathbf{h}_{ab}} = \frac{\det(\mathbf{h}_{ab} - \eta_{ab})\left(\mathbf{h}^{ab} - \eta^{ab}\right)\delta(\mathbf{h}_{ab} - \eta_{ab})}{\det(\mathbf{h}_{ab})\,\mathbf{h}^{ab}\delta\mathbf{h}_{ab}} \tag{16.197}$$

$$= \frac{\det(\mathbf{h}_{ab} - \eta_{ab})}{\det\mathbf{h}_{ab}}\left(1 - \frac{1}{D}\eta^{ab}\mathbf{h}_{ab}\right)\left(1 - \frac{\delta\eta_{ab}}{\delta\mathbf{h}_{ab}}\right) \tag{16.198}$$

$$= \frac{\det(\mathbf{h}_{ab} - \eta_{ab})}{\det\mathbf{h}_{ab}}\left(1 - \frac{1}{D}\eta_{ab}\mathbf{h}^{ab}\right), \tag{16.199}$$

where we have applied the well-known relations: the Jacobi formula for differentiating determinant $\delta\det\mathbf{P} = \det\mathbf{P}\mathbf{P}^T\delta\mathbf{P}$, the Cauchy determinant formula $\det\mathbf{PQ} = \det\mathbf{P}\det\mathbf{Q}$ and the natural property of induced metric $\mathbf{h}_{ab}\mathbf{h}^{ab} = D$, one receives finally the following result

$$\frac{\delta p_k}{\delta h} = \frac{1}{h}\left(1 - \frac{1}{D}\frac{\sum_\eta \exp\left(-\mathrm{Tr}\eta_{ab}{}^k\right)\frac{\eta_{ab}\mathbf{h}^{ab}\det\eta_{ab}}{\det(\mathbf{h}_{ab} - \eta_{ab})}}{\sum_\eta \exp\left(-\mathrm{Tr}\eta_{ab}{}^k\right)\frac{\det\eta_{ab}}{\det(\mathbf{h}_{ab} - \eta_{ab})}}\right), \tag{16.200}$$

where $h = \det \mathbf{h}_{ab}$ and $p_k = \mathrm{Tr}\mathbf{h}_{ab}{}^k$ is, in general, manifestly non-zero. In this manner, in the light of the established fact $\frac{\delta p_k}{\delta h} \neq 0$, and the Newton identities for the matrix invariants $kc_k = p_k - \sum_{i=1}^{k-1} c_i p_{k-i}$ implies also $\frac{\delta c_k}{\delta h} \neq 0$.

Interestingly, in the large N limit one has also $D \ggg 1$, and for this reason the formula (16.200) becomes much simpler

$$\frac{\delta p_k}{\delta h} = \frac{1}{h}, \tag{16.201}$$

and, consequently, in such a limit one has the solution

$$\mathrm{Tr}\mathbf{h}_{ab}{}^k = \ln \det \mathbf{h}_{ab} + \ln G_\infty, \tag{16.202}$$

where G_∞ is the integration constant. After application of the identity $\exp \mathrm{Tr}\mathbf{P} = \det \exp \mathbf{P}$ the solution (16.202) leads to the transcendental equation for induced metric $\mathbf{h}_{ab}$

$$\exp \mathbf{h}_{ab}{}^k = G_\infty \mathbf{h}_{ab}. \tag{16.203}$$

Interestingly, in the situation when the constant vanishes identically $G_\infty = 0$ and the variation of the gauge vector $\delta\theta^a$ is non-zero, the solutions of the equation (16.203) are identical with the matrices η_{ab} satisfying the equation $\int \exp \mathbf{h}_{ab}{}^k \delta\theta^a = 0$ and leading to the Dirac delta function (16.186) and all its ramifications presented in this section. In this manner, such an identification gives unambiguous significance for these matrices

Conclusion. *In the large N limit and for the identically vanishing constant $G_\infty = 0$, induced metric has the values defined by the matrices η_{ab}.*

When $G_\infty \neq 0$ then the crucial problem for definition of the large N limit behaviour of the gauge field theory is to find solutions to the transcendental equation for induced metric (16.203).

There are certain particularly interesting situations described by the trace-determinant formula (16.194). Such a case is, for instance, the Euclidean root, *i.e.* $\eta_{ab} \equiv \delta_{ab}$, where δ_{ab} is the $D \times D$ dimensional unit matrix. Then one has the following relation

$$\mathrm{Tr}\mathbf{h}^k = D + \ln|\det(\mathbf{h}_{ab} - \delta_{ab})|. \tag{16.204}$$

Another important case is when there is only one root η_{ab}. In such a situation, one has

$$\mathrm{Tr}\mathbf{h}^k = \mathrm{Tr}{\eta_{ab}}^k - \ln|\det\eta_{ab}| + \ln|\det(\mathbf{h}_{ab}-\eta_{ab})|. \tag{16.205}$$

Let us consider the trace-determinant formula (16.194) in some detail for the situation in which N is a finite natural number. Applying the elementary relation

$$\det(\mathbf{h}_{ab}-\eta_{ab}) = \det\left(\mathbf{h}_{ab}\left(1-\frac{\eta_{ab}}{\mathbf{h}_{ab}}\right)\right) = \det\mathbf{h}_{ab}\det\left(1-\frac{1}{D}\eta_{ab}\mathbf{h}^{ab}\right), \tag{16.206}$$

where we have used the Cauchy determinant formula $\det\mathbf{PQ} = \det\mathbf{P}\det\mathbf{Q}$ and the feature of induced metric $\mathbf{h}_{ab}\mathbf{h}^{ab} = D$, it is easy to notice that the trace-determinant formula (16.194) can be rewritten in the following form

$$\mathrm{Tr}{\mathbf{h}_{ab}}^k = \ln|h| - \ln\left|\sum_\eta \exp\left(-\mathrm{Tr}{\eta_{ab}}^k\right)\frac{\det\eta_{ab}}{1-\frac{1}{D}\eta_{ab}\mathbf{h}^{ab}}\right|, \tag{16.207}$$

and after using the Taylor power series expansion for $\frac{1}{D}\left|\eta_{ab}\mathbf{h}^{ab}\right| < 1$

$$\frac{1}{1-\frac{1}{D}\eta_{ab}\mathbf{h}^{ab}} = \sum_{n=0}^{\infty}\left(\frac{1}{D}\eta_{ab}\mathbf{h}^{ab}\right)^n \tag{16.208}$$

one receives the expansion formula

$$\mathrm{Tr}{\mathbf{h}_{ab}}^k = \ln|h| - \ln\left|\sum_\eta \exp\left(-\mathrm{Tr}{\eta_{ab}}^k\right)\det\eta_{ab}\sum_{n=0}^{\infty}\left(\frac{1}{D}\eta_{ab}\mathbf{h}^{ab}\right)^n\right|. \tag{16.209}$$

Similarly, in the situation $\frac{1}{D}\left|\eta_{ab}\mathbf{h}^{ab}\right| > 1$ the suitable Taylor power series expansion is

$$\frac{1}{1-\frac{1}{D}\eta_{ab}\mathbf{h}^{ab}} = -\frac{\frac{D}{\eta_{ab}\mathbf{h}^{ab}}}{1-\frac{D}{\eta_{ab}\mathbf{h}^{ab}}} = -\frac{D}{\eta_{ab}\mathbf{h}^{ab}}\sum_{n=0}^{\infty}\left(\frac{D}{\eta_{ab}\mathbf{h}^{ab}}\right)^n, \tag{16.210}$$

and for this reason one receives

$$\mathrm{Tr}\mathbf{h}_{ab}{}^{k} = \ln|h| - \ln\left|\sum_{\eta}\exp\left(-\mathrm{Tr}\eta_{ab}{}^{k}\right)\det\eta_{ab}\sum_{n=1}^{\infty}\left(\frac{D}{\eta_{ab}\mathbf{h}^{ab}}\right)^{n}\right|. \qquad (16.211)$$

The fundamental problem in the quantum gravity model is the relation between any gauge field $\mathbf{A}_{\mu}^{a}$ and induced space metric $\mathbf{h}_{ab}$. Let us take into account the definition of the Faddeev–Popov determinant

$$\frac{1}{\det\mathbf{h}_{ab}} = \int[\delta\theta]\delta\left(\int\mathbf{h}_{ab}\delta\theta^{b}\right), \qquad (16.212)$$

the following definition of the induced space metric

$$\mathbf{h}_{ab} = \frac{\delta F_{a}[\mathbf{A}_{\mu}^{\theta^{a}}(x)]}{\delta\theta^{b}(x)}, \qquad (16.213)$$

and the infinitesimal gauge transformation

$$\mathbf{A}_{\mu}^{\theta^{a}} = \mathbf{A}_{\mu}^{a} + \delta_{c}^{a}f_{ab}{}^{c}\theta^{b}\mathbf{A}_{\mu}^{a} + \frac{1}{g}\partial_{\mu}\theta^{a} + O[\theta^{2}]. \qquad (16.214)$$

The gauge group hypersurface is given by any restriction

$$F_{a}[\mathbf{A}_{\mu}^{\theta^{a}}] = 0, \qquad (16.215)$$

and, as an example, we shall discuss here all the situations

$$F_{a}[\mathbf{A}_{\mu}^{\theta^{a}}] = F_{a}\left(\mathbf{A}_{\mu}^{\theta^{a}}(x), \partial^{\mu}\mathbf{A}_{\mu}^{\theta^{a}}(x), x\right), \qquad (16.216)$$

where

$$\partial^{\mu}\mathbf{A}_{\mu}^{\theta^{a}} = \partial^{\mu}\mathbf{A}_{\mu}^{a} + \delta_{c}^{a}f_{ab}{}^{c}\partial^{\mu}\theta^{b}\mathbf{A}_{\mu}^{a} + \delta_{c}^{a}f_{ab}{}^{c}\theta^{b}\partial^{\mu}\mathbf{A}_{\mu}^{a} + \frac{1}{g}\Box\theta^{a} + O[\theta^{2}]. \quad (16.217)$$

It should be emphasized that the gauge fixing in the form (16.216) describes a number of gauges used in modern particle physics. These are particularly: the Coulomb (radiation) gauge, the Lorenz gauge, the Landau gauge, the Feynman gauge, the Fermi–Feynman gauge, the 't Hooft gauges, the Poincaré (Line, point, multipolar) gauge, the Fock–Schwinger gauge, the Gervais–Neveu gauge, Maximal Abelian Gauge

(MAG), background-field gauge, quantum gauge, the Nielsen–Kallosh gauge, all axial gauges: the Weyl (temporal, Hamiltonian) gauge, the Arnowitt–Fickler gauge, Space-Cone gauge, Light-Cone gauge, unitary gauge and others (For review of gauges Cf. *e.g.* Ref. [944]). In all these situations one can receive the induced space metric in the following form

$$\mathbf{h}_{ab}(x) = \frac{1}{V_{D+1}}\int d^{D+1}y\left(\frac{\delta F_a[\mathbf{A}_\mu^{\theta^a}(x)]}{\delta \mathbf{A}_\mu^{\theta^d}(y)}\frac{\delta \mathbf{A}_\mu^{\theta^d}(y)}{\delta\theta^b(x)} + \frac{\delta F_a[\mathbf{A}_\mu^{\theta^a}(x)]}{\delta\partial^{(y)\mu}\mathbf{A}_\mu^{\theta^d}(y)}\frac{\delta\partial^{(y)\mu}\mathbf{A}_\mu^{\theta^d}(y)}{\delta\theta^b(x)}\right.$$
$$\left. + \frac{\delta F_a[\mathbf{A}_\mu^{\theta^a}(x)]}{\delta x_\mu}\frac{\delta x_\mu}{\delta y_\nu}\frac{\delta y_\nu}{\delta\theta^b(y)}\right), \tag{16.218}$$

where $V_{D+1} = \int d^{D+1}y$ is the space-time volume. Deriving the variations

$$\delta\mathbf{A}_\mu^{\theta^a} = \delta\mathbf{A}_\mu^a + \delta_c^a f_{ab}{}^c\delta\theta^b\mathbf{A}_\mu^a + \delta_c^a f_{ab}{}^c\theta^b\delta\mathbf{A}_\mu^a + \frac{1}{g}\partial_\mu\delta\theta^a + O[\theta^2], \tag{16.219}$$

$$\delta\partial^\mu\mathbf{A}_\mu^{\theta^a} = \delta\partial^\mu\mathbf{A}_\mu^a + \delta_c^a f_{ab}{}^c\partial^\mu\delta\theta^b\mathbf{A}_\mu^a + \delta_c^a f_{ab}{}^c\partial^\mu\theta^b\delta\mathbf{A}_\mu^a + \delta_c^a f_{ab}{}^c\delta\theta^b\partial^\mu\mathbf{A}_\mu^a$$
$$+\delta_c^a f_{ab}{}^c\theta^b\delta\partial^\mu\mathbf{A}_\mu^a + \frac{1}{g}\Box\delta\theta^a + O[\theta^2], \tag{16.220}$$

and applying the derivatives

$$\frac{\delta\theta^a(y)}{\delta\theta^b(x)} = \delta_b^a\delta^{(D+1)}(y-x), \tag{16.221}$$

$$\frac{\delta x_\mu}{\delta y_\nu} = \delta_\mu^\nu\delta^{(D+1)}(x-y) = \delta_\mu^\nu\delta^{(D+1)}(y-x), \tag{16.222}$$

$$\frac{\delta\mathbf{A}_\mu^{\theta^d}(y)}{\delta\theta^b(x)} = \delta_b^d\left(\delta_c^b f_{db}{}^c\mathbf{A}_\mu^d(y) + \frac{1}{g}\partial_\mu^{(y)}\right)\delta^{(D+1)}(y-x), \tag{16.223}$$

$$\frac{\delta\partial^{(y)\mu}\mathbf{A}_\mu^{\theta^d}(y)}{\delta\theta^b(x)} =$$
$$\delta_b^d\left(\delta_c^b f_{db}{}^c\mathbf{A}_\mu^d(y)\partial^{(y)\mu} + \delta_c^b f_{db}{}^c\partial^{(y)\mu}\mathbf{A}_\mu^d(y) + \frac{1}{g}\Box^{(y)}\right)\delta^{(D+1)}(y-x), \tag{16.224}$$

the induced metric (16.218) takes the following form

$$\begin{aligned}\mathbf{h}_{ab}(x) &= \frac{1}{V_{D+1}}\int d^{D+1}y\left[\frac{\delta F_a[\mathbf{A}^{\theta^a}_\mu(x)]}{\delta \mathbf{A}^{\theta^b}_\mu(y)}\left(\delta^b_c f_{db}{}^c\mathbf{A}^d_\mu(y)+\frac{1}{g}\partial^{(y)}_\mu\right)\right.\\ &+ \frac{\delta F_a[\mathbf{A}^{\theta^a}_\mu(x)]}{\delta\partial^{(y)\mu}\mathbf{A}^{\theta^b}_\mu(y)}\left(\delta^b_c f_{db}{}^c\mathbf{A}^d_\mu(y)\partial^{(y)\mu}+\delta^b_c f_{db}{}^c\partial^{(y)\mu}\mathbf{A}^d_\mu(y)+\frac{1}{g}\square^{(y)}\right)\\ &+ \left.\frac{\partial^{(x)\mu}F_a[\mathbf{A}^{\theta^a}_\mu(x)]}{\partial^{(y)\mu}\theta^b(y)}\right]\delta^{(D+1)}(y-x), \qquad (16.225)\end{aligned}$$

so that after application of the integral identities with the Dirac delta function one receives the relations

$$\int d^{D+1}y\frac{\delta F_a[\mathbf{A}^{\theta^a}_\mu(x)]}{\delta \mathbf{A}^{\theta^b}_\mu(y)}\mathbf{A}^d_\mu(y)\delta^{(D+1)}(y-x) = \frac{\delta F_a[\mathbf{A}^{\theta^a}_\mu]}{\delta \mathbf{A}^{\theta^b}_\mu}\mathbf{A}^d_\mu, \qquad (16.226)$$

$$\int d^{D+1}y\frac{\delta F_a[\mathbf{A}^{\theta^a}_\mu(x)]}{\delta \mathbf{A}^{\theta^b}_\mu(y)}\partial^{(y)}_\mu\delta^{(D+1)}(y-x) = -\square\frac{\delta F_a[\mathbf{A}^{\theta^a}_\mu]}{\delta\partial^\mu \mathbf{A}^{\theta^b}_\mu}, \qquad (16.227)$$

$$\int d^{D+1}y\frac{\delta F_a[\mathbf{A}^{\theta^a}_\mu(x)]}{\partial^{(y)\mu}\delta \mathbf{A}^{\theta^b}_\mu(y)}\mathbf{A}^d_\mu(y)\partial^{(y)\mu}\delta^{(D+1)}(y-x) = -\partial^\mu\left(\frac{\delta F_a[\mathbf{A}^{\theta^a}_\mu]}{\delta\partial^\mu \mathbf{A}^{\theta^b}_\mu}\mathbf{A}^d_\mu\right), \qquad (16.228)$$

$$\int d^{D+1}y\frac{\delta F_a[\mathbf{A}^{\theta^a}_\mu(x)]}{\delta\partial^{(y)\mu}\mathbf{A}^{\theta^b}_\mu(y)}\partial^{(y)\mu}\mathbf{A}^d_\mu(y)\delta^{(D+1)}(y-x) = \frac{\delta F_a[\mathbf{A}^{\theta^a}_\mu]}{\delta\partial^\mu \mathbf{A}^{\theta^b}_\mu}\partial^\mu\mathbf{A}^d_\mu, \qquad (16.229)$$

$$\int d^{D+1}y\frac{\delta F_a[\mathbf{A}^{\theta^a}_\mu(x)]}{\delta\partial^{(y)\mu}\mathbf{A}^{\theta^b}_\mu(y)}\square^{(y)}\delta^{(D+1)}(y-x) = \square\frac{\delta F_a[\mathbf{A}^{\theta^a}_\mu]}{\delta\partial^\mu \mathbf{A}^{\theta^b}_\mu}, \qquad (16.230)$$

$$\int d^{D+1}y\frac{\partial^{(x)\mu}F_a[\mathbf{A}^{\theta^a}_\mu(x)]}{\partial^{(y)\mu}\theta^b(y)}\delta^{(D+1)}(y-x) = \frac{\partial^\mu F_a[\mathbf{A}^{\theta^a}_\mu]}{\partial^\mu\theta^b}, \qquad (16.231)$$

where in the right hand side we denoted $\square = \square^{(x)}$ and $\partial^\mu = \partial^{(x)\mu}$ and all quantities are expressed in the variable x. Applying the differentiation by parts

$$\partial^\mu\left(\frac{\delta F_a[\mathbf{A}^{\theta^a}_\mu]}{\delta\partial^\mu \mathbf{A}^{\theta^b}_\mu}\mathbf{A}^d_\mu\right) = \partial^\mu\frac{\delta F_a[\mathbf{A}^{\theta^a}_\mu]}{\delta\partial^\mu \mathbf{A}^{\theta^b}_\mu}\mathbf{A}^d_\mu + \frac{\delta F_a[\mathbf{A}^{\theta^a}_\mu]}{\delta\partial^\mu \mathbf{A}^{\theta^b}_\mu}\partial^\mu\mathbf{A}^d_\mu, \qquad (16.232)$$

and collecting all contributions together, one receives finally the induced metric in the following form

$$\mathbf{h}_{ab} = \delta^b_c f_{db}{}^c \mathfrak{L}^\mu_b \left(\mathbf{A}^\theta\right) F_a[\mathbf{A}^{\theta^a}_\mu]\mathbf{A}^d_\mu + \frac{\partial^\mu F_a[\mathbf{A}^{\theta^a}_\mu]}{\partial^\mu \theta^b}. \tag{16.233}$$

where we have introduced the suitable Euler–Lagrange operator

$$\mathfrak{L}^\mu_b \left(\mathbf{A}^\theta\right) = \frac{\delta}{\delta \mathbf{A}^{\theta^b}_\mu} - \partial^\mu \frac{\delta}{\delta \partial^\mu \mathbf{A}^{\theta^b}_\mu}. \tag{16.234}$$

Now the determinant of the induced matrix (16.233), which is the global dimension in our theory, can be established easily

$$h = \det\left(\delta^b_c f_{db}{}^c \mathfrak{L}^\mu_b \left(\mathbf{A}^\theta\right) F_a[\mathbf{A}^{\theta^a}_\mu]\mathbf{A}^d_\mu + \frac{\partial^\mu F_a[\mathbf{A}^{\theta^a}_\mu]}{\partial^\mu \theta^b}\right). \tag{16.235}$$

One sees now that the induced metric (16.233) is highly non-trivial. Namely, in the case when $F_a[\mathbf{A}^{\theta^a}_\mu]$ is a kind of Lagrangian, *i.e.* satisfies the Euler–Lagrange equations of motion

$$\frac{\delta F_a[\mathbf{A}^{\theta^a}_\mu]}{\delta \mathbf{A}^{\theta^b}_\mu} - \partial^\mu \frac{\delta F_a[\mathbf{A}^{\theta^a}_\mu]}{\delta \partial^\mu \mathbf{A}^{\theta^b}_\mu} = 0, \tag{16.236}$$

the induced metric (16.233) is much more simpler

$$\mathbf{h}_{ab} = \frac{\partial^\mu F_a[\mathbf{A}^{\theta^a}_\mu]}{\partial^\mu \theta^b}, \tag{16.237}$$

and the global dimension

$$h = \det \frac{\partial^\mu F_a[\mathbf{A}^{\theta^a}_\mu]}{\partial^\mu \theta^b}. \tag{16.238}$$

Let us call such specific spaces *Lagrangian quantum spaces*. In such a situation, it is clear that the gauge fixing restriction $F_a[\mathbf{A}^{\theta^a}_\mu] = 0$ can be added to the total Lagrangian of any Yang–Mills theory.

In this manner, we have proven that making use of the Faddeev–Popov determinant, the fundamental method of the Yang–Mills theories, allows to show that, in general, all the matrix invariants c_k of induced space metric $\mathbf{h}$ are functionals of its determinant $h = \det \mathbf{h}$, *i.e.* the global dimension in the model of quantum gravity presented in the previous chapters. It should be emphasized that such a situation, in general, must not be a general property of induced space metric $\mathbf{h}$. However, according to the idea presented in this section the induced space metric (16.225) is entirely formed by the Yang–Mills connection $\mathbf{A}_\mu$ governed by a suitable gauge field theory and, consequently, we propose to call such spaces by the name *quantum gauge spaces*, and the corresponding model of quantum gravity by the name *gauge quantum gravity*. In this manner, our model of higher-dimensional quantum gravity, in fact, describes the gauge quantum gravity of higher-dimensional quantum gauge spaces.

It should be emphasized that gauge quantum gravity in itself is an interesting model of quantum gravity. First of all, in the light of the fact that the induced metric of space is expressed solely through a gauge field, after quantization of a gauge field the induced metric is automatically quantized. The second important fact is that gauge quantum gravity inherits many features of the Yang–Mills theories, including the problems like *e.g.* Gribov ambiguity. We have shown that essentially the Gribov region of the gauge field theories is defined by the positiveness of the induced metric. Such a situation creates the natural possibility of existence of the Gribov copies in gauge quantum gravity.

D Space Dimension: Relation to String Theory

The quantum field-theoretical formulation of higher-dimensional quantum gravity, presented in this chapter, can be efficiently applied to the estimation of the dimension D of an induced space by making use of the parameters of embedded geometry, the energy density of Matter fields, and the wave functional solving the model of higher-dimensional quantum geometrodynamics. It can be seen by straightforward and rather non-tedious algebraic manipulation that the evolution equation of the quantum geometrodynamics

$$\left[\frac{\delta^2}{\delta h^2} - \frac{{}^{(D)}R - 2\Lambda - 2\kappa\rho[h]}{2\kappa^2 D^2 (D-2)}\frac{1}{h}\right]\Psi(h) = 0, \tag{16.239}$$

can be easily transformed to the equivalent equation for the space dimension D

$$D^2(D-2) = \frac{1}{2\kappa^2}\left({}^{(D)}R - 2\Lambda - 2\kappa\rho\right)\left(\frac{h}{\Psi}\frac{\delta^2\Psi}{\delta h^2}\right)^{-1} \equiv \Delta, \tag{16.240}$$

which can be solved immediately. For the case of $\Delta > 0$ the real solution to the equation (16.240) is

$$\begin{aligned} D &= \frac{2}{3} + \frac{1}{3}\left(8 + \frac{27}{2}\Delta - \frac{3}{2}\sqrt{3}\sqrt{32\Delta + 27\Delta^2}\right)^{1/3} + \\ &+ \frac{1}{3}\left(8 + \frac{27}{2}\Delta + \frac{3}{2}\sqrt{3}\sqrt{32\Delta + 27\Delta^2}\right)^{1/3}, \end{aligned} \tag{16.241}$$

and, consequently, in such a situation the physical values of the space dimension are

$$D > 2 \longrightarrow D = 3, 4, 5, \text{ etc.} \tag{16.242}$$

This is the relevant result and unambiguously establishes the following lower bound

$$\Delta \geqslant 9. \tag{16.243}$$

In the manifestly non-physical case $\Delta < 0$ one receives the solution

$$\begin{aligned} D &= \frac{2}{3} + \frac{1}{3}\left(8 - \frac{27}{2}|\Delta| - \frac{3}{2}\sqrt{3}\sqrt{-32|\Delta| + 27\Delta^2}\right)^{1/3} + \\ &+ \frac{1}{3}\left(8 - \frac{27}{2}|\Delta| + \frac{3}{2}\sqrt{3}\sqrt{-32|\Delta| + 27\Delta^2}\right)^{1/3}. \end{aligned} \tag{16.244}$$

Consequently, in such a situation, the physical value of the dimension of the classical space is

$$D \in (-\infty, 2)/\{0\} \longrightarrow D = 1. \tag{16.245}$$

However, such a value unambiguously establishes $\Delta = -1$, what is physically irrelevant result.

In the case of the identically trivial situation $\Delta = 0$, the space dimension has only two possible values of the space dimension

$$D = 0, 2 \tag{16.246}$$

and only the case $D = 2$ can be the physically relevant value of the space dimension. However, this case is inconsistent with the model of quantum gravity.

From our analysis arises an essential question: what is the physical significance of the specific combination $\Delta = D^3 - 2D^2$? In the light of the established lower bound (16.243) one can suggest that Δ is a dimension of a certain *effective space* and, therefore, $\Delta + 1$ is the dimension of the *effective space-time*. In such a situation, in the light of the spirit of string theory, one can suggest that such an effective space-time is the space-time in which strings and superstrings live. Because in the case of strings the upper bound for the space-time dimension is 26, the dimension of the effective space is at most

$$\Delta_{strings} = 25. \tag{16.247}$$

When supersymmetry is taken into account, *i.e.* for the case of the superstrings domination, then the upper bound for space-time dimensionality has a slightly different value – 11. In such a situation, one receives the maximal value of the dimension of the effective space

$$\Delta_{superstrings} = 10. \tag{16.248}$$

Such a state of affairs leads to the non-trivial ramification: the minimal value of the dimension D of embedded space must be equal to 3. In this manner, the cases of 0-, 1- and 2- dimensional classical spaces are naturally excluded from our considerations. Such a scenario of the effective space allows to establish the clear and unambiguous relationship between the higher-dimensional quantum gravity and string theory

Conclusion. *The consistent ramification of the existence of the higher-dimensional quantum gravity in the classical space-time is the emergent string theory in the effective space-time.*

The minimal dimension of the effective space-time, following from the lower bound (16.243), would be 10. Interestingly, in the case of strings $\Delta \leqslant 25$ the upper bound for the space dimension is

$$D_{strings} \lessapprox 3.7643, \tag{16.249}$$

which in the light of the lower bound (16.242) and the fact that the space dimension D should be a natural number leads to the ramification that the value 4 is the stable dimension of the classical space-time.

Similarly, in the case of superstrings $\Delta \leqslant 11$ one has the following upper bound

$$D_{superstrings} \lessapprox 3.0647. \tag{16.250}$$

The both received upper bounds (16.249) and (16.250) can be interpreted as the syndrome of the existence of comparatively small values of the fluctuation to the stable space dimension – 3.

Such a situation is apparently fully consistent with the method of dimensional regularization, which was applied by Gerardus 't Hooft and Martinus Justinus Godefridus Veltman [945] to elucidate problem of the renormalization of gauge field theories through the efficient regularization of the divergent loop integrals in the evaluation of the Feynman diagrams, and deserved the 1999 Nobel Prize in Physics. In such a regularization procedure the dimension of space-time is taken *ad hoc* as $4+\varepsilon$ where ε, the fluctuation to the space dimension, in the regularization procedure goes to the limit value 0^{-}. In this manner, in the light of the dimensional regularization the space dimensionality can be written by the formula

$$D = 3 + \varepsilon, \tag{16.251}$$

where ε is the fluctuation of the space dimension. In such a scenario one can receive the following values of the fluctuation to the space dimension

$$\varepsilon_{strings} \lessapprox 0.7643, \tag{16.252}$$

$$\varepsilon_{superstrings} \lessapprox 0.0647, \tag{16.253}$$

and, therefore, it can be seen straightforwardly that the supersymmetry significantly restricts fluctuations of the space dimensionality

$$\frac{\varepsilon_{superstrings}}{\varepsilon_{strings}} \approx 0.0846. \tag{16.254}$$

There arises a subsequent essential question: what is the value of the dimension Δ of the effective space for which the dimension of the classical space changes diametrically? If one applies the method of dimensional regularization then the dimension of space is $D = 3 + \varepsilon$, and one can put straightforwardly such a relation in the equation (16.241). In such a situation, the fluctuation of space dimension can be established in terms of the dimension of the effective space as follows

$$\begin{aligned}\varepsilon &= -\frac{7}{3}+\frac{1}{3}\left(8+\frac{27}{2}\Delta-\frac{3}{2}\sqrt{3}\sqrt{32\Delta+27\Delta^2}\right)^{1/3}+\\&+\frac{1}{3}\left(8+\frac{27}{2}\Delta+\frac{3}{2}\sqrt{3}\sqrt{32\Delta+27\Delta^2}\right)^{1/3}.\end{aligned}\qquad(16.255)$$

The relationship between the dimension of the effective space and the dimensional regularization parameter ε can be also easily derived

$$\Delta=(3+\varepsilon)^3-2(3+\varepsilon)^2=\varepsilon^3+7\varepsilon^2+15\varepsilon+9,\qquad(16.256)$$

and now it can be seen by simple straightforward calculation that the change in the dimension of the classical space by the value $\varepsilon=1$ corresponds to the following change in the dimension of the effective space

$$\Delta=32.\qquad(16.257)$$

This is a comparatively big value, which means that even a comparatively small change in the dimension of the classical space results in a much more significant change in the dimension of the effective space. Such a situation leads to highly non-trivial ramifications, among others the classical space-time of dimension 5 corresponds to the effective space-time of dimension 33. This is equivalent to the statement that when the dimension of the effective space-time is limited to $\Delta\gg 32$ then the fluctuation of the dimension of classical space-time is $\varepsilon\gg 1$. In this manner, the fluctuation ε of the classical space-time dimension is much less than 1 if and only if the dimension Δ of the effective space-time is much less than 32.

Let us take into account now the gauge quantum gravity, investigated in the previous section, in the context of the dimensional regularization. In such a situation, it is clear that the number of colors is

$$N=\sqrt{D+1}=\sqrt{4+\varepsilon},\qquad(16.258)$$

so that for comparatively small fluctuations $\varepsilon\ll 1$ one can take only the two first terms of the Taylor power series

$$N=2+\frac{\varepsilon}{4}+O(\varepsilon^2),\qquad(16.259)$$

and, consequently, one can establish the gauge group of the Yang–Mills theory generating the gauge quantum gravity

$$G = SU\left(2+\frac{\varepsilon}{4}\right). \tag{16.260}$$

It is good to see now the relation between N and the dimension of the effective space-time. It is easy to see that

$$\Delta = N^2(N^4+7)-(5N^4+3), \tag{16.261}$$

so that in the large-N limit the leading terms are

$$\Delta = N^6\left(1+\frac{7}{N^4}\right)-N^4\left(5+\frac{3}{N^4}\right)\approx N^6-5N^4, \tag{16.262}$$

whereas for small value of N the approximative behaviour is

$$\Delta = 7N^2-3. \tag{16.263}$$

Let us analyze briefly the dependence between N and the dimension Δ of the effective space. Applying the relation $N=\sqrt{D+1}$ within the formula (16.241, one receives

$$\begin{aligned} N &= \left[\frac{5}{3}+\frac{1}{3}\left(8+\frac{27}{2}\Delta-\frac{3}{2}\sqrt{3}\sqrt{32\Delta+27\Delta^2}\right)^{1/3}+\right.\\ &+\left.\frac{1}{3}\left(8+\frac{27}{2}\Delta+\frac{3}{2}\sqrt{3}\sqrt{32\Delta+27\Delta^2}\right)^{1/3}\right]^{1/2}. \end{aligned} \tag{16.264}$$

For small values of the dimension of the effective space Δ the formula (16.264) behaves as

$$N_{\Delta\ll 1} = \sqrt{3}+\frac{\Delta}{8\sqrt{3}}+O(\Delta^3), \tag{16.265}$$

whereas for large values of Δ its behaviour is

$$N_{\Delta\gg 1} = \Delta^{1/6}+O\left(\frac{1}{\Delta^{1/6}}\right). \tag{16.266}$$

Taking into account the lower bounds (16.249) and (16.250), making a straightforward calculation, one can see that for the case of strings the number of colors is bounded by

$$N_{strings} \lesssim 2.1827 \tag{16.267}$$

whereas for the case of superstings the bound is slightly different

$$N_{superstrings} \lesssim 2.0161. \tag{16.268}$$

In this manner, in the most general situation, gauge quantum gravity is fully consistent with the concept of effective space-time if and only if the number of colors N has one of the values

$$N = 1, 2, \tag{16.269}$$

i.e. the dimension of the classical space-time is 0 or 3. The case of 0-dimensional space is rather non-physical, so the full consistency is established for $N = 2$ and $D = 3$. In this manner, at such a consistency, the dimension of the effective space is $\Delta = 9$. In such a situation, the gauge group of the Yang–Mills theory is established unambiguously as $G = SU(2)$.

E Discussion

In this chapter we have presented the consistent model of higher dimensional quantum gravity, which arises from the global optimization of higher dimensional quantum geometrodynamics and formulation of quantum gravity as a quantum field theory in the Fock space which differs from string theory, but possesses certain straightforward relations to string theory. In fact, this result in the case of four-dimensional space-times has been established in the previous chapters of this book. In this chapter, however, we have generalized the results to the dimension four to the case of $D+1$-dimensional space-times having Lorentzian signature $(1, D)$. We have performed also another reduction of the Wheeler–DeWitt equation, in which *all* matrix invariants of induced metric of the D-dimensional embedded space were involved. Such a procedure has modified the results by constant multipliers only, and perhaps is a more adequate formulation of quantum gravity.

It is clear that the presented construction in itself is highly non-trivial, because allows to express quantum gravity as a quantum field theory formulated in the Fock space of creation and annihilation operators. Moreover, the theory is based on comparatively simple mathematics which is given by the one-dimensional Dirac equation. The quanta of gravity, gravitons, in such an approach are identified with axions and possess natural supersymmetry.

We have showed for the case of the four-dimensional space-time that the global one-dimensionality conjecture leads to the generally solvable model of quantum gravity. This model has been straightforwardly solved for a number of particular space-times solving the Einstein field equations. Similar ramifications can be preformed for the case of the model of higher dimensional quantum gravity, because this model is based on the minor generalization of the general idea formulated for the four-dimensional situation. It is worth stressing that particularly interesting solutions, which could be studied in further research, are higher dimensional generalizations of black holes and higher dimensional cosmological solutions.

In this manner, the model of quantum gravity presented in this chapter possesses a strong natural significance for quantum gravity phenomenology. This is caused by the fact that the quantum field theory creates the most natural possible background to generate empirically verifiable results, and gives the non-trivial philosophical interpretation of the model of quantum gravity. Naturally, there are further perspectives for the development of the presented approach to quantum gravity based on the reductive approach to quantum geometrodynamics. In our opinion, such a research line is the highly promising way to the consistent and constructive theorizing of quantized gravitational fields.

Part IV

Addendum

Photon Mass & Wave-Particle Duality

Introduction

One of the most intriguing ideas of modern theoretical physics is existence of the photon mass. Recall that the belief about the non-zero value of the photon mass in deeply rooted in experimental data, and was deduced in early 1970s by A. Mazet, C. Imbert, and S. Huard [297] by observations of total internal reflection to test the Goos–Hänchen effect [298] of the beam shift. Soon after this essential analysis, L. De Broglie and J.-P. Vigier [299] discussed this contradiction in the context of the quantum theory of radiation and generalized it to the quantum theory of massive spin-1 photons.

Still, the question of the non-zero value of the photon mass is one of the most important and actual problem of physics. The possibility of the non-zero photon mass was suggested in the investigation by Bass and Schrödinger [309] performed in the mid-1950s. In early 1980s H. Georgi, P. Ginsparg, and S.L. Glashow [311] suggested that the non-zero photon mass is the justification for Cosmic Microwave Background Radiation. Another good example is the analysis performed by R. Lakes [308], in which the photon mass is analyzed in the framework of the Proca equation. The diverse approaches to the problem of the non-zero photon mass were recently reviewed by A.S. Goldhaber and M.M. Nieto [310].

Today the problem of the non-zero photon mass is also often discussed, for the most recent studies see *e.g.* the papers in Ref. [946]. The question about the non-zero value of the photon mass straightforwardly arises from the experimental data, which manifestly demonstrate that the photon mass is small but detectable. Interestingly, in common understanding the concept of the non-zero photon mass can not be described within the framework based on the standard considerations of the wave-particle duality, *i.e.* by involving the fundamental ideas based on Special Relativity and Quantum Theory.

In this brief analysis, we shall present the concise and consistent proposal for constructive explanation of the non-zero photon mass. The construction is based solely on the most fundamental arguments following from the wave-particle duality. The non-trivial modification is the contribution due to the constant frequency field, which can be interpreted as the presence of the Zero-Point Frequency field. The results of this section were originally presented in the paper [947] of the author.

Non-Zero Photon Mass

In Special Relativity, the energy E of a particle possessing the mass m and moving with a three-momentum $\vec{p} = [p_x, p_y, p_z]$ is described by the Einstein energy-momentum relation

$$E^2 = m^2c^4 + p^2c^2, \tag{16.270}$$

where $p = \sqrt{p_x^2 + p_y^2 + p_z^2}$ is the value of three-momentum of a particle. The formula (16.270) can be treated as the definition of the rest mass m

$$m^2 = \frac{E^2}{c^4} - \frac{p^2}{c^2}, \tag{16.271}$$

when values of energy and momentum of a moving particle are known from experimental data. Recall that the wave-particle duality establishes the physical equivalence between the description in terms of energy E and momentum p appropriate for particles and the description in the language of frequency f and involves wavelength λ suitable for waves. The first relation is the Planck energy-frequency relation

$$E = hf, \tag{16.272}$$

and the second one is the De Broglie relation

$$p = \frac{h}{\lambda}. \tag{16.273}$$

Their empirical legitimacy has been established by a number of experiments. Applying these fundamental relations within the definition (16.271), one receives the formula

$$m^2 = \frac{h^2}{c^4}\left(f^2 - \frac{c^2}{\lambda^2}\right). \tag{16.274}$$

It is evidently seen that if one considers the non-zero value of the photon mass, then one should have the condition

$$f^2 - \frac{c^2}{\lambda^2} > 0. \tag{16.275}$$

Such a conclusion is manifestly wrong in the light of the standard formula for frequency

$$f = \frac{c}{\lambda}, \tag{16.276}$$

which is usually applied in description of wave phenomena. It suggests, however, that for the consistent and constructive explanation of the non-zero photon mass within the wave-particle duality, the celebrated formula (16.276) should be modified.

Let us assume that the frequency f of the photon is, in general, given by the Laurent series expansion in wavelength λ

$$f(\lambda) = \sum_{n=-\infty}^{\infty} a_n(\lambda - \lambda_0)^n, \tag{16.277}$$

where a_n are some coefficients having the dimension $[a_n] = T^{-1}L^{-n}$, which are given by the Cauchy integral formula

$$a_n = \frac{1}{2\pi i}\oint_\gamma \frac{f(\lambda)}{(\lambda - \lambda_0)^{n+1}} d\lambda, \tag{16.278}$$

where γ is counterclockwise around a closed, rectifiable path which contains no self-intersections, enclosing λ_0 and lies in an annulus A in which $f(\lambda)$ is analytic. For more detailed discussion Cf. *e.g.* Ref. [933]. In this manner, we suppose that, in general, the wavelength is a complex number

$$\lambda = \lambda_R + i\lambda_I, \tag{16.279}$$

where λ_R is the real part and λ_I is the imaginary part of λ. The wavelength λ_0 can be treated as a free parameter, but is a certain specific wavelength. In such a situation, the speed of light in vacuum $a_{-1} = c$ is the value of the residue of the function $f(\lambda)$, *i.e.*

$$c = \frac{1}{2\pi i} \oint_{\gamma} f(\lambda) d\lambda. \tag{16.280}$$

Let us consider the particular situation

$$f = f_0 + \frac{c}{\lambda}, \tag{16.281}$$

where f_0 is a certain reference constant frequency, which can be interpreted as the Zero-Point Frequency field. It can be seen by straightforward and easy calculation that in such a situation one has

$$f^2 - \frac{c^2}{\lambda^2} = f_0^2 + 2 f_0 \frac{c}{\lambda}, \tag{16.282}$$

and for this reason the rest mass of a particle becomes

$$m^2 = \frac{h^2}{c^4} \left(2 f_0 \frac{c}{\lambda} + f_0^2 \right). \tag{16.283}$$

Let us apply now the following formula

$$m^2 = m_0^2 + \delta m^2, \tag{16.284}$$

in which the background term

$$m_0^2 = \frac{h^2}{c^4} f_0^2, \tag{16.285}$$

is the square of the Zero-Point mass following from using the Einstein mass-energy relation $E = mc^2$ and the Planck energy-frequency relation $E = hf$. The correction δm^2 to this background contribution has the form

$$\delta m^2 = \frac{h^2}{c^4} \frac{2 f_0 c}{\lambda}. \tag{16.286}$$

If one presents the background frequency in the following form

$$f_0 = \frac{c^2}{h} m_0, \tag{16.287}$$

then the mass of a particle can be rewritten in the form

$$m = m_0\sqrt{1+\frac{2\ell_0}{\lambda}}, \tag{16.288}$$

which for small values of the wavelength λ in comparison to the scale ℓ_0 behaves like

$$m \approx m_0\left(1+\frac{\ell_0}{\lambda}\right). \tag{16.289}$$

here ℓ_0 is the Compton wavelength of the mass m_0

$$\ell_0 = \frac{h}{m_0 c}. \tag{16.290}$$

Interestingly, when value of the background frequency f_0 is small, say $f_0 = E_P/h$, where $E_P = \sqrt{\frac{\hbar c^5}{G}}$ is the Planck energy, then the term $\frac{h^2}{c^4}f_0^2$ has the following very small value

$$m_0^2 \approx M_P^2 \approx 4.73\cdot 10^{-16}\text{kg}^2. \tag{16.291}$$

In such a situation, the correction (16.286) is

$$\delta m^2 = 4\pi M_P^2\frac{\ell_P}{\lambda} \approx \frac{9.62084\cdot 10^{-50}\text{kg}^2\cdot\text{m}}{\lambda}. \tag{16.292}$$

In other words, when the wavelength of a particle is

$$\lambda \ll 4\pi\ell_P, \tag{16.293}$$

then the correction is larger then the contribution due to the background Zero-Point Frequency field. In such a situation, the mass of a particle is given by the formula

$$m = M_P\sqrt{1+\frac{4\pi\ell_P}{\lambda}} \approx M_P\left(1+\frac{2\pi\ell_P}{\lambda}\right). \tag{16.294}$$

Discussion

The wave-particle duality-based construction presented in this paper consistently explains the particular case of the photon. In such a situation, the mass formula (16.288) leads to clear and unambiguous explanation of the non-zero value of the photon mass, which up to the constant multiplier is the reciprocal of its wavelength. In the unmodified scenario in such a situation the rest photon mass is exactly equal to zero, and this case is described by Special Relativity in which the Lorentz symmetry works. However, the construction proposed above does not lead to this classical result, and photon mass is manifestly non-zero in the framework of the wave-particle duality, so that the Lorentz symmetry is violated manifestly.

The presented approach is strictly based on the Laurent series, which is the most general possible series expansion, the relation between frequency and wavelength certainly works for the photon if value of its mass lies below the measured threshold determined by experimental data. The dispersion relation, which has been proposed in this section, would be essential over a large distance, like for the galaxies which are very far from the terrestrial observer point of view.

It is worth stressing that standardly, a complex value of frequency is characteristic for a quantum particle in a tunneling potential. In this manner, in our scenario the massive photon is such a particle. The scenario indicates that the zero term or the residue term in the Laurent series corresponds to the propagation of photon in vacuum, but the physical meaning of the another contribution to the series is presently unclear.

Navier–Stokes Equations

The Millennium Problem

One of the pillars of fluid mechanics and one of the most mysterious and unsolved riddles of theoretical and mathematical physics is the problem of existence of smooth solutions to the system of partial differential equations given by the Navier–Stokes equations, first derived and studied by Claude-Louis Navier in 1827 [948] and George Gabriel Stokes in 1845 [949], and the continuity equation. Although the Navier–Stokes equations are more than 180 years old, still nobody found the general solutions to them and proved that in the case of $n > 2$ dimensional space smooth solutions to the Navier–Stokes equations exist in general. The importance of the problem is underlined by the wide range of applications of the Navier–Stokes equations in science and technology (Cf. Ref. [950]). Moreover, if one takes into account the point of view that the fluids are the solutions to the Navier–Stokes equations then it is clear that in $n = 3$ space there are smooth solutions to these equations. Despite such intuitive integrability, the mathematical procedure for construction of smooth solutions to the Navier–Stokes equations is still a great riddle.

The Navier–Stokes equations in the $n+1$ dimensional Minkowski space-time have the following form

$$\rho\left(\dot{v}_i + v_j\partial^j v_i\right) = -\partial_i p + f_i + \eta\triangle v_i + \left(\zeta + \frac{1}{3}\eta\right)\partial_i(\partial^j v_j), \qquad (16.295)$$

where $v_i = v_i(x,t)$ is the flow velocity n-vector function, $\rho = \rho(x,t)$ is the fluid density scalar function, $p = p(x,t)$ is the pressure scalar function, $f_i = f_i(x,t)$ is the body force n-vector function, and η and ζ are the so-called coefficients of viscosity which for most cases are assumed to be constant parameters. For more detailed introduction in the case of the $n=3$ dimensional space see *e.g.* Ref. [951]. Here we take into account the Newton convention of time derivatives, where a dotted quantity, like *e.g.* $\dot{f}$, is the time derivative of f and more dots under a quantity means higher order time derivatives. Similarly as in the previous chapters of this book, we take into account the Einstein summation convention introduced in the context of General Relativity. This says that if one meets an index twice in a single term, then it means summation of this term over all values of such an index. In the case of the Navier–Stokes equations (16.295) one has the following notation

$$v_j\partial^j v_i = \sum_{j=1}^{n} v_j\partial^j v_i, \tag{16.296}$$

$$\partial^j v_j = \sum_{j=1}^{n} \partial^j v_j, \tag{16.297}$$

$$\triangle v_i = \partial_j\partial^j v_i = \sum_{j=1}^{n} \partial_j\partial^j v_i. \tag{16.298}$$

The problem to solve is the system of equations given by the Navier–Stokes equations (16.300) taken together with the continuity equation

$$\dot{\rho} + \partial^i(\rho v_i) = 0. \tag{16.299}$$

For the case of the so-called incompressible viscous fluid, *i.e.* when the velocity vector is divergence-free $\partial^i v_i = 0$, the Navier–Stokes equations take the following form

$$\rho\left(\dot{v}_i + v_j\partial^j v_i\right) = \nu\rho\triangle v_i - \partial_i p + f_i, \tag{16.300}$$

where $\nu = \dfrac{\eta}{\rho}$, called the kinematic viscosity, is assumed to be a positive constant parameter. Here we shall consider such a situation.

It should be emphasized that to solve the problem there must be at least one globally controlled quantity. In usual situations it is assumed that there exists a positive constant $E < \infty$ such that for all $t \geqslant 0$ the total kinetic energy is globally controlled

$$\int_{\mathbb{R}^n} \frac{\rho|v|^2}{2} d^n x < E, \tag{16.301}$$

where $d^n x$ is a measure on $\mathbb{R}^n$, and

$$V_n = \int_{\mathbb{R}^n} d^n x < \infty \qquad (16.302)$$

is the finite constant volume of the n-dimensional Euclidean space.

The problem is to solve the system of partial differential equations (16.300) and (16.299) with respect to the pressure p and the velocity v_i which should be both smooth and globally defined functions, for given fluid density ρ, kinematic viscosity v and the body force f_i. In the most general situation the system of equations (16.300) and (16.299) is able to generate an equivalent system of equations for the spatial evolution of pressure p and both the spatial and time evolutions of the velocity vector field v_i. However, the time evolution of pressure p can be extracted by using a supplemental relationship between pressure p and fluid density ρ, like a thermodynamic equation of state.

Note that the problem of existence and smoothness of the solutions to the Navier–Stokes equations is one of the seven Millennium Prize Problems stated by the Clay Mathematics Institute [952] in 2000. In the Official Problem Description for this problem, stated by Charles Louis Fefferman [953], the attention is restricted to the case of constant fluid density with an initial-time condition for the velocity given by any smooth function on $\mathbb{R}^n$. Physically reasonable solutions belong to the space of continuous functions $C^\infty(\mathbb{R}^n \times [0,\infty))$. The problem is to establish whether such smooth physically reasonable solutions exist for the Navier–Stokes equations in general. C.L. Fefferman asks for a proof of one of the four statements: two for the solutions on $\mathbb{R}^3$ and, *alternatively*, two for the solutions on the torus $\mathbb{R}^3/\mathbb{Z}_3$. In the case of the $\mathbb{R}^3$ problem it is required to solve one of the subproblems. The first problem, referred by C.L. Fefferman as (A), is the existence and smoothness of the solutions, *i.e.* global regularity, to the Navier–Stokes equations for the initial datum of the vector $u_i = \rho v_i$,

$$u_i(x,0) = u_i^\circ(x). \qquad (16.303)$$

The problem is to prove that for the identically zero body forces, *i.e.* $f_i(x,t) = 0$, there exist smooth solutions. The second problem to solve, referred by C.L. Fefferman as (C), is breakdown of smoothness of the solutions to the Navier–Stokes equations when both $u_i^\circ(x) \in C^\infty(\mathbb{R}^3)$ and

$f_i(x,t) \in \left[C^\infty\left(\mathbb{R}^3 \times [0,\infty)\right)\right]^3$. Similar requirements are asked to prove in the case of the $\mathbb{R}^3/\mathbb{Z}_3$ problem. The subproblem referred as (B) requires to show that for identically vanishing body forces and smooth divergence-free $u_i^\circ(x)$ satisfying the periodicity condition

$$u_i^\circ(x+e_j) = u_i^\circ(x), \tag{16.304}$$

where e_j is j^{th} unit vector in $\mathbb{R}^n$ and $1 \leqslant j \leqslant n$, there are smooth solutions (p, u_i) such that the periodicity condition

$$u_i(x+e_j,t) = u_i(x,t), \tag{16.305}$$

is satisfied. The subproblem referred as (D) requires to show that for smooth body forces satisfying the periodicity condition

$$f_i(x+e_j) = f(x), \tag{16.306}$$

and smooth divergence-free $u_i^\circ(x)$ satisfying the periodicity condition (16.304), there exist no smooth solutions (p, u_i).

The equations (16.300), (16.299), and (16.303) form the Cauchy problem for the Navier–Stokes equations. Why global regularity, *i.e.* existence of smooth solutions, of $n = 3$ Navier–Stokes equations is worthy of a million dollars prize when similar problems are usually soluble for plenty of another equations? The most essential problem is the control of the solution by using the globally defined quantities or, in other words, to preserve global regularity to the Cauchy problem of nonlinear partial differential equations. There are three possible strategies to solve the problem: find exact and explicit solution or transform the Navier–Stokes equations to exactly and explicitly soluble equations, find a new globally controlled quantities, or find a new method leading to globally smooth solutions. For more precise statement see the detailed discussion by Terence Chi-Shen Tao [954].

In this section we focus attention on the general situation given by the system of partial differential equations (16.300) and (16.299) with the initial datum (16.303) and the global condition (16.301), and then consider the particular case given by the divergence-free velocity vector, constant fluid density and constant kinetic energy. We shall show that there are possible two simple strategies which both allow to define the evolution of the velocity vector in space and time, so that the problem of existence of smooth solutions to the Navier–Stokes equations in the most general situation can be solved unambiguously. The special case can be then similarly considered. In the Conclusion we give consistent answers to all the statements made by C.L. Fefferman [953].

Smooth Solutions

Let us present step-by-step the solution to the global regularity problem, *i.e.* the problem of existence of the smooth solutions, for the Navier–Stokes equations. Our strategy will be based on exact and explicit derivation of the equations for the velocity vector v_i and the pressure p of a fluid. We take into account the physically reasonable assumption that the fluid density ρ is non-zero everywhere.

Continuity Equation

First, let us write explicitly the continuity equation (16.299)

$$\dot{\rho}+v_i\partial^i\rho+\rho\partial^i v_i=0, \tag{16.307}$$

which allows to establish the divergence of velocity field

$$\partial^i v_i=-\frac{\partial^i\rho}{\rho}v_i-\frac{\dot{\rho}}{\rho}, \tag{16.308}$$

and subsequently calculate the following relations

$$\partial^j(\partial^i v_i) = -\partial^j\left(\frac{\dot{\rho}}{\rho}\right)-\partial^j\left(\frac{\partial^i\rho}{\rho}\right)v_i-\frac{\partial^i\rho}{\rho}\partial^j v_i, \tag{16.309}$$

$$\triangle(\partial^i v_i) = -\triangle\left(\frac{\dot{\rho}}{\rho}\right)-\triangle\left(\frac{\partial^i\rho}{\rho}\right)v_i-2\partial_j\left(\frac{\partial^i\rho}{\rho}\right)\partial^j v_i-\frac{\partial^i\rho}{\rho}\triangle v_i. \tag{16.310}$$

Differentiation of the continuity equation (16.307) with respect to the time coordinate gives

$$\ddot{\rho}+v_i\partial^i\dot{\rho}+\dot{v}_i\partial^i\rho+\dot{\rho}\partial^i v_i+\rho\partial^i\dot{v}_i=0, \tag{16.311}$$

which after making use of the divergence (16.308) and the elementary differential identities

$$\left(\frac{\dot{\rho}}{\rho}\right)^{\cdot} = \frac{\ddot{\rho}}{\rho} - \frac{\dot{\rho}^2}{\rho^2}, \tag{16.312}$$

$$\partial^i\left(\frac{\dot{\rho}}{\rho}\right) = \frac{\partial^i\dot{\rho}}{\rho} - \frac{\dot{\rho}}{\rho}\frac{\partial^i\rho}{\rho}. \tag{16.313}$$

yields the following relation

$$\partial^i\dot{v}_i = -\left(\frac{\dot{\rho}}{\rho}\right)^{\cdot} - \partial^i\left(\frac{\dot{\rho}}{\rho}\right)v_i - \frac{\partial^i\rho}{\rho}\dot{v}_i. \tag{16.314}$$

Finally, differentiation of the continuity equation (16.299) with respect to coordinate x_j gives the result

$$\partial^j\dot{\rho} + \partial^j\partial^i\rho v_i + 2\partial^i\rho\partial^j v_i + \rho\partial^j(\partial^i v_i) = 0, \tag{16.315}$$

which, after taking into account the relationship (16.309), gives the possibility to construct the equation

$$\partial^i\rho\left(\partial^j v_i + \frac{\partial^j\rho}{\rho}v_i\right) + \frac{\partial^j\rho}{\rho}\dot{\rho} = 0. \tag{16.316}$$

Application of the elementary identity $\partial^j\rho = \delta_i^j\partial^i\rho$ within the last term of the equation (16.316) gives

$$\partial^i\rho\left(\partial^j v_i + \frac{\partial^j\rho}{\rho}v_i + \frac{\dot{\rho}}{\rho}\delta_i^j\right) = 0, \tag{16.317}$$

and, consequently, leads to the differential equation for the spatial evolution of the velocity vector field

$$\partial^j v_i = -\frac{\partial^j\rho}{\rho}v_i - \frac{\dot{\rho}}{\rho}\delta_i^j. \tag{16.318}$$

Making use of the relation (16.318), after certain algebraic manipulations, one can establish

$$\triangle v_i = \partial_j(\partial^j v_i) = \left(2\frac{\partial^j\rho}{\rho}\frac{\partial_j\rho}{\rho} - \frac{\triangle\rho}{\rho}\right)v_i + 2\frac{\dot{\rho}}{\rho}\frac{\partial_i\rho}{\rho} - \frac{\partial_i\dot{\rho}}{\rho}, \tag{16.319}$$

and, in the similar way, the relation (16.309) can be rewritten in the following form

$$\partial^j(\partial^i v_i) = \left(2\frac{\partial^i\rho}{\rho}\frac{\partial^j\rho}{\rho} - \frac{\partial^i\partial^j\rho}{\rho}\right)v_i + 2\frac{\dot{\rho}}{\rho}\frac{\partial^j\rho}{\rho} - \frac{\partial^j\dot{\rho}}{\rho}. \tag{16.320}$$

Equation for Pressure

Differentiation of the Navier–Stokes equations (16.300) with respect to coordinate x_i and making some elementary algebraic manipulations gives the relation

$$\partial^i\rho\left(\dot v_i+v_j\partial^j v_i-\nu\triangle v_i\right)+\rho\left(\partial^i\dot v_i+\partial^i v_j\partial^j v_i+v_j\partial^j(\partial^i v_i)-\nu\triangle(\partial^i v_i)\right)$$
$$=\partial^i\left(f_i-\partial_i p\right), \tag{16.321}$$

which after making use of the relations (16.309), (16.310), and (16.314) takes the form

$$\partial^i v_j\partial^j v_i+2\nu\partial_j\left(\frac{\partial^i\rho}{\rho}\right)\partial^j v_i-2v_j\partial^j\left(\frac{\dot\rho}{\rho}\right)-v_j\partial^j\left(\frac{\partial^i\rho}{\rho}\right)v_i$$
$$+\nu\triangle\left(\frac{\partial^i\rho}{\rho}\right)v_i-\left(\frac{\dot\rho}{\rho}\right)^{\cdot}+\nu\triangle\left(\frac{\dot\rho}{\rho}\right)=\frac{1}{\rho}\partial^i\left(f_i-\partial_i p\right). \tag{16.322}$$

Let us apply the equation (16.318) to the equation (16.322). Calculating the contributions

$$\partial^i v_j\partial^j v_i = \frac{\partial^i\rho\partial^j\rho}{\rho^2}v_iv_j+2\frac{\dot\rho}{\rho}\frac{\partial^i\rho}{\rho}v_i+n\left(\frac{\dot\rho}{\rho}\right)^2, \tag{16.323}$$

$$2\nu\partial_j\left(\frac{\partial^i\rho}{\rho}\right)\partial^j v_i = -2\nu\partial_j\left(\frac{\partial^i\rho}{\rho}\right)\frac{\partial^j\rho}{\rho}v_i-2\nu\frac{\dot\rho}{\rho}\partial_i\left(\frac{\partial^i\rho}{\rho}\right), \tag{16.324}$$

where $n=\delta^j_j=\delta^j_i\delta^i_j$ is the dimension of space, and using these results within the equation (16.322), one receives finally the differential equation

$$\frac{1}{\rho}\partial^i\left(f_i-\partial_i p\right)=\left(2\frac{\partial^i\rho\partial^j\rho}{\rho^2}-\frac{\partial^i\partial^j\rho}{\rho}\right)v_iv_j$$
$$+\left(\nu\triangle\left(\frac{\partial^i\rho}{\rho}\right)-2\nu\partial_j\left(\frac{\partial^i\rho}{\rho}\right)\frac{\partial^j\rho}{\rho}+4\frac{\dot\rho}{\rho}\frac{\partial^i\rho}{\rho}-2\frac{\partial^i\dot\rho}{\rho}\right)v_i$$
$$+(n+1)\left(\frac{\dot\rho}{\rho}\right)^2-\frac{\ddot\rho}{\rho}+\nu\triangle\left(\frac{\dot\rho}{\rho}\right)-2\nu\frac{\dot\rho}{\rho}\partial_i\left(\frac{\partial^i\rho}{\rho}\right). \tag{16.325}$$

which can be integrated straightforwardly

$$f_i-\partial_i p=C_i(t)+\int_0^x\left[\left(2\frac{\partial^k\rho\partial^j\rho}{\rho}-\partial^k\partial^j\rho\right)v_kv_j\right.$$
$$+\left(\nu\rho\triangle\left(\frac{\partial^k\rho}{\rho}\right)-2\nu\partial_j\left(\frac{\partial^k\rho}{\rho}\right)\partial^j\rho+4\dot{\rho}\frac{\partial^k\rho}{\rho}-2\partial^k\dot{\rho}\right)v_k$$
$$\left.+(n+1)\rho\left(\frac{\dot{\rho}}{\rho}\right)^2-\ddot{\rho}+\nu\rho\triangle\left(\frac{\dot{\rho}}{\rho}\right)-2\nu\dot{\rho}\partial_k\left(\frac{\partial^k\rho}{\rho}\right)\right]dx_i', \tag{16.326}$$

where $C_i(t)$ is a smooth vector function on $[[0,\infty)]^3$. Some algebraic manipulations within the equation (16.326) lead to the differential equation for the spatial evolution of the pressure

$$\partial_i p=f_i-C_i(t)-\int_0^x\left[\left(2\frac{\partial^k\rho\partial^j\rho}{\rho}-\partial^k\partial^j\rho\right)v_kv_j\right.$$
$$+\left(\nu\rho\triangle\left(\frac{\partial^k\rho}{\rho}\right)-2\nu\partial_j\left(\frac{\partial^k\rho}{\rho}\right)\partial^j\rho+4\dot{\rho}\frac{\partial^k\rho}{\rho}-2\partial^k\dot{\rho}\right)v_k$$
$$\left.+(n+1)\rho\left(\frac{\dot{\rho}}{\rho}\right)^2-\ddot{\rho}+\nu\rho\triangle\left(\frac{\dot{\rho}}{\rho}\right)-2\nu\dot{\rho}\partial_k\left(\frac{\partial^k\rho}{\rho}\right)\right]dx_i', \tag{16.327}$$

which can be integrated immediately with the result

$$p(x,t)=C(t)-C_i(t)x^i+\int_0^x f_i dx'^i-\int_0^x dx'^i\int_0^{x'}\left[\left(2\frac{\partial^k\rho\partial^j\rho}{\rho}-\partial^k\partial^j\rho\right)v_kv_j\right.$$
$$+\left(\nu\rho\triangle\left(\frac{\partial^k\rho}{\rho}\right)-2\nu\partial_j\left(\frac{\partial^k\rho}{\rho}\right)\partial^j\rho+4\dot{\rho}\frac{\partial^k\rho}{\rho}-2\partial^k\dot{\rho}\right)v_k$$
$$\left.+(n+1)\rho\left(\frac{\dot{\rho}}{\rho}\right)^2-\ddot{\rho}+\nu\rho\triangle\left(\frac{\dot{\rho}}{\rho}\right)-2\nu\dot{\rho}\partial_k\left(\frac{\partial^k\rho}{\rho}\right)\right]dx_i'', \tag{16.328}$$

where $C(t)$ is a smooth function on $[0,\infty)$. It is easy to see that the functions $C(t)$ and $C_i(t)$ satisfy

$$C(t) = p(0,t), \tag{16.329}$$
$$C_i(t) = f_i(0,t)-\partial_i p(x,t)|_{x=0}. \tag{16.330}$$

Equation for Velocity Field

Now it can be seen straightforwardly that making use of the equations (16.318), (16.319) and (16.326) within the equation (16.300) leads to the equation for the time evolution of the velocity vector

$$\begin{aligned}\rho\dot{v}_i - \partial^j\rho v_i v_j - \dot{\rho}v_i = \nu\left(2\frac{\partial^j\rho\partial_j\rho}{\rho} - \triangle\rho\right)v_i + 2\nu\dot{\rho}\frac{\partial_i\rho}{\rho} - \nu\partial_i\dot{\rho}\\ +C_i(t) + \int_0^x\left[\left(2\frac{\partial^k\rho\partial^j\rho}{\rho} - \partial^k\partial^j\rho\right)v_k v_j\right.\\ +\left(\nu\rho\triangle\left(\frac{\partial^k\rho}{\rho}\right) - 2\nu\partial_j\left(\frac{\partial^k\rho}{\rho}\right)\partial^j\rho + 4\dot{\rho}\frac{\partial^k\rho}{\rho} - 2\partial^k\dot{\rho}\right)v_k\\ \left.+(n+1)\rho\left(\frac{\dot{\rho}}{\rho}\right)^2 - \ddot{\rho} + \nu\rho\triangle\left(\frac{\dot{\rho}}{\rho}\right) - 2\nu\dot{\rho}\partial_k\left(\frac{\partial^k\rho}{\rho}\right)\right]dx_i', \quad (16.331)\end{aligned}$$

which after introducing the variable $u_i = \rho v_i$ and making use of the differential relation $\dot{u}_i = \dot{\rho}v_i + \rho\dot{v}_i$ can be transformed into the form of an integro-differential equation

$$\begin{aligned}\dot{u}_i = \frac{\partial^j\rho}{\rho^2}u_i u_j + \nu\left(2\frac{\partial^j\rho\partial_j\rho}{\rho^2} - \frac{\triangle\rho}{\rho}\right)u_i + 2\frac{\dot{\rho}}{\rho}u_i + 2\nu\dot{\rho}\frac{\partial_i\rho}{\rho} - \nu\partial_i\dot{\rho}\\ +\int_0^x\left[(n+1)\rho\left(\frac{\dot{\rho}}{\rho}\right)^2 - \ddot{\rho} + \nu\rho\triangle\left(\frac{\dot{\rho}}{\rho}\right) - 2\nu\dot{\rho}\partial_k\left(\frac{\partial^k\rho}{\rho}\right)\right]dx_i'\\ +C_i(t) + \int_0^x\left[2\frac{v_j\partial^j\rho}{\rho}\frac{\partial^k\rho}{\rho} - \frac{v_j\partial^j\partial^k\rho}{\rho} + \nu\triangle\left(\frac{\partial^k\rho}{\rho}\right)\right.\\ \left.-2\nu\partial_j\left(\frac{\partial^k\rho}{\rho}\right)\frac{\partial^j\rho}{\rho} + 4\frac{\dot{\rho}}{\rho}\frac{\partial^k\rho}{\rho} - 2\frac{\partial^k\dot{\rho}}{\rho}\right]u_k dx_i', \quad (16.332)\end{aligned}$$

In the light of the Cauchy problem for the Navier–Stokes equations (16.303), the equation (16.332) is fully correct. Namely, for such a first order in the time derivative integro-differential equation we need only one condition in the form (16.303).

Divergence-Free Velocity

Let us focus our attention on the condition of a divergence-free velocity vector, *i.e.* $\partial^i v_i = 0$, which is exactly stated as the Millennium Prize Problem by C.L. Fefferman in the official description [953]. In such a situation, the continuity equation takes the following form

$$\dot{\rho} + v_i \partial^i \rho = 0, \tag{16.333}$$

and in the light of the relation (16.320) one has the condition

$$\left(2\frac{\partial^i \rho}{\rho}\frac{\partial^j \rho}{\rho^2} - \frac{\partial^i \partial^j \rho}{\rho^2}\right) u_i + 2\frac{\dot{\rho}}{\rho}\frac{\partial^j \rho}{\rho} - \frac{\partial^j \dot{\rho}}{\rho} = 0, \tag{16.334}$$

which allows to simplify the integro-differential equation (16.332)

$$\begin{aligned}\dot{u}_i = &\left(\frac{u_j \partial^j \rho}{\rho^2} + 2\frac{\dot{\rho}}{\rho}\right) u_i + C_i(t)\\ &+ \int_0^x \left[(n+1)\rho\left(\frac{\dot{\rho}}{\rho}\right)^2 - \ddot{\rho} + \nu\rho\triangle\left(\frac{\dot{\rho}}{\rho}\right) - 2\nu\dot{\rho}\partial_k\left(\frac{\partial^k \rho}{\rho}\right)\right] dx'_i\\ &+ \int_0^x \left[\nu\triangle\left(\frac{\partial^k \rho}{\rho}\right) - 2\nu\partial_j\left(\frac{\partial^k \rho}{\rho}\right)\frac{\partial^j \rho}{\rho} + 2\frac{\dot{\rho}}{\rho}\frac{\partial^k \rho}{\rho} - \frac{\partial^k \dot{\rho}}{\rho}\right] u_k dx'_i. \end{aligned} \tag{16.335}$$

Let us consider now the equation (16.318). Multiplication of both sides of this relation by the velocity vector v^i gives

$$v^i \partial^j v_i = -\frac{\partial^j \rho}{\rho} v^2 - \frac{\dot{\rho}}{\rho} v^j, \tag{16.336}$$

where $v^2 = v_i v^i$ is the square of the velocity vector field, which after taking into account the elementary differential relations

$$\frac{1}{2}\partial^j v^2 = v^i \partial^j v_i, \tag{16.337}$$

$$\partial^j \frac{\rho v^2}{2} = \frac{\partial^j \rho}{2} v^2 + \frac{\rho}{2}\partial^j v^2, \tag{16.338}$$

takes the following form

$$\partial^j \frac{\rho v^2}{2} = -\frac{\partial^j \rho}{2} v^2 - \dot{\rho} v^j, \tag{16.339}$$

or after multiplication of both sides by $-\left(\frac{\rho v^2}{2}\right)^{-1}$

$$-\partial^j \ln\left|\frac{\rho v^2}{2}\frac{V_n}{E_0}\right| = \frac{\partial^j \rho}{\rho} + \frac{\dot{\rho} v^j}{\frac{\rho v^2}{2}}, \tag{16.340}$$

where $\frac{E_0}{V_n}$ is the integration constant.

Let us take into account the condition (16.301). This is naturally satisfied for the following specific choice

$$\frac{\rho v^2}{2} = \frac{E_0}{V_n}, \tag{16.341}$$

and leads to the following bound

$$E_0 < E. \tag{16.342}$$

which is satisfied for some, arbitrarily large, constant E. In the light of the condition (16.341) the relation (16.340) leads to the equation

$$\frac{\partial^j \rho}{\rho} = -\frac{V_n}{E_0}\dot{\rho} v^j, \tag{16.343}$$

which after multiplication by $\frac{\rho v_j}{2}$ and minor algebraic manipulations gives the result

$$\frac{u_j \partial^j \rho}{\rho} = -2\dot{\rho}. \tag{16.344}$$

The relation (16.343) is helpful in derivation of the relation

$$\frac{\partial^k \partial^j \rho}{\rho} = \frac{\partial^k \rho \partial^j \rho}{\rho^2} + \frac{V_n}{E_0}\left(\frac{\dot{\rho}}{\rho}\partial^j \rho - \partial^j \dot{\rho}\right) v^k + \frac{V_n}{E_0}\frac{\dot{\rho}^2}{\rho}\delta^{jk}, \tag{16.345}$$

which allows to derive another interesting formula

$$\begin{aligned} \frac{\partial^k \triangle \rho}{\rho} &= \left(\frac{V_n}{E_0}\frac{\dot{\rho}^2}{\rho} - \frac{\triangle \rho}{\rho}\right)\frac{\partial^k \rho}{\rho} + \frac{V_n}{E_0}\left(\frac{\dot{\rho}}{\rho}\triangle \rho - \triangle \dot{\rho}\right) v^k \\ &+ \frac{V_n}{E_0}\frac{\dot{\rho}}{\rho}\left(3\partial^k \dot{\rho} - 2\frac{\dot{\rho}}{\rho}\partial^k \rho\right). \end{aligned} \tag{16.346}$$

Another relation, which will useful in further analysis, is

$$\frac{\partial^k \dot{\rho}}{\rho} = \frac{\dot{\rho}}{\rho}\frac{\partial^k \rho}{\rho} - \frac{V_n}{E_0}\left(\dot{\rho} v^k\right)^{\cdot}. \tag{16.347}$$

The helpfulness of the relations (16.344), (16.345), (16.346), and (16.347) can be straightforwardly seen in the explicit analysis of the third line of the expression (16.369)

$$\begin{aligned} &\left[\nu\triangle\left(\frac{\partial^k\rho}{\rho}\right) - 2\nu\partial_j\left(\frac{\partial^k\rho}{\rho}\right)\frac{\partial^j\rho}{\rho} + 2\frac{\dot{\rho}}{\rho}\frac{\partial^k\rho}{\rho} - \frac{\partial^k\dot{\rho}}{\rho}\right]u_k = \\ &= \left(3\nu\frac{\partial_j\rho\partial^j\rho}{\rho^2} + \nu\frac{\triangle\rho}{\rho} + 2\frac{\dot{\rho}}{\rho}\right)\frac{u_k\partial^k\rho}{\rho} \\ &+ \left(\nu\frac{\partial^k\triangle\rho}{\rho} - 4\nu\frac{\partial^k\partial^j\rho}{\rho}\frac{\partial_j\rho}{\rho} - \frac{\partial^k\dot{\rho}}{\rho}\right)u_k, \end{aligned} \tag{16.348}$$

which after some manipulations involving the following calculations

$$\frac{u_k\partial^k\dot{\rho}}{\rho} = -\frac{\dot{\rho}^2}{\rho} - 2\ddot{\rho}, \tag{16.349}$$

$$\nu\frac{u_k\partial^k\triangle\rho}{\rho} = 2\nu\left[2\frac{\dot{\rho}}{\rho}\triangle\rho - \triangle\dot{\rho} - \frac{V_n}{E_0}\left(3\dot{\rho}\ddot{\rho} + \frac{1}{2}\frac{\dot{\rho}^3}{\rho}\right)\right], \tag{16.350}$$

$$\nu\frac{u_k\partial^k\partial^j\rho}{\rho}\frac{\partial_j\rho}{\rho} = -2\nu\left(\frac{\partial^j\dot{\rho}\partial_j\rho}{\rho} + \frac{V_n}{E_0}\frac{\dot{\rho}^3}{\rho}\right), \tag{16.351}$$

can be transformed to the following form

$$\begin{aligned} &\left[\nu\triangle\left(\frac{\partial^k\rho}{\rho}\right) - 2\nu\partial_j\left(\frac{\partial^k\rho}{\rho}\right)\frac{\partial^j\rho}{\rho} + 2\frac{\dot{\rho}}{\rho}\frac{\partial^k\rho}{\rho} - \frac{\partial^k\dot{\rho}}{\rho}\right]u_k = 2\ddot{\rho} - 3\frac{\dot{\rho}^2}{\rho} \\ &+2\nu\left[3\frac{\dot{\rho}}{\rho}\left(\triangle\rho - \frac{\partial_j\rho\partial^j\rho}{\rho}\right) - \triangle\dot{\rho} + 4\frac{\partial_j\rho\partial^j\dot{\rho}}{\rho} + \frac{V_n}{E_0}\left(\frac{7}{2}\frac{\dot{\rho}^3}{\rho} - 3\dot{\rho}\ddot{\rho}\right)\right]. \end{aligned} \tag{16.352}$$

The final expression, which should be presented explicitly is

$$\rho\triangle\left(\frac{\dot{\rho}}{\rho}\right) - 2\dot{\rho}\partial_k\left(\frac{\partial^k\rho}{\rho}\right) = \triangle\dot{\rho} - 2\frac{\partial_j\rho\partial^j\dot{\rho}}{\rho} - 2\frac{\dot{\rho}}{\rho}\left(\triangle\rho - \frac{3}{2}\frac{\partial_j\rho\partial^j\rho}{\rho}\right). \tag{16.353}$$

In this manner, applying some elementary algebraic manipulations, the integro-differential equation (16.369) is eventually reduced to the first order partial differential equation

$$\dot{u}_i = C_i(t) + \int_0^x \left[(n-2)\frac{\dot{\rho}^2}{\rho} + \ddot{\rho} \right] dx_i' + \nu \int_0^x \left[6\frac{\partial_j \rho \partial^j \dot{\rho}}{\rho} - \triangle\dot{\rho} + \frac{\dot{\rho}}{\rho}\left(4\triangle\rho - 3\frac{\partial_j \rho \partial^j \rho}{\rho}\right) + \frac{V_n}{E_0}\left(7\frac{\dot{\rho}^3}{\rho} - 6\dot{\rho}\ddot{\rho}\right)\right] dx_i', \tag{16.354}$$

which can be integrated immediately

$$\begin{aligned} u_i(x,t) &= u_i^\circ(x) + \int_0^t \left[f_i(0,t') - \partial^i p(x,t')\big|_{x=0} \right] dt' \\ &+ \int_0^x \left[\int_0^t \left((n-2)\frac{\dot{\rho}^2}{\rho} + \ddot{\rho} \right) dt' \right] dx_i' \\ &+ \nu \int_0^x \left[\int_0^t \left(6\frac{\partial_j \rho \partial^j \dot{\rho}}{\rho} - \triangle\dot{\rho} + \frac{\dot{\rho}}{\rho}\left(4\triangle\rho - 3\frac{\partial_j \rho \partial^j \rho}{\rho}\right) \right.\right. \\ &\left.\left. + \frac{V_n}{E_0}\left(7\frac{\dot{\rho}^3}{\rho} - 6\dot{\rho}\ddot{\rho}\right)\right) dt' \right] dx_i', \end{aligned} \tag{16.355}$$

where $u_i^\circ(x) = u_i(x,0)$ is the initial datum of the vector $u_i = \rho v_i$, and we have used the relation (16.330) for the function $C_i(t)$.

Explicit Using of Continuity Equation

Slightly different results can be obtained as a result of explicit application of the continuity equation (16.333). Namely, this equation can be rewritten in the following form

$$\partial^k \rho = -\dot{\rho}\frac{v^k}{v^2}, \tag{16.356}$$

where $v^2 = v_j v^j$ is square of the velocity vector. The form (16.356) of the continuity equation is useful for derivation of some important relationships. First of all, by direct differentiation with respect to time, one obtains

$$\partial^k \dot{\rho} = \frac{\dot{\rho}\dot{v}^k}{v^2} - \frac{\ddot{\rho}v^k}{v^2}. \tag{16.357}$$

and by direct differentiation with respect to coordinate vector one receives

$$\partial^k \partial^j \rho = -\left(\frac{\dot{\rho}}{\rho}\partial^j \rho + \partial^j \dot{\rho}\right)\frac{v^k}{v^2} - \frac{\dot{\rho}^2}{\rho}\frac{\delta^{jk}}{v^2}, \tag{16.358}$$

and by using the identity $\triangle = \partial_j \partial^j$ also the formula

$$\partial^k \triangle \rho = -\left(\triangle \dot{\rho} + 2\frac{\partial_j \dot{\rho} \partial^j \rho}{\rho} + \frac{\dot{\rho}}{\rho}\triangle \rho + 2\frac{\dot{\rho}^3}{\rho^2 v^2}\right)\frac{v^k}{v^2} - \frac{\dot{\rho}}{\rho v^2}\left(3\partial^k \dot{\rho} + 2\frac{\dot{\rho}}{\rho}\partial^k \rho\right). \tag{16.359}$$

Helpfulness of the continuity equation (16.356) and the subsequent relations (16.357), (16.358) and (16.359) can be easily seen in construction of the following relations

$$\frac{u_k \partial^k \rho}{\rho} = -\dot{\rho}, \tag{16.360}$$

$$\frac{u_k \partial^k \dot{\rho}}{\rho} = \frac{1}{2}\dot{\rho}\frac{(\rho v^2)^{\cdot}}{\rho v^2} - \frac{1}{2}\frac{\dot{\rho}^2}{\rho} - \ddot{\rho}, \tag{16.361}$$

$$\frac{u_k \partial^k \partial^j \rho}{\rho}\frac{\partial_j \rho}{\rho} = \frac{1}{\rho v^2}\frac{\dot{\rho}^3}{\rho} - \frac{\dot{\rho}}{\rho}\frac{\partial_j \rho \partial^j \rho}{\rho} - \frac{\partial_j \rho \partial^j \dot{\rho}}{\rho}, \tag{16.362}$$

$$\frac{u_k \partial^k \triangle \rho}{\rho} = \frac{1}{\rho v^2}\left(2\dot{\rho}\ddot{\rho} + \frac{3}{2}\frac{\dot{\rho}^3}{\rho} - \frac{3}{2}\dot{\rho}^2\frac{(\rho v^2)^{\cdot}}{\rho v^2}\right) - \triangle \dot{\rho} - 2\frac{\partial_j \dot{\rho} \partial^j \rho}{\rho} - \frac{\dot{\rho}}{\rho}\triangle \rho, \tag{16.363}$$

which are useful in reduction of the third line of the expression (16.369)

$$\left[v\triangle\left(\frac{\partial^k \rho}{\rho}\right) - 2v\partial_j\left(\frac{\partial^k \rho}{\rho}\right)\frac{\partial^j \rho}{\rho} + 2\frac{\dot{\rho}}{\rho}\frac{\partial^k \rho}{\rho} - \frac{\partial^k \dot{\rho}}{\rho}\right]u_k = \tag{16.364}$$
$$\left(3v\frac{\partial_j \rho \partial^j \rho}{\rho^2} + v\frac{\triangle \rho}{\rho} + 2\frac{\dot{\rho}}{\rho}\right)\frac{u_k \partial^k \rho}{\rho} + \left(v\frac{\partial^k \triangle \rho}{\rho} - 4v\frac{\partial^k \partial^j \rho}{\rho}\frac{\partial_j \rho}{\rho} - \frac{\partial^k \dot{\rho}}{\rho}\right)u_k,$$

which after some manipulations takes the following form

$$\left[v\triangle\left(\frac{\partial^k \rho}{\rho}\right) - 2v\partial_j\left(\frac{\partial^k \rho}{\rho}\right)\frac{\partial^j \rho}{\rho} + 2\frac{\dot{\rho}}{\rho}\frac{\partial^k \rho}{\rho} - \frac{\partial^k \dot{\rho}}{\rho}\right]u_k =$$
$$\ddot{\rho} - \frac{3}{2}\frac{\dot{\rho}^2}{\rho} - \frac{1}{2}\dot{\rho}\frac{(\rho v^2)^{\cdot}}{\rho v^2} + v\left[2\frac{\partial_j \dot{\rho} \partial^j \rho}{\rho} + \frac{\dot{\rho}}{\rho}\frac{\partial_j \rho \partial^j \rho}{\rho} - 2\frac{\dot{\rho}}{\rho}\triangle \rho - \triangle \dot{\rho}\right.$$
$$\left. + \frac{1}{\rho v^2}\left(2\dot{\rho}\ddot{\rho} - \frac{5}{2}\frac{\dot{\rho}^3}{\rho} - \frac{3}{2}\dot{\rho}^2\frac{(\rho v^2)^{\cdot}}{\rho v^2}\right)\right]. \tag{16.365}$$

The final expression, which should be presented explicitly, is

$$\rho\triangle\left(\frac{\dot\rho}{\rho}\right)-2\dot\rho\partial_k\left(\frac{\partial^k\rho}{\rho}\right)=\triangle\dot\rho-2\frac{\partial_j\rho\partial^j\dot\rho}{\rho}-2\frac{\dot\rho}{\rho}\left(\triangle\rho-\frac{3}{2}\frac{\partial_j\rho\partial^j\rho}{\rho}\right). \tag{16.366}$$

By application of the relations (16.365) and (16.366), taking into account the specific condition of constant kinetic energy

$$\frac{\rho v^2}{2}=\frac{E_0}{V_n}, \tag{16.367}$$

and making several elementary algebraic manipulations, the integro-differential equation (16.369) becomes the first order inhomogeneous partial differential equation

$$\begin{aligned}\dot u_i=\frac{\dot\rho}{\rho}u_i+C_i(t)+\frac{2n-1}{2}\int_0^x\frac{\dot\rho^2}{\rho}dx_i' \\ +4\nu\int_0^x\frac{\dot\rho}{\rho}\left[\frac{V_n}{4E_0}\left(\rho\ddot\rho-\frac{5}{4}\dot\rho^2\right)+\frac{\partial_j\rho\partial^j\rho}{\rho}-\triangle\rho\right]dx_i',\end{aligned} \tag{16.368}$$

which can be solved by an elementary method of the theory of differential equations. The solution is

$$\begin{aligned}u_i(x,t)=u_i^\circ(x)\exp\left(\int_0^t\frac{\dot\rho}{\rho}dt'\right)+ \\ \exp\left(\int_0^t\frac{\dot\rho}{\rho}dt'\right)\int_0^t dt'\exp\left(-\int_0^{t'}\frac{\dot\rho}{\rho}dt''\right)\Bigg\{f_i(0,t)-\partial^i p(x,t)\big|_{x=0} \\ +\int_0^x dx_i'\frac{\dot\rho}{\rho}\left[\frac{2n-1}{2}\dot\rho+4\nu\left(\frac{V_n}{4E_0}\left(\rho\ddot\rho-\frac{5}{4}\dot\rho^2\right)+\frac{\partial_j\rho\partial^j\rho}{\rho}-\triangle\rho\right)\right]\Bigg\}.\end{aligned} \tag{16.369}$$

Conclusion

The highly non-trivial situation, two different types of solutions obtained by making use of two different integration methods, leads to the conclusion that the strategy based on the equation (16.318), which is the result arising from the full continuity equation (16.308, is more appropriate than the strategy based on explicit using the continuity equation (16.356) arising from appropriate reduction by using the divergence-free velocity vector condition. However, such a situation is only seeming, because if the divergence-free velocity vector condition is satisfied then there is only one solution of the continuity equation for the fluid density. The solution is a constant density

$$\rho(x,t) = \rho_0, \tag{16.370}$$

for which the obtained solutions for the pressure and velocity vector, in both integration strategies presented above, coincide

$$u_i(x,t) = u_i^\circ(x) + \int_0^t \left[f_i(0,t') - \partial_i p(x,t')\big|_{x=0} \right] dt', \tag{16.371}$$

$$p(x,t) = p(0,t) - \left[f_i(0,t) - \partial_i p(x,t)|_{x=0} \right] x^i + \int_0^x f_i(x',t) dx'^i, \tag{16.372}$$

where $p(0,t)$ and $\partial_i p(x,t)|_{x=0}$ are the initial data for the pressure and for correctness must be given *ad hoc*, *i.e.* integration does not determine them automatically but demands their existence. Interestingly, the solutions (16.371) and (16.372) do not depend on the viscosity parameter ν, which means that they are true for the case of the Euler equation, *i.e.* $\nu = 0$. In spite of such situation, it can be seen by straightforward and easy analysis that (16.371) and (16.372) solve the Navier–Stokes equations with constant fluid density and constant kinetic energy.

In this way, we can make the following true conclusions

1. The Cauchy problem for the Navier–Stokes equations, defined by C.L. Fefferman's [953] as the Millennium Prize Problem, is well defined.

2. Smooth solutions, for which total kinetic energy is the globally controlled quantity (16.301), always exist in any dimensionality n of the Euclidean space, including $n = 3$. The considered case

$$\frac{\rho v^2}{2} = \frac{E_0}{V_n}, \tag{16.373}$$

does not change the generality of the integration procedure presented in this section. For the most general choice

$$\frac{\rho v^2}{2} = \frac{\mathscr{E}(x,t)}{V_n}, \tag{16.374}$$

where $\mathscr{E}(x,t) \in C^\infty[\mathbb{R}^n \times [0,\infty)]$ satisfying the relation

$$\frac{1}{V_n}\int_{\mathbb{R}^n} \mathscr{E}(x,t)d^n x < E < \infty, \tag{16.375}$$

which arises from the condition (16.301), any of the two strategies still lead to the smooth solutions (16.371) and (16.372). The restriction (16.375) means that for any time t the space-average of the energy $\mathscr{E}(x,t)$ is globally controlled.

3. For the case of identically zero body forces, *i.e.* $f_i = 0$, smooth solutions (p,u_i) always exist and have the form

$$u_i(x,t) = u_i^\circ(x) - \int_0^t \partial_i p(x,t')\big|_{x=0} dt', \tag{16.376}$$

$$p(x,t) = p(0,t) + \partial_i p(x,t)|_{x=0} x^i \tag{16.377}$$

4. The divergence-free initial datum of velocity field $u_i(x,0) = u_i^\circ(x)$ must be existent for correctness of the integration method.

5. For the case of non-zero body forces $f_i(x,t)$ which belong to space $[C^\infty(\mathbb{R}\times[0,\infty))]^3$, smooth solutions exist if and only if the integrals

$$\int_0^t f_i(0,t')dt', \tag{16.378}$$

$$\int_0^x f_i(x',t)dx'^i, \tag{16.379}$$

are smooth functions. Surely, this is not clear whether there are smooth functions which arise as derivatives of non-smooth functions. Even if there exists such a function then the breakdown of smoothness to the Navier–Stokes equations is existent when both $u_i^\circ(x)$ and $f_i(x,t)$ are smooth functions. Such a situation could be true for any dimension n of Euclidean space, including $n=3$, *i.e.* $\mathbb{R}^3$.

In this way, we have solved the two problems presented by C.L. Fefferman [953]: (A) existence and smoothness of Navier–Stokes solutions on $\mathbb{R}^3$, and (C) breakdown of Navier–Stokes solutions on $\mathbb{R}^3$.

The problem of the torus $\mathbb{R}^3/\mathbb{Z}_3$ is also soluble within the framework presented in this section. For such a situation, C.L. Fefferman requires to give proof of two statements.

The first statement, referred as (B), says that for $\nu > 0$, $n = 3$, and smooth divergence-free $u_i^\circ(x)$ satisfying

$$u_i^\circ(x + e_j) = u_i^\circ(x), \tag{16.380}$$

where e_j is j^{th} unit vector in $\mathbb{R}^n$ and $1 \leqslant j \leqslant n$, and $f_i(x,t)$ being identically zero, there are smooth solutions to the Navier–Stokes equations satisfying the mentioned conditions and

$$u_i(x + e_j) = u_i(x), \tag{16.381}$$

This statement can be proven straightforwardly, by using the established solutions

$$u_i(x,t) = u_i^\circ(x) - \int_0^t \partial_i p(x,t')\big|_{x=0} dt', \tag{16.382}$$

$$p(x,t) = p(0,t) + \partial_i p(x,t)|_{x=0} x^i. \tag{16.383}$$

Easy verification shows that the periodicity condition (16.381) is satisfied if and only if

$$\int_0^t \partial_i p(x + e_j, t')\big|_{x=0} dt' = \int_0^t \partial_i p(x,t')\big|_{x=0} dt', \tag{16.384}$$

which is satisfied always for

$$p(x + e_j, t) = p(x,t) + p_0(t), \tag{16.385}$$

where $p_0(t)$ is a smooth function on $[0, \infty)$. The relation (16.385) means that pressure should be a quasiperiodic function in space coordinates. Straightforward calculation shows that

$$\begin{aligned} p(x + e_j, t) &= p(e_j, t) + \partial_j p(x + e_j, t)\big|_{x=0} (x^j + e^j) \\ &= p(0,t) + p_0(t) + \partial_j p(x,t)\big|_{x=0} (x^j + e^j) = p(x,t) + p_0(t), \end{aligned} \tag{16.386}$$

if and only if the following condition holds

$$\partial_j p(x,t)\big|_{x=0} e^j = 0. \tag{16.387}$$

The second problem given by Fefferman, referred as (D), is to show that for $n = 3$, $\nu > 0$, and smooth $u_i^\circ(x)$ satisfying the condition (16.380) and $f_i(x,t)$ satisfying the condition

$$f_i(x + e_j, t) = f_i(x,t), \tag{16.388}$$

there is no smooth solutions satisfying (16.381). Straightforward verification

$$u_i(x+e_j,t) = u_i^\circ(x+e_j) + \int_0^t \left[f_i(e_j,t') - \partial_i p(x+e_j,t')\big|_{x=0}\right] dt'$$

$$= u_i^\circ(x) + \int_0^t \left[f_i(0,t') - \partial_i p(x+e_j,t')\big|_{x=0}\right] dt' = u_i(x,t), \tag{16.389}$$

$$p(x+e_j,t) = p(e_j,t) - \left[f_j(e_j,t) - \partial_j p(x+e_j,t)\big|_{x=0}\right](x^j+e^j)$$

$$+ \int_0^{x+e_j} f_i(x',t)dx'^i = p(0,t) + p_0(t) - \left[f_j(0,t) - \partial_j p(x,t)\big|_{x=0}\right](x^j+e^j)$$

$$+ \int_0^x f_i(x',t)dx'^i + \int_x^{x+e_j} f_i(x',t)dx'^i = p(x,t), \tag{16.390}$$

shows that periodic smooth solutions exist if and only if the condition (16.384) holds and moreover

$$p_0(t) + \partial_j p(x,t)\big|_{x=0} e^j + \int_x^{x+e_j} f_i(x',t)dx'^i = 0. \tag{16.391}$$

The condition (16.391), however, can not be satisfied in the light of the fact that integral of a smooth function must not be a smooth function. This shows that in such a situation breakdown of smoothness of the solutions to the Navier–Stokes equations takes place.

In this manner, we have proven, for the specific situation given by the divergence-free velocity vector, constant fluid density, and constant kinetic energy, that all the statements referred by C.L. Fefferman [953] as (A), (B), (C) and (D) constitute a reasonable leeway to find the solution to the "Navier–Stokes Equations" Millennium Prize Problem of the Clay Mathematics Institute. The most general situation, in which both fluid density and kinetic energy are not constants, needs to make use of more complicated elementary calculations which can be performed in the way analogous to the analysis presented for the specific case, but we shall not present the details in this book.

Epilogue

> There are, at present, fundamental problems in theoretical physics (. . .) the solution of which (. . .) will presumably require a more drastic revision of our fundamental concepts than any that have gone before. Quite likely, these changes will be so great that it will be beyond the power of human intelligence to get the necessary new ideas by direct attempts to formulate the experimental data in mathematical terms. The theoretical worker in the future will, therefore, have to proceed in a more direct way. The most powerful method of advance that can be suggested at present is to employ all the resources of pure mathematics in attempts to perfect and generalize the mathematical formalism that forms the existing basis of theoretical physics, and after each success in this direction, to try to interpret the new mathematical features in terms of physical entities.
>
> *Paul Adrien Maurice Dirac*

This book, in my primary intention, has been written to present the constructive description of the physical reality. The philosophical background for such a description was the consistent realization of the fusion of two fundamental concepts of Antiquity: Æther, the idea of Aristotelian natural philosophy, and Multiverse which was initiated in the Epicurean philosophy and has straightforward influences in the Western tradition of Hinduism and Islam. The proposed approach in itself creates the unified point of view which I have proposed to call by the name *Æthereal Multiverse*. There is evidently an opportunity and a necessity to apply such an approach straightforwardly to the concrete problems of theoretical and mathematical physics. The ramifications of such a strategy can be confronted with experimental and observational data. Only such an approach guarantees full physical consistency of *Æthereal Multiverse* with Nature.

I believe that *Æthereal Multiverse* describes very accurately the actual state of affairs in Nature at comparatively small scales. In my opinion the energetic regime governed by the Planck scale, the specific scale in which quantum physics meets classical physics, is most adequate place for the existence of the effects arising due to the fusion of the concepts of Æther and Multiverse. However, such aesthetics is only my personal belief about cosmology and, therefore, there is a justified need for empirical verification for its all ramifications. Perhaps, the scope of applicability of the *Æthereal Multiverse* approach is much wider than I think presently, but it is equally possible that the region of applicability is completely different than I have suggested. Other possible candidates for finding out the new physics, which exists at comparatively small scales, include also the Compton scale or, introduced in this book, the Compton–Planck scale.

Anyway, this monograph presented the consistent and constructive development of certain fundamental knowledge of theoretical physics which has now received the status of the established facts. We have proven definitely that the straightforward philosophical reasoning involving the concepts of both Æther and Multiverse can be productively performed and applied to the consistent description of the new physics. The ontological spirit of the modal realism, which is the ideological foundation for our all considerations, has been resulted in the fruitful and quite fashionable development of these two fundamental ideas of ancient natural philosophy. At the first glance, such a state of affairs looks to be old-fashioned or manifestly fossilizing the old ideas of physics rather than constructing the foundations of new physics. The crucial point of the presented approach is the constructiveness and consistency of the applied theoretical methods based on the method of analogy with the well-established physical knowledge. The central methodological background for our deductions is the principle of simplicity, which enabled us to lay the foundations for new physics from the established knowledge.

From the philosophical point of view, the *Æthereal Multiverse* approach has been proven an obvious and highly non-trivial thing. We have showed, namely, that it is possible to efficiently and consistently join the natural philosophy systems of Aristotelian and Epicurean–Islamic–Hinduistic philosophies in one unified and extraordinary productive approach to description of the physical reality. This in itself is a paradoxical situation because these ancient philosophies have been understood as considerably different and manifestly divergent ideological systems, which are impossible to be unified in one productive description. Such a situation leads me to the strong conviction that the new physics can be effectively formulated by manifestly untypical and unorthodox configurations of the well-known philosophical systems and the well-established knowledge. In other words, it is my strong belief that the new physics is the only sophisticated epistemological structure which is hidden in philosophical interpretation of mathematical formalism of the old physics. The old knowledge should be reinterpreted and supplemented by new ramifications. However, this hidden structure can be straightforwardly extracted if and only if both unorthodox and rigorous methodology is applied unwaveringly. For instance, the identification method, which has been frequently applied in our considerations, is the only straightforward purely logical ramification of the more general method of philosophical interpretation, which plays the fundamental role in all natural sciences.

In this monograph I have presented certain results obtained in my earlier research work. The full collection of these results has been entirely revised and improved. In the light of the fallible nature of human reasoning I was forced to reject from these general epistemological and ontological foundations for the new physics some parts of my work which resulted from usual technical mistakes. However, I have complemented the gap by a number of new developments arising from the revised and definitely correct part of my research results. In spite of these efforts I have a deep conviction that all the applied changes and necessary modifications eventually allowed me to establish the physical truth arising from the *Æthereal Multiverse* approach. This level of discussion leads me to the opinion that this approach is the universal scheme of reasoning which, in the best possible way, is able to effectively describe the new physics and, above all, in itself creates the consistent logical background for the new physics. At the first glance such a logic can be difficult to accept or even impossible to apply. However, it has been shown in this book that the constructive and consistent application of such a scheme leads to the elegant picture of the physical reality.

The scenario arising from the *Æthereal Multiverse* approach is, in my opinion, an essentially novel proposal for joining description of theoretical physics and cosmology because, as a whole is a certain efficient application of both the methods and the philosophical background of relativity theory and quantum theory which have been definitely confirmed as the fundamental theories of physics. Their possible deformations and modifications arise from the most recent experimental data of high energy physics and particle astrophysics. I would be extraordinary satisfied if *Æthereal Multiverse* would turn out to be a productive approach to theoretical physics and cosmology. The crucial question is, however, whether arbitrary mathematical creations can be applied to the consistent description of the physical reality. This book showed that *certain ideas* allow to construct such a consistent approach. In general, the results of any mathematical reasoning applied to making any realistic physical scenarios should be verified by an empirical way, because the physical significance of any theory is established by the physical truth resulting from this theory. It should be emphasized, however, that constructive failures are always much more valuable than nonconstructive successes. It is most important to learn a lesson from and verify ramifications of the theory. In such a state of affairs I remember the words of Wolfgang Pauli: *This is not even wrong.*

I have been interested in phenomena involving complexity, diversity and evolution since I was a young boy.

Murray Gell–Mann

Bibliography

[1] R. Carroll, *On the Emergence Theme of Physics* (World Scientific, 2010)

[2] P.A. Schilpp, *Albert Einstein: Philosopher-Scientist* (MJF Books, 1970);
J. Mehra, *Einstein, Physics and Reality* (World Scientific, 1999);
A. Pais, *'Subtle is the Lord...'. The Science and the Life of Albert Einstein* (Oxford University Press, 2005)

[3] W.D. Ross and J.A. Smith (Eds.), *The Works of Aristotle* Twelve volumes (Clarendon Press, 1908-1952)

[4] T. Birch (Ed.), *The Works of the Honourable Robert Boyle*, 6 vols. (London 1672)

[5] I. Newton, *Opticks: or a Treatise of the Reflexions, Refractions, Inflections, and Colours of Light* (1704)

[6] G.L. Le Sage, *The Newtonian Lucretius*, in *The Le Sage Theory fo Gravitation* (Translated by C.G. Abbot, with introductory note by S.P. Langley), pp. 139-160 (Annual Report of the Board of Regents of the Smithsonian Institution, 1898);
T. Thomson, *Biographical Account of M. Le Sage*, Annals of Philosophy **11**, pp. 241-252 (Baldwin, 1818);
M.R. Edwards, *Pushing Gravity: New Perspectives on Le Sage's Theory of Gravitation* (C. Roy Keys, 2002)

[7] W. Thomson, Phil. Mag. **45**, 321 (1873);
W. Thomson, *Mathematical and Physical Papers* (Cambridge University Press, 1882-1911);

W. Thomson, *Collected Papers in Physics and Engineering* (Cambridge University Press, 1912);
Lord Kelvin, *Baltimore Lectures on Molecular Dynamics and the Wave Theory of Light* (C.J. Clay and Sons, 1904)

[8] J.C. Maxwell, Phil. Mag. **21** p. 161, 281, 338 (1861); Phil. Mag. **22**, p. 12, 85 (1861);
J.C. Maxwell, *Ether*, Encyclopædia Britannica (9th ed.) **8**, pp. 569-572 (1878);
J.C. Maxwell, *A Treatise on Electricity and Magnetism* (Clarendeon Press, 1873)

[9] A. Fresnel, Ann. Chim. Phys. **9**, 57 (1818);
H. Fizeau, C. R. Acad. Sci. **33**, 349 (1851); Phil. Mag. 4th Ser. **2**, 568 (1851); Ann. Chim. Phys. Ser. 3 **57**, 385 (1859); Phil. Mag. 4th Ser. **19**, 245 (1860);
T.R. Birks, *On Matter and Ether or the Secret Laws of Physical Change* (MacMillan, 1862);
S.T. Preston, *Physics of the Ether* (E. & F.N. Spon, 1875);
P.G. Tait, *Lectures on Some Recent Advances in Physical Science with a Special Lecture on Force* (MacMillan, 1876);
W. Barlow, *New Theories of Matter and of Force* (Sampson Low, 1885);
J.S. Russell, *The Wave and Translation in the Oceans of Water, Air, and Ether* (Trubner, 1885);
H.W. Watson and S.H. Burbury, *The Mathematical Theory of Electricity and Magnetism Vol. 1 Electrostatics & Vol. 2 Magnetism and Electrodynamics* (Clarendon, 1885 & 1889);
A.E. Dolbear, *Matter, Ether, and Motion. The Factors and Relations of Physical Science* (Lee and Shepard, 1892);
J. Tyndall, *New Fragments* (D. Appleton, 1896);
A.E. Dolbear, *Modes of Motion. Mechanical Concepts of Physical Phenomena* (Lee and Shepard, 1897);
J. Larmor, Phil. Trans. Roy. Soc. **190**, 205 (1897);
D. Mendeleeff, *Principles of Chemistry* (P.F. Collier and Son, 1897);
E. Wiechert, *Grundlagen der Elektrodynamik* (Teubner, 1899);
J. Larmor, *Aether and Matter. A Development of the Dynamical*

Relations of the Aether to Material Systems on the Basis of the Atomic Constitution of Matter Including a Discussion of the Influence of the Earth's Motion on Optical Phenomena (Cambridge University Press, 1900);
N.E. Gilbert, *Some Experiments Upon The Relations Between Ether, Matter, and Electricity*. Dissertation submitted to the Board of University Studies of the John Hopkins University for the Degree of Doctor of Philosophy (1901);
O. Reynolds, *On an Inversion of Ideas as to the Structure of the Universe* (Cambridge University Press, 1902);
W.G. Hooper, *Aether and Gravitation* (Chapman and Hall, 1903);
G. Adam, *Electricity. The Chemistry of Ether* (Whitaker & Ray, 1904);
J.E. Gore, *Studies in Astronomy* (Chatto & Windus, 1904);
D. Mendeleeff, *An Attempt Towards a Chemical Conception of the Ether* (Longmans, 1904);
G. Mie, *Moleküle, Atome, Weltäther* (Teubner, 1904);
J.J. Thomson, *Electricity and Matter* (Charles Scribner's Sons, 1904);
A.M. Clerke, *Modern Cosmogonies* (Adam and Charles Black, 1905);
W.A. Shenstone, *The New Physics and Chemistry. A Series of Popular Essays on Physical and Chemical Subjects* (Smith, Elder & Co., 1906);
S. Arrhenius, *Theories of Chemistry*. Being Delivered at the University of California, in Berkeley (Longmans, 1907);
O. Lodge, *Modern Views of Electricity* (MacMillan, 1907);
O. Lodge, *The Ether of Space* (Harper & Brothers, 1909);
G.W. De Tunzelmann, *A Treatise on Electrical Theory and the Problem of the Universe* (Griffin, 1910);
J. Larmor, *Aether*, in Encyclopædia Britannica (11th ed.) (1911);
F. Harris, *Gravitation* (Longmans, 1912);
W.W.R. Ball, *Mathematics Recreations and Essays* (6th ed., MacMillan, 1914);
M. Erwin, *The Universe and the Atom. The Ether Constitution, Creation and Structure of Atoms, Gravitation, and Electricity, Kinetically Explained* (Constable, 1915);

G.H. Darwin, *Scientific Papers* (Cambridge University Press, 1907-1916);
J.H. Jeans, *The Mathematical Theory of Electricity and Magnetism* (Cambridge University Press, 1911);
W.J. Spillman, *A theory of gravitation and related phenomena* (The New Era, 1915);
F.W. Very, *The Luminiferous Ether. (I) Its Relation to the Electron and to a Universal Interstellar Medium (II) Its Relation to the Atom* (The Four Seas, 1919);
B. Harrow, *From Netwon to Einstein. Changing Conceptions of the Universe* (Van Nostrand, 1920);
R.A. Sampson, *On Gravitation and Relativity* being the Halley Lecture delivered on June 12, 1920 (Claredon, 1920);
W. Bragg, *Electrons & Ether Waves.* Being the twenty-third Robert Boyle lecture on 11th May 1921 (Oxford University Press, 1921);
P.A. Campbell, *A Non-Euclidean Theory of Matter and Electricity* (Cambridge University Press, 1921);
E. Cunningham, *Relativity, The Electron Theory and Gravitation* (Longmans, 1921);
L. Page, *An Introduction to Electrodynamics From the Standpoint of the Electron Theory* (Ginn, 1922);
H.S. Redgrove, *Alchemy: Ancient and Modern* (William Rider & Son, 1922);
T.J.J. See, *Electrodynamic Wave-Theory of Physical Forces Vol. 1 & Vol. 2 New theory of the Aether* (Thos. P. Nichols & Son, 1917 & 1922)

[10] E.T. Whittaker, *A History of the Theories of Aether and Electricity* Vol. 1 The classical theories & Vol. 2 The modern theories 1900-1926 (2nd ed., Thomas Nelson and Sons, 1951)

[11] A.A. Michelson, Am. J. Sci. **22**, 120 (1881);
A.A. Michelson and E.W. Morley, Am. J. Sci. **34**, 333 (1887)

[12] A.A. Michelson, *Light Waves and Their Uses* (The University of Chicago Press, 1903)

[13] H.A. Lorentz, Zittingsverlag akad. v. Wet. **1**, 74 (1892);
H.A. Lorentz, *Versuch einer Theorie der elektrischen und optischen Erscheinungen in bewegten Körpern* (E.J. Brill, 1895);
H.A. Lorentz, *The Theory of Electrons and its Applications to the Phenomena of Light and Radiant Heat*. A Course of Lectures Delivered in Columbia University, New York, in March and April 1906 (2nd ed., Teubner, 1916);
H.A. Lorentz, *Problems on Modern Physics*. A Course of Lectures Delivered in the California Institute of Technology (Ginn, 1927);
H.A. Lorentz, *Lectures on Theoretical Physics. Vols. I-III*. Delivered at the University of Leiden (MacMillan, 1927-1931)

[14] P. Zeeman, Proc. Roy. Acad. Amsterdam **17**, 445 (1914); Proc. Roy. Acad. Amsterdam **18**, 398 (1915)

[15] O. Heaviside, The Electrician p. 23, 83, 147, 458 (1888-1889) & Electrical papers **2**, p. 490, 504 (1894); Phil. Mag. **27**(167), 324 (1889)

[16] G.F. FitzGerald, Letters to the Editor, Science **13**, 390 (1889)

[17] F.T. Trouton and H.R. Noble, Phil. Trans. Royal Soc. **A202**, 165 (1903);
F.T. Trouton and A.O. Rankine, Proc. Roy. Soc. **80** (420) (1908);
M.G. Sagnac, C. R. Acad. Sci. **157**, p. 708, 710, 1410 (1913);
M.G. Sagnac, J. Phys. Radium **4**(5), 177 (1914);
A.A. Michelson, Astrophys. J. **61**, 137 (1925);
A.A. Michelson, H.G. Gale, and F. Pearson, Astrophys. J. **61**, 139 (1925);
R.J. Kennedy and E.M. Thorndike, Phys. Rev. **42**(2), 400 (1932);
G.W. Hammar, Phys. Rev. **48**(5), 462 (1935)

[18] W. Pauli, *Theory of Relativity* (Pergamon Press, 1958)

[19] M. von Laue, Münchener Sitzungsb., 405 (1911)

[20] E.W. Morley and D. Miller, Proc. Am. Acad. Arts & Sci. **41**, 321 (1905); Science **21**(531), 339 (1905); Science **25**, 525 (1907);
D. Miller, Phys. Rev. **19**, 407 (1922); Proc. Nat. Acad. Sci. **11**, 306 (1925); Phys. Rev. **27**(6), 812 (1926); Science **63**, 433 (1926);

Astrophys. J. **68**, 341 (1928); J. Roy. Ast. Soc. Canada **24**, 82 (1930); Science **77**(2007), 587 (1933); Phys. Rev. **43**, 1054 (1933); Rev. Mod. Phys. **5**(2), 203 (1933); Nature **133**, 162 (1934);
G. Joos and D. Miller, Phys. Rev. **45**, 114 (1934)

[21] L. Silberstein, Science Suppl. - Science News **62**(1596), 8 (1925);
C.T. Chase, A repetition of the Trouton-Noble ether drift experiment. Master's thesis, California Instutute of Technology (1926);
H. Mineur, J. Roy. Ast. Soc. Canada **21**, 206 (1927);
A.A. Michelson, H.G. Gale, and F. Pearson, Nature **115**, 566 (1925);
A.A. Michelson, F.G. Pease, and F. Pearson, Nature **123**, 88 (1929); J. Optical Soc. Am. **18**, 181 (1929); Astrophys. J. **82**, 26 (1935)
Miller Challenges Einstein: Explains Ether Drift Research and Function of Interferometer, The Case Alumns, 10 (December 1929);
W.F.G. Swann, Phys. Rev. **35**, 336 (1930);
W.L. Laurence, *New Evidence held to Upset Einstein: Formula Based on Overlooked Optics Law Offered Here by Prof. Cartmel. Existence of Ether Seen. Difference in Speed of Light Observed From Earth Urged as Relativity Challenge.*, in New York Times, p. 1 (February 23, 1936);
H. Fletcher, Nat. Acad. Sci. **23**, 60 (1943);
W. Reich, *Cosmic Superimposition* (Orgone Institute Press, 1951), republished as *Ether, God, and Devil: Cosmic Superimposition* (Farrar, Strauss & Giroux, 1973);
G. Szekeres, Phys. Rev. **104**, 1791 (1956);
L. Swenson, *The Ethereal Aether: A History of the Michelson-Morley-Miller Aether-Drift Experiments* (U. Texas Press, 1972);
J. DeMeo, *Premilinary Analysis of Changes in Kansas Weather Coincidental to Experimental Operations with a Reich Cloudbuster*, Thesis, University of Kansas, Lawrence, Geography-Meteorology Dept., 1979; chapter in *The Orgone Accumulator Handbook* (Natural Energy, 1989); Pulse of the Planet **1**(2), 3 (1989); in *Proceedings of 72nd Annual Meeting of American Association for the Advancement of Science.*

Northern Arizona Univ., Flagstaff, Arizona, 2-6 June 1996, pp. 41-42 (SW & Rocky Mountain Division, 1996); Pulse of the Planet **5**, 138 (2002); Pulse of the Planet **5**, 114 (2002); in C. Whitney (Ed.), *Proceedings of the Natural Philosophy Alliance* **1**(1), pp. 15-20 (Spring 2004); James De Meo's Research Website: Orgone Biophysical Research Lab/Saharasia http://www.orgonlab.org/index.htm;
M. Allais, *L'Anisotropie de l'espace: La nécessaire révision de certains postulats des théories contemporaines* (Clément Juglar, 1997); in *21st Century Science and Technology*, pp. 26-34 (Spring, 1998); Pulse of the Planet **5**, 132 (2002); C. R. Acad. Sci. **327**(IIb), p. 1405, 1411 (1999); C. R. Acad. Sci. **1**(IV), 1205 (2000); Pulse of the Planet **5**, 132 (2002);
H. Munera, Apeiron **5** (1-2), 37 (1998); Ann. de la Fond. Louis de Broglie **27**(3), 463 (2002);
Yu.M. Galaev, Radiophysics and Electronics **5**(1), 119 (2000); Space-time and Substance **2**(5(10)), 211 (2001); Space-time and Substance **3**(5(15)), 207 (2002) R.P. Crease, *Finding the flaw in falsifiability*, Physics World (December 2002);
C. Lämmerzahl, H. Dittus, A. Peters, and S. Schiller, Class. Quantum Grav. **18**, 2499 (2002);
M. Consoli and E. Costanzo, e-Print: arXiv:astro-ph/0311576; Nuovo Cim. B **119**, 393 (2004), e-Print: arXiv:gr-qc/0406065; e-Print: arXiv:gr-qc/0511160; e-Print: arXiv:gr-qc/0604009; e-Print: arXiv:0710.5613[gr-qc]
M. Consoli and L. Pappalardo, e-Print: arXiv:0912.0103[gr-qc];
C.M.L. de Aragao, M. Consoli, and A. Grillo, e-Print: arXiv:gr-qc/0507048; e-Print: arXiv:0509066;
T.J. Roberts, e-Print: arXiv:physics/0608238 [physics.gen-ph];
R.T. Cahill, Apeiron **11**(1), 53 (2004); Prog. Phys., 60 (July 2006); Prog. Phys., 73 (October 2006); Prog. Phys., 63 (October 2007); e-Print: arXiv:0804.0039 [physics.gen-ph];
R.T. Cahill and F. Stokes Prog. Phys., 103 (April 2008);
J. Leach, A.J. Wright, J.B. Götte, J.M. Girkin, L. Allen, S. Franke-Arnold, S.M. Barnett, and M.J. Padgett, Phys. Rev. Lett. **100**, 153902 (2008)

[22] R.S. Shankland, S.W. McCuskey, F.C. Leone, and G. Kuerti, Rev.

Mod. Phys. **27**, 167 (1955)

[23] R.A. Muller, Scientific American **238**, 64 (1978)

[24] L. Kostro, *Einstein and the Ether* (Apeiron, 2000)

[25] L. Randall, *Warped Passages: Unraveling the Mysteries of the Universe's Hidden Dimensions* (Harper Perennial, 2006)

[26] A. Einstein, Ann. Phys. (Leipzig) **17**, 891 (1905)

[27] M. Laue, Ann. Phys. (Leipzig) **328**, 989 (1907)

[28] A. Einstein, Phys. Zeits. **10**, 817 (1909); Naturwissenschaften **6**(48), 697 (1918);
A. Einstein, *Ether and the Theory of Relativity*, in A. Einstein, *Sidelights on Relativity*, pp. 3-24 (Methuen, 1922);
A. Einstein, *Über den Äther*, Verhandl. der Schweizerischen Naturforsch. Gesellsch., pp. 85-93 (1924)

[29] A. Einstein and L. Infeld, *The Evolution of Physics. The Growth of Ideas from Early Concepts to Relativity and Quanta* (Simon and Schuster, 1938)

[30] E. Cartan, Ann. Sci. Ec. Norm. Sup. **40**, 325 (1923); Ann. Sci. Ec. Norm. Sup. **41**, 1 (1924)

[31] A.M. Trautman, in B. Hoffmann (Ed.), *Perspectives in Geometry and Relativity: Essays in Honor of Václav Hlavatý*, pp. 413–425 (Indiana University Press, 1966); in S. Deser and K.W. Ford (Eds.), *Lectures on General Relativity: Brandeis Summer Institute in Theoretical Physics, 1964, Vol. I*, pp. 1-248 (Prentice-Hall, 1965)

[32] C.W. Misner, K.S. Thorne, and J.A. Wheeler, *Gravitation* (W.H. Freeman, 1973)

[33] P.C.W. Davies and J.R. Brown (Eds.), *The Ghost in the Atom: A Discussion of the Mysteries oFf Quantum Physics* (Cambridge University Press, 1993)

[34] J.S. Bell, *Speakable and Unspeakable in Quantum Mechanics: Collected Papers on Quantum Philosophy* (Cambridge University Press, 1987)

[35] R.P. Feynman, R.B. Leighton, and M. Sands, *The Feynman Lectures on Physics Vol. II Mainly electromagnetism and matter*, Sec. 12-7 (Addison-Wesley, 1965)

[36] J.A. Wheeler and R.P. Feynman, Rev. Mod. Phys. **11**(3), 425 (1949)

[37] P.A.M. Dirac, Nature **168**, 906 (1951); Proc. Roy. Soc. (London) A **209**, 291 (1951); Sci. Month. **78**, 142 (1954)

[38] L. Infeld, Nature **169**, 702 (Jan. 31, 1952);
P.A.M. Dirac, Nature **169**, 702 (Feb. 16, 1952)

[39] M. Planck, Verhandl. Deutschen Phys. Gesell. **2**, 237 (1900)

[40] A. Einstein and O. Stern, Ann. Phys. (Leipzig) **14**, 489 (1913)

[41] J.A. Wheeler, *Geometrodynamics* (Academic Press, 1962)

[42] H. Aspden, Phys. Lett. **85A**, 411 (1981)

[43] H.E. Puthoff, Phys. Rev. D **35**, 3266 (1987)

[44] H.E. Puthoff, Phys. Rev. A **39**, 2333 (1989)

[45] A.D. Sakharov, Sov. Phys. - Dokl. **12**, 1040 (1968)

[46] H.E. Puthoff, Phys. Rev. A **40**, 4857 (1989); Errata and Comments, Phys. Rev. **A44**, 3382, 3385 (1991).

[47] B. Haisch, A. Rueda, and H.E. Puthoff, Phys. Rev. A **49**, 678 (1994)

[48] B.G. Sidharth, *When the Universe Took a U-turn* (World Scientific, 2010);
B.G. Sidharth, *The Thermodynamic Universe. Exploring the Limits of Physics* (World Scientific, 2008);
B.G. Sidharth, *The Universe of Fluctuations. The Architecure of Space-time and the Universe* (Springer, 2005)

[49] M. Gomes, J.R. Nascimento, A.Yu. Petrov, and A.J. Da Silva, Phys. Rev. **D81**, 045018 (2010);
W. Donnelly and T. Jacobson, Phys. Rev. **D82**, 064032 (2010);
S.M. Carroll, T.R. Dulaney, M.I. Gresham, and H. Tam, Phys. Rev. **D79**, 065011 (2009); Phys. Rev. **D79**, 065012 (2009);
S.M. Carroll and H. Tam, Phys. Rev. **D78**, 044047 (2008);
B. Withers, Class. Quantum Grav. **26**, 225009 (2009);
X. Kuang and Y. Ling, JCAP 0910, 024 (2009);
A. Chatrabhuti, P. Patcharamaneepakorn, and P. Wongjun, JHEP 0908, 019 (2009);
C. Armendariz-Picon and A. Diez-Tejedor, JCAP 0912, 018 (2009);
E.V. Linder and R.J. Scherrer, Phys. Rev. **D80**, 023008 (2009);
D.-C. Dai, R. Matsuo, and G. Starkman, Phys. Rev. **D78**, 104004 (2008);
C. Bonvin, R. Durrer, P.G. Ferreira, G. Starkman and T.G. Zlosnik, Phys. Rev. **D77**, 024037 (2008);
Y. Xie and T.-Y. Huang, Phys. Rev. **D77**, 124049 (2008);
T. Tamaki and U. Miyamoto, Phys. Rev. **D77**, 024026 (2008);
T. Jacobson, PoSQG-Ph, 020 (2007);
C. Eling, T. Jacobson, and M.C. Miller, Phys. Rev. **D76**, 042003 (2007);
D. Garfinkle, C. Eling, and T. Jacobson, Phys. Rev. **D76**, 024003 (2007);
R.A. Konoplya and A. Zhidenko, Phys. Lett. **B644**, 186 (2007); Phys. Lett. **B648**, 236 (2007);
T.G. Zlosnik, P.G. Ferreira, and G.D. Starkman, Phys. Rev. **D75**, 044017 (2007);
C. Eling, Phys. Rev. **D73**, 084026 (2006); Phys. Rev. **D76**, 084033 (2007);
C. Eling and T. Jacobson, Phys. Rev. **D74**, 084027 (2006); Class. Quantum Grav. **23**, 5625-5642 (2006); Class. Quantum Grav. **23**, 5643 (2006);
B.Z. Foster, Phys. Rev. **D73**, 024005 (2006); Phys. Rev. **D73**, 104012 (2006);
B.Z. Foster and T. Jacobson, Phys. Rev. **D73**, 064015 (2006);
M. Levin and X.-G. Wen, Phys. Rev. **D73**, 035122 (2006)

[50] T.A. Jacobson and R. Parentani, Scientific American Magazine (December 2005)

[51] Epicurus & Epicurean Philosophy, http://www.epicurus.net;
R. Geer, *Epicurus: Letters, Principal Doctrines, and Vatican Sayings* (Prentice Hall, 1997);
E.M. O'Connor, *The Essential Epicurus: Letters, Principal Doctrines, Vatican Sayings, and Fragments* (Prometheus Books, 1993);
H. Usener, *Epicurea* (Irvington, 1987);
N.W. DeWitt, *Epicurus and His Philosophy* (2nd ed., University of Minnesota Press, 1964);
R.D. Hicks, *Stoic and Epicurean* (Charles Scribner's Sons, 1925)

[52] A.Y. Ali, *The Meaning of the Holy Qur'an* (10th ed., Amana Publications, 2002); *The Qur'an: Text, Translation & Commentary (English and Arabic Edition)* (Tahrike Tarsile Qur'an, 1987)

[53] A. Setia, Islam & Science **2**(2), 161 (Winter 2004), http://findarticles.com/p/articles/mi_m0QYQ/is_2_2/ai_n9532826/

[54] P.N. Sinha, *A Study of the Bhagavata Purana or Esoteric Hinduism* (Benares, 1901);
V. Mani, *Puranic Encyclopaedia: A Comprehensive Dictionary with Special Reference to the Epic and Purānic Literature* (1st English ed., Motilal Banarsidass, 1975)

[55] W. James, *The Varieties of Religious Experience: A Study in Human Nature*. Being the Clifford Lectures on Natural Religion delivered at Edinburgh in 1901-1902 (Longmans, Gree & Co., 1909);
W. James, *A Pluralistic Universe* Hibbert Lectures 1908 (University of Nebraska Press, 1909);
William James: Writings 1902-1910 (Library of America, 1987)

[56] L. Wittgenstein, *Tractatus Logico-Philosophicus* (Kegan Paul, Trench and Trübner, 1922)

[57] P. Geach and M. Black (Eds.), *Translations from the Philosophical Writings of Gottlob Frege* (3rd ed., Blackwell, 1980)

[58] R. Suszko, in R. Parikh (Ed.), *Logic Colloquium: symposium on logic held at Boston, 1972-73*, pp. 169-239 (Lect. Notes Math. **453**, Springer, 1975);
R. Suszko, Studia Logica **27**, 7 (1971);
R. Suszko, Notre Dame Journal of Formal Logic **9**, 7 (1968);
R. Suszko, Analele Universitatii Bucuresti, Acta Logica **11**, 105 (1968)

[59] M. Omyła, Studies in Logic, Grammar, and Rhetoric **10**(23), 21 (2007);
M. Omyła, Philosophy of Science **14**(4(56)), 49 (2006);
M. Omyła, Philosophical Education **40**, 19 (2005);
M. Omyła (Ed.), *Logical Ideas of Roman Suszko*, (Warsaw University, 2001);
M. Omyła, J. Symbolic Logic **55**, 422 (1990);
M. Omyła, Ruch Filozoficzny XLVI(1), 27 (1989);
M. Omyła,*An Outline Non-Fregean Logic* (PWN-Polish Scientific Publishers, 1986);
M. Omyła, Language and Ontology, 195 (1982)

[60] B. Wolniewicz, Bulletin of the Section of Logic **34**, 227 (2005);
B. Wolniewicz, Logica Trianguli **6**, 69 (2003);
B. Wolniewicz, Logica Trianguli **5**, 89 (2001);
B. Wolniewicz, Logica Trianguli **4**, 69 (2000);
B. Wolniewicz, *Logic and Metaphysics. Studies in Wittgenstein's Ontology of Facts* (Polish Semiotic Society, 1999);
B. Wolniewicz, Bulletin of the Section of Logic **22**, 33 (1993);
B. Wolniewicz, in H.A. Lewis (Ed.), *Peter Geach: Philosophical Encounters*, pp. 77-81 (Kluwer, 1991);
B. Wolniewicz, Bulletin of the Section of Logic **18**, 105 (1989);
B. Wolniewicz, Bulletin of the Section of Logic **18**, 2 (1989);
B. Wolniewicz, Bulletin of the Section of Logic **15**, 132 (1986);
B. Wolniewicz, *Ontology of Situations: Foundations and Applications* (PWN-Polish Scientific Publishers, 1985);
B. Wolniewicz, Bulletin of the Section of Logic **13**, 1 (1984);
B. Wolniewicz, Bulletin of the Section of Logic **13**, 21 (1984);
B. Wolniewicz, Bulletin of the Section of Logic **13**, 255 (1984);
B. Wolniewicz, Bulletin of the Section of Logic **12**, 21 (1983);

B. Wolniewicz, Studia Logica **42**, 269 (1983);
B. Wolniewicz, Studia Logica **41**, 381 (1982);
B. Wolniewicz, Bulletin of the Section of Logic **11**, 134 (1982);
B. Wolniewicz, Bulletin of the Section of Logic **11**, 84 (1982);
B. Wolniewicz, Bulletin of the Section of Logic **10**, 17 (1981);
B. Wolniewicz, Bulletin of the Section of Logic **9**, 115 (1980);
B. Wolniewicz, Bulletin of the Section of Logic **9**, 57 (1980);
B. Wolniewicz, in C. Diamond (Ed.), *Intention and Intentionality: Essays in Honour of Professor G.E.M. Anscombe*, pp. 165-178 (Cornell University Press, 1979);
B. Wolniewicz, Bulletin of the Section of Logic **8**, 16 (1979);
B. Wolniewicz, Bulletin of the Section of Logic **7**, 143 (1978);
B. Wolniewicz, Dialectics and Humanism **5**, 171 (1978);
B. Wolniewicz, in E. D'Angelo, D. DeGrood, and D. Riepe (Eds.), *Contemporary East European philosophy. Vol. 3*, pp. 231-243 (Spartacus Books, 1971);
B. Wolniewicz, Theoria **36**, 161 (1970);
B. Wolniewicz, in R.S. Cohen and M.W. Wartofsky (Eds.), *Boston Studies in the Philosophy of Science. Vol. IV*, pp. 208-210 (Reidel Publishing Company 1969)

[61] J. Hawranek and J. Zygmunt, Bulletin of the Section of Logic **2**(3), 117 (1983);
H. Draskovicova, T.S. Fofanova, and V.I. Igoshin, *Ordered Sets and Lattices II* (American Mathematical Society, 1992)

[62] D. Lewis, *Counterfactuals* (Rev. printing, Blackwell, 1986);
D. Lewis, *On the Plurality of Worlds* (Blackwell, 1986);
D. Lewis, *Parts of Classes* (Blackwell, 1991)

[63] S. Kripke, *Naming and Necessity* (Harvard University Press, 1980)

[64] A. Tarski, Phil. Phenomen. Research **4**, 341 (1944)

[65] C. Wright, *Truth and Objectivity* (Harvard University Press, 1992)

[66] M. Lynch, *Truth as One and Many* (Oxford University Press, 2009);

M. Lynch, *True to Life: Why Truth Matters* (MIT Press, 2004);
M. Lynch, *Truth in Context: An Essay on Pluralism and Objectivity* (MIT Press, 1998)

[67] H.N. Goodman, *Languages of Art: An Approach to a Theory of Symbols* (2nd ed., Hackett, 1976);
H.N. Goodman, *The Structure of Appearance* (3rd ed., Reidel, 1977);
H.N. Goodman, *Ways of Worldmaking* (Hackett, 1978);
H.N. Goodman, *Fact, Fiction, and Forecast* (4th ed., Harvard University Press, 1983);
H.N. Goodman, *A Study of Qualities*, Dissertation at Harvard University (Garland, 1990)

[68] H. Everett III, *The Theory of the Universal Wavefunction*, PhD thesis, Princeton University (1957);
H. Everett III, Rev. Mod. Phys. **29**, 454 (1957)

[69] S. Osnaghi, F. Freitas and O. Freire Jr., Stud. Hist. Phil. Mod. Phys. **40**(2), 97 (2009)

[70] M. Born, Zeit. Physik **37**(12), 863 (1926)

[71] B.S. DeWitt, Physics Today **23**(9), 30 (September 1970)

[72] B.S. DeWitt, in C.M. DeWitt and J.A. Wheeler (Eds.), *Battelle Rencontres 1967. Lectures in Mathematics and Physics*, pp. 318-332 (W.A. Benjamin, 1968);
B.S. DeWitt, in B. D'Espagnat (Ed.), *Proceddings of the International School of Physics "Enrico Fermi", Course IL: Foundations of Quantum Mechanics*, pp. 211-262 (Academic Press, 1971)

[73] B.S. DeWitt and R.N. Graham, Am. J. Phys. **39**, 724 (1971)

[74] B.S. DeWitt and N. Graham (Eds.), *The Many-Worlds Interpretation of Quantum Mechanics: A Fundamental Exposition by Hugh Everett, III, with Papers by J.A. Wheeler, B.S. DeWitt, L.N. Cooper and D. Van Vechten, and N. Graham* (Princeton University Press, 1973)

[75] J.A. Wheeler and K.W. Ford, *Geons, Black Holes, and Quantum Foam: A Life in Physics* (W.W. Norton, 1998)

[76] J.A. Wheeler and W.H. Zurek (Eds.), *Quantum Theory and Measurement* (Princeton University Press, 1983)

[77] P. Byrne, Scientific American Magazine (December 2007);
P. Byrne, *The Many Worlds of Hugh Everett III: Multiple Universes, Mutual Assured Destruction, and the Meltdown of a Nuclear Family* (Oxford University Press, 2010)

[78] A.M. Gleason, J. Math. Mech. **6**, 885 (1957)

[79] J.B. Hartle, Am. J. Phys. **36**(8), 704 (1968)

[80] J.B. Hartle and S.W. Hawking, Phys. Rev. D **28**(12), 2960 (1983)

[81] A.D. Linde, Phys. Lett. B **129**, 177 (1983); Mod. Phys. Lett. A **1**, 81 (1986); Phys. Lett. B **175**, 395 (1986); Scientific American Magazine (March 1998)
A.D. Linde, *Inflation and Quantum Cosmology* (Academic Press, 1990);
A.D. Linde, *Particle Physics and Inflationary Cosmology* (Harwood Academic Publishers, 1990)

[82] F.J. Tipler, in R. Penrose and C.J. Isham (Eds.), *Quantum Concepts in Space and Time*, pp. 204-214 (Clarendon, 1986)

[83] J.R. Gribbin, *In Search of Schrödinger's Cat: Quantum Physics and Reality* (Bantam, 1984);
J.R. Gribbin, *In Search of the Multiverse: Parallel Worlds, Hidden Dimensions, and the Ultimate Quest for the Frontiers of Reality* (John Wiley & Sons, 2010)

[84] M. Lockwood, *Mind, Brain, & the Quantum* (Basil Blackwell, 1989)

[85] M. Gell-Mann and J.B. Hartle, in W.H. Zurek (Ed.), *Complexity, Entropy and the Physics of Information*, pp. 425-459 (Addison-Wesley, 1990)

[86] D.Z. Albert, *Quantum Mechanics and Experience* (Harvard University Press, 1992)

[87] R. Penrose, *Shadows of the Mind* (Oxford University Press, 1994)

[88] D.J. Chalmers, *The Conscious Mind* (Oxford University Press, 1996)

[89] D. Deutsch, *The Fabric of Reality: The Science of Parallel Universes and Its Implications* (Penguin Press, 1997)

[90] M. Kaku, *Hyperspace: A Scientific Odyssey Through Parallel Universes, Time Warps, and the Tenth Dimension* (Oxford University Press, 1994);
M. Kaku, *Parallel Worlds: The Science of Alternative Universes and Our Future in the Cosmos* (Allen Lane, 2005)

[91] R. Plaga, Found. Phys. **27**, 559 (1997)

[92] J.A. Barrett, *The Quantum Mechanics of Minds and Worlds* (Oxford University Press, 1999)

[93] D. Deutsch, Int. J. Theor. Phys. **24**(1), 1 (1985); Proc. Roy. Soc. London A **440**, 97 (1985); Proc. Roy. Soc. London A **445**, 3129 (1999); e-Print: arXiv:quant-ph/0104033

[94] D. Page, AIP Conf. Proc. **493**, 225 (1999), e-Print: arXiv:gr-qc/0001001

[95] L. Polley, e-Print: arXiv:quant-ph/9906124; e-Print: arXiv:quant-ph/0102113

[96] W.H. Zurek, Phys. Rev. A **71**, 052105 (2005)

[97] W.H. Zurek, Nature Physics **5**, 181 (2009); Phys. Rev. Lett. **90**, 120404 (2003)

[98] D. Wallace, in A. Ney and D.Z. Albert (Eds.), *The Wave Function: Essays in the Metaphysics of Quantum Mechanics* (Oxford University Press, forthcoming); in R. Batterman (ed.), The Oxford Handbook of Philosophy of Physics (Oxford University Press, forthcoming); in S. Saunders, J. Barrett, A. Kent, and D. Wallace

(Eds.), *Many Worlds? Everett, Quantum Theory, and Reality*, pp. 227-262 (Oxford University Press, 2010); Stud. Hist. Phil. Mod. Phys. **38**, 311 (2007); Brit. J. Phil. Sci. **57**, 655 (2006); Stud. Hist. Phil. Mod. Phys. **34**, 415 (2003); Stud. Hist. Phil. Mod. Phys. **34**, 86 (2003); Stud. Hist. Phil. Mod. Phys. **33**, 637 (2002); e-Print: arXiv:quant-ph/0211104;

[99] S. Saunders, Proc. Roy. Soc. London A **460**, 1771 (2004); e-Print: arXiv:quant-ph/0412194

[100] S. Saunders and D. Wallace, Brit. J. Phil. Sci. **59**, 293 (2008); Brit. J. Phil. Sci. **59**, 315 (2008)

[101] D. Parfit, *Reasons and Persons* (Oxford University Press, 1986)

[102] A. Kent, Int. J. Mod. Phys. A **5**, 1745 (1990), e-Print: arXiv: gr-qc/9703089; in S. Saunders, J. Barrett, A. Kent, and D. Wallace (Eds.), *Many Worlds? Everett, Quantum Theory, and Reality*, pp. 307-354 (Oxford University Press, 2010)

[103] N.P. Landsman, in F. Weinert, K. Hentschel, D. Greenberger and B. Falkenburg (Eds.), *Compendium of Quantum Physics*, pp. 6-9 (Springer, 2008)

[104] H. Janssen, *Reconstructing Reality: Environment-induced decoherence, the measurement problem, and the emergence of definiteness in quantum mechanics. A critical assessment* (Master Thesis, Radboud University Nijmegen, 2008)

[105] M. Gardner, *Are Universes Thicker Than Blackberries?: Discourses on Gödel, Magic Hexagrams, Little Red Riding Hood, and Other Mathematical and Pseudoscientific Topics* (W.W. Norton, 2004)

[106] L. Smolin, *The Life of the Cosmos* (Oxford University Press, 1999);
L. Smolin, *Three Roads to Quantum Gravity* (Basic Books, 2001)

[107] B. Greene, *The Elegant Universe: Superstrings, Hidden Dimensions, and the Quest for the Ultimate Theory* (Vintage Books, 2000);

B. Greene, *The Fabric of the Cosmos: Space, Time, and the Texture of Reality* (Alfred A. Knopf, 2004)

[108] C. Bruce, *Schrödinger's Rabbits. The Many Worlds of Quantum* (Joseph Henry Press, 2004)

[109] B. Carr (Ed.), *Universe or Multiverse?* (Cambridge University Press 2007)

[110] F.J. Dyson, *Imagined Worlds* (Harvard University Press, 1998)

[111] S.J. Dick, *Many Worlds: the new universe, extraterrestrial life, and the theological implications* (Templeton Foundation Press, 2000)

[112] M. Tegmark, Nature **448**, 23 (5 July 2007); in J.D. Barrow, P.C.W. Davies, and C.L. Harper (Eds.), *Science and Ultimate Reality: Quantum Theory, Cosmology, and Complexity*, pp. 459-491 (Cambridge University Press, 2004); Scientific American Magazine (May 2003); Fortsch. Phys. **46**, 855 (1998); Ann. Phys. N.Y. **270**, 1 (1998)

[113] V. Allori, *Fundamental Physical Theories: Mathematical Structures Grounded on a Primitive Ontology* (PhD Dissertation, The State University of New Jersey, 2007);
V. Allori, *On the Metaphysics of Quantum Mechanics*, in S. Lebihan (Ed.), *La philosophie de la physique: d'aujourd'hui a demain* (Editions Vuibert, forthcoming);
V. Allori, *Primitive Ontology and the Structure of Fundamental Physical Theories*, in A. Ney and D.Z. Albert (Eds.), *The Wave Function: Essays in the Metaphysics of Quantum Mechanics* (Oxford University Press, forthcoming)

[114] V. Allori, S. Goldstein, R. Tumulka, and N. Zanghì, Br. J. Philos. Sci. **62**(1), 1 (2011)

[115] M.S. Turner, Scientific American Magazine (August 2009)

[116] A. Jenkins and G. Perez, Scientific American Magazine (January 2010)

[117] J. Feng and M. Trodden, Scientific American Magazine (November 2010)

[118] G.F.R. Ellis, Scientific American Magazine (August 2011)

[119] A. Vilenkin and M. Tegmark, Scientific American Web Exclusives (July 2011)

[120] L.A. Glinka, *Natural Emergence Scheme: (Very) Early Universe as Static Multiverse of Superfluid Fermi-Bose Superstrings*, presentation within GEOSET, http://geoset.fsu.edu/L1.swf

[121] L.A. Glinka, Prespacetime Journal **1**(9), 1395 (November 2010) e-Print: viXra:1011.0007

[122] L.A. Glinka, Prespacetime Journal **2**(5), 733 (May 2011) e-Print: viXra:1106.0001

[123] P.E. Gibbs, Prespacetime Jounral **2**(5), 761 (May 2011)

[124] J.D. Barrow and F.J. Tipler, *The Anthropic Cosmological Principle* (Oxford University Press, 1985)

[125] F.J. Tipler, *The Physics of Immortality. Modern Cosmology, God and the Resurrection of the Dead* (Anchor Books, 1994)

[126] M. Gardner, The New York Review of Books **23**(8), 22 (May 1986)

[127] P.C.W. Davies, Int. J. Astrobiology **2**(2), 115 (2003)

[128] P.C.W. Davies, *Cosmic Jackpot: Why Our Universe is Just Right for Life* (Houghton-Mifflin, 2007)

[129] A.H. Guth, *The Inflationary Universe: The Quest For A New Theory Of Cosmic Origins* (Perseus Books, 1997)

[130] G. Gamow, Phys. Rev. **74**(4), 505 (1948);
G. Gamow, Nature **162**(4122), 680 (1948);
R.A. Alpher and R.C. Herman, Phys. Rev. **74**(12), 1737 (1948)

[131] A.G. Doroshkevich and I.D. Novikov, Sov. Phys. - Dokl. **9**(23), 4292 (1964)

[132] A.A. Penzias and R.W. Wilson, Astrophys. J.**142**, 419 (1965)

[133] R.H. Dicke, P.J.E. Peebles, P.G. Roll, and D.T. Wilkinson, Astrophys. J. **142**, 414 (1965)

[134] A.H. Guth, Phys. Rev. D **23**(2), 347 (1981)

[135] A.A. Starobinsky, Phys. Lett. B **91**, 99 (1980)

[136] D. Kazanas, Astrophys. J. **241**, L59 (1980)

[137] K. Sato, Phys. Lett. B **33**, 66 (1981)

[138] M.B. Einhorn and K. Sato, Nucl. Phys. B **180**(3), 385 (1981)

[139] A. Linde, Phys. Lett. B **108**(6), 389 (1982)

[140] A. Albrecht and P. Steinhardt, Phys. Rev. Lett. **48**(17), 1220 (1982)

[141] V.F. Mukhanov and G.V. Chibisov, JETP Lett. **33**, 532 (1981);
V.F. Mukhanov and G.V. Chibisov, Sov. Phys. JETP **56**, 258 (1982)

[142] G.W. Gibbons, S.W. Hawking, and S.T.C. Siklos (Eds.), *The Very Early Universe: Proceedings of the Nuffield Workshop, Cambridge, 21 June to 9 July, 1982* (Cambridge University Press, 1983)

[143] S.W. Hawking, Phys. Lett. B **115**, 295 (1982)

[144] A.A. Starobinsky, Phys. Lett. B **117**, 175 (1982)

[145] A.H. Guth and S.-Y. Pi, Phys. Rev. Lett. **49**(15), 1110 (1982)

[146] J.M. Bardeen, P. Steinhardt and M.S. Turner, Phys. Rev. D **28**, 679 (1983)

[147] L. Susskind, *The Cosmic Landscape: String Theory and the Illusion of Intelligent Design* (Little, Brown, 2005);
L. Susskind, e-Print: arXiv:hep-th/0302219

[148] A. Vilenkin, Phys. Rev. D **27**, 2848 (1983);
A. Vilenkin, *Many Worlds in One: The Search for Other Universes* (Hill and Wang, 2006)

[149] S. Weinberg, Phys. Rev. Lett. **59**, 2607 (1987)

[150] G. 't Hooft, e-Print: arXiv:gr-qc/9310026

[151] S.W. Hawking, *Black Holes and Baby Universes and Other Essays* (Bantam, 1994)

[152] M.J. Rees, *Before the Beginning: Our Universe and Others* (Addison-Wesley, 1997);
M.J. Rees, *Just Six Numbers: The Deep Forces That Shape The Universe* (Basic Books, 2001)

[153] J.D. Bekenstein, Scientific American, 59 (August 2003)

[154] S. Weinberg, *Dreams of a Final Theory: The Scientist's Search for the Ultimate Laws of Nature* (Vintage, 1994)

[155] B. Greene, *The Hidden Reality: Parallel Universes and the Deep Laws of the Cosmos* (Knopf, 2011)

[156] P. Steinhardt, *It's a Matter of Time*, Edge 2006 annual question, available at http://www.edge.org/q2006/q06_7.html

[157] L. Smolin, *The Trouble With Physics: The Rise of String Theory, the Fall of a Science, and What Comes Next* (Houghton Miffin Harcourt, 2006)

[158] P. Woit, *Not Even Wrong: The Failure of String Theory & the Continuing Challange to to Unify the Laws of Physics* (Basic Books, 2006)

[159] J.A. Nieto, Adv. Theor. Math. Phys. **8**, 177 (2004); Adv. Theor. Math. Phys. **10**, 747 (2006); J. Math. Phys. **45**, 285 (2004);
J.A. Nieto and M.C. Marín, J. Math. Phys. **41**, 7997 (2000); Int. J. Mod. Phys. A **18**, 5261 (2003);
C. Castro and J.A. Nieto, Int. J. Mod. Phys. A **22**, 2021 (2007)

[160] A. Einstein, Phys. Zeits. **15**, 176 (1914); Sitzungsber. Preuss. Akad. Wiss. Berlin **2**, 1030 (1914); Sitzungsber. Preuss. Akad. Wiss. Berlin **44**, 778 (1915); Sitzungsber. Preuss. Akad. Wiss. Berlin **46**, 799 (1915); Sitzungsber. Preuss. Akad. Wiss. Berlin

47, 831 (1915); Sitzungsber. Preuss. Akad. Wiss. Berlin **48**, 844 (1915); Ann. Phys. (Leipzig) **49**, 769 (1916); Sitzungsber. Preuss. Akad. Wiss. Berlin **2**, 1111 (1916); Sitzungsber. Preuss. Akad. Wiss. Berlin **1**, 142 (1917); Ann. Phys (Leipzig) **55**, 241 (1918); Sitzungsber. Preuss. Akad. Wiss. Berlin **1**, 349 (1919); Math. Ann. **7**, 99 (1926)

[161] B. Riemann, Gött. Abh. Ges. Wiss. **13**, 1 (1867)

[162] E. Cartan, *Leçons sur la Géométrie des Espaces de Riemann* (Gauthier-Villars, 1928)

[163] D. Hilbert, Königl. Gesell. d. Wiss. Göttingen, Nachr., Math.-Phys. Kl. **27**, 395 (1915); Königl. Gesell. d. Wiss. Göttingen, Nachr., Math.-Phys. Kl. **61**, 53 (1917)

[164] A. Palatini, Rend. Circ. Mat. Palermo **43**, 203 (1919)

[165] A.A. Friedmann, Zeits. f. Physik **10**, 377 (1922); Zeits. f. Physik **21**, 326 (1924);
G. Lemaître, J. Math. Phys. **4**, 188 (1925); Phys. Rev. **25**, 903 (1925); Bull. Astron. Inst. Netherlands **5**, N200, 273 (1930); Monthly Not. Roy. Astron. Soc. **91**, 490 (1931); Nature **127**, No. 3210, 706 (1931); Ann. Soc. Sci. Brux. **A47**, 49 (1927); Ann. Soc. Sci. Brux. **A53**, 51 (1933);
H.P. Robertson, Phil. Mag. **5**, 835 (1928); Proc. Nat. Acad. Sci. **15**, 822 (1929); Science **76**, 221 (1932); Rev. Mod. Phys. **5**, 62 (1933); Astrophys. J. **82**, 248 (1935); Astrophys. J. **83**, 187 (1936); Astrophys. J. **83**, 257 (1936);
A.G. Walker, Proc. Lon. Math. Soc. 2 **42**, 90, (1937)

[166] A. Einstein, *The Meaning of Relativity* (6th ed., Routledge, 2003)

[167] A.L. Zelmanov, Dokl. Acad. Nauk USSR **227(1)**, 78 (1976)

[168] A. Adem, J. Leida, and Y. Ruan, *Orbifolds and Stringy Topology* (Cambridge University Press, 2007)

[169] P.A.M. Dirac, *Lectures on Quantum Mechanics* (Belfer Graduate School of Science, Yeshiva University Press, 1964); Phys. Rev. **114**, 924 (1959); Phys. Rev. Lett. **2**, 368 (1959); Proc. Roy. Soc.

(London) A **246**, 326 (1958); Proc. Roy. Soc. (London) A **246**, 333 (1958); Can. J. Math. **2**, 129 (1950)

[170] R. Arnowitt, S. Deser and C.W. Misner, in L. Witten (Ed.), *Gravitation: an introduction to current research*, pp. 227-265 (John Wiley & Sons, 1961)

[171] S. Chandrasekhar, *The Mathematical Theory of Black Holes* (Clarendon Press, 1983);
E. Poisson, *A Relativist's Toolkit. The Mathematics of Black-Hole Mechanics*. (Cambridge University Press, 2004)

[172] R.P. Feynman, F.B. Morinigo, and W.G. Wagner, *Feynman Lectures on Gravitation* (Addison-Wesley, 1995)

[173] S. Weinberg, *Cosmology* (Oxford University Press, 2008);
O. Grøn and S. Hervik, *Einstein's General Relativity. With Modern Applications in Cosmology.* (Springer, 2007);
J. Plebański and A. Krasiński, *An Introduction to General Relativity and Cosmology* (Cambridge University Press, 2006);
P. Hoyng, *Relativistic Astrophysics and Cosmology: A Primer* (Springer, 2006);
T.-P. Cheng, *Relativity, Gravitation, and Cosmology: A Basic Introduction* (Oxford University Press, 2005);
S. Carroll, *Space-time and Geometry. An Introduction to General Relativity* (Addison-Wesley, 2004);
J.B. Hartle, *Gravity. An Introduction to Einstein's General Relativity* (Addison-Wesley, 2003);
A. Liddle, *An Introduction to Modern Cosmology* (2nd, John Wiley & Sons, 2003);
T. Padmanabhan, *Theoretical Astrophysics Volume III: Galaxies and Cosmology* (Cambridge University Press, 2002);
B. Ryden, *Introduction to Cosmology* (Addison-Wesley, 2002);
E.F. Taylor and J.A. Wheeler, *Exploring Black Holes. Introduction to General Relativity* (Addison-Wesley, 2000);
R. d'Inverno, *Introducing Einstein's Relativity* (Clarendon Press, 1998);
A. Krasiński, *Inhomogeneous Cosmological Models* (Cambridge University Press, 1997);

T. Padmanabhan, *Cosmology and Astrophysics Through Problems* (Cambridge University Press, 1996);
M. Lachièze-Rey, *Cosmology: A First Course* (Cambridge University Press, 1995);
L.D. Landau and E.M. Lifshitz, *Course of Theoretical Physics. Volume 2: The Classical Theory of Fields* (4th ed., Butterworth Heinemann, 1994);
P.J.E. Peebles, *Principles of Physical Cosmology* (Princeton University Press, 1993);
E.W. Kolb and M.S. Turner, *Early Universe* (Addison-Wesley, 1989);
M. Berry, *Principles of Cosmology and Gravitation* (Cambridge University Press, 1976);
V. Fock, *The Theory of Space, Time, and Gravitation* (2nd rev. ed., Pergamon Press, 1964)

[174] D. Lüst and S. Theisen, *Lectures on String Theory* (Lect. Notes Phys. **346**, Springer, 1989)

[175] J.A. Wheeler, Ann. Phys. N.Y. **2**, 604 (1957); *Geometrodynamics* (Academic Press, 1962); in C. DeWitt and B. DeWitt (Eds.), *Relativity, Groups, and Topology. Lectures Delivered at Les Houches During the 1963 Session of the Summer School of Theoretical Physics*, pp. 317-501 (Gordon and Breach, 1964); in C.M. DeWitt and J.A. Wheeler (Eds.), *Battelle Rencontres 1967 Lectures in Mathematics and Physics* (W.A. Benjamin, 1968), pp. 242-308; *Einsteins Vision* (Springer, 1968); in R.P. Gilbert and R. Newton (Eds.), *Analytic Methods in Mathematical Physics*, pp. 335-378 (Gordon and Breach, 1970)

[176] B.S. DeWitt, Phys. Rev. **160**, 1113 (1967)

[177] V.A. Fock, Z. Phys. **75**, 622 (1932)

[178] N.N. Bogoliubov and D.V. Shirkov, *Introduction to the Theory of Quantized Fields* (3rd English ed., John Wiley & Sons, 1980)

[179] J. von Neumann, Math. Ann. **104**, 570 (1931);
H. Araki and E.J. Woods, J. Math. Phys. **4**, 637 (1963)

[180] N.N. Bogoliubov, Sov. Phys. JETP **7**, 41 (1958)

[181] J.-P. Blaizot and G. Ripka, *Quantum Theory of Finite Systems* (Massachusetts Institute of Technology Press, 1986)

[182] M. Lemoine, J. Martin, and P. Peter (Eds.), *Inflationary Cosmology* (Springer, 2008);
V. Mukhanov, *Physical Foundations of Cosmology* (Cambridge University Press, 2005);
A.R. Liddle and D.H. Lyth, *Cosmological Inflation and Large-Scale Structure* (Cambridge University Press, 2000)

[183] V.F. Mukhanov, H.A. Feldman and R.H. Brandenberger, Phys. Rep. **215**, 203 (1992)

[184] J.A. Bardeen, Phys. Rev. D **22**, 1882 (1980)

[185] J.S. Schwinger, Phys. Rev. **82**, 664 (1951)

[186] J. Martin, in M. Lemoine, J. Martin, and P. Peter (Eds.), *Inflationary Cosmology*, pp. 193-242 (Springer, 2008)

[187] J. Martin, Braz. J. Phys. **34**, 1307 (2004)

[188] G. Jug and B.N. Shalaev, J. Phys. A **32**, 7249 (1999)

[189] I.M. Suslov, JETP Lett. **74**, 191 (2001)

[190] N.W. Ashcroft and N.D. Mermin, *Solid State Physics* (Harcourt College Publishers, 1976);
H. Haken, *Quantum Field Theory of Solids: An Introduction* (Elsevier, 1983);
G.D. Mahan, *Many-Particle Physics* (2nd ed., Plenum Press, 1990);
C. Kittel, *Introduction to Sold State Physics* (8th ed., John Wiley & Sons, 2005)

[191] E. Schrödinger, Sitzungsber. Preuss. Akad. Wiss., Phys.-math. Kl. **14**, 296 (1930)

[192] P.S. Laplace, Mém. Acad. Roy. Sci. Paris 1773, **VII**, 37 (1776), reprinted in *Œvres complétes de Laplace VIII*, pp. 201-275 (Gauthier-Villars, 1891)

[193] J.H. Poincaré, C. R. Acad. Sci. Paris **140**, 1504 (1905), reprinted in *Œuvres de Henri Poincaré IX*, pp. 489-493 (Gauthier-Villars, 1916);
J.H. Poincaré, Rev. Gen. Sci. Pures et Appl. **19**, 386 (1908), reprinted in *Œuvres de Henri Poincaré IX*, pp. 551-586 (Gauthier-Villars, 1916)

[194] A. Einstein, Sitzungsber. Preuss. Akad. Wiss. **1**, 688 (1916)

[195] A. Einstein, Sitzungsber. Preuss. Akad. Wiss. **1**, 154 (1918)

[196] T. Levi-Civita, Rend. della Reale Accad. dei Lincei **26**, 381 (1917)

[197] A.S. Eddington, Proc. Roy. Soc. (London) A **102**(716), 268 (1922)

[198] D. Kennefick, in H. Goenner, J. Renn, J. Ritter, and T. Sauer (Eds.), *The Expanding Worlds of General Relativity (Einstein Studies, Volume* **7**), pp. 207-234 (Birkhäuser, 1999);
D. Kennefick, Physics Today **48**(9), 43 (September 2005);
D. Kennefick, *Traveling at the Speed of Thought: Einstein and the Quest for Gravitational Waves* (Princeton University Press, 2007);

[199] A. Einstein and N. Rosen, J. Franklin Inst. **223**(1), 43 (1937)

[200] M. Born, *The Einstein Born Letters*, letter No. 71 (MacMillan, 1971)

[201] G. Beck, Zeits. f. Physik **33**, 713 (1925)

[202] L. Infeld, *Quest: The Evolution of a Physicist* (Doubleday, Doran & Co., 1941); *Quest: An Autobiography* (AMS Chelsea Publishing Company, 1965/1980/2006)

[203] H.W. Brinkmann, Math. Ann. **18**, 119 (1925)

[204] O.R. Baldwin and G.B. Jeffery, Proc. Roy. Soc. (London) A **111**(757), 95 (1926)

[205] N. Rosen, Phys. Zeits. Sow. **12**, 366 (1937)

[206] A. Einstein, L. Infeld and B. Hoffmann, Ann. Math. **39**(1), 65 (1938)

[207] N. Hu, Proc. Roy. Irish Acad. **51A**, 87 (1947)

[208] L. Infeld and A.E. Scheidegger, Canad. J. Math. **3**, 195 (1951)

[209] A.Z. Petrov, Uch. Zapiski Kazan Gos. Univ. **144**, 55 (1954); Doklady Akad. Nauk SSSR **105**, 905 (1955)

[210] N. Rosen, in A. Mercier and M. Kervaire (Eds.), *Jubilee of Relativity Theory: Proceedings of the Fiftieth Anniversary Conference, Bern, 11-16 July 1955*, pp. 171-175 (Helv. Phys. Acta Suppl. **4**, Birkhäuser-Verlag, 1955)

[211] G.C. McVittie, J. Rat. Mech. Anal. **4**, 201 (1955)

[212] F.A.E. Pirani, Math. Reviews **16**, 1165 (1955)

[213] J.N. Goldberg, Phys. Rev. **99**, 1873 (1955)

[214] G.C. McVittie, *General Relativity and Cosmology* (Chapman and Hall, 1956)

[215] F.A.E. Pirani, Bull. Acad. Pol. Sci. **III**(5), 143 (1956)

[216] F.A.E. Pirani, Phys. Rev. **105**, 1089 (1957)

[217] H. Bondi, Nature **179**(4569), 1072 (1957)

[218] J. Weber and J.A. Wheeler, Rev. Mod. Phys. **29**(3), 509 (1957)

[219] N. Rosen, Phys. Rev. **110**, 291 (1958)

[220] R.K. Sachs and P.G. Bergmann, Phys. Rev. **112**, 674 (1958)

[221] A.M. Trautman, Bull. Acad. Pol. Sci. **6**, 407 (1958); Bull. Acad. Pol. Sci. **6**, 627 (1958);

[222] H. Bondi, F.A.E. Pirani and I. Robinson, Proc. Roy. Soc. (London) A **251**, 519 (1959)

[223] R.P. Kerr, Nuovo Cim. **13**, 673 (1959)

[224] J. Stachel, *Energy Flow in Cylindrical Gravitational Waves* (Master thesis, Stevens Institute of Technology, 1959)

[225] A. Peres, Nuovo Cim. **11**, 644 (1959); Nuovo Cim. **15**, 351 (1960)

[226] L. Infeld and J.F. Plebański, *Motion and Relativity* (Pergamon Press, 1960)

[227] J. Weber, Phys. Rev. **117**, 306 (1960)

[228] B. Bertotti and J. Plebański, Ann. Physics **11**, 169 (1960)

[229] R. Sachs, Proc. Roy. Soc. (London) A **264**, 309 (1961)

[230] J.N. Goldberg and R.K. Sachs. Acta Phys. Polon. Suppl. **22**, 13 (1962)

[231] P. Havas and J.N. Goldberg, Phys. Rev. **128**, 398 (1962)

[232] E.T. Newman and R. Penrose, J. Math. Phys. **3**, 566 (1962)

[233] H. Bondi, M.G.J. van der Burg, and A.W.K. Metzner, Proc. Roy. Soc. (London) A **269**, 21 (1962)

[234] R.K. Sachs, Proc. Roy. Soc. (London) A **270**, 103 (1962)

[235] E. Newman and T. Unti, J. Math. Phys. **3**, 892 (1962)

[236] H. Bondi, in L. Infeld (Ed.), *Relativistic Theories of Gravity: Proceedings of conference held in Warsaw and Jabłonna, July 1962*, pp. 120-121 (Pergamon Press/Gauthier-Villars/P.W.N. Scientific Publishing House, 1964)

[237] I. Robinson and A. Trautman, Proc. Roy. Soc. (London) **265**, 463 (1962)

[238] M.E. Gertsenshtein, Soviet Physics JETP **14**(1), 84 (1962)

[239] R. Penrose, Phys. Rev. Lett. **10**, 66 (1963)

[240] F. Dyson, in A.G.W. Cameron (Ed.), *Interstellar Communications: A Collection of Reprints and Original Contributions*, pp. 115-120 (W.A. Benjamin, 1963)

[241] W.A. Fowler, Rev. Mod. Phys **36**, 545 (1964)

[242] I. Robinson, Ivor, A. Schild and E.L. Schucking, *Quasi-Stellar Sources and Gravitational Collapse: Proceedings of the First Texas Symposium on Relativistic Astrophysics* (University of Chicago Press, 1965)

[243] S. Chandrasekhar, Astrophys. J. **142**, 1488 (1965)

[244] P. Szekeres, J. Math. Phys. **7**, 751 (1966)

[245] F.I. Cooperstock, Phys. Rev. **163**, 1368 (1967)

[246] R.A. Isaacson, Phys. Rev. **166**, 1263 (1968); Phys. Rev. **166**, 1272 (1968)

[247] W.L. Burke, *The Coupling of Gravitational Radiation to Nonrelativistic Sources*, PhD thesis (Caltech, 1969)

[248] W.L. Burke and K.S. Thorne, in M. Carmeli, S.I. Fickler and L. Witten (Eds.), *Relativity: Proceedings of the Relativity Conference in the Midwest, held at Cincinnati, Ohio, June 2-6, 1969*, pp. 209-228 (Plenum Press, 1970)

[249] J. Weber, Phys. Rev. Lett. **22**, 1320 (1969)

[250] S. Chandrasekhar and F.P. Esposito, Astrophys. J. **160**, 153 (1970)

[251] P. Havas, in A. Lichnerowicz and Y. Choquet-Bruhat (Eds.), *Ondes et Radiation Gravitationelles: Actes du Colloques Internationaux du C.N.R.S. no. 220, Paris, 1973*, pp. 383-392 (Editions du CNRS, 1974)

[252] T. Damour and R. Ruffini, C. R. Acad. Sci. Paris Sér. **A279**, 971 (1974)

[253] R.A. Hulse and J.H. Taylor, Astrophys. J. **195**, L51 (1975)

[254] J. Ehlers, A. Rosenblum, J.N. Goldberg and P. Havas, Astrophys. J. **160**, 153 (1976); Astrophys. J. **208**, L77 (1976)

[255] B.K. Berger, D.M. Eardley, and D.W. Olson, Phys. Rev. D **16**, 3086 (1977)

[256] J. Ehlers, *Isolated Gravitating systems in General Relativity: Proceedings of the International School of Physics "Enrico Fermi", Course LXVII* (North Holland, 1979)

[257] J.H. Taylor and P.M. McCulloch, Ann. N. Y. Acad. Sci. **336**(1), 442 (1980)

[258] K.S. Thorne, Rev. Mod. Phys. **52**, 299 (1980)

[259] M. Walker and C.M. Will, Astrophys. J. **242**, L129 (1980)

[260] A. Rosenblum, Phys. Lett. A **81**(1), 1 (1981)

[261] F.I. Cooperstock and D.W. Hobill, Gen. Rel. Grav. **14**, 361 (1982)

[262] A. Rosenblum, J. Phys. A: Math. Gen. **16**, 2751 (1983)

[263] T. Damour, Phys. Rev. Lett. **51**, 1019 (1983)

[264] T. Futamase, Phys. Rev. D **28**, 2373 (1983)

[265] J.L. Anderson, Phys. Rev. D **36**, 2301 (1987)

[266] J. Ehlers, A. R. Prasanna, and R. A. Breuer, Class. Quantum Grav. **4**, 253 (1987)

[267] I. Robinson, Class. Quantum Grav. **6**, 1863 (1989)

[268] B.F. Schutz, Class. Quantum Grav. **6**, 1761 (1989)

[269] J.B. Griffiths, *Colliding Plane Waves in General Relativity* (Clarendon Press, 1991)

[270] F.I. Cooperstock, Found. Phys. **22**, 1011 (1992)

[271] A. Abramovici et al, Science **256**, 325 (1992)

[272] R. Gómez and J. Winicour, Phys. Rev. D **45**, 2776 (1992)

[273] M. Efroimsky, Class. Quantum Grav. **9**, 2601 (1992)

[274] N. Rosen and K.S. Virbhadra, Gen. Rel. Grav. **25**, 429 (1993)

[275] L. Blanchet and G. Schäfer, Class. Quantum Grav. **10**, 2699 (1993)

[276] D.E. Neville, Class. Quantum Grav. **10**, 2223 (1993)

[277] B.F. Schutz, Class. Quantum Grav. **10**, S135 (1993)

[278] L. Blanchet and B.S. Sathyaprakash, Class. Quantum Grav. **11**, 2807 (1994)

[279] T. Damour and B.R. Iyer, Class. Quantum Grav. **11**, 1353 (1994)

[280] G.A. Alekseev and J.B. Griffiths, Class. Quantum Grav. **13**, L13 (1996)

[281] L. Blanchet, B.R. Iyer, C.M. Will, and A.G. Wiseman, Class. Quantum Grav. **13**, 575 (1996)

[282] R. Maartens, G.F.R. Ellis, and S.T.C. Siklos, Class. Quantum Grav. **14**, 1927 (1997)

[283] R.A. Matzner et al, Class. Quantum Grav. **14**, L21 (1997)

[284] L.E. Simone, S.W. Leonard, E. Poisson, and C.M. Will, Class. Quantum Grav. **14**, 237 (1997)

[285] H. van Elst and G.F.R. Ellis, Phys. Rev. D **59**, 024013 (1999)

[286] J. Weber, *General Relativity And Gravitational Waves* (Dover, 2004)

[287] M. Maggiore, *Gravitational Waves. Volume 1 Theory and Experiments* (Oxford University Press, 2007)

[288] P. Jaranowski and A. Królak, *Analysis of Gravitational-Wave Data* (Cambridge University Press, 2009)

[289] J.D.E. Craighton and W.G. Anderson, *Gravitational-Wave Physics and Astronomy: An Introduction to Theory, Experiment and Data Analysis* (Wiley-VCH, 2011)

[290] R.M.L. Baker, AIP Conf. Proc. **699**, 1106 (2004); AIP Conf. Proc. **699**, 1146 (2004); AIP Conf. Proc. **699**, 1093 (2004); AIP Conf. Proc. **746**, 1306 (2005); Astron. Nachr. **327**(7), 710 (2006); AIP Conf. Proc. **880**, 1017 (2007); AIP Conf. Proc. **1103**, 591 (2009);

AIP Conf. Proc. **1103**, 548 (2009); AIP Conf. Proc. **1208**, 501 (2010);
R.M.L. Baker and C.S. Black, AIP Conf. Proc. **1103**, 582 (2009);
R.M.L. Baker, E.W. Davis, and R.C. Woods, AIP Conf. Proc. **746**, 1315 (2005);
R.M.L. Baker and F.Y. Li, AIP Conf. Proc. **746**, 1271 (2005);
R.M.L. Baker, F.Y. Li, and R. Li, AIP Conf. Proc. **813**, 1352 (2006);
R.M.L. Baker, G.V. Stephenson, and F.Y. Li, AIP Conf. Proc. **969**, 1036 (2008); AIP Conf. Proc. **969**, 1045 (2008);
R.M.L. Baker, R.C. Woods, and F.Y. Li, AIP Conf. Proc. **813**, 1280 (2006);
A.W. Beckwith, AIP Conf. Proc. **1208**, 477 (2010); AIP Conf. Proc. **1103**, 571 (2009); AIP Conf. Proc. **1103**, 564 (2009); AIP Conf. Proc. **969**, 1018 (2008); AIP Conf. Proc. **969**, 1091 (2008);
G. Fontana and R.M.L. Baker, AIP Conf. Proc. 813, **1249** (2006); AIP Conf. Proc. **880**, 1156 (2007);
C. Harper and G. Stephenson, AIP Conf. Proc. **880**, 1083 (2007);
F.Y. Li and R.M.L. Baker, Int. J. Mod. Phys. B **21**, 3274 (2007);
F.Y. Li, R.M.L. Baker, and Z. Fang, AIP Conf. Proc. **880**, 1139 (2007);
F.Y. Li, R.M.L. Baker, Z. Fang, G.V. Stephenson and Z. Chen, Eur. Phys. J. C **56**, 407 (2008);
F.Y. Li and Y. Nan, Chinese Phys. Lett. **26**(5), 050402 (2009);
F.Y. Li and M.-X. Tang, Acta Phys. Sin. **6**, 321 (1997);
F.Y. Li, M.X. Tang, J. Luo, and Y.C. Li, Phys. Rev. D **62**, 044018 (2000);
F.Y. Li, M.-X. Tang and P. Zhao, Acta Phys. Sin. **41**, 1919 (1992);
F.Y. Li and N. Yang, China Phys. Lett. **21**(11), 2113 (2004);
F.Y. Li, N. Yang, Z. Fang, R.M.L. Baker, G.V. Stephenson and H. Wen, Phys. Rev. D. **80**, 060413 (2009);
G.V. Stephenson, AIP Conf. Proc. **746**, 1264 (2005); AIP Conf. Proc. **1103**, 532 (2009); AIP Conf. Proc. **1103**, 542 (2009);
R.C. Woods, R.M.L. Baker, F.Y. Li, G.V. Stephenson, E.W. Davis and A.W. Beckwith, J. Mod. Phys. **2**(6), 498 (2011)

[291] V.F. Mukhanov, *Massive Gravity*, talk at *Quantum Theory and Gravitation, Eidgenössische Technische Hochschule, Zürich, Switzerland, June 14-24, 2011*

[292] E.M. Lifshitz, J. Phys. (USSR) **10**, 116 (1946)

[293] E. Bertschinger, in R. Schaeffer, J. Silk, M. Spiro, and J. Zinn-Justin (Eds.), *Cosmology and Large Scale Structure. Proc. Les Houches Summer School, Session LX*, pp. 273-347 (Elsevier Science, 1996)

[294] M.J.G. Veltman, in R. Balian and J. Zinn-Justin (Eds.), *Methods in Field Theory. Proc. Les Houches, Session XXVIII*, pp. 265-328 (North Holland, 1976)

[295] F. Englert and R. Brout, Phys. Rev. Lett. **13**(9), 321 (1964),
P.W. Higgs, Phys. Rev. Lett. **13**(16), 508 (1964),
G.S. Guralnik, C.R. Hagen, and T.W.B. Kibble, Phys. Rev. Lett. **13**(20), 585 (1964).

[296] Y. Nambu, Phys. Rev. **117**, 648 (1960);
J. Goldstone, Nuovo Cim. **19**, 154 (1961).

[297] A. Mazet, C. Imbert, and S. Huard, C. R. Acad. Sci. Ser. B **273**, 592 (1971)

[298] F. Goos and M. Hänchen, Ann. Phys. (Leipzig) **1**, 333 (1947)

[299] L. De Broglie and J.-P. Vigier, Phys. Rev. Lett. **28**(1), 1001 (1972)

[300] E.S. Abers and B.W. Lee, Phys. Rep. **9C**, 1 (1973);
T.-P. Cheng and L.-F. Li, *Gauge theory of elementary particle physics* (Clarendon Press, 1988)

[301] C. Itzykson and J.-B. Zuber, *Quantum Field Theory* (McGraw-Hill, 1980);
M. Peskin and D. Schroeder, *An Introduction to Quantum Field Theory* (Westview Press, 1995)

[302] V.L. Ginzburg and L.D. Landau, Zh. Eksp. Teor. Fiz. **20**, 1064 (1950)

[303] A.A. Abrikosov, Sov. Phys. JETP **5**, 1174 (1957)

[304] H.B. Nielsen and P. Olesen, Nucl. Phys. **B61**, 45 (1973)

[305] M.N. Chernodub, L. Faddeev, A.J. Niemi, JHEP **0812**, 014 (2008);
A. Bashir, Y. Concha-Sanchez, R. Delbourgo, and M.E. Tejeda-Yeomans, Phys. Rev. D **80**, 045007 (2009);
S.P. Kim and H.K. Lee, J. Korean Phys. Soc. **54**, 2605 (2009); Phys. Rev. D **76**, 125002 (2007)
C.Bernicot and J.-Ph.Guillet, JHEP **0801**, 059 (2008);
R.P. Malik and B.P. Mandal, Pramana **72**, 805 (2009);
D.S. Irvine, M.E. Carrington, G.Kunstatter, and D. Pickering, Phys. Rev. D **64**, 045015 (2001);
H. Kleinert and B. Van den Bossche, Nucl. Phys. B **632**, 51 (2002)

[306] A.B. Arbuzov, L.A. Glinka, and V.N. Pervushin, in *QUARKS-2008: Proceedings of the 15th International Seminar on High Energy Physics, Sergiev Posad, Russia, May 23-29, 2008*; e-Print: arXiv:0805.3075 [hep-ph]; e-Print: arXiv:0705.4672v6 [hep-ph]

[307] G. J. Troup, J. L. A. Francey, R. G. Turner, and A. Tirkel, Phys. Rev. Lett. **28**, 1540 (1972)

[308] R. Lakes, Phys. Rev. Lett. **80**(9), 1826 (1998)

[309] L. Bass and E. Schrödinger, Proc. R. Soc. London A **232**, 1 (1955)

[310] A.S. Goldhaber and M.M. Nieto, Rev. Mod. Phys. **43**, 277 (1971); Rev. Mod. Phys. **82**, 939 (2010)

[311] H. Georgi, P. Ginsparg, and S.L. Glashow, Nature (London) **306**, 765 (1983)

[312] Y. Aharonov and D. Bohm, Phys. Rev. **115**(3), 485 (1959)

[313] L.A. Glinka, Apeiron **16**, 147 (2009)

[314] W. Pauli, Ann. Inst. H. Poincaré **6**, 109 (1936)

[315] J.D. Bjorken and S.D. Drell, *Relativistic Quantum Mechanics* (McGraw-Hill, 1964);
W. Greiner, *Relativistic Quantum Mechanics. Wave Equations* (3rd ed., Springer, 2000);
M. Peskin and D. Schroeder, *An Introduction to Quantum Field Theory* (Westview Press, 1995);

M. Kaku, *Quantum Field Theory. A Modern Introduction* (Oxford University Press, 1993)

[316] P.A.M. Dirac, Proc. Roy. Soc. A **117**, 610 (1928); Proc. Roy. Soc. A **126**, 360 (1929)

[317] H.S. Snyder, Phys. Rev. **71**, 38 (1947); Phys. Rev. **72**, 68 (1947)

[318] M.V. Battisti and S. Meljanac, Phys. Rev. **D82**, 024028 (2010); Phys. Rev. **D79**, 067505 (2009),
S. Meljanac, D. Meljanac, A. Samsarov, and M. Stojic, Mod. Phys. Lett. **A25**, 579 (2010); e-Print: arXiv:0909.1706[math-ph]

[319] D.M. Mattingly, L. Maccione, M. Galaverni, S. Liberati, and G. Sigl, JCAP **1002**, 007 (2010),
L. Maccione, A.M. Taylor, D.M. Mattingly and S. Liberati, JCAP **0904**, 022 (2009),
L. Maccione, A.M. Taylor, D.M. Mattingly and S. Liberati, Int. J. Mod. Phys. **D18**, 1621 (2009),
B.G. Sidharth, Found. Phys. **38**, 89 (2008)

[320] W. Greiner, *Relativistic Quantum Mechanics* (3rd ed., Springer, 2000)

[321] M.E. Peskin and D.V. Schroeder, *An Introduction to Quantum Field Theory* (Addison-Wesley, 1995)

[322] C. Kiefer, *Quantum Gravity* (2nd ed., Oxford University Press, 2007)

[323] T.D. Lee, *Particle Physics and Introduction to Field Theory* (Harwood Academic Press, 1981); Phys. Lett. **122B**, 217 (1983)

[324] L. Bergström and A. Goobar, *Cosmology and Particle Astrophysics* (2nd. ed., Springer, 2006),
D.H. Perkins, *Particle Astrophysics* (Oxford University Press, 2003),
H.V. Klapdor-Kleingrothaus and K. Zuber, *Particle Astrophysics* (Rev. ed., IOP Publishing, 2000)

[325] A.H. Compton, Phys. Rev. **21(5)**, 483 (1923)

[326] R. Ferraro, *Einstein's Space-Time. An Introduction o Special and General Relativity* (Springer, 2007),
W. Rindler, *Relativity. Special, General, and Cosmological* (2nd ed., Oxford University Press, 2006),
P.M. Schwarz and J.H. Schwarz, *Special Relativity. From Einstein to Strings* (Cambridge University Press, 2004),
S. Gasiorowicz, *Quantum Physics* (3rd ed., John Wiley & Sons, 2003),
P.A. Tipler, *Modern Physics* (10th ed., Worth Publishers, 1992)

[327] M.K. Transtrum, *Quantum Mechanics in Snyder Space*, Honor's thesis advised by J.-F.S. Van Huele, Brigham Young University (2006)

[328] P.E. Gibbs, *Event-Symmetric Space-Time* (Weburbia Publication 1998), e-Print: viXra:0911.0042, electronic copies available at http://www.weburbia.com/press/esst.htm

[329] B.G. Sidharth, Int. J. Mod. Phys. E **19**, 79 (2010); e-Print: arXiv:0902.3342 [physics.gen-ph]; e-Print: arXiv:0903.4722 [physics.gen-ph]; e-Print: arXiv:0904.3639 [physics.gen-ph]; e-Print: arXiv:0911.1228 [physics.gen-ph]; e-Print: arXiv:1003.2330 [physics.gen-ph]; e-Print: arXiv:1008.2491 [physics.gen-ph]; e-Print: arXiv:1110.0929 [physics.gen-ph]

[330] L.A. Glinka, Apeiron **17**(4), 223 (2010); Apeiron **17**(4), 243 (2010)

[331] B.S. Lakshmi, Apeiron **16**(4), 485 (2009)

[332] C.V. Aditya and B.G. Sidharth, e-Print: arXiv:1010.0540 [physics.gen-ph]

[333] R. Andriambololona and Ch. Rakotonirina, EJTP **8**(25), 177 (2011)

[334] A. Connes, *Noncommutative Geometry* (Academic Press, 1994);
G. Dito, Lett. Math. Phys. **48**, 307 (1999);
N. Seiberg and E. Witten, JHEP **09**, 032 (1999);
D.J. Gross, A. Hashimoto, and N. Itzhaki, Adv. Theor. Math. Phys. **4**, 893 (2000);

A.S. Cattaneo and G. Felder, Prog. Theor. Phys. Suppl. **144**, 38 (2001);
N. Bel Baraka, Int. J. Theor. Phys. **41**(4), 737 (2002);
G. Dito and D. Sternheimer, Lect. Math. Theor. Phys. **1**, 9 (2002);
M.R. Douglas and N.A. Nekrasov, Rev. Mod. Phys. **73**, 977 (2002);
G. Fiore, M. Maceda, and J. Madore, J. Math. Phys. **43**, 6307 (2002);
R.J. Szabo, Phys. Rep. **378**, 207 (2003);
M. Chaichian, K. Nishijima and A. Tureanu, Phys. Lett. B **568**, 146 (2003);
L. Alvarez-Gaume and M. A. Vazquez-Mozo, Nucl. Phys. B **668**, 293 (2003);
A. Berard and H. Mohrbach, Phys. Rev. D **69**, 127701 (2004);
A. Das and J. Frenkel, Phys. Rev. D **69**, 065017 (2004);
M. Chaichian, M. N. Mnatsakanova, K. Nishijima, A. Tureanu, and Yu. A. Vernov, e-Print: arXiv:hep-th/0402212;
D.H.T. Franco and C.M.M. Polito, J. Math. Phys. **46**, 083503 (2005);
M. Chaichian, P.P. Kulish, K. Nshijima, and A. Tureanu, Phys. Lett. B **604**, 98 (2004);
C.D. Fosco and G. Torroba, Phys. Rev. D **71**, 065012 (2005);
C.-S. Chu, K. Furuta, and T. Inami, Int. J. Mod. Phys. A **21**, 67 (2006);
B. Schroer, Ann. Phys. N.Y. **319**, 92 (2005); Ann. Phys. N.Y. **321**, 435 (2006);
O.W. Greenberg, Phys. Rev. D **73**, 045014 (2006);
M.A. Soloviev, Theor. Math. Phys. **147**, 660 (2006); Theor. Math. Phys. **153**, 1351 (2007);
E. Harikumar and V.O. Rivelles, Class. Quantum Grav. **23**, 7551 (2006);
G. Fiore and J. Wess, Phys. Rev. D **75**, 105022 (2007);
M. Chaichian, M. N. Mnatsakanova, A. Tureanu, and Yu. S. Vernov, JHEP **0809**, 125 (2008);
M.V. Battisti and S. Meljanac, Phys. Rev. D **79**, 067505 (2009); Phys. Rev. D **82**, 024028 (2010);
M. Daszkiewicz, Mod. Phys. Lett. A **24**, 1325 (2009); Acta Phys. Polon. B **41**, 1881 (2010); Acta Phys. Polon. B **41**, 1889 (2010);

Mod. Phys. Lett. A **25**, 1059 (2010);
A.H. Chamseddine and A. Connes, Forts. Phys. **58**, 553 (2010);
D. Kolodrubetz and M. Marcolli, Phys. Lett. B **693**, 166 (2010);
E. Brown and R.B. Mann, Phys. Lett. B **694**(4-5), 440 (2010)

[335] M. Kontsevich, in G. Dito and D. Sternheimer (Eds.), *Proceedings of Euroconference Moshe Flato (CMF 2000), Dijon, France, 11-13 Sep 2000*, pp. 157-216 (Lett. Math. Phys. **66**(3), 2003)

[336] P.A.M. Dirac, *The Principles of Quantum Mechanics* (Clarendon Press, 1958)

[337] M.A. Markov, Prog. Theor. Phys. Suppl. **E65**, 85 (1965); Sov. Phys. JETP **24**, 584 (1967)

[338] V.G. Kadyshevsky, Sov. Phys. JETP **14**, 1340(1962); Nucl. Phys. B **141**, 477 (1978); in W. Beiglböck, A. Böhm, and E. Takasugi (Eds.), *Group Theoretical Methods in Physics: Seventh International Colloquium and Integrative Conference on Group Theory and Mathematical Physics, Held in Austin, Texas, September 11–16, 1978*, pp. 114-124 (Lect. Notes Phys. **94**, Springer, 1978); PEPAN **11**(1), 5 (1980);
V.G. Kadyshevsky and M. D. Mateev, Phys. Lett. B **106**, 139 (1981); Nuovo Cim. A **87**(3), 324 (1985);
M.V. Chizhov, A.D. Donkov, V.G. Kadyshevsky, and M.D. Mateev, Nuovo Cim. A **87**(3), 350 (1985); Nuovo Cim. A **87**(4), 373 (1985);
V.G. Kadyshevsky, Phys. Part. Nucl. **29**(3), 227 (1998);
V.G. Kadyshevsky, M.D. Mateev, V.N. Rodionov, and A.S. Sorin, Dokl. Phys. **51**(6), 287 (2006); CERN-TH/2007-150, e-Print: arXiv:0708.4205 [hep-ph]

[339] V.N. Rodionov, e-Print: arXiv:0903.4420 [hep-ph]

[340] A.E. Chubykalo, V.V. Dvoeglazov, D.J. Ernst, V.G. Kadyshevsky, and Y.S. Kim (Eds.), *Lorentz Group, CPT and Neutrinos: Proceedings of the International Workshop, Zacatecas, Mexico, 23-26 June 1999* (World Scientific, 2000)

[341] B.G. Sidharth, Found. Phys. **38**, 89 (2008); Found. Phys. **38**, 695 (2008)

[342] B.G. Sidharth, Int. J. Mod. Phys. E **14**, 1 (2005)

[343] B.G. Sidharth, Int. J. Mod. Phys. E **14**, 927 (2005)

[344] M. Kaku and J.T. Thompson, *Beyond Einstein: The Cosmic Quest for the Theory of the Universe* (Oxford University Press, 1997); G. Fraser, *Antimatter: The Ultimate Mirror* (Cambridge University Press, 2000)

[345] L. Maccione, S. Liberati, and G. Sigl, DESY 10039, e-Print: arXiv:1003.5468 [astro-ph.HE];
D.M. Mattingly, L. Maccione, M. Galaverni, S. Liberati, and G. Sigl, JCAP **1002**, 007 (2010);
S. Liberati and L. Maccione, Ann. Rev. Nucl. Part. Sci. **59**, 245 (2009);
L. Maccione, A.M. Taylor, D.M. Mattingly, and S. Liberati, JCAP **0904**, 022 (2009); L. Maccione, S. Liberati, A. Celotti, J.G. Kirk, and P. Ubertini, Phys. Rev. D **78**, 103003 (2008)

[346] B.G. Sidharth, *Private communication*, March-May 2009

[347] B.G. Sidharth, e-Print: arXiv:0811.4541 [physics.gen-ph]; e-Print: arXiv:0902.3342 [physics.gen-ph]

[348] G. Amelino-Camelia and J. Kowalski-Glikman (Eds.), *Planck Scale Effects in Astrophysics and Cosmology* (Lect. Notes Phys. **669**, Springer, 2005);
C. Callender and N. Huggett (Eds.), *Physics Meets Philosophy at the Planck Scale: Contemporary Theories in Quantum Gravity* (Cambridge University Press, 2004);
R.L. Amoroso, G. Hunter, N. Kafatos, and J.-P. Vigier (Eds.), *Gravitation and Cosmology: From the Hubble Radius to the Planck Scale. Proceedings of a Symposium in Honour of the* 80th *Birthday of Jean-Pierre Vigier* (Kluwer Academic Publishers, 2002)

[349] I.J.R. Aitchinson, *An Informal Introduction to Gauge Field Theories* (Cambridge University Press, 1984);
F. Halzen and A.D. Martin, *Quarks and Leptons: An Introductory Course in Modern Particle Physics* (John Wiley & Sons, 1984);

P.D.B. Collins, A.D. Martin, and E.J. Squires, *Particle Physics and Cosmology* (John Wiley & Sons, 1989);
G. Kane, *Modern Elementary Particle Physics: The Fundamental Particles and Forces?* (Addison-Wesley, 1993);
J.F. Donoghue, E. Golowich, and B.R. Holstein, *Dynamics of The Standard Model* (Cambridge University Press, 1996);
S. Weinberg, *The Quantum Theory of Fields. Vol. II Modern Applications* (Cambridge University Press, 1996);
B.R. Martin and G. Shaw, *Particle Physics* (2nd ed., John Wiley & Sons, 1997);
P.H. Frampton, *Gauge Field Theories* (2nd ed., John Wiley & Sons, 2000);
S. Pokorski, *Gauge Field Theories* (2nd ed., Cambridge Univeristy Press, 2000);
V. Rubakov, *Classical Theory of Gauge Fields* (Princeton University Press, 2002);
I.J.R. Aitchinson and A.J.G. Hey, *Gauge Theories in Particle Physics. Vols. I & II* (3rd ed., IOP Publishing, 2003-2004);
M. Guidry, *Gauge Field Theories: An Introduction with Applications* (Wiley-VCH, 2004)

[350] W. Greiner, S. Schramm, and E. Stein, *Quantum Chromodynamics* (3rd ed., Springer, 2007)

[351] Y. Fukuda, et al, Phys. Rev. Lett. **81**(6), 1158 (1998)

[352] G. ’t Hooft, Phys. Lett. B **198**, 61 (1987)

[353] C. Giunti and C.W. Kim, *Fundamentals of Neutrino Physics and Astrophysics* (Oxford University Press, 2007);
M. Lemoine and G. Sigl (Eds.), *Physics and Astrophysics of Ultra-High-Energy Cosmic Rays* (Lect. Notes Phys. **576**, Springer, 2001);
F.W. Stecker and S.T. Scully, New J. Phys. **11**, 085003 (2009)

[354] OPERA Collaboration (T. Adam, N. Agafonova, A. Aleksandrov et al), *Measurement of the neutrino velocity with the OPERA detector in the CNGS beam*, e-Print: arXiv:1109.4897v1 [hep-ex]

[355] E. Cartlidge, *BREAKING NEWS: Error Undoes Faster-Than-Light Neutrino Results*, available at http://news.sciencemag.org/scienceinsider/2012/02/breaking-news-error-undoes-faster.html;
E.S. Reich, *Flaws found in faster-than-light neutrino measurement*, avialable at http://www.nature.com/news/flaws-found-in-faster-than-light-neutrino-measurement-1.10099

[356] S. Weinberg, *The Quantum Theory of Fields. Vol. I Foundations* (Cambridge University Press, 1995)

[357] C.N. Yang and R. Mills, Phys. Rev. **96**(1), 191 (1954)

[358] S.L. Glashow, Nucl. Phys. **22**, 579 (1961); Rev. Mod. Phys. **52**, 539 (1980),
S. Weinberg, Phys. Rev. Lett. **19**, 1264 (1967); Rev. Mod. Phys. **52**, 515 (1980),
A Salam, in N. Svartholm (Ed.), *Proceedings of the 8th Nobel Symposium on 'Elementary Particle Theory, Relativistic Groups and Analyticity', Stockholm, Sweden, 1968*, pp. 367-377 (Almquist and Wiksells Forlag AB, 1969); Rev. Mod. Phys. **52**, 525 (1980),
S.M. Bilenky and J. Hošek, Phys. Rep. **90**(2), 73 (1982),
E.A. Paschos, *Electroweak Theory* (Cambridge University Press, 2007)

[359] T.D. Lee and C.N. Yang, Phys. Rev. **104**, 254 (1956); Phys. Rev. **106**, 1371 (1957)

[360] M. Gell-Mann, Phys. Rev. **125**, 1067 (1962)

[361] M. Gell-Mann, Phys. Lett. **8**, 214 (1964)

[362] G. Zweig, CERN-8182-TH-401 (1964); CERN-8419-TH-412 (1964)

[363] H. Fritzsch, M. Gell-Mann, and H. Leutwyler, Phys. Lett. B **47**, 365 (1973)

[364] D.J. Gross and F. Wilczek, Phys. Rev. Lett. **30**, 1343 (1973); Phys. Rev. D **8**, 3633 (1973); Phys. Rev. D **9**, 980 (1974);

H.D. Politzer, Phys. Rev. Lett. **30**, 1346 (1973); Phys. Rep. **14**, 129 (1974)

[365] M. Ikeda, S. Ogawa, and Y. Ohnuki, Prog. Theor. Phys. **22**, 715 (1959)

[366] S. Sakata, Prog. Theor. Phys. **16**, 686 (1956)

[367] Y. Ne'eman, Nucl. Phys. **26**(2), 222 (1961)

[368] N. Cabibo, Phys. Rev. Lett. **10**, 531 (1963)

[369] J.D. Bjorken and S.L. Glashow, Phys. Lett. **11**, 255 (1964)

[370] M. Kaku, Phys. Rev. D **13**, 2881 (1976)

[371] J. Santamarina, Acta Phys. Pol. **310**(6), 533 (1979)

[372] A.A. Belavin, A.M. Polyakov, A.S. Schwartz, and Yu.S. Tyupkin, Phys. Lett. B **59**(1), 85 (1975)

[373] H. Georgi and S.L. Glashow, Phys. Rev. Lett. **32**, 438 (1974)

[374] S. Barr, Phys. Lett. B **112**, 219 (1982)

[375] J.P. Derendinger, J.E. Kim, D.V. Nanopoulos, Phys. Lett. B **139**, 170 (1984)

[376] A. Hartanto and L.T. Handoko, Phys. Rev. D **71**, 095013 (2005), e-Print: arXiv:hep-ph/0504280v3

[377] S.K. Yun, Phys. Rev. D **18**, 3472 (1978)

[378] F. Gürsey and L.A. Radicati, Phys. Rev. Lett. **13**, 173 (1964)

[379] B. Sakita, Phys. Rev. **136** B, 1756 (1964)

[380] Z. Tavartkiladze, Phys. Lett. B **392**, 360 (1997), e-Print: arXiv:hep-ph/9610441

[381] J.E. Kim, Phys. Rev. D **23**, 27067 (1981)

[382] C.W. Kim and C. Roiesnel, Phys. Lett. B **93**(3), 343 (1980)

[383] J.E. Kim and H.S. Song, Phys. Rev. D **23**, 2102 (1981); Phys. Rev. D **25**, 2996 (1982)

[384] N.G. Deshpande and P.D. Mannheim, Phys. Lett. B **94**(3), 355 (1980); Phys. Rev. D **24**, 2923 (1981)

[385] K.S. Soh, P.Y. Pac, and S.K. Kim, Phys. Rev. D **27**, 1597 (1983)

[386] P.H. Frampton, Phys. Lett. B **89**, 352 (1980)

[387] Y. Fujimoto and P. Sodano, Phys. Rev. D **23**, 1657 (1981)

[388] S. Dimopoulos, P.H. Frampton, H. Georgi, and M.B. Wise, Phys. Lett. B **117**, 185 (1982)

[389] T. Kibble, J. Phys. A **9**, 1387 (1976)

[390] E. Witten, Phys. Lett. B **105**, 267 (1981)

[391] P.H. Frampton and J.E. Kim, Phys. Lett. B **120**(4-6), 349 (1983)

[392] P.H. Frampton, T. Matsuki and W.W. Wada, Phys. Rev. D **27**, 2204 (1983)

[393] H.R. Pagels, Phys. Rev. D **27**, 2299 (1983)

[394] P.H. Frampton and S.L. Glashow, Phys. Lett. B **131**(4-6), 340 (1983)

[395] K. Kang, C.K. Kim and J.K. Kim, Phys. Rev. D **33**, 260 (1986)

[396] J.E. Kim, Phys. Rev. D **21**, 1687 (1980)

[397] J.E. Kim, Phys. Rev. D **24**, 3007 (1981)

[398] A. De Rujula, H. Georgi, and S.L. Glashow, in K. Kang, H. Fried, and P. Frampton (Eds.), *Grand Unification. Proceedings, 5th Workshop, Providence, USA, APRIL 12-14, 1984*, p. 88 (World Scientific, 1984)

[399] K.S. Babu, X.-G. He, and S. Pakvasa, Phys. Rev. D **33**, 763 (1986), Pre-print: PRINT-85-0525 (HAWAII), UH-511-557-85 (July 1985)

[400] H. Nishimura and A. Okunishi, Phys. Lett. B **209**, 307 (1988)

[401] S. Willenbrock, Phys. Lett. B **561**, 130 (2003)

[402] C.D. Carone and J.M. Conroy, Phys. Rev. D **70**, 075013 (2004)

[403] A. Demaria and R. R. Volkas, Phys. Rev. D **71**, 105011 (2005)

[404] J. Sayre, S. Wiesenfeldt, and S. Willenbrock, Phys. Rev. D **73**, 035013 (2006)

[405] C. Cauet, H. Pas, and S. Wiesenfeldt, Phys. Rev. D **83**, 093008 (2011)

[406] N. Arkani-Hamed, A.G. Cohen, and H. Georgi, Phys. Rev. Lett. **86**, 4757 (2001)

[407] C.T. Hill, S. Pokorski, and J. Wang, Phys. Rev. D **64**, 105005 (2001)

[408] H.-C. Cheng, C.T. Hill, S. Pokorski, and J. Wang, Phys. Rev. D **64**, 065007 (2001)

[409] G. 't Hooft, Nucl. Phys. B **72**(3), 461 (1974)

[410] G. 't Hooft, Nucl. Phys. B **35**, 167 (1971)

[411] H. Georgi, *Lie Algebras in Particle Physics. From Isospin to Unified Theories* (2nd ed., Westview Press, 1999)

[412] J. Baez and J. Huerta, Bull. Am. Math. Soc. **47**, 483 (2010), e-Print: arXiv:0904.1556 [hep-th]

[413] L.D. Faddeev and V.N. Popov, Phys. Lett. B **25**, 29 (1967)

[414] V.N. Gribov, Nucl. Phys. B **139**, 1 (1978)

[415] C. Becchi, A. Rouet and R. Stora, Phys. Lett. B **52**, 344 (1974); Commun. Math. Phys. **42**, 127 (1975); Ann. Phys. N.Y. **98**(2), 287 (1976)

[416] I.V. Tyutin, Preprint of P.N. Lebedev Physical Institute **39** (1975), e-Print: arXiv:0812.0580 [hep-th]

[417] S. Gupta, Proc. Phys. Soc. A **63**(267), 681 (1950)

[418] K. Bleuler, Helv. Phys. Acta **23**(5), 567 (1950)

[419] M. Dehghani, Braz. J. Phys. **39**(3), 559 (2009)

[420] I.M. Singer, Commun. Math. Phys. **60**, 7 (1978)

[421] M.F. Atiyah and J.D.S. Jones, Commun. Math. Phys. **61**, 97 (1978)

[422] D. Zwanziger, Nucl. Phys. B **209**, 336 (1982); Nucl. Phys. B **323**, 513 (1989); Nucl. Phys. B **321**, 591 (1989)

[423] J.A. Gracey, Eur. Phys. J. C **70**, 451 (2010), e-Print: arXiv:1010.1104 [hep-th]; Phys. Rev. D **82**, 085032 (2010), e-Print: arXiv:1009.3889 [hep-th]; PoS QCD-TNT **09**, 016 (2009); JHEP **1002**, 009 (2010), e-Print: arXiv:0909.3411 [hep-th]; Braz. J. Phys. **37**, 226 (2007), e-Print: arXiv:hep-th/0701139; Nucl. Phys. Proc. Suppl. **160**, 122 (2006); JHEP **0605**, 052 (2006), e-Print: arXiv:hep-ph/0605077,
K.-I. Kondo, PoS QCD-TNT **09**, 021 (2009), e-Print: arXiv:0911.2880 [hep-th]; e-Print: arXiv:0905.1899 [hep-th]; Phys. Lett. B **678**, 322 (2009), e-Print: arXiv:0904.4897 [hep-th],
A. Cucchieri and T. Mendes, PoS QCD-TNT **09**, 026 (2009), e-Print: arXiv:1001.2584 [hep-lat],
M.A.L. Capri, A.J. Gomez, M.S. Guimaraes, V.E.R. Lemes, S.P. Sorella, and D.G. Tedesco, Phys. Rev. D **82**, 105019 (2010), e-Print: arXiv:1009.4135 [hep-th],
M.A.L. Capri, D. Dudal, V.E.R. Lemes, R.F. Sobreiro, S.P. Sorella, R. Thibes, and H. Verschelde, Eur. Phys. J. C **52**, 459 (2007), e-Print: arXiv:0705.3591 [hep-th],
D. Dudal, S.P. Sorella, and N. Vandersickel, e-Print: arXiv:1105.3371 [hep-th],
D. Dudal, N. Vandersickel, H. Verschelde, and S.P. Sorella, PoS QCD-TNT **09**, 012 (2009), e-Print: arXiv:0911.0082 [hep-th],
D. Dudal and N. Vandersickel, Phys. Lett. B **700**, 369 (2011), e-Print: arXiv:1010.3927 [hep-th],
D. Dudal, N. Vandersickel, L. Baulieu, S.P. Sorella, M.S.

Guimaraes, M.Q. Huber, O. Oliveira, and D. Zwanziger, PoS LC **2010**, 021 (2010), e-Print: arXiv:1009.5846 [hep-th],
D. Dudal, O. Oliveira, and N. Vandersickel, Phys. Rev. D **81**, 074505 (2010), e-Print: arXiv:1002.2374 [hep-lat],
D. Dudal, S.P. Sorella, N. Vandersickel, Eur. Phys. J. C **68**, 283 (2010), e-Print: arXiv:1001.3103 [hep-th],
D. Dudal, S.P. Sorella, N. Vandersickel, and H. Verschelde, JHEP **0908**, 110 (2009), e-Print: arXiv:0906.4257 [hep-th],
D. Dudal, J.A. Gracey, S.P. Sorella, N. Vandersickel, and H. Verschelde, Phys. Rev. D **78**, 125012 (2008), e-Print: arXiv:0808.0893 [hep-th]; Phys. Rev. D **78**, 065047 (2008), e-Print: arXiv:0806.4348 [hep-th],
D. Dudal, S.P. Sorella, N. Vandersickel, and H. Verschelde, Phys. Rev. D **77**, 071501 (2008), e-Print: arXiv:0711.4496 [hep-th],
M.Q. Huber, R. Alkofer, and S.P. Sorella, AIP Conf. Proc. **1343**, 158 (2011), e-Print: arXiv:1010.4802 [hep-th],
S.P. Sorella, D. Dudal, M.S. Guimaraes, and N. Vandersickel, PoS FACESQCD 022 (2010), e-Print: arXiv:1102.0574 [hep-th S. Upadhyay and B.P. Mandal, Europhys. Lett. **93**, 31001 (2011), e-Print: arXiv:1101.5448 [hep-th],
N. Vandersickel, D. Dudal, H. Verschelde, and S.P. Sorella, PoS QCD-TNT **09**, 048 (2009), e-Print: arXiv:0910.2653 [hep-th],
N. Vandersickel, D. Dudal, O. Oliveira, and S.P. Sorella, AIP Conf. Proc. **1343**, 155 (2011), e-Print: arXiv:1010.4448 [hep-th],
N. Vandersickel, e-Print: arXiv:1104.1315 [hep-th],
N. Vandersickel, D. Dudal, and S.P. Sorella, PoS FACESQCD 044 (2010), e-Print: arXiv:1102.0866 [hep-th]

[424] J. Manjavidze, A. Sissakian and V. Voronyuk, e-Print: arXiv:hep-th/0402214

[425] J. Manjavidze, Phys. Part. Nucl. **30**, 49 (1999), e-Print: arXiv:hep-ph/9802318,
J. Manjavidze and A. Sissakian, Phys. Rept. **346**, 1 (2001)

[426] J. Manjavidze and A. Sissakian, J. Math. Phys. **42**, 4158 (2001), e-Print: arXiv:hep-ph/0104298

[427] S.P. Sorella, L. Baulieu, D. Dudal, M.S. Guimaraes, M.Q. Huber, N. Vandersickel, and D. Zwanziger, AIP Conf. Proc. **1361**, 272 (2011), e-Print: arXiv:1003.0086 [hep-th],
M.A.L. Capri, A.J. Gomez, M.S. Guimaraes, V.E.R. Lemes, S.P. Sorella, and D.G. Tedesco, Phys. Rev. D **83**, 105001 (2011), e-Print: arXiv:1102.5695 [hep-th],
F. Canfora, A. Giacomini, and J. Oliva, e-Print: arXiv:1109.0909 [hep-th],
L.C.Q. Vilar, O.S. Ventura, and V.E.R. Lemes, e-Print: arXiv:1108.3305 [hep-th],
M.A.L. Capri, A.J. Gomez, M.S. Guimaraes, V.E.R. Lemes, and S.P. Sorella, J. Phys. A **43**, 245402 (2010), e-Print: arXiv:1002.1659 [hep-th],
J. Greensite, Phys. Rev. D **81**, 114011 (2010), e-Print: arXiv:1001.0784 [hep-th],
B. Holdom, Phys. Rev. D **82**, 105001 (2010), e-Print: arXiv: 1003.5410 [hep-th]; Phys. Rev. D **79**, 085013 (2009), e-Print: arXiv:0901.0497 [hep-ph]; Phys. Rev. D **78**, 125030 (2008), e-Print: arXiv:0712.2379 [hep-ph],
M.A.L. Capri, A.J. Gomez, V.E.R. Lemes, R.F. Sobreiro, and S.P. Sorella, Phys. Rev. D **79**, 025019 (2009), e-Print: arXiv:0811.2760 [hep-th],
K.-I. Kondo, e-Print: arXiv:0909.4866 [hep-th],
P. Boucaud, J.P. Leroy, A. Le Yaouanc, J. Micheli, O. Pene, and J. Rodriguez-Quintero, Phys. Rev. D **80**, 094501 (2009), e-Print: arXiv:0909.2615 [hep-ph],
S.P. Sorella, Phys. Rev. D **80**, 025013 (2009), e-Print: arXiv:0905.1010 [hep-th],
A.A. Slavnov, Theor. Math. Phys. **161**, 1497 (2009), e-Print: arXiv:0902.1847 [hep-th],
D. Dudal, S.P. Sorella, N. Vandersickel, and H. Verschelde, Phys. Lett. B **680**, 377 (2009), e-Print: arXiv:0808.3379 [hep-th],
H. Reinhardt and P. Watson, Phys. Rev. D **79**, 045013 (2009), e-Print: arXiv:0808.2436 [hep-th],
K. Langfeld, T. Heinzl, A. Ilderton, M. Lavelle, and D. McMullan, PoS CONFINEMENT **8**, 047 (2008), e-Print: arXiv:0812.2418 [hep-lat],

G. Kelnhofer, J. Math. Phys. **49**, 052302 (2008), e-Print: arXiv:0707.1186 [hep-th],
J.A. Gracey, in W. Janke and A. Pelster (Eds.), *Proceedings of 9th International Conference On Path Integrals: New Trends And Perspectives (PI07), 23-28 September 2007, Dresden, Germany*, pp. 167-172 (World Scientific, 2008), e-Print: arXiv:0711.3622 [hep-th],
T. Heinzl, K. Langfeld, M. Lavelle, and D. McMullan, Phys. Rev. D **76**, 114510 (2007), e-Print: arXiv:0705.2718 [hep-lat],
A. Ilderton, M. Lavelle, and D. McMullan, JHEP **0703**, 044 (2007), e-Print: arXiv:hep-th/0701168,
A. Ilderton, in B. Mueller, M.A. Rotondo and C.-I. Tan (Eds.), *Proceedings of the Ninth Workshop on Non-Perturbative Quantum Chromodynamics 'Institut Astrophysique de Paris, June 4-8, 2007* (SLAC Electronic Conference Proceedings Archive C0706044, 2007), e-Print: arXiv:0709.1671 [hep-th],
D. Dudal, M.A.L. Capri, J.A. Gracey, V.E.R. Lemes, R.F. Sobreiro, S.P. Sorella, R. Thibes, and H. Verschelde, Braz. J. Phys. **37**, 320 (2007), e-Print: arXiv:hep-th/0609160 [hep-th],
M.A.L. Capri, V.E.R. Lemes, R.F. Sobreiro, S.P. Sorella, and R. Thibes, Phys. Rev. D **74**, 105007 (2006), e-Print: arXiv:hep-th/0609212,
R.F. Sobreiro and S.P. Sorella, JHEP **0506**, 054 (2005), e-Print: arXiv:hep-th/0506165; e-Print: arXiv:hep-th/0504095,
G. Esposito, D.N. Pelliccia and F. Zaccaria, Int. J. Geom. Meth. Mod. Phys. **1**, 423 (2004), e-Print: arXiv:hep-th/0404240,
Ch. Fleischhack, Commun. Math. Phys. **234**, 423 (2003), e-Print: arXiv:math-ph/0007001,
A.G. Williams, Prog. Theor. Phys. Suppl. **151**, 154 (2003), e-Print: arXiv:hep-lat/0304003, Nucl. Phys. Proc. Suppl. **109** A, 141 (2002), e-Print: arXiv:hep-lat/0202010,
J.F. Grotowski and P.P. Schirmer, Commun. Math. Phys. **216**, 179 (2001),
M. Faber, J. Greensite, and S. Olejnik, Phys. Rev. D **64**, 034511 (2001), e-Print: arXiv:hep-lat/0103030,
K. Haller, Int. J. Mod. Phys. A **16**, 2789 (2001), e-Print: arXiv:hep-th/0004205,

C. Ewerz, Nucl. Phys. Proc. Suppl. **90**, 130 (2000), e-Print: arXiv:hep-ph/0008229,

F. Bruckmann, T. Heinzl, A. Wipf, and T. Tok, Nucl. Phys. B **584**, 589 (2000), e-Print: arXiv:hep-th/0001175,

T. Heinzl, in W. Lucha and K.M. Maung (Eds.), *Proceedings of 4th International Conference On Quark Confinement And The Hadron Spectrum, 3-8- July 2000, Vienna, Austria*, pp. 249-251 (World Scientific, 2001); Nucl. Phys. Proc. Suppl. **54** A, 194 (1997), e-Print: arXiv:hep-th/9609055,

V.M. Villanueva, J. Govaerts, J.-L. Lucio-Martinez, J. Phys. A **33**, 4183 (2000), e-Print: arXiv:hep-th/9909033,

T.G. Kovacs and E.T. Tomboulis, Phys. Lett. B **463**, 104 (1999), e-Print: arXiv:hep-lat/9905029,

C. Becchi, S. Giusto and C. Imbimbo, in R. Casalbuoni, R. Giachetti, V. Tognetti, R. Vaia, and P. Verrucchi (Eds.), *Proceedings of 6th International Conference On Path Integrals From PeV To TeV: 50 Years From Feynman's Paper (PI 98), 25-29 August 1998, Florence, Italy*, pp. 36-43 (World Scientific, 1999), e-Print: arXiv:hep-th/9811018,

M. Schaden, in H.M. Fried and B. Mueller (Eds.), *Proceedings of 4th Workshop On Quantum Chromodynamics, 1-6 June 1998, Paris, France*, pp. 418-431 (World Scientific, 1999), e-Print: arXiv:hep-th/9810162,

N. Duchting, S.V. Shabanov, and T. Strobl, Nucl. Phys. B **538**, 485 (1999), e-Print: arXiv:hep-th/9808151,

T.D. Lee, in A.V. Smilga (Ed.), *Proceedings of 3rd Workshop on Continuous Advances in QCD (QCD 98), 16-19 April 1998, Minneapolis*, pp. 16-19 (World Scientific, 1998),

Y. Hosotani, K. Saririan, and B. Tekin, in A.V. Smilga (Ed.), *Proceedings of 3rd Workshop on Continuous Advances in QCD (QCD 98), 16-19 April 1998, Minneapolis*, pp. 19-27 (World Scientific, 1998),

M. Testa, Phys. Lett. B **429**, 349 (1998), e-Print: arXiv:hep-lat/9803025

F.G. Scholtz and S.V. Shabanov, Ann Phys. N.Y. **263**, 119 (1998), e-Print: arXiv:hep-th/9701051,

A. Cucchieri and T. Mendes, Nucl. Phys. Proc. Suppl. **63**, 841

(1998), e-Print: arXiv:hep-lat/9710040,
R. Zucchini, Commun. Math. Phys. **185**, 723 (1997), e-Print: arXiv:hep-th/9607003,
R. Friedberg, T.D. Lee, Y. Pang, and H.C. Ren, Ann. Phys. N.Y. **246**, 381 (1996),
K. Fujikawa, Nucl. Phys. B **468**, 355 (1996), e-Print: arXiv:hep-th/9510111; Nucl. Phys. B **223**, 218 (1983); Prog. Theor. Phys. **61**, 627 (1979); e-Print: arXiv:hep-th/9511124,
N. Kalogeropoulos, J. Math. Phys. **35**, 587 (1994), e-Print: arXiv:hep-th/9305092,
D. McMullan, Commun. Math. Phys. **160**, 431 (1994),
F.G. Scholtz and G.B. Tupper, Phys. Rev. D **48**, 1792 (1993),
P. van Baal and R.E. Cutkosky, e-Print: arXiv:hep-lat/9208027,
P. van Baal, Nucl. Phys. B **369**, 259 (1992); e-Print: arXiv:hep-th/9711070,
G. Jungman, Mod. Phys. Lett. A **7**, 849 (1992),
G. Dell'Antonio and D. Zwanziger, Commun. Math. Phys. **138**, 291 (1991),
C. Parrinello and G. Jona-Lasinio, Phys. Lett. B **251**, 175 (1990),
H. Kanno, Lett. Math. Phys. **19**, 249 (1990),
R. Parthasarathy, Pramana **32**, 563 (1989); Lett. Math. Phys. **15**, 179 (1988),
M.E. Fabbrichesi, Phys. Lett. B **205**, 521 (1988),
Z. Jaskólski, J. Math. Phys. **29**, 1034 (1988),
G. Siopsis, Phys. Lett. B **187**, 351 (1987); Phys. Lett. B **136**, 175 (1984),
H. Neuberger, Phys. Lett. B **183**, 337 (1987),
L.J. Carson, Nucl. Phys. B **266**, 357 (1986),
F. Nill, in T. Dorlas, N.M. Hugenholtz, and M. Winnink (Eds.), *Statistical Mechanics and Field Theory: Mathematical Aspects: Proceedings of the International Conference on the Mathematical Aspects of Statistical Mechanics and Field Theory, Held in Groningen, the Netherlands, August 26-30, 1985*, pp. 325-328 (Lect. Notes Phys. **257**, Springer, 1986); MPI-PAE/PTh 39/85 (June 1985),
R.E. Cutkosky, Phys. Rev. D **30**, 447 (1984),
M. Horibe, A. Hosoya, and J. Sakamoto, Prog. Theor. Phys. **70**,

1636 (1983),
E. Malec, J. Math. Phys. **23**, 21 (1982),
A. Niemi, Nucl. Phys. B **189**, 115 (1981); Phys. Lett. B **91**, 411 (1980),
G. Ghika and M. Visinescu, Phys. Rev. D **21**, 1538 (1980),
T. Maskawa and H. Nakajima, Prog. Theor. Phys. **63**, 641 (1980),
T. Maskawa, in S. Homma, M. Kawaguchi, and H. Miyazawa (Eds.), *Proceedings of 19th International Conference On High Energy Physics, 23-30 August 1978, Tokyo*, pp. 500-502 (Phys. Soc. Japan, 1979),
H.J. de Vega, Phys. Rev. D **19**, 3072 (1979),
F.S. Henyey, Phys. Rev. D **20**, 1460 (1979),
K. Shigemoto, Lett. Nuovo Cim. **24**, 495 (1979),
A.K. Das and M. Kaku, Nuovo Cim. B **50**, 303 (1979),
P. Hirschfeld, Nucl. Phys. B **157**, 37 (1979),
A.P. Balachandran, H.S. Mani, R. Ramachandran, and P. Sharan, Phys. Rev. Lett. **40**, 988 (1978),
Y. Takahashi and M. Kobayashi, Phys. Lett. B **78**, 241 (1978),
M.B. Halpern and J. Koplik, Nucl. Phys. B **132**, 239 (1978),
J.P. Greensite, Phys. Rev. D **18**, 3842 (1978)

[428] A. Mallios, *Modern Differential Geometry in Gauge Theories. Volume II Yang-Mills Fields* (Birkhäuser, 2009),
M. Asorey and P.K. Mitter, Ann. Inst. Henri Poincaré **45**(1), 61 (1986),
M. Daniel and C.M. Viallet, Rev. Mod. Phys. **52**, 175 (1980),
M.F. Atiyah, *Geometry of Yang-Mills Fields* (Accademia Nazionale Dei Lincei Scuola Normale Superiore, 1979)

[429] T.T. Wu and C.N. Yang, in H. Mark and S. Fernbach (Eds.), *Properties of Matter Under Unusual Conditions*, pp. 344-354 (Interscience, 1968)

[430] P.A.M. Dirac, Proc. Roy. Soc. (London) A **133**, 60 (1931)

[431] T.T. Wu and C.N. Yang, Phys. Rev. D **12**, 3845 (1975); Nucl. Phys. B **107**(3), 365 (1976)

[432] K.G. Wilson, Phys. Rev. D **10**(8), 2445 (1974)

[433] A.J. Buras, J. Ellis, M.K. Gaillard, and D.V. Nanopoulos, Nucl. Phys. B **135**, 66 (1978)

[434] A. Guth and S. Tye, Phys. Rev. Lett. **44**(10), 631 (1980); Erratum: Phys. Rev. Lett. **44**(10), 963 (1980);
M.B. Einhorn, Phys. Rev. D **21**, 3295 (1980)

[435] Ya. B. Zel'dovich and M. Yu. Khlopov, Phys. Lett. B **79**, 239 (1978);
J. Preskill, Phys. Rev. Lett. **43**(19), 1365 (1979)

[436] P. Curie, Bull. Séances Soc. Fr. Phys., pp. 76-77 (1894)

[437] P. Curie, J. Phys. (Paris) **3**, 393 (1894)

[438] P. Curie, J. Phys. (Paris) **3**, 415 (1894)

[439] Z. Fang et al, Science **302**(5642), 92 (2003)

[440] C. Castelnovo, R. Moessner, and S.L. Sondhi, Nature **451**, 42 (2008)

[441] S.T. Bramwell et al, Nature **461**, 956 (2009), e-Print: arXiv:0907.0956 [cond-mat.other]

[442] T. Fennell et al, Science Express **326**(5951), 415 (2009)

[443] M.J.P. Gingras, Science **326**(5951), 375 (2009)

[444] D.J.P. Morris et al, Science Express **326**(5951), 411 (2009)

[445] L.A. Mol et al, J. App. Phys. **106**(6), 063913 (2009)

[446] G. Möller and R. Moessner, Phys. Rev. B **80**, 140409(R) (2009)

[447] H. Kadowaki et al, J. Phys. Soc. Japan **78**, 103706 (2009), e-Print: arXiv:0908.3568 [cond-mat.dis-nn]

[448] X.-L. Qi, R. Li, J. Zang, and S.-C. Zhang, Science **323**(5918), 1184 (2009)

[449] C. Castelnovo, ChemPhysChem **11**, 557 (2010)

[450] C. Castelnovo, R. Moessner, and S.L. Sondhi, e-Print: arXiv:1107.5482 [cond-mat.str-el]

[451] G.-W. Chern and O. Tchernyshyov, e-Print: arXiv:1109.0275 [cond-mat.str-el]

[452] S.R. Giblin et al, Nature Physics **7**, 252 (2011)

[453] L.D.C. Jaubert and P.C.W. Holdsworth, J. Phys.: Cond. Mat. **23**, 164222 (2011)

[454] B. Klemke, *Thermal Properties of Dysprosium-Titanate in the Spin Ice State*, PhD Dissertation, Institut für Festkörperphysik, Technischen Universität Berlin, 2011 http://opus.kobv.de/tuberlin/volltexte/2011/2942/

[455] S. Ladak et al, New J. Phys. **13**, 023023 (2011)

[456] E. Mengotti, Nature Physics **7**, 68 (2011)

[457] H.D. Zhou et al, Nature Communications **2**, 478 (2011)

[458] M.J. Harris et al, Phys. Rev. Lett. **79**(13), 2554 (1997)

[459] P. Debye and E. Hückel, Phys. Zeits. **24**, 185 (1923)

[460] L. Onsager, J. Chem. Phys. **2**, 599 (1934)

[461] M. Wien, Ann. Phys. (Leipzig) **85**, 795 (1928); Phys. Zeits. **29**, 751 (1928); Ann. Phys. (Leipzig) **1**, 400 (1929); Phys. Zeits. **32**, 545 (1931)

[462] G. 't Hooft, Nucl. Phys. B **79**, 276 (1974), Pre-print: CERN-TH-1876, May 1974, 13 pages

[463] A.M. Polyakov, JETP Lett. **20**, 194 (1974)

[464] A.M. Polyakov, Sov. Phys. JETP **41**(6), 988 (1975)

[465] E.V. Prokhvatilov and V.A. Franke, Yad. Fiz. **24**, 856 (1976);
F.A. Shaposnik, Lett. Nuovo Cim. **17**, 245 (1976);
E. Cremmer, F. Schaposnik, and Joel Scherk, Phys. Lett. B **65**, 78 (1976)

[466] H. Weyl, Math. Zeits. **23**, 271 (1925)

[467] É. Cartan, *Sur la structure des groupes de transformations finis et continus (Thèse)* (Nony, 1894); reprinted in *Oeuvres Complètes I*, pp. 137-288 (Gauthier-Villars, 1952; Editions du CNRS, 1984)

[468] E. van der Meer, *Magnetic Monopoles in Gauge Field Theories. On the realization of global symmetries on moduli spaces*,
Master Thesis at Institute for Theoretical Physics of University of Amsterdam, supervised by F.A. Bais (1997)

[469] P. Goddard and D.I. Olive, Rept. Prog. Phys. **41**, 1357 (1978)

[470] M.K. Prasad and C.M. Sommerfield, Phys. Rev. Lett. **35**, 760 (1975)

[471] E.B. Bogomol'nyi, Sov. J. Nucl. Phys. **24**, 449 (1976)

[472] J. Schwinger, Phys. Rev. **144**, 1087 (1966);
J. Schwinger, Phys. Rev. **173**, 1536 (1968);
J. Schwinger, Science **165**(3895), 757 (1969)

[473] D. Zwanziger, Phys. Rev. **176**, 1480 (1968); Phys. Rev. **176**, 1489 (1968)

[474] J.J. Aubert et al, Phys. Rev. Lett. **33**(23), 1404 (1974)

[475] J.E. Augustin et al, Phys. Rev. Lett. **33**, 1406 (1974)

[476] I. Bialynicki-Birula and Z. Bialynicka-Birula, Phys. Rev. D **3**(10), 2410 (1971)

[477] I. Bialynicki-Birula, Phys. Rev. D **3**(10), 2413 (1971)

[478] E. Witten, Phys. Lett. B **86**, 283 (1979)

[479] B. Julia and A. Zee, Phys. Rev. D **11**, 2227 (1975)

[480] G. 't Hooft, *50 Years of Yang-Mills Theory* (World Scientific, 2005);
Ya. Shnir, *Magnetic Monopoles* (Springer, 2005);
N.Manton and P.M. Sutcliffe, *Topological Solitons* (Cambridge

University Press, 2004);
M.A. Shifman, *Instantons in Gauge Theories* (World Scientific, 1994);
N.S. Craigie (Ed.), *Theory and Detection of Magnetic Monopoles in Gauge Theories: A Collected Set of Lecture Notes* (World Scientific, 1986);
S. Coleman, *Aspects of Symmetry* (Cambridge University Press, 1985);
R.A. Carrigan and W.P. Trower, *Magnetic Monopoles* (NATO ASI B Series **102**, Plenum Press, 1983);
P. Rossi, Phys. Rep. **86**(6), 317 (1982)

[481] K.M. Milton, Rept. Prog. Phys. **69**, 1637 (2006);
F.A. Bais and J. Striet, Phys. Lett. B **540**, 319 (2002);
F.A. Bais and B.J. Schroers, Nucl. Phys. B **512**, 250 (1998);
D. Singleton, Int. J. Theo. Phys. **34**(12), 2453 (1995);
D. Wilkinson and F.A. Bais, Phys. Rev. D **19**, 2410 (1979);
E.A. Olszewski, e-Print: arXiv:1003.1165 [hep-th];
D. Tong, e-Print: arXiv:hep-th/0509216

[482] J.R. Klauder (Ed.), *Magic without magic: John Archibald Wheeler* (Freeman, 1972)

[483] C.J. Isham, R. Penrose, and D.W. Sciama (Eds.), *Quantum Gravity. An Oxford symposium* (Oxford University Press, 1975)

[484] R. Balian and J. Zinn-Justin (Eds.), *Methods in Field Theory. Les Houches, École D'Été De Physique Théorique. Session XXVIII* (North-Holland, 1976)

[485] C.J. Isham, R. Penrose, and D.W. Sciama (Eds.), *Quantum Gravity 2. A Second Oxford Symposium* (Oxford University Press, 1981)

[486] S.M. Christensen (Ed.), *Quantum Theory of Gravity. Essays in honor of the 60th birthday of Bryce S. DeWitt* (Adam Hilger, 1984)

[487] M.A. Markov and P.C. West (Eds.), *Quantum Gravity. Proceedings of the Second Seminar, Oct 13-15, 1981, Moscow, USSR* (World Scientific, 1984)

[488] M.A. Markov, V.A. Berezin, and V.P. Frolov (Eds.), *Quantum Gravity. Proceedings of the Third Seminar, Oct 23-25, 1984, Moscow, USSR* (World Scientific, 1985)

[489] R. Penrose and C.J. Isham (Eds.), *Quantum Concepts in Space and Time* (Oxford University Press, 1986)

[490] T. Padmanabhan and J.V. Narlikar, *Gravity, Gauge Theories and Quantum Cosmology* (Kluwer Academic Publishers, 1986)

[491] H.J. de Vega and N. Sánchez (Eds.), *Field Theory, Quantum Gravity, and Strings. Proceedings of a Seminar Series Held at DAPHE, Observatoire de Meudon, and LPTHE, Université Pierre et Marie Curie, Paris, Between October 1984 and October 1985* (Lect. Notes Phys. **246**, Springer, 1986)

[492] H.J. de Vega and N. Sánchez (Eds.), *Field Theory, Quantum Gravity and Strings II Proceedings of a Seminar Series Held at DAPHE, Observatoire de Meudon, and LPTHE, Université Pierre et Marie Curie, Paris, Between October 1985 and October 1986* (Lect. Notes Phys. **280**, Springer, 1987)

[493] M.A. Markov, V.A. Berezin, and V.P. Frolov (Eds.), *Quantum Gravity. Proceedings of the Fourth Seminar, May 25-29, 1987, Moscow, USSR* (World Scientific, 1988)

[494] J. Audretsch and V. de Sabbata (Eds.), *Quantum Mechanics in Curved Space–Time* (Plenum Press, 1990)

[495] A. Ashtekar and J. Stachel (Eds.), *Conceptual Problems of Quantum Gravity* (Birkhäuser, 1991)

[496] S. Coleman, J.B. Hartle, T. Piran, and S. Weinberg (Eds.), *Quantum Cosmology and Baby Universes* (World Scientific, 1991)

[497] M.A. Markov, V.A. Berezin, and V.P. Frolov (Eds.), *Quantum Gravity. Proceedings of the Fifth Seminar, May 28-June 1, 1990, Moscow, USSR* (World Scientific, 1991)

[498] I.L. Buchbinder, S.D. Odintsov, and I.L. Shapiro, *Effective Action in Quantum Gravity* (IOP Publishing, 1992)

[499] D.J. Gross, T. Piran, and S. Weinberg (Eds.), *Two Dimensional Quantum Gravity and Random Surfaces* (World Scientific, 1992)

[500] M. Henneaux and C. Teitelboim, *Quantization of Gauge Systems* (Princeton University Press, 1992)

[501] M.C. Bento, O. Bertolami, J.M. Mourão, and R.F. Picken (Eds.), *Classical and Quantum Gravity* (World Scientific, 1993)

[502] G.W. Gibbons and S.W. Hawking (Eds.), *Euclidean Quantum Gravity* (World Scientific, 1993)

[503] J.C. Baez (Ed.), *Knots and Quantum Gravity* (Clarendon Press, 1994)

[504] J. Ehlers and H. Friedrich (Eds.), *Canonical Gravity: From Classical to Quantum* (Springer, 1994)

[505] G. Esposito, *Quantum Gravity, Quantum Cosmology and Lorentzian Geometries* (Springer, 1994)

[506] E. Prugovečki, *Principles of Quantum General Relativity* (World Scientific, 1995)

[507] P.D. D'Eath, *Supersymmetric Quantum Cosmology* (Cambridge University Press, 1996)

[508] R. Gambini and J. Pullin, *Loops, Knots, Gauge Theories and Quantum Gravity* (Cambridge University Press, 1996)

[509] P.G. Bergmann, V. De Sabbata, and H.J. Treder (Eds.), *Quantum Gravity: International School of Cosmology and Gravitation XIV Course: 80th Birthday Dedication to Peter G. Bergmann, held 11-19 May, 1995 in Erice, Italy* (World Scientific, 1996)

[510] G. Esposito, A.Yu. Kamenshchik, and G. Pollifrone, *Euclidean Quantum Gravity on Manifolds with Boundary* (Springer, 1997)

[511] G. 't Hooft, A. Jaffe, G. Mack, P.K. Mitter, and R. Stora (Eds.), *Quantum Fields and Quantum Space Time* (Plenum Press, 1997)

[512] P. Fré, V. Gorini, G. Magli, and U. Moschella, *Classical and Quantum Black Holes* (IOP Publishing, 1999)

[513] I.G. Avramidi, *Heat Kernel and Quantum Gravity* (Springer, 2000)

[514] J. Kowalski-Glikman (Ed.), *Towards Quantum Gravity. Proceedings of the XXXV International Winter School on Theoretical Physics, Held in Polanica, Poland, 2-11 February 1999* (Lect. Notes Phys. **541**, Springer, 2000)

[515] B.N. Kursunoglu, S.L. Mintz, and A. Perlmutter (Eds.), *Quantum Gravity, Generalized Theory of Gravitation and Superstring Theory-Based Unification* (Kluwer Academic Publishers, 2002)

[516] S. Carlip, *Quantum Gravity in 2+1 Dimensions* (Cambridge University Press, 2003)

[517] G.W. Gibbons, E.P.S. Shellard, and S.J. Rankin (Eds.), *The Future of Theoretical Physics and Cosmology* (Cambridge University Press, 2003)

[518] D. Giulini, C. Kiefer, and C. Lämmerzahl (Eds.), *Quantum Gravity. From Theory To Experimental Search* (Lect. Notes Phys. **631**, Springer, 2003)

[519] C. Rovelli, *Quantum Gravity* (Cambridge University Press, 2004)

[520] A. Gomberoff and D. Marolf (Eds.), *Lectures on Quantum Gravity* (Springer, 2005)

[521] D. Rickles, S. French, and J. Saatsi (Eds.), *The Structural Foundations of Quantum Gravity* (Clarendon Press, 2006)

[522] B. Carr (Ed.), *Universe of Multiverse?* (Cambridge University Press, 2007)

[523] B. Fauser, J. Tolksdorf, and E. Zeidler (Eds.), *Quantum Gravity. Mathematical Models and Experimental Bounds* (Birkhäuser, 2007)

[524] D. Gross, M. Henneaux, and A. Sevrin (Eds.), *The Quantum Structure of Space and Time* (World Scientific, 2007)

[525] V.F. Mukhanov and S. Winitzki, *Introduction to Quantum Effects in Gravity* (Cambridge University Press, 2007)

[526] T. Thiemann, *Modern Canonical Quantum General Relativity* (Cambridge University Press, 2007)

[527] H.W. Hamber, *Quantum Gravitation: The Feynman Path Integral Approach* (Springer, 2009)

[528] D. Oriti, *Approaches to Quantum Gravity. Toward a New Understanding of Space, Time, and Matter* (Cambridge University Press, 2009)

[529] M. Bojowald, *Canonical Gravity and Applications: Cosmology, Black Holes, and Quantum Gravity* (Cambridge University Press, 2010)

[530] J. Murugan, A. Weltman, and G.F.R. Ellis, *Foundations of Space and Time: Reflections on Quantum Gravity* (Cambridge University Press, 2011)

[531] L.A. Glinka, Oriental J. Phys. **3**(1), 1 (2011); Grav. Cosmol. **15**(4), 317 (2009); in B.G. Sidharth, F. Honsell, O. Mansutti, K. Sreenivasan, and A. De Angelis (Eds.), *Frontiers of Fundamental and Computational Physics. 9th International Symposium, Udine and Trieste, Italy 7-9 January 2008*, pp. 94-99 (AIP Conf. Proc. **1018**, 2008); in E. Ivanov and S. Fedoruk (Eds.), *Supersymmetries and Quantum Symmetries (SQS'07): Proceedings of International Workshop, held in Dubna, Russia, July 30 - August 4, 2007*, pp. 406-411 (Dubna JINR 2008); SIGMA 3, 087 (2007); e-Print: arXiv:0711.1380 [gr-qc]

[532] S.W. Hawking, Comm. Math. Phys. **43**, 199 (1975); Phys. Rev. D **13**, 191 (1976); Phys. Rev. D **14**, 2460 (1976); Comm. Math. Phys. **55**, 133 (1977); Phys. Rev. D **18**, 1747 (1978); Pont. Acad. Sci. Scri. Varia **48**, 563 (1982); Ph ys. Lett. B **134**, 403 (1984); Nucl. Phys. B **239**, 257 (1984); Phys. Rev. D **32**, 259 (1985); Phys. Rev. D **32**, 2489 (1985) Phys. Rev. D **37**, 904 (1988); Phys. Rev. D **53**, 3099 (1996); Phys. Scr. T **117**, 49 (2005); Phys. Rev. D **72**, 084013 (2005); in S.W. Hawking and W. Israel (Eds.),

General Relativity: An Einstein centenary survey, pp. 746-785 (Cambridge University Press 1979); in M. Levy and S. Deser (Eds.), *Recent Developments in Gravitation. Cargese 1978*, pp. 145-175 (Plenum Press 1979); in B.S. DeWitt and R. Stora (Eds.), *Relativity, Groups, and Topology II*, pp. 333-381 (Elsevier 1984); in H.J. de Vega and N. Sánchez (Eds.), *Field Theory, Quantum Gravity, and Strings. Proceedings of a Seminar Series Held at DAPHE, Observatoire de Meudon, and LPTHE, Université Pierre et Marie Curie, Paris, Between October 1984 and October 1985*, pp. 1-46 (Springer 1986); in S.W. Hawking and W. Israel (Eds.), *Three hundred years of gravitation*, pp. 631-652 (Cambridge University Press 1987); in J.J. Halliwell, J. Perez-Marcader, and W.H. Zurek (Eds.), *Physical Origins of Time Asymmetry*, pp. 346-356 (Cambridge University Press 1992); *Hawking on the Big Bang and Black Holes* (World Scientific, 1993)

[533] G.W. Gibbons and S.W. Hawking, Phys. Rev. D **15**, 2752 (1977)

[534] J.B. Hartle and S.W. Hawking, Phys. Rev. D **28**, 2960 (1983)

[535] S.W. Hawking and J.C. Luttrell, Phys. Lett. B **143**, 83 (1984); Nucl. Phys. B **247**, 250 (1984)

[536] S.W. Hawking and Z.C. Wu, Phys. Lett. B **151**, 15 (1985)

[537] J.J. Halliwell and S.W. Hawking, Phys. Rev. D **31**, 1777 (1985)

[538] S.W. Hawking and D. Page, Nucl. Phys. B **264**, 185 (1986); Nucl. Phys. B **298**, 789 (1988); Phys. Rev. D **42**, 2655 (1992)

[539] S.W. Hawking, R. Laflamme, and G.W. Lyons, Phys. Rev. D **47**, 5342 (1993)

[540] R. Bousso and S.W. Hawking, Phys. Rev. D **52**, 5660 (1995)

[541] S.W. Hawking and S.F. Ross, Phys. Rev. D **52**, 5862 (1995)

[542] M.J. Cassidy and S.W. Hawking, Phys. Rev. D **57**, 2372 (1998)

[543] S.W. Hawking and C.J. Hunter, Phys. Rev. D **59**, 044025 (1999)

[544] S.W. Hawking, T. Hertog, and H.S. Reall, Phys. Rev. D **62**, 043501 (2000); Phys. Rev. D **63**, 083504 (2001)

[545] S.W. Hawking, T. Hertog, and N. Turok, Phys. Rev. D **62**, 063502 (2000)

[546] S.W. Hawking and T. Hertog, Phys. Rev. D **66**, 123509 (2002); Phys. Rev. D **73**, 123527 (2006)

[547] J.B. Hartle, S.W. Hawking, and T. Hertog, Phys. Rev. Lett. **100**, 201301 (2008); Phys. Rev. D **77**, 123537 (2008)

[548] M. Kriele, *Space-time. Foundations of General Relativity and Differential Geometry* (Lect. Notes Phys. **59**, Springer, 1999)

[549] P. Petersen, *Riemannian Geometry* (2nd ed., Grad. Texts Math. **171**, Springer, 2006)

[550] C.F. Gauss, Comm. Soc. Reg. Sci. Gött. Rec. **6**, 99 (1828)

[551] D. Codazzi, Ann. Mat. Pura Appl. **2**, 101 (1868-1869)

[552] A. Hanson, T. Regge, and C. Teitelboim, *Constrained Hamiltonian Systems* (Contributi del Centro Linceo Interdisciplinare di Scienze Matematiche e loro Applicazioni, n. **22**, Accademia Nazionale dei Lincei, 1976)

[553] B. DeWitt, *The Global Approach to Quantum Field Theory, Vols. 1-2* (Int. Ser. Monogr. Phys. **114**, Clarendon Press, 2003)

[554] J.F. Nash, Ann. Math. **56**, 405 (1952); Ann. Math. **63**, 20 (1956)

[555] A. Kowalczyk, Bull. Acad. Polon. Sci. Ser. Sci. Math. **28**, 385 (1981)

[556] S. Masahiro, *Nash Manifolds* (Lect. Notes Math. **1269**, Springer 1987)

[557] M. Günther, Ann. Global Anal. Geom. **7**, 69 (1989); Math. Nachr. **144**, 165 (1989)

[558] J.W. York, Phys. Rev. Lett. **28**, 1082 (1972)

[559] S. Weinberg, *Gravitation and Cosmology. Principles and Applications of the General Theory of Relativity* (John Wiley & Sons, 1972);
R.K. Sachs and H. Wu, *General Relativity for Mathematicians* (Springer, 1977);
B.F. Schutz, *Geometrical Methods of Mathematical Physics* (Cambridge University Press, 1980);
M. Carmeli, *Classical Fields: General Relativity and Gauge Theory* (John Wiley & Sons, 1982);
B. O'Neill, *Semi-Riemannian Geometry with Applications to Relativity* (Academic Press, 1983);
R.M. Wald, *General Relativity* (University of Chicago, 1984);
N. Straumann, *General Relativity and Relativistic Astrophysics* (Springer, 1984);
F. De Felice and C.J.S. Clarke, *Relativity on Curved Manifolds* (Cambridge University Press, 1990);
L.P. Hughston and K.P. Tod, *An Introduction to General Relativity* (Cambridge University Press, 1994);
J. Stewart, *Advanced General Relativity* (Cambridge University Press, 1996);
H. Stephani, D. Kramer, M. A. H. MacCallum, C. Hoenselaers, and E. Herlt, *Exact Solutions of Einstein's Field Equations* (2nd ed., Cambridge University Press, 2003);
G.S. Hall, *Symmetries and Curvature Structure in General Relativity* (World Scientific, 2004);
N.M.J. Woodhouse, *General Relativity* (Springer, 2006);
L. Ryder, *Introduction to General Relativity* (Cambridge University Press, 2009)

[560] H. Goldstein, Ch. Poole, and J. Safko, *Classical Mechanics* (3rd ed., Pearson, 2002)

[561] A. Peres, Nuovo Cim. **26**, 53 (1962)

[562] L.D. Faddeev, Usp. Fiz. Nauk **136**, 435 (1982)

[563] B.S. DeWitt, in M. Carmeli, S.I. Fickler, and L. Witten (Eds.), *Relativity. Proceedings of the Relativity Conference in the*

Midwest held at Cincinnati, Ohio, June 2-6, 1969, pp. 359-374 (Plenum Press, 1970)

[564] A.E. Fischer, Ann. Global Anal. Geom. **14**, 263 (1996); Fields Inst. Comm. **7**, 107 (1996); in M. Gotay, J. Marsden, and V. Moncrief (Eds.), *Mathematical Aspects of Classical Field Theory, proceedings of the AMS-IMS-SIAM Joint Summer Research Conference, Seattle, Washington, July 20-26, 1991*, pp. 331-366 (Contemporary Mathematics **132**, AMS, 1992); J. Math. Phys. **27**, 718 (1986); Gen. Rel. Grav. **15**, 1191 (1983)

[565] O. Pekonen, J. Geom. Phys. **4** 493 (1987)

[566] D. Giulini, Gen. Rel. Grav. **41**, 785 (2009); in H. Falcke and F.W. Hehl (Eds.), *The Galactic Black Hole: Lectures on General Relativity and Astrophysics*, pp. 178-206 (IOP Publishing, 2003); in H.-P. Breuer and F. Petruccione (Eds.), *Relativistic Quantum Measurement and Decoherence: Proceedings of Workshop on Relativistic Quantum Measurement and Decoherence, 9-10 Apr 1999, Naples, Italy*, pp. 67-92 (Lect. Notes Phys. **559**, Springer, 2000); in F. Hehl, C. Kiefer, and R. Metzler (Eds.), *Black Holes: Theory and Observation*, pp. 224–243 (Lect. Notes Phys. **514**, Springer, 1998); Helv. Phys. Acta **69**, 333 (1996); Helv. Phys. Acta **68**, 438 (1995); Helv. Phys. Acta **68**, 87 (1995); Phys. Rev. D **5**, 5630 (1995); Comm. Math. Phys. **148**, 353 (1992)

[567] A.E. Fischer and V.E. Moncrief, in S. Cotsakis and G.W. Gibbons (Eds.), *Global Structure and Evolution in General Relativity: Proceedings of the First Samos Meeting on Cosmology, Geometry and Relativity, held at Karlovassi, Samos, Greece, September 5-7, 1994*, pp. 111–173 (Lect. Notes Phys. **460**, Springer, 1996)

[568] N. Steenrod, *The Topology of Fibre Bundles* (Princeton University Press, 1951);
R. Sulanke and P. Wintgen, *Differentialgeometrie und Faserbündel* (VEB Deutscher Verlag der Wissenschaften, 1972)

[569] A.E. Fischer, in M. Carmeli, S.I. Fickler, and L. Witten (Eds.), *Relativity. Proceedings of the Relativity Conference in the Midwest held at Cincinnati, Ohio, June 2-6, 1969*, pp. 303-359 (Plenum Press, 1970)

[570] D. Giulini and C. Kiefer, Phys. Lett. A **193**, 21 (1994)

[571] K.V. Kuchař, J. Math. Phys. **13**, 768 (1972);
C. Teitelboim, Lett. Nuovo Cim. **3**, 397 (1972);
S.A. Hojman, K. Kuchař, and C. Teitelboim, Nature (London), Phys. Sci. **97**, 245 (1973);
K.V. Kuchař, J. Math. Phys. **15**, 708 (1974)

[572] F. Barbero, Phys. Rev. D **51**(10), 5507 (1995);
G. Immirzi, Class. Quantum Grav. **14**(10), L117 (1997)

[573] D. Witt, J. Math. Phys. **27**(2), 573 (1986)

[574] J. Friedman and R. Sorkin, Phys. Rev. Lett. **44**, 1100 (1980)

[575] C.J. Isham, in J.J. Duff and C.J. Isham (Eds.), *Quantum Structure of Space and Time. Proceedings of the Nuffield Workshop, August 3–21 1981, Imperial College London*, pp. 37–52 (Cambridge University Press, 1982)

[576] R. Sorkin, in P.G. Bergmann and V. De Sabbata (Eds.), *Topological Properties and Global Structure of Space–Time* (NATO Advanced Study Institutes Series B**138**, Kluwer Academic Publishers, 1986), p. 249; in S. De Filippo, M. Marinaro, G. Marmo, and G. Vilasi (Eds.), *Geometrical and Algebraic Aspects of Nonlinear Field Theory*, pp. 201–218 (Elsevier, 1989)

[577] F. Dowker and R. Sorkin, in R.C. Hilborn and G.M. Tino (Eds.), *Spin-Statistics Connections and Commutation Relations: Experimental Tests and Theoretical Implications*, pp. 205–218 (American Institute of Physics, 2000); Class. Quantum Grav. **15**, 1153 (1998);
C. Aneziris, A.P. Balachandran, M. Bourdeau, S. Jo, T.R. Ramadas and R.D. Sorkin, Int. J. Mod. Phys. A **14**(20), 5459 (1989); Mod. Phys. Lett. A **4**(4), 331 (1989)

[578] C.T.C. Wall, *Surgery on Compact Manifolds* (2nd ed., edited and with a foreword by A.A. Ranicki, Mathematical Surveys and Monographs **69**, American Mathematical Society, 1999);
J. Roe, *Index Theory, Coarse Geometry, and Topology of Manifolds* (American Mathematical Society, 1996);
A.A. Kosinski, *Differential Manifolds* (Pure and Applied Mathematics **138**, Academic Press, 1993)

[579] G. Perelman, e-Print: arXiv:math/0303109v1 [math.DG]

[580] J.W. Milnor, Am. J. Math. **84**(1), 1 (1962)

[581] D.I. Fouxe-Rabinovitch, Rec. Math. **8**(50), 265 (1940); Rec. Math. **9**(51), 297 (1941)

[582] D. McCullough and A. Miller, Mem. Am. Math. Soc. **61**(344) (1986)

[583] N.D. Gilbert, Proc. Lond. Math. Soc. **54**, 115 (1987)

[584] D. Giulini, in B. Fauser, J. Tolksdorf, and E. Zeidler (Eds.), *Quantum Gravity. Mathematical Models and Experimental Bounds*, pp. 161-201 (Birkhäuser, 2007)

[585] J. Lelong-Ferrand, Mém. Cl. Sci. Acad. Roy. Belg. **39**(5), 3 (1971)

[586] R. Geroch, J. Math. Phys. **13**, 394 (1972);
P. Breitenlohner and D. Maison, Ann. Inst. Henri Poincaré A **46**(2), 215 (1987)

[587] R.A. Matzner and Ch.W. Misner, Phys. Rev. **154**, 1229 (1967)

[588] J. Ehlers, *Konstruktionen und Charakterisierung von Lösungen der Einsteinschen Gravitationsfeldgleichungen*, (PhD Thesis, University of Hamburg, 1957);
R. Geroch, J. Math. Phys. **12**(6), 918 (1971);
M. Mars, Class. Quantum Grav. **18**(4), 719 (2001)

[589] P.J. Hilton and S. Wylie, *Homology Theory: An Introduction to Algebraic Topology* (Cambridge University Press, 1965);
S.-T. Hu, *Homology Theory: A First Course in Algebraic Topology*

(Holden-Day, 1966);
J.W. Milnor, Bull. Amer. Math. Soc. **72**, 358 (1966);
M.M. Cohen, *A Course in Simple-Homotopy Theory* (Springer, 1973);
H. Seifert and W. Threlfall, *A Textbook of Topology* (Pure Appl. Math. **89**, Translated from the German edition of 1934, Academic Press, 1980);
M. Audin, *The Topology of Torus Action on Symplectic Manifolds* (Birkhäuser, 1991);
G.E. Bredon, *Topology and Geometry* (Graduate Texts in Mathematics **139**, Springer, 1993);
J. Stillwell, *Classical Topology and Combinatorial Group Theory* (2nd ed., Springer, 1993);
S.P. Novikov, *Topology I: General Survey* (Springer, 1995);
I.M. James, *Handbook of Algebraic Topology* (North-Holland, 1995);
W.P. Thurston, *Three-Dimensional Geometry and Topology* (Princeton University Press, 1997);
I.M. James, *History of Topology* (North-Holland, 1999);
A. Hatcher, *Algebraic Topology* (Cambridge University Press, 2002);
A.V. Bolsinov and A.T. Fomenko, *Integrable Hamiltonian Systems: Geometry, Topology, Classification* (Chapman & Hall/CRC, 2004)

[590] H. Tietze, Monatsh. Math. Phys. **19**, 1 (1908)

[591] J.W. Alexander II, Trans. Amer. Math. Soc. **20**, 339 (1919)

[592] W. Threlfall and H. Seifert, Math. Ann. **104**, 1 (1930)

[593] J.H.C. Whitehead, Ann. Math. **42** (5), 1197 (1941)

[594] K. Reidemeister, Abh. Math. Sem. Univ. Hamburg **11**, 102 (1935); J. Reine. Angew. Math. **173**, 164 (1935)

[595] W. Franz, J. Reine Angew. Math. **173**, 245 (1935)

[596] G. De Rham, Mat. Sb., **1**(5), 737 (1936); Comment. Math. Helv. **12**, 191 (1939); Ann. Inst. Fourier, Grenoble **2**, 51 (1950)

[597] G. de Rham, M. Kervaire, and S. Maumary, *Torsion et type simple d'homotopie* (Lect. Notes Math. **48**, Springer, 1967)

[598] K. Reidemeister, Bull. Amer. Math. Soc. **56**, 297 (1950)

[599] C.P. Rourke and B.J. Sanderson, *Introduction to Piecewise-Linear Topology* (Springer, 1972)

[600] E.J. Brody, Ann. Math. **71**, 163 (1960)

[601] D.B. Ray and I.M. Singer, Adv. Math. **7**(2), 145 (1971); Ann. Math. **98**(1), 154 (1973); Proc. Symp. Pure Math. **23**, 167 (1973)

[602] J. Cheeger, PNAS **74**(7), 2651 (1977); Ann. of Math. **109**(2), 259 (1979)

[603] W. Müller, Adv. Math. **28**(3), 233 (1978)

[604] J.H. Przytycki and A. Yasuhara, Geom. Ded. **98**(1) (2003)

[605] P. Salvatore and R. Longoni, Riccardo, Topology **44**, 375 (2005)

[606] V. Turaev, *Torsions of 3-Dimensional Manifolds* (Progress in Mathematics **208**, Birkhäuser 2002);
L.I. Nicolaescu, *The Reidemeister Torsion of 3-Manifolds* (de Gruyter Studies in Mathematics **30**, Walter de Gruyter & Co. 2003);
G. Mislin, in I.M. James (Ed.), *Handbook of Algebraic Topology*, Chapter 26 (Elsevier Science, 1995)

[607] M. Rueff, Compositio Math. **6**, 161 (1938)

[608] U. Brehm and J. Świątkowski, Sonderforschungsbereich **288**, Preprint No. **59** (1993)

[609] H. Seifert, Acta Math. **60**, 147 (1933);
P. Orlik, *Seifert Manifolds* (Lect. Notes Math. **291**, Springer 1972)

[610] E.E. Moise, Ann. Math. **56**, 96 (1952);
E.E. Moise, *Geometric Topology in Dimensions 2 and 3* (Springer, 1977)

[611] Yu.B. Rudyak, e-Print: arXiv:math/0105047v1 [math.AT]; A. Ranicki (Ed.), *The Hauptvermutung Book* (K-Monographs in Mathematics **1**, Kluwer Academic Publishers, 1996); D.P. Sullivan, Bull. Amer. Math. Soc. **73**, 59 (1967); J.W. Milnor, Ann. Math. **74**(2), 575 (1961); R.H. Bing, Ann. Math. **69**, 37 (1959)

[612] K.V. Kuchař, in B. Hu, M. Ryan, and C. Vishveshvara (Eds.), *Directions in General Relativity*, pp. 201-221 (Cambridge University Press, 1993); in G. Kunstatter, D. Vincent, and J. Williams (Eds.), *Proceedings of the 4th Canadian Conference on General Relativity and Relativistic Astrophysics*, pp. 211-314 (World Scientific, 1992)

[613] S. Tomonaga, Prog. Theor. Phys. **1**, 27 (1946); Prog. Theor. Phys. **2**, 101 (1946)

[614] J. Schwinger, Phys. Rev. **74**, 1439 (1948)

[615] J.D. Brown and K.V. Kuchař, Phys. Rev. D **51**, 5600 (1995)

[616] W.G. Unruh, Phys. Rev. D **40**, 104852 (1989)

[617] D.A. Craig and P. Singh, Phys. Rev. D **82**, 12352 (2010); AIP Conf. Proc. **1232**, 275 (2010)

[618] A. Kleinschmidt and H. Nicolai, Int. J. Mod. Phys. D **19**(14), 2305 (2010)

[619] L.A. Glinka, Grav. Cosmol. **16**(1), 7 (2010); Concepts Phys. **6**, 19 (2009); New Adv. Phys. **2**, 1 (2008); e-Print: arXiv:0804.3516 [gr-qc]

[620] C. Kiefer, Gen. Rel. Grav. **41**, 877 (2009); in J. Kowalski-Glikman (Ed.), *Towards Quantum Gravity: Proceedings of the XXXV International Winter School on Theoretical Physics, Held in Polanica, Poland, 2-11 February 1999*, pp. 158-187 (Lect. Notes Phys. **541**, Springer 2000); in J. Ehlers and H. Friedrich (Eds.), *Canonical Gravity: From Classical to Quantum. Proceedings of the 117th WE Heraeus Seminar Held at Bad Honnef, Germany, 13-17 September 1993*,

pp. 170–212 (Springer, 1994); Ann. Phys. (Leipzig) **15**, 129 (2006)

[621] H. Yoshino and M. Shibata, Phys. Rev. D **80**, 084025 (2009)

[622] L. Brewin, Phys. Rev. D **80**, 084030 (2009); Gen. Rel. Grav. **39**, 521 (2007); Class. Quantum Grav. **15**, 2427 (1998)

[623] M. Alcubierre, *Introduction to 3+1 Numerical Relativity* (International Series of Monographs on Physics **140**, Oxford University Press, 2008)

[624] T. Damour and H. Nicolai, Int. J. Mod. Phys. D **17**, 525 (2008)

[625] P. Gusin, Phys. Rev. D **77**, 066017 (2008)

[626] J.F. Barbero G., AIP Conf. Proc. **1023**, 3 (2008)

[627] S. Zonetti and G. Montani, Int. J. Mod. Phys. A **23**, 1240 (2008)

[628] F. Cianfrani and G. Montani, Int. J. Mod. Phys. A **23**, 1149 (2008)

[629] E. Castellanos, A. Camacho, and J.I. Rivas, AIP Conf. Proc. **977**, 202 (2008)

[630] J.F. Barbero G. and E.J.S. Villasenor, Phys. Rev. D **77**, 121502 (2008)

[631] S. Carlip, Class. Quantum Grav. **25**, 154010 (2008)

[632] M.S. El Naschie, Chaos Solitons Fractals **36**, 808 (2008)

[633] G. Montani and F. Cianfrani, Class. Quantum Grav. **25**, 065007 (2008)

[634] R. Benini and G. Montani, Int. J. Mod. Phys. A **23**, 1244 (2008)

[635] R. Garattini, J. Phys. A **41**, 164057 (2008); Nucl. Phys. Proc. Suppl. **57**, 316 (1997)

[636] I.Ya. Aref'eva and I. Volovich, Int. J. Geom. Meth. Mod. Phys. **05**, 641 (2008)

[637] W. Nelson and M. Sakellariadou, Phys. Lett. B **661**, 37 (2008)

[638] J.A. Isenberg, Int. J. Mod. Phys. D **17**, 265 (2008); in N.T. Bishop and S.D. Maharaj (Eds.), *General Relativity and Gravitation: Proceedings of the 16th International Conference*, (World Scientific, 2002); in J.P. Harnard and S. Shnider (Eds.), *Geometrical and Topological Methods in Gauge Theories: Proceedings of the Canadian Mathematical Society Summer Research Institute, McGill University, Montreal, September 3-8, 1979*, pp. 223-248 (Ann. Phys. N.Y. **129**, Springer, 1980)

[639] D. Giulini and C. Kiefer, in I.-O. Stamatescu and E. Seiler (Eds.), *Approaches to Fundamental Physics: An Assessment of Current Theoretical Ideas*, pp. 131-150 (Lect. Notes Phys. **721**, Springer, 2007); Class. Quantum Grav. **12**, 403 (1995)

[640] C. Lämmerzahl, in B. Fauser, J. Tolksdorf, and E. Zeidler (Eds.), *Quantum Gravity: Mathematical Models and Experimental Bounds*, pp. 15–39 (Birkhaüser, 2007)

[641] A.Y. Kamenshchik, C. Kiefer, and B. Sandhöfer, Phys. Rev. D **76**, 064032 (2007)

[642] Ch. Soo, Class. Quantum Grav. **24**, 1547 (2007)

[643] M. Thibeault and C. Simeone, Int. J. Mod. Phys. D **16**, 1303 (2007)

[644] R. Carroll, Theor. Math. Phys. **152**, 904 (2007)

[645] L. Hardy, J. Phys. A **40**, 3081 (2007)

[646] D. Oriti, in B. Fauser, J. Tolksdorf, and E. Zeidler (Eds.), *Quantum Gravity. Mathematical Models and Experimental Bounds*, pp. 101-126 (Birkhäuser, 2007); Braz. J. Phys. **35**(2B), 481 (2005)

[647] I. Rodnianski, in M. Sanz-Solé, J. Soria-de Diego, J.L. Verona-Malumbres, and J.M. Verdera-Melenchón (Eds.), *Proceedings of the International Congress of Mathematicians: Madrid, Spain, August 22-30, 2006. Vol 3*, pp. 421-442 (European Mathematical Society, 2007)

[648] Y. Ma, Front. Phys. China **1**, 125, (2006)

[649] D. Rickles, in D. Rickles, S. French, and J. Saatsi (Eds.), *The Structural Foundations of Quantum Gravity*, pp. 152-195 (Clarendon Press, 2006); Stud. Hist. Phil. Mod. Phys. **36**, 691 (2005)

[650] J. Mattingly, in J. Eisenstaedt and A. Kox (Eds.), *The Universe of General Relativity, Vol. 11 of Einstein Studies*, pp. 327–338 (Birkhaüser, 2006)

[651] A.B. Henriques, Gen. Rel. Grav. **38**, 1645 (2006)

[652] D. Maxwell, J. Reine, Angew. Math. **590**, 1 (2006)

[653] M.P. Dąbrowski, C. Kiefer, and B. Sandhöfer, Phys. Rev. D **74**, 044022 (2006)

[654] R. Bartnik and J. Isenberg, Class. Quantum Grav. **23**, 2559 (2006)

[655] C. Kiefer, J. Müller-Hill, and C. Vaz, Phys. Rev. D **73**, 044025 (2006)

[656] S. Klainerman and I. Rodnianski, Geom. Funct. Anal. **16**(1), 164 (2006); Invent. Math. **159**, 437 (2005)

[657] N. Pinto-Neto, Found. Phys. **35**, 577 (2005)

[658] M.J.W. Hall, Gen. Rel. Grav. **37**, 1505 (2005)

[659] C. Kiefer, T. Lück, and P. Moniz, Phys. Rev. D **72**, 045006 (2005)

[660] T.P. Shestakova and C. Simeone, Grav. Cosmol. **10**(3(39)), 161 (2004); Grav. Cosmol. **10**(4(40)), 257 (2004)

[661] A.P. Gentle, N.D. George, W.A. Miller, and A. Kheyfets, Int. J. Mod. Phys. A **19**(10), 1609 (2004)

[662] T-Y. Cao, Stud. Hist. Phil. Mod. Phys. **32**(2), 181 (2004)

[663] Y. Choquét-Bruhat, Class. Quantum Grav. **21**, S127 (2004)

[664] T. Kubota, T. Ueno, and N. Yokoi, Phys. Lett. B **579**, 200 (2004)

[665] K. Meissner, Class. Quantum Grav. **21**, 5245 (2004)

[666] J. Butterfield and C.J. Isham, in C. Callender and N. Huggett (Eds.), *Physics Meets Philosophy at the Planck Scale: Contemporary Theories in Quantum Gravity*, pp. 33-89 (Cambridge University Press, 2004); Found. Phys. **30**, 1707 (2000); in J. Butterfield (Ed.), *The Arguments of Time*, pp. 111-168 (Oxford University Press, 1999)

[667] L. Andersson and V. Moncrief, in P.T. Chruściel and H. Friedrich (Eds.), *The Einstein Equations and the Large Scale Behavior of Gravitational Fields: 50 Years of the Cauchy Problem in General Relativity*, pp. 299–330 (Birkhäuser, 2004); Ann. Inst. H. Poincaré **4**, 1 (2003)

[668] S. Weinstein, in C. Callender and N. Huggett (Eds.), *Physics Meets Philosophy at the Planck Scale: Contemporary Theories in Quantum Gravity*, pp. 90-100 (Cambridge University Press, 2004); in T. Piran (Ed.), *Proceedings of the 8th Marcel Grossmann Meeting on Recent Developments in Theoretical and Experimental General Relativity, Gravitation and Relativistic Field Theories (MG 8), Jerusalem, Israel, 22-27 June 1997*, pp. 875-877 (World Scientific, 1999)

[669] C.J. Isham, Adv. Theor. Math. Phys. **8**, 797 (2004); Adv. Theor. Math. Phys. **7**, 807 (2004); Adv. Theor. Math. Phys. **7**, 331 (2003); in G.W. Gibbons, E.P.S. Shellard, and S.J. Rankin (Eds.), *The Future of Theoretical Physics and Cosmology: Celebrating Stephen Hawking's 60th Birthday*, pp. 384-408 (Cambridge University Press, 2003); in M. Francaviglia, G. Longhi, L. Lusanna and E. Sorace (Eds.), *Proceedings of the 14th International Conference on General Relativity and Gravitation*, pp. 167-209 (World Scientific, 1997); Class. Quantum Grav. **13**, A5 (1996); in J. Ehlers and H. Friedrich (Eds.), *Canonical Gravity: From Classical to Quantum. Proceedings of the 117th WE Heraeus Seminar Held at Bad Honnef, Germany, 13-17 September 1993*, pp. 1-21 (Springer, 1994); in L.A. Ibort and

M.A. Rodriguez (Eds.), *Integrable Systems, Quantum Groups and Quantum Field Theories*, pp. 157-288 (Kluwer Academic Publishers, 1993); in M.A. del Olmo, M. Santander, and J. Mateos-Guilarte (Eds.),*Group Theoretical Methods in Physics: Proceedings of the XIX International Colloquium, held in Salamanca, Spain, 29 June - 5 July 1992*, pp. 157-288 (1992); in H. Mitter and H. Gausterer (Eds.), *Proceedings of 30th Internationale Universitätswochen für Kernphysik: Recent Aspects of Quantum Fields*, pp. 123-229 (Lect. Notes Phys. **396**, Springer, 1991); in A. Ashtekar and J.J. Stachel (Eds.), *2nd Osgood Hill Conference: Festschrift Paul Dirac*, pp. 351-400 (Birkhäuser, 1991); in H.C. Lee (Ed.), *Proceedings of 1989 Banff NATO ASI: Physics, Geometry and Topology*, pp. 129-190 (Plenum Press, 1990); Nucl. Phys. B Proc. Suppl. **6**, 349, (1989); in M.A.H. MacCallum (Ed.), *Proceedings of 11th International Conference on General Relativity and Gravitation, 6-12 Jul 1986, Stockholm*, pp. 99-149 (Cambridge University Press, 1987); in A.T. Davies and D.G. Sutherland (Eds.), *Proceedings of 28th Scottish Universities Summer School In Physics: Supersymmetry and Supergravity, 28 Jul-17 Aug 1985, Edinburgh, Scotland* (SUSSP Publications **28**, Edinburgh University Press, 1986), p. 0001; in S.M. Christensen (Ed.), *Quantum Theory of Gravity: Essays in honor of the 60th birthday of Bryce S. De Witt*, pp. 299-314 (Adam Hilger, 1984); in B.S. DeWitt and R. Stora (Eds.), *Relativity, Groups, and Topology II*, pp. 1059-1290 (Elsevier 1984); Proc. Roy. Soc. Lond. A **368**, 33 (1979); Proc. Roy. Soc. Lond. A **351**, 209 (1976); in C.J. Isham, R. Penrose, and D.W. Sciama (Eds.), *Quantum Gravity: an Oxford Symposium*, pp. 1-77 (Clarendon Press, 1975)

[670] S. Klainerman and F. Nicolo, *The Evolution Problem in General Relativity* (Prog. Math. Phys. **25**, Birkhäuser, 2003)

[671] H. Lindblad and I. Rodnianski, C. R. Math. Acad. Sci. Paris **336**(11), 901 (2003)

[672] J. Isenberg, R. Mazzeo, and D. Pollack, Ann. Inst. H. Poincaré **4**, 369 (2003)

[673] E. Anderson, J. Barbour, B. Foster, and N. 'O Murchadha, Class. Quantum Grav. **20**, 1571 (2003)

[674] R.J. Gleiser, C.N. Kozameh, and F. Parisi, Class. Quantum Grav. **20**, 4375 (2003)

[675] D. Christodoulou, in V.G. Gurzadyan, R.T. Jantzen, and R. Ruffini (Eds.), *Proceedings of the Ninth Marcel Grossmann Meeting on General Relativity, University of Rome "La Sapienza", July 2-8, 2000*, pp. 44–54 (World Scientific, 2003); Ann. Math. **149**, 183 (1999); Comm. Pure Appl. Math. **46**, 1131 (1993); Phys. Rev. Lett. **25**, 1596 (1970)

[676] P. Hájíček, in D. Giulini, C. Kiefer, and C. Lämmerzahl (Eds.), *Quantum Gravity. From Theory To Experimental Search*, pp. 255–299 (Lect. Notes Phys. **631**, Springer, 2003)

[677] M.J.W. Hall, K. Kumar, and M. Reginatto, J. Phys A: Math. Gen. **36**, 9779 (2003)

[678] J.B. Hartle, in G.W. Gibbons, E.P.S. Shellard, and S.J. Rankin (Eds.), *The Future of Theoretical Physics and Cosmology: Celebrating Stephen Hawking's 60th Birthday*, pp. 38-50, 615-620 (Cambridge University Press, 2003)

[679] G. Gibbons, in G.W. Gibbons, E.P.S. Shellard, and S.J. Rankin (Eds.), *The Future of Theoretical Physics and Cosmology: Celebrating Stephen Hawking's 60th Birthday*, pp. 351-372 (Cambridge University Press, 2003)

[680] F. Dowker, in G.W. Gibbons, E.P.S. Shellard, and S.J. Rankin (Eds.), *The Future of Theoretical Physics and Cosmology: Celebrating Stephen Hawking's 60th Birthday*, pp. 436-452 (Cambridge University Press, 2003)

[681] D. Page, in G.W. Gibbons, E.P.S. Shellard, and S.J. Rankin (Eds.), *The Future of Theoretical Physics and Cosmology: Celebrating Stephen Hawking's 60th Birthday*, pp. 621-648 (Cambridge University Press, 2003)

[682] A. Vilenkin, in G.W. Gibbons, E.P.S. Shellard, and S.J. Rankin (Eds.), *The Future of Theoretical Physics and Cosmology: Celebrating Stephen Hawking's 60th Birthday*, pp. 649-666 (Cambridge University Press, 2003)

[683] J. Halliwell, in G.W. Gibbons, E.P.S. Shellard, and S.J. Rankin (Eds.), *The Future of Theoretical Physics and Cosmology: Celebrating Stephen Hawking's 60th Birthday*, pp. 675-692 (Cambridge University Press, 2003)

[684] N. Pinto-Neto and E.S. Santini, Gen. Rel. Grav. **34**, 505 (2002); Phys. Rev. D **59**, 123517 (1999)

[685] J.A. Belinchon, Int. J. Mod. Phys. D **11**, 527 (2002)

[686] M. Castagnino, G. Catren, and R. Ferraro, Class. Quantum Grav. **19**, 4729 (2002)

[687] A.E. Fischer and V.E. Moncrief, in P. Holmes, P.K. Newton, and A. Weinstein (Eds.), *Geometry, Dynamics, and Mechanics: 60th Birthday Volume for J.E. Marsden*, pp. 1-60 (Springer, 2002); in V. De Alfaro, J. Nelson, M. Cadoni, M. Cavaglia, and A.T. Filippov (Eds.), *Constrained Dynamics and Quantum Gravity 1999: Abstracts of the Third Meeting on Constrained Dynamics and Quantum Gravity, QC99, Villasimius, Italy, 13-17 September 1999*, pp. 83-102 (Nucl. Phys. B Proc. Suppl. **88**, North-Holland, 2000); in B. Fiedler, K. Groger, and J. Sprekels (Eds.), *Equadiff 99: International Conference on Differential Equations, Berlin, August 17, 1999. Vol. 1*, pp. 279-282 (World Scientific, 1999); Class. Quantum Grav. **16**, L79 (1999); in S. Cotsakis and G.W. Gibbons (Eds.), *Mathematical and Quantum Aspects of Relativity, held at Pythagoreon, Samos, greece, 31 August - 4 Septmeber 1998*, pp. 70-101 (Lect. Notes Phys. **537**, Springer, 1998); in J. Nelson (Ed.), *Proceedings of the Second meeting on Constrained Dynamics and Quatum Gravity QG96, Santa Margherita Ligure, Italy, 17 September 1996*, pp. 142-161 (Nucl. Phys. B Proc. Suppl. **57**, North-Holland, 1997); Gen. Rel. Grav. **28**, 221 (1996); Gen. Rel. Grav. **28**, 207 (1996); in M. Flato, R. Kerner, and A. Lichnerowicz (Eds.), Physics on

Manifolds, Proceedings of the International Colloquium in honour of Yvonne Choquet-Bruhat, Paris, June 3-5, 1992, pp. 11-151 (Kluwer Academic Publishers, 1994);

[688] G.F.R. Ellis, in S. Bonometto, V. Gorini, and U. Moschella (Eds.), *Modern Cosmology*, pp. 108-158 (Institite of Physics, 2002)

[689] M. Kenmoku, H. Kubotani, E. Takasugi, and Y. Yamazaki, Prog. Theor. Phys. **105**, 897 (2001); Phys. Rev. D **59**, 124004 (1999)

[690] S. Biswas, A. Shaw, and D. Biswas, Int. J. Mod. Phys. D **10**, 585 (2001)

[691] H. Yamazaki and T. Hara, Prog. Theor. Phys. **106**, 323 (2001)

[692] S. Carlip, Rep. Prog. Phys. **64**, 885 (2001)

[693] M. Anderson, Comm. Math. Phys. **222**, 533 (2001)

[694] S. Goldstein and S. Teufel, in C. Callender and N. Huggett (Eds.), *Physics Meets Philosophy at the Planck Scale*, pp. 275–289 (Cambridge University Press, 2001)

[695] H. Friedrich and A. Rendall, in B.G. Schmidt (Ed.), *Einstein's Field Equations and Their Physical Implications: Selected Essays in Honour of Jürgen Ehlers*, pp. 127-223 (Lect. Notes Phys. **540**, Springer, 2000)

[696] A. Kheyfets and W.A. Miller, Int. J. Mod. Phys. A **15**, 4125 (2000); Phys. Rev. D **51**, 493 (1995)

[697] M.A. Ahmed, Nuovo Cim. B **115**, 1127 (2000)

[698] M.L. Filchenkov, Russ. Phys. J. **43**, 921 (2000)

[699] K.S. Stelle, Nucl. Phys. B Proc. Suppl. **88**, 3 (2000)

[700] G.Lifschytz and V. Periwal, JHEP **0004**, 026 (2000)

[701] Y. Choquét-Bruhat, J. Isenberg, and J.W. York, Jr., Phys. Rev. D **61**, 084034 (2000)

[702] M. Kenmoku, Grav. Cosmol. **5**, 289 (1999)

[703] B.S. DeWitt, in T. Piran (Ed.), *Proceedings of the 8th Marcel Grossmann Meeting on Recent Developments in Theoretical and Experimental General Relativity, Gravitation and Relativistic Field Theories (MG 8), Jerusalem, Israel, 22-27 June 1997*, pp. 6-25 (World Scientific, 1999); in R.E. Allen (Ed.), *Relativity, Particle Physics And Cosmology: Proceedings of Richard Arnowitt Fest: A Symposium On Supersymmetry And Gravitation, 5-7 Apr 1998, College Station, Texas*, pp. 70-92 (World Scientific, 1999); in S.W. Hawking and W. Israel (Eds.), *General Relativity: An Einstein centenary survey*, pp. 680-745 (Cambridge University Press, 1979); Phys. Rep. **19**, 295 (1975); Gen. Rel. Grav. **1**, 181 (1970); *Dynamical Theory of Groups and Fields* (Gordon and Breach, 1965); in L. Witten (Ed.), *Gravitation: An Introduction to Current Research*, pp. 266–381 (John Wiley & Sons, 1962); Phys. Rev. **162**, 1195 (1967); Phys. Rev. **162**, 1239 (1967)

[704] K. Kuchař, in J. Butterfield (Ed.), *The Arguments of Time*, pp. 169-195 (Oxford University Press, 1999); in M. Francaviglia, G. Longhi, L. Lusanna and E. Sorace (Eds.), *Proceedings of the 14th International Conference on General Relativity and Gravitation*, pp. 511-514 (World Scientific, 1997); Phys. Rev. D **50**, 3961 (1994); in R. Gleiser, C. Kozameh and O. Moreschi (Eds.), *General Relativity and Gravitation 1992: Proceedings of the Thirteenth International Conference on General Relativity and Gravitation, held Huerta Grande, Cordoba, 28 June-4 July, 1992*, pp. 119-150 (IOP Publishing, 1993); Phys. Rev. D **43**, 3332 (1991); Phys. Rev. D **39**, 2263 (1989); Cont. Math. **71**, 285 (1988); Phys. Rev. D **34**, 3044 (1986); Phys. Rev. D **34**, 3031 (1986); J. Math. Phys. **22**, 2640 (1981); in W. Israel (Ed.), *Relativity, Astrophysics and Cosmology: Proceedings of The Summer School held 14-26 August, 1972 at The Banff Centre, Banff, Alberta*, pp. 238-288 (Kluwer Academic Publishers, 1973); J. Math. Phys. **13**, 768 (1972); Phys. Rev. D **4**, 955 (1971); J. Math. Phys. **11**, 3322 (1970)

[705] J.W. Norbury, Eur. J. Phys. **19**, 143 (1998)

[706] A.O. Barvinsky and C. Kiefer, Nucl. Phys. B **526**, 509 (1998)

[707] V.A. Berezin, A. Boyarsky, and A.Yu. Neronov, Phys. Rev. D **57**, 1118 (1998)

[708] A. Mostafazadeh, J. Math. Phys. **39**, 4499 (1998)

[709] T. Brotz, Phys. Rev. D **57**, 2349 (1998)

[710] H. Lu, J. Maharana, S. Mukherji, and C.N. Pope, Phys. Rev. D **57**, 2219 (1998)

[711] J. Makela and P. Repo, Phys. Rev. D **57**, 4899 (1998)

[712] R. Gleiser, O. Nicasio, R. Price, and J. Pullin, Phys. Rev. D **57**, 3401 (1998)

[713] T. Horiguchi, Nuovo Cim. B **113**, 429 (1998); Nuovo Cim. B **112**, 1227 (1997); Nuovo Cim. B **112**, 1107 (1997); Nuovo Cim. B **111**, 293 (1996); Nuovo Cim. B **111**, 85 (1996); Nuovo Cim. B **111**, 49 (1996); Nuovo Cim. B **111**, 165 (1996); Nuovo Cim. B **110**, 839 (1995); Mod. Phys. Lett. A **9**, 1429 (1994); Mod. Phys. Lett. A **8**, 777 (1993); Phys. Rev. D **48**, 5764 (1993)

[714] A. Błaut and J. Kowalski-Glikman, Phys. Lett. B **406**, 33 (1997)

[715] J.W. Barrett and L. Crane, Class. Quantum Grav. **14**, 2113 (1997); J. Math. Phys. **39**, 3296 (1998)

[716] R. Sorkin, Int. J. Theor. Phys. **36**, 2759 (1997)

[717] J. Isenberg and J. Park, Class. Quantum Grav. **14**, A189 (1997)

[718] M.D. Pollock, Int. J. Mod. Phys. D **6**, 91 (1997); Mod. Phys. Lett. A **12**, 2057 (1997); Int. J. Mod. Phys. D **4**, 305 (1995); Int. J. Mod. Phys. D **3**, 579 (1994); Int. J. Mod. Phys. A **7**, 4149 (1992)

[719] R. Parentani, Phys. Rev. D **56**, 4618 (1997)

[720] M.A. Scheel, T.W. Baumgarte, G.B. Cook, S.L. Shapiro, and S.A. Teukolsky, Phys. Rev. d **56**, 6320 (1997)

[721] A. Ambjørn, M. Carfora, and A. Marzouli, *The Geometry of Dynamical Triangulations* (Springer, 1997); in F. David,

P. Ginsparg and J. Zinn-Justin (Eds.), *Fluctuating Geometries in Statistical Mechanics and Field Theory: Proceedings of the Les Houches Summer School*, pp. 77–195 (North-Holland, 1996)

[722] T.W. Baumgarte, G.B. Cook, M.A. Scheel, S.L. Shapiro, and S.A. Teukolsky, Phys. Rev. Lett. **79**, 1182 (1997); Phys. Rev. D **54**, 4849 (1996)

[723] J. Louko and S.N. Winters-Hilt, Phys. Rev. D **54**, 2647 (1996)

[724] G. Lifschytz, S.D. Mathur, and M. Ortiz, Phys. Rev. D **53**, 766 (1996)

[725] R. Penrose, Gen. Rel. Grav. **28**, 581 (1996)

[726] P. Hübner, Phys. Rev. D **53**, 701 (1996)

[727] S.R. Brandt and E. Seidel, Phys. Rev. D **54**, 1403 (1996)

[728] J. Kowalski-Glikman and K. Meissner, Phys. Lett. B **376**, 48 (1996)

[729] J. Kowalski-Glikman, in B. Jancewicz, J.T. Sobczyk, and J. Lukierski (Eds.), *From Field Theory to Quantum Groups. Birthday volume dedicated to Jerzy Lukierski*, pp. 229-242 (World Scientific, 1996)

[730] D.L. Wiltshire, in B. Robson, N. Visvanathan and W.S. Woolcock (eds.), *Cosmology: The Physics of the Universe: Proceedings of the 8th Physics Summer School, Australian National University, Canberra, Australia, 16 January-3 February, 1995*, pp. 473-531 (World Scientific, 1996)

[731] B. Robson, in B. Robson, N. Visvanathan and W.S. Woolcock (eds.), *Cosmology: The Physics of the Universe: Proceedings of the 8th Physics Summer School, Australian National University, Canberra, Australia, 16 January-3 February, 1995*, pp. 473–531 (World Scientific, 1996)

[732] M. Perry, in G.S. Hall and J.R. Pulham, *General Relativity: Proceedings of the Forty Sixth Scottish Universities Summer*

School in Physics, Aberdeen, July 1995, pp. 377-406 (NATO ASI/SUSSP Publications/IOP Publishing, 1996)

[733] N.T. Bishop, R. Gomez, P.R. Holvorcem, R.A. Matzner, P. Papadopoulos, and J. Winicour, Phys. Rev. Lett. **76**, 4303 (1996)

[734] T.W. Baumgarte, S.L. Shapiro, and S.A. Teukolsky, Astrophys. J. **458**, 680 (1996); Astrophys. J. **443**, 717 (1995)

[735] M. Varadarajan, Phys. Rev. D **52**, 7080 (1995)

[736] S.P. Kim, Phys. Rev. D **52**, 3382 (1995)

[737] N.P. Landsman, Class. Quantum Grav. **12**, L119 (1995)

[738] J. Feinberg and Y. Peleg, Phys. Rev. D **52**, 1988 (1995)

[739] T. Horiguchi, K. Maeda, and M. Sakamoto, Phys. Lett. B **344**, 105 (1995)

[740] F. Embacher, Grav. Cosmol. **1**, 46 (1995)

[741] C. Rovelli and L. Smolin, Phys. Rev. D **52**, 5743 (1995); in B.R. Lyer (Ed.), *Highlights in Gravitation and Cosmology* (Cambridge University Press, 1988)

[742] A. Ashtekar, in B. Julia and J. Zinn-Justin (Eds.), *Les Houches, Session LVII, 1992: Gravitation and Quantizations*, pp. 181-283 (Elsevier, 1995)

[743] A.M. Abrahams, A. Andreson, Y. Choquét-Bruhat, and J.W. York, Jr., Phys. Rev. Lett. **75**, 3377 (1995)

[744] S. Carlip, Class. Quantum Grav. **11**, 31 (1994)

[745] P. Mansfield, Nucl. Phys. B **418**, 113 (1994)

[746] E. Adi and S. Solomon, Phys. Lett. B **336**, 152 (1994)

[747] A. Ishikawa, Phys. Rev. D **50**, 2609 (1994)

[748] J.E. Lidsey, Class. Quantum Grav. **11**, 1211 (1994)

[749] A.M. Abrahams, G.B. Cook, S.L. Shapiro, and. S.A. Teukolsky, Phys. Rev. D **49**, 5153 (1994)

[750] J.B. Barbour, Class. Quantum Grav. **11**, 2853 (1994); Class. Quantum Grav. **11**, 2875 (1994); in R. Penrose and C.J. Isham (Eds.), *Quantum Concepts in Space and Time*, pp. 236-246 (Oxford University Press, 1986)

[751] A.M. Abrahams and C.R. Evans, Phys. Rev. Lett. **70**, 2980 (1993); Phys. Rev. D **37**, 318 (1988)

[752] G. Hayward and K. Wong, Phys. Rev. D **47**, 4778 (1993); Phys. Rev. D **46**, 620 (1992)

[753] M.W. Choptuik, Phys. Rev. Lett. **70**, 9 (1993)

[754] S. Capozziello and R. de Ritis, Int. J. Mod. Phys. D **2**, 373 (1993)

[755] N. Pinto-Neto and A.F. Velasco, Gen. Rel. Grav. **25**, 991 (1993)

[756] J. Maekelae, Phys. Rev. D **48**, 1679 (1993)

[757] S. Abe, Phys. Rev. D **47**, 718 (1993)

[758] H. Anada, Y. Mizumoto, and T. Kitazoe, Mod. Phys. Lett. A **8**, 45 (1993); Mod. Phys. Lett. A **8**, 1065 (1993)

[759] D.S. Salopek, J.M. Stewart, and J. Parry, Phys. Rev. D **48**, 719 (1993)

[760] H.-J. Matschull, Class. Quantum Grav. **10**, L149 (1993)

[761] T. Hori, Prog. Theor. Phys. **90**, 743 (1993)

[762] P.D. D'Eath, S.W. Hawking, and O. Obregon, Phys. Lett. B **300**, 44 (1993)

[763] N. Pinto-Neto and A. F. Velasco, Gen. Rel. Grav. **25**(10), 991 (1993)

[764] S.L. Shapiro and S.A. Teukolsky, Astrophys. J. **45**, 2739 (1992); Astrophys. J. **307**, 575 (1986); Astrophys. J. **335**, 199 (1980)

[765] E. Seidel and W.-M. Suen, Phys. Rev. Lett. **69**, 1845 (1992)

[766] S. Chakraborty, Int. J. Theor. Phys. **31**, 289 (1992)

[767] F.D. Mazzitelli, Phys. Rev. D **46**, 4758 (1992)

[768] K. Shimizu and S. Wada, Int. J. Mod. Phys. A **7**, 1627 (1992)

[769] C.L. Stone and K.V. Kuchař, Class. Quantum Grav. **9**, 757 (1992)

[770] M. Ferraris, M. Francaviglia, and I. Sinicco, Nuovo Cim. B **107**, 11 (1992)

[771] G. Gorelik, in J. Eisenstaedt (Ed.), *Studies in the History of General Relativity, Vol. 3 of Einstein Studies*, pp. 364–379 (Birkhaüser, 1992)

[772] M.J. Duncan, in P. Langacker and M. Cvetic (Eds.), *Proceedings: Testing The Standard Model - TASI-90: Theoretical Advanced Study Inst. In Elementary Particle Physics, 3-29 Jun 1990, Boulder, Colorado*, pp. 743-770 (World Scientific, 1991)

[773] G. Giampieri, Phys. Lett. B **261**, 411 (1991)

[774] J.P. Dias and M. Figueira Ann. Poincare Phys. Theor. **54**, 17 (1991)

[775] M.J. Duncan and L.G. Jensen, Nucl. Phys. B **361**, 695 (1991)

[776] Z. Bern, S.K. Blau, and E. Mottola, Phys. Rev. D **33**, 1212 (1991)

[777] P.O. Mazur, Phys. Lett. B **262**, 405 (1991)

[778] G.T. Horowitz, Class. Quantum Grav. **8**, 587 (1991); Phys. Rev. D **31**, 1169 (1985)

[779] C. Kiefer and T.P. Singh, Phys. Rev. D **44**, 1067 (1991)

[780] K.V. Kuchař and C.G. Torre, Phys. Rev. D **43**, 419 (1991); Phys. Rev. D **44**, 3116 (1991)

[781] P. Hájíček and K.V. Kuchař Phys. Rev. D **41**, 1091 (1990)

[782] G.B. Cook and J.W. York, Phys. Rev. D **41**, 1077 (1990)

[783] T. Padmanabhan and T.P. Singh Class. Quantum Grav. **7**, 411 (1990)

[784] A.O. Barvinsky, Phys. Lett. B **241**, 201 (1990)

[785] W. Fischler, D. Morgan, and J. Polchinski, Phys. Rev. D **42**, 4042 (1990)

[786] G.B. Cook, *Initial Data for the Two-Body Problem of General Relativity*, PhD Thesis, University of North Carolina (1990)

[787] C.R. Evans, L.S. Finn, and D.W. Hobill (Eds.), *Frontiers in Numerical Relativity* (Cambridge University Press, 1989)

[788] J.W. York, Jr., in C.R. Evans, L.S. Finn, and D.W. Hobill (Eds.), *Frontiers in Numerical Relativity*, pp. 89-109 (Cambridge University Press, 1989); in F.J. Tipler (Ed.), *Essays in General Relativity*, pp. 39-58 (Academic Press, 1980); in L. Smarr (Ed.), *Sources of Gravitational Radiation*, pp. 83-126 (Cambridge University Press, 1979); Ann. Inst. H. Poincaré **21**, 319 (1974); J. Math. Phys. **14**, 456 (1973); Phys. Rev. Lett. **26**, 1656 (1971)

[789] A. Vilenkin, Phys. Rev. D **39**, 1116 (1989)

[790] S. Weinberg, Rev. Mod. Phys. **61**, 1 (1989)

[791] V.D. Ivashchuk, V.N. Melnikov and A.I. Zhuk Nuovo Cim. B **104**, 575 (1989)

[792] W. Fishler, I. Klebanov, J. Polchinski, and L. Susskind, Nucl. Phys. B **327**, 157 (1989)

[793] E. Alvarez, Rev. Mod. Phys. **61** (3), 561 (1989)

[794] S.B. Giddings and A. Strominger, Nucl. Phys. B **321**, 481 (1989)

[795] A. Hosoya and M. Morikawa, Phys. Rev. D **39**, 1123 (1989)

[796] M. McGuigan, Phys. Rev. D **39**, 2229 (1989); Phys. Rev. D **38**, 3031 (1988)

[797] P.J. Schinder, S.A. Bludman, and T. Piran, Phys. Rev. D **37**, 2722 (1988)

[798] V.A. Rubakov, Phys. Lett. B **214**, 503 (1988)

[799] L. Smolin, in J.A. Isenberg (Ed.), *Mathematics and General Relativity: Proceedings of the AMS-IMS-SIAM Joint Summer Research Conference held June 22—28,1986 with support from the National Science Foundation*, pp. 55-98 (Contemporary Mathematics **71**, AMS, 1988)

[800] T. Jacobson, in J.A. Isenberg (Ed.), *Mathematics and General Relativity: Proceedings of the AMS-IMS-SIAM Joint Summer Research Conference held June 22—28,1986 with support from the National Science Foundation*, pp. 99-104 (Contemporary Mathematics **71**, AMS, 1988)

[801] J.L. Friedman and I. Jack, Phys. Rev. D **37**, 3495 (1988)

[802] N.T. Bishop, Gen. Rel. Grav. **20** 573 (1988); Gen. Rel. Grav. **16**, 589 (1984); Gen. Rel. Grav. **14**, 717 (1982)

[803] H.C. Tsamis and R.P. Woodard, Phys. Rev. D **36**, 3641 (1987)

[804] M. Gleiser, R. Holman, and N.P. Neto, Nucl. Phys. B **294**, 1164 (1987), FERMILAB-Pub-87/73-A

[805] J. Thornburg, Class. Quantum Grav. **4**, 1119 (1987)

[806] C.R. Evans, in M.P. Ulmer (Ed.), *Proceedings of the 13th Texas Symposium on Relativistic Astrophysics*, pp. 152-156 (World Scientific, 1987); in J.M. Centrella (Ed.), *Dynamical Spacetimes and Numerical Relativity*, pp. 3-39 (Cambridge University Press, 1986); *A Method for Numerical Relativity: Simulation of Axisymmetric Gravitational Collapse and Gravitational Radiation Generation*, PhD thesis (University of Texas at Austin, 1984)

[807] K.S. Thorne, in S.W. Hawking and W. Israel (Eds.), *Three hundred years of gravitation*, pp. 330-458 (Cambridge University Press 1987)

[808] A. Anderson and B. DeWitt, Found. Phys. **16**, 91 (1986)

[809] M.W. Choptuik and W.G. Unruh, Gen. Rel. Grav. **18**, 813 (1986)

[810] H. Friedrich, Comm. Math. Phys. **107**, 587 (1986),; J. Geom. Phys. **3**, 101 (1986)

[811] L.Z. Fang and M. Li, Phys. Lett. B **169**, 28 (1986)

[812] A.O. Barvinsky and V.N. Ponomarev, Sov. Phys. J. **29**(3), 187 (1986); Phys. Lett. B **167**, 289 (1986)

[813] W.Z. Chao, Phys. Rev. D **31**, 3079 (1985)

[814] R. Brandenburger, Rev. Mod. Phys. **57**, 1 (1985)

[815] D.N. Page, Phys. Rev. D **32**, 2496 (1985)

[816] P. Amsterdamski, Phys. Rev. D **31**, 3073 (1985)

[817] C.J. Isham and K.V. Kuchař, Ann. Phys. N.Y. **164**, 288 (1985); Ann. Phys. N.Y. **164**, 316 (1985)

[818] T. Nakamura, Y. Kojima, and K. Oohara, Phys. Lett. A **107**, 452 (1985); Phys. Lett. A **106**, 235 (1984)

[819] C.J. Isham and A.C. Kakas, Class. Quantum Grav. **1**, 621 (1984); Class. Quantum Grav. **1**, 633 (1984)

[820] A.D. Linde, Rep. Prog. Phys. **47**, 925 (1984)

[821] J. Bowen, J. Rauber, and J.W. York, Jr., Class. Quantum Grav. **1**, 591 (1984)

[822] A.D. Kulkarni, J. Math. Phys. **25**, 1028 (1984)

[823] A.D. Kulkarni, L.C. Shepley, and J.W. York, Jr., Phys. Lett. A **96**, 228 (1983)

[824] N. Deruelle and T. Piran (Eds.), *Gravitational Radiation* (North-Holland, 1983)

[825] R. Schoen and S.T. Yau, Comm. Math. Phys. **90**, 575 (1983); Comm. Math. Phys. **65**, 45 (1979)

[826] J.M. Bowen, Gen. Rel. Grav. **14**, 1183 (1982); Gen. Rel. Grav. **11**, 227 (1979)

[827] J.W. York, Jr. and T. Piran, in R. Matzner and L. Shepley (Eds.), *Spacetime and Geometry: The Alfred Schild Lectures*, pp. 147-176 (University of Texas Press, 1982)

[828] M. Cantor and A.D. Kulkarni, Phys. Rev. D **25**, 2521 (1982)

[829] J. Isenberg and J.E. Marsden, Phys. Rep. **89**, 179 (1982)

[830] D.N. Page and C.D. Geilker, Phys. Rev. Lett. **47**, 979 (1981)

[831] A.E. Fischer, J.E. Marsden, and V.E. Moncrief, Ann. Inst. H. Poincaré **33**, 147 (1980); in F.J. Tipler (Ed.), *Essays in General Relativity*, pp. 79-96 (Academic Press, 1980)

[832] J.M. Bowen and J.W. York, Jr., Phys. Rev. D **21**, 2047 (1980)

[833] J.A. Isenberg and J.M. Nester, in A. Held (Ed.), *General Relativity and Gravitation. One Hundred Years After the Birth of Albert Einstein*, pp. 23-97 (Plenum Press, 1980)

[834] P.G. Bergmann and A.B. Komar, in A. Held (Ed.), *General Relativity and Gravitation. One Hundred Years After the Birth of Albert Einstein*, pp. 227-254 (Plenum Press, 1980); Int. J. Theor. Phys. **5**, 15 (1972); in *Recent Developments in General Relativity. A collection of papers dedicated to Leopold Infeld*, pp. 31-46 (Pergamon Press/PWN-Polish Scientific Publishers, 1962)

[835] Y. Choquét-Bruhat and J.W. York, Jr., in A. Held (Ed.), *General Relativity and Gravitation. One Hundred Years After the Birth of Albert Einstein*, pp. 99-172 (Plenum Press, 1980)

[836] V.G. Lapchinsky and V.A. Rubakov, Acta Phys. Pol. B **10**, 1041 (1979)

[837] D. Christodoulou, M. Francaviglia, and W.M. Tulczyjew, Gen. Rel. Grav. **10**, 567 (1979)

[838] A.E. Fischer and J.E. Marsden, in J. Ehlers (Ed.), *Isolated Gravitating Systems in General Relativity: Proceedings of The International School Of Physics "Enrico Fermi" Course LXVII*,

pp. 396-456 (North-Holland, 1979); in S.W. Hawking and W. Israel (Eds.), *General Relativity: An Einstein centenary survey*, pp. 138-211 (Cambridge University Press 1979); Can. J. Math. **39**(1), 193 (1977); Gen. Rel. Grav. **7**, 915 (1976); Sympos. Math. **XIV**, 193 (1974); Gen. Rel. Grav. **5**, 73 (1974); Bull. Amer. Math. Soc. **80**, 479 (1974); Bull. Amer. Math. Soc. **79**, 997 (1973); Gen. Rel. Grav. **4**, 309 (1973); in D. Farnsworth, J. Fink, J. Porter, and A. Thompson (Eds.), *Methods of Local and Global Differential Geometry in General Relativity: Proceedings of the Regional Conference on Relativity held at the University of Pittsburgh, Pittsburgh, Pennsylvania, July 13–17, 1970*, pp. 176-188 (Lect. Notes Phys. **14**, Springer, 1972); J. Math. Phys. **13**, 546 (1972); Comm. Math. Phys. **28**, 1 (1972)

[839] D. Christodoulou and M. Francaviglia, in J. Ehlers (Ed.), *Isolated Gravitating Systems in General Relativity: Proceedings of The International School Of Physics "Enrico Fermi" Course LXVII*, pp. 480-495 (North-Holland, 1979)

[840] Y. Choquét-Bruhat, A.E. Fischer and J.E. Marsden, in J. Ehlers (Ed.), *Isolated Gravitating Systems in General Relativity: Proceedings of The International School Of Physics "Enrico Fermi" Course LXVII*, pp. 322-395 (North-Holland, 1979)

[841] L. Smarr (Ed.), *Sources of Gravitational Radiation* (Cambridge University Press, 1979); in M.D. Papagiannis (Ed.), *Eighth Texas Symposium on Relativistic Astrophysics*, pp. 569-604 (The New York Academy of Sciences, 1977)

[842] M. Francaviglia, Riv. Nuovo Cim. **1**, 1303 (1978)

[843] H.-S. Tsao, Phys. Lett. B **68**, 79 (1977)

[844] K. Eppley, Phys. Rev. D **16**, 1609 (1977)

[845] D. Karmer, Acta Phys. Pol. B **7**, 117 (1976)

[846] S.A. Hojman, K. Kuchař, and C. Teitelboim, Nature (London), Ann. Phys. N.Y. **96**, 88 (1976)

[847] N. Ó Murchada and J.W. York, Jr., Gen. Rel. and Grav. **7**, 257 (1976); Phys. Rev. D **10**, 437 (1974); Phys. Rev. D **10**, 2345 (1974); J. Math. Phys. **14**, 1551 (1973)

[848] M.A.H. MacCallum, in C.J. Isham, R. Penrose, and D.W. Sciama (Eds.), *Quantum Gravity: an Oxford Symposium*, pp. 174-208 (Clarendon Press, 1975)

[849] P.B. Gilkey, J. Diff. Geom. **10**, 601 (1975); Proc. Symp. Pure. Math. **27**, 265 (1975)

[850] A. Čadež, Ann. Phys. N.Y. **83**, 449 (1974)

[851] A. Ashtekar and R. Geroch, Rep. Prog. Phys. **37**, 1211 (1974)

[852] T. Regge and C. Teitelboim, Ann. Phys. NY **88**, 286, (1974); Phys. Lett. B **53**, 101 (1974)

[853] J. Tarski, Ann. Inst. H. Poincaré **20**(1), 95 (1974); Ann. Inst. H. Poincaré **17**, 171 (1972); Ann. Inst. H. Poincaré **17**, 313 (1972); Ann. Inst. H. Poincaré **11**, 331 (1969); J. Math. Phys. **7**, 560 (1966)

[854] G. 't Hooft, Nucl. Phys. B **62**, 444 (1973); in M. Lévy and S. Deser (Eds.), *Recent Developments in Gravitation, Proceedings from Cargèse 1978, Vol. B44 of NATO Advanced Study Institutes Series*, pp. 323–345 (Kluwer Academic Publishers, 1979)

[855] C. Teitelboim, Phys. Rev. D **25**, 3159 (1982); in A. Held (Ed.), *General Relativity and Gravitation. One Hundred Years After the Birth of Albert Einstein*, pp. 195-226 (Plenum Press, 1980); Phys. Lett. B **56**, 376 (1975); Ann. Phys. N.Y. **79**(2), 542 (1973); Phys. Rev. D **8**, 3266 (1973); Phys. Rev. D **5**, 2941 (1972)

[856] C.J. Isham, Abdus Salam, and J.A. Strathdee, Phys. Lett. B **46**, 407 (1973)

[857] F. Lund, Phys. Rev. D **8**, 3247 (1973)

[858] F. Estabrook, H. Wahlquist, S. Christensen, B. DeWitt, L. Smarr, and E. Tsiang, Phys. Rev. D **7**, 2814 (1973)

[859] S.W. Hawking and G.F.R. Ellis, *The Large Scale Structure of Space-Time* (Cambridge University Press, 1973)

[860] B.K. Berger, D.M. Chitre, V.E. Moncrief, and Y. Nutku, Phys. Rev. D **5**, 2467 (1972)

[861] R. Geroch, J. Math. Phys. **13**, 956 (1972); J. Math. Phys. **8**, 782 (1967)

[862] M. Ryan, *Hamiltonian Cosmology* (Springer, 1972)

[863] V. Moncrief and C. Teitelboim, Phys. Rev. D **6**, 966 (1972)

[864] V. Moncrief, Phys. Rev. D **5**, 277 (1972)

[865] G.W. Gibbons, Comm. Math. Phys. **27**, 87 (1972)

[866] G.W. Gibbons and B.F. Schutz, Mon. Not. Roy. Astr. Soc. **159**, 41P (1972)

[867] V.K. Patodi, J. Diff. Geom. **5**, 233 (1971)

[868] J.R. Klauder and E.W. Aslaksen, Phys. Rev. D **2**, 272 (1970)

[869] D.R. Brill and R.H. Gowdy, Rep. Prog. Phys. **33**, 413 (1970)

[870] Y. Choquét-Bruhat, Séminaire Jean Leray **2**, 1 (1969-1970)

[871] C.W. Misner, in M. Carmeli, S.I. Fickler, and L. Witten (Eds.), *Relativity. Proceedings of the Relativity Conference in the Midwest held at Cincinnati, Ohio, June 2-6, 1969*, pp. 55-79 (Plenum Press, 1970); Phys. Rev. Lett. **22**, 1071 (1969); Phys. Rev. **186**, 1319 (1969); Phys. Rev. **186**, 1328 (1969); Ann. Phys. N.Y. **24**, 102 (1963); Phys. Rev. **118**, 1110 (1960); Rev. Mod. Phys. **29**, 497 (1957)

[872] U.H. Gerlach, Phys. Rev. **177**, 1929 (1969)

[873] A. Peres, Phys. Rev. **171**, 1335 (1968)

[874] H.P. McKean and I.M. Singer, J. Diff. Geom. **1**, 43 (1967)

[875] A.B. Komar, Phys. Rev. **153**, 1385 (1967); Phys. Rev. **164**, 1595 (1967)

[876] P.G. Bergmann, in M. Carmeli, S.I. Fickler, and L. Witten (Eds.), *Relativity. Proceedings of the Relativity Conference in the Midwest held at Cincinnati, Ohio, June 2-6, 1969*, pp. 43-53 (Plenum Press, 1970); Phys. Rev. **144**, 1078 (1966); Rev. Mod. Phys. **33**, 510 (1961); Rev. Mod. Phys. **29**, 352 (1957); Nuovo Cim. **3**, 1177 (1956)

[877] M. May and R.H. White, Phys. Rev. **141**, 1232 (1966)

[878] H. Leutwyler, Nuovo Cim. **42**, 159 (1966); Phys. Rev. **134**, B1155 (1964)

[879] S. Hahn and R.W. Lindquist, Ann. Phys. N.Y. **29**, 304 (1964)

[880] R.W. Lindquist, J. Math. Phys. **4**, 938 (1963)

[881] D.R. Brill and R.W. Lindquist Phys. Rev. **131**, 471 (1963)

[882] R.F. Beierlein, D.H. Sharp, and J.A. Wheeler, Phys. Rev. **126**, 1864 (1962)

[883] S. Mandelstam, Proce. Roy. Soc. London A **270**(1342), 346 (1962)

[884] R. Arnowitt, S. Deser, and C.W. Misner, Phys. Rev. **122**, 997 (1961); Phys. Rev. **121**, 1556 (1961); Ann. Phys. N.Y. **11**, 116, (1960); J. Math. Phys. **1**, 434 (1960); Phys. Rev. Lett. **4**, 375 (1960); Nuovo Cim. **15**, 487 (1960); Phys. Rev. **120**, 313 (1960); Phys. Rev. **118**, 1100 (1960); Phys. Rev. **117**, 1595 (1960); Phys. Rev. **116**, 1322 (1959); Phys. Rev. **113**, 745 (1959)

[885] J.L. Anderson, Phys. Rev. **114**, 1182 (1959)

[886] P.W. Higgs, Phys. Rev. Lett. **3**, 66 (1959); Phys. Rev. Lett. **1**, 373 (1958)

[887] A. Lichnerowicz, *Théories relativistes de la gravitation* (Masson, 1955)

[888] F.A.E. Pirani and A. Schild, Phys. Rev. **79**, 986 (1950)

[889] A. Ashtekar, Phys. Rev. Lett. **57**, 2244 (1986); Phys. Rev. D **36**, 1587 (1987)

[890] C. Rovelli and L. Smolin, Nucl. Phys. B **331**, 80 (1990)

[891] T. Jacobson and L. Smolin, Nucl. Phys. B **299**, 295 (1988);
C. Rovelli and L. Smolin, Phys. Rev. Lett. **61**, 1155 (1988); Nucl. Phys. B **133**, 90 (1990); Nucl. Phys. B **442**, 593 (1995); Nucl. Phys. B **456**, 753 (1995);
A. Ashtekar, *New Perspectives in Canonical Gravity* (Bibliopolis, 1988);
A. Ashtekar and R.S. Tate, *Non-perturbative Canonical Gravity* (World Scientific, 1991);
C. Rovelli, Class. Quantum Grav. **6**, 911 (1989); Nucl. Phys. B **331**, 80 (1990); Class. Quantum Grav. **8**, 297 (1991); in A.I. Janis and J.R. Porter (Eds.), *Recent Advances in General Relativity: Essays in Honor of Ted Newman*, pp. 85-102 (Birkhäuser, 1992); in N. Dadhich and J. Narlikar (Eds.), *Gravitation and Relativity: At the Turn of the Millenium. Proceedings of the GR-15 Conference, held at IUCAA, Pune, India, December 16–21, 1997*, pp. 281–331 (Inter-University Centre for Astronomy and Astrophysics, 1998);
A. Ashtekar, C. Rovelli and L. Smolin, Phys. Rev. D **44**, 1740 (1991);
A. Ashtekar and J. Lewandowski, Class. Quantum Grav. **14**, A55 (1997); Class. Quantum Grav. **21**, R53 (2004)

[892] M. Bojowald, Class. Quantum Grav. **17**, 1489 (2000); Class. Quantum Grav. **17**, 1509 (2000); Class. Quantum Grav. **18**, 1055 (2001); Class. Quantum Grav. **18**, 1071 (2001); Class. Quantum Grav. **18**, L109 (2001); Phys. Rev. D **64**, 084018 (2001); Phys. Rev. Lett. **87**, 121301 (2001); Phys. Rev. Lett. **86**, 5227 (2001); Phys. Rev. Lett. **89**, 261301 (2002); Class. Quantum Grav. **19**, 5113 (2002); Class. Quantum Grav. **19**, 2717 (2002); Class. Quantum Grav. **20**, 2595 (2003); J. Phys. Conf. Ser. **24**, 77 (2005); Class. Quantum Grav. **23**, 987 (2006); Nature Phys. **3**(8), 523 (2007); Gen. Rel. Grav. **40**, 2659 (2008); Phys. Rev. Lett. **100**, 221301 (2008); Living Rev. Rel. **11**, 4 (2008); Class. Quantum Grav. **26**, 075020 (2009); AIP Conf. Proc. **1196**, 62 (2009); AIP Conf. Proc. **1256**, 66 (2010)

[893] A. Ashtekar and J. Lewandowski, in J. Baez (Ed.), *Knots and Quantum Gravity*, pp. 21-61 (Oxford University Press, 1994)

[894] A. Ashtekar and C. Isham, Class. Quantum Grav. **9**, 1433 (1992)

[895] R. Giles, Phys. Rev. D **24**, 2160 (1981)

[896] A. Ashtekar and J. Lewandowski, J. Math. Phys. **36**, 2170 (1995)

[897] M. Bojowald, D. Mulryne, W. Nelson, and R. Tavakol, Phys. Rev. D **82**, 124055 (2010);
M. Bojowald, J.D. Reyes, R. Tibrewala, Phys. Rev. D **80**, 084002 (2009);
M. Bojowald, G.M. Hossain, M. Kagan, S. Shankaranarayanan, Phys. Rev. D **78**, 063547 (2008); Phys. Rev. D **79**, 043505 (2009), Phys. Rev. D **82**, 109903 (2010);
M. Bojowald and J.D. Reyes, Class. Quantum Grav. **26**, 035018 (2009);
M. Bojowald and R. Das, Class. Quantum Grav. **25**, 195006 (2008); Phys. Rev. D **75**, 123521 (2007);
M. Bojowald, T. Harada, and R. Tibrewala, Phys. Rev. D **78**, 064057 (2008);
M. Bojowald, R. Das, and R.J. Scherrer, Phys. Rev. D **77**, 084003 (2008);
M. Bojowald and G.M. Hossain, Phys. Rev. D **77**, 023508 (2008);
M. Bojowald, H. Hernandez, and A. Skirzewski, Phys. Rev. D **76**, 063511 (2007);
M. Bojowald, D. Cartin, and G. Khanna, Phys. Rev. D **76**, 064018 (2007);
M. Bojowald, H. Hernandez, M. Kagan, and A. Skirzewski, Phys. Rev. D **75**, 064022 (2007);
M. Bojowald and A. Rej, Class. Quantum Grav. **22**, 3399 (2005);
M. Bojowald and H.A. Morales-Tecotl, in N. Breton, J.L. Cervantes-Cota, and M. Salgado (Eds.), *Proceedings of 5th Mexican School On Gravitation And Mathematical Physics: The Early Universe And Observational Cosmology (DGFM 2002), 24-29 Nov 2002, Playa del Carmen, Quintana Roo, Mexico*, pp. 421-462 (Lect. Notes Phys. **646**, Springer, 2004);
A. Ashtekar, M.Campiglia, and A. Henderson, Phys. Rev. D **82**,

124043 (2010);
A. Ashtekar, M. Bojowald, and J. Lewandowski, Adv. Theor. Math. Phys. **7**, 233 (2003);
A. Ashtekar, W. Kaminski, and J. Lewandowski. Phys. Rev. D **79**, 064030 (2009);
A. Henderson, C. Rovelli, F. Vidotto, and E. Wilson-Ewing, Class. Quantum Grav. **28**, 025003 (2011);
M. Domagala, K. Giesel, W. Kaminski, and J. Lewandowski, Phys. Rev. D **82**, 104038 (2010);
W. Kaminski, M. Kisielowski, and J. Lewandowski, Class. Quantum Grav. **27**, 095006 (2010);
W. Kaminski, J. Lewandowski, and T. Pawlowski, Class. Quantum Grav. **26**, 035012 (2009); Class. Quantum Grav. **26**, 245016 (2009);
Ł. Szulc, Phys. Rev. D **78**, 064035 (2008);
I. Agullo, G.J.F. Barbero, E.F. Borja, J. Diaz-Polo, and E.J.S. Villasenor, Phys. Rev. D **82**, 084029 (2010); Phys. Rev. D **80**, 084006 (2009);
D.-W. Choiu and L.-F. Li, Phys. Rev. D **80**, 043512 (2009);
D.-W. Chiou and M. Geiller, Phys. Rev. D **82**, 064012 (2010);
A. Corichi, T. Vukasinac, and J.A. Zapata, AIP Conf. Proc. **1256**, 172 (2010);
C. Beetle and J. Engle, Class. Quantum Grav. **27**, 235024 (2010);
C. Rovelli and S. Speziale, Phys. Rev. D **82**, 044018 (2010);
Y. Ma, C. Soo, and J. Yang, Phys. Rev. D **81**, 124026 (2010);
L. Fatibene, M. Ferraris, and M. Francaviglia, Class. Quantum Grav. **27**, 185016 (2010);
S. Mercuri, PoS ISFTG, 016 (2009);
R. Gambini and J. Pullin, J. Phys. Conf. Ser. **189**, 012034 (2009); Adv. Sci. Lett. **2**, 251 (2009); Class. Quantum Grav. **26**, 035002 (2009);
H. Sahlmann, Class. Quantum Grav. **27**, 225007 (2010);
J. Brunnemann and D. Rideout, Class. Quantum Grav. **27**, 205008 (2010);
M. Han, Class. Quantum Grav. **27**, 215009 (2010);
F. Cianfrani and G. Montani, Phys. Rev. D **80**, 084045 (2009);
J.F. Barbero G., J. Phys. Conf. Ser. **175**, 012005 (2009);

A. Perez, AIP Conf. Proc. **1132**, 386 (2009);
Ch.-A. Li, J.-J. Jiang, and J.-Q. Su, Sci. China G **52**, 1179 (2009);
A. DeBenedictis, Can. J. Phys. **87**, 255 (2009);
J.M. Garcia-Islas, Can. J. Phys. **88**, 223 (2010);
S.J. Gates, Jr., S.V. Ketov, and N. Yunes, Phys. Rev. D **80**, 065003 (2009);
M. Bojowald, J.D. Reyes, and R. Tibrewala, Phys. Rev. D **80**, 084002 (2009);
D. Mamone and C. Rovelli, Class. Quantum Grav. **26**, 245013 (2009);
J. Grain and A. Barrau, Phys. Rev. Lett. **102**, 081301 (2009);
J. Engle, Class. Quantum Grav. **27**, 035003 (2010);
F. Cianfrani and G. Montani, Phys. Rev. Lett. **102**, 091301 (2009);
M. Bojowald and J.D. Reyes, Class. Quantum Grav. **26**, 035018 (2009);
S. Speziale, Adv. Sci. Lett. **2**, 280 (2009);
C, Perini, C. Rovelli, S. Speziale, Phys. Lett. B **682**, 78 (2009);
E. Bianchi, Nucl. Phys. B **807**, 591 (2009);
K. Giesel and T. Thiemann, Class. Quantum Grav. **27**, 175009 (2010); Class. Quantum Grav.**23**, 5667 (2006); Class. Quantum Grav. **23**, 5693 (2006);
J. Brunnemann and T. Thiemann, Class. Quantum Grav. **23**, 1429 (2006); Class. Quantum Grav. **23**, 1395 (2006); Class. Quantum Grav. **23**, 1289 (2006);
B. Bahr and T. Thiemann, Class. Quantum Grav. **26**, 235022 (2009); Class. Quantum Grav. **26**, 045011 (2009); Class. Quantum Grav. **26**, 045012 (2009); Class. Quantum Grav. **24**, 2109 (2007);
B. Dittrich and T. Thiemann, J. Math. Phys. **50**, 012503 (2009); Class. Quantum Grav. **23**, 1143 (2006); Class. Quantum Grav. **23**, 1143 (2006); Class. Quantum Grav. **23**, 1089 (2006); Class. Quantum Grav. **23**, 1067 (2006); Class. Quantum Grav. **23**, 1025 (2006);
T. Thiemann, in D. Oriti (Ed.), *Approaches to Quantum Gravity. Toward a New Understanding of Space, Time, and Matter*, pp. 235-252 (Cambridge University Press, 2009); Int. J. Mod. Phys.

A **23**, 8, 1113 (2008); in I.-O. Stamatescu and E. Seiler (Eds.), *Approaches to Fundamental Physics: An Assessment of Current Theoretical Ideas*, pp. 185-263 (Lect. Notes Phys. **721**, Springer, 2007); Class. Quantum Grav. **23**, 2211 (2006); in D. Giulini, C. Kiefer, and C. Lämmerzahl (Eds.), *Quantum Gravity. From Theory To Experimental Search*, pp. 41-135 (Lect. Notes Phys. **631**, Springer, 2003);
A. Ashtekar, Gen. Rel. Grav. **41**, 707 (2009);
C. Fleischhack, in B. Fauser, J. Tolksdorf, and E. Zeidler (Eds.), *Quantum Gravity. Mathematical Models and Experimental Bounds*, pp. 203-220 (Birkhäuser, 2007);
A. Perez, in D. Oriti (Ed.), *Approaches to Quantum Gravity. Toward a New Understanding of Space, Time, and Matter*, pp. 272-289 (Cambridge University Press, 2009)

[898] I. M. Gelfand and M. A. Naimark, Rec. Math. [Mat. Sbornik] N.S. **12**(54), 197 (1943)

[899] I. Segal, Trans. Amer. Math. Soc. **61**, 69 (1947); Bull. Amer. Math. Soc. **53**, 73 (1947)

[900] R. d'Inverno, in G.S. Hall and J.R. Pulham (Eds.), *General Relativity: Proceedings of the Forty Sixth Scottish Universities Summer School in Physics, Aberdeen, July 1995*, pp. 331-376 (NATO ASI/SUSSP Publications/IOP Publishing, 1996);
B. Brügmann, Ann. Phys. (Leipzig) **9**, 227 (2000);
L. Lehner, Class. Quantum Grav. **18**, R25 (2001); e-Print: arXiv:gr-qc/0106072;
U. Sperhake, *Non-linear numerical Schemes in General Relativity*, PhD thesis, University of Southampton (2001), e-Print: arXiv:gr-qc/0201086;
C. Bona and C. Palenzuela-Luque, *Elements of Numerical Relativity: From Einstein's Equations to Black Hole Simulations* (Springer, 2005);
J. Frauendiener, D.J.W. Giulini, and V.Perlick (Eds.), *Analytical and Numerical Approaches to Mathematical Relativity: With a Foreword by Roger Penrose* (Springer, 2006);
E. Gourgoulhon, e-Print: arXiv:gr-qc/0703035.

[901] G. Darmois, Mém. Sci. Math. **25**, 1 (1927)

[902] A. Lichnerowicz, *Sur certains problèmes globaux relatifs au système des équations d'Einstein (Thèse)* (Actual. Sci. Ind. **833**, Hermann, 1939)

[903] A. Lichnerowicz, J. Math. Pures Appl. **23**, 37 (1944)

[904] Y. Fourès-Bruhat, Acta Math. **88**, 141 (1952)

[905] Y. Choquét-Bruhat, J. Rat. Mech. Anal. **5**, 951 (1956)

[906] J.W. York, Phys. Rev. Lett. **28**, 1082 (1972); J. Math. Phys. **13**, 125 (1972)

[907] N. Ó Murchadha and J.W. York, Phys. Rev. D **10**, 428 (1974)

[908] L. Smarr, Ann. N.Y. Acad. of Sci. **302**, 569 (1977)

[909] L. Smarr and J.W. York, Jr., Phys. Rev. D **17**, 2529 (1978); Phys. Rev. D **17**, 1945 (1978)

[910] J.M. Bardeen and T. Piran, Phys. Rep. **96**, 206 (1983)

[911] T. Nakamura, Prog. Theor. Phys. **65**, 1876 (1981)

[912] R.F. Stark and T. Piran, Phys. Rev. Lett. **55**, 891 (1985)

[913] T. Nakamura, K. Oohara, and Y. Kojima, Prog. Theor. Phys. Suppl. **90**, 1 (1987)

[914] C. Bona and J. Massó, Phys. Rev. D **40**, 1022 (1989); Phys. Rev. Lett. **68**, 1097 (1992); Phys. Rev. D **38**, 2419 (1988)

[915] Y. Choquét-Bruhat and J.W. York, Jr., C. R. Acad. Sci. Paris Sér. I **321**, 1089, (1995)

[916] L.E. Kidder, M.A. Scheel, and S.A. Teukolsky, Phys. Rev. D **64**, 064017 (2001)

[917] M. Shibata and T. Nakamura, Phys. Rev. D **52**, 5428 (1995)

[918] T.W. Baumgarte and S.L. Shapiro, Phys. Rev. D **59**, 024007 (1999)

[919] J.W. York, Phys. Rev. Lett. **82**, 1350 (1999)

[920] M. Shibata, Prog. Theor. Phys. **101**, 1199 (1999); Phys. Rev. D **60**, 104052 (1999)

[921] S.A. Hayward, Phys. Rev. D **61**, 101503 (2000)

[922] S. Bonazzola, E. Gourgoulhon, P. Grandclément, and J. Novak, Phys. Rev. D **70**, 104007 (2004)

[923] H. Stephani, *Relativity: An Introduction to Special and General Relativity* (3rd ed., Cambridge University Press, 2004)

[924] A.L. Besse, *Einstein Manifolds* (Springer, 1987)

[925] E. Calabi, Ann. Sci. Ec. Norm. Sup. **12**, 268 (1979); Proc. Internat. Congress. Math. Amsterdam **2**, 206 (1954); in *Algebraic Geometry and Topology. A symposium in honor of S. Lefschetz*, pp. 78-89 (Princeton University Press, 1957); Trans. Amer. Math. Soc. **67**, 401 (1958)
S.T. Yau, Enseignement Math. 33, 109 (1986); in W.A Bardeen and A.R While (Eds.), *Proceedings of Argonne Symposium on Anomalies, Geometry, Topology*, pp. 395–406 (World Scientific, 1985); Comm. Pure and Appl. Math. **31**(3), 339 (1979); Amer. J. Math. **100**, 197 (1978); Proc. Nat. Acad. Sci. USA **74**, 1798 (1977); in H-D. Cao and S.-H. Yau (Eds.), *Surveys in Differential Geometry, Vol. 13, Geometry, Analysis, and Algebraic Geometry: Forty Years of the Journal of Differential Geometry*, pp. 277-318 (Surv. Differ. Geom. **13**, International Press, 2009)

[926] E. Calabi, Ann. Ecol. Norm. Sup. **12**, 269 (1979);
A. Beauville, J. Diff. Geom. **18**, 755 (1983)

[927] M.H. Emam, Class. Quantum Grav. **27**, 163001 (2010); C.M. Hull, U. Lindstrom, M. Rocek, R. von Unge, and M. Zabzine, JHEP **1008**, 060 (2010); H. Lu and Z.-L. Wang, J. Geom. Phys. **60**, 1741 (2010); T. Eguchi and K. Hikami, Lett. Math. Phys. **92**, 269 (2010); H. Lu, Y. Pang and Z.-L. Wang, Class. Quantum Grav. **27**, 155018 (2010); Y.-H. He, S.-J. Lee, and A. Lukas, JHEP **1005**, 071 (2010); J. Fu, Z. Wang, and D. Wu Math. Res. Lett. **17**, 887

(2010); A.D. Popov, Nucl. Phys. B **828**, 594 (2010); R. Heluani and M. Zabzine, Adv. Math. **223**, 1815 (2010); I. Biswas and B. McKay J. Geom. Phys. **60**, 661 (2010); A.Q. Velez and A. Boer, Comm. Math. Phys. **297**, 597 (2010); M. Kreuzer, Ukr. J. Phys. **55**, 613 (2010); F. Loran and H. Soltanpanahi, Adv. Theor. Math. Phys. **13**, 637 (2009); M. Eto, T. Fujimori, S.B. Gudnason, M.Nitta, and K. Ohashi, Nucl. Phys. B **815**, 495 (2009); H. Jockers and M. Soroush, Comm. Math. Phys. **290**, 249 (2009); Nucl. Phys. B **821**, 535 (2009); A.S. Haupt, A. Lukas, K.S. Stelle, JHEP **0905**, 069 (2009); S. Rollenske, R.P. Thomas, Journal of Topology **2**, 405 (2009); Y.-G. Oh and K. Zhu, Asian J. Math. **13**, 323 (2009); B. Haghighat and A. Klemm, JHEP **0901**, 029 (2009); V. Tosatti, J. Eur. Math. Soc. **11**, 755 (2009); F.P. Correia, JHEP **0912**, 004 (2009); M. Kreuzer, Fortsch. Phys. **57**, 625 (2009); H. Ooguri and M. Yamazaki, Phys. Rev. Lett. **102**, 161601 (2009); Comm. Math. Phys. **292**, 179 (2009); E. Palti, JHEP **0904**, 099 (2009); S.P. de Alwis Phys. Lett. B **675**, 377 (2009); R.S. Garavuso, M. Kreuzer, and A. Noll JHEP **0903**, 007 (2009); M.C. Brambilla, Rev. Mat. Compl. **22**, 1 (2009); F. Witt, Rend. Sem. Mat. Univ. Politec. Torino **66**, 1 (2008); K. Ueda and M. Yamazaki, JHEP **0812**, 045 (2008); V. Braun, T. Brelidze, M.R. Douglas, and B.A. Ovrut, JHEP**0807**, 120 (2008); JHEP **0805**, 080 (2008); A. Ricco, Int. J. Mod. Phys. A **23**, 2187 (2008); L. Covi, M. Gomez-Reino, C. Gross, J. Louis, G.A. Palma, and C.A. Scrucca JHEP **0806**, 057 (2008); S. McReynolds, Mod. Phys. Lett. A **23**, 1841 (2008); S. Mizoguchi, Phys. Lett. B **669**, 352 (2008); N. Halmagyi and T. Okuda, JHEP **0803**, 028 (2008); M. Cvetic and T. Weigand, Phys. Rev. Lett. **100**, 251601 (2008); M. Cicoli, J.P. Conlon, and F. Quevedo, JHEP **0801**, 052 (2008); A. Misra and P. Shukla, Nucl. Phys. B **799**, 165 (2008); M.R. Douglas, R.L. Karp, S. Lukic, and R. Reinbacher, J. Math. Phys. **49**, 032302 (2008); C. Doran, B. Greene, S. Judes, Comm. Math. Phys. **280**, 675 (2008); L. Grant and K. Narayan, Class. Quantum Grav. **25**, 045010 (2008); V. Tosatti, B. Weinkove, S.-T. Yau, Proc. London Math. Soc. **97**, 401 (2008); A. Klemm and R. Pandharipande, Comm. Math. Phys. **281**, 621 (2008); V. Bouchard and R. Donagi, Comm. Numb. Theor. Phys. **2**, 1 (2008); A. Klemm and M. Marino, Comm. Math. Phys. **280**, 27 (2008);

H. Fang, Z. Lu, and K.-I. Yoshikawa, J. Diff. Geom. **80**(2), 175 (2008); O. Iyama and I. Reiten, Am. J. Math. **130**(4), 1087 (2008); S. Cynk and C. Meyer, Rocky Mountain J. Math. **38**(6), 1937 (2008); Int. J. Math. **18**(3), 331 (2007); P. Grange and S. Schafer-Nameki JHEP **0710**, 052 (2007); O.A.P. Mac Conamhna, Phys. Rev. D **76**, 106010 (2007); R. Sriharsha, JHEP **0703**, 095 (2007); D. Conti, J. Geom. Phys. **57**(12), 2483 (2007); J. Gomis and T. Okuda, JHEP **0707**, 005 (2007); JHEP **0702**, 083 (2007); A. Fayyazuddin, Class. Quantum Grav. **24**, 3151 (2007); D.Tsimpis, JHEP **0703**, 099 (2007); R. D'Auria, S. Ferrara, and M. Trigiante, Nucl. Phys. B **780**, 28 (2007); W. Chen, M. Cvetic, H. Lu, C.N. Pope, and J.F. Vazquez-Poritz, Nucl. Phys. B **785**, 74 (2007); B. Andreas and G. Curio, J. Geom. Phys. **57**, 2249 (2007); R. Ahl Laamara, A. Belhaj, L.B. Drissi, and E.H. Saidi, Nucl. Phys. B **776**, 287 (2007); P.S. Aspinwall, J. Math. Phys. **48**, 082304 (2007); J.P. Conlon, D. Cremades, and F. Quevedo, JHEP **0701**, 022 (2007); D.-E. Diaconescu, A. Garcia-Raboso, R.L. Karp, and K. Sinha Adv. Theor. Math. Phys. **11**, 471 (2007); S. Bellucci, S. Krivonos, and A. Shcherbakov, Phys. Lett. B **645**, 299 (2007); T. Oota and Y. Yasui, Phys. Lett. B **639**, 54 (2006); R. Blumenhagen, S. Moster, and T. Weigand, Nucl. Phys. B **751**, 186 (2006); S. Okada, Int. Math. Res. Not. **2006**, 58743 (2006); S. Bellucci, S. Ferrara, A. Marrani, and A. Yeranyan, Riv. Nuovo Cim. **29**(5), 1 (2006); P. Kaura and A. Misra, Fortsch. Phys. **54**, 1109 (2006); D.-E. Diaconescu, A. Garcia-Raboso, and K. Sinha, JHEP **0606**, 058 (2006); R. Ahl Laamara, A. Belhaj, L.B. Drissi, and E.H. Saidi, J. Phys. A **39**, 5965 (2006); M. Aganagic, D. Jafferis, and N. Saulina, JHEP **0612**, 018 (2006); U. Bruzzo and A. Ricco, Lett. Math. Phys. **76**, 57 (2006); B. Forbes and M. Jinzenji, JHEP **0603**, 061 (2006); T. Eguchi and Y. Tachikawa, JHEP **0601**,100 (2006); Z.Lu and X. Sun, Comm. Math. Phys. **261**, 297 (2006); J. Inst. Math. Jussieu **3**(2), 185 (2004); T. Bridgeland, Comm. Math. Phys. **266**, 715 (2006); A. Fayyazuddin, T.Z. Husain, and I. Pappa, Phys. Rev. D **73**, 126004 (2006); N. Halmagyi, A. Sinkovics, and P. Sulkowski, JHEP **0601**, 040 (2006); T. Pantev and E. Sharpe, Nucl. Phys. B **733**, 233 (2006); L.N. Lipatov, A. Sabio Vera, V.N. Velizhanin, and G.G. Volkov, Int. J. Mod. Phys. A **21**, 2953 (2006);

M.B. Schulz, JHEP **0605**, 023 (2006); M. Schuett, Canad. Math. Bull. **49**(2), 296 (2006); Coll. Math. **55**(2), 219 (2004); J. Bryan and R. Pandharipande, Geom. Topol. Monogr. **8**, 97 (2006); H. Jockers and J. Louis, Nucl. Phys. B **705**, 167 (2005); T.W. Grimm, Fortsch. Phys. **53**, 1179 (2005); H. Jockers, Fortsch. Phys. **53**, 1087 (2005); P. Koerber, JHEP **0508**, 099 (2005); D. Berenstein, C.P. Herzog, P. Ouyang, and S. Pinansky, JHEP **0509**, 084 (2005); A. Bilal and S. Metzger, JHEP **0508**, 097 (2005); V. Balasubramanian, P. Berglund, J.P. Conlon, and F. Quevedo, JHEP **0503**, 007 (2005); T.W. Grimm and J. Louis, Nucl. Phys. B **718**, 153 (2005); T. Kimura, Nucl. Phys. B **711**, 163 (2005); M. Rocek and N. Wadhwa, Adv. Theor. Math. Phys. **9**, 315 (2005); A. Belhaj, L.B. Drissi, J. Rasmussen, E.H. Saidi, and A. Sebbar, J. Phys. A **38**, 6405 (2005); B. Andreas and D.H. Ruiperez, Adv. Theor. Math. Phys. **9**, 253 (2005); A. Klemm, M. Kreuzer, E. Riegler, and E. Scheidegger, JHEP **0505**, 023 (2005); T. Okuda, JHEP **0503**, 047 (2005); M. Arai, M. Nitta, and N. Sakai, Prog. Theor. Phys. **113**, 657 (2005); D. Mülsch and B. Geyer, Int. J. Geom. Meth. Mod. Phys. **2**, 409 (2005); A. Misra and A.Nanda, Fortsch. Phys. **53**, 246 (2005); D. Huybrechts, Int. J. Math. **16**, 13 (2005); A. Misra, Int. J. Mod. Phys. A **20**, 2059 (2005); Fortsch. Phys. **52**, 831 (2004) B. Banos and A. Swann, Class. Quantum Grav. **21**, 3127 (2004); M. Hssaini, M. Kessabi, B. Maroufi, and M.B.Sedra, Afr. J. Math. Phys. **1**, 301 (2004); J.P. Conlon and F. Quevedo, JHEP **0410**, 039 (2004); U.H. Danielsson, M.E. Olsson, and M. Vonk, JHEP **0411**, 007 (2004); V. Bouchard, B. Florea, and M. Marino, JHEP **0412**, 035 (2004); T. Eguchi and Y. Sugawara, JHEP **0405**, 014 (2004); T.W. Grimm and J. Louis, Nucl. Phys. B **699**, 387 (2004); G. Volkov, Int. J. Mod. Phys. A **19**, 4835 (2004); M. Lynker, R. Schimmrigk, and S. Stewart, Nucl. Phys. B **700**, 463 (2004); V. Braun, B.A. Ovrut, T. Pantev, and R. Reinbacher, JHEP **0412**, 062 (2004); M. Grana, T.W. Grimm, H. Jockers, and J. Louis, Nucl. Phys. B **690**, 21 (2004); A. Giryavets, S. Kachru, P.K. Tripathy, and S.P. Trivedi, JHEP **0404**, 003 (2004); H. Lu, C.N. Pope, and K.S. Stelle, JHEP **0407**, 072 (2004); L. Jarv, T. Mohaupt, and F. Saueressig, JCAP **0402**, 012 (2004); F. Ferrari, Adv. Theor. Math. Phys. **7**, 619 (2004); R. Donagi, B.A. Ovrut,

T. Pantev, and R. Reinbacher, JHEP **0401**, 022 (2004); S. Govindarajan and J. Majumder, Pramana **62**, 711 (2004); M. Grana, R. Minasian, M. Petrini, and A. Tomasiello, Compt. Ren. Phys. **5**, 979 (2004); JHEP **0408**, 046 (2004); A.A. Malykh, Y. Nutku, M.B. Sheftel, J. Phys. A **36**, 10023 (2003); M.L. Barberis, Math. Phys. Anal. Geom. **6**, 1 (2003); T. Johnsen and A.L. Knutsen, Comm. Algebra **31**(8), 3917 (2003); L.R. Huiszoon and K. Schalm, JHEP **0311**, 019 (2003); D. Kastor, JHEP **0307**, 040 (2003); M.Dunajski and L.J. Mason, J. Math. Phys. **44**, 3430 (2003); Comm. Math. Phys. **213**, 641 (2000); W.-D. Ruan, J. Sympl. Geom. **1**(3), 435 (2002); M. Dunajski and P. Tod, Differ. Geom. Appl. **14**, 39 (2001); M. Cvetic, G.W. Gibbons, H. Lu, and C.N. Pope, Nucl. Phys. B **617**, 151 (2001); Z.Lu, J. Geom. Anal. **11**(1), 103 (2001); J. Geom. Anal. **11**(4), 635 (2001); G. Grantcharov and Y.S. Poon, Comm. Math. Phys. **213**, 19 (2000); G. Papadopoulos and J. Gutowski, Nucl. Phys. B **551**, 650 (1999); J.P. Gauntlett, G.W. Gibbons, G. Papadopoulos, and P.K. Townsend, Nucl. Phys. B **500**, 133 (1997); L. Rozansky and E. Witten, Selecta Math. **3**, 401 (1997); B. Greene, e-Print: arXiv:hep-th/9702155

[928] W.K. Clifford, Amer. J. Math. **1**, 350 (1878);
R. Brauer and H. Weyl, Amer. J. Math. **57**, 425 (1935);
É. Cartan, *Théorie des spineurs* (Actualités Scientifiques et Industrielles, No. **643** et 701, Hermann, 1938);
C. Chevalley, *The Algebraic Theory of Spinors* (Columbia University Press, 1954);
C.T.C. Wall, J. Reine Ang. Math. **213**, 187 (1963);
M.F. Atiyah, R. Bott, and A. Shapiro, Topology **3** (Suppl. 1), 3 (1964);
M.M. Postnikov, *Lie Groups and Lie Algebras* (Mir, 1986)
P. Budinich and A. Trautman, *The Spinorial Chessboard* (Trieste Notes in Physics, Springer, 1988);
J.E. Gilbert and M.A.M. Murray, *Clifford algebras and Dirac operators in harmonic analysis* (Cambridge University Press, 1991);
A. Trautman and K. Trautman, J. Geom. Phys. **15**, 1 (1994);
I.R. Porteous, *Clifford Algebras and The Classical Groups* (Cambridge Studies in Advanced Mathematics vol. **50**,

Cambridge University Press, 1995);
A. Trautman, Contemporary Math. **203**, 3 (1997);
P. Lounesto, *Clifford Algebras and Spinors* (2nd ed., London Mathematical Society, Lect. Notes Series **286**, Cambridge University Press, 2001);
V.V. Fernández, A.M. Moya, and W.A. Rodrigues Jr., Adv. Appl. Clifford Alg. **11**, 1 (2001);
J.C. Baez, Bull. Amer. Math. Soc. (N. S.) **39**, 145 (2002);
A. Tarutman, in J.-P. Françoise, G.L. Naber and Tsou S.T. (Eds.), *Encyclopedia of Mathematical Physics, Vol. 1*, pp. 518-530 (Elsevier, 2006);
P. Anglès, *Conformal Groups in Geometry and Spin Structures* (Birkhäuser, 2008);
J. Gallier, e-Print: arXiv: 0805.0311 [math.GM]

[929] Yu.A. Brychkov, *Handbook of Special Functions: Derivatives, Integrals, Series, and Other Formulas* (CRC Press, 2008);
S. Kanemitsu and H. Tsukada, *Vistas of Special Functions* (World Scientific, 2007);
G.E. Andrews, R. Askey, and R. Roy, *Special Functions* (Cambridge University Press 1999);
A.F. Nikiforov and V.B. Uvarov, *Special Functions of Mathematical Physics: A Unified Introduction with Applications* (Birkhäuser, 1988);
Z.X. Wang and D.R. Guo, *Special Functions* (World Scientific, 1989);
A. Wawrzyńczyk, *Group Representations and Special Functions* (Kluwer Academic Publishers/PWN-Polish Scientific Publishers, 1984);
W.W. Bell, *Special Functions for Scientists and Engineers* (D. Van Nostrand, 1968);
W. Miller, Jr., *Lie Theory and Special Functions* (Academic Press, 1968);
N.N. Lebedev, *Special Functions and Their Applications* (Prentice-Hall, 1965)

[930] B.P. Rynne and M.A. Youngson, *Linear Functional Analysis* (2nd ed., Springer, 2008);

V.N. Hansen, *Functional Analysis: Entering Hilbert Space* (World Scientific, 2006);
K. Saxe, *Beginning Functional Analysis* (Springer, 2002);
K. Atkinson and W. Han, *Theoretical Numerical Analysis - A Functional Analysis Framework* (Springer, 2001);
E. Kreyszig, *Introductory Functional Analysis with Applications* (John Wiley & Sons, 1978);
R.F. Curtain and A.J. Pritchard, *Functional Analysis in Modern Applied Mathematics* (Academic Press, 1977);
P. Roman, *Some Modern Mathematics for Physicists and Other Outsiders. An Introduction to Algebra, Topology, and Functional Analysis* Vols. 1 & 2 (Pergamon Press, 1975);
S.K. Berberian, *Lectures in Functional Analysis and Operator Theory* (Springer, 1974);
W. Rudin, *Functional Analysis* (McGraw-Hill, 1973)

[931] H. Lebesgue, *Leçons sur l'intégraion et le recherche des fonctions primitives* (Gauthier-Villars, 1904; 2nd ed. 1928); Ann. Fac. Sci. Univ. Toulouse Ser. 3 **1**, 25 (1909);
N. Dunford and J.T. Schwartz, *Linear Operators* (Wiley, 1958)

[932] I. Białynicki-Birula and Z. Białynicka-Birula, *Quantum Electrodynamics* (Pergamon Press/PWN-Polish Scientific Publishers, 1975)

[933] L.V. Ahlfors, *Complex Analysis. An Introduction To The Theory Of Analytic Functions Of One Complex Variable* (McGraw-Hill, 1966);
R. Nevanlinna and V. Paatero, *Introduction to Complex Analysis* (Addison-Wesley, 1969);
W. Rudin, *Real and Complex Analysis* (3rd ed., McGraw-Hill, 1987);
T.W. Gamelin, *Complex Analysis* (Springer, 2001);
M. Ablowitz and A.S. Fokas, *Complex Variables: Introduction and Applications* (2nd ed., Cambridge University Press, 2003);
E. Freitag and R. Busam, *Complex Analysis* (Springer, 2005)

[934] A. Messiah, *Quantum Mechanics* (North-Holland, 1962);
L.I. Schiff, *Quantum Mechanics* (3rd ed., Mc-Graw-Hill, 1968);

E. Merzbacher, *Quantum Mechanics* (2nd ed., John Wiley & Sons, 1970);
J.E.G. Farine, *Quantum Theory of Scattering Processes* (Pergamon Press, 1973);
W.O. Amrein, J.M. Jauch, and K.B. Sinha, *Scattering Theory in Quantum Mechanics* (W.A. Benjamin, 1977);
R.L. Liboff, *Introductory Quantum Mechanics* (Addison-Wesley, 1980);
A. Galindo and P. Pascual, *Quantum Mechanics* (Springer, 1990);
L.D. Landau and E.M. Lifshitz, *Course of Theoretical Physics. Volume 3: Quantum Mechanics: Non-Relativistic Theory* (3rd ed., Pergamon Press, 1991);
I. Białynicki-Birula, M. Cieplak, and J. Kamiński, *Theory of Quanta* (Oxford University Press, 1992);
J.J. Sakurai, *Modern Quantum Mechanics* (Addison-Wesley, 1994);
R. Shankar, *Quantum Mechanics* (2nd ed., Plenum Press, 1994);
D.J. Griffits, *Introduction to Quantum Mechanics* (Prentice Hill, 1995);
H.A. Bethe and R. Jackiw, *Intermediate Quantum Mechanics* (3rd ed., Westview, 1997);
W. Greiner, *Quantum Mechanics: An Introduction* (4th ed., Springer, 2001);
J. Schwinger, *Quantum Mechanics: Symbolism of Atomic Measurements* ed. by B.-G. Englert (Springer, 2001);
J.-L. Basdevant and J. Dalibard, *Quantum Mechanics* (Springer, 2002);
K. Gottfried and T.-M. Yan, *Quantum Mechanics: Fundamentals* (2nd ed., Springer, 2003);
A.C. Phillips, *Introduction to Quantum Mechanics* (John Wiley & Sons, 2003);
C. Cohen-Tannoudji, B. Diu, and F. Laloe, *Quantum Mechanics* (Wiley, 2006);
F. Schwabl, *Quantum Mechanics* (4th ed., Springer, 2007);
G. Auletta, M. Fortunato, and G. Parisi, *Quantum Mechanics* (Cambridge University Press, 2009);
N. Zettili, *Quantum Mechanics: Concepts and Applications* (John

Wiley & Sons, 2009)

[935] M. Bôcher, *An Introduction to the Study of Integral Equations* (Cambridge University Press, 1909);
D. Hilbert, *Grundzüge einer allgemeinen Theorie der linearen Integralgleischungen* (Teubner, 1912)
T. Lalesco, *Introduction a la théorie des équations intégrales. Avec une préface de É. Picard* (A. Hermann et Fils, 1912);
V. Volterra, *Leçons sur les équations intégrales et les équations intégro-différentielles* (Gauthier-Villars, 1913); *Theory of Functionals and of Integral and Integro-Differential Equations* (Blackie & Son, 1930); Rend. Accad. Lincei **5**, 177 (1896); Rend. Accad. Lincei **5**, 289 (1896); Ann. Mat. Pura Appl. Ser. 2 **25**, 139 (1897);
A. Kneser, *Die Integralgleichungen und ihre Anwendungen in der mathematischen Physik* (Vieweg, 1922);
H. Hochstadt, *Integral Equations* (John Wiley & Sons, 1973);
M.L. Krasnov, *Integral Equations: Introduction to the theory* (in Russian) (Nauka, 1975);
A.B. Mingarelli, *Volterra–Stieltjes Integral Equations and Generalized Ordinary Differential Expressions* (Lect. Notes Math. **989**, Springer-Verlag, 1983);
D. Porter and D.S.G. Stirling, *Integral equations: A practical treatment, from spectral theory to applications* (Cambridge University Press, 1990);
A.D. Polyanin and A.V. Manzhirov, *Handbook of Integral Equations* (2nd ed., Chapman & Hall/CRC, 2008)

[936] D.N. Zubarev, V. Morozov, and G. Röpke, *Statistical Mechanics of Nonequilibrium Processes* (John Wiley & Sons, 1996)

[937] K. Huang, *Statistical Mechanics* (2nd ed., John Wiley & Sons, 1987)

[938] R. Hagedorn, Suppl. Nuovo Cim. **3**, 147 (1965); Suppl. Nuovo Cim. **6**, 169 (1968)

[939] J. Ambjørn, J. Jurkiewicz, and R. Loll, in D. Oriti (Ed.), *Approaches to Quantum Gravity. Toward a New Understanding*

of Space, Time, and Matter, pp. 341-359 (Cambridge University Press, 2009); Int. J. Mod. Phys. D **17**, 2515 (2009); Contemp. Phys. **47**, 103 (2006); Phys. Rev. Lett. **95**, 171301 (2005); Phys. Lett. B **607**, 205 (2005); Phys. Rev. D **72**, 064014 (2005); Phys. Lett. B **581**, 255 (2004); Phys. Rev. Lett. **93**, 131 (2004); Nucl. Phys. Proc. Suppl. **106**, 980 (2002); Nucl. Phys. B **610**, 347 (2001); Phys. Rev. D **64**, 044011 (2001); Nucl. Phys. Proc. Suppl. **94**, 689 (2001); Phys. Rev. Lett. **85**, 924 (2000)

[940] J. Ambjorn, A. Goerlich, J. Jurkiewicz, and R. Loll, Phys. Rev. D **78**, 063544 (2008); Phys. Rev. Lett. **100**, 091304 (2008)

[941] B.F.L. Ward, Int. J. Mod. Phys. D **17**, 627 (2008)

[942] T. Padmanabhan, Phys. Rev. D **28**, 745 (1983)

[943] G. Birkhoff and S. Mac Lane, *A Survey of Modern Algebra* (4th ed., Macmillan, 1977);
L. Childs, *A Concrete Introduction to Higher Algebra* (Springer, 1979);
L.C. Grove, *Algebra* (Academic Press, 1983);
A. Kurosh, *Higher Algebra* (5th print., Mir Publishers, 1988);
S.H. Friedberg, A.J. Insel, and L.E. Spence, *Linear Algebra* (2nd ed., Prentice Hall, 1989);
M.L. Curtis, *Abstract Linear Algebra* (Springer, 1990);
J.A. Gallian, *Contemporary Abstract Algebra* (2nd ed., D.C. Heath and Company, 1990);
F. Gantmacher, *Matrix Theory* (2nd ed., American Mathematical Society, 1990);
B. Fine and G. Rosenberger, *The Fundamental Theorem of Algebra* (Springer, 1997);
J.B. Fraleigh, *A First Course in Abstract Algebra* (7th ed., Addison-Wesley, 2003);
T.W. Hungerford, *Algebra* (Graduate Texts in Mathematics **73**, Springer, 2003);
D.S. Dummit and R.M. Foote, *Abstract Algebra* (3rd ed., John Wiley & Sons, 2004);
W.J. Gilbert and W.K. Nicholson, *Modern Algebra with*

Applications (2nd ed., John Wiley & Sons, 2004);
S. Barnard, *Higher Algebra* (Read Books, 2008)

[944] W. Siegel, *Fields*, e-Print: arXiv:hep-th/9912205

[945] G. 't Hooft and M. Veltman, Nucl. Phys. B **44**, 189 (1972)

[946] Q.W. Wang and X.F. Lu, Commun. Theor. Phys. **43**, 704 (2005);
B. Altschul, Phys. Rev. D **73**, 036005 (2006);
L.B. Okun, Acta Phys. Polon. B **37**, 565 (2006);
A. Accioly, J. Helayël-Neto, and E. Scatena, Phys. Rev. D **82**, 065026 (2010); Int. J. Mod. Phys. D **19**, 2393 (2010)

[947] L.A. Glinka, NeuroQuantology **10**(1), 11 (2012)

[948] C.L.M.H. Navier, Mém. Acad. Sci. Inst. Fr. **6**(2), 389 (1827)

[949] G.G. Stokes, Trans. Camb. Phil. Soc. **8**, 287 (1845)

[950] O.A. Ladyzhenskaya, *The Mathematical Theory of Viscous Incompressible Flow* (Mathematics and Its Applications **2**, 2nd ed., Gordon and Breach, 1969);
R. Salvi (Ed.), *The Navier–Stokes Equations: Theory and Numerical Methods* (Lecture Notes in Pure and Applied Mathematics Volume **223**, Marcel Dekker, 2002);
R.Kh. Zeytounian, *Theory and Applications of Viscous Fluid Flows* (Springer Verlag, 2004);
C. Foias, O. Manley, R. Rosa and R. Temam, *Navier–Stokes Equations and Turbulence* (Encyclopedia of Mathematics and its Applications **83**, Cambridge University Press, 2004);
G.P. Galdi, *An Introduction to the Mathematical Theory of the Navier–Stokes Equations: Steady-State Problems* (Springer Monographs in Mathematics, 2nd ed., Springer, 2011)

[951] L.D. Landau and E.M. Lifshitz, *Course of Theoretical Physics. Volume 6: Fluid Mechanics* (2nd ed, Pergamon Press, 1987)

[952] Clay Mathematics Institute, *Navier–Stokes Equation*, http://www.claymath.org/millennium/Navier-Stokes_Equations/

[953] C.L. Fefferman, *Existence and Smoothness of The Navier–Stokes Equation*, http://www.claymath.org/millennium/Navier-Stokes_Equations/navierstokes.pdf

[954] T. Tao, *Why global regularity for Navier–Stokes is hard*, http://terrytao.wordpress.com/2007/03/18/why-global-regularity-for-navier-stokes-is-hard

Index